T0234552

Lecture Notes in Computer Science　　13433

Founding Editors

Gerhard Goos
Karlsruhe Institute of Technology, Karlsruhe, Germany

Juris Hartmanis
Cornell University, Ithaca, NY, USA

Editorial Board Members

Elisa Bertino
Purdue University, West Lafayette, IN, USA

Wen Gao
Peking University, Beijing, China

Bernhard Steffen ⓘ
TU Dortmund University, Dortmund, Germany

Moti Yung ⓘ
Columbia University, New York, NY, USA

More information about this series at https://link.springer.com/bookseries/558

Linwei Wang · Qi Dou · P. Thomas Fletcher ·
Stefanie Speidel · Shuo Li (Eds.)

Medical Image Computing and Computer Assisted Intervention – MICCAI 2022

25th International Conference
Singapore, September 18–22, 2022
Proceedings, Part III

Springer

Editors
Linwei Wang
Rochester Institute of Technology
Rochester, NY, USA

Qi Dou ⓘ
Chinese University of Hong Kong
Hong Kong, Hong Kong

P. Thomas Fletcher ⓘ
University of Virginia
Charlottesville, VA, USA

Stefanie Speidel ⓘ
National Center for Tumor Diseases
(NCT/UCC)
Dresden, Germany

Shuo Li ⓘ
Case Western Reserve University
Cleveland, OH, USA

ISSN 0302-9743   ISSN 1611-3349 (electronic)
Lecture Notes in Computer Science
ISBN 978-3-031-16436-1   ISBN 978-3-031-16437-8 (eBook)
https://doi.org/10.1007/978-3-031-16437-8

© The Editor(s) (if applicable) and The Author(s), under exclusive license
to Springer Nature Switzerland AG 2022
This work is subject to copyright. All rights are reserved by the Publisher, whether the whole or part of the material is concerned, specifically the rights of translation, reprinting, reuse of illustrations, recitation, broadcasting, reproduction on microfilms or in any other physical way, and transmission or information storage and retrieval, electronic adaptation, computer software, or by similar or dissimilar methodology now known or hereafter developed.
The use of general descriptive names, registered names, trademarks, service marks, etc. in this publication does not imply, even in the absence of a specific statement, that such names are exempt from the relevant protective laws and regulations and therefore free for general use.
The publisher, the authors, and the editors are safe to assume that the advice and information in this book are believed to be true and accurate at the date of publication. Neither the publisher nor the authors or the editors give a warranty, expressed or implied, with respect to the material contained herein or for any errors or omissions that may have been made. The publisher remains neutral with regard to jurisdictional claims in published maps and institutional affiliations.

This Springer imprint is published by the registered company Springer Nature Switzerland AG
The registered company address is: Gewerbestrasse 11, 6330 Cham, Switzerland

Preface

We are pleased to present the proceedings of the 25th International Conference on Medical Image Computing and Computer-Assisted Intervention (MICCAI) which – after two difficult years of virtual conferences – was held in a hybrid fashion at the Resort World Convention Centre in Singapore, September 18–22, 2022. The conference also featured 36 workshops, 11 tutorials, and 38 challenges held on September 18 and September 22. The conference was also co-located with the 2nd Conference on Clinical Translation on Medical Image Computing and Computer-Assisted Intervention (CLINICCAI) on September 20.

MICCAI 2022 had an approximately 14% increase in submissions and accepted papers compared with MICCAI 2021. These papers, which comprise eight volumes of Lecture Notes in Computer Science (LNCS) proceedings, were selected after a thorough double-blind peer-review process. Following the example set by the previous program chairs of past MICCAI conferences, we employed Microsoft's Conference Managing Toolkit (CMT) for paper submissions and double-blind peer-reviews, and the Toronto Paper Matching System (TPMS) to assist with automatic paper assignment to area chairs and reviewers.

From 2811 original intentions to submit, 1865 full submissions were received and 1831 submissions reviewed. Of these, 67% were considered as pure Medical Image Computing (MIC), 7% as pure Computer-Assisted Interventions (CAI), and 26% as both MIC and CAI. The MICCAI 2022 Program Committee (PC) comprised 107 area chairs, with 52 from the Americas, 33 from Europe, and 22 from the Asia-Pacific or Middle East regions. We maintained gender balance with 37% women scientists on the PC.

Each area chair was assigned 16–18 manuscripts, for each of which they were asked to suggest up to 15 suggested potential reviewers. Subsequently, over 1320 invited reviewers were asked to bid for the papers for which they had been suggested. Final reviewer allocations via CMT took account of PC suggestions, reviewer bidding, and TPMS scores, finally allocating 4–6 papers per reviewer. Based on the double-blinded reviews, area chairs' recommendations, and program chairs' global adjustments, 249 papers (14%) were provisionally accepted, 901 papers (49%) were provisionally rejected, and 675 papers (37%) proceeded into the rebuttal stage.

During the rebuttal phase, two additional area chairs were assigned to each rebuttal paper using CMT and TPMS scores. After the authors' rebuttals were submitted, all reviewers of the rebuttal papers were invited to assess the rebuttal, participate in a double-blinded discussion with fellow reviewers and area chairs, and finalize their rating (with the opportunity to revise their rating as appropriate). The three area chairs then independently provided their recommendations to accept or reject the paper, considering the manuscript, the reviews, and the rebuttal. The final decision of acceptance was based on majority voting of the area chair recommendations. The program chairs reviewed all decisions and provided their inputs in extreme cases where a large divergence existed between the area chairs and reviewers in their recommendations. This process resulted in the acceptance of a total of 574 papers, reaching an overall acceptance rate of 31% for MICCAI 2022.

In our additional effort to ensure review quality, two Reviewer Tutorials and two Area Chair Orientations were held in early March, virtually in different time zones, to introduce the reviewers and area chairs to the MICCAI 2022 review process and the best practice for high-quality reviews. Two additional Area Chair meetings were held virtually in July to inform the area chairs of the outcome of the review process and to collect feedback for future conferences.

For the MICCAI 2022 proceedings, 574 accepted papers were organized in eight volumes as follows:

- Part I, LNCS Volume 13431: Brain Development and Atlases, DWI and Tractography, Functional Brain Networks, Neuroimaging, Heart and Lung Imaging, and Dermatology
- Part II, LNCS Volume 13432: Computational (Integrative) Pathology, Computational Anatomy and Physiology, Ophthalmology, and Fetal Imaging
- Part III, LNCS Volume 13433: Breast Imaging, Colonoscopy, and Computer Aided Diagnosis
- Part IV, LNCS Volume 13434: Microscopic Image Analysis, Positron Emission Tomography, Ultrasound Imaging, Video Data Analysis, and Image Segmentation I
- Part V, LNCS Volume 13435: Image Segmentation II and Integration of Imaging with Non-imaging Biomarkers
- Part VI, LNCS Volume 13436: Image Registration and Image Reconstruction
- Part VII, LNCS Volume 13437: Image-Guided Interventions and Surgery, Outcome and Disease Prediction, Surgical Data Science, Surgical Planning and Simulation, and Machine Learning – Domain Adaptation and Generalization
- Part VIII, LNCS Volume 13438: Machine Learning – Weakly-supervised Learning, Machine Learning – Model Interpretation, Machine Learning – Uncertainty, and Machine Learning Theory and Methodologies

We would like to thank everyone who contributed to the success of MICCAI 2022 and the quality of its proceedings. These include the MICCAI Society for support and feedback, and our sponsors for their financial support and presence onsite. We especially express our gratitude to the MICCAI Submission System Manager Kitty Wong for her thorough support throughout the paper submission, review, program planning, and proceeding preparation process – the Program Committee simply would not have be able to function without her. We are also grateful for the dedication and support of all of the organizers of the workshops, tutorials, and challenges, Jianming Liang, Wufeng Xue, Jun Cheng, Qian Tao, Xi Chen, Islem Rekik, Sophia Bano, Andrea Lara, Yunliang Cai, Pingkun Yan, Pallavi Tiwari, Ingerid Reinertsen, Gongning Luo, without whom the exciting peripheral events would have not been feasible. Behind the scenes, the MICCAI secretariat personnel, Janette Wallace and Johanne Langford, kept a close eye on logistics and budgets, while Mehmet Eldegez and his team from Dekon Congress & Tourism, MICCAI 2022's Professional Conference Organization, managed the website and local organization. We are especially grateful to all members of the Program Committee for their diligent work in the reviewer assignments and final paper selection, as well as the reviewers for their support during the entire process. Finally, and most importantly, we thank all authors, co-authors, students/postdocs, and supervisors, for submitting and presenting their high-quality work which made MICCAI 2022 a successful event.

We look forward to seeing you in Vancouver, Canada at MICCAI 2023!

September 2022

Linwei Wang
Qi Dou
P. Thomas Fletcher
Stefanie Speidel
Shuo Li

Organization

General Chair

Shuo Li — Case Western Reserve University, USA

Program Committee Chairs

Linwei Wang	Rochester Institute of Technology, USA
Qi Dou	The Chinese University of Hong Kong, China
P. Thomas Fletcher	University of Virginia, USA
Stefanie Speidel	National Center for Tumor Diseases Dresden, Germany

Workshop Team

Wufeng Xue	Shenzhen University, China
Jun Cheng	Agency for Science, Technology and Research, Singapore
Qian Tao	Delft University of Technology, the Netherlands
Xi Chen	Stern School of Business, NYU, USA

Challenges Team

Pingkun Yan	Rensselaer Polytechnic Institute, USA
Pallavi Tiwari	Case Western Reserve University, USA
Ingerid Reinertsen	SINTEF Digital and NTNU, Trondheim, Norway
Gongning Luo	Harbin Institute of Technology, China

Tutorial Team

Islem Rekik	Istanbul Technical University, Turkey
Sophia Bano	University College London, UK
Andrea Lara	Universidad Industrial de Santander, Colombia
Yunliang Cai	Humana, USA

Clinical Day Chairs

Jason Chan	The Chinese University of Hong Kong, China
Heike I. Grabsch	University of Leeds, UK and Maastricht University, the Netherlands
Nicolas Padoy	University of Strasbourg & Institute of Image-Guided Surgery, IHU Strasbourg, France

Young Investigators and Early Career Development Program Chairs

Marius Linguraru	Children's National Institute, USA
Antonio Porras	University of Colorado Anschutz Medical Campus, USA
Nicole Rieke	NVIDIA, Deutschland
Daniel Racoceanu	Sorbonne University, France

Social Media Chairs

Chenchu Xu	Anhui University, China
Dong Zhang	University of British Columbia, Canada

Student Board Liaison

Camila Bustillo	Technische Universität Darmstadt, Germany
Vanessa Gonzalez Duque	Ecole centrale de Nantes, France

Submission Platform Manager

Kitty Wong	The MICCAI Society, Canada

Virtual Platform Manager

John Baxter	INSERM, Université de Rennes 1, France

Program Committee

Ehsan Adeli	Stanford University, USA
Pablo Arbelaez	Universidad de los Andes, Colombia
John Ashburner	University College London, UK
Ulas Bagci	Northwestern University, USA
Sophia Bano	University College London, UK
Adrien Bartoli	Université Clermont Auvergne, France
Kayhan Batmanghelich	University of Pittsburgh, USA

Hrvoje Bogunovic	Medical University of Vienna, Austria
Ester Bonmati	University College London, UK
Esther Bron	Erasmus MC, the Netherlands
Gustavo Carneiro	University of Adelaide, Australia
Hao Chen	Hong Kong University of Science and Technology, China
Jun Cheng	Agency for Science, Technology and Research, Singapore
Li Cheng	University of Alberta, Canada
Adrian Dalca	Massachusetts Institute of Technology, USA
Jose Dolz	ETS Montreal, Canada
Shireen Elhabian	University of Utah, USA
Sandy Engelhardt	University Hospital Heidelberg, Germany
Ruogu Fang	University of Florida, USA
Aasa Feragen	Technical University of Denmark, Denmark
Moti Freiman	Technion - Israel Institute of Technology, Israel
Huazhu Fu	Agency for Science, Technology and Research, Singapore
Mingchen Gao	University at Buffalo, SUNY, USA
Zhifan Gao	Sun Yat-sen University, China
Stamatia Giannarou	Imperial College London, UK
Alberto Gomez	King's College London, UK
Ilker Hacihaliloglu	University of British Columbia, Canada
Adam Harrison	PAII Inc., USA
Mattias Heinrich	University of Lübeck, Germany
Yipeng Hu	University College London, UK
Junzhou Huang	University of Texas at Arlington, USA
Sharon Xiaolei Huang	Pennsylvania State University, USA
Yuankai Huo	Vanderbilt University, USA
Jayender Jagadeesan	Brigham and Women's Hospital, USA
Won-Ki Jeong	Korea University, Korea
Xi Jiang	University of Electronic Science and Technology of China, China
Anand Joshi	University of Southern California, USA
Shantanu Joshi	University of California, Los Angeles, USA
Bernhard Kainz	Imperial College London, UK
Marta Kersten-Oertel	Concordia University, Canada
Fahmi Khalifa	Mansoura University, Egypt
Seong Tae Kim	Kyung Hee University, Korea
Minjeong Kim	University of North Carolina at Greensboro, USA
Baiying Lei	Shenzhen University, China
Gang Li	University of North Carolina at Chapel Hill, USA

Xiaoxiao Li	University of British Columbia, Canada
Jianming Liang	Arizona State University, USA
Herve Lombaert	ETS Montreal, Canada
Marco Lorenzi	Inria Sophia Antipolis, France
Le Lu	Alibaba USA Inc., USA
Klaus Maier-Hein	German Cancer Research Center (DKFZ), Germany
Anne Martel	Sunnybrook Research Institute, Canada
Diana Mateus	Centrale Nantes, France
Mehdi Moradi	IBM Research, USA
Hien Nguyen	University of Houston, USA
Mads Nielsen	University of Copenhagen, Denmark
Ilkay Oksuz	Istanbul Technical University, Turkey
Tingying Peng	Helmholtz Zentrum Muenchen, Germany
Caroline Petitjean	Université de Rouen, France
Gemma Piella	Universitat Pompeu Fabra, Spain
Chen Qin	University of Edinburgh, UK
Hedyeh Rafii-Tari	Auris Health Inc., USA
Tammy Riklin Raviv	Ben-Gurion University of the Negev, Israel
Hassan Rivaz	Concordia University, Canada
Michal Rosen-Zvi	IBM Research, Israel
Su Ruan	University of Rouen, France
Thomas Schultz	University of Bonn, Germany
Sharmishtaa Seshamani	Allen Institute, USA
Feng Shi	United Imaging Intelligence, China
Yonggang Shi	University of Southern California, USA
Yang Song	University of New South Wales, Australia
Rachel Sparks	King's College London, UK
Carole Sudre	University College London, UK
Tanveer Syeda-Mahmood	IBM Research, USA
Qian Tao	Delft University of Technology, the Netherlands
Tolga Tasdizen	University of Utah, USA
Pallavi Tiwari	Case Western Reserve University, USA
Mathias Unberath	Johns Hopkins University, USA
Martin Urschler	University of Auckland, New Zealand
Maria Vakalopoulou	University of Paris Saclay, France
Harini Veeraraghavan	Memorial Sloan Kettering Cancer Center, USA
Satish Viswanath	Case Western Reserve University, USA
Christian Wachinger	Technical University of Munich, Germany
Hua Wang	Colorado School of Mines, USA
Hongzhi Wang	IBM Research, USA
Ken C. L. Wong	IBM Almaden Research Center, USA

Fuyong Xing	University of Colorado Denver, USA
Ziyue Xu	NVIDIA, USA
Yanwu Xu	Baidu Inc., China
Pingkun Yan	Rensselaer Polytechnic Institute, USA
Guang Yang	Imperial College London, UK
Jianhua Yao	Tencent, China
Zhaozheng Yin	Stony Brook University, USA
Lequan Yu	University of Hong Kong, China
Yixuan Yuan	City University of Hong Kong, China
Ling Zhang	Alibaba Group, USA
Miaomiao Zhang	University of Virginia, USA
Ya Zhang	Shanghai Jiao Tong University, China
Rongchang Zhao	Central South University, China
Yitian Zhao	Chinese Academy of Sciences, China
Yefeng Zheng	Tencent Jarvis Lab, China
Guoyan Zheng	Shanghai Jiao Tong University, China
Luping Zhou	University of Sydney, Australia
Yuyin Zhou	Stanford University, USA
Dajiang Zhu	University of Texas at Arlington, USA
Lilla Zöllei	Massachusetts General Hospital, USA
Maria A. Zuluaga	EURECOM, France

Reviewers

Alireza Akhondi-asl	Manas Nag
Fernando Arambula	Tianye Niu
Nicolas Boutry	Seokhwan Oh
Qilei Chen	Theodoros Pissas
Zhihao Chen	Harish RaviPrakash
Javid Dadashkarimi	Maria Sainz de Cea
Marleen De Bruijne	Hai Su
Mohammad Eslami	Wenjun Tan
Sayan Ghosal	Fatmatulzehra Uslu
Estibaliz Gómez-de-Mariscal	Fons van der Sommen
Charles Hatt	Gijs van Tulder
Yongxiang Huang	Dong Wei
Samra Irshad	Pengcheng Xi
Anithapriya Krishnan	Chen Yang
Rodney LaLonde	Kun Yuan
Jie Liu	Hang Zhang
Jinyang Liu	Wei Zhang
Qing Lyu	Yuyao Zhang
Hassan Mohy-ud-Din	Tengda Zhao

Yingying Zhu
Yuemin Zhu
Alaa Eldin Abdelaal
Amir Abdi
Mazdak Abulnaga
Burak Acar
Iman Aganj
Priya Aggarwal
Ola Ahmad
Seyed-Ahmad Ahmadi
Euijoon Ahn
Faranak Akbarifar
Cem Akbaş
Saad Ullah Akram
Tajwar Aleef
Daniel Alexander
Hazrat Ali
Sharib Ali
Max Allan
Pablo Alvarez
Vincent Andrearczyk
Elsa Angelini
Sameer Antani
Michela Antonelli
Ignacio Arganda-Carreras
Mohammad Ali Armin
Josep Arnal
Md Ashikuzzaman
Mehdi Astaraki
Marc Aubreville
Chloé Audigier
Angelica Aviles-Rivero
Ruqayya Awan
Suyash Awate
Qinle Ba
Morteza Babaie
Meritxell Bach Cuadra
Hyeon-Min Bae
Junjie Bai
Wenjia Bai
Ujjwal Baid
Pradeep Bajracharya
Yaël Balbastre
Abhirup Banerjee
Sreya Banerjee

Shunxing Bao
Adrian Barbu
Sumana Basu
Deepti Bathula
Christian Baumgartner
John Baxter
Sharareh Bayat
Bahareh Behboodi
Hamid Behnam
Sutanu Bera
Christos Bergeles
Jose Bernal
Gabriel Bernardino
Alaa Bessadok
Riddhish Bhalodia
Indrani Bhattacharya
Chitresh Bhushan
Lei Bi
Qi Bi
Gui-Bin Bian
Alexander Bigalke
Ricardo Bigolin Lanfredi
Benjamin Billot
Ryoma Bise
Sangeeta Biswas
Stefano B. Blumberg
Sebastian Bodenstedt
Bhushan Borotikar
Ilaria Boscolo Galazzo
Behzad Bozorgtabar
Nadia Brancati
Katharina Breininger
Rupert Brooks
Tom Brosch
Mikael Brudfors
Qirong Bu
Ninon Burgos
Nikolay Burlutskiy
Michał Byra
Ryan Cabeen
Mariano Cabezas
Hongmin Cai
Jinzheng Cai
Weidong Cai
Sema Candemir

Qing Cao
Weiguo Cao
Yankun Cao
Aaron Carass
Ruben Cardenes
M. Jorge Cardoso
Owen Carmichael
Alessandro Casella
Matthieu Chabanas
Ahmad Chaddad
Jayasree Chakraborty
Sylvie Chambon
Yi Hao Chan
Ming-Ching Chang
Peng Chang
Violeta Chang
Sudhanya Chatterjee
Christos Chatzichristos
Antong Chen
Chao Chen
Chen Chen
Cheng Chen
Dongdong Chen
Fang Chen
Geng Chen
Hanbo Chen
Jianan Chen
Jianxu Chen
Jie Chen
Junxiang Chen
Junying Chen
Junyu Chen
Lei Chen
Li Chen
Liangjun Chen
Liyun Chen
Min Chen
Pingjun Chen
Qiang Chen
Runnan Chen
Shuai Chen
Xi Chen
Xiaoran Chen
Xin Chen
Xinjian Chen

Xuejin Chen
Yuanyuan Chen
Zhaolin Chen
Zhen Chen
Zhineng Chen
Zhixiang Chen
Erkang Cheng
Jianhong Cheng
Jun Cheng
Philip Chikontwe
Min-Kook Choi
Gary Christensen
Argyrios Christodoulidis
Stergios Christodoulidis
Albert Chung
Özgün Çiçek
Matthew Clarkson
Dana Cobzas
Jaume Coll-Font
Toby Collins
Olivier Commowick
Runmin Cong
Yulai Cong
Pierre-Henri Conze
Timothy Cootes
Teresa Correia
Pierrick Coupé
Hadrien Courtecuisse
Jeffrey Craley
Alessandro Crimi
Can Cui
Hejie Cui
Hui Cui
Zhiming Cui
Kathleen Curran
Claire Cury
Tobias Czempiel
Vedrana Dahl
Tareen Dawood
Laura Daza
Charles Delahunt
Herve Delingette
Ugur Demir
Liang-Jian Deng
Ruining Deng

Yang Deng
Cem Deniz
Felix Denzinger
Adrien Depeursinge
Hrishikesh Deshpande
Christian Desrosiers
Neel Dey
Anuja Dharmaratne
Li Ding
Xinghao Ding
Zhipeng Ding
Ines Domingues
Juan Pedro Dominguez-Morales
Mengjin Dong
Nanqing Dong
Sven Dorkenwald
Haoran Dou
Simon Drouin
Karen Drukker
Niharika D'Souza
Guodong Du
Lei Du
Dingna Duan
Hongyi Duanmu
Nicolas Duchateau
James Duncan
Nicha Dvornek
Dmitry V. Dylov
Oleh Dzyubachyk
Jan Egger
Alma Eguizabal
Gudmundur Einarsson
Ahmet Ekin
Ahmed Elazab
Ahmed Elnakib
Amr Elsawy
Mohamed Elsharkawy
Ertunc Erdil
Marius Erdt
Floris Ernst
Boris Escalante-Ramírez
Hooman Esfandiari
Nazila Esmaeili
Marco Esposito
Théo Estienne

Christian Ewert
Deng-Ping Fan
Xin Fan
Yonghui Fan
Yubo Fan
Chaowei Fang
Huihui Fang
Xi Fang
Yingying Fang
Zhenghan Fang
Mohsen Farzi
Hamid Fehri
Lina Felsner
Jianjiang Feng
Jun Feng
Ruibin Feng
Yuan Feng
Zishun Feng
Aaron Fenster
Henrique Fernandes
Ricardo Ferrari
Lukas Fischer
Antonio Foncubierta-Rodríguez
Nils Daniel Forkert
Wolfgang Freysinger
Bianca Freytag
Xueyang Fu
Yunguan Fu
Gareth Funka-Lea
Pedro Furtado
Ryo Furukawa
Laurent Gajny
Francesca Galassi
Adrian Galdran
Jiangzhang Gan
Yu Gan
Melanie Ganz
Dongxu Gao
Linlin Gao
Riqiang Gao
Siyuan Gao
Yunhe Gao
Zeyu Gao
Gautam Gare
Bao Ge

Rongjun Ge
Sairam Geethanath
Shiv Gehlot
Yasmeen George
Nils Gessert
Olivier Gevaert
Ramtin Gharleghi
Sandesh Ghimire
Andrea Giovannini
Gabriel Girard
Rémi Giraud
Ben Glocker
Ehsan Golkar
Arnold Gomez
Ricardo Gonzales
Camila Gonzalez
Cristina González
German Gonzalez
Sharath Gopal
Karthik Gopinath
Pietro Gori
Michael Götz
Shuiping Gou
Maged Goubran
Sobhan Goudarzi
Alejandro Granados
Mara Graziani
Yun Gu
Zaiwang Gu
Hao Guan
Dazhou Guo
Hengtao Guo
Jixiang Guo
Jun Guo
Pengfei Guo
Xiaoqing Guo
Yi Guo
Yuyu Guo
Vikash Gupta
Prashnna Gyawali
Stathis Hadjidemetriou
Fatemeh Haghighi
Justin Haldar
Mohammad Hamghalam
Kamal Hammouda

Bing Han
Liang Han
Seungjae Han
Xiaoguang Han
Zhongyi Han
Jonny Hancox
Lasse Hansen
Huaying Hao
Jinkui Hao
Xiaoke Hao
Mohammad Minhazul Haq
Nandinee Haq
Rabia Haq
Michael Hardisty
Nobuhiko Hata
Ali Hatamizadeh
Andreas Hauptmann
Huiguang He
Nanjun He
Shenghua He
Yuting He
Tobias Heimann
Stefan Heldmann
Sobhan Hemati
Alessa Hering
Monica Hernandez
Estefania Hernandez-Martin
Carlos Hernandez-Matas
Javier Herrera-Vega
Kilian Hett
David Ho
Yi Hong
Yoonmi Hong
Mohammad Reza Hosseinzadeh Taher
Benjamin Hou
Wentai Hou
William Hsu
Dan Hu
Rongyao Hu
Xiaoling Hu
Xintao Hu
Yan Hu
Ling Huang
Sharon Xiaolei Huang
Xiaoyang Huang

Yangsibo Huang
Yi-Jie Huang
Yijin Huang
Yixing Huang
Yue Huang
Zhi Huang
Ziyi Huang
Arnaud Huaulmé
Jiayu Huo
Raabid Hussain
Sarfaraz Hussein
Khoi Huynh
Seong Jae Hwang
Ilknur Icke
Kay Igwe
Abdullah Al Zubaer Imran
Ismail Irmakci
Benjamin Irving
Mohammad Shafkat Islam
Koichi Ito
Hayato Itoh
Yuji Iwahori
Mohammad Jafari
Andras Jakab
Amir Jamaludin
Mirek Janatka
Vincent Jaouen
Uditha Jarayathne
Ronnachai Jaroensri
Golara Javadi
Rohit Jena
Rachid Jennane
Todd Jensen
Debesh Jha
Ge-Peng Ji
Yuanfeng Ji
Zhanghexuan Ji
Haozhe Jia
Meirui Jiang
Tingting Jiang
Xiajun Jiang
Xiang Jiang
Zekun Jiang
Jianbo Jiao
Jieqing Jiao

Zhicheng Jiao
Chen Jin
Dakai Jin
Qiangguo Jin
Taisong Jin
Yueming Jin
Baoyu Jing
Bin Jing
Yaqub Jonmohamadi
Lie Ju
Yohan Jun
Alain Jungo
Manjunath K N
Abdolrahim Kadkhodamohammadi
Ali Kafaei Zad Tehrani
Dagmar Kainmueller
Siva Teja Kakileti
John Kalafut
Konstantinos Kamnitsas
Michael C. Kampffmeyer
Qingbo Kang
Neerav Karani
Turkay Kart
Satyananda Kashyap
Alexander Katzmann
Anees Kazi
Hengjin Ke
Hamza Kebiri
Erwan Kerrien
Hoel Kervadec
Farzad Khalvati
Bishesh Khanal
Pulkit Khandelwal
Maksim Kholiavchenko
Ron Kikinis
Daeseung Kim
Jae-Hun Kim
Jaeil Kim
Jinman Kim
Won Hwa Kim
Andrew King
Atilla Kiraly
Yoshiro Kitamura
Stefan Klein
Tobias Klinder

Lisa Koch
Satoshi Kondo
Bin Kong
Fanwei Kong
Ender Konukoglu
Aishik Konwer
Bongjin Koo
Ivica Kopriva
Kivanc Kose
Anna Kreshuk
Frithjof Kruggel
Thomas Kuestner
David Kügler
Hugo Kuijf
Arjan Kuijper
Kuldeep Kumar
Manuela Kunz
Holger Kunze
Tahsin Kurc
Anvar Kurmukov
Yoshihiro Kuroda
Jin Tae Kwak
Francesco La Rosa
Aymen Laadhari
Dmitrii Lachinov
Alain Lalande
Bennett Landman
Axel Largent
Carole Lartizien
Max-Heinrich Laves
Ho Hin Lee
Hyekyoung Lee
Jong Taek Lee
Jong-Hwan Lee
Soochahn Lee
Wen Hui Lei
Yiming Lei
Rogers Jeffrey Leo John
Juan Leon
Bo Li
Bowen Li
Chen Li
Hongming Li
Hongwei Li
Jian Li

Jianning Li
Jiayun Li
Jieyu Li
Junhua Li
Kang Li
Lei Li
Mengzhang Li
Qing Li
Quanzheng Li
Shaohua Li
Shulong Li
Weijian Li
Weikai Li
Wenyuan Li
Xiang Li
Xingyu Li
Xiu Li
Yang Li
Yuexiang Li
Yunxiang Li
Zeju Li
Zhang Li
Zhiyuan Li
Zhjin Li
Zi Li
Chunfeng Lian
Sheng Lian
Libin Liang
Peixian Liang
Yuan Liang
Haofu Liao
Hongen Liao
Ruizhi Liao
Wei Liao
Xiangyun Liao
Gilbert Lim
Hongxiang Lin
Jianyu Lin
Li Lin
Tiancheng Lin
Yiqun Lin
Zudi Lin
Claudia Lindner
Bin Liu
Bo Liu

Chuanbin Liu
Daochang Liu
Dong Liu
Dongnan Liu
Fenglin Liu
Han Liu
Hao Liu
Haozhe Liu
Hong Liu
Huafeng Liu
Huiye Liu
Jianfei Liu
Jiang Liu
Jingya Liu
Kefei Liu
Lihao Liu
Mengting Liu
Peirong Liu
Peng Liu
Qin Liu
Qun Liu
Shenghua Liu
Shuangjun Liu
Sidong Liu
Tianrui Liu
Xiao Liu
Xingtong Liu
Xinwen Liu
Xinyang Liu
Xinyu Liu
Yan Liu
Yanbei Liu
Yi Liu
Yikang Liu
Yong Liu
Yue Liu
Yuhang Liu
Zewen Liu
Zhe Liu
Andrea Loddo
Nicolas Loménie
Yonghao Long
Zhongjie Long
Daniel Lopes
Bin Lou

Nicolas Loy Rodas
Charles Lu
Huanxiang Lu
Xing Lu
Yao Lu
Yuhang Lu
Gongning Luo
Jie Luo
Jiebo Luo
Luyang Luo
Ma Luo
Xiangde Luo
Cuong Ly
Ilwoo Lyu
Yanjun Lyu
Yuanyuan Lyu
Sharath M S
Chunwei Ma
Hehuan Ma
Junbo Ma
Wenao Ma
Yuhui Ma
Anderson Maciel
S. Sara Mahdavi
Mohammed Mahmoud
Andreas Maier
Michail Mamalakis
Ilja Manakov
Brett Marinelli
Yassine Marrakchi
Fabio Martinez
Martin Maška
Tejas Sudharshan Mathai
Dimitrios Mavroeidis
Pau Medrano-Gracia
Raghav Mehta
Felix Meissen
Qingjie Meng
Yanda Meng
Martin Menten
Alexandre Merasli
Stijn Michielse
Leo Milecki
Fausto Milletari
Zhe Min

Tadashi Miyamoto
Sara Moccia
Omid Mohareri
Tony C. W. Mok
Rodrigo Moreno
Kensaku Mori
Lia Morra
Aliasghar Mortazi
Hamed Mozaffari
Pritam Mukherjee
Anirban Mukhopadhyay
Henning Müller
Balamurali Murugesan
Tinashe Mutsvangwa
Andriy Myronenko
Saad Nadeem
Ahmed Naglah
Usman Naseem
Vishwesh Nath
Rodrigo Nava
Nassir Navab
Peter Neher
Amin Nejatbakhsh
Dominik Neumann
Duy Nguyen Ho Minh
Dong Ni
Haomiao Ni
Hannes Nickisch
Jingxin Nie
Aditya Nigam
Lipeng Ning
Xia Ning
Sijie Niu
Jack Noble
Jorge Novo
Chinedu Nwoye
Mohammad Obeid
Masahiro Oda
Steffen Oeltze-Jafra
Ayşe Oktay
Hugo Oliveira
Sara Oliveira
Arnau Oliver
Emanuele Olivetti
Jimena Olveres

Doruk Oner
John Onofrey
Felipe Orihuela-Espina
Marcos Ortega
Yoshito Otake
Sebastian Otálora
Cheng Ouyang
Jiahong Ouyang
Xi Ouyang
Utku Ozbulak
Michal Ozery-Flato
Danielle Pace
José Blas Pagador Carrasco
Daniel Pak
Jin Pan
Siyuan Pan
Yongsheng Pan
Pankaj Pandey
Prashant Pandey
Egor Panfilov
Joao Papa
Bartlomiej Papiez
Nripesh Parajuli
Hyunjin Park
Sanghyun Park
Akash Parvatikar
Magdalini Paschali
Diego Patiño Cortés
Mayank Patwari
Angshuman Paul
Yuchen Pei
Yuru Pei
Chengtao Peng
Jialin Peng
Wei Peng
Yifan Peng
Matteo Pennisi
Antonio Pepe
Oscar Perdomo
Sérgio Pereira
Jose-Antonio Pérez-Carrasco
Fernando Pérez-García
Jorge Perez-Gonzalez
Matthias Perkonigg
Mehran Pesteie

Jorg Peters
Terry Peters
Eike Petersen
Jens Petersen
Micha Pfeiffer
Dzung Pham
Hieu Pham
Ashish Phophalia
Tomasz Pieciak
Antonio Pinheiro
Kilian Pohl
Sebastian Pölsterl
Iulia A. Popescu
Alison Pouch
Prateek Prasanna
Raphael Prevost
Juan Prieto
Federica Proietto Salanitri
Sergi Pujades
Kumaradevan Punithakumar
Haikun Qi
Huan Qi
Buyue Qian
Yan Qiang
Yuchuan Qiao
Zhi Qiao
Fangbo Qin
Wenjian Qin
Yanguo Qin
Yulei Qin
Hui Qu
Kha Gia Quach
Tran Minh Quan
Sandro Queirós
Prashanth R.
Mehdi Rahim
Jagath Rajapakse
Kashif Rajpoot
Dhanesh Ramachandram
Xuming Ran
Hatem Rashwan
Daniele Ravì
Keerthi Sravan Ravi
Surreerat Reaungamornrat
Samuel Remedios

Yudan Ren
Mauricio Reyes
Constantino Reyes-Aldasoro
Hadrien Reynaud
David Richmond
Anne-Marie Rickmann
Laurent Risser
Leticia Rittner
Dominik Rivoir
Emma Robinson
Jessica Rodgers
Rafael Rodrigues
Robert Rohling
Lukasz Roszkowiak
Holger Roth
Karsten Roth
José Rouco
Daniel Rueckert
Danny Ruijters
Mirabela Rusu
Ario Sadafi
Shaheer Ullah Saeed
Monjoy Saha
Pranjal Sahu
Olivier Salvado
Ricardo Sanchez-Matilla
Robin Sandkuehler
Gianmarco Santini
Anil Kumar Sao
Duygu Sarikaya
Olivier Saut
Fabio Scarpa
Nico Scherf
Markus Schirmer
Alexander Schlaefer
Jerome Schmid
Julia Schnabel
Andreas Schuh
Christina Schwarz-Gsaxner
Martin Schweiger
Michaël Sdika
Suman Sedai
Matthias Seibold
Raghavendra Selvan
Sourya Sengupta

Carmen Serrano
Ahmed Shaffie
Keyur Shah
Rutwik Shah
Ahmed Shahin
Mohammad Abuzar Shaikh
S. Shailja
Shayan Shams
Hongming Shan
Xinxin Shan
Mostafa Sharifzadeh
Anuja Sharma
Harshita Sharma
Gregory Sharp
Li Shen
Liyue Shen
Mali Shen
Mingren Shen
Yiqing Shen
Ziyi Shen
Luyao Shi
Xiaoshuang Shi
Yiyu Shi
Hoo-Chang Shin
Boris Shirokikh
Suprosanna Shit
Suzanne Shontz
Yucheng Shu
Alberto Signoroni
Carlos Silva
Wilson Silva
Margarida Silveira
Vivek Singh
Sumedha Singla
Ayushi Sinha
Elena Sizikova
Rajath Soans
Hessam Sokooti
Hong Song
Weinan Song
Youyi Song
Aristeidis Sotiras
Bella Specktor
William Speier
Ziga Spiclin

Jon Sporring
Anuroop Sriram
Vinkle Srivastav
Lawrence Staib
Johannes Stegmaier
Joshua Stough
Danail Stoyanov
Justin Strait
Iain Styles
Ruisheng Su
Vaishnavi Subramanian
Gérard Subsol
Yao Sui
Heung-Il Suk
Shipra Suman
Jian Sun
Li Sun
Liyan Sun
Wenqing Sun
Yue Sun
Vaanathi Sundaresan
Kyung Sung
Yannick Suter
Raphael Sznitman
Eleonora Tagliabue
Roger Tam
Chaowei Tan
Hao Tang
Sheng Tang
Thomas Tang
Youbao Tang
Yucheng Tang
Zihao Tang
Rong Tao
Elias Tappeiner
Mickael Tardy
Giacomo Tarroni
Paul Thienphrapa
Stephen Thompson
Yu Tian
Aleksei Tiulpin
Tal Tlusty
Maryam Toloubidokhti
Jocelyne Troccaz
Roger Trullo

Chialing Tsai
Sudhakar Tummala
Régis Vaillant
Jeya Maria Jose Valanarasu
Juan Miguel Valverde
Thomas Varsavsky
Francisco Vasconcelos
Serge Vasylechko
S. Swaroop Vedula
Roberto Vega
Gonzalo Vegas Sanchez-Ferrero
Gopalkrishna Veni
Archana Venkataraman
Athanasios Vlontzos
Ingmar Voigt
Eugene Vorontsov
Xiaohua Wan
Bo Wang
Changmiao Wang
Chunliang Wang
Clinton Wang
Dadong Wang
Fan Wang
Guotai Wang
Haifeng Wang
Hong Wang
Hongkai Wang
Hongyu Wang
Hu Wang
Juan Wang
Junyan Wang
Ke Wang
Li Wang
Liansheng Wang
Manning Wang
Nizhuan Wang
Qiuli Wang
Renzhen Wang
Rongguang Wang
Ruixuan Wang
Runze Wang
Shujun Wang
Shuo Wang
Shuqiang Wang
Tianchen Wang

Tongxin Wang
Wenzhe Wang
Xi Wang
Xiangdong Wang
Xiaosong Wang
Yalin Wang
Yan Wang
Yi Wang
Yixin Wang
Zeyi Wang
Zuhui Wang
Jonathan Weber
Donglai Wei
Dongming Wei
Lifang Wei
Wolfgang Wein
Michael Wels
Cédric Wemmert
Matthias Wilms
Adam Wittek
Marek Wodzinski
Julia Wolleb
Jonghye Woo
Chongruo Wu
Chunpeng Wu
Ji Wu
Jianfeng Wu
Jie Ying Wu
Jiong Wu
Junde Wu
Pengxiang Wu
Xia Wu
Xiyin Wu
Yawen Wu
Ye Wu
Yicheng Wu
Zhengwang Wu
Tobias Wuerfl
James Xia
Siyu Xia
Yingda Xia
Lei Xiang
Tiange Xiang
Deqiang Xiao
Yiming Xiao

Hongtao Xie
Jianyang Xie
Lingxi Xie
Long Xie
Weidi Xie
Yiting Xie
Yutong Xie
Fangxu Xing
Jiarui Xing
Xiaohan Xing
Chenchu Xu
Hai Xu
Hongming Xu
Jiaqi Xu
Junshen Xu
Kele Xu
Min Xu
Minfeng Xu
Moucheng Xu
Qinwei Xu
Rui Xu
Xiaowei Xu
Xinxing Xu
Xuanang Xu
Yanwu Xu
Yanyu Xu
Yongchao Xu
Zhe Xu
Zhenghua Xu
Zhoubing Xu
Kai Xuan
Cheng Xue
Jie Xue
Wufeng Xue
Yuan Xue
Faridah Yahya
Chaochao Yan
Jiangpeng Yan
Ke Yan
Ming Yan
Qingsen Yan
Yuguang Yan
Zengqiang Yan
Baoyao Yang
Changchun Yang

Chao-Han Huck Yang
Dong Yang
Fan Yang
Feng Yang
Fengting Yang
Ge Yang
Guanyu Yang
Hao-Hsiang Yang
Heran Yang
Hongxu Yang
Huijuan Yang
Jiawei Yang
Jinyu Yang
Lin Yang
Peng Yang
Pengshuai Yang
Xiaohui Yang
Xin Yang
Yan Yang
Yifan Yang
Yujiu Yang
Zhicheng Yang
Jiangchao Yao
Jiawen Yao
Li Yao
Linlin Yao
Qingsong Yao
Chuyang Ye
Dong Hye Ye
Huihui Ye
Menglong Ye
Youngjin Yoo
Chenyu You
Haichao Yu
Hanchao Yu
Jinhua Yu
Ke Yu
Qi Yu
Renping Yu
Thomas Yu
Xiaowei Yu
Zhen Yu
Pengyu Yuan
Paul Yushkevich
Ghada Zamzmi

Ramy Zeineldin
Dong Zeng
Rui Zeng
Zhiwei Zhai
Kun Zhan
Bokai Zhang
Chaoyi Zhang
Daoqiang Zhang
Fa Zhang
Fan Zhang
Hao Zhang
Jianpeng Zhang
Jiawei Zhang
Jingqing Zhang
Jingyang Zhang
Jiong Zhang
Jun Zhang
Ke Zhang
Lefei Zhang
Lei Zhang
Lichi Zhang
Lu Zhang
Ning Zhang
Pengfei Zhang
Qiang Zhang
Rongzhao Zhang
Ruipeng Zhang
Ruisi Zhang
Shengping Zhang
Shihao Zhang
Tianyang Zhang
Tong Zhang
Tuo Zhang
Wen Zhang
Xiaoran Zhang
Xin Zhang
Yanfu Zhang
Yao Zhang
Yi Zhang
Yongqin Zhang
You Zhang
Youshan Zhang
Yu Zhang
Yubo Zhang
Yue Zhang

Yulun Zhang
Yundong Zhang
Yunyan Zhang
Yuxin Zhang
Zheng Zhang
Zhicheng Zhang
Can Zhao
Changchen Zhao
Fenqiang Zhao
He Zhao
Jianfeng Zhao
Jun Zhao
Li Zhao
Liang Zhao
Lin Zhao
Qingyu Zhao
Shen Zhao
Shijie Zhao
Tianyi Zhao
Wei Zhao
Xiaole Zhao
Xuandong Zhao
Yang Zhao
Yue Zhao
Zixu Zhao
Ziyuan Zhao
Xingjian Zhen
Haiyong Zheng
Hao Zheng
Kang Zheng
Qinghe Zheng
Shenhai Zheng
Yalin Zheng
Yinqiang Zheng
Yushan Zheng
Tao Zhong
Zichun Zhong
Bo Zhou
Haoyin Zhou
Hong-Yu Zhou
Huiyu Zhou
Kang Zhou
Qin Zhou
S. Kevin Zhou
Sihang Zhou

Tao Zhou
Tianfei Zhou
Wei Zhou
Xiao-Hu Zhou
Xiao-Yun Zhou
Yanning Zhou
Yaxuan Zhou
Youjia Zhou
Yukun Zhou
Zhiguo Zhou
Zongwei Zhou
Dongxiao Zhu
Haidong Zhu
Hancan Zhu

Lei Zhu
Qikui Zhu
Xiaofeng Zhu
Xinliang Zhu
Zhonghang Zhu
Zhuotun Zhu
Veronika Zimmer
David Zimmerer
Weiwei Zong
Yukai Zou
Lianrui Zuo
Gerald Zwettler
Reyer Zwiggelaar

Outstanding Area Chairs

Ester Bonmati   University College London, UK
Tolga Tasdizen   University of Utah, USA
Yanwu Xu   Baidu Inc., China

Outstanding Reviewers

Seyed-Ahmad Ahmadi   NVIDIA, Germany
Katharina Breininger Friedrich-Alexander-Universität
                      Erlangen-Nürnberg, Germany
Mariano Cabezas   University of Sydney, Australia
Nicha Dvornek   Yale University, USA
Adrian Galdran   Universitat Pompeu Fabra, Spain
Alexander Katzmann   Siemens Healthineers, Germany
Tony C. W. Mok   Hong Kong University of Science and
                      Technology, China
Sérgio Pereira   Lunit Inc., Korea
David Richmond   Genentech, USA
Dominik Rivoir   National Center for Tumor Diseases (NCT)
                      Dresden, Germany
Fons van der Sommen   Eindhoven University of Technology,
                      the Netherlands
Yushan Zheng   Beihang University, China

Honorable Mentions (Reviewers)

Chloé Audigier   Siemens Healthineers, Switzerland
Qinle Ba   Roche, USA

Meritxell Bach Cuadra   University of Lausanne, Switzerland
Gabriel Bernardino   CREATIS, Université Lyon 1, France
Benjamin Billot   University College London, UK
Tom Brosch   Philips Research Hamburg, Germany
Ruben Cardenes   Ultivue, Germany
Owen Carmichael   Pennington Biomedical Research Center, USA
Li Chen   University of Washington, USA
Xinjian Chen   Soochow University, Taiwan
Philip Chikontwe   Daegu Gyeongbuk Institute of Science and
                               Technology, Korea
Argyrios Christodoulidis   Centre for Research and Technology
                               Hellas/Information Technologies Institute,
                               Greece
Albert Chung   Hong Kong University of Science and
                               Technology, China
Pierre-Henri Conze   IMT Atlantique, France
Jeffrey Craley   Johns Hopkins University, USA
Felix Denzinger   Friedrich-Alexander University
                               Erlangen-Nürnberg, Germany
Adrien Depeursinge   HES-SO Valais-Wallis, Switzerland
Neel Dey   New York University, USA
Guodong Du   Xiamen University, China
Nicolas Duchateau   CREATIS, Université Lyon 1, France
Dmitry V. Dylov   Skolkovo Institute of Science and Technology,
                               Russia
Hooman Esfandiari   University of Zurich, Switzerland
Deng-Ping Fan   ETH Zurich, Switzerland
Chaowei Fang   Xidian University, China
Nils Daniel Forkert   Department of Radiology & Hotchkiss Brain
                               Institute, University of Calgary, Canada
Nils Gessert   Hamburg University of Technology, Germany
Karthik Gopinath   ETS Montreal, Canada
Mara Graziani   IBM Research, Switzerland
Liang Han   Stony Brook University, USA
Nandinee Haq   Hitachi, Canada
Ali Hatamizadeh   NVIDIA Corporation, USA
Samra Irshad   Swinburne University of Technology, Australia
Hayato Itoh   Nagoya University, Japan
Meirui Jiang   The Chinese University of Hong Kong, China
Baoyu Jing   University of Illinois at Urbana-Champaign, USA
Manjunath K N   Manipal Institute of Technology, India
Ali Kafaei Zad Tehrani   Concordia University, Canada
Konstantinos Kamnitsas   Imperial College London, UK

Pulkit Khandelwal	University of Pennsylvania, USA
Andrew King	King's College London, UK
Stefan Klein	Erasmus MC, the Netherlands
Ender Konukoglu	ETH Zurich, Switzerland
Ivica Kopriva	Rudjer Boskovich Institute, Croatia
David Kügler	German Center for Neurodegenerative Diseases, Germany
Manuela Kunz	National Research Council Canada, Canada
Gilbert Lim	National University of Singapore, Singapore
Tiancheng Lin	Shanghai Jiao Tong University, China
Bin Lou	Siemens Healthineers, USA
Hehuan Ma	University of Texas at Arlington, USA
Ilja Manakov	ImFusion, Germany
Felix Meissen	Technische Universität München, Germany
Martin Menten	Imperial College London, UK
Leo Milecki	CentraleSupelec, France
Lia Morra	Politecnico di Torino, Italy
Dominik Neumann	Siemens Healthineers, Germany
Chinedu Nwoye	University of Strasbourg, France
Masahiro Oda	Nagoya University, Japan
Sebastian Otálora	Bern University Hospital, Switzerland
Michal Ozery-Flato	IBM Research, Israel
Egor Panfilov	University of Oulu, Finland
Bartlomiej Papiez	University of Oxford, UK
Nripesh Parajuli	Caption Health, USA
Sanghyun Park	DGIST, Korea
Terry Peters	Robarts Research Institute, Canada
Theodoros Pissas	University College London, UK
Raphael Prevost	ImFusion, Germany
Yulei Qin	Tencent, China
Emma Robinson	King's College London, UK
Robert Rohling	University of British Columbia, Canada
José Rouco	University of A Coruña, Spain
Jerome Schmid	HES-SO University of Applied Sciences and Arts Western Switzerland, Switzerland
Christina Schwarz-Gsaxner	Graz University of Technology, Austria
Liyue Shen	Stanford University, USA
Luyao Shi	IBM Research, USA
Vivek Singh	Siemens Healthineers, USA
Weinan Song	UCLA, USA
Aristeidis Sotiras	Washington University in St. Louis, USA
Danail Stoyanov	University College London, UK

Ruisheng Su	Erasmus MC, the Netherlands
Liyan Sun	Xiamen University, China
Raphael Sznitman	University of Bern, Switzerland
Elias Tappeiner	UMIT - Private University for Health Sciences, Medical Informatics and Technology, Austria
Mickael Tardy	Hera-MI, France
Juan Miguel Valverde	University of Eastern Finland, Finland
Eugene Vorontsov	Polytechnique Montreal, Canada
Bo Wang	CtrsVision, USA
Tongxin Wang	Meta Platforms, Inc., USA
Yan Wang	Sichuan University, China
Yixin Wang	University of Chinese Academy of Sciences, China
Jie Ying Wu	Johns Hopkins University, USA
Lei Xiang	Subtle Medical Inc, USA
Jiaqi Xu	The Chinese University of Hong Kong, China
Zhoubing Xu	Siemens Healthineers, USA
Ke Yan	Alibaba DAMO Academy, China
Baoyao Yang	School of Computers, Guangdong University of Technology, China
Changchun Yang	Delft University of Technology, the Netherlands
Yujiu Yang	Tsinghua University, China
Youngjin Yoo	Siemens Healthineers, USA
Ning Zhang	Bloomberg, USA
Jianfeng Zhao	Western University, Canada
Tao Zhou	Nanjing University of Science and Technology, China
Veronika Zimmer	Technical University Munich, Germany

Mentorship Program (Mentors)

Ulas Bagci	Northwestern University, USA
Kayhan Batmanghelich	University of Pittsburgh, USA
Hrvoje Bogunovic	Medical University of Vienna, Austria
Ninon Burgos	CNRS - Paris Brain Institute, France
Hao Chen	Hong Kong University of Science and Technology, China
Jun Cheng	Institute for Infocomm Research, Singapore
Li Cheng	University of Alberta, Canada
Aasa Feragen	Technical University of Denmark, Denmark
Zhifan Gao	Sun Yat-sen University, China
Stamatia Giannarou	Imperial College London, UK
Sharon Huang	Pennsylvania State University, USA

Anand Joshi	University of Southern California, USA
Bernhard Kainz	Friedrich-Alexander-Universität Erlangen-Nürnberg, Germany and Imperial College London, UK
Baiying Lei	Shenzhen University, China
Karim Lekadir	Universitat de Barcelona, Spain
Xiaoxiao Li	University of British Columbia, Canada
Jianming Liang	Arizona State University, USA
Marius George Linguraru	Children's National Hospital, George Washington University, USA
Anne Martel	University of Toronto, Canada
Antonio Porras	University of Colorado Anschutz Medical Campus, USA
Chen Qin	University of Edinburgh, UK
Julia Schnabel	Helmholtz Munich, TU Munich, Germany and King's College London, UK
Yang Song	University of New South Wales, Australia
Tanveer Syeda-Mahmood	IBM Research - Almaden Labs, USA
Pallavi Tiwari	University of Wisconsin Madison, USA
Mathias Unberath	Johns Hopkins University, USA
Maria Vakalopoulou	CentraleSupelec, France
Harini Veeraraghavan	Memorial Sloan Kettering Cancer Center, USA
Satish Viswanath	Case Western Reserve University, USA
Guang Yang	Imperial College London, UK
Lequan Yu	University of Hong Kong, China
Miaomiao Zhang	University of Virginia, USA
Rongchang Zhao	Central South University, China
Luping Zhou	University of Sydney, Australia
Lilla Zollei	Massachusetts General Hospital, Harvard Medical School, USA
Maria A. Zuluaga	EURECOM, France

Contents – Part III

Computer Aided Diagnosis

Breast Imaging

Multi-view Local Co-occurrence and Global Consistency Learning Improve Mammogram Classification Generalisation

Yuanhong Chen[1]([✉]), Hu Wang[1], Chong Wang[1], Yu Tian[1], Fengbei Liu[1], Yuyuan Liu[1], Michael Elliott[2], Davis J. McCarthy[2], Helen Frazer[2,3], and Gustavo Carneiro[1]

[1] Australian Institute for Machine Learning, The University of Adelaide, Adelaide, Australia
{yuanhong.chen,hu.wang,chong.wang,yu.tian,fengbei.liu,yuyuan.liu, gustavo.carneiro}@adelaide.edu.au
[2] St. Vincent's Institute of Medical Research, Melbourne, Australia
{melliott,dmccarthy}@svi.edu.au
[3] St. Vincent's Hospital Melbourne, Melbourne, Australia
helen.frazer@svha.org.au

Abstract. When analysing screening mammograms, radiologists can naturally process information across two ipsilateral views of each breast, namely the cranio-caudal (CC) and mediolateral-oblique (MLO) views. These multiple related images provide complementary diagnostic information and can improve the radiologist's classification accuracy. Unfortunately, most existing deep learning systems, trained with globally-labelled images, lack the ability to jointly analyse and integrate global and local information from these multiple views. By ignoring the potentially valuable information present in multiple images of a screening episode, one limits the potential accuracy of these systems. Here, we propose a new multi-view global-local analysis method that mimics the radiologist's reading procedure, based on a global consistency learning and local co-occurrence learning of ipsilateral views in mammograms. Extensive experiments show that our model outperforms competing methods, in terms of classification accuracy and generalisation, on a large-scale private dataset and two publicly available datasets, where models are exclusively trained and tested with global labels.

Keywords: Deep learning · Supervised learning · Mammogram classification · Multi-view

---

Supplementary Information The online version contains supplementary material available at https://doi.org/10.1007/978-3-031-16437-8_1.

© The Author(s), under exclusive license to Springer Nature Switzerland AG 2022
L. Wang et al. (Eds.): MICCAI 2022, LNCS 13433, pp. 3–13, 2022.
https://doi.org/10.1007/978-3-031-16437-8_1

1   Introduction

Breast cancer is the most common cancer worldwide and the fifth leading cause of cancer related death [27]. Early detection of breast cancer saves lives and population screening programs demonstrate reductions in mortality [11]. One of the best ways of improving the chances of survival from breast cancer is based on its early detection from screening mammogram exams [20]. A screening mammogram exam contains two ipsilateral views of each breast, namely bilateral craniocaudal (CC) and mediolateral oblique (MLO), where radiologists analyse both views in an integrated manner by searching for global architectural distortions and local masses and calcification. Furthermore, radiologists tend to be fairly adaptable to images from different machines and new domains with varying cancer prevalence. Some automated systems have been developed [1,4,22,24,32] but none have achieved accuracy or generalisability in low cancer prevalent screening populations that is at the same level as radiologists [5]. We argue that for systems to reach the accuracy and generalisability of radiologists, they will need to mimic the manual analysis explained above.

Global-local analysis methods combine the analysis of the whole image and local patches to jointly perform the classification. Shen et al. [23,24] propose a classifier that relies on concatenated features by a global network (using the whole image) and a local network (using image patches). Another global-local approach shows state-of-the-art (SOTA) classification accuracy with a two-stage training pipeline that first trains a patch classifier, which is then fine-tuned to work with a whole image classification [22]. Although these two models [22,23] show remarkable performance, they do not explore cross-view information and depend on relatively inaccurate region proposals, which can lead to sub-optimal solutions. To exploit cross-view mammograms, previous approaches process the two views without trying to find the inter-relationship between lesions from both views [1]. Alternatively, Ma et al. [13] implement a relation network [9] to learn the inter-relationships between the region proposals, which can be inaccurate and introduce noise that may lead to a poor learning process, and the lack of global-local analysis may lead to sub-optimal performance. MommiNet-v2 [32] is a system focused on the detection and classification of masses from mammograms that explores cross-view information from ipsilateral and bilateral mammographic views, but its focus on the detection and classification of masses limits its performance on global classification that also considers other types of lesions, such as architectural distortions and calcification.

The generalisation of mammogram classification methods is a topic that has received little attention, with few papers showing methods that can be trained in one dataset and tested in another one. A recent paper exposes this issue and presents a meta-repository of mammogram classification models, comparing five models on seven datasets [26]. Importantly, these models are not fine-tuned to any of the datasets. Testing results based on area under the receiver operating characteristic curve (AUC) show that the GMIC model [24] is about 20% better than Faster R-CNN [18] on the NYU reader study test set [31], but Faster R-CNN outperforms GMIC on the INBreast dataset [14] by 4%, and GLAM [12] reaches 85.3% on the NYU test set but only gets 61.2% on INBreast and 78.5% on CMMD [15]. Such experiments show the importance of assessing the ability of models to generalise to testing sets from different populations and with images produced by different machines, compared with the training set.

In this paper, we introduce the Multi-View local Co-occurrence and global Consistency Learning (BRAIxMVCCL) model which is trained in an end-to-end manner for a joint global and local analysis of ipsilateral mammographic views. BRAIxMVCCL has a novel global consistency module that is trained to penalise differences in feature representation from the two views, and also to form a global cross-view representation. Moreover, we propose a new local co-occurrence learning method to produce a representation based on the estimation of the relationship between local regions from the two views. The final classifier merges the global and local representations to output a prediction. To summarise, the **key contributions of the proposed BRAIxMVCCL** for multi-view mammogram classification are: 1) a novel global consistency module that produces a consistent cross-view global feature, 2) a new local co-occurrence module that automatically estimates the relationship between local regions from the two views, and 3) a classifier that relies on global and local features from the CC and MLO mammographic views. We evaluate BRAIxMVCCL on a large-scale (around 2 million images) private dataset and on the publicly available INBreast [14] and CMMD [15] datasets without model fine-tuning to test for generalisability. In all problems, we only use the global labels available for the training and testing images. Our method outperforms the SOTA on all datasets in classification accuracy and generalisability.

2 Proposed Method

We assume the availability of a multi-view mammogram training set that contains weakly labelled ipsilateral views (i.e., CC and MLO) of each breast, denoted by $\mathcal{D} = \{\mathbf{x}_i^m, \mathbf{x}_i^a, \mathbf{y}_i\}_{i=1}^{|\mathcal{D}|}$, where $\mathbf{x} \in \mathcal{X} \subset \mathbb{R}^{H \times W}$ represents a mammogram of height H and width W, $\mathbf{x}_i^m$ represents the main view, $\mathbf{x}_i^a$ is the auxiliary view ($m, a \in \{CC, MLO\}$, interchangeably), and $\mathbf{y}_i \in \mathcal{Y} = \{0, 1\}$ denotes the label ($1 =$ cancer and $0 =$ non-cancer). The testing set is similarly defined.

2.1 Multi-view Local Co-occurrence and Global Consistency Learning

The proposed BRAIxMVCCL model, depicted in Fig. 1, consists of a backbone feature extractor denoted as $\mathbf{u}^m = f_\phi^B(\mathbf{x}^m)$ and $\mathbf{u}^a = f_\phi^B(\mathbf{x}^a)$, a global consistency module $\mathbf{z}^G = f_\eta^G(\mathbf{u}^m, \mathbf{u}^a)$ that forms the global representation from the two views, a local co-occurrence module $\mathbf{z}^m, \mathbf{z}^a = f_\gamma^L(\mathbf{u}^m, \mathbf{u}^a)$ that explores the cross-view feature relationships at local regions, and the cancer prediction classifier $\tilde{\mathbf{y}} = p_\theta(\mathbf{y} = 1|\mathbf{x}^m, \mathbf{x}^a) = f_\psi(\mathbf{z}^G, \mathbf{z}^m, \mathbf{z}^a) \in [0, 1]$, where $p_\theta(\mathbf{y} = 0|\mathbf{x}^m, \mathbf{x}^a) = 1 - p_\theta(\mathbf{y} = 1|\mathbf{x}^m, \mathbf{x}^a)$. The BRAIxMVCCL parameter

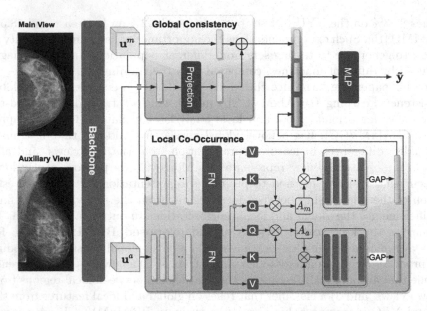

Fig. 1. BRAIxMVCCL takes two mammographic views (main and auxiliary) and uses a backbone model to extract the main and auxiliary features $\mathbf{u}^m$ and $\mathbf{u}^a$, where the main components are: 1) a global consistency module that learns a projection from the auxiliary to the main view, and then combines the main and projected auxiliary features to produce a global representation; 2) a local co-occurrence module that models the local semantic relationships between the two views to produce local representations; and 3) a fusion of the local and global representations to output the prediction $\tilde{\mathbf{y}}$. GAP stands for global average pooling and MLP for multi-layer perception.

$\theta = \{\phi, \eta, \gamma, \psi\} \in \Theta$ represents all module parameters and is estimated with the binary cross entropy loss (BCE) and a global consistency loss, as follows

$$\theta^* = \arg\min_\theta \frac{1}{\mathcal{D}} \sum_{(\mathbf{x}_i^m, \mathbf{x}_i^a, \mathbf{y}_i) \in \mathcal{D}} \ell_{bce}(\mathbf{y}_i, \mathbf{x}_i^m, \mathbf{x}_i^a) + \ell_{sim}(\mathbf{x}_i^m, \mathbf{x}_i^a), \qquad (1)$$

where $\ell_{bce}(\mathbf{y}_i, \mathbf{x}_i^m, \mathbf{x}_i^a) = -\big(\mathbf{y}_i \log p_\theta(\mathbf{y}_i|\mathbf{x}_i^m, \mathbf{x}_i^a) + (1 - \mathbf{y}_i) \log(1 - p_\theta(\mathbf{y}_i|\mathbf{x}_i^m, \mathbf{x}_i^a))$ and $\ell_{sim}(.)$ denotes the global consistency loss defined in (2). Below, we explain how we train each component of BRAIxMVCCL.

When analysing mammograms, radiologists focus on global features (e.g., architectural distortions), local features (e.g., masses and calcification), where this analysis depends on complementary information present in both views [6]. To emulate the multi-view global analysis, our **global consistency module (GCM)** is trained to achieve two objectives: 1) the global representations of both mammographic views should be similar to each other, and 2) the complementarity of both views has to be effectively characterised. Taking the backbone features $\mathbf{u}^m, \mathbf{u}^a \in \mathbb{R}^{\widehat{H} \times \widehat{W} \times D}$, where $\widehat{H} < H$ and $\widehat{W} < W$, we first compute the features $\mathbf{g}^m, \mathbf{g}^a \in \mathbb{R}^D$ with global max pooling. To penalise differences between the two global representations, we define the two mapping functions $\tilde{g}^m = f^{a \to m}(g^a)$ and $\tilde{g}^a = f^{m \to a}(g^m)$ that can transform one view feature into another view feature. We maximise the similarity between features for training the global consistency module, as follows:

$$\ell_{sim}(\mathbf{x}_i^m, \mathbf{x}_i^a) = -0.5(c(\tilde{g}_i^m, g_i^m) + c(\tilde{g}_i^a, g_i^a)), \tag{2}$$

where $c(\mathbf{x}_1, \mathbf{x}_2) = \frac{\mathbf{x}_1 \cdot \mathbf{x}_2}{\max(\|\mathbf{x}_1\|_2 \cdot \|\mathbf{x}_2\|_2, \epsilon)}$ with $\epsilon > 0$ being a factor to avoid division by zero. The output from the global consistency module is formed with a skip connection [8], as in $\mathbf{z}^G = g^m + \tilde{g}^m$ to not only improve model training, but also to characterise the complementary of both views.

The **local co-occurrence module (LCM)** aims to build feature vectors from the two views by analysing long-range interactions between samples from both views. Starting from the backbone features $\mathbf{u}^m$ and $\mathbf{u}^a$, we transform $\mathbf{u}^m$ into the main local feature matrix $\mathbf{U}^m = [\mathbf{u}_1^m, \mathbf{u}_2^m, ..., \mathbf{u}_D^m]$, with $\mathbf{u}_d^m \in \mathbb{R}^{\widehat{H}\widehat{W}}$ and similarly for the auxiliary local feature matrix $\mathbf{U}^a$, where $\mathbf{U}^m, \mathbf{U}^a \in \mathbb{R}^{\widehat{H}\widehat{W} \times D}$. To estimate the local co-occurrence of lesions between the mammographic views, we propose the formulation of a cross-view feature representation with:

$$\widetilde{\mathbf{U}}^m = f^A(\mathbf{U}^a, \mathbf{U}^m, \mathbf{U}^m), \text{ and } \widetilde{\mathbf{U}}^a = f^A(\mathbf{U}^m, \mathbf{U}^a, \mathbf{U}^a), \tag{3}$$

where $\widetilde{\mathbf{U}}^m, \widetilde{\mathbf{U}}^a \in \mathbb{R}^{\widehat{H}\widehat{W} \times D'}$, $f^A(\mathbf{Q}, \mathbf{K}, \mathbf{V}) = f^{mlp}\left(f^{mha}(\mathbf{Q}, \mathbf{K}, \mathbf{V})\right)$, with $f^{mlp}(.)$ representing a multi-layer perceptron and the multi-head attention (MHA) [3,29] is defined as

$$f^{mha}(\mathbf{Q}, \mathbf{K}, \mathbf{V}) = softmax\left(\frac{(\mathbf{Q}\mathbf{W}_q)(\mathbf{K}\mathbf{W}_k^T)}{\sqrt{D}}\right)(\mathbf{V}\mathbf{W}_v)\right), \tag{4}$$

where $\mathbf{W}_q, \mathbf{W}_k, \mathbf{W}_v \in \mathbb{R}^{D \times D'}$ are the linear layers.

The view-wise local features for each view is obtained with $\mathbf{z}^m = GAP(\widetilde{\mathbf{U}}^m)$ (similarly for $\mathbf{z}^a$), where $\mathbf{z}^m, \mathbf{z}^a \in \mathbf{R}^{D'}$, and GAP represents the global average pooling. Finally, the integration of the global and local modules is obtained via a concatenation fusion, where we concatenate $\mathbf{z}^G, \mathbf{z}^m, \mathbf{z}^a$ before applying the last MLP classification layer, denoted by $f_\psi(\mathbf{z}^G, \mathbf{z}^m, \mathbf{z}^a)$, to estimate $p_\theta(\mathbf{y}|\mathbf{x}^m, \mathbf{x}^a)$.

3   Experimental Results

3.1   Datasets

Private Mammogram Dataset. The ADMANI (Annotated Digital Mammograms and Associated Non-Image data) dataset is collected from several breast screening clinics from 2013 to 2017 and is composed of ADMANI-1 and ADMANI-2. ADMANI-1 contains 139,034 exams (taken from 2013 to 2015) with 5,901 cancer cases (containing malignant findings) and 133,133 non-cancer cases (with benign lesions or no findings). This dataset is split 80/10/10 for train/validation/test in a patient-wise manner. ADMANI-2 contains 1,691,654 cases with 5,232 cancer cases and 1,686,422 non-cancer cases and is a real world, low cancer prevalence screening population cohort. Each exam on ADMANI-1,2 has two views (CC and MLO) per breast produced by one of the following manufactures: SIEMENS, HOLOGIC, FUJIFILM Corporation, Philips Digital Mammography Sweden AB, KONICA MINOLTA, GE MEDICAL SYSTEMS, Philips Medical Systems and Agfa.

Public Mammogram Datasets. The Chinese Mammography Database (CMMD) [15] contains 5,202 mammograms from 1,775 patients with normal, benign or malignant biopsy-confirmed tumours[1]. Following [26], we use the entire CMMD dataset for testing, which contains 2,632 mammograms with malignant findings and 2,568 non-malignant mammograms. All images were produced by GE Senographe DS mammography system [15]. The *INBreast* dataset [14] contains 115 exams from Centro Hospitalar de S. João in Portugal. Following [26], we use the official testing set that contains 31 biopsy confirmed exams, with 15 malignant exams and 16 non-malignant exams. Unlike the meta-repository [26], we did not evaluate on DDSM [25] since it is a relatively old dataset (released in 1997) with low image quality (images are digitised from X-ray films), and we could not evaluate on OPTIMAM [7], CSAW-CC [2] and NYU [31] datasets since the first two have restricted access (only partnering academic groups can gain partial access to the data [26]), and the NYU dataset [31] is private.

3.2 Implementation Details

We pre-process each image to remove text annotations and background noise outside the breast region, then we crop and pad the pre-processed images to fit into the target image size 1536×768 to avoid distortion. During data loading process, we resize the input image pairs to 1536×768 pixels and flip all images so that the breast points toward the right hand size of the image. To improve training generalisation, we use data augmentation based on random vertical flipping and random affine transformation. BRAIxMVCCL uses EfficientNet-b0 [28], initialized with ImageNet [19] pre-trained weights, as the backbone feature extractor. Our training relies on Adam optimiser [10] using a learning rate of 0.0001, weight decay of 10^{-6}, batch size of 8 images and 20 epochs. We use ReduceLROnPlateau to dynamically control the learning rate reduction based on the BCE loss during model validation, where the reduction factor is set to 0.1. Hyper-parameters are estimated using the validation set. All experiments are implemented with Pytorch [16] and conducted on an NVIDIA RTX 3090 GPU (24 GB). The training takes 23.5 h on ADMANI-1. The testing takes 42.17 ms per image. For the fine-tuning of GMIC [24] on ADMANI-1,2 datasets, we load the model checkpoint (trained on around 0.8M images from NYU training set) from the official github repository[2] and follow hyper-parameter setting from the author's github[3].

---

[1] Similarly to the meta-repository in [26], we remove the study #D1-0951 as the pre-processing failed in this examination.

[2] https://github.com/nyukat/GMIC.

[3] https://github.com/nyukat/GMIC/issues/4.

For fine-tuning DMV-CNN, we load the model checkpoint from the official repository[4] and fine-tune the model with using the Adam optimiser [10] with learning rate of 10^{-6}, a minibatch of size 8 and weight decay of 10^{-6}. The classification results are assessed with the area under the receiver operating characteristic curve (AUC-ROC) and area under the precision-recall curve (AUC-PR).

3.3 Results

Table 1 shows the testing AUC-ROC results on ADMANI-1,2. We calculate the **image-level** bootstrap AUC with 2,000 bootstrap replicates, from which we present the mean and the lower and upper bounds of the 95% confidence interval (CI). Our model shows significantly better results than the SOTA model GMIC [24] with a 5.89% mean improvement. We evaluate the generalisation of the models trained on ADMANI-1 on the large-scale dataset ADMANI-2. It is worth mentioning that the ADMANI-2 dataset contains images produced by machines not used in training (i.e., GE MEDICAL SYSTEMS, Philips Medical Systems and Agfa) and by a different image processing algorithm for the SIEMENS machine. On ADMANI-2, our model achieves a mean AUC-ROC of 92.60%, which is 9.52% better than the baseline EfficientNet-b0 [28], 15.47% better than DMV-CNN and 8.21% comparing with GMIC. According to the CI values, our results are significantly better than all other approaches.

Table 3 shows the AUC-ROC and AUC-PR results of our model trained on ADMANI-1 on two publicly available datasets, namely INBreast [14] and CMMD [15], following the evaluation protocol in [26]. We report 95% bootstrap confidence intervals with 2,000 bootstrap replicates based on **breast-level** evaluation[5]. Our BRAIxMVCCL (single) achieves better results than other approaches, surpassing the mean AUC-ROC result by GMIC (single) by 6.8% on INBreast and 2.7% on CMMD, and also outperforming GMIC (top-5 ensemble) by 1.4% on INBreast and 2.1% on CMMD, even though our BRAIxMVCCL is not an ensemble method. We show a qualitative visual comparison between GMIC [24] and our BRAIxMVCCL in Fig. 2 where results demonstrates that our model can correctly classify challenging cancer and non-cancer cases missed by GMIC.

3.4 Ablation Study

To evaluate the effectiveness of each component of our model, we show an ablation study based on testing AUC-ROC results in Table 2. The study starts with the backbone model EfficientNet-b0 [28], which achieves 88.61% (shown in first row of Table 1). Taking the representations extracted by the backbone model obtained from the two views and fusing them by concatenating the representations, and passing them through the classification layer (column 'Fusion' in

---

[4] https://github.com/nyukat/breast_cancer_classifier.
[5] The breast-level result is obtained by averaging the predictions from both views.

Table 1. Image-level mean and 95% CI AUC-ROC results for ADMANI datasets.

Methods	ADMANI-1 (test set)	ADMANI-2 (whole set)
EfficientNet-b0 [28]	0.886 (0.873–0.890)	0.838 (0.828–0.847)
DMV-CNN [22]	0.799 (0.780–0.817)	0.771 (0.766–0.777
GMIC [24]	0.889 (0.870–0.903)	0.844 (0.835–0.853)
BRAIxMVCCL	**0.948 (0.937–0.953)**	**0.926 (0.922–0.930)**

Table 2. Ablation study of key components of our BRAIxMVCCL model.

Fusion	SA	LCM	GCM	Testing AUC
✓				0.9088
✓	✓			0.9183
✓		✓		0.9368
✓			✓	0.9241
✓	✓		✓	0.9353
✓		✓	✓	**0.9478**

Table 3. Testing results show breast-level estimates of AUC-ROC and AUC-RP on INBreast [14] and CCMD [15]. Results are reported with 95% CI calculated from bootstrapping with 2,000 replicates. Results from other methods are obtained from [26].

Dataset	AUC	End2end [21] (DDSM)	End2end [21] (INbreast)	Faster [17] R-CNN	DMV-CNN [22]	GMIC [24] (single)	GMIC [24] (top-5 ensemble)	GLAM [12]	BRAIxMVCCL (single)
INBreast	ROC	0.676 (0.469-0.853)	0.977 (0.931-1.000)	0.969 (0.917-1.000)	0.802 (0.648-0.934)	0.926 (0.806-1.000)	**0.980 (0.940-1.000)**	0.612 (0.425-0.789)	0.994 (0.985-1.000)
	PR	0.605 (0.339-0.806)	0.955 (0.853-1.000)	0.936 (0.814-1.000)	0.739 (0.506-0.906)	0.899 (0.726-1.000)	**0.957 (0.856-1.000)**	0.531 (0.278-0.738)	0.986 (0.966-1.000)
CMMD†	ROC	0.534 (0.512-0.557)	0.449 (0.428-0.473)	0.806 (0.789-0.823)	0.740 (0.720-0.759)	0.825 (0.809-0.841)	**0.831 (0.815-0.846)**	0.785 (0.767-0.803)	0.852 (0.840-0.863)
	PR	0.517 (0.491-0.544)	0.462 (0.438-0.488)	0.844 (0.828-0.859)	0.785 (0.764-0.806)	0.854 (0.836-0.869)	**0.859 (0.842-0.875)**	0.818 (0.798-0.837)	0.876 (0.864-0.887)

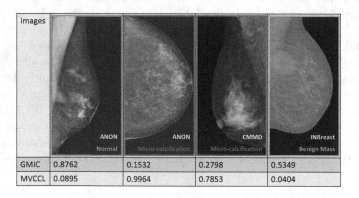

Fig. 2. Qualitative visual comparison of the classification of challenging cancer (red border) and non-cancer (green border) cases produced by GMIC and our BRAIxMVCCL on ADMANI, CMMD and INBreast datasets. The numbers below each image indicate the cancer probability prediction by each model. (Color figure online)

Table 2) increases the performance to 90.88% (first row in Table 2). To demonstrate the effectiveness of our local co-occurrence module (LCM), we compare it against a self-attention (SA) module [29,30], which integrates cross-view features and performs the self-attention mechanism. We show that our proposed LCM is 1.97% better than SA. The model achieves 92.41% with our global consistency module (GCM) and can be further improved to 93.53% by using a self-attention (SA) module [29, 30] to combine the cross-view global features. Finally, replacing SA by LCM further boosts the result by 1.25%. Our proposed model achieves a testing AUC of 94.78% which improves the baseline model by 6.17%.

4   Conclusion

We proposed a novel multi-view system for the task of high-resolution mammogram classification, which analyses global and local information of unregistered ipsilateral views. We developed a global consistency module to model the consistency between global feature representations and a local co-occurrence module to model the region-level correspondence in a weakly supervised scenario. We performed extensive experiments on large-scale clinical datasets and also evaluated the model on cross-domain datasets. We achieved SOTA classification performance and generalisability across datasets. In future work, we plan to study partial multi-view learning to resolve the issue of dependence on multi-view data.

Acknowledgement. This work is supported by funding from the Australian Government under the Medical Research Future Fund - Grant MRFAI000090 for the Transforming Breast Cancer Screening with Artificial Intelligence (BRAIx) Project. We thank the St Vincent's Institute of Medical Research for providing the GPUs to support the numerical calculations in this paper.

References

1. Carneiro, G., Nascimento, J., Bradley, A.P.: Automated analysis of unregistered multi-view mammograms with deep learning. IEEE Trans. Med. Imaging **36**(11), 2355–2365 (2017)
2. Dembrower, K., Lindholm, P., Strand, F.: A multi-million mammography image dataset and population-based screening cohort for the training and evaluation of deep neural networks-the cohort of screen-aged women (CSAW). J. Digit. Imaging **33**(2), 408–413 (2020)
3. Dosovitskiy, A., et al.: An image is worth 16 × 16 words: transformers for image recognition at scale. arXiv preprint arXiv:2010.11929 (2020)
4. Frazer, H.M., Qin, A.K., Pan, H., Brotchie, P.: Evaluation of deep learning-based artificial intelligence techniques for breast cancer detection on mammograms: results from a retrospective study using a BreastScreen Victoria dataset. J. Med. Imaging Radiat. Oncol. **65**(5), 529–537 (2021)
5. Freeman, K., et al.: Use of artificial intelligence for image analysis in breast cancer screening programmes: systematic review of test accuracy. bmj **374** (2021)
6. Hackshaw, A., Wald, N., Michell, M., Field, S., Wilson, A.: An investigation into why two-view mammography is better than one-view in breast cancer screening. Clin. Radiol. **55**(6), 454–458 (2000)
7. Halling-Brown, M.D., et al.: Optimam mammography image database: a large-scale resource of mammography images and clinical data. Radiol.: Artif. Intell. **3**(1), e200103 (2020)

8. He, K., Zhang, X., Ren, S., Sun, J.: Deep residual learning for image recognition. In: Proceedings of the IEEE Conference on Computer Vision and Pattern Recognition, pp. 770–778 (2016)

9. Hu, H., Gu, J., Zhang, Z., Dai, J., Wei, Y.: Relation networks for object detection. In: Proceedings of the IEEE Conference on Computer Vision and Pattern Recognition, pp. 3588–3597 (2018)

10. Kingma, D.P., Ba, J.: Adam: a method for stochastic optimization. arXiv preprint arXiv:1412.6980 (2014)

11. Lauby-Secretan, B., et al.: Breast-cancer screening-viewpoint of the IARC working group. N. Engl. J. Med. **372**(24), 2353–2358 (2015)

12. Liu, K., Shen, Y., Wu, N., Chlkedowski, J., Fernandez-Granda, C., Geras, K.J.: Weakly-supervised high-resolution segmentation of mammography images for breast cancer diagnosis. arXiv preprint arXiv:2106.07049 (2021)

13. Ma, J., Li, X., Li, H., Wang, R., Menze, B., Zheng, W.S.: Cross-view relation networks for mammogram mass detection. In: 2020 25th International Conference on Pattern Recognition (ICPR), pp. 8632–8638. IEEE (2021)

14. Moreira, I.C., Amaral, I., Domingues, I., Cardoso, A., Cardoso, M.J., Cardoso, J.S.: INbreast: toward a full-field digital mammographic database. Acad. Radiol. **19**(2), 236–248 (2012)

15. Nolan, T.: The Chinese mammography database (CMMD) (2021). https://wiki.cancerimagingarchive.net/pages/viewpage.action?pageId=70230508. Accessed 21 Aug 2021

16. Paszke, A., et al.: Pytorch: an imperative style, high-performance deep learning library. In: Advances in Neural Information Processing Systems, vol. 32, pp. 8026–8037 (2019)

17. Ren, S., He, K., Girshick, R., Sun, J.: Faster R-CNN: towards real-time object detection with region proposal networks. In: Advances in Neural Information Processing Systems, vol. 28 (2015)

18. Ribli, D., Horváth, A., Unger, Z., Pollner, P., Csabai, I.: Detecting and classifying lesions in mammograms with deep learning. Sci. Rep. **8**(1), 1–7 (2018)

19. Russakovsky, O., et al.: ImageNet large scale visual recognition challenge. Int. J. Comput. Vis. **115**(3), 211–252 (2015)

20. Selvi, R.: Breast Diseases: Imaging and Clinical Management. Springer, Cham (2014)

21. Shen, L.: End-to-end training for whole image breast cancer diagnosis using an all convolutional design. arXiv preprint arXiv:1711.05775 (2017)

22. Shen, L., Margolies, L.R., Rothstein, J.H., Fluder, E., McBride, R., Sieh, W.: Deep learning to improve breast cancer detection on screening mammography. Sci. Rep. **9**(1), 1–12 (2019)

23. Shen, Y., et al.: Globally-aware multiple instance classifier for breast cancer screening. In: Suk, H.-I., Liu, M., Yan, P., Lian, C. (eds.) MLMI 2019. LNCS, vol. 11861, pp. 18–26. Springer, Cham (2019). https://doi.org/10.1007/978-3-030-32692-0_3

24. Shen, Y., et al.: An interpretable classifier for high-resolution breast cancer screening images utilizing weakly supervised localization. Med. Image Anal. **68**, 101908 (2021)

25. Smith, K.: CBIS-DDSM (2021). https://wiki.cancerimagingarchive.net/display/Public/CBIS-DDSM. Accessed 21 Aug 2021

26. Stadnick, B., et al.: Meta-repository of screening mammography classifiers. arxiv:2108.04800 (2021)

27. Sung, H., et al.: Global cancer statistics 2020: GLOBOCAN estimates of incidence and mortality worldwide for 36 cancers in 185 countries. CA: Cancer J. Clin. **71**(3), 209–249 (2021)
28. Tan, M., Le, Q.: EfficientNet: rethinking model scaling for convolutional neural networks. In: International Conference on Machine Learning, pp. 6105–6114. PMLR (2019)
29. Vaswani, A., et al.: Attention is all you need. In: Advances in Neural Information Processing Systems, pp. 5998–6008 (2017)
30. Wang, X., Girshick, R., Gupta, A., He, K.: Non-local neural networks. In: Proceedings of the IEEE Conference on Computer Vision and Pattern Recognition, pp. 7794–7803 (2018)
31. Wu, N., et al.: The NYU breast cancer screening dataset v1.0. Technical report, New York University (2019). https://cs.nyu.edu/~kgeras/reports/datav1.0.pdf
32. Yang, Z., et al.: MommiNet-v2: mammographic multi-view mass identification networks. Med. Image Anal. 102204 (2021)

Knowledge Distillation to Ensemble Global and Interpretable Prototype-Based Mammogram Classification Models

Chong Wang[1]([✉]), Yuanhong Chen[1], Yuyuan Liu[1], Yu Tian[1], Fengbei Liu[1], Davis J. McCarthy[2], Michael Elliott[2], Helen Frazer[3], and Gustavo Carneiro[1]

[1] Australian Institute for Machine Learning, The University of Adelaide,
Adelaide, Australia
chong.wang@adelaide.edu.au
[2] St. Vincent's Institute of Medical Research, Melbourne, Australia
[3] St. Vincent's Hospital Melbourne, Melbourne, Australia

Abstract. State-of-the-art (SOTA) deep learning mammogram classifiers, trained with weakly-labelled images, often rely on global models that produce predictions with limited interpretability, which is a key barrier to their successful translation into clinical practice. On the other hand, prototype-based models improve interpretability by associating predictions with training image prototypes, but they are less accurate than global models and their prototypes tend to have poor diversity. We address these two issues with the proposal of BRAIxProtoPNet++, which adds interpretability to a global model by ensembling it with a prototype-based model. BRAIxProtoPNet++ distills the knowledge of the global model when training the prototype-based model with the goal of increasing the classification accuracy of the ensemble. Moreover, we propose an approach to increase prototype diversity by guaranteeing that all prototypes are associated with different training images. Experiments on weakly-labelled private and public datasets show that BRAIxProtoPNet++ has higher classification accuracy than SOTA global and prototype-based models. Using lesion localisation to assess model interpretability, we show BRAIxProtoPNet++ is more effective than other prototype-based models and post-hoc explanation of global models. Finally, we show that the diversity of the prototypes learned by BRAIxProtoPNet++ is superior to SOTA prototype-based approaches.

Keywords: Interpretability · Explainability · Prototype-based model · Mammogram classification · Breast cancer diagnosis · Deep learning

Supplementary Information The online version contains supplementary material available at https://doi.org/10.1007/978-3-031-16437-8_2.

© The Author(s), under exclusive license to Springer Nature Switzerland AG 2022
L. Wang et al. (Eds.): MICCAI 2022, LNCS 13433, pp. 14–24, 2022.
https://doi.org/10.1007/978-3-031-16437-8_2

1   Introduction

Deep learning models [13,14] have shown promising performance in many medical imaging applications (e.g., mammography [18], radiology [7], diagnostics [12], ophthalmology [4,5]), even achieving accuracy as high as human radiologists [21]. Regardless of the encouraging performance of these models trained with weakly-labelled images, the limited interpretability of their predictions remains a barrier to their successful translation into clinical practice [16]. Recently, some studies use post-hoc explanations (e.g., Grad-CAM [17]) to highlight image regions associated with model predictions. However, such highlighted classification-relevant image regions are often not reliable and insufficient for interpretability [16].

There is a growing interest in the development of effective interpretable methods for medical image classifiers trained with weakly-labelled images. Khakzar et al. [10] train a chest X-ray classifier with perturbed adversarial samples to form more reliable class activation maps (CAM) [17]. Ilanchezian et al. [9] introduce BagNets for interpretable gender classification in retinal fundus images, which can reveal how local image evidence is integrated into global image decisions. Chen et al. [3] present ProtoPNet that learns a set of class-specific prototypes, and a test image is classified by evaluating its similarity to these prototypes, which provides a unique understanding of the inner workings of the model. Furthermore, XProtoNet [11] learns prototypes with disease occurrence maps for interpretable chest X-ray classification. To make the model focus on clinically interpretable features, additional signals (e.g., nuclei and fat droplets) are provided to supervise the attention map of a global biopsy image classifier [23]. In general, the above approaches are either based on poorly interpretable post-hoc explanations from highly accurate global classifiers or achieve good interpretability from less accurate local (e.g., prototype-based) classifiers.

In this paper, we present BRAIxProtoPNet++, a novel, accurate, and interpretable mammogram classification model. BRAIxProtoPNet++ ensembles a highly accurate global classification model with an interpretable ProtoPNet model, with the goal of achieving better classification accuracy than both global and ProtoPNet models and satisfactory interpretability. This goal is achieved by distilling the knowledge from the global model when training the ProtoPNet. Furthermore, BRAIxProtoPNet++ increases the prototype diversity from ProtoPNet, with a new prototype selection strategy that guarantees that the prototypes are associated with a diverse set of training images. To summarise, **our contributions are:** 1) a new approach to add interpretability to accurate global mammogram classifiers; 2) a new global and prototype-based ensemble model, named BRAIxProtoPNet++, trained with knowledge distillation to enable effective interpretability and higher classification accuracy than both the global and prototype-based models; and 3) improved prototype diversity of BRAIxProtoPNet++ compared to previous prototype-based models [3]. Experimental results on weakly-supervised private and public datasets [2] show that BRAIxProtoPNet++ improves classification accuracy and exhibits promising interpretability results compared to existing non-interpretable global classifiers and recently proposed interpretable models.

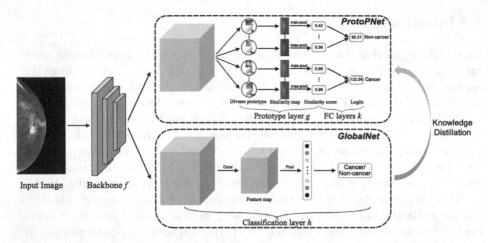

Fig. 1. The architecture of our proposed BRAIxProtoPNet++, consisting of a shared CNN backbone, a global classifier GlobalNet, and an interpretable ProtoPNet that distils the knowledge from GlobalNet, to form an accurate ensemble classifier, and maximises the prototype diversity during training.

2   Proposed Method

In this section, we introduce our proposed method that relies on the weakly-labelled dataset $\mathcal{D} = \{(\mathbf{x}, \mathbf{y})_i\}_{i=1}^{|\mathcal{D}|}$, where $\mathbf{x} \in \mathbb{R}^{H \times W}$ represents the mammogram of size $H \times W$, and $\mathbf{y} \in \{0, 1\}^2$ denotes a one-hot representation of the image label (e.g., cancer versus non-cancer).

2.1   BRAIxProtoPNet++

Our proposed BRAIxProtoPNet++ with knowledge distillation and diverse prototypes, depicted in Fig. 1, takes an accurate global image classifier (GlobalNet), trained with the weakly-labelled training set $\mathcal{D}$, and integrates it with the interpretable ProtoPNet model. ProtoPNet achieves classification by comparing local image parts with learned prototypes, which tends not to be as accurate as the holistic classification by GlobalNet. This happens because even if local abnormalities are crucial for identifying breast cancer, information from the whole mammogram (e.g., lesions spreading in different spatial locations, or contrast between healthy and abnormal regions) may help to reach accurate classification. Hence, to improve ProtoPNet's accuracy, we propose to distill the knowledge of GlobalNet to train ProtoPNet, using our new knowledge distillation (KD) loss function. Another limitation of ProtoPNet is the poor diversity of the learned prototypes that can negatively impact classification accuracy and model interpretability. We address this issue with a new prototype selection approach to increase the prototype diversity of the original ProtoPNet.

Model. BRAIxProtoPNet++ comprises a CNN backbone (e.g., DenseNet [8] or EfficientNet [20]) represented by $\mathbf{X} = f_{\theta_f}(\mathbf{x})$, where $\mathbf{X} \in \mathbb{R}^{\frac{H}{32} \times \frac{W}{32} \times D}$, and θ_f denotes the backbone parameters. The classification layer of the GlobalNet is denoted by $\tilde{\mathbf{y}}^G = h_{\theta_h}(\mathbf{X})$, where $\tilde{\mathbf{y}}^G \in [0,1]^2$ is the model prediction. The ProtoPNet is denoted by $\tilde{\mathbf{y}}^L = k_{\theta_k}(g_{\theta_g}(\mathbf{X}))$, where $\tilde{\mathbf{y}}^L \in [0,1]^2$ is the model prediction, $g_{\theta_g}(\cdot)$ represents the prototype layer, and $k_{\theta_k}(\cdot)$ the fully connected (FC) layers. The prototype layer has M learnable class-representative prototypes $\mathcal{P} = \{\mathbf{p}_m\}_{m=1}^M$, with $M/2$ prototypes for each class and $\mathbf{p}_m \in \mathbb{R}^D$, which are used to form similarity maps $\mathbf{S}_m(h,w) = e^{\frac{-||\mathbf{X}(h,w)-\mathbf{p}_m||_2^2}{T}}$, where $h \in \{1, ..., \frac{H}{32}\}$ and $w \in \{1, ..., \frac{W}{32}\}$ denote spatial indexes in similarity maps, and T is a temperature factor. The prototype layer $g_{\theta_g}(\cdot)$ outputs M similarity scores from max-pooling $\left\{ \max_{h,w} \mathbf{S}_m(h,w) \right\}_{m=1}^M$, which are fed to $k_{\theta_k}(\cdot)$ to obtain classification result.

Training. BRAIxProtoPNet++ is trained by minimising the following objective:

$$\ell_{PPN++}(\mathcal{D}, \theta_f, \theta_g, \theta_k, \theta_h, \mathcal{P}) = \ell_{PPN}(\mathcal{D}, \theta_f, \theta_g, \theta_k, \mathcal{P}) + \\ \alpha \ell_{CEG}(\mathcal{D}, \theta_f, \theta_h) + \beta \ell_{KD}(\mathcal{D}, \theta_f, \theta_g, \theta_k, \mathcal{P}), \quad (1)$$

where α and β are weighting hyper-parameters, $\ell_{PPN}(\mathcal{D}, \theta_f, \theta_g, \theta_k, \theta_h)$ is the ProtoPNet loss defined in (3), $\ell_{CEG}(\mathcal{D}, \theta_f, \theta_h)$ represents the cross-entropy loss to train θ_f and θ_h for GlobalNet using label $\mathbf{y}$ and prediction $\tilde{\mathbf{y}}^G$, and

$$\ell_{KD}(\mathcal{D}, \theta_f, \theta_g, \theta_k, \mathcal{P}) = \frac{1}{|\mathcal{D}|} \sum_{i=1}^{|\mathcal{D}|} \max(0, (\mathbf{y}_i)^\top (\tilde{\mathbf{y}}_i^G) - (\mathbf{y}_i)^\top (\tilde{\mathbf{y}}_i^L) + \omega) \quad (2)$$

is our proposed knowledge distillation (KD) loss, with $(\mathbf{y}_i)^\top (\tilde{\mathbf{y}}_i^G)$ and $(\mathbf{y}_i)^\top (\tilde{\mathbf{y}}_i^L)$ denoting the predicted probability of the labelled class from the GlobalNet and ProtoPNet, and $\omega > 0$ representing a pre-defined margin to control ProtoPNet's confidence gain. Our novel KD loss in (2) is designed to distill the knowledge [1] from GlobalNet to ProtoPNet to increase the classification accuracy of ProtoPNet and enable a better ensemble classification using both models.

The ProtoPNet loss introduced in (1) is defined by:

$$\ell_{PPN}(\mathcal{D}, \theta_f, \theta_g, \theta_k, \mathcal{P}) = \ell_{CEL}(\mathcal{D}, \theta_f, \theta_g, \theta_k, \mathcal{P}) + \\ \lambda_1 \ell_{CT}(\mathcal{D}, \theta_f, \theta_g, \mathcal{P}) + \lambda_2 \max(0, \gamma - \ell_{SP}(\mathcal{D}, \theta_f, \theta_g, \mathcal{P})), \quad (3)$$

where λ_1, λ_2, and γ denote hyper-parameters, $\ell_{CEL}(\mathcal{D}, \theta_f, \theta_g, \theta_k, \mathcal{P})$ is the cross-entropy loss between the label $\mathbf{y}$ and the ProtoPNet output $\tilde{\mathbf{y}}^L$, and

$$\ell_{CT}(\mathcal{D}, \theta_f, \theta_g, \mathcal{P}) = \frac{1}{|\mathcal{D}|} \sum_{i=1}^{|\mathcal{D}|} \min_{\mathbf{p}_m \in \mathcal{P}_{\mathbf{y}_i}} \min_{\mathbf{z} \in \mathbf{X}_i} ||\mathbf{z} - \mathbf{p}_m||_2^2, \quad (4)$$

$$\ell_{SP}(\mathcal{D}, \theta_f, \theta_g, \mathcal{P}) = \frac{1}{|\mathcal{D}|} \sum_{i=1}^{|\mathcal{D}|} \min_{\mathbf{p}_m \notin \mathcal{P}_{\mathbf{y}_i}} \min_{\mathbf{z} \in \mathbf{X}_i} ||\mathbf{z} - \mathbf{p}_m||_2^2, \quad (5)$$

where we abuse the notation to represent $\mathbf{z} \in \mathbb{R}^D$ as one of the $\frac{H}{32} \times \frac{W}{32}$ feature vectors of $\mathbf{X}_i = f_{\theta_f}(\mathbf{x}_i)$, and $\mathcal{P}_{\mathbf{y}_i} \subset \mathcal{P}$ is the set of prototypes with class label $\mathbf{y}_i$. For each training image, the cluster loss in (4) encourages the input image to have at least one local feature close to one of the prototypes of its own class, while the separation loss in (5) ensures all local features to be far from the prototypes that are not from the image's class.

Note in (3) that compared with the original ProtoPNet [3], we introduce the hinge loss [22] on $\ell_{SP}(\cdot)$ to impose a margin that mitigates the risk of overfitting. After each training epoch, we update each prototype $\mathbf{p}_m$ to be represented by the nearest latent feature vector $\mathbf{z}$ from all training images of the same class. Specifically, we replace $\mathbf{p}_m$ with the nearest feature vector $\mathbf{z}$, as in:

$$\mathbf{p}_m \leftarrow \arg \min_{\mathbf{z} \in \mathbf{X}_{i \in \{1, \dots, |\mathcal{D}|\}}} ||\mathbf{z} - \mathbf{p}_m||_2^2. \tag{6}$$

One practical limitation in [3] is that there is no guarantee of diversity among prototypes. Here, we enforce prototype diversity in $\mathcal{P}$ by ensuring that we never have the same training image used for updating more than one prototype in (6). This is achieved with the following 2-step algorithm: 1) for each prototype $\mathbf{p}_m$, compute the distances to all training images of the same class, and sort the distances in ascending order; and 2) select prototypes, where for the first prototype $\mathbf{p}_1$, we choose its nearest image and record the image index indicating that the image has been used, then for the second prototype $\mathbf{p}_2$, we do the same operation, but if the selected image has been used by previous prototypes (e.g., $\mathbf{p}_1$), we will skip this image and use one of the next nearest images to $\mathbf{p}_2$. The prototype selection stage is performed sequentially until all prototypes are updated by different training images. We show in the experiments that this greedy prototype selection strategy improves the prototype diversity of ProtoPNet.

Testing. The final prediction of BRAIxProtoPNet++ is obtained by averaging the GlobalNet and ProtoPNet predictions $\tilde{\mathbf{y}}^G$ and $\tilde{\mathbf{y}}^L$, and the interpretability is reached by showing the prototypes $\mathbf{p}_m \in \mathcal{P}$ that produced the largest max-pooling score, together with the corresponding similarity map $\mathbf{S}_m$.

3 Experimental Results

3.1 Dataset

The experiments are performed on a private large-scale breast screening mammogram ADMANI (Annotated Digital Mammograms and Associated Non-Image data) dataset. It contains high-resolution (size of 5416×4040) 4-view mammograms (L-CC, L-MLO, R-CC, and R-MLO) with diagnosis outcome per view (i.e., cancer and no cancer findings). The dataset has 20592 (3262 cancer, 17330 non-cancer) training images and 22525 (806 cancer, 21719 non-cancer) test images, where there is no overlap of patient data between training and test sets. In the test set, 410 cancer images have lesion annotations labelled by experienced radiologists for evaluating cancer localisation. We also use the public

Chinese Mammography Database (CMMD) [2] to validate the generalisation performance of BRAIxProtoPNet++. CMMD consists of 5200 (2632 cancer, 2568 non-cancer) 4-view test mammograms with size 2294×1914 pixels.

3.2   Experimental Setup

The BRAIxProtoPNet++ is implemented in Pytorch [15]. The model is trained using Adam optimiser with an initial learning rate of 0.001, weight decay of 0.00001, and batch size of 16. The hyper-parameters in (1)–(3) are set using simple general rules (e.g., small values for $\omega, \lambda_1, \lambda_2, \alpha, \beta$ should be close to 1, and $\gamma \gg 1$), but model results are relatively robust to a large range of their values (for the experiments below, we have: $\alpha = 1, \beta = 0.5, \omega = 0.2, \lambda_1 = 0.1, \lambda_2 = 0.1, \gamma = 10$). For the two datasets, images are pre-processed using the Otsu threshold algorithm to crop the breast region, which is subsequently resized to $H = 1536, W = 768$. The feature size $D = 128$ in (5), and the temperature parameter $T = 128$. The number of prototypes $M = 400$ (200 for cancer class and 200 for non-cancer class). We use EfficientNet-B0 [20] and DenseNet-121 [8] as SOTA backbones. The training of BRAIxProtoPNet++ is divided into three stages: 1) training of backbone and GlobalNet, 2) training of ProtoPNet with a frozen backbone and GlobalNet, and 3) fine-tuning of the whole framework. Data augmentation techniques (e.g., translation, rotation, and scaling) are used to improve generalisation. All experiments are conducted on a machine with AMD Ryzen 9 3900X CPU, 2 GeForce RTX 3090 GPUs, and 32 GB RAM. The training of BRAIxProtoPNet++ takes about 28 h, and the average testing time is about 0.0013 s per image.

Classification accuracy is assessed with the area under the receiver operating characteristic curve (AUC). To evaluate model interpretability, we measure the area under the precision recall curve (PR-AUC) for the cancer localisation on test samples. To evaluate prototype diversity, we calculate the mean pairwise cosine distance and $L2$ distance between learned prototypes from the same class.

3.3   Results

We compare our proposed method with the following models: EfficientNet-B0 [20], DenseNet-121 [8], Sparse MIL [24], GMIC [18], and ProtoPNet [3]. For all these models, we use the publicly available codes provided by the papers. EfficientNet-B0 and DenseNet-121 are non-interpretable models. Sparse MIL can localise lesions by dividing a mammogram into regions that are classified using multiple-instance learning with a sparsity constraint. For a fair comparison, we use EfficientNet-B0 as the backbone for Sparse MIL. GMIC uses a global module to select the most informative regions of an input mammogram, then it relies on a local module to analyse those selected regions, it finally employs a fusion module to aggregate the global and local outputs for classification. All methods above, and our BRAIxProtoPNet++, are trained on the training set from ADMANI, and tested on the ADMANI testing set and the whole CMMD dataset.

Table 1. AUC results on ADMANI and CMMD datasets. The best result is in bold.

Methods			Test AUC	
			ADMANI	CMMD
DenseNet-121 [8]			88.54	82.38
EfficientNet-B0 [20]			89.62	76.41
Sparse MIL [24]			89.75	81.33
GMIC [18]			89.98	81.03
ProtoPNet (DenseNet-121) [3]			87.12	80.23
ProtoPNet (EfficientNet-B0) [3]			88.30	79.61
Ours (DenseNet-121)	w/o KD	ProtoPNet	87.32	80.09
		GlobalNet	88.45	82.42
		Ensemble	88.87	82.50
	w/KD	ProtoPNet	88.35	80.67
		GlobalNet	88.61	82.52
		Ensemble	89.54	**82.65**
Ours (EfficientNet-B0)	w/o KD	ProtoPNet	88.63	79.01
		GlobalNet	90.11	76.50
		Ensemble	90.18	80.45
	w/KD	ProtoPNet	89.55	79.86
		GlobalNet	90.12	76.47
		Ensemble	**90.68**	81.65

Table 1 shows the test AUC results of all methods on ADMANI and CMMD datasets. For our BRAIxProtoPNet++, we present the classification results of the ProtoPNet and GlobalNet branches independently, and their ensemble result to show the importance of combining the classification results of both branches. We also show results with (w/KD) and without (w/o KD) distilling the knowledge from GlobalNet to train the ProtoPNet branch. Our best result is achieved with the ensemble model trained with KD, which reaches SOTA results on ADMANI and CMMD datasets. Note that original ProtoPNet's AUC [3] is worse than the non-interpretable global classifiers EfficientNet-B0 and DenseNet-121. However, the application of KD to the original ProtoPNet provides substantial AUC improvement, showing the importance of KD. It is observed that using DenseNet-121 as backbone exhibits better generalisation results on CMMD than using EfficientNet-B0, which means that DenseNet-121 is more robust against domain shift [6]. For the GMIC model, we note a discrepancy on the CMMD result on Table 1 (AUC = 81.03) and the published result in [19] (AUC = 82.50). This is explained by the different training set and input image setup used by GMIC in [19], so to enable a fair comparison, we present the result by GMIC with the same experimental conditions as all other methods in the Table.

Figure 2 (a) displays the learned non-cancer and cancer prototypes and their source training images. We can see that the cancer prototypes come from regions

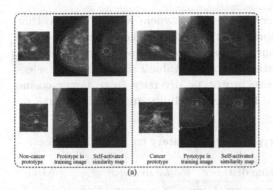
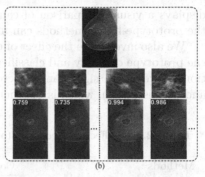

Fig. 2. (a) Examples of non-cancer (left) and cancer prototypes (right) from BRAIx-ProtoPNet++. (b) The interpretable classification. First row: test image. Second row: top-2 activated non-cancer (left) and cancer (right) prototypes. Third row: similarity maps with the max-pooling score for classification.

containing cancerous visual biomarkers (e.g., malignant mass) which align with radiologists' criterion for breast cancer diagnosis, while non-cancer prototypes are from normal breast tissues or benign regions. Figure 2 (b) shows the interpretable reasoning of BRAIxProtoPNet++ on a cancerous test image. We can see that our model classifies the image as cancer because the lesion present in the image looks more like the cancer prototypes than the non-cancer ones.

We also evaluate model interpretability by assessing the cancer localisation. The cancer regions are predicted by applying a threshold of 0.5 on the Grad-CAM (EfficientNet-B0 and DenseNet-121), malignant map (Sparse MIL), salience map (GMIC), and similarity map with the top-1 activated cancer prototype (ProtoP-Net and BRAIxProtoPNet++). For all models, we exclude images with classification probability less than 0.1 since they are classified as non-cancer. When computing PR-AUC, we threshold several values (from 0.05 to 0.5) for the intersection over union (IoU) between predicted cancer region and ground-truth cancer mask to obtain a series of PR-AUC values, as shown in Fig. 3. We can see that our BRAIxProtoPNet++ consistently achieves superior cancer localisation performance over the other methods under different IoU thresholds. Figure 4

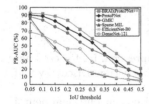
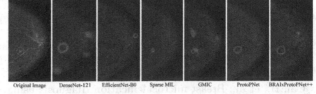

Fig. 3. PR-AUC in different IoU thresholds.

Fig. 4. Visual results of cancer localisation. Yellow circle in the original image indicates cancer region. (Color figure online)

displays a visual comparison of cancer localisation, where we can observe that the prototype-based methods can more accurately detect the cancer region.

We also investigate the effect of our proposed prototype selection strategy on the prototype diversity and classification accuracy. Table 2 shows that our selection strategy can significantly increase prototype diversity (note larger cosine and $L2$ distances), which is beneficial for interpretability and classification.

Table 2. The effect of greedy prototype selection strategy on ADMANI dataset.

Methods	Cosine distance		$L2$ distance		AUC
	Non-cancer	Cancer	Non-cancer	Cancer	
ProtoPNet w/o greedy selection	0.034	0.061	0.805	0.827	88.11
ProtoPNet w/ greedy selection	0.074	0.094	1.215	1.712	88.30

4   Conclusion

In this paper, we presented BRAIxProtoPNet++ to realise accurate mammogram classification with effective prototype-based interpretability. Our approach has been designed to enable the integration of prototype-based interpretable model to any highly accurate global mammogram classifier, where we distil the knowledge of the global model when training the prototype-based model to increase the classification accuracy of the ensemble. We also proposed a method to increase the diversity of the learned prototypes. Experimental results on private and public datasets show that BRAIxProtoPNet++ has SOTA classification and interpretability results. One potential limitation in our BRAIxProtoPNet++ is that the learned prototypes are class-specific, we will explore to learn class-agnostic prototypes for mammogram classification in the future work.

Acknowledgement. This work was supported by funding from the Australian Government under the Medical Research Future Fund - Grant MRFAI000090 for the Transforming Breast Cancer Screening with Artificial Intelligence (BRAIx) Project, and the Australian Research Council through grants DP180103232 and FT190100525.

References

1. Buciluă, C., Caruana, R., Niculescu-Mizil, A.: Model compression. In: Proceedings of the 12th ACM SIGKDD International Conference on Knowledge Discovery and Data Mining, pp. 535–541 (2006)
2. Cai, H., et al.: Breast microcalcification diagnosis using deep convolutional neural network from digital mammograms. Comput. Math. Methods Med. **2019** (2019)
3. Chen, C., Li, O., Tao, D., Barnett, A., Rudin, C., Su, J.K.: This looks like that: deep learning for interpretable image recognition. In: Advances in Neural Information Processing Systems, vol. 32, pp. 8930–8941 (2019)

4. Fang, L., Cunefare, D., Wang, C., Guymer, R.H., Li, S., Farsiu, S.: Automatic segmentation of nine retinal layer boundaries in oct images of non-exudative AMD patients using deep learning and graph search. Biomed. Opt. Express **8**(5), 2732–2744 (2017)
5. Fang, L., Wang, C., Li, S., Rabbani, H., Chen, X., Liu, Z.: Attention to lesion: lesion-aware convolutional neural network for retinal optical coherence tomography image classification. IEEE Trans. Med. Imaging **38**(8), 1959–1970 (2019)
6. He, X., et al.: Sample-efficient deep learning for Covid-19 diagnosis based on CT scans. IEEE Trans. Med. Imaging 1–10 (2020)
7. Hermoza, R., Maicas, G., Nascimento, J.C., Carneiro, G.: Region proposals for saliency map refinement for weakly-supervised disease localisation and classification. In: Martel, A.L., et al. (eds.) MICCAI 2020. LNCS, vol. 12266, pp. 539–549. Springer, Cham (2020). https://doi.org/10.1007/978-3-030-59725-2_52
8. Huang, G., Liu, Z., Van Der Maaten, L., Weinberger, K.Q.: Densely connected convolutional networks. In: Proceedings of the IEEE Conference on Computer Vision and Pattern Recognition, pp. 4700–4708 (2017)
9. Ilanchezian, I., Kobak, D., Faber, H., Ziemssen, F., Berens, P., Ayhan, M.S.: Interpretable gender classification from retinal fundus images using BagNets. In: de Bruijne, M., et al. (eds.) MICCAI 2021. LNCS, vol. 12903, pp. 477–487. Springer, Cham (2021). https://doi.org/10.1007/978-3-030-87199-4_45
10. Khakzar, A., Albarqouni, S., Navab, N.: Learning interpretable features via adversarially robust optimization. In: Shen, D., et al. (eds.) MICCAI 2019. LNCS, vol. 11769, pp. 793–800. Springer, Cham (2019). https://doi.org/10.1007/978-3-030-32226-7_88
11. Kim, E., Kim, S., Seo, M., Yoon, S.: XProtoNet: diagnosis in chest radiography with global and local explanations. In: Proceedings of the IEEE/CVF Conference on Computer Vision and Pattern Recognition, pp. 15719–15728 (2021)
12. Kleppe, A., Skrede, O.J., De Raedt, S., Liestøl, K., Kerr, D.J., Danielsen, H.E.: Designing deep learning studies in cancer diagnostics. Nat. Rev. Cancer **21**(3), 199–211 (2021)
13. Krizhevsky, A., Sutskever, I., Hinton, G.E.: ImageNet classification with deep convolutional neural networks. In: Advances in Neural Information Processing Systems, vol. 25, pp. 1097–1105 (2012)
14. LeCun, Y., Bengio, Y., Hinton, G.: Deep learning. Nature **521**(7553), 436–444 (2015)
15. Paszke, A., et al.: Pytorch: an imperative style, high-performance deep learning library. In: Advances in Neural Information Processing Systems, vol. 32 (2019)
16. Rudin, C.: Stop explaining black box machine learning models for high stakes decisions and use interpretable models instead. Nat. Mach. Intell. **1**(5), 206–215 (2019)
17. Selvaraju, R.R., Cogswell, M., Das, A., Vedantam, R., Parikh, D., Batra, D.: Grad-CAM: visual explanations from deep networks via gradient-based localization. In: Proceedings of the IEEE International Conference on Computer Vision, pp. 618–626 (2017)
18. Shen, Y., et al.: An interpretable classifier for high-resolution breast cancer screening images utilizing weakly supervised localization. Med. Image Anal. **68**, 101908 (2021)
19. Stadnick, B., et al.: Meta-repository of screening mammography classifiers. arXiv preprint arXiv:2108.04800 (2021)

20. Tan, M., Le, Q.: EfficientNet: rethinking model scaling for convolutional neural networks. In: International Conference on Machine Learning, pp. 6105–6114. PMLR (2019)
21. Wu, N., et al.: Deep neural networks improve radiologists' performance in breast cancer screening. IEEE Trans. Med. Imaging **39**(4), 1184–1194 (2019)
22. Xing, H.J., Ji, M.: Robust one-class support vector machine with rescaled hinge loss function. Pattern Recogn. **84**, 152–164 (2018)
23. Yin, C., Liu, S., Shao, R., Yuen, P.C.: Focusing on clinically interpretable features: selective attention regularization for liver biopsy image classification. In: de Bruijne, M., et al. (eds.) MICCAI 2021. LNCS, vol. 12905, pp. 153–162. Springer, Cham (2021). https://doi.org/10.1007/978-3-030-87240-3_15
24. Zhu, W., Lou, Q., Vang, Y.S., Xie, X.: Deep multi-instance networks with sparse label assignment for whole mammogram classification. In: Descoteaux, M., Maier-Hein, L., Franz, A., Jannin, P., Collins, D.L., Duchesne, S. (eds.) MICCAI 2017. LNCS, vol. 10435, pp. 603–611. Springer, Cham (2017). https://doi.org/10.1007/978-3-319-66179-7_69

Deep is a Luxury We Don't Have

Ahmed Taha[✉], Yen Nhi Truong Vu, Brent Mombourquette,
Thomas Paul Matthews, Jason Su, and Sadanand Singh

WhiteRabbit.AI, Santa Clara, CA 95054, USA
ahmdtaha.us@gmail.com

Abstract. Medical images come in high resolutions. A high resolution
is vital for finding malignant tissues at an early stage. Yet, this resolution
presents a challenge in terms of modeling long range dependencies. Shallow transformers eliminate this problem, but they suffer from quadratic
complexity. In this paper, we tackle this complexity by leveraging a linear
self-attention approximation. Through this approximation, we propose
an efficient vision model called **HCT** that stands for **H**igh resolution
Convolutional **T**ransformer. HCT brings transformers' merits to high
resolution images at a significantly lower cost. We evaluate HCT using
a high resolution mammography dataset. HCT is significantly superior
to its CNN counterpart. Furthermore, we demonstrate HCT's fitness for
medical images by evaluating its effective receptive field. Code available
at https://bit.ly/3ykBhhf.

Keywords: Medical images · High resolution · Transformers

1 Introduction

Medical images have high spatial dimensions (*e.g.*, 6 million pixels). This poses
a challenge in terms of modeling long range spatial dependencies. To learn these
dependencies, one can build a deeper network by stacking more layers. Yet,
this fails for two reasons: (1) the computational cost of each layer is significant
because of the high resolution input; (2) stacking convolutional layers expands
the *effective* receptive field sublinearly [19]. This means that a huge number of
layers is needed. For high resolution inputs, a deep network is a luxury.

Compared to convolutional neural networks (CNNs), transformers [1,30] are
superior in terms of modeling long range dependencies. Yet, transformers are
computationally expensive due to the attention layers' quadratic complexity.
Furthermore, transformers are data hungry [7]. This limits the utility of transformers in medical images which are both high resolution and scarce. In this
paper, we tackle these challenges and propose a **H**igh resolution **C**onvolutional
Transformer (HCT). Through HCT, we achieve statistically significant better

---

A. Taha and Y. N. T. Vu—Equal Contribution.

Supplementary Information The online version contains supplementary material
available at https://doi.org/10.1007/978-3-031-16437-8_3.

© The Author(s), under exclusive license to Springer Nature Switzerland AG 2022
L. Wang et al. (Eds.): MICCAI 2022, LNCS 13433, pp. 25–35, 2022.
https://doi.org/10.1007/978-3-031-16437-8_3

Table 1. Effective receptive field [19] (ERF) evaluation for GMIC [25] and HCT. The first row shows the ERF using 100 breast images (left and right) randomly sampled. The second and third rows show the ERF using 100 right and left breast images randomly sampled, respectively. To highlight the ERF difference, we aggregate the ERF across images' rows and columns. The GMIC's ERF (blue curves) is highly concentrated around the center pixel and has a Gaussian shape. In contrast, the HCT's ERF (orange curves) is less concentrated around the center and spreads *dynamically* to the breasts' locations without explicit supervision. Finally, the square-root of ERF (Sqrt EFR) highlights the difference between the GMIC and HCT's ERF. These high resolution images are best viewed on a screen.

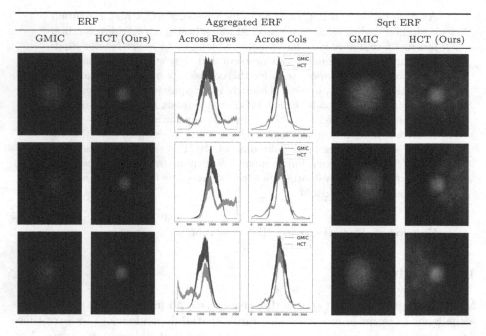

performance and emphasize the following message: Keep an eye on your network's effective receptive field (ERF).

Table 1 compares the rigid ERF of a CNN architecture (*e.g.,* GMIC [25]) versus the flexible ERF of HCT. Table 1 evaluates the ERF three times (rows): (I) using $N = 100$ random breast images (left and right), (II) using $N = 100$ random right breast images, and (III) using $N = 100$ random left breast images. The GMIC's ERF is rigid and follows a Gaussian distribution. This aligns with Luo *et al.* [19] findings. In contrast, HCT's ERF is dynamic.

Thanks to the self-attention layers, HCT focuses *dynamically* on regions based on their contents and not their spatial positions. The HCT's ERF is less concentrated on the center but spans both horizontally and vertically according to the image's content. For instance, the second row in Table 1 shows how the aggregated ERF of HCT is shifted toward both right (Across Rows) and top (Across Columns) region, *i.e.,* where a right breast is typically located. This emphasizes the value of dynamic attention in transformers.

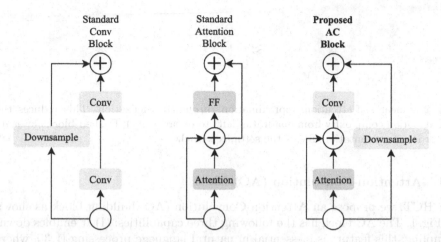

Fig. 1. Neural networks' building blocks. (Left) A standard convolutional block for vision models with spatial downsampling capability. (Center) A standard attention block for language models with long range attention capability. (Right) Our Attention-Convolutional (AC) block with both spatial downsampling and long range attention capabilities. In the AC block, the conv layer both reduces the spatial resolution and increases the number of channels. Batchnorm and activation (*e.g.,* RELU) layers are omitted for visualization purposes. The AC block's pseudocode is provided in the appendix.

Transformers are ubiquitous in medical imaging applications. They have been used for classification [11,21,25,28,37], segmentation [10,14,29,32,35,39], and image denoising [20,40]. Yet, this recent literature leverages low resolution inputs to avoid the computational cost challenge. Conversely, HCT is designed for high resolution input, *i.e.,* 2–3K pixels per dimension. This is a key difference between our paper and recent literature. In summary, our key contributions are

1. We propose an efficient high resolution transformer-based model, HCT, for medical images (Sect. 2). We demonstrate HCT's superiority using a high resolution mammography dataset (Sect. 3).
2. We emphasize the importance of dynamic receptive fields for mammography datasets (Table 1 and Sect. 4). This finding is important for medical images where a large portion of an image is irrelevant (*e.g.,* an empty region).

2   HCT: High Resolution Convolutional Transformer

In this section, we propose a network building block that combines both self-attention and convolution operations (Sect. 2.1). Then, we leverage a linear self-attention approximation to mitigate the attention layer's quadratic complexity (Sect. 2.2). Finally, we introduce HCT, a convolutional transformer architecture designed for high resolution images (Sect. 2.3).

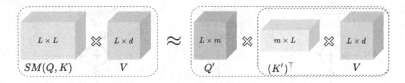

Fig. 2. Linear self-attention approximation through kernelization. This reduces the self-attention complexity from quadratic (left) to linear (right). Dashed-blocks indicate the order of computation. SM is the softmax kernel.

2.1   Attention-Convolution (AC) Block

For HCT, we propose an **A**ttention-**C**onvolution (AC) building block as shown in Fig. 1. The AC block has the following three capabilities: (I) It enables down-sampling; this feature is inessential in natural language processing [1,30] where the number of output and input tokens are equal. Yet, for 2D/3D images, it is important to downsample the spatial resolution as the network's depth increases; (II) The proposed AC block supports different input resolutions. This allows our model to pretrain on small image patches, then finetune on high reso-lution images. Furthermore, the AC block supports various input resolutions during inference; (III) The AC block consumes and produces spatial feature maps. Thus, it integrates seamlessly into architectures designed for classifica-tion [11,21,25,26,37], detection [18,26], and segmentation [10,14,29,32,35,39].

To capitalize on these three capabilities, it is vital to reduce the attention layer's computational cost. Section 2.2 illustrates how we achieve this.

2.2   Efficient Attention

Designing efficient transformers is an active area of research [27]. To avoid the quadratic complexity of attention layers, we leverage the Fast Attention Via pos-itive Orthogonal Random features approach (FAVOR+), a.k.a. Performers [4]. In this section, we (1) present an overview of Performers, and (2) compare Per-formers with other alternatives (*e.g.*, Reformer [15]).

Choromanski *et al.* [4] regard the attention mechanism through kerneliza-tion, *i.e.*, $\mathbb{K}(x,y) = \phi(x)\phi(y)^\top$, where a kernel $\mathbb{K}$ applied on $x, y \in \mathbb{R}^d$ can be approximated using random feature maps $\phi : \mathbb{R}^d \to \mathbb{R}_+^m$. To avoid the quadratic complexity $(L \times L)$ of $SM(Q, K)$, it is rewritten into $\phi(Q)\phi(K)^\top = Q'(K')^\top$ where L is the number of input tokens, SM is the softmax kernel applied on the tokens' queries Q and Keys K. Concretely, the softmax kernel is approximated using random feature maps ϕ defined as follows

$$\phi(x) = \frac{h(x)}{\sqrt{m}} \left(f_1(w_1^\top x), ..., f_1(w_m^\top x), ..., f_l(w_1^\top x), ..., f_l(w_m^\top x) \right), \qquad (1)$$

where functions $f_1, ..., f_l : \mathbb{R} \to \mathbb{R}$, $h : \mathbb{R}^d \to \mathbb{R}$, and vectors $w_1, ..., w_m \overset{iid}{\sim} \mathcal{D}$ for some distribution $\mathcal{D} \in \mathcal{P}(\mathbb{R}^d)$. Figure 2 illustrates how Performers change the order of computation to avoid the attention's layer quadratic complexity.

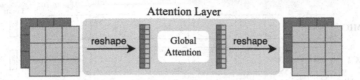

Fig. 3. The proposed self-attention layer flattens the input feature maps before applying global self-attention, then reshapes the result into spatial feature maps. Accordingly, this layer integrates seamlessly in various vision architectures designed for different tasks (*e.g.*, classification or detection).

Choromanski *et al.* [4] sampled the vectors $w_1, ..., w_m$ from an isotropic distribution $\mathcal{D}$ and entangled them to be exactly orthogonal. Then, the softmax kernel is approximated using $l = 1$, $f_1 = \exp$, and $h(x) = \exp(-\frac{||x||^2}{2})$. This formula is denoted as Performer-Softmax. Then, Choromanski *et al.* [4] proposed another stabler random feature maps, Performer-RELU, by setting $l = 1$, $f_1 = \text{RELU}$, and $h(x) = 1$. The simplicity of this kernel approximation is why we choose Performers over other alternatives.

Performers are simpler than other linear approximations. For instance, a Performer leverages standard layers (*e.g.*, linear layers). Thus, it is simpler than Reformer [15] which leverages reversible layers [12]. A Performer supports global self-attention, *i.e.*, between all input tokens. This reduces the number of hyperparameters compared to block-wise approaches such as shifted window [17], block-wise [23], and local-attention [22]. Furthermore, a Performer makes no assumption about the maximum number of tokens while Linformer [31] does. Accordingly, Performer supports different input resolutions without bells and whistles. Basically, Performers make the least assumptions while being relatively simple. By leveraging Performers in HCT, we promote HCT to other applications beyond our mammography evaluation setup.

2.3   The HCT Architecture

In this section, we describe the HCT architecture and how the attention layer – in the AC block – operates on spatial data, *e.g.*, 2D/3D images. We pick a simple approach that introduces the least hyperparameters. Since self-attention layers are permutation-invariant, we simply flatten the input 3D feature maps from $\mathbb{R}^{W \times H \times C}$ into a sequence of 1D tokens, *i.e.*, $\mathbb{R}^{WH \times C}$ as shown in Fig. 3. This approach supports various input formats (*e.g.*, volumetric data) as well as different input resolutions. Furthermore, this approach integrates seamlessly into other vision tasks such as detection and segmentation. Despite its simplicity and merits, the attention layer (Fig. 3) brings one last challenge to be tackled.

Attention layers lack inductive bias. Accordingly, transformers are data hungry [7] while labeled medical images are scarce. Compared to a convolutional layer, the global self-attention layer (Fig. 3) is more vulnerable to overfitting and optimization instability. To tackle these challenges, Xiao *et al.* [34] have promoted early convolutional layers to increase optimization stability and improve

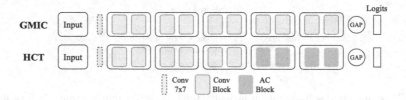

Fig. 4. Two mammography classification architectures: GMIC [25] and our HCT. Both architectures are ResNet-22 variants. While GMIC is a pure CNN architecture, HCT is a convolutional transformer-based architecture. GAP denotes global average pooling.

Table 2. Statistics of the OPTIMAM mammography dataset.

	Train	Val	Test
Malignant images (Positive)	8,923	1,097	1,195
Non-malignant images (Negative)	49,453	6,169	6,163
Total	58,376	7,266	7,358

generalization performance. Similar observations have been echoed in recent literature [5,6,33]. Accordingly, we integrate our AC block into a variant ResNet-22 architecture. This variant, dubbed GMIC [25], has been proposed for breast cancer classification. Thus, GMIC serves both as a backbone for architectural changes and as a baseline for quantitative evaluation.

Figure 4 presents both GMIC and HCT architectures, both ResNet-22 variants. Compared to the standard ResNet-22, both architectures employ large strides at early convolutional layers. A large stride (*e.g.*, $s = 2$) may mitigate but not eliminate the computational cost of high resolution inputs. Both architectures have an initial conv-layer with a 7×7 kernel followed by five residual stages. Each stage has two blocks and each block has two layers (*e.g.*, conv or attention) as shown in Fig. 1. GMIC uses the standard convolutional block (Fig. 1, left), while HCT uses the AC block (Fig. 1, right). Our experiments (Sect. 3) explore other architecture variants as baselines for quantitative evaluation.

3   Experiments

Dataset: We evaluate HCT using a high resolution mammography dataset. This dataset comes from the National Health Service OPTIMAM database [13]. The dataset contains 148,448 full-field digital mammography (FFDM) images from 38,189 exams and 17,322 patients. The dataset is divided randomly into train/val/test splits at the patient level with a ratio of 80:10:10. An image is labeled malignant (positive) if it has a malignant biopsy in the corresponding breast within 12 months of the screening exam. An image is labeled non-malignant (negative) if it has either a benign biopsy within 12 months of the screening exam or 24 months of non-malignant imaging follow-up. Images that violate the above criteria are excluded from the dataset. The validation split is used for hyperparameter tuning. Table 2 presents OPTIMAM statistics.

Table 3. Quantitative evaluation on OPTIMAM. We report both the number of parameters (millions) and the architecture's performance. Performance is reported using AUC and their 95% confidence intervals (CI) are in square brackets. We evaluate both the small patch and the full image models. The linear-approximation column denotes the linear approximation method employed. E.g., Per-Softmax stands for Performer-Softmax. The CI is constructed via percentile bootstrapping with 10,000 bootstrap iterations.

Architecture	Params	Linear approx.	Patch	Image
			Adam Optimizer	
GMIC	2.80	–	96.13 [95.43, 96.78]	85.04 [83.74, 86.36]
HCT (**ours**)	1.73	Nyström	**96.41 [95.76, 97.01]**	84.83 [83.49, 86.14]
HCT (**ours**)	1.73	Per-Softmax	96.35 [95.68, 96.97]	**86.64 [85.38, 87.86]**
HCT (**ours**)	1.73	Per-RELU	96.34 [95.66, 96.97]	86.29 [85.02, 87.54]
			Adam + ASAM Optimizer	
GMIC	2.80	–	96.29 [95.62, 96.92]	86.58 [85.34, 87.80]
HCT (**ours**)	1.73	Nyström	96.65 [96.02, 97.23]	86.73 [85.49, 87.95]
HCT (**ours**)	1.73	Per-Softmax	96.68 [96.05, 97.26]	87.39 [86.14, 88.59]
HCT (**ours**)	1.73	Per-RELU	**96.73 [96.09, 97.32]**	**88.00 [86.80, 89.18]**

Baselines: We evaluate HCT using two Performer variants: **Performer-RELU** and **Performer-Softmax**. We also evaluate HCT with **Nyströmformer** [36], *i.e.*, another linear attention approximation. The Nyströmformer has a hyper-parameter q, the number of landmarks. We set $q = \max(W, H)$, where W and H are the width and height of the feature map, respectively. Finally, we evaluate HCT against a pure CNN architecture: **GMIC**. GMIC [25] is an established benchmark [9] for high resolution mammography. To evaluate all architectures on a single benchmark, GMIC denotes the first module *only* from Shen *et al.* [25]. This eliminates any post-processing steps, *e.g.*, extracting ROI proposals.

Technical Details: All networks are initialized randomly, pretrained on small patches (512×512), then finetuned on high resolution images (3328×2560). This is a common practice in mammography literature [9,18]. All models are trained with the cross entropy loss. For the patch model, we use a cosine learning rate decay with an initial learning rate $lr = 6e^{-4}$, batch size $b = 160$, and $e = 80$ epochs. For the image model, we also use the cosine learning rate decay, but with $lr = 2e^{-5}$, $b = 32$, and $e = 80$. All patch models are trained on a single 2080Ti GPU while image models are trained on a single A6000 GPU.

We evaluate HCT using both Adam and the adaptive sharpness-aware minimization [8,16] (ASAM) optimizers. ASAM simultaneously minimizes the loss value and the loss sharpness. This improves generalization [8] especially for transformer-based models [3]. We set the ASAM hyperparameter $\rho = 0.05$. We used a public implementation of Performer[1]. We use the default Performer hyper-parameters [4], *i.e.*, $n_h = 8$ heads, $p = 0.1$ dropout rate, $m = \frac{d}{n_h} \log \frac{d}{n_h}$, where

---

[1] https://github.com/lucidrains/performer-pytorch.

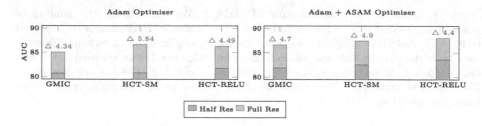

Fig. 5. Quantitative evaluation using half (1664 × 1280) and full (3328 × 2560) resolution inputs. $\triangle$ denotes the *absolute* improvement margin achieved on the full resolution by the same architecture. This highlights the advantage of high resolution inputs when looking for malignant tissues at an early stage.

d is the number of channels. We used standard augmentation techniques (*e.g.,* random flipping and Gaussian noise).

Results: Table 3 presents a classification evaluation on OPTIMAM. We report both the area under the receiver operating characteristic curve (AUC) and the 95% confidence interval. HCT with Performers outperforms GMIC consistently. RELU has been both promoted empirically by Choromanski *et al.* [4] and defended theoretically by Schlag *et al.* [24]. Similarly, we found Performer-RELU superior to Performer-Softmax when the ASAM optimizer was used. With the vanilla Adam optimizer, we attribute Performer-RELU's second best – and not the best – performance to stochastic training noise.

4   Ablation Study

In this section, we compute the effective receptive field (ERF) of both GMIC and HCT. Through this qualitative evaluation, we demonstrate HCT's fitness for medical images. Finally, we evaluate the performance of HCT using half and full resolution inputs.

Effective Receptive Field Study: We follow Luo *et al.* [19] procedure to compute the ERF. Specifically, we feed an input image to our model. Then, we set the gradient signal to one at the center pixel of the output feature map, and to zero everywhere else. Then, we back-propagate this gradient through the network to get the input's gradients. We repeat this process using $N = 100$ randomly sampled breast images. Finally, we set the ERF to the mean of these N inputs' gradients. Table 1 presents the ERF of both GMIC and HCT with Performer-RELU. HCT is superior in terms of modeling long range dependencies.

Resolution Study: To emphasize the importance of high resolution inputs for medical images, we train GMIC and HCT using both half and full resolutions inputs. Figure 5 delivers an evaluation of all models: GMIC, HCT with Performance-RELU (HCT-RELU), and HCT with Performer-Softmax (HCT-SM). The full resolution models beat the half resolution models with a significant margin (+4% AUC).

Discussion: Compared to a low resolution input, a high resolution input will always pose a computational challenge. As technology develops, we will have better resources (*e.g.*, GPUs), but image-acquisition quality will improve as well. HCT is a ResNet-22 variant, a shallow architecture compared to ResNet-101/152 and recent architectural development [2, 38]. For high resolution inputs, a deep network is a luxury.

5 Conclusion

We have proposed HCT, a convolutional transformer for high resolution inputs. HCT leverages linear attention approximation to avoid the quadratic complexity of vanilla self-attention layers. Through this approximation, HCT boosts performance, models long range dependencies, and maintains a superior effective receptive field.

References

1. Bahdanau, D., Cho, K., Bengio, Y.: Neural machine translation by jointly learning to align and translate. In: ICLR (2015)
2. Brown, T., et al.: Language models are few-shot learners. In: NeurIPS (2020)
3. Chen, X., Hsieh, C.J., Gong, B.: When vision transformers outperform ResNets without pretraining or strong data augmentations. In: ICLR (2022)
4. Choromanski, K., et al.: Rethinking attention with performers. arXiv preprint arXiv:2009.14794 (2020)
5. Cordonnier, J.B., Loukas, A., Jaggi, M.: On the relationship between self-attention and convolutional layers. arXiv preprint arXiv:1911.03584 (2019)
6. d'Ascoli, S., Touvron, H., Leavitt, M., Morcos, A., Biroli, G., Sagun, L.: ConViT: improving vision transformers with soft convolutional inductive biases. In: ICML (2021)
7. Dosovitskiy, A., et al.: An image is worth 16×16 words: transformers for image recognition at scale. In: ICLR (2021)
8. Foret, P., Kleiner, A., Mobahi, H., Neyshabur, B.: Sharpness-aware minimization for efficiently improving generalization. arXiv preprint arXiv:2010.01412 (2020)
9. Frazer, H.M., Qin, A.K., Pan, H., Brotchie, P.: Evaluation of deep learning-based artificial intelligence techniques for breast cancer detection on mammograms: results from a retrospective study using a BreastScreen Victoria dataset. J. Med. Imaging Radiat. Oncol. (2021)
10. Gao, Y., Zhou, M., Metaxas, D.N.: UTNet: a hybrid transformer architecture for medical image segmentation. In: de Bruijne, M., et al. (eds.) MICCAI 2021. LNCS, vol. 12903, pp. 61–71. Springer, Cham (2021). https://doi.org/10.1007/978-3-030-87199-4_6
11. Geras, K.J., et al.: High-resolution breast cancer screening with multi-view deep convolutional neural networks. arXiv preprint arXiv:1703.07047 (2017)
12. Gomez, A.N., Ren, M., Urtasun, R., Grosse, R.B.: The reversible residual network: backpropagation without storing activations. In: NeurIPS (2017)
13. Halling-Brown, M.D., et al.: Optimam mammography image database: a large-scale resource of mammography images and clinical data. Radiol.: Artif. Intell. (2020)

14. Karimi, D., Vasylechko, S.D., Gholipour, A.: Convolution-free medical image segmentation using transformers. In: de Bruijne, M., et al. (eds.) MICCAI 2021. LNCS, vol. 12901, pp. 78–88. Springer, Cham (2021). https://doi.org/10.1007/978-3-030-87193-2_8

15. Kitaev, N., Kaiser, Ł., Levskaya, A.: Reformer: the efficient transformer. arXiv preprint arXiv:2001.04451 (2020)

16. Kwon, J., Kim, J., Park, H., Choi, I.K.: ASAM: adaptive sharpness-aware minimization for scale-invariant learning of deep neural networks. In: ICML (2021)

17. Liu, Z., et al.: Swin transformer: hierarchical vision transformer using shifted windows. In: ICCV (2021)

18. Lotter, W., et al.: Robust breast cancer detection in mammography and digital breast tomosynthesis using an annotation-efficient deep learning approach. Nat. Med. **27**, 244–249 (2021)

19. Luo, W., Li, Y., Urtasun, R., Zemel, R.: Understanding the effective receptive field in deep convolutional neural networks. In: NeurIPS (2016)

20. Luthra, A., Sulakhe, H., Mittal, T., Iyer, A., Yadav, S.: Eformer: edge enhancement based transformer for medical image denoising. In: ICCVW (2021)

21. Matsoukas, C., Haslum, J.F., Söderberg, M., Smith, K.: Is it time to replace CNNs with transformers for medical images? In: ICCVW (2021)

22. Parmar, N., et al.: Image transformer. In: ICML (2018)

23. Qiu, J., Ma, H., Levy, O., Yih, S.W.T., Wang, S., Tang, J.: Blockwise self-attention for long document understanding. arXiv preprint arXiv:1911.02972 (2019)

24. Schlag, I., Irie, K., Schmidhuber, J.: Linear transformers are secretly fast weight programmers. In: ICML (2021)

25. Shen, Y., et al.: Globally-aware multiple instance classifier for breast cancer screening. In: Suk, H.-I., Liu, M., Yan, P., Lian, C. (eds.) MLMI 2019. LNCS, vol. 11861, pp. 18–26. Springer, Cham (2019). https://doi.org/10.1007/978-3-030-32692-0_3

26. Shen, Y., et al.: An interpretable classifier for high-resolution breast cancer screening images utilizing weakly supervised localization. Med. Image Anal. (2020)

27. Tay, Y., Dehghani, M., Bahri, D., Metzler, D.: Efficient transformers: a survey. arXiv preprint arXiv:2009.06732 (2020)

28. van Tulder, G., Tong, Y., Marchiori, E.: Multi-view analysis of unregistered medical images using cross-view transformers. In: de Bruijne, M., et al. (eds.) MICCAI 2021. LNCS, vol. 12903, pp. 104–113. Springer, Cham (2021). https://doi.org/10.1007/978-3-030-87199-4_10

29. Valanarasu, J.M.J., Oza, P., Hacihaliloglu, I., Patel, V.M.: Medical transformer: gated axial-attention for medical image segmentation. In: de Bruijne, M., et al. (eds.) MICCAI 2021. LNCS, vol. 12901, pp. 36–46. Springer, Cham (2021). https://doi.org/10.1007/978-3-030-87193-2_4

30. Vaswani, A., et al.: Attention is all you need. In: NeurIPS (2017)

31. Wang, S., Li, B.Z., Khabsa, M., Fang, H., Ma, H.: Linformer: self-attention with linear complexity. arXiv preprint arXiv:2006.04768 (2020)

32. Wang, W., Chen, C., Ding, M., Yu, H., Zha, S., Li, J.: TransBTS: multimodal brain tumor segmentation using transformer. In: de Bruijne, M., et al. (eds.) MICCAI 2021. LNCS, vol. 12901, pp. 109–119. Springer, Cham (2021). https://doi.org/10.1007/978-3-030-87193-2_11

33. Wu, H., et al.: CVT: introducing convolutions to vision transformers. In: ICCV (2021)

34. Xiao, T., Dollar, P., Singh, M., Mintun, E., Darrell, T., Girshick, R.: Early convolutions help transformers see better. In: NeurIPS (2021)

35. Xie, Y., Zhang, J., Shen, C., Xia, Y.: CoTr: efficiently bridging CNN and transformer for 3D medical image segmentation. In: de Bruijne, M., et al. (eds.) MICCAI 2021. LNCS, vol. 12903, pp. 171–180. Springer, Cham (2021). https://doi.org/10.1007/978-3-030-87199-4_16
36. Xiong, Y., et al.: Nystromformer: a nystrom-based algorithm for approximating self-attention. In: AAAI (2021)
37. Yang, Z., et al.: MommiNet-v2: mammographic multi-view mass identification networks. Med. Image Anal. (2021)
38. Yuan, L., et al.: Florence: a new foundation model for computer vision. arXiv preprint arXiv:2111.11432 (2021)
39. Zhang, Y., Liu, H., Hu, Q.: TransFuse: fusing transformers and CNNs for medical image segmentation. In: de Bruijne, M., et al. (eds.) MICCAI 2021. LNCS, vol. 12901, pp. 14–24. Springer, Cham (2021). https://doi.org/10.1007/978-3-030-87193-2_2
40. Zhang, Z., Yu, L., Liang, X., Zhao, W., Xing, L.: TransCT: dual-path transformer for low dose computed tomography. In: de Bruijne, M., et al. (eds.) MICCAI 2021. LNCS, vol. 12906, pp. 55–64. Springer, Cham (2021). https://doi.org/10.1007/978-3-030-87231-1_6

PD-DWI: Predicting Response to Neoadjuvant Chemotherapy in Invasive Breast Cancer with Physiologically-Decomposed Diffusion-Weighted MRI Machine-Learning Model

Maya Gilad[1]([✉])[iD] and Moti Freiman[2][iD]

[1] Efi Arazi School of Computer Science, Reichman University, Herzliya, Israel
MayaEsther.Gilad@post.runi.ac.il
[2] Faculty of Biomedical Engineering, Technion, Haifa, Israel

Abstract. Early prediction of pathological complete response (pCR) following neoadjuvant chemotherapy (NAC) for breast cancer plays a critical role in surgical planning and optimizing treatment strategies. Recently, machine and deep-learning based methods were suggested for early pCR prediction from multi-parametric MRI (mp-MRI) data including dynamic contrast-enhanced MRI and diffusion-weighted MRI (DWI) with moderate success. We introduce PD-DWI (https://github.com/TechnionComputationalMRILab/PD-DWI), a physiologically decomposed DWI machine-learning model to predict pCR from DWI and clinical data. Our model first decomposes the raw DWI data into the various physiological cues that are influencing the DWI signal and then uses the decomposed data, in addition to clinical variables, as the input features of a radiomics-based XGBoost model. We demonstrated the added-value of our PD-DWI model over conventional machine-learning approaches for pCR prediction from mp-MRI data using the publicly available Breast Multi-parametric MRI for prediction of NAC Response (BMMR2) challenge. Our model substantially improves the area under the curve (AUC), compared to the current best result on the leaderboard (0.8849 vs. 0.8397) for the challenge test set. PD-DWI has the potential to improve prediction of pCR following NAC for breast cancer, reduce overall mp-MRI acquisition times and eliminate the need for contrast-agent injection.

Keywords: Breast DWI · Response to neoadjuvant chemotherapy · Machine-learning · Radiomics

1 Introduction

Breast cancer is the most prevalent cancer and is one of the leading causes of cancer mortality worldwide [2,3,8]. Neoadjuvant chemotherapy (NAC), a preoperative standard-of-care for invasive breast cancer, has improved breast cancer

© The Author(s), under exclusive license to Springer Nature Switzerland AG 2022
L. Wang et al. (Eds.): MICCAI 2022, LNCS 13433, pp. 36–45, 2022.
https://doi.org/10.1007/978-3-031-16437-8_4

treatment effectiveness and has been associated with better survival rates [3, 20, 21]. Pathological complete response (pCR), the absence of residual invasive disease in either breast or axillary lymph nodes, is used to assess patient response to NAC. While notable pCR rates have been reported even in the most aggressive tumors, approximately 20% of breast cancers are resistant to NAC; hence, early response prediction of pCR might enable efficient selection and, when needed, alteration of treatment strategies and improve surgical planning [4,5,19,20].

Multi-parametric MRI (mp-MRI) approaches were proposed to address the urgent need for early prediction of pCR. Commonly, dynamic contrast-enhanced magnetic resonance imaging (DCE-MRI) and diffusion-weighted MRI (DWI) are used in mp-MRI assessment of pCR due to their higher sensitivity to tissue architecture compared to anatomical MRI [1,10,16,19,23]. Recently, radiomics and high-dimensional image features approaches were applied to predict pCR after NAC from mp-MRI data of breast cancer patients. Joo et al. [14] and Duanmu et al. [7] introduced deep-learning models to fuse high-dimensional mp-MRI image features, including DCE-MRI and DWI, and clinical information to predict pCR after NAC in breast cancer. Huang et al. [13] and Liu et al. [17] used radiomics features extracted from mp-MRI data for pCR prediction.

However, the mp-MRI data represents an overall aggregation of various physiological cues influencing the acquired signal. Specifically, DWI data with the computed apparent diffusion coefficient (ADC) map is known to represent an overall aggregation of both pure diffusion, inversely correlated with cellular density, and pseudo-diffusion, correlated with the collective motion of blood water molecules in the micro-capillary network [1,21,22]. As pCR is characterized by reduced cellular density, and therefore an increase in the presence of pure diffusion on one hand, and reduced blood flow in the capillaries and therefore reduced pseudo-diffusion on the other hand, the naïve aggregation of these cues by the ADC map may eliminate important features for pCR prediction [1,21]. While separation between the diffusion and pseudo-diffusion components can be achieved by acquiring DWI data with multiple b-values and fitting the "Intravoxel incoherent motion" (IVIM) model to the DWI data [1,16,21], lengthy acquisition times preclude clinical utilization [1,9].

In this work, we present PD-DWI, a physiologically decomposed DWI model to predict pCR solely from clinical DWI data. Our model first decomposes the DWI data into pseudo-diffusion ADC and the pseudo-diffusion fraction maps representing the different physiological cues influencing the DWI signal using linear approximation. Then, we use these maps, in addition to clinical variables, as the input features of a radiomics-based XGBoost model for pCR prediction.

We demonstrated the added-value of our PD-DWI model over conventional machine-learning approaches for pCR prediction from mp-MRI data using the publicly available "Breast Multiparametric MRI for prediction of NAC Response" (BMMR2) challenge[1]. Our model substantially improves the area under the curve (AUC) compared to current best result on the leaderboard (0.8849 vs. 0.8397) on the challenge test set.

---

[1] https://qin-challenge-acrin.centralus.cloudapp.azure.com/competitions/2.

Our approach does not require lengthy DWI data acquisition with multiple b-values and eliminates the need for Gadolinium-based contrast agent injection and DCE-MRI acquisition for early pCR prediction following NAC for invasive breast cancer patients.

2   Method

2.1   Patient Cohort

We used the BMMR2 challenge dataset. This dataset was curated from the ACRIN 6698 multi-center study [6,18,19]. The dataset includes 191 subjects from multiple institutions and has been pre-divided into training and test sub-groups by challenge organizers (60%–40% split, stratified by pCR outcome).

All subjects had longitudinal mp-MRI studies, including standardized DWI and DCE-MRI scans, at three time points: T0 (pre-NAC), T1 (3 weeks NAC), and T2 (12 weeks NAC). DWI data was acquired with a single-shot echo planar imaging sequence with parallel imaging (reduction factor, two or greater); fat suppression; a repetition time of greater than 4000 ms; echo time minimum; flip angle, 90°; field of view, 300–360 mm; acquired matrix, 128 × 128 to 192 × 192; in-plane resolution, 1.7–2.8 mm; section thickness, 4–5 mm; and imaging time, 5 or fewer minutes. Diffusion gradients were applied in three orthogonal directions by using diffusion weightings (b-values) of 0, 100, 600, and 800 s/mm². No respiratory triggering or other motion compensation methods were used [19].

DWI whole-tumor manual segmentation and DCE-MRI functional tumor segmentation were provided for each time point, in addition to overall ADC and Signal Enhancement Ratio (SER) maps. Non-imaging clinical data included demographic data (age, race), 4-level lesion type, 4-level hormonal receptor HR/HER2 status, 3-level tumor grade, and MRI measured longest diameter (cm) at T0.

Reference histopathological pCR outcome (0/1) was defined as no residual invasive disease in either breast or axillary lymph nodes on the basis of postsurgical histopathologic examination performed by the ACRIN 6698 institutional pathologists who were blinded to mp-MRI data [19]. The reference pCR labels were available to challenge participants only for training sub-groups.

2.2   Physiological Decomposition of Clinical Breast DWI

In the framework of DWI, the random displacement of individual water molecules in the tissue results in signal attenuation in the presence of magnetic field encoding gradient pulses. This attenuation increases with the ADC and the degree of sensitization-to-diffusion of the MRI pulse sequence (b-value), taking the form of a mono-exponential decay with the b-value:

$$s_i = s_0 \exp\left(-b_i ADC\right) \tag{1}$$

where s_i is the signal at b-value b_i and s_0 is the signal without sensitizing the diffusion gradients. While diffusion and blood micro-circulation are two entirely different physical phenomena, randomness results from the collective motion of blood water molecules in the micro-capillary network that may be depicted as a "pseudo-diffusion" process that influences the DWI signal decay for low b-values $(0\text{--}200\,\text{s/mm}^2)$ [9].

The overall DWI signal attenuation can be produced with the bi-exponential IVIM signal decay model proposed by Le-Bihan [15]:

$$s_i = s_0 \left(F \exp\left(-b_i \left(D^* + D\right)\right) + (1 - F) \exp\left(-b_i D\right) \right) \tag{2}$$

where D is the pure diffusion coefficient, D^* is the pseudo-diffusion coefficient and F is the pseudo-diffusion fraction.

Direct physiological decomposition of the BMMR2 challenge DWI data, by fitting the IVIM model to the DWI data, is not possible due to the limited number of b-values used during the acquisition. Instead, we used a linear approximation of the pseudo-diffusion and pseudo-diffusion fraction components as follows. We first calculated an ADC map reflecting mostly pseudo-diffusion (ADC_{0-100}) with Eq. 1 by using only DWI data acquired at low b-values $(\leq 100\,\text{s/mm}^2)$, overall ADC map using the entire DWI data (ADC_{0-800}), and a pseudo-diffusion-free ADC map $(ADC_{100-800})$ using only DWI data acquired at high b-values (i.e. $\geq 100\,\text{s/mm}^2$). Then, we calculate the pseudo-diffusion fraction map (F) which represents the relative contribution of the pseudo-diffusion cue to the overall DWI signal [12].

Figure 1 presents the averaged DWI signal over the region of interest for two representative patients from the BMMR2 challenge training set along with the ADC_{0-100}, ADC_{0-800}, $ADC_{100-800}$, and the pseudo-diffusion fraction (F) parameter data. The reduced slope of ADC_{0-100} and F parameter data reflect reduced pseudo-diffusion and pseudo-diffusion fraction which represent a better pCR to NAC.

Figure 2 presents the DWI data along with the overall ADC map (ADC_{0-800}) and its physiological decomposition to the pseudo-diffusion ADC (ADC_{0-100}) and the pseudo-diffusion fraction (F).

2.3 Machine Learning Model

Our PD-DWI model was composed of physiological maps extracted from the DWI data in combination with clinical data. Figure 3 presents the overall model architecture. The acquired DWI data at each time-point is decomposed into the different physiological cues reflecting pseudo-diffusion (ADC_{0-100}) and pseudo-diffusion-fraction (F). 3D Radiomics features were extracted from the segmented tumor in ADC and F maps using the PyRadiomics software package [11].

We integrated clinical data into our model using several methods. By modeling the 4-level hormonal receptor status as two binary features (HR status and HER2 status) we allowed the model to leverage differences in radiomic features between tumor types. By converting the 3-level tumor grade to an ordinal categorical arrangement (Low - 1, Intermediate - 2, and High - 3) we allowed the model to use the relation between different severity levels. One patient in the

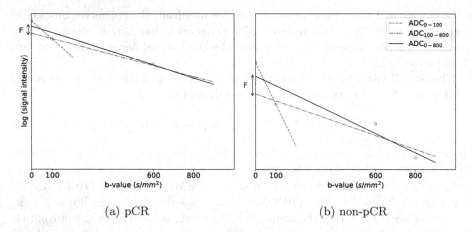

Fig. 1. DWI signal decay as a function of the b-value along with the different ADC models of representative pCR and Non pCR patients.

training set lacked a tumor grade and was imputed as High using Most-Frequent imputation strategy. We applied One-Hot-Encoding on race and lesion type data.

The total number of extracted radiomic and clinical features was relatively high in comparison to the BMMR2 challenge dataset size. Thus, a subset of 100 features with the highest ANOVA F-values was selected in order to minimize model over-fitting and increase model robustness. Ultimately, an XGBoost classifier was trained on selected features and predicted the pCR probability of a given patient.

2.4 Evaluation Methodology

Relation Between DWI Signal Attenuation Decay and pCR Prediction. We first explored the relation between signal attenuation decay and pCR prediction. We calculated additional ADC maps using various subsets of the b-values. Specifically, we calculated the additional following maps: $ADC_{100-800}$ using b-values 100–800, ADC_{0-800} using all available b-values, and finally ADC_{0-100} using b-values 0 and 100.

We assessed the impact of each component in our model by creating four additional models as follows. Three models were created using ADC_{0-100}, $ADC_{100-800}$ and ADC_{0-800}, without the F map at all. An additional model was created with only the F map.

We assessed the performance of our PD-DWI model in comparison to the other DWI-based models based on the AUC metric achieved on the BMMR2 challenge test set using all available time-points (i.e. pre-NAC, early-NAC, and mid-NAC). Since pCR outcomes of the test set were not available to challenge participants, the AUC metrics of evaluated models were calculated by BMMR2 challenge organizers upon submission of pCR predictions for all patients in the test set.

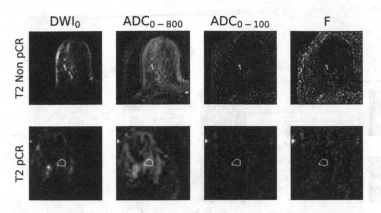

Fig. 2. DWI data of pCR and non-PCR patients at T2 (mid-NAC). The white contour represents the tumor segmentation used for analysis as given by challenge organizers.

Early Prediction of pCR over the Course of NAC. We determined the added-value of our PD-DWI model in predicting pCR during the course of NAC in comparison to a baseline model that was built using the same pipeline, but include all imaging data available as part of the BMMR2 challenge dataset (ADC_{0-800}+SER). Specifically, the baseline model used both an ADC_{0-800} map computed from the DWI data and a SER map computed from the DCE-MRI data by the challenge organizers.

We used the AUC computed by the challenge organizers on the test as the evaluation metric.

Implementation Details. All models were implemented[2] with python 3.8 and xgboost 1.5.1, scikit-learn 1.0.2, pandas 1.3.5 and pyradiomics 3.0.1 packages. All models were trained on the BMMR2 challenge training set. We removed one patient from the training set due to a missing DWI image with b-value $= 100 \, s/mm^2$.

We performed hyper-tuning of the number of features k that were fed to the XGBoost classifier. For XGBoost classifier, parameters hyper-tuning was focused on *min-child-weight*,*max-depth*, and *subsample* parameters. By assigning *min-child-weight* with values larger than 1 (default) and limiting *max-depth* value, we guaranteed that each node in the boosted tree represents a significant share of the samples and also that the total number of nodes in the boosted tree is limited. The combination of both assures the boosted tree is based on a low number of features. Using the *subsample* parameter we added another layer of mitigation against model-overfit by randomly selecting a new subset of samples on each training iteration. To avoid over-fitting, parameters hyper-tuning was performed with K-fold cross-validation strategy using the training set.

We handled label imbalance (70% non-pCR, 30% pCR) by adjusting XGBoost's *scale-pos-weight* parameter based on the proportion of pCR and non-

---

[2] https://github.com/TechnionComputationalMRILab/PD-DWI.

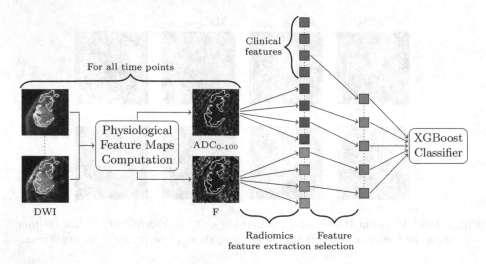

Fig. 3. Model architecture: The acquired DWI data at each time-point is decomposed into the different physiological cues. 3D Radiomics features are extracted from the segmented tumor in ADC and F maps and categorical clinical data are concatenated. Then a feature selection process is applied to select the most informative features. Finally, the selected features are fed into an XGBoost classifier.

pCR samples in the dataset. Hence, we encouraged XGBoost to correct errors on pCR samples despite their smaller share in dataset.

3   Results

Relation Between DWI Signal Attenuation Decay and pCR Prediction. Table 1 summarizes the comparison between all five model settings. Our PD-DWI model achieved a higher AUC score than the ADC-only models and the F-only model. This result suggests changes in both cellular density and blood flow in the micro-capillary network as a response to NAC. Amongst the ADC-only models, ADC_{0-800} had the highest AUC score, followed by ADC_{0-100}. AUC scores of $ADC_{100-800}$ and F-only model scores were relatively similar. With the release of test set labels, we evaluated the statistical significance of PD-DWI model's improved performance. PD-DWI model achieved the best Cohen's Kappa (κ) score whilst ADC_{0-800} was second to last. PD-DWI model's improved performance was statistically significant compared to F-only model (Sensitivity test, 0.65, $p < 0.05$) and baseline model (Sensitivity test, 0.6, $p < 0.05$).

Table 1. Models prediction performance comparison at time-point T2 (mid-NAC).

	AUC	F_1	κ
Baseline	0.8065	0.5714	0.4038
ADC_{0-100}-only	0.8581	0.7391	0.6215
$ADC_{100-800}$-only	0.8465	0.6667	0.494
ADC_{0-800}-only	0.8781	0.6531	0.4823
F-only	0.8423	0.6522	0.4953
PD-DWI	**0.8849**	**0.7391**	**0.6215**

Table 2. BMMR2 Challenge performance (https://qin-challenge-acrin.ce ntralus.cloudapp.azure.com/competit ions/2\#results) comparison at time-point T2 (mid-NAC).

	AUC
Pre-challenge benchmark	0.78
DKFZ Team	0.8031
IBM - BigMed	0.8380
PennMed CBIG	0.8397
PD-DWI	**0.8849**

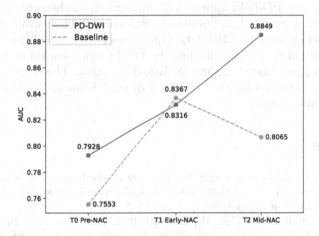

Fig. 4. Model performance at the different phases of NAC treatment.

Table 2 presents a comparison between the results obtained by our PD-DWI model and the current BMMR2 challenge leaderboard at time-point T2 (mid-NAC). Our PD-DWI model outperformed the best score in the challenge. To the best of our knowledge, our model has the best pCR prediction performance for the BMMR2 dataset.

Early Prediction of pCR over the Course of NAC. We evaluated the added-value of the PD-DWI model for early prediction of pCR following NAC compared to the baseline model. Figure 4 summarizes model performance as a function of the different time points of NAC treatment. As expected both models' pCR predictions improve given additional imaging data acquired during the course of the treatment. Specifically, our PD-DWI model consistently outperforms the baseline model in both T0 and T2. While our model performed about the same on T1, our approach does not require contrast injection and DCE-MRI acquisition. These findings suggest that changes in pseudo-diffusion and pseudo-diffusion fraction are more evident compared to changes in overall ADC values and SER values during the course of the treatment. Finally, our PD-DWI model performance using only pre-NAC data achieved a higher pCR prediction performance compared to the BMMR2 benchmark which used all time-points as provided by challenge organizers (0.7928 vs 0.78).

4   Conclusions

We introduced PD-DWI, a physiologically decomposed DWI machine-learning model, to predict pCR following NAC in invasive breast cancer. Our model accounts for the different physiological cues associated with pCR as reflected by the DWI signal rather than using aggregated information by means of the ADC map. The proposed PD-DWI approach demonstrated a substantial improvement in pCR prediction over the best published results on a publicly available challenge for pCR prediction from mp-MRI data. Our model is based solely on clinical DWI data, thus eliminating the need for lengthy DWI acquisition times, Gadolinium-based contrast agent injections, and DCE-MRI imaging. The proposed approach can be directly extended for the prediction of pCR following NAC in additional oncological applications.

References

1. Baltzer, P., et al.: Diffusion-weighted imaging of the breast-a consensus and mission statement from the EUSOBI International Breast Diffusion-Weighted Imaging working group. Eur. Radiol. **30**(3), 1436–1450 (2020)
2. Banaie, M., Soltanian-Zadeh, H., Saligheh-Rad, H.R., Gity, M.: Spatiotemporal features of DCE-MRI for breast cancer diagnosis. Comput. Methods Programs Biomed. **155** (2018). https://doi.org/10.1016/j.cmpb.2017.12.015
3. Bhushan, A., Gonsalves, A., Menon, J.U.: Current state of breast cancer diagnosis, treatment, and theranostics. Pharmaceutics **13** (2021). https://doi.org/10.3390/pharmaceutics13050723
4. Cain, E.H., Saha, A., Harowicz, M.R., Marks, J.R., Marcom, P.K., Mazurowski, M.A.: Multivariate machine learning models for prediction of pathologic response to neoadjuvant therapy in breast cancer using MRI features: a study using an independent validation set. Breast Cancer Res. Treat. **173**(2), 455–463 (2018). https://doi.org/10.1007/s10549-018-4990-9
5. Chen, X., Chen, X., Yang, J., Li, Y., Fan, W., Yang, Z.: Combining dynamic contrast-enhanced magnetic resonance imaging and apparent diffusion coefficient maps for a radiomics nomogram to predict pathological complete response to neoadjuvant chemotherapy in breast cancer patients. J. Comput. Assisted Tomogr. **44** (2020). https://doi.org/10.1097/RCT.0000000000000978
6. Clark, K., et al.: The Cancer Imaging Archive (TCIA): maintaining and operating a public information repository. J. Digit. Imaging **26**(6), 1045–1057 (2013). https://doi.org/10.1007/s10278-013-9622-7

7. Duanmu, H., et al.: Prediction of pathological complete response to neoadjuvant chemotherapy in breast cancer using deep learning with integrative imaging, molecular and demographic data. In: Martel, A.L., et al. (eds.) MICCAI 2020. LNCS, vol. 12262, pp. 242–252. Springer, Cham (2020). https://doi.org/10.1007/978-3-030-59713-9_24
8. Ferlay, J., et al.: Cancer statistics for the year 2020: an overview. Int. J. Cancer **149** (2021). https://doi.org/10.1002/ijc.33588
9. Freiman, M., Voss, S.D., Mulkern, R.V., Perez-Rossello, J.M., Callahan, M.J., Warfield, S.K.: In vivo assessment of optimal b-value range for perfusion-insensitive apparent diffusion coefficient imaging. Med. Phys. **39** (2012). https://doi.org/10.1118/1.4736516
10. Gao, W., Guo, N., Dong, T.: Diffusion-weighted imaging in monitoring the pathological response to neoadjuvant chemotherapy in patients with breast cancer: a meta-analysis. World J. Surgical Oncol. **16** (2018). https://doi.org/10.1186/s12957-018-1438-y
11. Griethuysen, J.J.V., et al.: Computational radiomics system to decode the radiographic phenotype. Cancer Res. **77** (2017). https://doi.org/10.1158/0008-5472.CAN-17-0339
12. Gurney-Champion, O.J., et al.: Comparison of six fit algorithms for the intra-voxel incoherent motion model of diffusion-weighted magnetic resonance imaging data of pancreatic cancer patients. PLoS ONE **13**(4), e0194590 (2018)
13. Huang, Y., et al.: Prediction of tumor shrinkage pattern to neoadjuvant chemotherapy using a multiparametric MRI-based machine learning model in patients with breast cancer. Front. Bioeng. Biotechnol. 558 (2021)
14. Joo, S., et al.: Multimodal deep learning models for the prediction of pathologic response to neoadjuvant chemotherapy in breast cancer. Sci. Rep. **11**(1), 1–8 (2021)
15. Le Bihan, D., Breton, E., Lallemand, D., Aubin, M., Vignaud, J., Laval-Jeantet, M.: Separation of diffusion and perfusion in intravoxel incoherent motion MR imaging. Radiology **168**(2), 497–505 (1988)
16. Liang, J., et al.: Intravoxel incoherent motion diffusion-weighted imaging for quantitative differentiation of breast tumors: a meta-analysis. Front. Oncol. **10** (2020). https://doi.org/10.3389/fonc.2020.585486
17. Liu, Z., et al.: Radiomics of multiparametric MRI for pretreatment prediction of pathologic complete response to neoadjuvant chemotherapy in breast cancer: a multicenter study. Clin. Cancer Res. **25**(12), 3538–3547 (2019)
18. Newitt, D.C., et al.: ACRIN 6698/I-SPY2 Breast DWI (2021). https://doi.org/10.7937/TCIA.KK02-6D95. https://wiki.cancerimagingarchive.net/x/lwH9Ag
19. Partridge, S.C., et al.: Diffusion-weighted MRI findings predict pathologic response in neoadjuvant treatment of breast cancer: the ACRIN 6698 multicenter trial. Radiology **289** (2018). https://doi.org/10.1148/radiol.2018180273. https://pubs.rsna.org/doi/full/10.1148/radiol.2018180273
20. Song, D., Man, X., Jin, M., Li, Q., Wang, H., Du, Y.: A decision-making supporting prediction method for breast cancer neoadjuvant chemotherapy. Front. Oncol. **10** (2021). https://doi.org/10.3389/fonc.2020.592556
21. Suo, S., et al.: J. Transl. Med. **19** (2021). https://doi.org/10.1186/s12967-021-02886-3
22. Woodhams, R., et al.: ADC mapping of benign and malignant breast tumors. Magn. Reson. Med. Sci. **4** (2005). https://doi.org/10.2463/mrms.4.35
23. Zhang, M., et al.: Multiparametric MRI model with dynamic contrast-enhanced and diffusion-weighted imaging enables breast cancer diagnosis with high accuracy. J. Magn. Reson. Imaging **49** (2019). https://doi.org/10.1002/jmri.26285

Transformer Based Multi-view Network for Mammographic Image Classification

Zizhao Sun[1], Huiqin Jiang[1]([✉]), Ling Ma[1], Zhan Yu[2], and Hongwei Xu[2]

[1] Zhengzhou University, Zhengzhou, China
{iehqjiang,ielma}@zzu.edu.cn
[2] The First Affiliated Hospital of Zhengzhou University, Zhengzhou, China

Abstract. Most of the existing multi-view mammographic image analysis methods adopt a simple fusion strategy: features concatenation, which is widely used in many features fusion methods. However, concatenation based methods can't extract cross view information very effectively because different views are likely to be unaligned. Recently, many researchers have attempted to introduce attention mechanism related methods into the field of multi-view mammography analysis. But these attention mechanism based methods still partly rely on convolution, so they can't take full advantages of attention mechanism. To take full advantage of multi-view information, we propose a novel pure transformer based multi-view network to solve the question of mammographic image classification. In our primary network, we use a transformer based backbone network to extract image features, a "cross view attention block" structure to fuse multi-view information, and a "classification token" to gather all useful information to make the final prediction. Besides, we compare the performance when fusing multi-view information at different stages of the backbone network using a novel designed "(shifted) window based cross view attention block" structure and compare the results when fusing different views' information. The results on DDSM dataset show that our networks can effectively use multi-view information to make judgments and outperform the concatenation and convolution based methods.

Keywords: Multi-view · Mammographic image classification · Transformer · Cross view attention

1 Introduction

Breast Cancer is the most common cancer among women, and early diagnosis can significantly improve patients' survival. Now, early screening of breast cancer patients mainly relies on mammography. When using mammography to screen a patient, the patient's each breast will be sampled from two different positions (Craniocaudal, CC and Mediolateral Oblique, MLO) respectively, producing four images, which can be expressed as L-CC, R-CC, L-MLO, and R-MLO.

Since the doctor makes a diagnosis based on careful observation of the four images shown in Fig. 1, most mammographic image classification methods are multi-view based

© The Author(s), under exclusive license to Springer Nature Switzerland AG 2022
L. Wang et al. (Eds.): MICCAI 2022, LNCS 13433, pp. 46–54, 2022.
https://doi.org/10.1007/978-3-031-16437-8_5

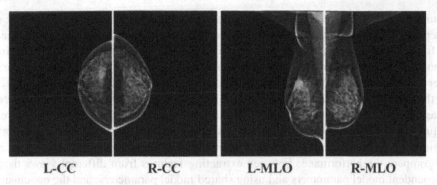

L-CC	**R-CC**	**L-MLO**	**R-MLO**

Fig. 1. The four images acquired by one mammography inspection.

method. However, most of the existing multi-view mammographic image analysis methods fuse multi-view information by simply concatenating features, which can't effectively extract cross view information because different views are likely to be unaligned and the fusions are usually conducted after pooling layers.

Recently, the methods using attention mechanism are widely studied. Some researchers have tried to use attention based methods to fuse multi-view information and obtain better performance than concatenation based multi-view methods. But all of these methods still partly rely on convolution, so they can't take full advantage of the attention mechanism.

To take full advantage of multi-view information, we propose a transformer based multi-view network for mammographic image classification. Our network use a transformer based backbone to extract image features and use "cross view attention block" after the last stage of the backbone to fuse multi-view information. A learnable "classification token" is employed to gather all useful information to make the prediction. The main contributions of our work are as follows:

1. We design a multi-view network based entirely on transformer architecture. The used "cross view attention block" can work better in a pure transformer style.
2. We introduce a learnable "classification token" into our network. This token can gather all useful information to make better prediction.
3. We design "(Shifted) Window based Cross View Attention Block". This structure can fuse cross view information anywhere in the network with low computational cost.

We conduct experiments on DDSM dataset [16], the results show that our network outperforms the concatenation and convolution based methods.

2   Related Work

Most multi-view mammographic image analysis methods fuse multi-view information using a simple strategy: feature concatenation. Sun et al. [1] use two independent CNN networks to extract CC view's and MLO view's features respectively and input the features into subsequent networks after concatenating them. Nasir et al. [2] use four independent CNN networks to extract four views' features respectively and concatenate the four views' features after global pooling layer, then the fused features are fed into a full connection (fc) layer to obtain the final results. Wu et al. [3] also concatenate the features after global pooling, but there are two differences between their work and reference [2]: (1) they use two successive fc layers after features concatenation so that the model can better learn the relationship between different views' features; (2) all CC views share a feature extraction network, and all MLO views share another feature extraction network. Besides, they compare the performance when using different fusion strategies. Li et al. [4] compare the performance between extracting features from different views using independent model parameters and using shared model parameters, and the conclusion is that the network can obtain better results when using shared model parameters to extract different views' features. Other networks also use feature concatenation strategy in different locations in models to fuse multi-view information (e.g., Liu et al. [10], Yan et al. [17], Yang et al. [8]). Because different views are likely to be unaligned, most concatenation based methods fuse multi-view information after global pooling layer. Therefore they lose much local information, which prevents them from taking full advantage of multi-view information. In our networks, we use "cross view attention block" to fuse multi-view information, which is more efficient than feature concatenation. And we explored where is the best position in the network to fuse multi-view information using designed "(Shifted) Window based Cross View Attention Block".

Recently, many researchers have attempted to introduce attention mechanism related methods into the field of multi-view mammography analysis. Zhao et al. [5] calculate a spatial attention map for the homogeneous views of the contralateral breasts and calculate a channel attention map for the inhomogeneous views of the ipsilateral breast. Then they conduct point-wise multiplication between each view and its two corresponding attention maps successively. Based on Relation Network [6], Ma et al. [7] design a Cross-View Relation Block structure to merge the visual and geometric information from another view's ROIs into the current view's ROIs. Yang et al. [8] propose a two-branch network: the first branch learn the relation between main view and inhomogeneous view of the same breast, using a Relation Block structure similar with [7]; another branch learn the relation between main view and homogeneous view of the contralateral breast, using the strategy of feature concatenation. Then they input these two feature maps into a feature fusion network to obtain the final result. Van et al. [9] propose the Cross-View Transformer, and they add this structure after the third stage of Resnet-18 to fuse information from CC view and MLO view. All of these attention mechanism based methods still partly rely on convolution, so they can't take full advantages of attention mechanism. Our network is based entirely on transformer and the results shows that the used "cross view attention block" can work better in a pure transformer style.

Transformer, proposed by Vaswani et al. [11], is a new structure based on self-attention mechanism. Dosovitskiy et al. [12] propose a network that is entirely based on transformer architecture. To use transformer in image classification task, they split an image into fixed-size patches, linearly embedd each of them, add an extra learnable "classification token", add position embeddings, and then feed the resulting sequence of vectors into a standard transformer encoder. In order to reduce the computational comlexity, Liu et al. [13] propose a hierarchical vision transformer model named Swin Transformer. Swin Transformer divide the image into small windows, conduct self-attention operations in each window instead of in the whole image, and solve the problem of information interaction between different windows through Shifted Window based Self-Attention. Tsai et al. [14] propose a Crossmodal Attention structure to learn the information between multi-modal data. This Crossmodal Attention structure is similar to the Cross-View Transformer proposed by van et al. [9] and cross view attention used in our network.

The above multi-view mammography analysis methods have great contributions, but they still have some limitations. Concatenation based methods lose much local information, so they can't use multi-view information effectively. Attention based methods can better learn cross view information, but all of them still partly rely on convolution, so they can't take full advantage of attention mechanism. And to our knowledge, no research has explored where is the best position in the network to fuse multi-view information. To address these limitations, we propose our network.

3 Methods

In this section, we describe three baseline models and our proposed models. All the multi-view models adopt three fusion strategies respectively, which is similar to [3]: (1) view-wise, only fuse the information from homogeneous views of contralateral breasts; (2) breast-wise, only fuse the information from inhomogeneous views of ipsilateral breast; (3) joint, fuse all views' information.

3.1 Baseline Models

There are two single-view baseline models: a CNN based model Resnet-50 [15] and the smallest swin transformer model Swin-T [13]. And there is a multi-view baseline model: Resnet-50 with feature concatenation after global pooling, whose structure is similar with reference [3].

3.2 Proposed Models

An overview of our proposed models is shown in Fig. 2. Our models first input four images of the same case simultaneously and feed them into four same backbone networks with shared weights. Then, a learnable "classification token" is concatenated to all feature vectors outputted from backbone networks respectively. To integrate a view's own information into the "classification token", the concatenated feature vectors is fed into a self-attention block. Then, models use cross view attention block to integrate cross-view information through different fusion strategies. Finally, the "classification tokens" is separated out and sent to the following layers to obtain the results.

Backbone. The backbone of our model is replaceable and its task is to extract features from images. It should be noted that if the backbone networks are CNN based networks, the feature maps (h, w, c) outputted from backbone networks should be flattened to feature vectors $(h \times w, c)$ before being fed into the following network.

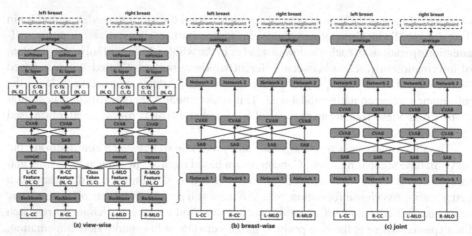

Fig. 2. Overview of the proposed models. There are some things to be explained: (1) the white-background elements are the data in different positions of the models, and some data are labeled with their dimensionality; the green-background elements are the components of the networks. (2) "SAB" means "Self Attention Block", "CVAB" means "Cross View Attention Block", "C-Tk" means "Classification Token", and "F" means "Features". (3) figure (a) is a complete structure of the view-wise model, but figure (b) and figure (c) leave out two parts of the structure for concision and the omitted parts are same as figure (a). (Color figure online)

Classification Token. Similar to ViT [12], we add a learnable classification token to the features sequence. And this token rather than the original image features is fed into the classification head to obtain the final result. The classification token can aggregate all useful information to obtain better results.

Self Attention Block (SAB). Our self attention block is similar to most transformer models. It consists of two layer norm (LN) layers, a multi-head self-attention (MSA) layer, and a 2-layer MLP with a GELU non-linearity in between. MSA is the standard version proposed in [11]. The self attention block is computed as:

$$z_{in} = x + E_{pos}$$
$$\hat{z} = MSA(LN(z_{in})) + z_{in} \qquad (1)$$
$$z_{out} = MLP(LN(\hat{z})) + \hat{z}$$

where E_{pos} denotes learnable 1D position embedding.

Cross View Attention Block (CVAB). The cross view attention block is similar to self attention block. Replace the MSA layer in self attention block with multi-head cross view attention (MCVA) layer, the self attention block becomes cross view attention block. The difference between self attention with cross view attention is: the Q, K, V matrixes of self attention are derived from the same view, but the Q matrix of cross view attention is derived from the target view α while the K matrix and the V matrix are derived from the source view β. The cross view attention can be expressed as $CVA_{\beta \rightarrow \alpha}$ when target view is α and source view is β.

(Shifted) Window Based Cross View Attention Block. Inserting the cross view attention block in the early stages of the backbone networks is difficult since it's computationally expensive to compute the attention on large feature maps' all pixels. To explore where is the best position to fuse multi-view information, we designed the window based cross view attention block (W-CVAB) and shifted window based cross view attention block (SW-CVAB). Replace all MSA operations in window based self attention block (W-SAB) and shifted window based self attention block (SW-SAB) proposed in reference [13] with MCVA operations, while guarantee that the two views are divided into windows and windows are shifted in the same way, the (S)W-SAB will become (S)W-CVAB. We use them in combination of SW-CVAB followed by W-CVAB.

With this structure, models can fuse cross-view information in every stage with an acceptable amount of computation. we compare the performance when fusing cross-view information at different stages of the model in Sect. 4.

4 Experiments and Results

4.1 Data

We choose all cancer cases in DDSM dataset as our dataset to evaluate the models and solve a binary classification problem to predict a breast is malignant or not malignant. Although these cases are labeled as cancer cases, most of them have cancer in just one breast. Therefore, the dataset we use includes both malignant and non-malignant images, and the positive and negative samples are roughly balanced.

Our dataset includes 891 cases and every case includes 4 images (L-CC, R-CC, L-MLO, R-MLO). So there are 3564 images, including 1753 malignant images and 1811 non-malignant images. And there are 1782 breasts, including 908 malignant breasts and 874 non-malignant breasts.

We use random cropping and random translation as data augmentation. We flip all images of right breasts horizontally and resize all images to 224×224 before sending them into models.

4.2 Setting

We employ an AdamW optimizer for 300 epochs using a step decay learning rate scheduler and 20 epochs of linear warm-up. We early stop the training when auc has not grown for 30 epochs and use the model with the best auc. An initial learning rate of 0.00001 and a weight decay of 0.01 are used. We employ 5-fold cross-validation to evaluate the models and report the mean of three runs. For all backbone networks, we use pre-trained weights on ImageNet.

Table 1. Comparison of fusing cross-view information at different stages. The single view version is marked with *. We report the mean and standard deviation of auc values over three runs. The best result is in bold font.

Method	AUC ± std.dev.
Last stage fusion	**0.840 ± 0.003**
Third stage fusion	0.837 ± 0.005
Second stage fusion	0.749 ± 0.003
Single view*	0.722 ± 0.001*
First stage fusion	0.712 ± 0.002
Before all stages fusion	0.666 ± 0.002
All stage fusion	0.782 ± 0.004

Table 2. Comparison between our primary models and baseline models. We report the mean and standard deviation of auc values over three runs. The best result in every fusion strategy is in bold font, and the best result in all models is marked with an underline.

Method	View-wise	Breast-wise	Joint
Resnet-50 (single view)	0.679 ± 0.002	0.679 ± 0.002	0.679 ± 0.002
Swin-T (single view)	0.722 ± 0.001	0.722 ± 0.001	0.722 ± 0.001
Resnet-50 (feature concatenation)	0.766 ± 0.002	0.675 ± 0.004	0.750 ± 0.006
Resnet-50 (last stage cva)	0.752 ± 0.003	0.681 ± 0.003	0.774 ± 0.005
Swin-T (feature concatenation)	0.834 ± 0.003	0.743 ± 0.002	0.835 ± 0.006
Swin-T (last stage cva)	0.840 ± 0.003	0.753 ± 0.003	0.842 ± 0.002
Swin-T (last stage cva, with C-Tk)	**0.843 ± 0.004**	**0.761 ± 0.002**	**0.846 ± 0.003**

4.3 Results and Analysis

Comparison of Fusing Cross-View Information at Different Stages. By   inserting the W-CVAB and SW-CVAB at different stages of backbones, the models can fuse cross view information at different stages. The results are shown in Table 1. From this set of results, it can be seen that models can achieve better results by fusing multi-view information at a higher layer of the models, and fusing multi-view information before the second stage has a negative impact on result. Therefore, our primary models fuse cross-view information after the last stage.

In this group of experiments, we all use the view-wise models with Swin-T back-bone and do not use classification token.

Comparison to Baseline Models. Table 2 shows the comparison between our primary models and other baseline models. In every fusion strategy, our model has the best result.

And from the last two lines we can see that all three types of models can benefit from using the learnable classification token.

In view-wise models, using cross view attention block at the last stage can improve the AUC by 0.118 when the backbone is Swin-T, but can only improve the AUC by 0.073 when the backbone is Resnet-50. We attribute this difference to the pure trans-former architecture when using Swin-T as the backbone. The cross view attention block can work better with transformer backbone. There is the same phenomenon in joint models.

In breast-wise models, using both methods to fuse multi-view information provides little AUC improvement when the backbone is Renset-50. But when choosing Swin-T as the backbone, using cross view attention block after the last stage can still provide 0.31 AUC improvement. We attribute this difference to the pure transformer architecture too.

The joint models are combinations of the view-wise models and breast-wise models, but they don't surpass the view-wise models very much. Combined with the poor results of breast-wise models, we believe that the inhomogeneous view of ipsilateral breast can't provide much useful information in this task.

5 Discussion and Conclusion

In this paper, We build a pure transformer based multi-view network. Our primary networks use a transformer based backbone network to extract images features, and use the cross-view attention block after the last stage of the backbone network to fuse cross-view information. Then our networks use a classification token to gather all useful information and make the final prediction using this token. Experiment results show that our networks have better results than the feature concatenation based multi-view networks and networks using attention mechanism but employing CNN back-bones.

There are some extended works we plan to do. The first one is to apply the network to a BI-RADS classification task and switch the network into a multi-task mode. The multi-task model may improve performance of every task compared to its single-task version. The second one is to try adding the classification token at the start of the whole network. It may be complicated when using Swin Transformer as the backbone because Swin Transformer conducts self-attention operations in each window and there are many windows in the early stage of Swin Transformer. Combining all windows' information may be difficult.

Acknowledgement. This research is supported by Zhengzhou collaborative innovation major special project (20XTZX11020).

References

1. Sun, L., Wang, J., Hu, Z., Xu, Y., Cui, Z.: Multi-view convolutional neural networks for mammographic image classification. IEEE Access **7**, 126273–126282 (2019)
2. Nasir Khan, H., Shahid, A.R., Raza, B., Dar, A.H., Alquhayz, H.: Multi-view feature fusion based four views model for mammogram classification using convolutional neural network. IEEE Access **7**, 165724–165733 (2019)

3. Wu, N., et al.: Deep neural networks improve radiologists' performance in breast cancer screening. IEEE Trans. Med. Imaging **39**, 1184–1194 (2020)

4. Li, C., et al.: Multi-view mammographic density classification by dilated and attention-guided residual learning. IEEE/ACM Trans. Comput. Biol. Bioinf. **18**, 1003–1013 (2021)

5. Zhao, X., Yu, L., Wang, X.: Cross-view attention network for breast cancer screening from multi-view mammograms. In: 2020 IEEE International Conference on Acoustics, Speech and Signal Processing (ICASSP), ICASSP 2020, pp. 1050–1054. IEEE, Barcelona (2020)

6. Hu, H., Gu, J., Zhang, Z., Dai, J., Wei, Y.: Relation networks for object detection. In: 2018 IEEE/CVF Conference on Computer Vision and Pattern Recognition (CVPR), pp. 3588–3597. IEEE, Salt Lake City (2018)

7. Ma, J., Li, X., Li, H., Wang, R., Menze, B., Zheng, W.-S.: Cross-view relation networks for mammogram mass detection. In: 2020 25th International Conference on Pattern Recognition (ICPR), pp. 8632–8638. IEEE, Milan (2021)

8. Yang, Z., et al.: MommiNet-v2: mammographic multi-view mass identification networks. Med. Image Anal. **73**, 102204 (2021)

9. van Tulder, G., Tong, Y., Marchiori, E.: Multi-view analysis of unregistered medical images using cross-view transformers. In: de Bruijne, M., et al. (eds.) MICCAI 2021. LNCS, vol. 12903, pp. 104–113. Springer, Cham (2021). https://doi.org/10.1007/978-3-030-87199-4_10

10. Liu, Y., et al.: Compare and contrast: detecting mammographic soft-tissue lesions with C^2-Net. Med. Image Anal. **71**, 101999 (2021)

11. Vaswani, A., et al.: Attention is all you need. In: Advances in Neural Information Processing Systems 30, pp. 5998–6008 (2017)

12. Dosovitskiy, A., et al.: An image is worth 16×16 words: transformers for image recognition at scale. In: ICLR (2021)

13. Liu, Z., et al.: Swin transformer: hierarchical vision transformer using shifted windows. In: ICCV (2021)

14. Tsai, Y.-H.H., Bai, S., Liang, P.P., Kolter, J.Z., Morency, L.-P., Salakhutdinov, R.: Multimodal transformer for unaligned multimodal language sequences. In: Proceedings of the 57th Annual Meeting of the Association for Computational Linguistics, pp. 6558–6569. Association for Computational Linguistics, Florence (2019)

15. He, K., Zhang, X., Ren, S., Sun, J.: Deep residual learning for image recognition. In: 2016 IEEE Conference on Computer Vision and Pattern Recognition (CVPR), pp. 770–778. IEEE, Las Vegas (2016)

16. Heath, M., Bowyer, K., Kopans, D., Kegelmeyer, P., Moore, R., Chang, K.: Current status of the digital database for screening mammography. In: Karssemeijer, N., Thijssen, M., Hendriks, J., van Erning, L. (eds.) Digital Mammography. Computational Imaging and Vision, vol. 13, pp. 457–460. Springer, Dordrecht (1998). https://doi.org/10.1007/978-94-011-5318-8_75

17. Yan, Y., Conze, P.-H., Lamard, M., Quellec, G., Cochener, B., Coatrieux, G.: Towards improved breast mass detection using dual-view mammogram matching. Med. Image Anal. **71**, 102083 (2021)

Intra-class Contrastive Learning Improves Computer Aided Diagnosis of Breast Cancer in Mammography

Kihyun You[1], Suho Lee[1,2], Kyuhee Jo[1,3], Eunkyung Park[1], Thijs Kooi[1], and Hyeonseob Nam[1(✉)]

[1] Lunit Inc., Seoul, Republic of Korea
{ukihyun,shlee,kjo,ekpark,tkooi,hsnam}@lunit.io
[2] Department of Data Science, Seoul National University of Science and Technology, Seoul, Republic of Korea
swlee@ds.seoultech.ac.kr
[3] Johns Hopkins University, Baltimore, MD, USA
kjo3@jh.edu

Abstract. Radiologists consider fine-grained characteristics of mammograms as well as patient-specific information before making the final diagnosis. Recent literature suggests that a similar strategy works for Computer Aided Diagnosis (CAD) models; multi-task learning with radiological and patient features as auxiliary classification tasks improves the model performance in breast cancer detection. Unfortunately, the additional labels that these learning paradigms require, such as patient age, breast density, and lesion type, are often unavailable due to privacy restrictions and annotation costs. In this paper, we introduce a contrastive learning framework comprising a Lesion Contrastive Loss (LCL) and a Normal Contrastive Loss (NCL), which jointly encourage models to learn subtle variations beyond class labels in a self-supervised manner. The proposed loss functions effectively utilize the multi-view property of mammograms to sample contrastive image pairs. Unlike previous multi-task learning approaches, our method improves cancer detection performance without additional annotations. Experimental results further demonstrate that the proposed losses produce discriminative intra-class features and reduce false positive rates in challenging cases.

Keywords: Mammography · Multi-task learning · Contrastive learning

1 Introduction

Mammography is the most common and cost-effective method for early detection of breast cancer—the second most common cancer in women worldwide [22].

Supplementary Information The online version contains supplementary material available at https://doi.org/10.1007/978-3-031-16437-8_6.

© The Author(s), under exclusive license to Springer Nature Switzerland AG 2022
L. Wang et al. (Eds.): MICCAI 2022, LNCS 13433, pp. 55–64, 2022.
https://doi.org/10.1007/978-3-031-16437-8_6

Computer aided diagnosis (CAD) systems have been used as a way to assist radiologists in reading millions of mammograms [11]. Thanks to the abundance of screening data and the recent advance in deep neural networks, CAD applications for mammography have been developed rapidly; some papers demonstrate performance comparable to its human counterpart [19,21].

Literature suggests that imitating how a radiologist reads a mammogram is a promising way to improve the performance of CAD models. For example, models that use multiple views (left and right or mediolateral oblique (MLO) and craniocaudal (CC)) in a mammographic exam outperform models that only work on a single view [12,15,17,25]. Furthermore, multi-task learning, where a model is trained to predict various radiological features such as the Breast Imaging-Reporting and Data System (BI-RADS) category [23], breast density, radiological subtype of a lesion as well as patient features like age, has shown to be effective in improving the performance of cancer detection [12,25]. However, this extra information is often unavailable due to privacy issues and huge annotation costs.

Contrastive learning is a training technique that can bypass this problem by learning useful features from the data themselves. Its goal is to pull similar examples (i.e., positive pairs) closer and push dissimilar examples (i.e., negative pairs) farther in the embedding space. Contrastive learning has recently surged in popularity for its success in self-supervised learning [4,8] and supervised representation learning [7,18,20]. In medical image analysis, contrastive learning has been used to pre-train a triage system for mammograms [2] and to improve domain generalization across multiple vendors [14]. Yan et al. [24] uses contrastive learning to match lesion patches generated from two different views of a patient as a part of multi-task learning for mammography. Chen et al. [3] exploits anatomical symmetry of body parts to detect fraction on X-ray by pixel-level contrastive learning with flipped images.

In this paper, we propose two contrastive losses, a Lesion Contrastive Loss (LCL) and a Normal Contrastive Loss (NCL), that leverage the multi-view nature of mammograms to explore rich radiological features beyond binary (cancer vs. non-cancer) class labels. The LCL attracts embeddings of different views of the same lesion and repels embeddings of different lesions. Similarly, the NCL attracts embeddings of bilateral normal breasts and repels embeddings of normal breasts from different patients. By contrasting samples that have the same class but are from distinct patients, our method allows the model to indirectly learn variations within the class, i.e., the intra-class variations. These losses are optimized jointly along with the image-level cancer classification loss via multi-task learning.

The main contributions of this paper are summarized as follows. 1) We introduce LCL and NCL that enable the model to learn auxiliary information without explicit annotations. To the best of our knowledge, our framework is the first approach in mammography that exploits contrastive learning to model intra-class variations in a self-supervised manner. 2) We show that our losses improve the overall cancer detection performance of vanilla baseline model as well as previous multi-task learning approaches, demonstrating the general applicability and

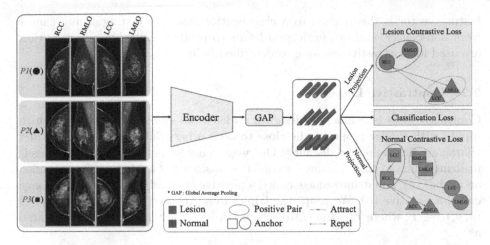

Fig. 1. Our contrastive learning framework consists of a Lesion Contrastive Loss (LCL) and a Normal Contrastive Loss (NCL). The LCL learns the similarity of a lesion seen at different views of a patient and the dissimilarity of lesions from different patients. The NCL, in contrast, utilizes parenchymal symmetry between both breasts of a normal patient and diversity of normal breasts from different patients.

efficacy of our method. 3) Extensive experiments on a large scale in-house dataset show that our method reduces false positive rates on challenging negative cases and produces features that are transferable to more fine-grained classification tasks. These results suggest that LCL and NCL combined generates features that more aptly reflect intra-class variations.

2   Method

Our work aims to exploit the four-view nature of mammograms (RCC, LCC, RMLO, and LMLO) to learn intra-class variations without using extra labels. Inspired by contrastive clustering [13], we introduce two contrastive losses— Lesion Contrastive Loss (LCL) and Normal Contrastive Loss (NCL)—that together encourage the model to explore intricate variability within class beyond the simple binary classification defined by the presence or absence of lesions. The overview of the proposed method is illustrated in Fig. 1.

2.1   Multi-task Learning

Following the success of multi-task learning in mammograms [12,24], our method is built upon a multi-task learning framework comprising three losses: an image-level classification loss, LCL and NCL. As shown in Fig. 1, an input mammogram is first fed into a fully-convolutional encoder, which consists of a pre-trained convolutional neural network followed by additional convolutional layers to further abstract the output feature maps. Global average pooling (GAP) is subsequently applied to reduce the feature maps into a 1-dimensional vector. The produced feature vector is then passed to a classification head for distinguishing cancer from non-cancer, and two projection heads to produce embedding vectors that are used for contrastive learning as described below.

2.2   Contrastive Losses

Contrastive loss is a function designed to learn an embedding space in which inputs with similar properties lie close to each other while inputs with distinct characteristics are placed distantly. One way to implement the contrastive loss is utilizing a unit of three images—a triplet—consisting of a reference image called an anchor (a), a matching image called a positive (p), and a non-matching image called a negative (n). We adopt the triplet loss proposed in [20]. Given triplets $(a, p, n) \in T$, where T is the entire set of triplets, the triplet loss is formulated as

$$L_{triplet} = \frac{1}{|T|} \sum_{(a,p,n)\in T} max(d(a,p)^2 - d(a,n)^2 + m, 0), \tag{1}$$

where m is a margin and $d(i, j)$ is the L2 distance between two images i and j on the embedding space. The triplet loss tries to keep the distance from the anchor (a) to the positive (p) smaller than the distance to the negative (n) by a margin m. We apply this formula to both LCL and NCL.

Lesion Contrastive Loss. Radiologists read mammograms not only by assessing the malignancy of a lesion (e.g. BI-RADS category) but also by describing various radiological features of the lesion such as the lesion type (e.g. mass, calcification, asymmetry, and distortion), shape (e.g. oval, round, and irregular), size, density, etc. These features are not captured when a model is trained merely with malignancy labels such as BI-RADS categories or biopsy results.

Motivated by this, we introduce the Lesion Contrastive Loss (LCL) that indirectly learns the various aspects of lesions by comparing breasts containing lesions in a self-supervised manner. In mammogram screening, lesions often in two different images called CC and MLO views. In LCL, the two images containing the same lesion (e.g. RCC and RMLO from a patient having cancer in the right breast) are pulled together, while images of different lesions (e.g. RCCs of two patients having cancer in their right breasts) are repelled from each other. Let an image containing a lesion be an anchor a. The ipsilateral image of the same lesion acts as a positive (p) and other patient's image that also contain a lesion (hence a different lesion) act as a negative (n) in Eq. 1

Normal Contrastive Loss. Normal breasts (i.e., breasts that contain neither a malignant nor benign lesion) exhibit radiological variations such as parenchymal patterns (breast density), unsuspicious benign patterns, surgical scars, and other textural characteristics. Literature indicates that these features may associate with the risk of cancer [1]. Additionally, a normal contralateral breast of a positive exam has a higher risk of developing cancer [10,26].

The normal Contrastive Loss (NCL) is designed to make better use of this latent variability among normal breasts. Based on the assumption that bilateral breasts are anatomically symmetric [3], the NCL draws together the images of left and right breasts of a normal patient (e.g. RCC and LCC of a normal patient) and pushes apart normal images from different patients (e.g. RCCs of two normal patients). Let an image without a lesion, i.e., a normal sample, to be an anchor (a). The corresponding view of its opposite breast acts as a positive (p), and an image of normal breast from a different patient acts as a negative (n).

2.3   Hard Negative Sampling

The proposed triplet strategy requires all four views of each case to be present in a single mini-batch. To this end, we draw batches on a case level. Furthermore, to ensure unbiased training, we maintain class balance by sampling an equal number of lesion (cases including a cancer or benign lesion) and non-lesion cases (cases not including a lesion) for each batch.

Naive sampling of triplets in a mini batch can be problematic because the negative pairs are often trivial. In other words, the negative sample is too distant from the anchor to enter the margin of the triplet loss and yield a significant training signal. To address this problem, we employ Hard Negative Mining (HNM) to select effective triplets that violate the margin constraint. Given an anchor image, we calculate pairwise L2 distances to all candidate negative images in the mini batch and select the five closest images as negatives. This not only accelerates training but also improves the final performance as demonstrated in ablation studies.

3   Experiments

3.1   Dataset and Metrics

We construct an in-house mammography dataset consisting of 88,753 exams of four-view full-field digital mammograms from 10 institutions in two countries; eight in South Korea and two in the USA. Among the exams, 11,276 are cancer positive confirmed by biopsy, 36,636 are benign confirmed by biopsy or at least one year of follow-up, and 40,841 are normal confirmed by at least one year of follow-up. The mammograms are taken using devices from three manufacturers: GE, Hologic and Siemens with ratio 48.9%, 46.1%, and 5.0% respectively. We sample a validation set of 2,943 (986 cancer, 963 benign, 994 normal) exams and a test set of 2,928 (953 cancer, 991 benign, and 984 normal)[1] exams to evaluate models.

We compute three metrics to evaluate the results: area under the receiver operating characteristic curve (AUROC), sensitivity at a fixed specificity (0.8), and specificity at a fixed sensitivity (0.8). The DeLong test [5] is used to generate confidence intervals for AUROC and a p-value when comparing two methods (Fig. 2).

---

[1] We randomly sampled 1,000 exams per category for each validation and test set, and a few outlier exams (e.g. breast implants) are excluded.

Table 1. Cancer diagnosis performance (AUROC with 95% confidence interval and p-value, sensitivity at specificity 0.8 and vice versa) of our method compared with three settings. Our method improves every baseline with a statistically significant margin.

Method	AUROC	p-value	Sens.	Spec.
Baseline	0.862 ± 0.010	-	0.760	0.742
+Ours	**0.878 ± 0.009**	<0.0001	**0.781**	**0.773**
Multi-task learning of [25]	0.867 ± 0.010	-	0.764	0.750
+Ours	**0.880 ± 0.009**	<0.0001	**0.780**	**0.775**
Multi-task learning of [12]	0.876 ± 0.009	-	0.782	0.778
+Ours	**0.885 ± 0.009**	<0.0001	**0.799**	**0.798**

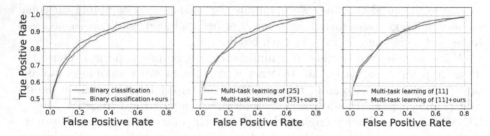

Fig. 2. ROC curves of each baseline and our method.

3.2   Implementation Detail

Our method is implemented in PyTorch, and trained with four GPUs of NVIDIA V100. We adopt ResNet-34 [9] pre-trained on ImageNet [6] as the backbone and use two additional convolution layers followed by global average pooling. A sigmoid classification layer is applied to predict a cancer score which is used to calculate binary cross-entropy loss. The projection layer for each of our contrastive losses is composed of 1×1 convolution and L2 normalization. We trained the models for 200K iterations with SGD where the learning rate, weight decay, and momentum are set to 0.005, 0.0001, and 0.9, respectively; the learning rate follows the cosine annealing learning rate scheduling [16].

During training, the input images are resized to 960×640 and the batch size is set to 128 (32 exams and 4 views per exam, as mentioned in Sect. 2.3) to fit our GPU memory constraint. To avoid overfitting, we augment images with various geometric transformations (translation, scaling, rotation, etc.) and photometric transformations (brightness, contrast, noise, etc.). For two contrastive losses, we use the triplet loss [20] with weight coefficient 5 and margin 0.2. We also investigate two more forms of contrastive loss: pair-wise contrastive loss [7] and InfoNCE loss [18]. Hyperparameters for these losses are specified in the ablation study in Sect. 3.3.

Table 2. Ablation analysis on the type of contrastive loss and the three components of our method. HNM means hard negative mining in Sect. 2.3.

Type of loss	LCL	NCL	HNM	AUROC	p-value	Sens.	Spec.
None (Baseline)				0.862 ± 0.010	-	0.760	0.742
Triplet [20]	✓			0.871 ± 0.010	0.0007	0.763	0.751
		✓		0.869 ± 0.010	0.0061	0.768	0.759
	✓	✓		0.874 ± 0.010	<0.0001	0.778	0.769
	✓	✓	✓	$\mathbf{0.878 \pm 0.010}$	<0.0001	**0.781**	**0.773**
Pairwise loss [7]	✓	✓	✓	0.873 ± 0.010	<0.0001	0.776	0.767
InfoNCE [18]	✓	✓	✓	0.876 ± 0.009	<0.0001	0.777	0.768

3.3 Result and Analysis

Performance Comparison. We evaluate the proposed loss functions by adding them to three settings: 1) a naive baseline with binary classification between cancer and non-cancer (i.e., benign and normal), 2) the multi-task learning setting proposed in MommiNet-v2 [25] which trains with BI-RADS categories as an additional regression task, and 3) the multi-task learning framework proposed in MVMT [12] which utilizes abundant extra information such as the BI-RADS category, breast density, radiological subtype (i.e., mass, micro-calcification, etc.), conspicuity, and patient age for additional tasks.

As shown in Table 1, adding auxiliary tasks using additional labels tends to improve the performance of the models. Notably, our unsupervised, label-free intra-class learning method outperforms supervised multi-task learning approaches in terms of the AUROC score. Furthermore, our method boosts the performance of previous multi-task approaches even further, which implies that LCL and NCL encourage models to learn orthogonal characteristics that are not captured in supervised multi-task learning methods.

Ablation Study. We perform an ablation analysis to demonstrate the contribution of each component (LCL, NCL, and HNM) and the choice of the specific implementation of contrastive loss. Table 2 shows that each part of the proposed framework contributes to the increased performance. Furthermore, our method achieves improvement regardless of the specific type of contrastive loss: triplet loss [20], pairwise loss [7] (weight 10 and margin 0.2), and InfoNCE [18] (weight 0.2 and temperature coefficient 2). The triplet loss yields the best performance.

Feature Analysis. We analyze the feature representation trained with our method by quantifying its capability to learn intra-class distinction. We train supervised linear classifiers on top of the trained features for two downstream tasks, lesion type (mass/asymmetry/distortion/micro-calcification) and breast density (A/B/C/D) predictions, and measure accuracy on the test set. Labels for these analyses were provided by board-certified radiologists on the in-house dataset. As shown in Table 3, our method significantly outperforms the baseline in both tasks with very low p-values, under 10^{-5}, which demonstrates that

Table 3. Linear classification accuracy (%) with trained feature representations. Results are shown with 95% confidence interval.

Method	Lesion type	Breast density
Baseline	53.9 ± 0.053	75.1 ± 0.030
+ Ours	**58.2 ± 0.055**	**78.1 ± 0.029**

Table 4. Subgroup analysis in terms of false positive rate (%) at sensitivity 0.8 with three different negative groups. Out method effectively reduces false positives in difficult negative groups.

Negatives	All	Biopsy-proven benign	Contralateral breast
Baseline	20.70	47.51	42.89
+ Ours	**16.41**	**40.33**	**35.51**

the learned representations effectively reflect intra-class variations among lesion breasts and normal breasts.

Subgroup Analysis. We perform a subgroup analysis on groups of data that have previously been prone to false positive predictions: (1) a set of biopsy-proven benign lesions and (2) a set of contralateral breasts (the opposite side to the breast diagnosed with biopsy-proven malignancy). The size of each negative subset is 181 and 949 cases respectively. A biopsy-proven benign lesion is not cancerous but is suspicious enough to be recalled for a pathological examination and can therefore be seen as a challenging negative sample. Furthermore, a breast that is contralateral to one that contains cancer does not only share anatomical similarity with its malignant counterpart such as texture and breast density, but also has a higher risk of developing cancer [10,26]. This, combined with the batch dependency in exam-based sampling, may undesirably cause models to associate contralateral breasts with the presence of cancer.

We show that the proposed loss functions mitigate false positive predictions for both subgroups. As shown in Table 4, our framework significantly reduces the false positive rate, demonstrating that learning intra-class variation helps handling ambiguous examples near the class boundary.

4 Conclusion

We presented an intra-class contrastive learning framework that learns auxiliary information without supervision. We proved the efficacy of learning an embedding space that reflects intra-class variations through extensive experiments with a total of 88,753 exams. Compared to supervised multi-task learning approaches, our LCL+NCL substantially improves the model performance without requiring additional labels. Furthermore, we showed that our losses help models to learn features that are transferable to more fine-grained classification tasks such as breast density and lesion subtype prediction. Lastly, our proposed method reduces false positive rate on radiologically challenging cases (biopsy-proved benign) and contextually challenging cases (contralateral breast of a cancerous breast).

References

1. Boyd, N.F., et al.: Mammographic density and the risk and detection of breast cancer. New Engl. J. Med. **356**(3), 227–236 (2007). https://doi.org/10.1056/NEJMoa062790, pMID: 17229950
2. Cao, Z., et al.: Supervised contrastive pre-training for mammographic triage screening models. In: de Bruijne, M., et al. (eds.) MICCAI 2021. LNCS, vol. 12907, pp. 129–139. Springer, Cham (2021). https://doi.org/10.1007/978-3-030-87234-2_13
3. Chen, H., et al.: Anatomy-aware Siamese network: exploiting semantic asymmetry for accurate pelvic fracture detection in X-ray images. CoRR abs/2007.01464 (2020). https://arxiv.org/abs/2007.01464
4. Chen, T., Kornblith, S., Norouzi, M., Hinton, G.: A simple framework for contrastive learning of visual representations (2020)
5. DeLong, E.R., DeLong, D.M., Clarke-Pearson, D.L.: Comparing the areas under two or more correlated receiver operating characteristic curves: a nonparametric approach. Biometrics 837–845 (1988)
6. Deng, J., Dong, W., Socher, R., Li, L.J., Li, K., Fei-Fei, L.: ImageNet: a large-scale hierarchical image database. In: 2009 IEEE Conference on Computer Vision and Pattern Recognition, pp. 248–255 (2009). https://doi.org/10.1109/CVPR.2009.5206848
7. Hadsell, R., Chopra, S., LeCun, Y.: Dimensionality reduction by learning an invariant mapping. In: 2006 IEEE Computer Society Conference on Computer Vision and Pattern Recognition (CVPR 2006), vol. 2, pp. 1735–1742 (2006). https://doi.org/10.1109/CVPR.2006.100
8. He, K., Fan, H., Wu, Y., Xie, S., Girshick, R.B.: Momentum contrast for unsupervised visual representation learning. CoRR abs/1911.05722 (2019). http://arxiv.org/abs/1911.05722
9. He, K., Zhang, X., Ren, S., Sun, J.: Deep residual learning for image recognition. CoRR abs/1512.03385 (2015). http://arxiv.org/abs/1512.03385
10. Hungness, E.S., et al.: Bilateral synchronous breast cancer: mode of detection and comparison of histologic features between the 2 breasts. Surgery **128**(4), 702–707 (2000)
11. Kim, H.E., et al.: Changes in cancer detection and false-positive recall in mammography using artificial intelligence: a retrospective, multireader study. Lancet Digit. Health **2**(3), e138–e148 (2020). https://doi.org/10.1016/S2589-7500(20)30003-0. https://www.sciencedirect.com/science/article/pii/S2589750020300030
12. Kyono, T., Gilbert, F.J., van der Schaar, M.: Multi-view multi-task learning for improving autonomous mammogram diagnosis. In: Doshi-Velez, F., et al. (eds.) Proceedings of the 4th Machine Learning for Healthcare Conference. Proceedings of Machine Learning Research, vol. 106, pp. 571–591. PMLR (2019). https://proceedings.mlr.press/v106/kyono19a.html
13. Li, Y., Hu, P., Liu, Z., Peng, D., Zhou, J.T., Peng, X.: Contrastive clustering. In: Proceedings of the AAAI Conference on Artificial Intelligence, vol. 35, no. 10, pp. 8547–8555 (2021). https://ojs.aaai.org/index.php/AAAI/article/view/17037

14. Li, Z., et al.: Domain generalization for mammography detection via multi-style and multi-view contrastive learning (2021)
15. Liu, Y., Zhang, F., Chen, C., Wang, S., Wang, Y., Yu, Y.: Act like a radiologist: towards reliable multi-view correspondence reasoning for mammogram mass detection. IEEE Trans. Pattern Anal. Mach. Intell. 1 (2021). https://doi.org/10.1109/TPAMI.2021.3085783
16. Loshchilov, I., Hutter, F.: SGDR: stochastic gradient descent with restarts. CoRR abs/1608.03983 (2016). http://arxiv.org/abs/1608.03983
17. Ma, J., Li, X., Li, H., Wang, R., Menze, B., Zheng, W.S.: Cross-view relation networks for mammogram mass detection. In: 2020 25th International Conference on Pattern Recognition (ICPR), pp. 8632–8638 (2021). https://doi.org/10.1109/ICPR48806.2021.9413132
18. van den Oord, A., Li, Y., Vinyals, O.: Representation learning with contrastive predictive coding. CoRR abs/1807.03748 (2018). http://arxiv.org/abs/1807.03748
19. Salim, M., et al.: External evaluation of 3 commercial artificial intelligence algorithms for independent assessment of screening mammograms. JAMA Oncol. 6(10), 1581–1588 (2020)
20. Schroff, F., Kalenichenko, D., Philbin, J.: FaceNet: a unified embedding for face recognition and clustering. CoRR abs/1503.03832 (2015). http://arxiv.org/abs/1503.03832
21. Sechopoulos, I., Teuwen, J., Mann, R.: Artificial intelligence for breast cancer detection in mammography and digital breast tomosynthesis: state of the art. In: Seminars in Cancer Biology, vol. 72, pp. 214–225. Elsevier (2021)
22. Siegel, R.L., Miller, K.D., Fuchs, H.E., Jemal, A.: Cancer statistics, 2022. CA: Cancer J. Clin. 72(1), 7–33 (2022). https://doi.org/10.3322/caac.21708. https://acsjournals.onlinelibrary.wiley.com/doi/abs/10.3322/caac.21708
23. Spak, D., Plaxco, J., Santiago, L., Dryden, M., Dogan, B.: BI-RADS® fifth edition: a summary of changes. Diagn. Int. Imaging 98(3), 179–190 (2017). https://doi.org/10.1016/j.diii.2017.01.001. https://www.sciencedirect.com/science/article/pii/S2211568417300013
24. Yan, Y., Conze, P.H., Lamard, M., Quellec, G., Cochener, B., Coatrieux, G.: Multitasking Siamese networks for breast mass detection using dual-view mammogram matching (2020). https://doi.org/10.1007/978-3-030-59861-7_32
25. Yang, Z., et al.: MommiNet-v2: mammographic multi-view mass identification networks. Med. Image Anal. 73, 102204 (2021). https://doi.org/10.1016/j.media.2021.102204. https://www.sciencedirect.com/science/article/pii/S1361841521002498
26. Yi, M., et al.: Predictors of contralateral breast cancer in patients with unilateral breast cancer undergoing contralateral prophylactic mastectomy. Cancer 115(5), 962–971 (2009)

Colonoscopy

BoxPolyp: Boost Generalized Polyp Segmentation Using Extra Coarse Bounding Box Annotations

Jun Wei[1,2,3], Yiwen Hu[1,2,3,6], Guanbin Li[7], Shuguang Cui[1,2,3], S. Kevin Zhou[1,4,5], and Zhen Li[1,2,3(✉)]

[1] School of Science and Engineering, The Chinese University of Hong Kong (Shenzhen), Shenzhen, China
lizhen@cuhk.edu.cn
[2] Shenzhen Research Institute of Big Data, Shenzhen, China
[3] The Future Network of Intelligence Institute, Shenzhen, China
[4] School of Biomedical Engineering and Suzhou Institute for Advanced Research, University of Science and Technology of China, Suzhou, China
[5] Institute of Computing Technology, Chinese Academy of Sciences, Beijing, China
[6] Institute of Urology, The Third Affiliated Hospital of Shenzhen University (Luohu Hospital Group), Shenzhen, China
[7] School of Computer Science and Engineering, Sun Yat-Sen University, Guangzhou, China

Abstract. Accurate polyp segmentation is of great importance for colorectal cancer diagnosis and treatment. However, due to the high cost of producing accurate mask annotations, existing polyp segmentation methods suffer from severe data shortage and impaired model generalization. Reversely, coarse polyp bounding box annotations are more accessible. Thus, in this paper, we propose a boosted **BoxPolyp** model to make full use of both accurate mask and extra coarse box annotations. In practice, box annotations are applied to alleviate the over-fitting issue of previous polyp segmentation models, which generate fine-grained polyp area through the iterative boosted segmentation model. To achieve this goal, a fusion filter sampling (FFS) module is firstly proposed to generate pixel-wise pseudo labels from box annotations with less noise, leading to significant performance improvements. Besides, considering the appearance consistency of the same polyp, an image consistency (IC) loss is designed. Such IC loss explicitly narrows the distance between features extracted by two different networks, which improves the robustness of the model. Note that our BoxPolyp is a plug-and-play model, which can be merged into any appealing backbone. Quantitative and qualitative experimental results on five challenging benchmarks confirm that our proposed model outperforms previous state-of-the-art methods by a large margin.

Keywords: Polyp segmentation · Colonoscopy · Colorectal cancer

J. Wei and Y. Hu—Equal Contributions.

© The Author(s), under exclusive license to Springer Nature Switzerland AG 2022
L. Wang et al. (Eds.): MICCAI 2022, LNCS 13433, pp. 67–77, 2022.
https://doi.org/10.1007/978-3-031-16437-8_7

1 Introduction

Colorectal Cancer (CRC) is one of the leading causes of malignant tumors death worldwide. As the precursor of CRC, colorectal polyps are the driver of CRC morbidity and mortality. Therefore, accurate polyp segmentation and diagnosis are of great significance to the survival of patients. Thanks to the evolution of computer technology, massive polyp segmentation models [1,4,6–8,17,20,23,24, 26,29] have been proposed and achieved remarkable performance.

However, these models are always plagued by data shortages and suffer from severe over-fitting issue. Since the popularity of U-Net [17] and FCN [13], most of polyp segmentation models [7,23] are based on convolutional neural networks (CNNs), which outperform traditional handcrafted ones but are data hungry. Unfortunately, accurate labeling of polyp masks is time-consuming and laborious, requiring pixel-by-pixel operation. Therefore, existing polyp segmentation datasets are relatively small. In particular, the widely adopted polyp training set [7,23] contains only 1,451 images, far from enough to feed a large capacity CNN model. Thus, models trained on this dataset exhibit the unstable performance and are sensitive to noise, which hampers the practical clinical usage.

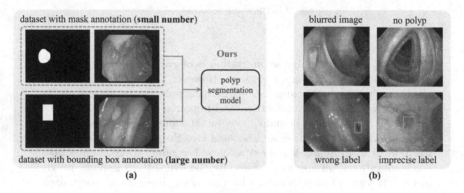

Fig. 1. (a) Our proposed polyp segmentation model using both accurate mask annotations and coarse bounding box ones. (b) Common annotation issues of the polyp detection dataset LDPolypVideo [14]. The second row shows the noisy annotations.

To provide effective clinical assistance, a generalized polyp segmentation model is urgently needed. In this paper, we struggle to achieve this goal using extra coarse bounding box annotations to expand the small segmentation dataset. Specifically, a large open-released polyp detection dataset LDPolypVideo [14] is adopted, which consists of 160 polyp video clips with 40,266 frames. Though all these images are labeled with only coarse bounding box annotations, they provide sufficient polyp appearance information and are much cheaper. Figure 1(a) shows the core idea of our method. Trained with a few images with mask annotations and a lot of images with bounding box annotations, a more generalized polyp segmentation model is achieved. However, directly applying the bounding box annotations of LDPolypVideo is suboptimal. Because the bounding box area contains many background pixels. Taking the bounding box area as polyp mask will bring a lot of noise. Besides, as shown in Fig. 1(b), LDPolypVideo contains many blurred images, images with no polyps, wrong labels and imprecise labels, which also will mislead the model training.

To make the most of the good parts of LDPolypVideo annotations and reduce the bad parts, we propose the novel BoxPolyp model which mainly consists of two modules: fusion filter sampling and image consistency loss. In practice, fusion filter sampling (FFS) aims to generate pseudo labels for high-confidence regions of each image in LDPolypVideo. By combining the raw bounding box annotations and the predicted masks (derived from the model trained on a small polyp segmentation dataset), FFS efficiently produces pixel-wise pseudo masks for deterministic regions. For uncertain regions, pseudo masks are inaccurate and therefore discarded. However, these discarded regions also contain valuable information. To fully explore these regions, we propose the image consistency (IC) loss instead of generating pseudo masks. IC loss applies two different networks to extract features from the same image and explicitly reduces the distance between features of the uncertain regions. By forcing feature alignment, our model could learn robust polyp feature representations, requiring no mask annotations.

In summary, our contributions are three-folds: (1) We are the first to boost a generalized polyp segmentation model through extra bounding box annotations. (2) We propose the fusion filter sampling to generate pseudo masks with less noise and design the image consistency loss to enhance the feature robustness of uncertain regions. (3) Our proposed BoxPolyp is a plug-and-play model, which can largely enhance polyp segmentation performance using different backbones.

2   Related Work

Traditional polyp segmentation models [20,28] are mostly based on low-level features (i.e., color, texture and boundary). But limited by the poor semantics, these models fail when dealing with complex scenarios. Recently, fully convolutional networks (FCN) [13] have been widely adopted for polyp segmentation and make great progress. For example, U-Net [17], U-Net++ [29] and ResUNet++ [11] use the encoder-decoder architecture to handle the segmentation tasks, which has become the standard paradigm for subsequent works. However, the polyp boundaries are not well handled by these methods. Afterwards, PsiNet [15], LODNet [5], PraNet [7], MSNet [27] and SFANet [8] force the model to learn the feature differences, which greatly enhances the model's perception for polyp boundaries and achieve the promising results.

Besides, ACSNet [24], HRENet [18] and CCBANet [16] pay more attention to context information. By adaptively aggregating multi-scale contexts, the ambiguity of local features will be reduced, thus leading to highly confident predictions. Unlike the above methods, SANet [23] deals with the polyp segmentation task in terms of data distribution. By eliminating the color bias of the image, SANet achieves robust performance gains in different scenarios. Furthermore, with the success of transformer in image processing, researchers start working on the long-distance dependency. For instance, PNSNet [12] uses a self-attention block to mine the temporal and spatial relations in polyp videos. Polyp-Pvt [6] directly introduces a transformer encoder to replace the widely used CNN backbones. Differently, Transfuse [25] combines both CNN and transformer to extract spatial correlation and global context. All these methods have achieved remarkable performance. But limited by training set, these models suffer from the overfitting issue. Therefore, we propose to use the cheap bounding box annotations to boost a generalized polyp segmentation model.

3   Method

Figure 2 depicts the whole framework of the proposed BoxPolyp segmentation model, consisting of two parts: fusion filter sampling and image consistency loss. Without special instructions, we use SANet [23] as our baseline model.

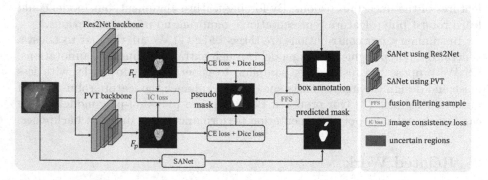

Fig. 2. The pipeline for our proposed BoxPolyp model. First, a SANet [23] trained on the small polyp segmentation dataset is used to predict the pixel-wise mask for each box-annotated image. Then, a FFS module combines the predicted mask and the box annotation to get the deterministic regions as pseudo labels. For regions of uncertainty, we propose the IC loss to reduce the distance between features extracted from two different backbones (*i.e.*, Res2Net [9] and PVT [22])

3.1   Fusion Filter Sampling

We integrate polyp detection dataset (*i.e.*, LDPolypVideo [14]) to enhance the polyp segmentation model. But LDPolypVideo is flawed in two ways. First, as shown in Fig. 1(b), there exists many wrongs and imprecise labels in LDPolypVideo, bringing noise for supervision. Second, bounding box annotations only provide coarse polyp contours and some background pixels are also included. Directly taking bounding box masks as pseudo labels will mislead the model. To solve the above issues, we propose fusion filter sampling (FFS) to generate pseudo masks with less noise interference.

Specifically, FFS filters out noise through object-level bounding box annotations and pixel-level pseudo masks, where object-level annotations weed out mislabeled or hard images and pixel-level masks filter out the background pixels in bounding box regions. For the object-level operation, given an image I, we first convert its bounding box annotations into a binary mask B, as shown in Fig. 3(a). Meanwhile, a pre-trained SANet [23] model (trained on small polyp segmentation dataset) is applied to get a coarse prediction P for I. Intuitively, if there is a big difference between B and P, I may be a hard sample or a mislabeled sample. In either case, I will be filtered out and not involved in the model training. Thus, the issues shown in Fig. 1(b) will be alleviated. In practice, we choose Dice $d = \frac{2BP}{B+P}$ to measure the difference between B and P. Only images with $d > 0.7$ will be selected out to minimize the impact of object-level wrong annotations. For the pixel-level operation, we combine the complementarity of B and P to refine the pseudo masks. Specifically, we choose pixels where both B and P are equal to 1 as the foreground F. Namely, $F = B \cap P$. Similarly, only pixels where both B and P are equal to 0 will be regarded as the background K. Namely, $K = (1 - B) \cap (1 - P)$. Other pixels belong to the uncertain regions, as shown in Fig. 3(a). During training, only F and K are involved in supervision, while uncertain regions are dealt with IC loss (described in Sect. 3.2). Through FFS, we maximize the utilization of bounding box annotations and minimize the potential noise interference.

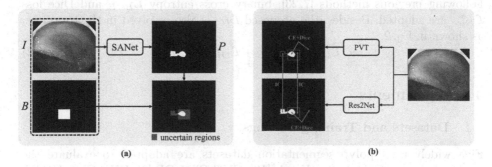

Fig. 3. (a) Refined pseudo mask generation using fusion filter sampling module, which consists of foreground, background and uncertain regions. (b) Different supervision for regions of certainty and regions of uncertainty.

3.2 Image Consistency (IC) Loss

By combining bounding box annotations and predicted masks, FFS module obtains deterministic foreground and background regions for supervision, as shown in Fig. 3(a). However, the regions of uncertainty are not supervised during training. Because no matter the box mask or the predicted mask is used as the pseudo label, it will bring a lot of noise which is harmful to the model generalization. In view of this, we propose the image consistency loss which mines supervisory information from the relationship between images, instead of pseudo labels.

Specifically, for each polyp image, we send it to two SANet models but with different backbone networks (*i.e.*, Res2Net [9] and PVT [22]), as shown in Fig. 2. Due to the different architectures (*i.e.*, CNN and Transformer), the features F_r and F_p extracted by Res2Net and PVT present different characteristics. Meanwhile, F_r and F_p come from the same image. They should have similar appearance. To bring the supervision for regions of uncertainty, we propose the IC loss to explicitly reduce the distance between F_r and F_p, as shown in Eq. 1.

$$\mathcal{L}_{IC} = \frac{\sum_{i,j}(F_r^{i,j} - F_p^{i,j})^2 \cdot U^{i,j}}{\sum_{i,j} U^{i,j}} \tag{1}$$

where i and j are the pixel indexes of polyp regions, U represents the mask of uncertain regions. Thus, IC loss focuses on regions without labels. Supervised by the IC loss, our model outputs more consistent predictions and greatly reduces the over-fitting risk.

3.3 Loss Function

Following previous methods [7,23], binary cross entropy $\mathcal{L}_{BCE}$ and Dice loss $\mathcal{L}_{Dice}$ are adopted. Besides, the proposed $\mathcal{L}_{IC}$ is also involved in the total loss, as shown in Eq. 2.

$$\mathcal{L}_{total} = \mathcal{L}_{BCE} + \mathcal{L}_{Dice} + \mathcal{L}_{IC} \tag{2}$$

4 Experiments

4.1 Datasets and Training Settings

Five widely used polyp segmentation datasets are adopted to evaluate the model performance, including Kvasir [10], CVC-ClinicDB [2], CVC-ColonDB [3], EndoScene [21] and ETIS [19]. For the comparability, we follow the same dataset partition as [7]. Besides, nine state-of-the-art methods are used for comparison, namely U-Net [17], U-Net++ [29], ResUNet [26], ResUNet++ [11], SFA [8], PraNet [7], SANet [23], MSNet [27] and Polyp-Pvt [6]. Pytorch is used to implement our BoxPolyp model. All input images are uniformly resized to 352×352. For data augmentation, random flip, random rotation and multi-scale training are adopted. The whole network is trained in an end-to-end way with a AdamW optimizer. Initial learning rate and batch size are set to 1e-4 and 16, respectively. We train the entire model for 80 epochs.

Table 1. Performance comparison with different polyp segmentation models. The red column represents the weighted average (wAVG) performance of different testing datasets. Below the dataset name is the image number of each dataset.

Methods	ColonDB		Kvasir		ClinicDB		EndoScene		ETIS		wAVG	
	380		100		62		60		196		798	
	Dice	IoU	Dice	IoU	Dice	IoU	Dice	IoU	Dice	IoU	Dice	IoU
U-Net	.512	.444	.818	.746	.823	.750	.710	.627	.398	.335	.561	.493
U-Net++	.483	.410	.821	.743	.794	.729	.707	.624	.401	.344	.546	.476
ResUNet	-	-	.791	-	.779	-	-	-	-	-	-	-
ResUNet++	-	-	.813	.793	.796	.796	-	-	-	-	-	-
SFA	.469	.347	.723	.611	.700	.607	.467	.329	.297	.217	.476	.367
PraNet	.712	.640	.898	.840	.899	.849	.871	.797	.628	.567	.741	.675
MSNet	.751	.671	.905	.849	.918	.869	.865	.799	.723	.652	.785	.714
SANet	.753	.670	.904	.847	.916	.859	.888	.815	.750	.654	.794	.714
Ours-Res2Net	.820	.741	.910	.857	.904	.849	.903	.835	.829	.742	.846	.771
Polyp-Pvt	.808	.727	.917	.864	.937	.889	.900	.833	.787	.706	.833	.760
Ours-Pvt	.819	.739	.918	.868	.918	.868	.906	.840	.842	.755	.851	.776

4.2 Quantitative Comparison

To prove the effectiveness of the proposed BoxPolyp, nine state-of-the-art models are used for comparison, as shown in Table 1. BoxPolyp surpasses previous methods by a large margin on the weighted average (wAVG) performace of five datasets, demonstrating the superior performance of the proposed methods. In addition, Fig. 4 shows the Dice values of the above models under different thresholds (used to binarize the mask). From these curves, we observe that Box-Polyp consistently outperforms other models, which proves its good capability for polyp segmentation.

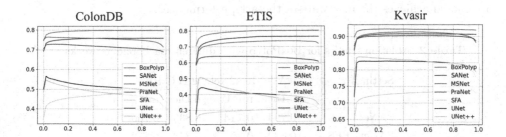

Fig. 4. Dice curves under different thresholds on three polyp datasets.

4.3 Visual Comparison

Figure 5 visualizes some predictions of different models. Compared with other counterparts, our method not only clearly highlights the polyp regions but also suppresses the background noise. Even for challenging scenarios, our model still handles well and generates accurate segmentation mask.

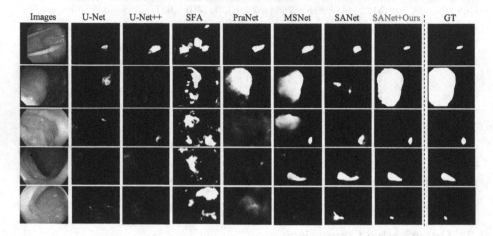

Fig. 5. Visual comparison between the proposed method and six state-of-the-art ones.

4.4 Ablation Study

To investigate the importance of each component in BoxPolyp, the weighted average (wAVG) performace is adopted. We evaluate the model on both Res2Net [9] and PVT [22] for ablation studies. As shown in Table 2, all proposed modules are beneficial for the final predictions. Combining all these modules, our model achieves the new state-of-the-art performance.

Table 2. Ablation studies for BoxPolyp with different backbone networks.

Settings	wAVG-Res2Net		wAVG-PVT	
	mDice	mIoU	mDice	mIoU
SANet	0.794	0.714	0.833	0.760
SANet+FFS	0.839	0.757	0.848	0.772
SANet+FFS+IC	0.846	0.771	0.851	0.776

5   Conclusion

Limited by the size of the dataset, existing polyp segmentation models are vulnerable to noise and suffer from over-fitting. For the first time, we leverage the cheap bounding box annotations to alleviate data shortage for a polyp segmentation task. Although coarse, these annotations can greatly improve the model generalization. It is achieved by the proposed FFS module and IC loss. In the future, we will explore the design of a weakly-supervised polyp segmentation model based on only bounding box annotations without masks.

Acknowledgement. This work is supported by the Guangdong Provincial Key Laboratory of Big Data Computing, The Chinese University of Hong Kong, Shenzhen, by NSFC-Youth 61902335, by Key Area R&D Program of Guangdong Province with grant No. 2018B030338001, by the National Key R&D Program of China with grant No. 2018YFB1800800, by Shenzhen Outstanding Talents Training Fund, by Guangdong Research Project No. 2017ZT07X152, by Guangdong Regional Joint Fund-Key Projects 2019B1515120039, by the NSFC 61931024 & 81922046, by helixon biotechnology company Fund and CCF-Tencent Open Fund.

References

1. Akbari, M., et al.: Polyp segmentation in colonoscopy images using fully convolutional network. In: 2018 40th Annual International Conference of the IEEE Engineering in Medicine and Biology Society (EMBC), pp. 69–72 (2018)
2. Bernal, J., Sánchez, F.J., Fernández-Esparrach, G., Gil, D., Rodríguez, C., Vilariño, F.: WM-DOVA maps for accurate polyp highlighting in colonoscopy: validation vs. saliency maps from physicians. Comput. Med. Imaging Graph. **43**, 99–111 (2015)
3. Bernal, J., Sánchez, J., Vilarino, F.: Towards automatic polyp detection with a polyp appearance model. Pattern Recogn. **45**(9), 3166–3182 (2012)
4. Brandao, P., et al.: Fully convolutional neural networks for polyp segmentation in colonoscopy. In: Medical Imaging 2017: Computer-Aided Diagnosis, vol. 10134, p. 101340F (2017)
5. Cheng, M., Kong, Z., Song, G., Tian, Y., Liang, Y., Chen, J.: Learnable oriented-derivative network for polyp segmentation. In: de Bruijne, M., et al. (eds.) MICCAI 2021. LNCS, vol. 12901, pp. 720–730. Springer, Cham (2021). https://doi.org/10.1007/978-3-030-87193-2_68
6. Dong, B., Wang, W., Fan, D.P., Li, J., Fu, H., Shao, L.: Polyp-pvt: polyp segmentation with pyramid vision transformers. arXiv preprint arXiv:2108.06932 (2021)
7. Fan, D.-P., et al.: PraNet: parallel reverse attention network for polyp segmentation. In: Martel, A.L., et al. (eds.) MICCAI 2020. LNCS, vol. 12266, pp. 263–273. Springer, Cham (2020). https://doi.org/10.1007/978-3-030-59725-2_26
8. Fang, Y., Chen, C., Yuan, Y., Tong, K.: Selective feature aggregation network with area-boundary constraints for polyp segmentation. In: Shen, D., et al. (eds.) MICCAI 2019. LNCS, vol. 11764, pp. 302–310. Springer, Cham (2019). https://doi.org/10.1007/978-3-030-32239-7_34
9. Gao, S., Cheng, M., Zhao, K., Zhang, X., Yang, M., Torr, P.H.S.: Res2net: a new multi-scale backbone architecture. IEEE Trans. Pattern Anal. Mach. Intell. **43**(2), 652–662 (2021)

10. Jha, D., et al.: Kvasir-SEG: a segmented polyp dataset. In: Ro, Y.M., et al. (eds.) MMM 2020. LNCS, vol. 11962, pp. 451–462. Springer, Cham (2020). https://doi.org/10.1007/978-3-030-37734-2_37

11. Jha, D., et al.: Resunet++: an advanced architecture for medical image segmentation. In: 2019 IEEE International Symposium on Multimedia (ISM), pp. 225–2255. IEEE (2019)

12. Ji, G.-P., et al.: Progressively normalized self-attention network for video polyp segmentation. In: de Bruijne, M., et al. (eds.) MICCAI 2021. LNCS, vol. 12901, pp. 142–152. Springer, Cham (2021). https://doi.org/10.1007/978-3-030-87193-2_14

13. Long, J., Shelhamer, E., Darrell, T.: Fully convolutional networks for semantic segmentation. In: Proceedings of the IEEE Conference on Computer Vision and Pattern Recognition, pp. 3431–3440 (2015)

14. Ma, Y., Chen, X., Cheng, K., Li, Y., Sun, B.: LDPolypVideo benchmark: a large-scale colonoscopy video dataset of diverse polyps. In: de Bruijne, M., et al. (eds.) MICCAI 2021. LNCS, vol. 12905, pp. 387–396. Springer, Cham (2021). https://doi.org/10.1007/978-3-030-87240-3_37

15. Murugesan, B., Sarveswaran, K., Shankaranarayana, S.M., Ram, K., Joseph, J., Sivaprakasam, M.: Psi-Net: shape and boundary aware joint multi-task deep network for medical image segmentation. In: 2019 41st Annual International Conference of the IEEE Engineering in Medicine and Biology Society (EMBC), pp. 7223–7226 (2019)

16. Nguyen, T.-C., Nguyen, T.-P., Diep, G.-H., Tran-Dinh, A.-H., Nguyen, T.V., Tran, M.-T.: CCBANet: cascading context and balancing attention for polyp segmentation. In: de Bruijne, M., et al. (eds.) MICCAI 2021. LNCS, vol. 12901, pp. 633–643. Springer, Cham (2021). https://doi.org/10.1007/978-3-030-87193-2_60

17. Ronneberger, O., Fischer, P., Brox, T.: U-Net: convolutional networks for biomedical image segmentation. In: Navab, N., Hornegger, J., Wells, W.M., Frangi, A.F. (eds.) MICCAI 2015. LNCS, vol. 9351, pp. 234–241. Springer, Cham (2015). https://doi.org/10.1007/978-3-319-24574-4_28

18. Shen, Y., Jia, X., Meng, M.Q.-H.: HRENet: a hard region enhancement network for polyp segmentation. In: de Bruijne, M., et al. (eds.) MICCAI 2021. LNCS, vol. 12901, pp. 559–568. Springer, Cham (2021). https://doi.org/10.1007/978-3-030-87193-2_53

19. Silva, J., Histace, A., Romain, O., Dray, X., Granado, B.: Toward embedded detection of polyps in WCE images for early diagnosis of colorectal cancer. Int. J. Comput. Assist. Radiol. Surg. 9(2), 283–293 (2014). https://doi.org/10.1007/s11548-013-0926-3

20. Tajbakhsh, N., Gurudu, S.R., Liang, J.: Automated polyp detection in colonoscopy videos using shape and context information. IEEE Trans. Med. Imaging 35(2), 630–644 (2015)

21. Vázquez, D., et al.: A benchmark for endoluminal scene segmentation of colonoscopy images. J. Healthc. Eng. 2017, 1–9 (2017)

22. Wang, W., et al.: PVTv2: improved baselines with pyramid vision transformer. Comput. Visual Media 8(3), 1–10 (2022). https://doi.org/10.1007/s41095-022-0274-8

23. Wei, J., Hu, Y., Zhang, R., Li, Z., Zhou, S.K., Cui, S.: Shallow attention network for polyp segmentation. In: de Bruijne, M., et al. (eds.) MICCAI 2021. LNCS, vol. 12901, pp. 699–708. Springer, Cham (2021). https://doi.org/10.1007/978-3-030-87193-2_66

24. Zhang, R., Li, G., Li, Z., Cui, S., Qian, D., Yu, Y.: Adaptive context selection for polyp segmentation. In: Martel, A.L., et al. (eds.) MICCAI 2020. LNCS, vol. 12266, pp. 253–262. Springer, Cham (2020). https://doi.org/10.1007/978-3-030-59725-2_25

25. Zhang, Y., Liu, H., Hu, Q.: TransFuse: fusing transformers and CNNs for medical image segmentation. In: de Bruijne, M., et al. (eds.) MICCAI 2021. LNCS, vol. 12901, pp. 14–24. Springer, Cham (2021). https://doi.org/10.1007/978-3-030-87193-2_2

26. Zhang, Z., Liu, Q., Wang, Y.: Road extraction by deep residual U-Net. IEEE Geosci. Remote Sens. Lett. 15(5), 749–753 (2018)

27. Zhao, X., Zhang, L., Lu, H.: Automatic polyp segmentation via multi-scale subtraction network. In: de Bruijne, M., et al. (eds.) MICCAI 2021. LNCS, vol. 12901, pp. 120–130. Springer, Cham (2021). https://doi.org/10.1007/978-3-030-87193-2_12

28. Zhou, S., et al.: A review of deep learning in medical imaging: image traits, technology trends, case studies with progress highlights, and future promises. Proc. IEEE 109(5), 820–838 (2020)

29. Zhou, Z., Rahman Siddiquee, M.M., Tajbakhsh, N., Liang, J.: UNet++: a nested U-Net architecture for medical image segmentation. In: Stoyanov, D., et al. (eds.) DLMIA/ML-CDS -2018. LNCS, vol. 11045, pp. 3–11. Springer, Cham (2018). https://doi.org/10.1007/978-3-030-00889-5_1

FFCNet: Fourier Transform-Based Frequency Learning and Complex Convolutional Network for Colon Disease Classification

Kai-Ni Wang[1], Yuting He[2], Shuaishuai Zhuang[3], Juzheng Miao[1], Xiaopu He[3], Ping Zhou[1], Guanyu Yang[2,4], Guang-Quan Zhou[1(✉)], and Shuo Li[5]

[1] School of Biological Science and Medical Engineering, Southeast University, Nanjing, China
`guangquan.zhou@seu.edu.cn`
[2] LIST, Key Laboratory of Computer Network and Information Integration (Southeast University), Ministry of Education, Nanjing, China
[3] The First Affiliated Hospital of Nanjing Medical University, Nanjing, China
[4] Centre de Recherche en Information Biomédicale Sino-Français (CRIBs), Strasbourg, France
[5] Department of Medical Biophysics, University of Western Ontario, London, ON, Canada

Abstract. Reliable automatic classification of colonoscopy images is of great significance in assessing the stage of colonic lesions and formulating appropriate treatment plans. However, it is challenging due to uneven brightness, location variability, inter-class similarity, and intra-class dissimilarity, affecting the classification accuracy. To address the above issues, we propose a **F**ourier-based **F**requency **C**omplex **N**etwork (FFCNet) for colon disease classification in this study. Specifically, FFC-Net is a novel complex network that enables the combination of complex convolutional networks with frequency learning to overcome the loss of phase information caused by real convolution operations. Also, our Fourier transform transfers the average brightness of an image to a point in the spectrum (the DC component), alleviating the effects of uneven brightness by decoupling image content and brightness. Moreover, the image patch scrambling module in FFCNet generates random local spectral blocks, empowering the network to learn long-range and local disease-specific features and improving the discriminative ability of hard samples. We evaluated the proposed FFCNet on an in-house dataset with 2568 colonoscopy images, showing our method achieves high performance outperforming previous state-of-the-art methods with an accuracy of 86.35% and an accuracy of 4.46% higher than the backbone. The project page with code is available at https://github.com/soleilssss/FFCNet.

Supplementary Information The online version contains supplementary material available at https://doi.org/10.1007/978-3-031-16437-8_8.

© The Author(s), under exclusive license to Springer Nature Switzerland AG 2022
L. Wang et al. (Eds.): MICCAI 2022, LNCS 13433, pp. 78–87, 2022.
https://doi.org/10.1007/978-3-031-16437-8_8

Keywords: Colon disease classification · Frequency learning ·
Complex convolutional network

1   Introduction

Accurate classification of early colon lesions plays an important role in diagnosis
and treatment [25]. Colorectal cancer is usually diagnosed at an advanced stage
due to insignificant early clinical symptoms, resulting in a high mortality rate
[5,10]. In clinical practice, colonoscopy is the most commonly used method to
diagnose colorectal lesions [1,17]. However, manual lesion classification is generally time-consuming and potentially subjective. Therefore, automated classification of colorectal lesions from colonoscopy images is critical in clinical analysis
because it: **1)** helps physicians determine the type of colonic disease; **2)** for-
mulates the most appropriate treatment options; **3)** compresses the duration of
colonoscopy [12].

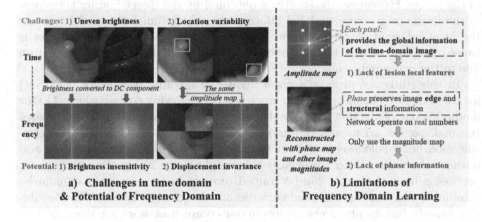

Fig. 1. Frequency-domain learning has potential and limitations in colon disease classification tasks. a) Potential: Spectrum is brightness-insensitive and displacement-
invariant. b) Limitations: Lack of image local information and phase information.

Existing research has achieved progress in classifying colon diseases [2,26],
but brightness imbalance and location variability remain intractable challenges
(Fig. 1a)). The unbalanced illumination from the endoscope probe induces apparent color and brightness differences even in normal images, degrading the classification performance and increasing the difficulty in the model generalization.
Some studies rely on data augmentation to resolve brightness imbalances [15].
Wei et al. designed the color exchange operation to force the model to focus
more on the target shape and structure by generating images of various colors
[23]. However, the significant differences in brightness and color of endoscopic
images prevent data augmentation from encompassing all distributions, resulting

in poor performance in some distinct regions. On the other hand, the locational variability of intestinal lesions is reflected in their appearance in various regions of the lumen wall. As a result, it is difficult for the network to learn all positional changes on small datasets and is prone to overfitting [11]. Several works applied transfer learning to alleviate the location variability issues by acquiring features from large datasets [13,22]. Nonetheless, different feature distributions between source and target domain in transfer learning can lead to insufficient adaptability in network training.

Frequency learning has great potential for its brightness insensitivity and displacement invariance, which is recognized to improve the ability to discriminate the colon diseases (Fig. 1a)) [18,24]. As we know, the DC component (only one value) in the spectrum represents the average brightness of the image [3]. This breaks the correlation between image content and brightness, prompting the model to focus more on the target shape and structure. The magnitude map has displacement invariance and remains unchanged with the spatial movement of the time-domain image, which avoids the influence of changes in the lesion location. Moreover, the phase map provides contour and structural information of the image. **However**, the spectrum obtained by the direct Fourier transform of the image only contains global information, which limits the model to learn the local information of the lesion. Besides, the lack of phase information is caused by the fact that the network only learns from the magnitude spectrum due to the real number operation (Fig. 1b)).

In this study, we propose a novel frequency learning framework (FFCNet) for colon disease classification. Our work has the following contributions: **1)** Our method is one of the first to study the automatic classification of colon diseases (normal, polyps, adenomas, cancers) in the whole process. This four-level classification helps colonoscopists to determine the type of lesions accurately and advance the clinical diagnosis of early colorectal cancer. **2)** For the first time, we propose a framework that can be trained directly in the frequency domain by combining complex convolutional networks and frequency learning. The convolution kernels, blocks, and architectures in our complex network are modified into complex operations to enable direct learning of the spectrum with complex numbers, thus avoiding the loss of phase features caused by real network operations. Moreover, the spectral brightness insensitivity and displacement invariance have the ability to resolve the uneven brightness and positional variability of time-domain images. **3)** We innovatively present an image patch scrambling module embedded in FFCNet to generate local spectrograms. Spectral blocks provide local features of lesions so that the model has stronger discriminative ability of inter-class similarity. Also, through random shuffling operations, spectral patches that appear at different locations in the image will reveal long-range information to the model. **4)** The proposed method is competitive against well-known CNN architectures in experiments. This work has also sparked discussions on how classical CNN architectures can exploit spatial and frequency features in solving real-world problems to improve performance.

2   Methodology

FFCNet (Fig. 2) is composed of a patch scrambling module and a frequency-domain complex network. The patch scrambling module (Sect. 2.1) obtains a complex spectrogram by slicing the time-domain image after the Discrete Fourier Transform (DFT) and then scrambling, which effectively aggregates local information and improves the learning ability of non-local information. The frequency-domain complex network (Sect. 2.2) is capable of handling complex numbers based on the original network architecture. Specifically, replacing convolution, ReLU, batch normalization (BN) with complex convolution, complex ReLU, and complex BN enables the network to calculate complex spectrum to extract richer feature information.

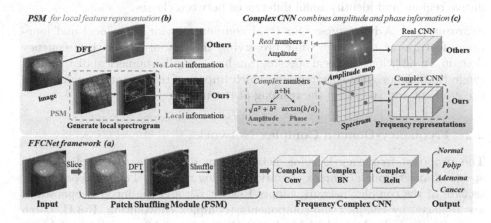

Fig. 2. FFCNet is an end-to-end architecture consisting of a patch scrambling algorithm and a frequency complex CNN. (a) An overview of the proposed method. (b) The patch scrambling module which is introduced in Sect. 2.1. (c) The frequency-domain complex CNN which is introduced in Sect. 2.2.

2.1   Patch Shuffling Module (PSM)

The proposed PSM transforming time-domain images into frequency-domain representations consists of image dicing, DFT, and random shuffling. Compared with the image spectrum without dicing, the network will only learn global information due to the spectral characteristics that a point in the frequency domain can affect the entire image, which neglects the necessary local features for colon classification. Hence, as shown in Fig. 2b), we perform DFT after slicing images to guide the network to focus on recognizable local features. On the other hand, the scrambled spectrum block further improves the long-distance feature learning of the model.

Given an input image I, we first uniformly partition the image into $K \times K$ patches denoted by matrix R. R_{ij} denotes an image patch where i and j are the horizontal and vertical indices, respectively ($1 \leq i \leq K$, $1 \leq j \leq K$). After the dicing, each block will be transformed to the frequency domain. Then, for each image block m, denoted as $f_m(x, y)$, with size $M \times N$, the DFT is computed according to the following expression:

$$F_m(u, v) = \sum_{x=0}^{M-1} \sum_{y=0}^{N-1} F_m(x, y) \, e^{-j2\pi(\frac{ux}{M} + \frac{vy}{N})} \tag{1}$$

Finally, the spectral patch will be randomly shuffled with probability p. Since the neatly arranged spectrum has been corrupted, in order to identify these randomly arranged spectral blocks, the classification network has to find discriminative regions and identify small differences between classes.

Summarized Advantages: Our PSM compensates for both local and long-range features of the image while preserving the advantages of the frequency domain. In addition, the local spectrogram has a smaller numerical distribution range than the original spectrogram, which improves the convergence degree of the gradient and speeds up the training of the model.

2.2 Frequency-Domain Complex Network

The proposed frequency learning network can directly learn complex-valued spectrograms through complex operations during training. The backbone architecture of the complex network adopts ResNet [7], and the internal operations are replaced by complex sub-components (complex convolution, ReLU, batch normalization). Each residual block consists of two $3*3$ convolutional layers and one connection path. Thus, the proposed network takes into account the advantages of frequency information features and the rich expressive power of complex operations.

Complex Convolution. In order to perform an operation equivalent to the traditional real-valued 2D convolution in the complex domain, the real part a and the imaginary part b of the complex matrix $P = a + bi$ in the spectrogram are respectively input into the network. Meanwhile, two sets of convolution kernels c and d are inserted to simulate the real and imaginary parts of the complex convolution kernel $Q = c + di$. Complex convolution can be expressed as:

$$P * Q = (a + bi) * (c + di) = \underbrace{(a * c - b * d)}_{real} + \underbrace{(a * d + b * c)}_{imaginary} i \tag{2}$$

where a, b, c, d are all real numbers.

Complex ReLU. The neural network relies on the ReLU function to introduce nonlinearity to promote the sparsity of the network. The ReLU function sets all negative values in the matrix to zero and does not change the remaining values.

The complex ReLU is the addition of the real and imaginary parts after applying ReLU respectively. Complex ReLU satisfies the Cauchy-Riemann equation when both the real and imaginary parts are strictly positive or strictly negative [21]. The specific formula is as follows:

$$ReLU\,(P) = \underbrace{ReLU\,(a)}_{real} + \underbrace{ReLU\,(b)i}_{imaginary} \tag{3}$$

Complex BN. BN is often employed to accelerate learning in neural networks. BN forcibly pulls the distribution of the input values of each layer of neural network back to a standard normal distribution with a mean of 0 and a variance of 1. For BN of complex numbers, it is unreasonable to translate and scale it so that it has a mean of 0 and a variance of 1. This normalization does not ensure that the variances of the real and imaginary parts are equal. It will be oval, possibly with high eccentricity. Hence, we treat it as a two-dimensional vector to change the data distribution.

Given a batch input x to compute the mean and variance, the normalized $\tilde{x}$ is expressed as:

$$\tilde{x} = \frac{(x - E\,(x))}{\sqrt{C}} \tag{4}$$

where c is the covariance matrix and $E\,(x)$ is the mean of the data. c is a 2×2 matrix represented as: where $R\,(x)$ and $I\,(x)$ represent the real and imaginary parts of $I\,(x)$, respectively. Similar to real number normalization, learnable reconstruction parameters γ and β are introduced to restore the feature distribution to be learned by the network. The difference is that the shift parameter is a complex parameter with two learnable components (real and imaginary). The scaling parameter is a 2×2 positive semi-definite matrix with only three degrees of freedom. There are three learnable components. The complex BN is defined as:

$$BN\,(x) = \gamma\left(\tilde{x}\right) + \beta \tag{5}$$

Summarized Advantages: Our elaborate complex CNN implements a full range of frequency-domain analysis, learning both magnitude and phase information in a unified architecture. Not only that, complex convolution, ReLU, and BN maintain the strength of easier optimization of complex numbers, and further improve the expressive ability of frequency features.

3   Experiments and Results

Experiment Protocol. 1) Datasets. The study included 3568 standard white-light endoscopic images including 865 normal, 843 polyps, 896 adenomas, and 964 cancers. We randomly split the dataset into training, validation and testing in a 6:2:2 ratio. **2) Settings.** We use version 1.1 of PyTorch [14] to perform all experiments on NVIDIA TITAN X (PASCAL) GPU machines and evaluate our

Table 1. FFCNet yields higher performance than different classical classification methods on each metric (%).

	Accuracy	Precision	Recall	F1-score
ResNet [7]	81.89	81.96	81.89	81.91
MobileNet [8]	81.11	81.14	81.11	81.07
EfficientNet [20]	84.44	84.76	84.44	84.55
DenseNet [9]	84.86	84.86	84.86	84.80
GoogLeNet [19]	85.42	85.72	85.42	85.52
CoAtNet [4]	85.93	86.08	85.93	85.94
Fast [4]	81.57	81.86	81.57	81.60
GFNet [16]	84.81	84.92	84.81	84.86
K-Space [6]	83.28	83.42	83.28	83.24
FFCNet	**86.35**	**86.61**	**86.35**	**86.44**

proposed method on the widely used classification backbone network ResNet-18. Input images are resized to a fixed size of 400×400. Random horizontal and vertical flips were applied for data augmentation. During training, we trained all network by SGD optimizers with a learning rate of 0.1 and a mini-batch size of 32 for 600 epochs. The probability p of patch shuffling was set to 0.3. During testing, data augmentation and patch scrambling algorithms were disabled. The diced spectrum of the original image was fed into the complex classification network for final prediction. **3) Evaluation metrics.** We evaluate the classification performance using four metrics: Accuracy, Precision, Recall and F1-score. More details are in our *Supplementary Material*.

Comparative Experiments Show the Superiority of Our FFCNet: The comparison of our network with several classical classification methods shows great potential for application in colonoscopy classification scenarios. Compared with other methods (Table 1), FFCNet has the highest accuracy (86.35%), precision (86.61%), recall (86.35%) and F1-score (86.44%). In particular, the accuracy of FFCNet is 4.46% higher than that of the backbone network ResNet, indicating that the frequency features provide a significant improvement to the architecture.

Comparisons with frequency, complex and transform networks also indicate the superiority of our FFCNet: a joint CNN and transformer network (CoAt-Net, 85.72%), a CNN network with added frequency modules (Fast, 81.57%), a transformer network with added frequency modules (GFNet, 84.81%), and other complex networks (K-Space, 83.28%). FFCNet outperforms them without requiring pre-training, indicating that the designed model is more efficient.

The results of the confusion matrix demonstrate the superiority of FFC-Net in discriminating similarity between classes. Our network obtains excellent performance in all categories Fig. 3. Not only that, the network achieves the highest accuracy in the most difficult to distinguish polyps (84.44%) and adenomas (79.53%), respectively. This finding is attributed to our PSM guiding the

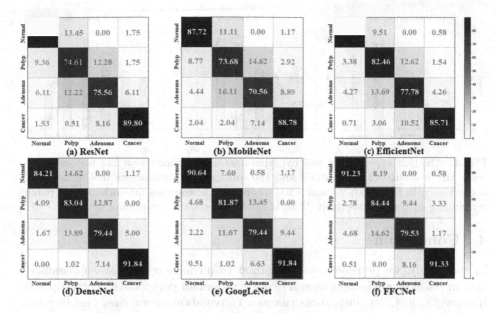

Fig. 3. The results of the confusion matrix illustrate that FFCNet obtains excellent performance in all categories. The numbers in the confusion matrix represent the percentage of predicted classes.

network to acquire disease-specific information by learning local and long-range features.

Ablation Experiments Demonstrate the Contribution of the Proposed Module: Figure 4a) shows the average accuracy of the ablation experiments quantitatively, demonstrating that each submodule contributes to the performance improvement. We first build a baseline model using magnitude maps with ResNet and gradually incorporate each of the submodules discussed in Sect. 3 into the baseline model. After adding PSM alone, local information is supplemented resulting in an accuracy improvement of 9.2% and 4.3%, compared to the baseline and baseline models with complex networks. Incorporating our proposed complex network to the baseline model improved by 3.9% verifies the ability of complex networks to mine phase information. The last two columns validate the importance of random shuffling significantly improving classification accuracy by learning distance information.

Hyperparameter Experiments Analyze the Superiority of the Architecture: We evaluate the effect of patch number and random scramble probability on the network (Fig. 4b), c)). The slicing operation brings local information of the image, so the accuracy of the network gradually increases as the number of slices grows. However, if the image patch is too small, it will destroy the advantages of frequency domain features and reduce model performance. Likewise, random shuffling allows the network to learn from information over long distances, yet the difficulty of network learning also increases. Therefore, we shuffle the image to stabilize the network performance with a certain probability.

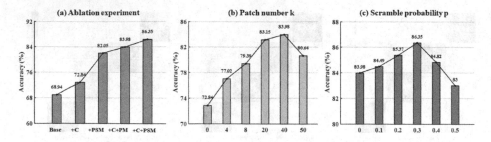

Fig. 4. Ablation experiments and hyperparameter experiments on the test set illustrate the necessity of submodules and the influence of parameters in the network, respectively. In figure (a), 'C' and 'PSM' represent the complex network and the patch scrambling module, respectively. 'PM' is the PSM removal scrambling.

4   Conclusion

In this study, we propose a novel method to advance colon disease classification in colonoscopy images from a frequency domain perspective. The proposed framework, FFCNet, introduces complex convolution operations, enabling the network to directly operate on complex spectra to obtain rich texture features and eliminate the influence of brightness imbalance. Furthermore, the patch scrambling algorithm we developed preprocesses the spectrogram so that the network can learn both long-range and local information. Finally, we compare the performance of the framework with other methods. The results show that our frequency-domain complex number framework is competitive with time-domain models in diagnosing colon diseases.

Acknowledgements. This work was supported by the National Key R&D Program Project (2018YFA0704102).

References

1. Bibbins-Domingo, K., et al.: Screening for colorectal cancer: US preventive services task force recommendation statement. JAMA **315**(23), 2564–2575 (2016)
2. Carneiro, G., Pu, L.Z.C.T., Singh, R., Burt, A.: Deep learning uncertainty and confidence calibration for the five-class polyp classification from colonoscopy. Med. Image Anal. **62**, 101653 (2020)
3. Chi, L., Jiang, B., Mu, Y.: Fast Fourier convolution. Adv. Neural Inf. Process. Syst. **33**, 4479–4488 (2020)
4. Dai, Z., Liu, H., Le, Q.V., Tan, M.: CoAtnNet: marrying convolution and attention for all data sizes. Adv. Neural Inf. Process. Syst. **34**, 3965–3977 (2021)
5. Elbediwy, A., et al.: Integrin signalling regulates YAP and TAZ to control skin homeostasis. Development **143**(10), 1674–1687 (2016)
6. Han, Y., Sunwoo, L., Ye, J.C.: k-Space deep learning for accelerated MRI. IEEE Trans. Med. Imaging **39**(2), 377–386 (2019)
7. He, K., Zhang, X., Ren, S., Sun, J.: Deep residual learning for image recognition. In: Proceedings of the IEEE Conference on Computer Vision and Pattern Recognition, pp. 770–778 (2016)

8. Howard, A.G., et al.: MobileNets: efficient convolutional neural networks for mobile vision applications. arXiv preprint arXiv:1704.04861 (2017)

9. Huang, G., Liu, Z., Van Der Maaten, L., Weinberger, K.Q.: Densely connected convolutional networks. In: Proceedings of the IEEE Conference on Computer Vision and Pattern Recognition, pp. 4700–4708 (2017)

10. Ladabaum, U., Dominitz, J.A., Kahi, C., Schoen, R.E.: Strategies for colorectal cancer screening. Gastroenterology **158**(2), 418–432 (2020)

11. Liu, X., Guo, X., Liu, Y., Yuan, Y.: Consolidated domain adaptive detection and localization framework for cross-device colonoscopic images. Med. Image Anal. **71**, 102052 (2021)

12. Mármol, I., Sánchez-de-Diego, C., Pradilla Dieste, A., Cerrada, E., Rodriguez Yoldi, M.J.: Colorectal carcinoma: a general overview and future perspectives in colorectal cancer. Int. J. Mol. Sci. **18**(1), 197 (2017)

13. Misawa, M., et al.: Artificial intelligence-assisted polyp detection for colonoscopy: initial experience. Gastroenterology **154**(8), 2027–2029 (2018)

14. Paszke, A., et al.: Pytorch: an imperative style, high- performance deep learning library. In: Advances in Neural Information Processing Systems 32 (2019)

15. Qadir, H.A., Shin, Y., Solhusvik, J., Bergsland, J., Aabakken, L., Balasingham, I.: Toward real-time polyp detection using fully CNNs for 2D Gaussian shapes prediction. Med. Image Anal. **68**, 101897 (2021)

16. Rao, Y., Zhao, W., Zhu, Z., Lu, J., Zhou, J.: Global filter networks for image classification. Adv. Neural Inf. Process. Syst. **34**, 980–993 (2021)

17. Rex, D.K., et al.: Colorectal cancer screening: recommendations for physicians and patients from the US multi-society task force on colorectal cancer. Gastroenterology **153**(1), 307–323 (2017)

18. Stuchi, J.A., Boccato, L., Attux, R.: Frequency learning for image classification. CoRR abs/2006.15476 (2020)

19. Szegedy, C., et al.: Going deeper with convolutions. In: Proceedings of the IEEE Conference on Computer Vision and Pattern Recognition, pp. 1–9 (2015)

20. Tan, M., Le, Q.: EfficientNet: rethinking model scaling for convolutional neural networks. In: International Conference on Machine Learning, pp. 6105–6114. PMLR (2019)

21. Trabelsi, C., et al.: Deep complex networks. CoRR abs/1705.09792 (2017)

22. Wang, Y., Feng, Z., Song, L., Liu, X., Liu, S.: Multiclassification of endoscopic colonoscopy images based on deep transfer learning. Comput. Math. Methods Med. **2021**, 1–21 (2021)

23. Wei, J., Hu, Y., Zhang, R., Li, Z., Zhou, S.K., Cui, S.: Shallow attention network for polyp segmentation. In: de Bruijne, M., et al. (eds.) MICCAI 2021. LNCS, vol. 12901, pp. 699–708. Springer, Cham (2021). https://doi.org/10.1007/978-3-030-87193-2_66

24. Xu, K., Qin, M., Sun, F., Wang, Y., Chen, Y.K., Ren, F.: Learning in the frequency domain. In: Proceedings of the IEEE/CVF Conference on Computer Vision and Pattern Recognition, pp. 1740–1749 (2020)

25. Zhang, R., et al.: Automatic detection and classification of colorectal polyps by transferring low- level CNN features from nonmedical domain. IEEE J. Biomed. Health Inform. **21**(1), 41–47 (2016)

26. Zhang, R., Zheng, Y., Poon, C.C., Shen, D., Lau, J.Y.: Polyp detection during colonoscopy using a regression-based convolutional neural network with a tracker. Pattern Recogn. **83**, 209–219 (2018)

Contrastive Transformer-Based Multiple Instance Learning for Weakly Supervised Polyp Frame Detection

Yu Tian[1,2,4]([✉]), Guansong Pang[3], Fengbei Liu[1], Yuyuan Liu[1], Chong Wang[1], Yuanhong Chen[1], Johan Verjans[1,2], and Gustavo Carneiro[1]

[1] Australian Institute for Machine Learning, University of Adelaide, Adelaide, Australia
[2] South Australian Health and Medical Research Institute, Adelaide, Australia
[3] Singapore Management University, Singapore, Singapore
[4] Harvard Medical School, Boston, MA, USA
ytian11@meei.harvard.edu

Abstract. Current polyp detection methods from colonoscopy videos use exclusively normal (i.e., healthy) training images, which i) ignore the importance of temporal information in consecutive video frames, and ii) lack knowledge about the polyps. Consequently, they often have high detection errors, especially on challenging polyp cases (e.g., small, flat, or partially visible polyps). In this work, we formulate polyp detection as a weakly-supervised anomaly detection task that uses video-level labelled training data to detect frame-level polyps. In particular, we propose a novel convolutional transformer-based multiple instance learning method designed to identify abnormal frames (i.e., frames with polyps) from anomalous videos (i.e., videos containing at least one frame with polyp). In our method, local and global temporal dependencies are seamlessly captured while we simultaneously optimise video and snippet-level anomaly scores. A contrastive snippet mining method is also proposed to enable an effective modelling of the challenging polyp cases. The resulting method achieves a detection accuracy that is substantially better than current state-of-the-art approaches on a new large-scale colonoscopy video dataset introduced in this work. Our code and dataset are available at https://github.com/tianyu0207/weakly-polyp.

Keywords: Polyp detection · Colonoscopy · Weakly-supervised learning · Video anomaly detection · Vision transformer

1 Introduction and Background

Colonoscopy has become a vital exam for colorectal cancer (CRC) early diagnosis. This exam targets the early detection of polyps (a precursor of colon cancer),

This work was supported by the Australian Research Council through grants DP180103232 and FT190100525.

© The Author(s), under exclusive license to Springer Nature Switzerland AG 2022
L. Wang et al. (Eds.): MICCAI 2022, LNCS 13433, pp. 88–98, 2022.
https://doi.org/10.1007/978-3-031-16437-8_9

which can improve survival rate by up to 95% [9,19,22,23,26]. During the procedure, doctors inspect the lower bowel with a scope to find polyps, but the quality of the exam depends on the ability of doctors to avoid mis-detections [19]. This can be alleviated by systems that automatically assist doctors detect frames containing polyps from colonoscopy videos. Nevertheless, accurate polyp detection is challenging due to the variable appearance, size and shape of colon polyps and their rare occurrence in an colonoscopy video.

One way to mitigate polyp detection challenges is with fully supervised training approaches, but given the expensive acquisition of fully labelled training sets, recent approaches have formulated the problem as an unsupervised anomaly detection (UAD) task [4,14,21,25]. These UAD methods [4,14,21,25] are trained with only normal training images and videos, and abnormal testing images and videos that contain polyps are detected as anomalous events. However, UAD approaches do not use training images or *snippets* (i.e., a set of consecutive video frames) containing polyps, so they are ineffective in recognising polyps of diverse characteristics, especially those that are small, partially visible, or irregularly shaped. As shown in a number of recent studies [16,17,20,22,24,29], incorporating some knowledge about anomalies into the training of anomaly detectors has improved the detection accuracy of hard anomalies. For example, weakly-supervised video anomaly detection (WVAD) [20,24,29] relies on video-level labelled data to train detection models. The video-level labels only indicate whether the whole video contains anomalies or not, which is easier to acquire than fully-labelled datasets with frame-level annotations. The WVAD formulation is yet to be explored in the detection of polyps from colonoscopy, but it is of utmost importance because colonoscopy videos are often annotated with video-level labels in real-world datasets.

Most existing WVAD methods [7,12,20,24,29,31,32] rely on multiple instance learning (MIL), in which all snippets in a normal video are treated as normal snippets, while each abnormal video is assumed to have at least one abnormal snippet. This approach can utilise video-level labels to train an anomaly-informed detector to find anomalous frames, but MIL methods often fail to select rare abnormal snippets in anomalous videos, especially the challenging abnormal snippets that have subtle visual appearance differences from the normal ones (e.g., small and flat colon polyps or frames with partially visible polyps–see Fig. 2). Consequently, they perform poorly in detecting these subtle anomalous snippets. Moreover, the WVAD methods above are trained on individual images, ignoring the important temporal dependencies in colonoscopy videos that can be explored for a more stable polyp detection performance.

In this paper, we introduce the first WVAD method specifically designed for detecting polyp frames from colonoscopy videos. Our method introduces a new contrastive snippet mining (CSM) algorithm to identify hard and easy normal and abnormal snippets. These snippets are further used to simultaneously optimise video and snippet-level anomaly scores, which effectively reduces detection errors, such as mis-classifying snippets with subtle polyps as normal ones, or normal snippets containing feces and water as abnormal ones. The exploration of global temporal dependency is also incorporated into our model with a transformer module, enabling a more stable anomaly classifier for colonoscopy videos. To resolve the poor modelling of local temporal dependency suffered by the transformer module [12,28], we also propose a convolutional transformer block to capture local correlations between neighbouring snippets. Our contributions are summarised as follows:

– To the best of our knowledge, this is the first work to tackle polyp detection from colonoscopy in a weakly supervised video anomaly detection manner.
– We propose a new transformer-based MIL framework that optimises anomaly scores in both snippet and video levels, resulting in more accurate anomaly scoring of polyp snippets.
– We introduce a new contrastive snippet mining (CSM) approach to identify hard and easy normal and abnormal snippets, where we pull the hard and easy snippets of the same class (i.e., normal or abnormal) together using a contrastive loss. This helps improve the robustness in detecting subtle polyp tissues and challenging normal snippets containing feces and water.
– We propose a new WVAD benchmark containing a large-scale diverse colonoscopy video dataset that combines several public colonoscopy datasets.

Our extensive empirical results show that our method achieves substantially better results than six state-of-the-art (SOTA) competing approaches on our newly proposed benchmark.

2   Method

Our method is trained with a set of weakly-labelled videos $\mathcal{D} = \{(\mathbf{F}_i, y_i)\}_{i=1}^{|\mathcal{D}|}$, where $\mathbf{F} \in \mathcal{F} \subset \mathbb{R}^{T \times D}$ represents pre-computed features (e.g., I3D [3]) of dimension D from T video snippets, and $y \in \mathcal{Y} = \{0, 1\}$ denotes the video-level annotation ($y_i = 0$ if $\mathbf{F}_i$ is a normal video and $y_i = 1$ otherwise), with each video being equally divided into a fixed number of snippets. Our method aims to learn a convolutional transformer MIL anomaly classifier for the T snippets, as in $r_{\theta,\phi} : \mathcal{F} \to [0,1]^T$, where this function is decomposed as $r_{\theta,\phi}(\mathbf{F}) = s_\phi(f_\theta(\mathbf{F}))$, with $f_\theta : \mathcal{F} \to \mathcal{X}$ being the transformer-based temporal feature encoder parameterised by θ (with $\mathcal{X} \subset \mathbb{R}^{T \times D}$) and $s_\phi : \mathcal{X} \to [0,1]^T$ denoting the MIL anomaly classifier, parameterised by ϕ, to optimise snippet-level anomaly scores.

2.1   Convolutional Transformer MIL Network

Motivated by the recent success of transformer architectures in analysing the global context of images [6] and videos [1], we propose to use a transformer to model the temporal information between the snippets of colonoscopy videos. Standard transformer without convolution [6] cannot learn the local structure between adjacent snippets, which is important for modelling local temporal relations because adjacent snippets are often highly correlated [20,25,29]. Hence, we

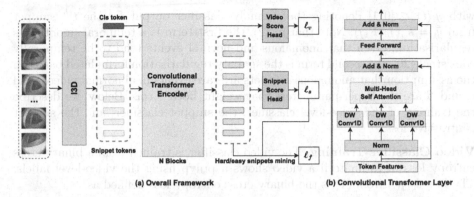

Fig. 1. (a) The architecture of our method consists an I3D [3] snippet feature extractor and a convolutional transformer MIL network. The I3D features are considered as snippet feature tokens to the transformer to predict snippet-wise anomaly scores using a snippet classifier. The Cls token is applied for a video classifier to predict if a video contains anomalies. The output features from the transformer are utilised to mine hard and easy snippets from normal and abnormal videos. The anomaly scores and hard/easy snippet representations are optimised by three proposed losses in (1). (b) The proposed temporal convolutional transformer layer replaces the linear projection with depthwise separable convolution (DW Conv1D) [5].

replace the linear token projection of the transformer by convolution operations. More specifically, we follow [28] and adopt the depth-wise separable 1D convolution [5] on the temporal dimension, as shown in Fig. 1(b). As shown in Fig. 1(a), the encoder comprises N convolutional transformer blocks that produce the final temporal feature representation $\mathbf{X} = f_\theta(\mathbf{F})$.

2.2 Transformer-Based MIL Training

The training of our model comprises a joint optimisation of a transformer-based temporal feature learning, a contrastive snippet mining (CSM) that is used to train a CSM-enabled MIL classifier, and a video-level classifier, with

$$\theta^*, \phi^*, \gamma^* = \arg\min_{\theta, \phi, \gamma} \ell_{cnt}(\mathcal{D}; \theta) + \ell_{snp}(\mathcal{D}; \theta, \phi) + \ell_{vid}(\mathcal{D}; \theta, \gamma) + \ell_{reg}(\mathcal{D}; \theta, \phi) \quad (1)$$

where $\ell_{cnt}(.)$ denotes a contrastive loss that uses the mined hard and easy normal and abnormal snippet features, $\ell_{snp}(.)$ is a loss function to train the snippet classifier $s_\phi(.)$ using the top k snippet-level anomaly scores from normal and abnormal videos, $\ell_{vid}(.)$ is a loss function to train the video classifier to predict whether the video contains anomalies, θ, ϕ and γ are respectively parameters of $\ell_{cnt}(.)$, $\ell_{snp}(.)$ and $\ell_{vid}(.)$, and the regularisation loss is defined by

$$\ell_{reg}(\mathcal{D}; \theta, \phi) = \sum_{(\mathbf{F}_i, y_i) \in \mathcal{D}} \alpha \left(\frac{1}{T} \sum_{t=2}^{T} (\tilde{y}_i(t) - \tilde{y}_i(t-1))^2 \right) + \beta \left(\frac{1}{T} \sum_{t=1}^{T} |\tilde{y}_i(t)| \right), \quad (2)$$

with $\tilde{y}_i(t) \in [0,1]$ denoting the anomaly classifier output for the tth snippet from $\tilde{y}_i = s_\phi(f_\theta(\mathbf{F}_i))$. Note that in (2), the first term is a temporal smoothness regularisation, given that anomalous and normal events tend to be temporally consistent [20], the second term is the sparsity regularisation formulated based on the assumption that anomalous snippets are rare events in abnormal videos, and α and β are the hyper-parameters that weight both terms. Below, we describe the training of the video-level classifier, the snippet classifier, and the snippet contrastive loss.

Video Classifier Training. The video classifier is trained from a binary cross entropy loss to estimate if a video shows a polyp using the video-level labels. The loss $\ell_{vid}(.)$ from (1) is the binary cross entropy loss defined as

$$\ell_{vid}(\mathcal{D}; \theta, \gamma) = - \sum_{(\mathbf{F}_i, y_i) \in \mathcal{D}} \left(y_i \log(v_\gamma(f_\theta(\mathbf{F}_i))) + (1 - y_i) \log(1 - v_\gamma(f_\theta(\mathbf{F}_i))) \right), \quad (3)$$

where $v_\gamma : \mathcal{X} \to [0,1]$ is the video level anomaly classifier parameterised by γ.

Snippet Classifier Training. The snippet classifier is optimised by training a top k ranking loss function using a set that contains the k snippets with the largest anomaly scores from $s_\phi(\mathbf{F})$ in (1). More specifically, we propose the following loss $\ell_{snp}(.)$ from (1) that maximises the separability between normal and abnormal videos:

$$\ell_{snp}(\mathcal{D}; \theta, \phi) = \sum_{\substack{(\mathbf{F}_i, y_i) \in \mathcal{D}, y_i=1 \\ (\mathbf{F}_j, y_j) \in \mathcal{D}, y_j=0}} \max\left(0, 1 - g_k(s_\phi(f_\theta(\mathbf{F}_i))) - g_k(s_\phi(f_\theta(\mathbf{F}_j)))\right), \quad (4)$$

where $g_k(.)$ returns the mean anomaly score from $s_\phi(.)$ of the top k snippets from a video [13,24].

Contrastive Snippet Mining. To make anomaly classification robust to hard normal and abnormal snippets, inspired by [31], we propose the following novel snippet contrastive loss:

$$\ell_{cnt}(\mathcal{D}; \theta) = \ell_c(\mathcal{D}^{HA}, \mathcal{D}^{EA}, \mathcal{D}^{EN}; \theta) + \ell_c(\mathcal{D}^{HN}, \mathcal{D}^{EN}, \mathcal{D}^{EA}; \theta), \quad (5)$$

where $\mathcal{D}^{HA}$ and $\mathcal{D}^{EA}$ represent sets of hard and easy abnormal snippets, while $\mathcal{D}^{HN}$ and $\mathcal{D}^{EN}$ denote sets of hard and easy normal snippets,

$$\ell_c(\mathcal{D}^{HA}, \mathcal{D}^{EA}, \mathcal{D}^{EN}; \theta) = \sum_{\mathbf{F}_i \in \mathcal{D}^{HA}, \mathbf{F}_j \in \mathcal{D}^{EA}} \log \frac{\exp[\frac{1}{\tau} f_\theta(\mathbf{F}_i)^\top f_\theta(\mathbf{F}_j)]}{\exp[\frac{1}{\tau} f_\theta(\mathbf{F}_i)^\top f_\theta(\mathbf{F}_j)] + \sum_{\mathbf{F}_m \in \mathcal{D}^{EN}} \exp[\frac{1}{\tau} f_\theta(\mathbf{F}_i)^\top f_\theta(\mathbf{F}_m)]},$$
$$(6)$$

and in a similar way we compute $\ell_c(\mathcal{D}^{HN}, \mathcal{D}^{EN}, \mathcal{D}^{EA}; \theta)$. The idea explored in (5) is to pull together easy and hard snippet features in $\mathcal{X}$ from the same class (normal or abnormal) and push apart features from different classes.

The selection of $\mathcal{D}^{HN}, \mathcal{D}^{EN}, \mathcal{D}^{HA}, \mathcal{D}^{EA}$ and their incorporation into our MIL learning framework is one key contribution of this work to address the poor detection accuracy of hard anomalous snippets in existing WVAD methods. Specifically, for abnormal videos, we first classify each of their T snippets with

$\hat{y}(t) = (\tilde{y}(t) > \epsilon)$, where $\tilde{y} = s_\phi(f_\theta(\mathbf{F}))$. We then identify the temporal edge snippets and missed pseudo abnormal snippets as hard anomalies $\mathcal{D}^{HA}$. For temporal edge detection, we use the erosion operator to subtract the original and eroded sequences and locate such transitional edge snippets, which are considered as hard anomalies (See Fig. 2 - temporal edge detection), and inserted into $\mathcal{D}^{HA}$. For locating the missed pseudo abnormal snippets, we assume that a subtle anomalous event (i.e., a small/flat polyp) happens in a region of K consecutive snippets when $\frac{R}{K}$ (majority) of them have $\hat{y}(t) = 1$, where K and R are respectively the hyper-parameters to control the temporal length of the pseudo abnormal region and the ratio of the minimum number of the abnormal pseudo snippets inside that region. The incorrectly predicted normal snippets inside abnormal regions (i.e., missed abnormal snippets in Fig. 2) are also inserted into $\mathcal{D}^{HA}$ as hard anomalies.

This hard anomaly selection process is motivated by the following two main observations: 1) subtle abnormal snippets from anomalous videos share similar characteristics to normal snippets (i.e., small and flat polyps) and consequently have low anomaly scores, and this can be easily identified from the adjacent abnormal snippets with higher anomaly scores since abnormal frames containing polyps are often contiguous; and 2) the transitional snippets between normal and abnormal events often contain noise such as water, endoscope pipe or partially visible polyps, so they are unreliable and can lead to inaccurate detection.

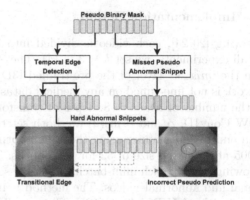

Fig. 2. Hard abnormal snippet mining algorithm to select temporal edge snippets and missed pseudo abnormal snippets. Those two types of hard anomalies represent: 1) transitional frames where polyps may be partially visible, or 2) subtle (i.e., small and flat) polyps that can lead to incorrect low anomaly scores.

Hard normal (HN) snippets (e.g., healthy frames containing water and feces) are collected by selecting the snippets with top k anomaly scores from normal videos since normal videos do not have any abnormalities, so the ones with incorrectly predicted higher scores can be deemed as hard normal. For easy snippet mining, we hypothesis that the snippets with the smallest k anomaly scores from normal videos and the snippets with top k anomaly scores from abnormal videos are easy normal (EN) and easy abnormal (EA).

3　Experiments and Results

3.1　Dataset

To form a real-world large-scale video polyp detection dataset, we collected colonoscopy videos from two widely used public datasets: Hyper-Kvasir [2] and LDPolypVideo [15]. The new dataset contains 61 normal videos without polyps and 102 abnormal videos with polyps for training, and 30 normal videos and 60 abnormal videos for testing. The videos in the training set have video-level labels and the videos in testing set contain frame-level labels. This dataset contains over one million frames and has diverse polyps with various sizes and shapes, making it one of the largest and most challenging colonoscopy datasets in the field. The dataset setup will be publicly available upon paper acceptance.

3.2　Implementation Details

Following [20,24], each video is divided into 32 video snippets, i.e., $T = 32$. For all experiments, we set $k = 3$ in (4). The 2048D input tokens are extracted from the 'mix_5c' layer of the pre-trained I3D [10] network. Note that the I3D network is not fine-tuned on any medical dataset. For the transformer block, we set the number of heads to 8, depth of transformer blocks to 12, and use a 3×1 DW Conv1D. α and β in (2) are both set to $5e - 4$. Our method is trained in an end-to-end manner using the Adam optimiser [11] with a weight decay of 0.0005 and a batch size of 32 for 200 epochs. The learning rate is set to 0.001. Following [20,24], each mini-batch consists of samples from 32 randomly selected normal and abnormal videos. The method is implemented in PyTorch [18] and trained with a NVIDIA 3090 GPU. The overall training times takes around 2.5 h, and the mean inference time takes 0.06s per frame – this time includes the I3D extraction time. For all baselines, we use the same I3D backbone and benchmark setup as ours.

3.3　Evaluation on Polyp Frame Detection

Baselines. We train six SOTA WVAD baselines: DeepMIL [20], GCN-Ano [32], CLAWS [30], AR-Net [27], MIST [7], and RTFM [24]. The same experimental setup as our approach is applied to these baselines for fair comparison.

Evaluation Measures. Similarly to previous papers [8,20], we use the frame-level area under the ROC curve (AUC) as the evaluation measure. Given that the AUC can produce optimistic results for imbalanced problems, such as anomaly detection, we follow [16,29] and use average precision (AP) as another evaluation measure. Larger AUC and AP values indicate better performance.

Quantitative Comparison. We show the quantitative comparison results in Table 1. Our model achieves the best 98.4% AUC and 86.6% AP and outperforms all six SOTA methods by a large margin. We obtain a maximum 10% and a minimum 2% AUC improvement, and a maximum 18% and a minimum 6%

Table 1. Comparison of frame-level AUC and AP performance with other SOTA WVADs on colonoscopy dataset using the same I3D feature extractor.

Method	Publication	AUC	AP
DeepMIL [20]	CVPR'18	89.41	68.53
GCN-Ano [32]	CVPR'19	92.13	75.39
CLAWS [30]	ECCV'20	95.62	80.42
AR-Net [27]	ICME'20	88.59	71.58
MIST [7]	CVPR'21	94.53	72.85
RTFM [24]	ICCV'21	96.30	77.96
Ours		**98.41**	**86.63**

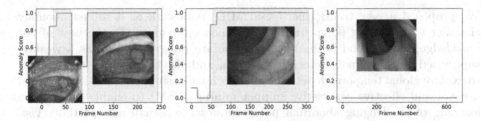

Fig. 3. Anomaly scores (orange curve) of our method on test videos. Pink areas indicate the labelled testing abnormal events. (Color figure online)

AP improvement over the second best approaches. Our method substantially surpasses the most recent WVAD approach RTFM [24] by 8% AP.

Qualitative Comparison. In Fig. 3, we show the anomaly scores produced by our model for test videos from our polyp detection dataset. As illustrated by the orange curves, our model can effectively produce small anomaly scores for normal snippets and large anomaly scores for abnormal snippets. Our model is also able to detect multiple anomalous events (e.g., videos with two polyp event occurrences - first figure in Fig. 3) in one video. Also, our model can also detect the subtle polyps (middle figure in Fig. 3).

3.4 Ablation Study

Table 2 shows the contribution of each component of our proposed method on the testing set. The baseline top-k MIL network, trained with ℓ_{snp}, achieves 92.8% AUC and 71.9% AP. Our method obtains a significant performance gain by adding the proposed convolutional transformer encoder (CTE). Adding the video classifier, represented by the loss $\ell_{vid}(.)$, boosts the performance by about 2% AUC and 3% AP. The proposed hard/easy snippet contrastive loss, denoted by the loss $\ell_{cnt}(.)$, further improve the performance (e.g., increasing AP by about 4%), indicating the effectiveness of addressing the hard anomaly issues.

Table 2. Ablation studies for polyp frame detection. The linear network with top-k MIL ranking loss is considered as the baseline, and CTE denotes the convolutional transformer encoder.

top-k (ℓ_{snp})	CTE	ℓ_{vid}	ℓ_{cnt}	AUC	AP
✓				92.88	71.96
✓	✓			94.92	79.56
✓	✓	✓		96.74	82.88
✓	✓	✓	✓	**98.41**	**86.63**

4   Conclusion

We proposed a new transformer-based MIL framework as a robust anomaly classifier for detecting polyp frames in colonoscopy videos. To the best of our knowledge, our method is the first to formulate polyp detection as a weakly-supervised video anomaly detection problem, and also to introduce transformer to explore global temporal dependency between video snippets. We also proposed a novel and effective contrastive snippet mining (CSM) to enable an effective learning of challenging abnormal polyp frames (i.e., small and partially visible polyps) and normal frames (i.e., water and feces). The resulting anomaly classifier showed SOTA results on our proposed large-scale colonoscopy dataset. Despite the remarkable performance on detecting polyp frames, our model may fail for online inference due to the transformer self-attention operation. We plan to further investigate the online self-attention techniques in future work.

References

1. Arnab, A., Dehghani, M., Heigold, G., Sun, C., Lučić, M., Schmid, C.: ViViT: a video vision transformer. In: Proceedings of the IEEE/CVF International Conference on Computer Vision, pp. 6836–6846 (2021)
2. Borgli, H., et al.: HyperKvasir, a comprehensive multi-class image and video dataset for gastrointestinal endoscopy. Sci. Data **7**(1), 1–14 (2020)
3. Carreira, J., Zisserman, A.: Quo vadis, action recognition? A new model and the kinetics dataset. In: Proceedings of the IEEE Conference on Computer Vision and Pattern Recognition, pp. 6299–6308 (2017)
4. Chen, Y., Tian, Y., Pang, G., Carneiro, G.: Deep one-class classification via interpolated gaussian descriptor. arXiv preprint arXiv:2101.10043 (2021)
5. Chollet, F.: Xception: deep learning with depthwise separable convolutions. In: Proceedings of the IEEE Conference on Computer Vision and Pattern Recognition, pp. 1251–1258 (2017)
6. Dosovitskiy, A., et al.: An image is worth 16x16 words: transformers for image recognition at scale. arXiv preprint arXiv:2010.11929 (2020)
7. Feng, J.C., Hong, F.T., Zheng, W.S.: MIST: multiple instance self-training framework for video anomaly detection. In: Proceedings of the IEEE/CVF Conference on Computer Vision and Pattern Recognition, pp. 14009–14018 (2021)

8. Gong, D., et al.: Memorizing normality to detect anomaly: memory-augmented deep autoencoder for unsupervised anomaly detection. In: ICCV, pp. 1705–1714 (2019)

9. Ji, G.-P., et al.: Progressively normalized self-attention network for video polyp segmentation. In: de Bruijne, M., et al. (eds.) MICCAI 2021. LNCS, vol. 12901, pp. 142–152. Springer, Cham (2021). https://doi.org/10.1007/978-3-030-87193-2_14

10. Kay, W., et al.: The kinetics human action video dataset. arXiv preprint arXiv:1705.06950 (2017)

11. Kingma, D.P., Ba, J.: Adam: a method for stochastic optimization. arXiv preprint arXiv:1412.6980 (2014)

12. Li, S., Liu, F., Jiao, L.: Self-training multi-sequence learning with transformer for weakly supervised video anomaly detection. In: Proceedings of the AAAI, Virtual 24 (2022)

13. Li, W., Vasconcelos, N.: Multiple instance learning for soft bags via top instances. In: Proceedings of the IEEE Conference on Computer Vision and Pattern Recognition, pp. 4277–4285 (2015)

14. Liu, Y., et al.: Photoshopping colonoscopy video frames. In: ISBI, pp. 1–5 (2020)

15. Ma, Y., Chen, X., Cheng, K., Li, Y., Sun, B.: LDPolypVideo benchmark: a large-scale colonoscopy video dataset of diverse polyps. In: de Bruijne, M. (ed.) MICCAI 2021. LNCS, vol. 12905, pp. 387–396. Springer, Cham (2021). https://doi.org/10.1007/978-3-030-87240-3_37

16. Pang, G., van den Hengel, A., Shen, C., Cao, L.: Toward deep supervised anomaly detection: reinforcement learning from partially labeled anomaly data. In: Proceedings of the 27th ACM SIGKDD Conference on Knowledge Discovery and Data Mining, pp. 1298–1308 (2021)

17. Pang, G., Shen, C., van den Hengel, A.: Deep anomaly detection with deviation networks. In: Proceedings of the 25th ACM SIGKDD International Conference on Knowledge Discovery and Data Mining, pp. 353–362 (2019)

18. Paszke, A., et al.: Pytorch: an imperative style, high-performance deep learning library. In: Wallach, H., Larochelle, H., Beygelzimer, A., d Alché-Buc, F., Fox, E., Garnett, R. (eds.) Advances in Neural Information Processing Systems 32, pp. 8024–8035. Curran Associates, Inc. (2019). http://papers.neurips.cc/paper/9015-pytorch-an-imperative-style-high-performance-deep-learning-library.pdf

19. Pu, L.Z.C.T., et al.: Computer-aided diagnosis for characterisation of colorectal lesions: a comprehensive software including serrated lesions. Gastrointest. Endosc. 92(4), 891–899 (2020)

20. Sultani, W., Chen, C., Shah, M.: Real-world anomaly detection in surveillance videos. In: Proceedings of the IEEE Conference on Computer Vision and Pattern Recognition, pp. 6479–6488 (2018)

21. Tian, Y., Liu, F., et al.: Self-supervised multi-class pre-training for unsupervised anomaly detection and segmentation in medical images. arXiv preprint arXiv:2109.01303 (2021)

22. Tian, Yu., Maicas, G., Pu, L.Z.C.T., Singh, R., Verjans, J.W., Carneiro, G.: Few-shot anomaly detection for polyp frames from colonoscopy. In: Martel, A.L., et al. (eds.) MICCAI 2020. LNCS, vol. 12266, pp. 274–284. Springer, Cham (2020). https://doi.org/10.1007/978-3-030-59725-2_27

23. Tian, Y., et al.: Detecting, localising and classifying polyps from colonoscopy videos using deep learning. arXiv preprint arXiv:2101.03285 (2021)

24. Tian, Y., Pang, G., Chen, Y., Singh, R., Verjans, J.W., Carneiro, G.: Weakly-supervised video anomaly detection with robust temporal feature magnitude learn-

ing. In: Proceedings of the IEEE/CVF International Conference on Computer Vision, pp. 4975–4986 (2021)

25. Tian, Yu., et al.: Constrained contrastive distribution learning for unsupervised anomaly detection and localisation in medical images. In: de Bruijne, M., et al. (eds.) MICCAI 2021. LNCS, vol. 12905, pp. 128–140. Springer, Cham (2021). https://doi.org/10.1007/978-3-030-87240-3_13

26. Tian, Y., et al.: One-stage five-class polyp detection and classification. In: 2019 IEEE 16th International Symposium on Biomedical Imaging (ISBI 2019), pp. 70–73. IEEE (2019)

27. Wan, B., Fang, Y., Xia, X., Mei, J.: Weakly supervised video anomaly detection via center-guided discriminative learning. In: 2020 IEEE International Conference on Multimedia and Expo (ICME), pp. 1–6 (2020)

28. Wu, H., et al.: CvT: introducing convolutions to vision transformers. arXiv preprint arXiv:2103.15808 (2021)

29. Wu, P., et al.: Not only look, but also listen: learning multimodal violence detection under weak supervision. In: Vedaldi, A., Bischof, H., Brox, T., Frahm, J.-M. (eds.) ECCV 2020. LNCS, vol. 12375, pp. 322–339. Springer, Cham (2020). https://doi.org/10.1007/978-3-030-58577-8_20

30. Zaheer, M.Z., Mahmood, A., Astrid, M., Lee, S.-I.: CLAWS: clustering assisted weakly supervised learning with normalcy suppression for anomalous event detection. In: Vedaldi, A., Bischof, H., Brox, T., Frahm, J.-M. (eds.) ECCV 2020. LNCS, vol. 12367, pp. 358–376. Springer, Cham (2020). https://doi.org/10.1007/978-3-030-58542-6_22

31. Zhang, C., Cao, M., Yang, D., Chen, J., Zou, Y.: Cola: weakly-supervised temporal action localization with snippet contrastive learning. In: Proceedings of the IEEE/CVF Conference on Computer Vision and Pattern Recognition, pp. 16010–16019 (2021)

32. Zhong, J.X., Li, N., Kong, W., Liu, S., Li, T.H., Li, G.: Graph convolutional label noise cleaner: train a plug-and-play action classifier for anomaly detection. In: Proceedings of the IEEE Conference on Computer Vision and Pattern Recognition, pp. 1237–1246 (2019)

Lesion-Aware Dynamic Kernel for Polyp Segmentation

Ruifei Zhang[1], Peiwen Lai[1], Xiang Wan[2,4], De-Jun Fan[3], Feng Gao[3], Xiao-Jian Wu[3], and Guanbin Li[1,2(✉)]

[1] School of Computer Science and Engineering, Sun Yat-sen University, Guangzhou, China
liguanbin@mail.sysu.edu.cn
[2] Shenzhen Research Institute of Big Data, Shenzhen, China
[3] The Sixth Affiliated Hospital, Sun Yat-sen University, Guangzhou, China
[4] Pazhou Lab, Guangzhou, China

Abstract. Automatic and accurate polyp segmentation plays an essential role in early colorectal cancer diagnosis. However, it has always been a challenging task due to 1) the diverse shape, size, brightness and other appearance characteristics of polyps, 2) the tiny contrast between concealed polyps and their surrounding regions. To address these problems, we propose a lesion-aware dynamic network (LDNet) for polyp segmentation, which is a traditional u-shape encoder-decoder structure incorporated with a dynamic kernel generation and updating scheme. Specifically, the designed segmentation head is conditioned on the global context features of the input image and iteratively updated by the extracted lesion features according to polyp segmentation predictions. This simple but effective scheme endows our model with powerful segmentation performance and generalization capability. Besides, we utilize the extracted lesion representation to enhance the feature contrast between the polyp and background regions by a tailored lesion-aware cross-attention module (LCA), and design an efficient self-attention module (ESA) to capture long-range context relations, further improving the segmentation accuracy. Extensive experiments on four public polyp benchmarks and our collected large-scale polyp dataset demonstrate the superior performance of our method compared with other state-of-the-art approaches. The source code is available at https://github.com/ReaFly/LDNet.

1 Introduction

Colorectal Cancer (CRC) is one of the most common cancer diseases around the world [17]. However, actually, most CRC starts from a benign polyp and gets progressively worse over several years. Thus, early polyp detection and removal make essential roles to reduce the incidence of CRC. In clinical practice, colonoscopy is a common examination tool for early polyp screening. An accurate and automatic polyp segmentation algorithm based on colonoscopy images can greatly support clinicians and alleviate the reliance on expensive labor, which is of great clinical significance.

© The Author(s), under exclusive license to Springer Nature Switzerland AG 2022
L. Wang et al. (Eds.): MICCAI 2022, LNCS 13433, pp. 99–109, 2022.
https://doi.org/10.1007/978-3-031-16437-8_10

However, accurate polyp segmentation still remains challenge due to the diverse but concealed characteristics of polyps. Early traditional approaches [12, 19] utilize hand-craft features to detect polyps, failing to cope with complex scenario and suffering from high misdiagnosis rate. With the advance of deep learning technology, plenty of CNN-based methods are developed and applied for polyp segmentation. Fully convolution network [11] is first proposed for semantic segmentation, and then its variants [1,3] also make a great breakthrough in the polyp segmentation task. UNet [16] adopts an encoder-decoder structure and introduces skip-connections to bridge each stage between encoder and decoder, supplying multi-level information to obtain a high-resolution segmentation map through successive up-sampling operations. UNet++ [27] introduces more dense and nested connections, aiming to alleviate the semantic difference of features maps between encoder and decoder. Recently, to better overcome the above mentioned challenges, some networks specially designed for polyp segmentation task have been proposed. For example, PraNet [6] adopts a reverse attention mechanism to mine finer boundary cues based on the initial segmentation map. ACSNet [23] adaptively selects and integrates both global contexts and local information, achieving more robust polyp segmentation performance. CCBANet [13] proposes the cascading context and the attention balance modules to aggregate better feature representation. SANet [21] designs the color exchange operation to alleviate the color diversity of polyps, and proposes a shallow attention module to select more useful shallow features, obtaining comparable segmentation results. However, existing methods mainly focus on enhancing the network's lesion representation from the view of feature selection [6,13,23] or data augmentation [21], and no attempts have been made to consider the structural design of the network from the perspective of improving the flexibility and adaptability of model feature learning, which limits their generalization.

To this end, we design a Lesion-aware Dynamic Network (LDNet) for the polyp segmentation task. Inspired by [9,24], we believe that a dynamic kernel can adaptively adjust parameters according to the input image, and thus achieving stronger feature exploration capabilities in exchange for better segmentation performance. Specifically, our unique kernel (also known as segmentation head) is dynamically generated basing on the global features of the input image, and generates one polyp segmentation prediction in each decoder stage. Accordingly, these segmentation results serve as clues to extract refined polyp features, which in turn update our kernel parameters with better lesion perception. For some complex polyp regions, the dynamic kernel generation and update mechanism we designed can step-wisely learn and mine discriminative regional features and gradually improve the segmentation results, enhancing the generalization of the model. Besides, we design two attention modules, *i.e.,* Efficient Self-Attention (ESA) and Lesion-aware Cross-Attention (LCA). The former is used to capture global feature relations, while the latter is designed to enhance feature contrast between lesions and other background regions, further improving the segmentation performance. In summary, the contributions of this paper mainly include three folds: (1) We design a lesion-aware dynamic network for polyp segmenta-

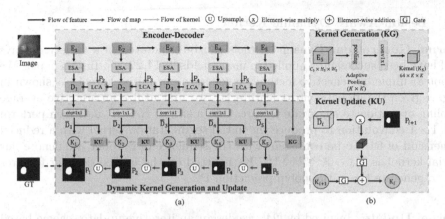

Fig. 1. (a) Overview of our LDNet. (b) Illustration of kernel generation and update.

tion. The introduction of a dynamic kernel generation and update mechanism endows the model with generalizability to discriminate polyp regions with diverse shapes, sizes, and appearances. (2) Our tailored ESA and LCA modules enhance the polyp feature representation, which helps to mine concealed polyps with low visual contrast. (3) Extensive experiments on four public polyp benchmarks and our collected large-scale polyp dataset demonstrate the effectiveness of our proposed method.

2   Methodology

The overview of our LDNet is shown in Fig. 1, which is a general encoder-decoder structure, incorporated with our designed dynamic kernel scheme and attention modules. The Res2Net [8] is utilized as our encoder, consisting of five blocks. The generated feature map of each block is denoted as $\{\mathbf{E}_i\}_{i=1}^5$. Accordingly, five-layer decoder blocks are adopted and their respective generated features are defined as $\{\mathbf{D}_i\}_{i=1}^5$. 1×1 convolution is utilized to unify the dimension of $\mathbf{D}_i$ to 64, denoted as $\bar{\mathbf{D}}_i$, which are adaptive to subsequent kernel update operations. In contrast to previous methods [6,7,21,23] with a static segmentation head, which is agnostic to the input images and remains fixed during the inference stage, we design a dynamic kernel as our segmentation head. The dynamic kernel is essentially a convolution operator used to produce segmentation result, but its parameters are initially generated by the global feature $\mathbf{E}_5$ of the input, and iteratively updated in the multi-stage decoder process based on the current decoder features $\bar{\mathbf{D}}_i$ and its previous segmentation result $\mathbf{P}_{i+1}$, which is employed to make a new prediction $\mathbf{P}_i$. For the convenience of expression, we denote the sequential updated kernels as $\{\mathbf{K}_i\}_{i=1}^5$. Each segmentation prediction is supervised by the corresponding down-sampled Ground Truth, and the prediction $\mathbf{P}_1$ of the last decoder stage is the final result of our model. We detail the dynamic kernel scheme and attention modules in the following sections.

2.1   Lesion-Aware Dynamic Kernel

Kernel Generation. Dynamic kernels can be generated in a variety of ways and have been successfully applied in many fields [9,14,22,24]. In this paper, We adopt a simple but effective method to generate our initial kernel. As shown in Fig. 1, given the global context feature $\mathbf{E}_5$, we first utilize an adaptive average pooling operation to aggregate features into a size of $K \times K$, and then perform one 1×1 convolution to produce the initial segmentation kernel with a reduced dimension of 64. To be consistent with the sequence of decoder, we denote our initial kernel as $\mathbf{K}_5 \in \mathbb{R}^{1 \times 64 \times K \times K}$. $\mathbf{K}_5$ is acted on the unified decoder features $\bar{\mathbf{D}}_5$ to generate the initial polyp prediction $\mathbf{P}_5$.

Kernel Update. Inspired by [24], we design an iterative update scheme based on the encoder-decoder architecture to improve our dynamic kernel. Given the i-th unified decoder features $\bar{\mathbf{D}}_i \in \mathbb{R}^{64 \times H_i \times W_i}$ and previous polyp segmentation result $\mathbf{P}_{i+1} \in \mathbb{R}^{1 \times H_{i+1} \times W_{i+1}}$, we first extract lesion features as:

$$\mathbf{F}_i = \sum^{H_i} \sum^{W_i} up_2(\mathbf{P}_{i+1}) \circ \bar{\mathbf{D}}_i, \tag{1}$$

where up_2 denotes up-sampling the prediction map by a factor of 2 to keep a same size with feature map. '$\circ$' represents the element-wise multiplication with broadcasting mechanism.

The essential operation of the kernel update is to integrate the lesion representations extracted by the current decoder features into previous kernel parameters. In this way, the kernel can not only perceive the lesion characteristics to be segmented in advance, but gradually incorporate multi-scale lesion information, thus enhancing its discrimination ability for polyps. Since the previous polyp prediction may be inaccurate, as in [24], we further utilize a gate mechanism to filter the noise in lesion features and achieve an adaptive kernel update. The formulation is:

$$\mathbf{K}_i = \mathbf{G}_i^F \circ \phi_1(\mathbf{F}_i) + \mathbf{G}_i^K \circ \phi_2(\mathbf{K}_{i+1}), \tag{2}$$

where ϕ_1 and ϕ_2 denote linear transformations. $\mathbf{G}_i^F$ and $\mathbf{G}_i^K$ are two gates, which are obtained by the element-wise multiplication between the variants of $\mathbf{F}_i$ and $\mathbf{K}_{i+1}$ followed by different linear transformation and Sigmoid function (σ), respectively:

$$\mathbf{G}_i = \phi_3(\mathbf{F}_i) \circ \phi_4(\mathbf{K}_{i+1}) \tag{3}$$

$$\mathbf{G}_i^K = \sigma(\phi_5(\mathbf{G}_i)), \mathbf{G}_i^F = \sigma(\phi_6(\mathbf{G}_i)) \tag{4}$$

The updated kernel $\mathbf{K}_i$ is acted on the specific decoder feature to make a new prediction $\mathbf{P}_i$. Both of them are sent to the $(i-1)$-th decoder stage to iteratively perform the above update scheme.

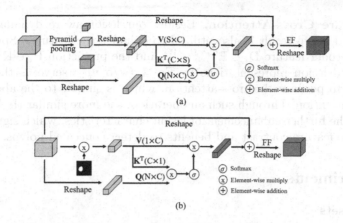

Fig. 2. (a) Illustration of ESA. (b) Illustration of LCA. FF denotes the feed-forward layer. We omit the residual addition between the input and output of FF for simplicity.

2.2 Attention Modules

Efficient Self-attention. Self-attention mechanism is first proposed in Transformer [20], and recently has played a significant role in many tasks [4,5] due to its strong long-range modeling capability, however is criticized for prohibitive computation and memory cost. To overcome these challenges, we borrow the idea from [26,28] and design our ESA module. As shown in Fig. 2, we follow the component of Transformer but replace the original self-attention with our ESA layer, followed by a feed-forward layer and a reshaping operation. We also perform a multi-head parallel scheme to further improve the performance. Specifically, given one encoder feature map $\mathbf{E}_i \in \mathbb{R}^{C_i \times H_i \times W_i}$, details of our ESA layer are formulated as follows:

$$\text{ESA}(\mathbf{E}_i) = \phi_o(\text{concat}(\text{head}^0, ..., \text{head}^n)), \tag{5}$$

$$\text{head}^j = \text{Attention}(\phi_q^j(\mathbf{Q}), \phi_k^j(\mathbf{K}), \phi_v^j(\mathbf{V})), \tag{6}$$

where ϕ_o, ϕ_q^j, ϕ_k^j, ϕ_v^j denote the linear projections, and n is the number of heads. $\mathbf{Q} \in \mathbb{R}^{N_i \times C_i}(N_i = H_i \times W_i)$ is reshaped from the $\mathbf{E}_i$. $\mathbf{K}, \mathbf{V} \in \mathbb{R}^{S \times C_i}$ are obtained by the pyramid pooling operation [26], which includes 1×1, 3×3, 5×5 adaptive average pooling to down-sample the feature map, followed by reshaping and concatenating operations. Thanks to such a sampling process, we utilize fewer representative global features to perform the standard attention [20], not only introducing global relations to original features, but significantly saving the computation overhead ($S = 1 \times 1 + 3 \times 3 + 5 \times 5 \ll N_i$). Attention($\cdot$) is formulated as:

$$\text{Attention}(\mathbf{q}, \mathbf{k}, \mathbf{v}) = \text{softmax}(\frac{\mathbf{q}\mathbf{k}^T}{\sqrt{d_k}})\mathbf{v}, \tag{7}$$

where d_k is the dimension of each head, equal to $\frac{C_i}{n}$.

Lesion-Aware Cross-Attention. Besides our lesion-aware dynamic kernel, the predicted polyp result is also utilized to enhance the features. Specifically, given the decoder feature $\mathbf{D}_i \in \mathbb{R}^{C_i \times H_i \times W_i}$ and the prediction $\mathbf{P}_i \in \mathbb{R}^{1 \times H_i \times W_i}$, the extracted lesion representations by Eq. 1 (w/o up_2) serve as the $\mathbf{K}$ and $\mathbf{V} \in \mathbb{R}^{1 \times C_i}$ to perform the cross-attention, which is similar to the above mentioned self-attention. Through such an operation, the more similar the region to the lesion, the further enhancement of lesion characteristics, which significantly improves the feature contrast and benefits to detect concealed polyps.

3   Experiments

3.1   Datasets

Public Polyp Benchmarks. We evaluate our proposed LDNet on four public polyp datasets, including Kvasir-SEG [10], CVC-ClinicDB [2], CVC-ColonDB [19] and ETIS [18]. Following the same setting in [6,21], we randomly select 80% images respectively from Kvasir-SEG and CVC-ClinicDB and fuse them together as our training set, 10% as validation set. The remaining data of Kvasir-SEG and CVC-ClinicDB, and other two unseen datasets are used for testing.

Our Collected Large-Scale Polyp Dataset. We also evaluate LDNet on our collected polyp dataset, which has 5175 images in total. This dataset is randomly split into 60% for training, 20% for validation, and the remaining for testing.

3.2   Implementation Details and Evaluation Metrics

Our method is implemented based on PyTorch framework [15] and runs on an NVIDIA GeForce RTX 2080 Ti GPU. We simply set $K = 1$ in the kernel generation and $n = 8$ in the multi-head attention mechanism. The SGD optimizer is utilized to train the model, with batch size of 8, momentum of 0.9 and weight decay of 10^{-5}. The initial learning rate is set to 0.001, and adjusted by a poly learning rate policy, which is $lr = lr_{init} \times (1 - \frac{epoch}{nEpoch})^{power}$, where $power = 0.9$, $nEpoch = 80$. All images are uniformly resized to 256×256. To avoid overfitting, data augmentations including random horizontal and vertical flips, rotation, random cropping are used in the training stage. A combination of Binary Cross-Entropy loss and Dice loss is used to supervise the training process.

As in [7,23], eight common metrics are adopted to evaluate polyp segmentation performance, including *Recall, Specificity, Precision, Dice Score, IoU for Polyp (IoUp), IoU for Background (IoUb), Mean IoU (mIoU)* and *Accuracy*.

3.3   Experiments on the Public Polyp Benchmarks

We compare our LDNet with several state-of-the-art methods, including UNet [16], ResUNet [25], UNet++ [27], ACSNet [23], PraNet [6], SANet [21], CCBANet [13], on the public polyp benchmarks. As shown in Table 1, our LDNet

Table 1. Comparison with other state-of-the-art methods on four benchmark datasets. The best three results are highlighted in red, green and blue, respectively.

	Methods	Rec	Spec	Prec	Dice	IoUp	IoUb	mIoU	Acc
Kvasir	UNet [16]	87.04	97.25	84.28	82.60	73.39	93.89	83.64	95.05
	ResUNet [25]	84.70	97.17	83.00	80.50	70.60	93.19	81.89	94.43
	UNet++ [27]	89.23	97.20	85.57	84.77	76.42	94.23	85.32	95.44
	ACSNet [23]	91.35	98.39	91.46	89.54	83.72	96.42	90.07	97.16
	PraNet [6]	93.90	97.33	89.87	90.32	84.55	95.98	90.26	96.75
	CCBANet [13]	90.71	98.04	91.02	89.04	82.82	96.21	89.52	97.02
	SANet [21]	92.06	98.20	91.14	89.92	83.97	96.54	90.26	97.18
	Ours	92.72	98.05	92.04	90.70	85.30	96.71	91.01	97.35
CVC- ClinicDB	UNet [16]	88.61	98.70	85.10	85.12	77.78	97.70	87.74	97.95
	ResUNet [25]	90.89	99.25	90.22	89.98	82.77	98.18	90.47	98.37
	UNet++ [27]	87.78	99.21	90.02	87.99	80.69	97.92	89.30	98.12
	ACSNet [23]	93.46	99.54	94.63	93.80	88.57	98.95	93.76	99.08
	PraNet [6]	95.22	99.34	92.25	93.49	88.08	98.92	93.50	99.05
	CCBANet [13]	94.89	99.22	91.39	92.83	86.96	98.79	92.87	98.93
	SANet [21]	94.74	99.41	92.88	93.61	88.26	98.94	93.60	99.07
	Ours	94.49	99.51	94.53	94.31	89.48	98.95	94.21	99.08
CVC-ColonDB	UNet [16]	63.05	98.00	68.01	56.40	47.32	94.51	70.92	94.84
	ResUNet [25]	59.91	98.06	65.29	54.87	44.31	93.77	69.04	94.06
	UNet++ [27]	63.49	98.59	77.79	60.77	52.64	95.19	73.92	95.48
	ACSNet [23]	77.38	99.26	81.72	75.51	67.38	96.16	81.77	96.32
	PraNet [6]	81.85	98.54	78.43	76.24	68.29	96.06	82.17	96.26
	CCBANet [13]	82.34	98.39	77.79	75.36	66.57	95.89	81.23	96.14
	SANet [21]	75.21	99.09	81.43	73.50	65.47	96.19	80.83	96.40
	Ours	83.46	98.49	81.15	78.43	70.58	96.21	83.39	96.48
ETIS	UNet [16]	47.33	96.36	48.05	34.81	28.38	94.72	61.55	94.90
	ResUNet [25]	49.12	97.21	56.85	38.65	30.54	95.27	62.90	95.43
	UNet++ [27]	55.52	95.40	59.14	40.91	33.86	93.87	63.87	94.07
	ACSNet [23]	78.31	98.44	68.81	69.44	60.96	97.78	79.37	97.89
	PraNet [6]	81.20	98.73	72.23	72.38	64.07	98.29	81.18	98.38
	CCBANet [13]	78.70	97.19	61.12	62.63	53.81	96.52	75.17	96.66
	SANet [21]	77.08	99.04	72.73	72.26	63.33	98.47	80.90	98.54
	Ours	82.83	98.44	72.07	74.37	66.50	98.01	82.26	98.10

achieves superior performance over other methods across four datasets on most metrics. In particular, on the two seen datasets, *i.e.,* Kvasir and CVC-ClinicDB, the proposed LDNet obtains the best *Dice* and *mIoU* scores, outperforming other methods. On the other two unseen datasets, the LDNet also shows strong generalization ability and achieves 78.43% and 74.37% *Dice* scores, 2.19% and

Table 2. Comparison with other state-of-the-art methods and ablation study on our collected dataset.

Methods	Rec	Spec	Prec	Dice	IoUp	IoUb	mIoU	Acc
UNet [16]	87.89	97.27	87.23	85.00	77.48	93.95	85.71	95.64
UNet++ [27]	89.88	97.43	88.18	86.92	79.88	94.56	87.26	96.21
ACSNet [23]	92.43	97.79	90.94	90.54	84.64	95.75	90.19	97.11
PraNet [6]	92.86	97.87	90.52	90.64	84.60	95.91	90.25	97.28
CCBANet [13]	91.91	97.79	91.32	90.39	84.36	95.73	90.04	97.10
SANet [21]	92.18	98.22	91.67	90.75	84.98	96.02	90.50	97.27
Ours	93.22	98.15	92.16	91.66	86.28	96.39	91.34	97.55
Baseline	92.02	97.03	87.75	88.30	81.26	94.95	88.11	96.54
Baseline+DK	92.22	97.58	90.41	89.88	83.86	95.53	89.70	96.92
Baseline+DK+ESAs	91.76	98.25	92.14	90.74	84.91	95.85	90.38	97.14

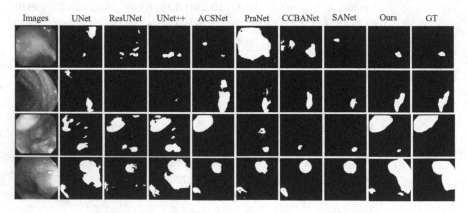

Images   UNet   ResUNet   UNet++   ACSNet   PraNet   CCBANet   SANet   Ours   GT

Fig. 3. Visual comparison of polyp segmentation results.

1.99% improvements over the second best approaches, further demonstrating the effectiveness of our approach. Some visualization examples are shown in Fig. 3.

3.4   Experiments on the Collected Large-Scale Polyp Dataset

On our collected large-scale polyp dataset, we compare the LDNet with UNet [16], UNet++ [27], ACSNet [23], PraNet [6], SANet [21] and CCBANet [13]. As shown in Table 2, our method again achieves the best performance, with a *Dice* of 91.66% and a *mIoU* of 91.34%, respectively.

3.5 Ablation Study

We conduct a series of ablation studies on our collected polyp dataset to verify the effectiveness of our designed dynamic kernel scheme and attention modules. Specifically, we utilize the traditional u-shape structure with a static segmentation head as our baseline, and gradually replace the static head with our designed dynamic kernels, then further add ESA and LCA modules, denoting as Baseline, Baseline+DK, Baseline+DK+ESAs and Ours respectively. As shown in Table 2, the introduction of the dynamic kernel significantly enhances the performance of the baseline, with a 1.58% improvement of *Dice* score. With the addition of our ESA and LCA modules, the scores of *Dice* and *mIoU* are further boosted by 0.86% and 0.68%, 0.92% and 0.96%, respectively.

4 Conclusion

In this paper, we propose the lesion-aware dynamic kernel (LDNet) for polyp segmentation, which is generated conditioned on the global information and updated by the multi-level lesion features. We believe that such a dynamic kernel can endow our model with more flexibility to attend diverse polyps regions. Besides, we also improve the feature representation and enhance the context contrast by two tailored attention modules, *i.e.*, ESA and LCA, which is beneficial for detecting concealed polyps. Extensive experiments and ablation studies demonstrate the effectiveness of our proposed method.

Acknowledgements. This work is supported in part by the Chinese Key-Area Research and Development Program of Guangdong Province (2020B0101350001), in part by the Guangdong Basic and Applied Basic Research Foundation (2020B 1515020048), in part by the National Natural Science Foundation of China (61976250), in part by the Guangzhou Science and technology project (20210202 0633), and is also supported by the Guangdong Provincial Key Laboratory of Big Data Computing, The Chinese University of Hong Kong, Shenzhen.

References

1. Akbari, M., et al.: Polyp segmentation in colonoscopy images using fully convolutional network. In: 2018 40th Annual International Conference of the IEEE Engineering in Medicine and Biology Society, pp. 69–72 (2018)
2. Bernal, J., Sánchez, F.J., Fernández-Esparrach, G., Gil, D., Rodríguez, C., Vilariño, F.: WM-DOVA maps for accurate polyp highlighting in colonoscopy: validation vs. saliency maps from physicians. Comput. Med. Imaging Graph. **43**, 99–111 (2015)
3. Brandao, P., et al.: Fully convolutional neural networks for polyp segmentation in colonoscopy. In: Medical Imaging 2017: Computer-Aided Diagnosis, vol. 10134, p. 101340F. International Society for Optics and Photonics (2017)
4. Carion, N., Massa, F., Synnaeve, G., Usunier, N., Kirillov, A., Zagoruyko, S.: End-to-end object detection with transformers. In: Vedaldi, A., Bischof, H., Brox, T., Frahm, J.-M. (eds.) ECCV 2020. LNCS, vol. 12346, pp. 213–229. Springer, Cham (2020). https://doi.org/10.1007/978-3-030-58452-8_13

5. Dosovitskiy, A., et al.: An image is worth 16x16 words: transformers for image recognition at scale. arXiv preprint arXiv:2010.11929 (2020)
6. Fan, D.-P., et al.: PraNet: parallel reverse attention network for polyp segmentation. In: Martel, A.L., et al. (eds.) MICCAI 2020. LNCS, vol. 12266, pp. 263–273. Springer, Cham (2020). https://doi.org/10.1007/978-3-030-59725-2_26
7. Fang, Y., Chen, C., Yuan, Y., Tong, K.: Selective feature aggregation network with area-boundary constraints for polyp segmentation. In: Shen, D., et al. (eds.) MICCAI 2019. LNCS, vol. 11764, pp. 302–310. Springer, Cham (2019). https://doi.org/10.1007/978-3-030-32239-7_34
8. Gao, S.H., Cheng, M.M., Zhao, K., Zhang, X.Y., Yang, M.H., Torr, P.: Res2net: a new multi-scale backbone architecture. IEEE Trans. Pattern Anal. Mach. Intell. **43**(2), 652–662 (2019)
9. He, J., Deng, Z., Qiao, Y.: Dynamic multi-scale filters for semantic segmentation. In: Proceedings of the IEEE/CVF International Conference on Computer Vision, pp. 3562–3572 (2019)
10. Jha, D., et al.: Kvasir-SEG: a segmented polyp dataset. In: Ro, Y.M., et al. (eds.) MMM 2020. LNCS, vol. 11962, pp. 451–462. Springer, Cham (2020). https://doi.org/10.1007/978-3-030-37734-2_37
11. Long, J., Shelhamer, E., Darrell, T.: Fully convolutional networks for semantic segmentation. In: Proceedings of the IEEE Conference on Computer Vision and Pattern Recognition, pp. 3431–3440 (2015)
12. Mamonov, A.V., Figueiredo, I.N., Figueiredo, P.N., Tsai, Y.H.R.: Automated polyp detection in colon capsule endoscopy. IEEE Trans. Med. Imaging **33**(7), 1488–1502 (2014)
13. Nguyen, T.-C., Nguyen, T.-P., Diep, G.-H., Tran-Dinh, A.-H., Nguyen, T.V., Tran, M.-T.: CCBANet: cascading context and balancing attention for polyp segmentation. In: de Bruijne, M., et al. (eds.) MICCAI 2021. LNCS, vol. 12901, pp. 633–643. Springer, Cham (2021). https://doi.org/10.1007/978-3-030-87193-2_60
14. Pang, Y., Zhang, L., Zhao, X., Lu, H.: Hierarchical dynamic filtering network for RGB-D salient object detection. In: Vedaldi, A., Bischof, H., Brox, T., Frahm, J.-M. (eds.) ECCV 2020. LNCS, vol. 12370, pp. 235–252. Springer, Cham (2020). https://doi.org/10.1007/978-3-030-58595-2_15
15. Paszke, A., et al.: Pytorch: an imperative style, high-performance deep learning library. In: Advances in Neural Information Processing Systems, pp. 8026–8037 (2019)
16. Ronneberger, O., Fischer, P., Brox, T.: U-Net: convolutional networks for biomedical image segmentation. In: Navab, N., Hornegger, J., Wells, W.M., Frangi, A.F. (eds.) MICCAI 2015. LNCS, vol. 9351, pp. 234–241. Springer, Cham (2015). https://doi.org/10.1007/978-3-319-24574-4_28
17. Siegel, R.L., Miller, K.D., Fuchs, H.E., Jemal, A.: Cancer statistics, 2022. CA Cancer J. Clin. **72**(1), 7–33 (2022)
18. Silva, J., Histace, A., Romain, O., Dray, X., Granado, B.: Toward embedded detection of polyps in WCE images for early diagnosis of colorectal cancer. Int. J. Comput. Assist. Radiol. Surg. **9**(2), 283–293 (2014)
19. Tajbakhsh, N., Gurudu, S.R., Liang, J.: Automated polyp detection in colonoscopy videos using shape and context information. IEEE Trans. Med. Imaging **35**(2), 630–644 (2015)
20. Vaswani, A., et al.: Attention is all you need. In: Advances in Neural Information Processing Systems 30 (2017)

21. Wei, J., Hu, Y., Zhang, R., Li, Z., Zhou, S.K., Cui, S.: Shallow attention network for polyp segmentation. In: de Bruijne, M., et al. (eds.) MICCAI 2021. LNCS, vol. 12901, pp. 699–708. Springer, Cham (2021). https://doi.org/10.1007/978-3-030-87193-2_66

22. Zhang, J., Xie, Y., Xia, Y., Shen, C.: DoDNet: learning to segment multi-organ and tumors from multiple partially labeled datasets. In: Proceedings of the IEEE/CVF Conference on Computer Vision and Pattern Recognition, pp. 1195–1204 (2021)

23. Zhang, R., Li, G., Li, Z., Cui, S., Qian, D., Yu, Y.: Adaptive context selection for polyp segmentation. In: Martel, A.L., et al. (eds.) MICCAI 2020. LNCS, vol. 12266, pp. 253–262. Springer, Cham (2020). https://doi.org/10.1007/978-3-030-59725-2_25

24. Zhang, W., Pang, J., Chen, K., Loy, C.C.: K-net: towards unified image segmentation. In: Advances in Neural Information Processing Systems 34 (2021)

25. Zhang, Z., Liu, Q., Wang, Y.: Road extraction by deep residual U-Net. IEEE Geosci. Remote Sens. Lett. **15**(5), 749–753 (2018)

26. Zhao, H., Shi, J., Qi, X., Wang, X., Jia, J.: Pyramid scene parsing network. In: Proceedings of the IEEE Conference on Computer Vision and Pattern Recognition, pp. 2881–2890 (2017)

27. Zhou, Z., Rahman Siddiquee, M.M., Tajbakhsh, N., Liang, J.: UNet++: a nested U-Net architecture for medical image segmentation. In: Stoyanov, D., et al. (eds.) DLMIA/ML-CDS -2018. LNCS, vol. 11045, pp. 3–11. Springer, Cham (2018). https://doi.org/10.1007/978-3-030-00889-5_1

28. Zhu, Z., Xu, M., Bai, S., Huang, T., Bai, X.: Asymmetric non-local neural networks for semantic segmentation. In: Proceedings of the IEEE/CVF International Conference on Computer Vision, pp. 593–602 (2019)

Stepwise Feature Fusion: Local Guides Global

Jinfeng Wang[1,2], Qiming Huang[1], Feilong Tang[1], Jia Meng[1], Jionglong Su[1(✉)], and Sifan Song[1,2(✉)]

[1] Xi'an Jiaotong-Liverpool University, Suzhou, China
[2] University of Liverpool, Liverpool, UK
Jionglong.Su@xjtlu.edu.cn, Sifan.Song19@student.xjtlu.edu.cn
https://github.com/Qiming-Huang/ssformer

Abstract. Colonoscopy, currently the most efficient and recognized colon polyp detection technology, is necessary for early screening and prevention of colorectal cancer. However, due to the varying size and complex morphological features of colonic polyps as well as the indistinct boundary between polyps and mucosa, accurate segmentation of polyps is still challenging. Deep learning has become popular for accurate polyp segmentation tasks with excellent results. However, due to the structure of polyps image and the varying shapes of polyps, it is easy for existing deep learning models to overfit the current dataset. As a result, the model may not process unseen colonoscopy data. To address this, we propose a new state-of-the-art model for medical image segmentation, the SSFormer, which uses a pyramid Transformer encoder to improve the generalization ability of models. Specifically, our proposed Progressive Locality Decoder can be adapted to the pyramid Transformer backbone to emphasize local features and restrict attention dispersion. The SSFormer achieves state-of-the-art performance in both learning and generalization assessment.

Keywords: Polyp segmentation · Deep learning · Generalization

1 Introduction

Colorectal cancer (CRC) is common cancer whose cancer risk may be reduced through early screening and removal of colon polyps [6,9]. However, accurate polyp segmentation is still a challenge due to the variable size and shape of polyps, as well as the indistinct boundaries between polyps and mucosa [6]. An accurate segmentation algorithm based on deep learning can effectively improve the accuracy and efficiency of polyp segmentation. Many image segmentation models based on the Convolutional Neural Networks (CNN) recently achieved excellent learning ability in several polyp segmentation benchmarks. [6,9,11,16,

---

J. Wang, Q. Huang and F. Tang—Contributed equally.

© The Author(s), under exclusive license to Springer Nature Switzerland AG 2022
L. Wang et al. (Eds.): MICCAI 2022, LNCS 13433, pp. 110–120, 2022.
https://doi.org/10.1007/978-3-031-16437-8_11

25] However, due to the top-down modeling method of the CNN model and the variability in the morphology of polyps but relatively simple structure of the polyps image, this model lacks generalization ability and is difficult to process unseen datasets. To improve the generalization ability of the deep learning model, we shall incorporate the Transformer architecture into the polyp segmentation task.

The Transformer [18] was initially proposed as a bottom-up model architecture in the natural language processing (NLP) community. Dosovitskiy *et al.*. proposed the Vision Transformer (ViT) [5] that achieved superior performance in image classification tasks. The Transformer is different from CNN which the weight parameters are trained in the kernel to extract and mix the features among elements in the receptive field. In contrast, the Transformer obtains similarities of all patch pairs through the dot product between the patch vectors to adaptively extract and mix features between all patches. This enables the Transformer to have an efficient global receptive field and reduces the inductive bias of the model. As a result, the Transformer has a more robust generalization ability than CNN and Multilayer Perceptron-like structures [12]. However, the low inductive bias and powerful global receptive field make it difficult for the Transformer model to capture task-specific critical local details adequately. In addition, with the deepening of the Transformer model, the global features are continuously mixed and converged [24], resulting in attention dispersion. These make it difficult for the Transformer model to accurately predict detailed information in the dense prediction task of semantic segmentation.

In order to achieve high generalization and accurate polyp automatic segmentation, a novel state-of-the-art (SOTA) medical image segmentation model, SSFormer, is proposed which uses a pyramid Transformer encoder [10,19,20,22] for excellent generalization and multi-scale feature processing capabilities. In our model, the Progressive Locality Decoder (PLD), based on a multi-stage feature aggregation structure, functions as the decoder. The multi-stage feature aggregation structure can enable features of different depths and expressive powers to guide each other, which we believe can address the problems of attention dispersion and underestimation of local features to improve the detail processing ability. Segformer [22] optimized the encoder of the pyramid structure of PVT [19] and proposed a multi-stage feature aggregation decoder, which predicts features of different scales and depths separately through simple upsampling and then parallel fusion. SETR [23] uses the traditional Transformer as the encoder and proposes an MLA decoder with a multi-stage feature aggregation structure. Their excellent performances demonstrate that the decoding method of multi-stage feature aggregation is beneficial to improving the performance of Transformer in dense prediction tasks. Our proposed PLD adopts a stepwise adaptive method to emphasise local features and integrate them into global features, making the fusion of features more efficient.

The main contributions of this paper are: 1) We introduce the pyramid Transformer architecture into the polyp segmentation task to increase the generalization ability of the neural network; 2) We propose a new decoder PLD suitable for

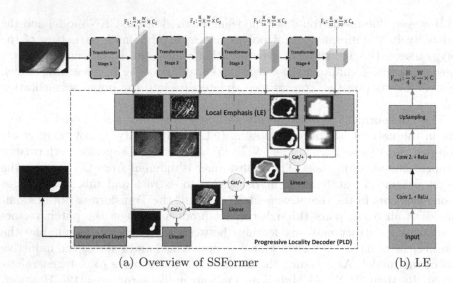

(a) Overview of SSFormer   (b) LE

Fig. 1. (a) The overview of SSFormer; (b) the structure of Local Emphasis module. In this figure, The lines with arrows and the feature maps next to them represent unemphasized features, local emphasized features, and fused features from top to bottom along the feature stream direction, respectively. The remainder of the PLD in Figure (a), excluding the Local Emphasis (LE), is the Stepwise Feature Aggregation (SFA). Feature fusion units can use concatenation (Cat) or addition (+) operations.

Transformer feature pyramids, which can smooth and effectively emphasise the local features in the Transformer to improve the detailed information processing ability of the neural network; 3) Our proposed SSFormer improves the SOTA performances of the ETIS benchmark, CVC-ClinicDB benchmark, and Kvasir benchmark by about 3%, 1.8%, and 1%, respectively. In addition, SSFormer has achieved state-of-the-art and superior performance in 2018 Data Science Bowl and ISIC-2018 benchmarks.

2 Methodology

2.1 Transformer Encoder

In order for our model to have enough generalization ability and multi-scale feature processing ability to carry out polyp segmentation, we use the Transformer based on the pyramid structure instead of CNN as the encoder. To this end, we adopt the encoder design of PVTv2 [20] and Segformer to construct the encoder. They both use the convolution operation to replace the PE operation of the traditional Transformer for consistency of spatial information, excellent performance and stability.

2.2 Aggregate Local and Global Features Stepwise (PLD)

Experiments [13,23] have demonstrated that the sufficiency of local features obtained in the shallow part of the Transformer directly affects the performance of the model. However, we believe that the existing Transformer model lacks local and detailed information processing ability to focus on critical detailed features (such as contour, veins and texture). As a result, this makes it difficult for the model to locate the more decisive local feature distribution (mucosa can be considered a distribution composed of local features such as unique veins and textures). We propose a novel multi-stage feature aggregation decoder PLD for feature pyramids to address this issue. Figure 1(a) shows that the PLD consists of the Local Emphasis (LE) module and the Stepwise Feature Aggregation (SFA) module. The experimental section compares PLD with other existing decoders with various encoders that can generate feature pyramids. We compare the attention distribution before the final prediction of several typical multi-stage feature aggregation decoders for Transformers. As demonstrated in Fig. 2(a), after PLD fuses multi-stage features, the prediction head can accurately focus on critical targets. In addition, our PLD can be used for other Pyramid Transformer encoders and can improve the model's accuracy. There is a further demonstration in Sect. 3.3.

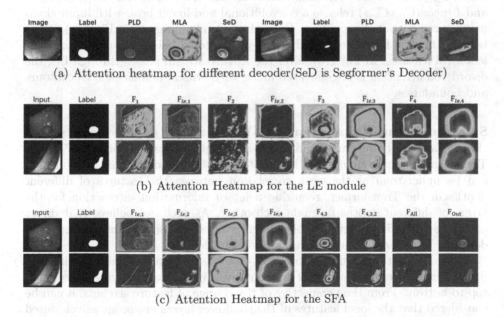

(a) Attention heatmap for different decoder(SeD is Segformer's Decoder)

(b) Attention Heatmap for the LE module

(c) Attention Heatmap for the SFA

Fig. 2. Attention heatmap of feature flow through the PLD process. Figure (a) shows that the LE module successfully focuses the model's attention on critical details. Figure (b) shows that the SFA structure effectively constrains the model's chaotic attention stepwise to fine critical regions.

Local Emphasis. In the Transformer, each patch in the image will mix the information of all other patches, even if their correlation is not high. After a large number of self-attention operations, the feature streams will converge, further exacerbating the attention dispersion or attention collapse [24]. Furthermore, we argue that the attention matrix in the self-attention mechanism can be viewed as a global non-preset convolution kernel. We designed the LE module using the local receptive field of the convolution kernel to increase the macro weights of the patches around the query patch to refocus attention on neighboring features thus reducing attention dispersion. In Fig. 1(b), the module consists of the convolution operators, activation functions, and a bilinear upsampling layer. We utilize the fixed receptive field of the convolution operator to mix the features of the adjacent patches of each patch, thereby increasing the associated weights of the adjacent patches to the center patch, thus emphasizing the local features of each patch. Since the feature types of the feature streams from different depths are different, we do not share the convolution weights for the feature streams at different levels in the feature pyramid. The formula for strengthening local features is as follows:

$$F_{le,i} = ReLU(Conv_i(C,C)(ReLU(Conv_i(C_i,C)(F_i)))), \qquad (1)$$

where $F_{le,i}$ refer to the local emphasized feature from stage i, $Conv_i(C_{in}, C_{out})$ and $Linear(C_{in}, C_{out})$ refer to a convolutional and linear layer with input channel C_{in} and output channel C_{out}. From the feature map given in Fig. 1(a), it can be seen that the LE can effectively clean up cluttered noises and emphasize critical local features. In Fig. 2(b), after the feature stream passes through LE, the disordered attention is re-condensed along with critical details such as contours and boundaries.

Stepwise Feature Aggregation (SFA). Experiments [13] have demonstrated that the amount of information interacted through residual connections [7] in the Transformer is more significant than that of the CNN model. This phenomenon can be understood as the weak correlation between the features of different depths in the Transformer, requiring a lot of information interaction for the layers of different depths to guide each other. As such, we believe that direct parallel aggregation of features of different stages with significant differences in depth in Transformer may generate an information gap.

In order for the feature aggregation to be as smooth as possible, the SFA progressively fuses the features of different levels in the feature pyramid from the top to bottom. From the perspective of the change of feature streams, it can be considered that the local features of the shallower layers are progressively fused into the global features of the deeper layer. This feature fusion method can reduce the information gap between the fused high-dimensional and low-dimensional features. As given in Fig. 2(c), local features gradually guide the attention of the model to critical regions in the SFA. In Fig. 1(a), the SFA consists of feature fusion units, linear fusion layers, and a linear prediction layer. The feature map of the fused structure in Fig. 1(a) (image with red border) shows that the SFA effectively incorporates local features into high-dimensional features and guides the feature stream into critical regions.

$$F_{i-1,i} = \begin{cases} Linear(2C,C)(Concat(F_{i-1},F_i)), \\ OR, \\ Linear(C,C)(Add(F_{i-1},F_i)), \end{cases} \tag{2}$$

Since the feature stream has the same shape after passing through the LE module, we can use concatenation or addition operation in the feature fusion unit as Eq. 2. In Table 4, we see that both perform equally well. Concatenation is the default in SSFormer.

2.3 Stepwise Segmentation Transformer

Based on the different encoder scales, we propose the SSFormer-S (Standard) and the SSFormer-L (Large) model. They achieve SOTA and competitive performance in several polyp segmentation benchmarks. Details are given in the experimental section. Moreover, SSFormer also achieved SOTA and competitive performance in ISIC-2018 and 2018 DATA Science Bowl.

3 Experiments

3.1 Experimental Setup

Dataset and Evaluation Matrix. Since the colon polyp segmentation task requires the model to have both accurate prediction and generalization capabilities, the performance of model on experimental and unseen benchmark datasets needs to be assessed separately. Therefore, following the experimental scheme of MSRF-Net [16], we train and test SSFormer on the Kavsir-SEG [8] and CVC-ClinicDB [1] benchmark datasets, respectively, to assess the accurate prediction and learning ability of models in the Kavsir-SEG and CVC-ClinicDB test set, respectively. In order to assess the generalization ability of SSFormer, we tested the model trained in Kavsir-SEG on CVC-ClinicDB and vice versa.

We refer to the experimental scheme of PraNet [6] and UACANet [9] that randomly extract 1450 images from the Kavsir and CVC-ClinicDB benchmark datasets to construct a training set (For fairness evaluation, we used the same training set as UACANet and PraNet), then test the model trained in this training set on the CVC-ColonDB [2] and ETIS [15] benchmark datasets. This test can demonstrate our model's accurate prediction and generalization ability in unseen datasets. Due to the variety of types and sizes of polyps in ETIS, it is the most challenging benchmark. The ISIC-2018 [4,17] and 2018 Data Science Bowl [3] benchmark datasets were also used in additional experiments. To unify the performance measures of the above two schemes, we only use mean Dice and mean IoU as evaluation matrices in our experimentation.

Implementation Details. We implement our model in PyTorch, which an NVIDIA TESLA A100 GPU accelerates. The AdamW optimizer is used with an initial learning rate of 0.0001, a decay rate of 0.1, and a decay period of 40 epochs. The training period is 200 epochs. Our loss function is the combined loss of Dice loss and BCE loss. During training, we resize the image to 352 × 352. We employ random flipping, scaling, rotation, and random dilation and erosion as data augmentation operations.

Table 1. The performance of the SOTA methods was trained and tested on the same benchmark dataset, used to assess learning ability, the scores in the table refer to [16]

Dataset	CVC-ClinicDB		Kvasir-SEG		ISIC-2018		2018 Data-Sci Bowl	
Methods	mDice	mIoU	mDice	mIoU	mDice	mIoU	mDice	mIoU
U-Net	0.9145	0.8654	0.8629	0.8176	0.8554	0.7847	0.9080	0.8314
U-Net++	0.8453	0.7559	0.7475	0.6313	0.8094	0.7288	0.7705	0.3010
Deeplabv3+	0.8897	0.8706	0.8965	0.8575	0.8772	0.8128	0.8857	0.8367
MSRF-Net	0.9420	**0.9043**	0.9217	**0.8914**	0.8824	0.8373	0.9224	0.8534
SSFormer-S	0.9268	0.8759	0.9261	0.8743	0.9195	0.8615	**0.9254**	**0.8652**
SSFormer-L	**0.9447**	0.8995	**0.9357**	0.8905	**0.9242**	**0.8675**	0.9230	0.8614

3.2   Results

Learning Ability. We split the CVC-ClinicDB and Kvasir benchmark datasets into 80% training set, 10% evaluation set and 10% test set according to the first scheme mentioned in Sect. 3.1. Table 1 demonstrates that our model improves the SOTA result by about 1.8% on the CVC-ClinicDB benchmark and about 1% on the Kvasir benchmark. These performances demonstrate the superior accurate prediction and learning abilities of SSFormer.

Furthermore, to assess the performance of SSFormer on other medical segmentation benchmarks, we conduct additional experiments on the ISIC-2018 and 2018 Data Science Bowl benchmark datasets. The results in Table 1 reveal that our model achieves the SOTA and excellent performance on two benchmarks, 2018 Data Science and ISIC-2018, respectively.

Generalization Ability. We test the SSFormer trained on the CVC-ClinicDB and Kvasir datasets on the Kvasir and CVC-ClinicDB benchmarks, respectively. As mentioned in Sect. 3.1, this test result can reflect the generalization ability of our model. In Table 2, our model achieves outstanding performance using this testing scheme. In addition, to further assess the generalization ability of SSFormer, we refer to the experimental scheme of PraNet, use the training set constructed from part of the Kvasir and CVC-ClinicDB datasets for training,

Table 2. Generalization Test 1

Train set	CVC-ClinicDB		Kvasir-SEG	
Test set	Kvasir-SEG		CVC-ClinicDB	
Methods	mDice	mIoU	mDice	mIoU
U-Net	0.6222	0.4588	0.7172	0.6133
U-Net++	0.5926	0.4564	0.4265	0.3345
Deeplabv3+	0.6746	0.5327	0.6509	0.5385
MSRF-Net	0.7575	0.6337	0.7921	0.6498
SSFormer-S	0.7790	0.6977	0.7966	0.7229
SSFormer-L	**0.8270**	**0.7348**	**0.8339**	**0.7573**

Table 3. Generalization Test 2

Train set	Kvasir & CVC-ClinicDB			
Test set	CVC-ColonDB		ETIS	
Method	mDice	mIoU	mDic	mIoU
UACANet-S	0.783	0.704	0.694	0.615
UACANet-L	0.751	0.678	0.766	0.689
CaraNet	0.773	0.689	0.747	0.672
PraNet	0.712	0.640	0.628	0.567
SSformer-S	0.772	0.697	0.767	0.698
SSformer-L	**0.802**	**0.721**	**0.796**	**0.720**

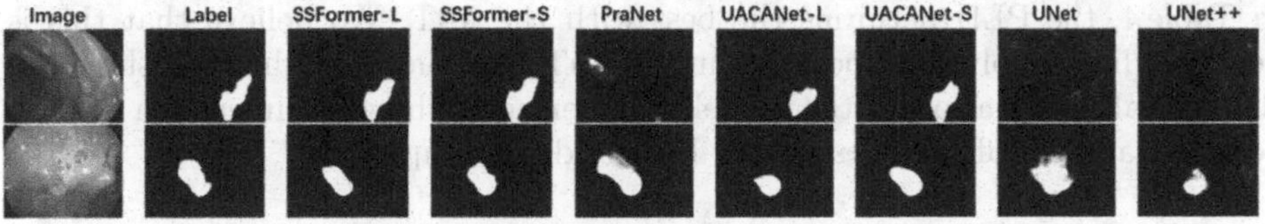

Fig. 3. Predicted results of different methods

and test the model on the CVC-ColonDB and ETIS benchmarks. The results in Table 3 demonstrate that our model significantly improves the SOTA performance (3%) in the most challenging ETIS and achieves superior performance in CVC-ColonDB. Figure 3 gives the prediction accuracy of our model on the ETIS benchmark. These results can prove that SSFormer has robust generalization and accurate prediction abilities. (The scores in Table 2 and Table 3 are obtained from [9,14,16,25]).

Table 4. Different encoder and decoder combinations performance with the same hyperparameter settings. The performance of different encoder and decoder combinations. The score is the performance of the model on the (CVC-ClinicDB, Kvasir) dataset group. (SeD is Segformer's Decoder, MiT is the Segformer's Encoder, and the CvT is proposed in [21])

Encoder\Decoder	MLA [23]	SeD [22]	PLD-Cat	PLD-Add
CvT [21]	0.898, 0.912	0.820, 0.889	0.912, 0.923	-
PvT [19]	0.809, 0.799	0.588, 0.618	0.828, 0.801	-
MiT [22]	0.907, 0.893	0.911, 0.903	0.916, 0.925	0.923, 0.897

Table 5. The effect of PLD components on model performance. The score in the table follow (mDice, mIOU).

Decoder\Dataset	Kvasir-SEG	ISIC-2018	2018 Data-Science Bowl
Without PLD	0.869, 0.918	0.855, 0.894	0.835, 0.904
LE	0.877, 0.925	0.860, 0.909	0.850, 0.915
SFA	0.885, 0.930	0.863, 0.918	0.858, 0.920
LE+SFA	**0.891, 0.936**	**0.868, 0.924**	**0.861, 0.923**

3.3 Ablation Study

In Table 4, the PLD performs the best with the MiT. We believe that this is because the convolution operation inside MiT can maintain the consistency of the spatial information of the model. Furthermore, the experiments in Table 5 demonstrate the effectiveness of the PLD and its components.

4 Conclusions

In this research, we propose a novel deep learning model SSFormer, with robust generalization and learning ability. These are critical for polyp segmentation. Furthermore, we find that our model also demonstrates powerful learning ability in ISIC-2018 and 2018 Data Science Bowl benchmarks in additional experiments. We believe that the SSFormer has great potential to improve deep learning performance in other medical image segmentation tasks. Furthermore, experiments demonstrate that our proposed local feature emphasis module effectively constrains the attention dispersion of Transformers. Therefore, our research can be further used to optimize the Transformer backbone network for the general computer vision community and high generalizability medical applications.

Acknowledgments. This work was supported by the Key Program Special Fund in XJTLU (KSF-A-22).

References

1. Bernal, J., Sánchez, F.J., Fernández-Esparrach, G., Gil, D., Rodríguez, C., Vilariño, F.: WM-DOVA maps for accurate polyp highlighting in colonoscopy: validation vs. saliency maps from physicians. Comput. Med. Imaging Graph. **43**, 99–111 (2015)
2. Bernal, J., Sánchez, J., Vilarino, F.: Towards automatic polyp detection with a polyp appearance model. Pattern Recogn. **45**(9), 3166–3182 (2012)
3. Caicedo, J.C., et al.: Nucleus segmentation across imaging experiments: the 2018 data science bowl. Nat. Methods **16**(12), 1247–1253 (2019)
4. Codella, N.C., et al.: Skin lesion analysis toward melanoma detection: a challenge at the 2017 international symposium on biomedical imaging (ISBI), hosted by the international skin imaging collaboration (ISIC). In: 2018 IEEE 15th International Symposium on Biomedical Imaging (ISBI 2018), pp. 168–172. IEEE (2018)

5. Dosovitskiy, A., et al.: An image is worth 16x16 words: transformers for image recognition at scale. arXiv preprint arXiv:2010.11929 (2020)
6. Fan, D.-P., et al.: PraNet: parallel reverse attention network for polyp segmentation. In: Martel, A.L., et al. (eds.) MICCAI 2020. LNCS, vol. 12266, pp. 263–273. Springer, Cham (2020). https://doi.org/10.1007/978-3-030-59725-2_26
7. He, K., Zhang, X., Ren, S., Sun, J.: Deep residual learning for image recognition. In: Proceedings of the IEEE Conference on Computer Vision and Pattern Recognition, pp. 770–778 (2016)
8. Jha, D., et al.: Kvasir-SEG: a segmented polyp dataset. In: Ro, Y.M., et al. (eds.) MMM 2020. LNCS, vol. 11962, pp. 451–462. Springer, Cham (2020). https://doi.org/10.1007/978-3-030-37734-2_37
9. Kim, T., Lee, H., Kim, D.: UACANet: uncertainty augmented context attention for polyp segmentation. In: Proceedings of the 29th ACM International Conference on Multimedia, pp. 2167–2175 (2021)
10. Li, G., Xu, D., Cheng, X., Si, L., Zheng, C.: SimViT: exploring a simple vision transformer with sliding windows (2021)
11. Lou, A., Guan, S., Loew, M.: CaraNet: context axial reverse attention network for segmentation of small medical objects. arXiv preprint arXiv:2108.07368 (2021)
12. Naseer, M.M., Ranasinghe, K., Khan, S.H., Hayat, M., Shahbaz Khan, F., Yang, M.H.: Intriguing properties of vision transformers. In: Advances in Neural Information Processing Systems 34 (2021)
13. Raghu, M., Unterthiner, T., Kornblith, S., Zhang, C., Dosovitskiy, A.: Do vision transformers see like convolutional neural networks? In: Advances in Neural Information Processing Systems 34 (2021)
14. Ronneberger, O., Fischer, P., Brox, T.: U-Net: convolutional networks for biomedical image segmentation. In: Navab, N., Hornegger, J., Wells, W.M., Frangi, A.F. (eds.) MICCAI 2015. LNCS, vol. 9351, pp. 234–241. Springer, Cham (2015). https://doi.org/10.1007/978-3-319-24574-4_28
15. Silva, J., Histace, A., Romain, O., Dray, X., Granado, B.: Toward embedded detection of polyps in WCE images for early diagnosis of colorectal cancer. Int. J. Comput. Assist. Radiol. Surg. 9(2), 283–293 (2014)
16. Srivastava, A., et al.: MSRF-Net: a multi-scale residual fusion network for biomedical image segmentation. arXiv preprint arXiv:2105.07451 (2021)
17. Tschandl, P., Rosendahl, C., Kittler, H.: The HAM10000 dataset, a large collection of multi-source dermatoscopic images of common pigmented skin lesions. Sci. Data 5(1), 1–9 (2018)
18. Vaswani, A., et al.: Attention is all you need. In: Advances in Neural Information Processing Systems 30 (2017)
19. Wang, W., et al.: Pyramid vision transformer: a versatile backbone for dense prediction without convolutions. In: Proceedings of the IEEE/CVF International Conference on Computer Vision, pp. 568–578 (2021)
20. Wang, W., et al.: PVT v2: improved baselines with pyramid vision transformer. Comput. Visual Media 8, 415–424 (2022). https://doi.org/10.1007/s41095-022-0274-8
21. Wu, H., et al.: CvT: introducing convolutions to vision transformers. In: Proceedings of the IEEE/CVF International Conference on Computer Vision, pp. 22–31 (2021)
22. Xie, E., et al.: SegFormer: simple and efficient design for semantic segmentation with transformers. In: Advances in Neural Information Processing Systems 34 (2021)

23. Zheng, S., et al.: Rethinking semantic segmentation from a sequence-to-sequence perspective with transformers. In: Proceedings of the IEEE/CVF Conference on Computer Vision and Pattern Recognition, pp. 6881–6890 (2021)
24. Zhou, D., et al.: DeepViT: towards deeper vision transformer. arXiv preprint arXiv:2103.11886 (2021)
25. Zhou, Z., Rahman Siddiquee, M.M., Tajbakhsh, N., Liang, J.: UNet++: a nested U-Net architecture for medical image segmentation. In: Stoyanov, D., et al. (eds.) DLMIA/ML-CDS -2018. LNCS, vol. 11045, pp. 3–11. Springer, Cham (2018). https://doi.org/10.1007/978-3-030-00889-5_1

Stay Focused - Enhancing Model Interpretability Through Guided Feature Training

Alexander C. Jenke[1]([✉])([iD]), Sebastian Bodenstedt[1], Martin Wagner[2],
Johanna M. Brandenburg[2], Antonia Stern[3], Lars Mündermann[3],
Marius Distler[4], Jürgen Weitz[4], Beat P. Müller-Stich[2], and Stefanie Speidel[1]

[1] Department for Translational Surgical Oncology, National Center for Tumor
Diseases (NCT), Partner Site Dresden, Dresden, Germany
{alexander.jenke,sebastian.bodenstedt,stefanie.speidel}@nct-dresden.de
[2] Department of General, Visceral and Transplantation Surgery,
University of Heidelberg, Heidelberg, Germany
[3] KARL STORZ SE & Co. KG, Tuttlingen, Germany
[4] Department of Visceral, Thoracic and Vascular Surgery, Faculty of Medicine,
University Hospital Carl Gustav Carus, TU Dresden, Dresden, Germany

Abstract. In computer-assisted surgery, artificial intelligence (AI) methods need to be interpretable, as a clinician has to understand a model's decision. To improve the visual interpretability of convolutional neural network, we propose to indirectly guide the feature development process of the model with augmented training data in which unimportant regions in an image have been blurred. On a public dataset, we show that our proposed training workflow results in better visual interpretability of the model and improves the overall model performance. To numerically evaluate heat maps, produced by explainable AI methods, we propose a new metric evaluating the focus with regards to a mask of the region of interest. Further, we are able to show that the resulting model is more robust against changes in the background by focusing the features onto the important areas of the scene and therefore improve model generalization.

Keywords: Explainable artificial intelligence · Surgical data science · Instrument presence detection · Computer-assisted surgery

1 Introduction

In computer-assisted surgery (CAS), the output of artificial intelligence (AI) algorithms needs to be interpretable, as a clinician has to understand a model's

Supplementary Information The online version contains supplementary material available at https://doi.org/10.1007/978-3-031-16437-8_12.

© The Author(s), under exclusive license to Springer Nature Switzerland AG 2022
L. Wang et al. (Eds.): MICCAI 2022, LNCS 13433, pp. 121–129, 2022.
https://doi.org/10.1007/978-3-031-16437-8_12

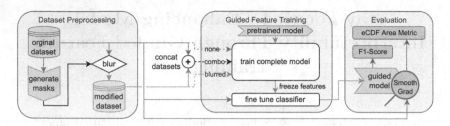

Fig. 1. GFT pipeline: a pretrained model is trained on original and/or modified data. Features are frozen and the classifier is fine-tuned on original data. The model takes single images as input and outputs a binary multilabel tool presence classification.

decision to be able to assess and use the information. The successful translation of CAS systems into the clinic depends on meaningful verification, human interpretability, and generalizability [8]. Interpretability can be achieved with explainable AI (XAI) methods. For convolutional neural networks (CNN), which are used for image processing tasks, usually visual explanations are used [2]. These highlight areas which influenced the CNN's decision and therefore mark important features in the image. Applying XAI to state-of-the-art AI architectures reveals, that these models often consider different features than expected. As preliminary experiments have shown, common models that determine if a certain instrument type is currently visible (instrument presence detection) lack focus on the instruments themselves and often decide based on features in the background. To tackle this issue, we present a novel approach for guided feature training (GFT), by augmenting the input data without modifying the model architecture. Thereby, we shift the model's focus onto the essential areas in the image without requiring large amounts of additional annotation.

The improved focus by GFT makes the models more robust against changing backgrounds and therefore improves generalizability. To numerically evaluate the focus improvement on the instruments, we propose a new metric, evaluating heat maps produced by XAI methods with regards to a mask of the region of interest. The source code is made public at https://gitlab.com/nct_tso_public/gft.

2   Methods

In this section, we introduce methods for guided feature training and propose a metric to numerically evaluate the model's focus, which is determined via SmoothGrad [10]. The established pipeline is shown in Fig. 1.

2.1   Dataset Prepossessing

A standard CNN, using the DeepLabV3 architecture [4] with a ResNet101 backbone and pretrained on COCO train2017 [6], serves as foundation for the proposed data preprocessing system. This architecture was selected due to its success in several Endoscopic Vision (EndoVis) segmentation challenges [1,3]. The ResNet101 [5] backbone showed the best performance in preliminary studies.

The model is trained for frame-based binary instrument segmentation using the SGD optimizer [12] and a OneCycle learn rate [11]. As the masks are intended to modify the images without affecting the instrument regions, the used binary cross entropy loss BCE is extended by a regularizing term FN to increase recall, accepting detriment of precision:

$$FN(x, \hat{x}) = -\frac{1}{n} \sum_{i=1}^{n} ((x_i - 0.5) - |(x_i - 0.5)| + (x_i - 1)) * \hat{x}_i \tag{1}$$

$$L(x, \hat{x}) = BCE(x, \hat{x}) + w * FN(x, \hat{x}) \tag{2}$$

where x is the model output, ranged from 0 to 1, $\hat{x}$ is the binary target, and w is a weighting factor to control the influence of the regularizing term. As the term in FN is multiplied with the binary target $\hat{x}$, it only applies to positive samples and the term strongly penalizes false negative predictions of the model.

The parts of a frame that don't contain instruments are blurred with the masks, with the goal that features belonging to background regions cannot be learned.

2.2 Guided Feature Training

In the guided feature training (GFT) block of the pipeline, we aim to train a multi-class classification model for frame-based instrument presence detection. The idea of GFT is to train different parts of the model using different combinations of the original and/or modified dataset without modifying the model's architecture. Therefore, GFT can be applied to all standard classification architectures.

In our study, we opted for a ResNet architecture [5] pretrained on ImageNet, as it was used by 7 of the 8 participants in the instrument classification task of the 2019 EndoVis Surgical Skill and Workflow Challenge achieving decent results [13]. Due to the limited size of our dataset, we decided to use the smaller ResNet18 instead of a ResNet50.

The model is trained for the frame-wise instrument presence detection task using the AdamW optimizer [7] and a OneCycle learn rate [11]. As shown in Fig. 1 the model is trained in two steps. In the first step, the model is trained on the modified dataset with or without the original dataset, depending on the selected guidance mode, to fine tune the features on the surgical video data. In the second step, the classifier part of the model is further trained on the original dataset to be able to optimally classify unmodified data based on the guided features.

In our study we investigate two different guidance modes, *blurred* and *combo* and compare those to a baseline without guidance (*none*). In the *none* mode only the original dataset is used during feature training. In the *blurred* mode only the modified dataset is used during feature training and in the *combo* mode the original and modified datasets are concatenated, creating a new set twice the size of the original.

2.3 Evaluation

SmoothGrad. The aim of GFT is to enforce the model's focus on the region of interest (ROI) which is essential to the classification task. In our case, the ROI includes the instruments to be detected. To be able to evaluate the focus we use SmoothGrad [10], which is a visual model-agnostic post-hoc XAI method. SmoothGrad averages multiple sensitivity maps, based on the gradient of the class score while applying noise to the input image. This produces a robust heat map for each class of the output label, describing the importance of each pixel of the input image onto the decision made. During evaluation, the heat maps of all predicted tools are combined using the maximum value for each pixel. Examples of SmoothGrad heat maps can be found in Fig. 4.

eCDF-Area Metric. The heat maps produced by SmoothGrad can be used to visually assess the model's focus on single examples. However, to evaluate the average performance of a model a numerical abstraction is needed. As the value range of the heat maps is not expressive and can only be used to compare different pixels within one map, no fixed threshold can be used to determine a set of important pixels. Further, existing metrics like the pointing game, introduced in [9,14], do not fully fulfill our requirements, as they only focus on the most important pixels while leaving the remaining distribution out of the evaluation.

To overcome this difficulty, we introduce the eCDF-Area metric. We aim to numerically score the focus of important pixels onto the ROI by calculating the area between the empirical cumulative distribution functions (eCDF) over the pixels in the ROI and the background (BG). Let X be the set of pixel values in the considered area and $eCDF_X(t) = |\{x|x \in X, x \leq t\}| \cdot |X|^{-1}$ the eCDF of the set X, where t is the running threshold value. Then the eCDF-Area metric can be defined as follows:

$$eCDFArea = \int eCDF_{BG}(t)dt - \int eCDF_{ROI}(t)dt \qquad (3)$$

The score results in a range of $[-1; 1]$, where 1 implies that pixels in the ROI and background respectively have the same value while the pixels in the ROI have a higher value. This would be a perfect focus distribution. The value of 0 implies a homogeneous distribution of pixel values into the two considered areas and -1 is the inverse distribution of 1 with the higher pixel values in the background.

Due to the use of an eCDF for each area respectively, the metric is invariant to the size of the ROI. As we do not expect a model to focus on the complete area of the ROI, we don't expect the metric to be close to 1 but rather between 0 and 1. The metric indicates an improvement in focus by achieving a higher value than a comparison model on the same evaluated frames.

For better interpretability, the pixels in the heat map are clipped to the 95th percentile, clipping the lowest and highest values. This does not change the relation between two eCDF-Areas, but only increases the observable change in metric values.

Fig. 2. Example of a fake image. The background (left) and tools (center) are combined into a fake image (right). Existing instruments in the background are blurred.

Fake Images. To further investigate the dependency of the trained models on the background, we evaluate on fake images. These images were created by masking instruments based on human segmentations and recombining them with different backgrounds, which were created by blurring out existing instruments and leaving only background information. An example is shown in Fig. 2. The model focus was evaluated using the eCDF-Area metric as well as the F1-Score on the same tools over varying backgrounds.

3   Results

To evaluate the proposed GFT, we perform three trials. The first trial serves as a proof of concept for GFT, where the models are trained using segmentation masks provided by human annotators. The second trial aims to prove the validity of the automatic mask generation approach without affecting the model performance or effects of GFT. The third trial scales up the GFT approach investigating the effects of GFT in its realistic application. All experiments were performed on an Nvidia GTX 1080.

3.1   Dataset

We evaluate the proposed GFT on the publicly available HeiChole dataset [13] containing instrument presence annotations and the HeiSurF dataset [3] which extends HeiChole by adding full scene segmentations annotated by 3 surgical experts. HeiSurF will be made publicly available at the time of journal publication of the challenge results.

Both datasets contain the same 33 videos of laparoscopic cholecystectomies of 3 centers with an average duration of 42.4 min each, of which 24 videos of 2 centers are publicly available. The data sets provide a frame-wise annotation of the visibility of 7 different instrument types (grasper, clipper, coagulation instruments, scissors, suction-irrigation, specimen bag, and stapler) of which 5 are used in our study. The remaining two were excluded due to rare occurrences (stapler) and being visually divergent (specimen bag). A full scene segmentation mask is provided for every 2 min of video, resulting in 466 frames, which in our work were binarized by only using the instrument annotation.

Table 1. Each line shows how the model performance compares for the different trials and guidance modes. The values shown are the F1-Score for each tool and the unweighted mean over all tools (μ) evaluated on the test data sampled with 1 fps. The best performances for each trial are in bold.

Trial	Guidance	Grasper	Clipper	Coagulation	Scissor	Suction	μ
Ground Truth Masks	None	**73%**	11%	73%	22%	32%	42%
	Blurred	72%	19%	74%	**30%**	50%	49%
	Combo	64%	**36%**	**76%**	28%	**54%**	**52%**
Automated Masks	None	**80%**	0%	79%	16%	55%	46%
	Blurred	68%	25%	78%	22%	56%	50%
	Combo	74%	**26%**	**84%**	**31%**	**67%**	**56%**
Larger Dataset (autom. Masks)	None	**88%**	**80%**	**92%**	47%	**77%**	76.7%
	Blurred	83%	75%	90%	47%	73%	74%
	Combo	87%	79%	90%	**54%**	76%	**77.3%**

For our first trial and the training of the segmentation model, the 466 frames with instrument segmentations were used. For the second trial, the videos were also sampled every 2 min, but with an offset of one minute and therefore selecting the frames in the middle of the segmented samples. This resulted in 457 frames being automatically masked by the DeepLabV3 model. For the third trial, the videos were sampled with 1 fps and automatically segmented by the DeepLabV3 even if a human segmentation was given.

The evaluation F1-Scores were calculated on 24144 non-public frames of the test data sampled at 1 fps and the DeepLabV3 was evaluated on the 249 segmented frames of the non-public test data. All frames were down scaled to a size of 480×320 pixels.

3.2   Instrument Segmentation

The DeepLabV3 was trained for 100 epochs using the SGD optimizer with a learning rate of 1×10^{-3}, a weight decay of 1×10^{-5}, a momentum of 0.9, and a batch size of 4. The BCE loss is weighted according to the number of positive samples of classes. The regularizing term was weighted with a factor $w = 5$. The model reached a Dice Score of 72%, Precision of 68%, and a Recall of 88%. As previously stated, we intentionally opted for a higher recall at the expense of precision using the regularizing term, as we aim for masks with complete tools.

3.3   Instrument Prediction

The classification models were trained in all trials for 100 epochs for feature development and classifier tuning each, using the AdamW optimizer with a learning rate of 1×10^{-4} and a batch size of 4. The BCE loss is weighted according to the number of positive samples of classes. Over all trials, the models with combo guidance resulted in the best average F1-Score compared to the other guidance modes. The models trained on the larger dataset reached clearly higher average F1-Scores of up to 77.3%. The detailed results of all trials with the different guidance modes broken down for all tools can be found in Table 1.

3.4 Model Interpretability

The model focus was evaluated using the eCDF Area, which was calculated for all frames of the used train data with instruments visible and a tool detected by the model. The distributions of the resulting values are shown in Fig. 3a. With no guidance, the models reach a median eCDF Area of 0.054 and 0.079 for the first and second trial respectively. The models trained with combo guidance reach 0.229 and 0.240 respectively. An example of the focus improvement is shown in Fig. 4.

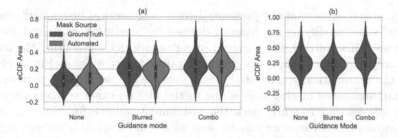

Fig. 3. Distributions of eCDF-Areas for (a) the first and second trial evaluated on the training data (b) the third trial evaluated on the fake images. Ground truth and automated masks show similar interpretability throughout the trials. GFT improves the interpretability in all three trials, while the effect is stronger in the first two trials.

Fig. 4. Example of SmoothGrad heatmaps, important areas are highlighted in yellow. As shown, GFT improves the focus as important pixels shift from the background onto the instrument. This is also numerically supported by the increased eCDF-Area metric. (Color figure online)

3.5 Fake Images

The models of the third trial were evaluated on fake images. The distributions of the eCDF Areas are shown in Fig. 3b. Further, the F1-Scores for each tool were calculated for each used background. The resulting distributions can be found in Fig. 5.

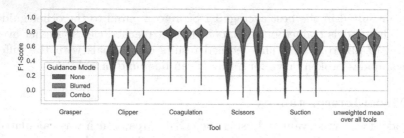

Fig. 5. Distribution of F1-Scores on the fake images over the different backgrounds for the guidance modes. Especially for the harder to learn tools (Clipper & Scissors) an improvement due to GFT can be observed.

4 Conclusion

We were able to show that the presence detection of surgical instruments can be improved when modifying the training dataset by blurring background areas in the frames. This blurring was achieved by automatically segmenting the instruments, only requiring a couple of segmented frames for training. As Fig. 3a clearly shows, the automatically generated masks provide the same GFT performance as human-annotated masks. This is due to the fact no exact segmentations are needed. Masks that roughly mark the instruments suffice for blurring. Automatic segmentation therefore can be seen as a valid approach for guiding features without requiring huge amounts of additional annotation, which alleviates the initially limiting requirement for masks in GFT.

As shown in Table 1, combo guidance strongly improves the prediction performance on tools which are poorly detected without guidance, like Clipper and Scissors. On the other hand, we find slight performance losses on good performing tools. However, throughout all experiments, the combo guidance mode reaches the highest average F1-Scores, while this effect is stronger on lower numbers of frames. Figure 5 shows that especially hard to learn tools in traditional training show a widely spread distribution of F1-Scores. This can be interpreted as that they are more sensitive to background changes. We were able to show that GFT shifts the focus away from the background. This increases robustness against changing backgrounds, as the more densely distributed F1-Scores in Fig. 3b show.

While GFT can not be seen as a XAI method itself, its primary aim is to influence the model features, guiding them explicitly to areas which a human expects to be highlighted when analyzing a classification prediction with visual XAI. In our trials GFT guided the focus of the model onto the instruments it was supposed to detect and thereby improved the interpretability. This was shown with the newly introduced eCDF-Area metric numerically evaluating the distribution of important pixels in a focus heat map in relation to a ROI. We argue GFT addresses an important part of the XAI research of improving model interpretability to meet humans expectations.

We see high potential in our proposed method improving state-of-the-art models, as we do not need to modify the model architecture and nearly no additional annotation is required. Additionally, we have observed stronger effects on smaller datasets. Therefore, future work on the potential of GFT to lower the required amount of data is needed. Our results on detecting tools were promising, however, translating GFT to more challenging scenes has to be researched. Here further work on the segmenting models is needed. Incorporating the classifier's focus might help to support the training of the segmenting model.

Acknowledgements. Funded by the Federal Ministry of Health, Germany (BMG), as part of the SurgOmics project.

References

1. Allan, M., et al.: 2018 robotic scene segmentation challenge, August 2020. arXiv:2001.11190 [cs]
2. Barredo Arrieta, A., et al.: Explainable Artificial Intelligence (XAI): concepts, taxonomies, opportunities and challenges toward responsible AI. Inf. Fusion **58**, 82–115 (2020)
3. Bodenstedt, S., et al.: Result presentation of endoscopic vision challenge 2021 - HeiChole surgical workflow analysis and full scene segmentation (HeiSurF) (2021). https://www.synapse.org/#!Synapse:syn25101790
4. Chen, L.C., Papandreou, G., Schroff, F., Adam, H.: Rethinking atrous convolution for semantic image segmentation, December 2017. arXiv:1706.05587 [cs]
5. He, K., Zhang, X., Ren, S., Sun, J.: Deep residual learning for image recognition, December 2015. arXiv:1512.03385 [cs]
6. Lin, T.Y., et al.: Microsoft COCO: common objects in context, February 2015. arXiv:1405.0312 [cs]
7. Loshchilov, I., Hutter, F.: Decoupled weight decay regularization, January 2019. arXiv:1711.05101 [cs, math]
8. Maier-Hein, L., et al.: Surgical data science - from concepts toward clinical translation. Med. Image Anal. **76**, 102306 (2022)
9. Petsiuk, V., Das, A., Saenko, K.: RISE: randomized input sampling for explanation of black-box models, September 2018. arXiv:1806.07421 [cs]
10. Smilkov, D., Thorat, N., Kim, B., Viégas, F., Wattenberg, M.: SmoothGrad: removing noise by adding noise, June 2017. arXiv:1706.03825 [cs, stat]
11. Smith, L.N., Topin, N.: Super-convergence: very fast training of neural networks using large learning rates, May 2018. arXiv:1708.07120 [cs, stat]
12. Sutskever, I., Martens, J., Dahl, G., Hinton, G.: On the importance of initialization and momentum in deep learning. In: International Conference on Machine Learning, pp. 1139–1147, May 2013
13. Wagner, M., et al.: Comparative validation of machine learning algorithms for surgical workflow and skill analysis with the HeiChole benchmark, September 2021. arXiv:2109.14956 [cs, eess]
14. Zhang, J., Bargal, S.A., Lin, Z., Brandt, J., Shen, X., Sclaroff, S.: Top-down neural attention by excitation backprop. Int. J. Comput. Vision **126**(10), 1084–1102 (2017). https://doi.org/10.1007/s11263-017-1059-x

On the Uncertain Single-View Depths in Colonoscopies

Javier Rodriguez-Puigvert[✉], David Recasens, Javier Civera,
and Ruben Martinez-Cantin

Universidad de Zaragoza, Zaragoza, Spain
{jrp,recasens,jcivera,rmcantin}@unizar.es

Abstract. Estimating depth information from endoscopic images is a prerequisite for a wide set of AI-assisted technologies, such as accurate localization and measurement of tumors, or identification of non-inspected areas. As the domain specificity of colonoscopies –deformable low-texture environments with fluids, poor lighting conditions and abrupt sensor motions– pose challenges to multi-view 3D reconstructions, single-view depth learning stands out as a promising line of research. Depth learning can be extended in a Bayesian setting, which enables continual learning, improves decision making and can be used to compute confidence intervals or quantify uncertainty for in-body measurements. In this paper, we explore for the first time Bayesian deep networks for single-view depth estimation in colonoscopies. Our specific contribution is two-fold: 1) an exhaustive analysis of scalable Bayesian networks for depth learning in different datasets, highlighting challenges and conclusions regarding synthetic-to-real domain changes and supervised vs. self-supervised methods; and 2) a novel teacher-student approach to deep depth learning that takes into account the teacher uncertainty.

Keywords: Single-view depth · Bayesian deep networks · Depth from monocular endoscopies

1 Introduction

Depth perception inside the human body is one of the cornerstones to enable automated assistance tools in medical procedures (e.g. virtual augmentations and annotations, accurate measurements or 3D registration of tools and interest regions) and, in the long run, the full automation of certain procedures and medical robotics. Monocular cameras stand out as very convenient sensors, as they are minimally invasive for in-vivo patients, but estimating depth from colonoscopy images is a challenge. Multi-view approaches are accurate and robust in many applications outside the body, e.g. [27], but assume certain rigidity, texture and illumination conditions that are not fulfilled in in-body images. Single-view 3D

Supplementary Information The online version contains supplementary material available at https://doi.org/10.1007/978-3-031-16437-8_13.

© The Author(s), under exclusive license to Springer Nature Switzerland AG 2022
L. Wang et al. (Eds.): MICCAI 2022, LNCS 13433, pp. 130–140, 2022.
https://doi.org/10.1007/978-3-031-16437-8_13

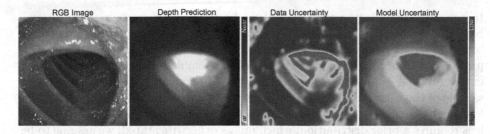

Fig. 1. Depth and uncertainty predictions for a colonoscopy image. Dark/bright colors stands for near/far depths, blue/red stands for low/high uncertainties. Note the higher uncertainties in darker and farther areas and in reflections. (Color figure online)

geometry is ill-posed, since infinite 3D scenes can explain a single 2D view [11]. In this last case, deep neural networks have shown impressive results in last years [5,7,9]. However, the vast majority of deep learning models lack any metric or intuition about their predictive accuracy. In a critical environment such as the inside of the human body, uncertainty quantification is essential. Specifically, in medical robotics it allows us to properly account for uncertainty in control and decision making, and in SLAM [4] to safely navigate inside the body. It also provides confidence intervals in in-body measurements (e.g., polyps), which is valuable for doctors to decide how to act. Uncertainty quantification is, in general, a must-have for robust, interpretable and safe AI systems. Bayesian deep learning perfectly combines the fields of deep learning and uncertainty quantification in a sound and grounded approach. However, for high-dimensional deep networks, accurate Bayesian inference is intractable. Only bootstrapping methods such as deep ensembles [16] have shown to produce well-calibrated uncertainties in many computer vision tasks at reasonable cost [10].

In this work, we address for the first time the use of Bayesian deep networks for depth prediction in colonoscopies. Figure 1 illustrates the predicted depth and uncertainties of one of our Bayesian model in a real colonoscopy image. Specifically, we first benchmark thoroughly, in synthetic and real data, supervised and self-supervised learning approaches in the colonoscopic domain. We demonstrate that Bayesian models trained on synthetic data can be transferred adequately to similar domains and we quantify the generalization of depth and uncertainty to real scenarios. Secondly, we propose a novel teacher-student method that models the uncertainty of the teacher in the loss, achieving state-of-the-art performance in a real colonoscopy domain.

2 Preliminaries and Related Work

Bayesian deep learning is a form of deep learning that performs probabilistic inference on deep network models. This enables uncertainty quantification for the model and the predictions. For high-dimensional deep networks, Bayesian inference is intractable and some approximate inference methods such as relying on variational inference or Laplace approximations might perform poorly.

In practice, sampling methods based on bootstrapping, such as deep ensembles [16], or Monte Carlo, like MC-dropout [8], have shown to be the most scalable, reliable and efficient approaches for depth estimation and other computer vision tasks [10]. In particular, deep ensembles have shown to perform extremely well even with a reduced number of samples, because each random sample of the network weight is optimized using a maximum a posteriori (MAP) loss $\mathcal{L}_{MAP} = \mathcal{L}_{LL} + \mathcal{L}_{prior}$, resulting in a high probability sample. The MAP loss requires a prior distribution, which unless otherwise stated, we assume to be a Gaussian distribution over the weights $\mathcal{L}_{prior} = ||\theta||^2$. For the data likelihood $\mathcal{L}_{LL}$, we use a loss function based on the Laplace distribution for which the predicted mean $\mu(x)$ and the predicted scale $\sigma(x)$ come from the network described in Sect. 3, with two output channels [15]. The variance associated with the scale term represents the uncertainty associated with the data, also called *aleatoric uncertainty* or $\sigma_a(x)$. Furthermore, in deep ensembles, the variance in the prediction from the multiple models of the ensemble is the uncertainty that is due to the lack of knowledge in the model, which is also called *epistemic uncertainty* or $\sigma_e(x)$. For example, data uncertainty might appear in poorly illuminated areas or with lack of texture, while model uncertainty arises from data that is different from the training dataset. Note that while the model uncertainty can be reduced with larger training datasets, data uncertainty is irreducible. Model uncertainty is particularly relevant to address domain changes. To illustrate that, in Sect. 6 we present results of models trained on synthetic data and tested on real data.

Single-View Depth Learning has demonstrated a remarkable performance recently. Some methods rely on accurate ground truth labels at training [5,7,30], which is not trivial in many application domains. Self-supervision without depth labels was achieved by enforcing multi-view photometric consistency during training [9,36,37]. In the medical domain, supervised depth learning was addressed by Visentini et al. [32] with autoencoders and by Shen et al. [29] with GANs, both using ground truth from phantom models. Other works based on GANs were trained with synthetic models [2,20,21,24], and Cheng et al. [3] added a temporal consistency loss. Self-supervised learning is a natural choice for endoscopies to overcome the lack of depth labels on the target domain [22,25,28]. Although depth or stereo are not common for in-vivo procedures, several works use them for training [12,19,35]. Others train in phantoms [31] or synthetic data [6,13], facing the risk of not generalizing to the target domain. In this paper we study the limits of such generalization. SfM supervision was addressed by Liu et al. [18] using siamese networks and by Widya et al. [34] using GANs. Note that none of these references address uncertainty quantification, which we cover in this work.Kendall et al. [15] combine epistemic and aleatoric uncertainty by using a MC Dropout approximation of the posterior distribution. This approach obtains pixel-wise depth and uncertainty predictions in a supervised setting. Ilg et al. [14] propose a multi-hypothesis network to quantify the uncertainty of optical flow. Traditional self-supervised losses to regress depth are limited due to the aleatoric uncertainty of input images [17]. Poggi et al. [23] address such problem by introducing a teacher-student architecture to learn depth and

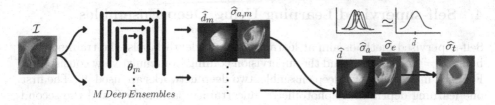

Fig. 2. Forward propagation of supervised deep ensembles. Our deep ensembles model a Gaussian distribution $N(\widehat{d}, \widehat{\sigma}_t^2)$. $\widehat{d}$ comes from averaging all ensembles depth output and $\widehat{\sigma}_t^2$ from joining data and model uncertainties

uncertainty. As key advantages, teacher-student architectures provide aleatoric uncertainty for depth and avoid photometric losses and pose regression networks, which are frequently unstable. In this work, we evaluate supervised and self-supervised approaches in colonoscopies images and propose a novel teacher-student approach that includes teacher uncertainty during training. Among the scalable Bayesian methods for single-view depth prediction, deep ensembles show the best calibrated uncertainty [23, 26] and hence we choose them as our model.

3 Supervised Learning Using Deep Ensembles

Let our dataset $\mathcal{D} = \{\{\mathcal{I}_1, d_1\}, \ldots, \{\mathcal{I}_N, d_N\}\}$ be composed by N samples, where each sample $i \in \{1, \ldots, N\}$ contains the input image $\mathcal{I}_i \in \{0, \ldots, 255\}^{w \times h \times 3}$ and per-pixel depth labels $d_i \in \mathbb{R}_{>0}^{w \times h}$. Regarding our network, we use an encoder-decoder architecture with skip connections, inspired by Monodepth2 [9], with two output layers. Thus, for every new image $\mathcal{I}$ the network predicts its pixel-wise depth $\widehat{d}(\mathcal{I}, \boldsymbol{\theta}) \in \mathbb{R}_{>0}^{w \times h}$ and data variance $\widehat{\sigma}_a^2(\mathcal{I}, \boldsymbol{\theta}) \in \mathbb{R}_{>0}^{w \times h}$. As commented in Sect. 2, we use a MAP loss, with $\mathcal{L}_{prior} = ||\boldsymbol{\theta}||^2$ and:

$$\mathcal{L}_{LL} = \frac{1}{w \cdot h} \sum_{j \in \Omega_i} \left(\frac{|| d[j] - \widehat{d}[j] \,||_1}{\widehat{\sigma}_a[j]} + \log \widehat{\sigma}_a[j] \right) \tag{1}$$

where $[\cdot]$ is the sampling operator and $j \in \Omega$ refers to the pixel coordinates in the image domain Ω. The per-pixel depth labels d can be obtained from ground truth depth d^{GT} or from SfM 3D reconstructions d^{SfM} [27]. A *deep ensemble model* is composed by M networks with weights $\{\theta_m\}_{m=1}^M$, each of them trained separately starting from different random seeds. We denote as $(\widehat{d}_m, \widehat{\sigma}_{a,m}^2)$ the output of the mth ensemble (see Fig. 2). We obtain the mean depth of the ensemble $\widehat{d}$ and its epistemic uncertainty $\widehat{\sigma}_e^2$ using the total mean and variance of the full model. The total uncertainty $\widehat{\sigma}_t^2 = \widehat{\sigma}_a^2 + \widehat{\sigma}_e^2$ combines the data $\widehat{\sigma}_a^2$ and model $\widehat{\sigma}_e^2$ uncertainties which results from the law of total variance.

$$\widehat{d} = \frac{1}{M} \sum_{m=0}^M \widehat{d}_m, \quad \widehat{\sigma}_a^2 = \frac{1}{M} \sum_{m=0}^M \widehat{\sigma}_{a,m}^2, \quad \widehat{\sigma}_e^2 = \frac{1}{M} \sum_{m=0}^M \left(\widehat{d} - \widehat{d}_m \right)^2 \tag{2}$$

4   Self-supervised Learning Using Deep Ensembles

Self-supervised methods aim at learning *without* depth labels, the training data being $\mathcal{D} = \{\mathcal{I}_1, \ldots, \mathcal{I}_N\}$ and the supervision coming from multi-view consistency. For each instance m of a deep ensemble, two deep networks are used [9]. The first one learning depth and a photometric uncertainty parameter $\widehat{u}$ and the second one learning to predict relative camera motion. We use a pseudo-likelihood for the loss function, that uses both networks for photometric consistency:

$$\mathcal{L}_{LL,m} = \frac{1}{w \cdot h} \sum_{j \in \Omega_i} \left(\frac{\mathcal{F}_p[j]}{\widehat{u}_m[j]} + \log \widehat{u}_m[j] \right) \tag{3}$$

where $\mathcal{F}_p$ is the photometric residual and $\widehat{u}_m$ an uncertainty prediction. The photometric residual $\mathcal{F}_p[j]$ of pixel j in a target image $\mathcal{I}_i$ is the minimum – between the warped images $\mathcal{I}_{i' \to i}$ from the previous and posterior images $\mathcal{I}_{i'}$ to the target one $\mathcal{I}_i$– of the sum of the photometric reprojection error and Structural Similarity Index Measure (SSIM) [33]:

$$\mathcal{F}_p[j] = \min((1 - \alpha)\|\mathcal{I}_i[j] - \mathcal{I}_{i' \to i}[j]\|_1 + \frac{\alpha}{2}(1 - \text{SSIM}(\mathcal{I}_i, \mathcal{I}_{i' \to i}, j)) \tag{4}$$

being $\alpha \in [0, 1]$ the relative weight of the addends; and $\mathcal{I}_i[j]$ and $\mathcal{I}_{i' \to i}[j] = \mathcal{I}_{i'}[j']$ the color values of pixel j of the target image $\mathcal{I}_i$ and of the warped image $\mathcal{I}_{i' \to i}$. To obtain this latter term, we warp every pixel j from the target image domain Ω_i to that of the source image $\Omega_{i'}$ using:

$$j' = \pi\left(\mathbf{R}_{i'i}\pi^{-1}(j, \widehat{d}_i[j]) + \mathbf{t}_{i'i}\right) \tag{5}$$

$\mathbf{R}_{i'i} \in SO(3)$ and $\mathbf{t}_{i'i} \in \mathbb{R}^3$ are the rotation and translation from Ω_i to $\Omega_{i'}$, and π and π^{-1} the projection and back-projection functions (3D point to pixel and vice versa). In this case, the prior loss also incorporates an edge-aware smoothness term $\mathcal{F}_s$, regularizing the predictions [9]. Thus, the prior term becomes $\mathcal{L}_{prior} = \|\theta\|^2 + \lambda_u \mathcal{F}_s[j]$, where λ_u calibrates the effect of the smoothness in terms of the reprojection uncertainty. This prior term is then combined to obtain $\mathcal{L}_{MAP}$ as described in Sect. 2. We obtain the ensemble prediction by model averaging as in the supervised case (Eq. 2). In this case, the data uncertainty for the depth prediction $\widehat{\sigma}_{a,m}^2$ cannot be extracted from the photometric uncertainty parameter $\widehat{u}$. Due to this, only model uncertainty will be considered in the experiments $(\widehat{\sigma}_t^2 = \widehat{\sigma}_e^2)$.

5   Teacher-Student with Uncertain Teacher

In the endoscopic domain, accurate depth training labels can only be obtained from RGB-D endoscopes (which are highly unusual) or synthetic data (that is affected by domain change). We propose the use a of a Bayesian teacher trained on synthetic colonoscopies that produces depth and uncertainty labels. The teacher's epistemic uncertainty allows us to overcome the domain gap automatically. Specifically, our novel teacher-student architecture models depth labels from the predictive posterior of the teacher $d \sim \mathcal{N}(\hat{d}, \sigma_T^2)$ (σ_T^2 is the total teacher variance). Thus, the likelihood must incorporate both the teacher and student distributions, which is used in the training loss. As before, the loss is based on a Laplacian likelihood

$$\mathcal{L}_{LL,m} = \frac{1}{w \cdot h} \sum_{j \in \Omega_i} \left(\frac{||\hat{d}_T[j] - \hat{d}[j]\,||_1}{\hat{\sigma}_m[j]} + \log \hat{\sigma}_m[j] \right) \tag{6}$$

where the per-pixel variance is the sum of the teacher predictive variance and the aleatoric one predicted by the student $\hat{\sigma}_m^2 = \hat{\sigma}_T^2 + \hat{\sigma}_{a,m}^2$. Our student is hence aware of the label reliability, which will be affected by the domain change.

6   Experimental Results

We present results in synthetic and real colonoscopies. Our first dataset is the one generated by Rau et al. [24], containing RGB images rendered from a 3D model of the colon in 15 different texture and illumination conditions. The second one, the EndoMapper dataset [1] contains real monocular colonoscopies. We evaluate each method using the following depth error metrics [5]: absolute relative error $1/w \cdot h \sum_{j \in \Omega_i} |d[j] - \hat{d}[j]|/\hat{d}[j]$, square relative error $1/w \cdot h \sum_{j \in \Omega_i} (d[j] - \hat{d}[j])^2/\hat{d}[j]$, root mean square error $(1/w \cdot h \sum_{j \in \Omega_i} (d[j] - \hat{d}[j])^2)^{1/2}$, rsme log $(1/w \cdot h \sum_{j \in \Omega_i} (\log d[j] - \log \hat{d}[j])^2)^{1/2}$ and $\delta < 1.25^i$: $1/w \cdot h \sum_{j \in \Omega_i} \max(d[j]/\hat{d}[j], \hat{d}[j]/d[j]) < 1.25^i$. We report also the Area Under the Calibration Error (AUCE) in terms of absolute uncertainty calibration [10]. Since our methods output a Gaussian distribution $\mathcal{N}(\hat{d}, \sigma^2)$ per pixel, we generate prediction intervals $\hat{d} \pm \phi^{-1}(\frac{p+1}{2})\sigma$ of confidence level $p \in [0, 1]$ being ϕ the CDF of the standard normal distribution. In a perfectly calibrated model, the proportion of pixels for which the prediction intervals covers the ground truth coincides the confidence level.

Synthetic Colon Dataset. We evaluate three training alternatives: GT (ground truth) depth supervision, SfM supervision and self-supervision. We use 6,550 images for training and 720 images for testing. We observed that training more than 18 networks per ensemble does not improve the performance significantly, so we use this number in our experiments. In SfM-related experiments, we use COLMAP [27]. Since d^{SfM} is up to scale, we compute a scale correction factor s_i per image $\mathcal{I}_i$ as follows: $s_i^{SfM} = \text{median}(d_i^{GT})/\text{median}(d_i^{SfM})$. This scale correction is also applied to predictions of self-supervised and supervised SfM models that are also up-to-scale. Table 1 shows the metrics for the depth error and its uncertainty.

Supervising a deep ensemble with d^{GT} labels achieves the best depth metrics. In terms of uncertainty, self-supervised and supervised with SfM are underconfident, in contrast to supervised with GT depth, which is overconfident and

Table 1. Depth and uncertainty metrics in the synthetic dataset. RMSE in mm.

Approach	Abs_{Rel}	Sq_{Rel}	RMSE	$RMSE_{log}$	$\delta < 1.25$	$\delta < 1.25^2$	$\delta < 1.25^3$	AUCE
Supervised GT	**0.050**	**0.335**	**2.996**	**0.102**	**0.978**	**0.993**	**0.997**	+0.190
Supervised SfM	0.172	2.568	7.409	0.269	0.852	0.939	0.962	**−0.116**
Self-supervised	0.179	1.774	7.601	0.243	0.792	0.938	0.972	−0.152

presents higher (worse) absolute AUCE. Figure 3 shows that the aleatoric uncertainty supervised by GT is high around the haustras and in dark areas. The epistemic uncertainty grows with the scene depth. The uncertainty supervised by SfM is high in areas where there are typically holes in SfM reconstructions. Similarly to models trained with GT, the aleatoric uncertainty is also visible in the haustras and the epistemic in the deepest areas. Photometric self-supervision tends to offer the worst performance.

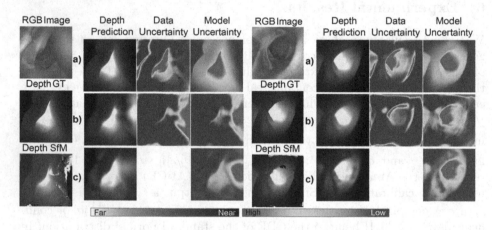

Fig. 3. Qualitative depth and uncertainty examples of (supervised learning, supervised learning SfM) and self supervised learning in synthetic images. a) Supervised GT, b) Supervised SfM and c) Self-supervised

EndoMapper Dataset. This experiment evaluates Bayesian depth networks in real colonoscopies. We use the model previously trained with synthetic ground truth depth ("Supervised GT") to analyse the effect of the domain change. In addition, we also present results from self-supervised training, a baseline teacher-student method [23] and our novel uncertain teacher approach. In real colonoscopies viewpoints change abruptly, images might be saturated or blurry, a considerable amount of liquid might appear and the colon itself produces significant occlusions. For this reasons, we remove images with partial or total visibility issues. We finally use 6,912 images out of the 14,400 images in the complete colonoscopic procedure. In order to obtain depth and uncertainty metrics, we create a 3D reconstruction of the colon using COLMAP (see supplementary

Table 2. Depth and uncertainty metrics in the EndoMapper dataset.

Approach	Abs$_{Rel}$	Sq$_{Rel}$	RMSE	RMSE$_{log}$	$\delta < 1.25$	$\delta < 1.25^2$	$\delta < 1.25^3$	AUCE
Supervised GT	0.240	0.644	2.595	0.308	0.645	0.898	0.962	−0.148
Self-supervised	0.371	1.260	4.603	0.431	0.417	0.721	0.886	−0.273
Teacher-student	0.234	0.600	2.532	0.301	0.657	0.903	0.963	−0.328
Uncertain teacher (ours)	**0.230**	**0.572**	**2.458**	**0.298**	**0.667**	**0.906**	**0.964**	**−0.129**

material). We also use 18-network ensembles for all methods. Table 2 shows the results. Our "Uncertain teacher" shows in general the smallest depth errors and the highest correlation between depth errors and predicted uncertainties.

For self-supervised methods, this real setting is challenging due to reflections, fluids and deformations, all of them aspects that are not considered in the photometric reprojection model of self-supervised losses. "Supervised GT" is affected by domain change, as it was trained on synthetic data. However, we observe that it successfully generalizes to the real domain and outperforms the self-supervised method. Based on this observation, we use synthetic supervision in the "Teacher-student" baseline and our "Uncertain teacher". In general, teacher-student depth metrics outperform the models trained with GT supervision in the synthetic domain and with self-supervision in the real domain. However, "Teacher-student" presents the worst AUCE metric, as the teacher uncertainty is not taken into account at training time. Our "Uncertain teacher" is the one presenting the best depth and uncertainty metrics, as it appropriately models the noise coming from domain transfer in the depth labels. Figure 4 shows qualitative results for the "Supervised GT", "Teacher-student" and "Uncertain teacher" models. Note that the data uncertainty captures light reflection and depth discontinuities in supervised learning. On the other hand, the model

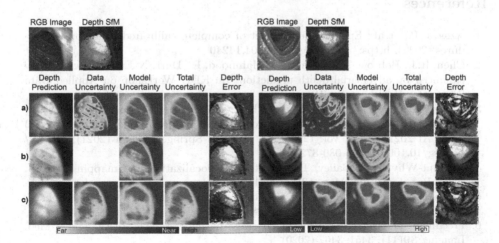

Fig. 4. Qualitative depth and uncertainty examples for a) supervised, b) self-supervised and c) uncertain teacher-student learning in real images.

uncertainty grows for the deeper areas. Observing these results, we can conclude that the domain change from synthetic to real colon images is not significant. Models trained on synthetic data generalize to real images and outperform models trained with self-supervision on the target domain, due to the challenges mentioned in the previous paragraphs.

7   Conclusions

All systems building on depth predictions from color images benefit from uncertainty estimates, in order to obtain robust, explainable and dependable assistance and decisions. In this paper, we have explored for the first time supervised and self-supervised approaches for depth and uncertainty single-view predictions in colonoscopies. From our experimental results, we extract several conclusions. Firstly, using ground truth depth as supervisory signal outperforms self-supervised learning and results in better calibrated models. Secondly, approaches based on photometric self-supervision and on SfM supervision coexist in the literature and there is a lack of analysis and results showing which type is more convenient. Thirdly, our experiments show that models trained in synthetic colonoscopies generalize to real colonoscopy images. Finally, we have proposed a novel teacher-student architecture that incorporates the teacher uncertainty in the loss, and have shown that it produces lower depth errors and better calibrated uncertainties than previous teacher-student architectures.

Acknowledgments. This work was supported by EndoMapper GA 863146 (EU-H2020), RTI2018-096903-B-I00, BES-2016-078426, PID2021-127685NB-I00 (FEDER/ Spanish Government), DGA-T45 17R/FSE (Aragón Government).

References

1. Azagra, P., et al.: Endomapper dataset of complete calibrated endoscopy procedures (2022). https://arxiv.org/abs/2204.14240
2. Chen, R.J., Bobrow, T.L., Athey, T., Mahmood, F., Durr, N.J.: SLAM endoscopy enhanced by adversarial depth prediction. In: KDD Workshop on Applied Data Science for Healthcare (2019)
3. Cheng, K., Ma, Y., Sun, B., Li, Y., Chen, X.: Depth estimation for colonoscopy images with self-supervised learning from videos. In: de Bruijne, M., et al. (eds.) MICCAI 2021. LNCS, vol. 12906, pp. 119–128. Springer, Cham (2021). https://doi.org/10.1007/978-3-030-87231-1_12
4. Durrant-Whyte, H., Bailey, T.: Simultaneous localization and mapping: Part I. IEEE Robot. Autom. Mag. **13**(2), 99–110 (2006)
5. Eigen, D., Puhrsch, C., Fergus, R.: Depth map prediction from a single image using a multi-scale deep network. In: NeurIPS (2014)
6. Freedman, D.: Detecting deficient coverage in colonoscopies. IEEE Trans. Med. Imaging **39**(11), 3451–3462 (2020)
7. Fu, H., Gong, M., Wang, C., Batmanghelich, K., Tao, D.: Deep ordinal regression network for monocular depth estimation. In: CVPR (2018)

8. Gal, Y., Ghahramani, Z.: Dropout as a Bayesian approximation: representing model uncertainty in deep learning. In: ICML (2016)
9. Godard, C., Mac Aodha, O., Firman, M., Brostow, G.J.: Digging into self-supervised monocular depth estimation. In: ICCV (2019)
10. Gustafsson, F.K., Danelljan, M., Schon, T.B.: Evaluating scalable Bayesian deep learning methods for robust computer vision. In: CVPR Workshops (2020)
11. Hartley, R., Zisserman, A.: Multiple View Geometry in Computer Vision, 2nd edn. Cambridge University Press, Cambridge (2003)
12. Huang, B., et al.: Self-supervised generative adversarial network for depth estimation in laparoscopic images. In: de Bruijne, M., et al. (eds.) MICCAI 2021. LNCS, vol. 12904, pp. 227–237. Springer, Cham (2021). https://doi.org/10.1007/978-3-030-87202-1_22
13. Hwang, S.J., Park, S.J., Kim, G.M., Baek, J.H.: Unsupervised monocular depth estimation for colonoscope system using feedback network. Sensors **21**(8), 2691 (2021)
14. Ilg, E., et al.: Uncertainty estimates and multi-hypotheses networks for optical flow. In: Ferrari, V., Hebert, M., Sminchisescu, C., Weiss, Y. (eds.) ECCV 2018. LNCS, vol. 11211, pp. 677–693. Springer, Cham (2018). https://doi.org/10.1007/978-3-030-01234-2_40
15. Kendall, A., Gal, Y.: What uncertainties do we need in Bayesian deep learning for computer vision? In: NeurIPS (2017)
16. Lakshminarayanan, B., Pritzel, A., Blundell, C.: Simple and scalable predictive uncertainty estimation using deep ensembles. In: NeurIPS (2017)
17. Li, Z., et al.: On the sins of image synthesis loss for self-supervised depth estimation. arXiv preprint arXiv:2109.06163 (2021)
18. Liu, X., et al.: Dense depth estimation in monocular endoscopy with self-supervised learning methods. IEEE Trans. Med. Imaging **39**(5), 1438–1447 (2019)
19. Luo, H., Hu, Q., Jia, F.: Details preserved unsupervised depth estimation by fusing traditional stereo knowledge from laparoscopic images. Healthc. Technol. Lett. **6**(6), 154 (2019)
20. Mahmood, F., Chen, R., Durr, N.J.: Unsupervised reverse domain adaptation for synthetic medical images via adversarial training. IEEE Trans. Med. Imaging **37**(12), 2572–2581 (2018)
21. Mahmood, F., Durr, N.J.: Deep learning and conditional random fields-based depth estimation and topographical reconstruction from conventional endoscopy. Med. Image Anal. **48**, 230–243 (2018)
22. Ozyoruk, K.B., et al.: EndoSLAM dataset and an unsupervised monocular visual odometry and depth estimation approach for endoscopic videos. Med. Image Anal. **71**, 102058 (2021)
23. Poggi, M., Aleotti, F., Tosi, F., Mattoccia, S.: On the uncertainty of self-supervised monocular depth estimation. In: CVPR (2020)
24. Rau, A., et al.: Implicit domain adaptation with conditional generative adversarial networks for depth prediction in endoscopy. Int. J. Comput. Assist. Radiol. Surg. **14**(7), 1167–1176 (2019). https://doi.org/10.1007/s11548-019-01962-w
25. Recasens, D., Lamarca, J., Fácil, J.M., Montiel, J., Civera, J.: Endo-depth-and-motion: reconstruction and tracking in endoscopic videos using depth networks and photometric constraints. IEEE Robot. Autom. Lett. **6**(4), 7225–7232 (2021)
26. Rodriguez-Puigvert, J., Martinez-Cantin, R., Civera, J.: Bayesian deep neural networks for supervised learning of single-view depth. IEEE Robot. Autom. Lett. **7**(2), 2565–2572 (2022)

27. Schonberger, J.L., Frahm, J.M.: Structure-from-motion revisited. In: CVPR (2016)
28. Sharan, L., et al.: Domain gap in adapting self-supervised depth estimation methods for stereo-endoscopy. Curr. Dir. Biomed. Eng. **6**(1), 1–5 (2020)
29. Shen, M., Gu, Y., Liu, N., Yang, G.Z.: Context-aware depth and pose estimation for bronchoscopic navigation. IEEE Robot. Autom. Lett. **4**(2), 732–739 (2019)
30. Song, M., Lim, S., Kim, W.: Monocular depth estimation using Laplacian pyramid-based depth residuals. IEEE Trans. Circuits Syst. Video Technol. **31**(11), 4381–4393 (2021)
31. Turan, M., et al.: Unsupervised odometry and depth learning for endoscopic capsule robots. In: IROS (2018)
32. Visentini-Scarzanella, M., Sugiura, T., Kaneko, T., Koto, S.: Deep monocular 3D reconstruction for assisted navigation in bronchoscopy. Int. J. Comput. Assist. Radiol. Surg. **12**(7), 1089–1099 (2017)
33. Wang, Z., Bovik, A.C., Sheikh, H.R., Simoncelli, E.P.: Image quality assessment: from error visibility to structural similarity. IEEE Trans. Image Process. **13**(4), 600–612 (2004)
34. Widya, A.R., Monno, Y., Okutomi, M., Suzuki, S., Gotoda, T., Miki, K.: Self-supervised monocular depth estimation in gastroendoscopy using GAN-augmented images. In: Medical Imaging 2021: Image Processing (2021)
35. Xu, K., Chen, Z., Jia, F.: Unsupervised binocular depth prediction network for laparoscopic surgery. Comput. Assist. Surg. **24**(sup1), 30–35 (2019)
36. Zhan, H., Garg, R., Saroj Weerasekera, C., Li, K., Agarwal, H., Reid, I.: Unsupervised learning of monocular depth estimation and visual odometry with deep feature reconstruction. In: CVPR (2018)
37. Zhou, T., Brown, M., Snavely, N., Lowe, D.G.: Unsupervised learning of depth and ego-motion from video. In: CVPR (2017)

Toward Clinically Assisted Colorectal Polyp Recognition via Structured Cross-Modal Representation Consistency

Weijie Ma[1], Ye Zhu[1], Ruimao Zhang[1](✉), Jie Yang[1], Yiwen Hu[1], Zhen Li[1], and Li Xiang[2]

[1] Shenzhen Research Institute of Big Data, The Chinese University of Hong Kong (Shenzhen), Shenzhen, China
weijiema@link.cuhk.edu.cn, zhangruimao@cuhk.edu.cn
[2] Longgang District People's Hospital of Shenzhen, Shenzhen, China

Abstract. The colorectal polyps classification is a critical clinical examination. To improve the classification accuracy, most computer-aided diagnosis algorithms recognize colorectal polyps by adopting Narrow-Band Imaging (NBI). However, the NBI usually suffers from missing utilization in real clinic scenarios since the acquisition of this specific image requires manual switching of the light mode when polyps have been detected by using White-Light (WL) images. To avoid the above situation, we propose a novel method to directly achieve accurate white-light colonoscopy image classification by conducting structured cross-modal representation consistency. In practice, a pair of multi-modal images, *i.e.* NBI and WL, are fed into a shared Transformer to extract hierarchical feature representations. Then a novel designed Spatial Attention Module (SAM) is adopted to calculate the similarities between class token and patch tokens for a specific modality image. By aligning the class tokens and spatial attention maps of paired NBI and WL images at different levels, the Transformer achieves the ability to keep both global and local representation consistency for the above two modalities. Extensive experimental results illustrate the proposed method outperforms the recent studies with a margin, realizing multi-modal prediction with a single Transformer while greatly improving the classification accuracy when only with WL images. Code is available at https://github.com/WeijieMax/CPC-Trans.

Keywords: Colorectal polyps classification · Multi-modal represntation learning · Transformer architecture

1 Introduction

Adenomatous polyp is considered to be the underlying cause of colorectal cancer (CRC) [23], which is the second lethal cancer and the third most commonly diagnosed malignancy [9]. Detection and resection of the polyps usually depend on colonoscopy. However, in clinical practice, the standard white-light (WL)

© The Author(s), under exclusive license to Springer Nature Switzerland AG 2022
L. Wang et al. (Eds.): MICCAI 2022, LNCS 13433, pp. 141–150, 2022.
https://doi.org/10.1007/978-3-031-16437-8_14

observation could only provide limited discriminative information between neo-plasticism and nonneoplastic colorectal polyps.

To improve classification accuracy, computer-aided diagnosis systems (CADx) are introduced, most of which adopt Narrow-Band Imaging (NBI), Blue Light Imaging (BLI) and other enhanced colonoscopy images [1,10]. For example, Fonolia et al. [6] proposed a CADx system for classifying colorectal polyps combining WL, BLI and Linked Colour Imaging (LCI) modalities, which achieved encouraging performance. Moreover, Franklin et al. [13] introduced a robust frame-level strategy using NBI sequences to learn a deep convolutional representation, which achieved an average classification accuracy of 90.79%.

Despite the enhanced imaging, endoscopists often rely on WL images before they change the light mode to detect the possible polyps, which means that the WL images may fail to capture discriminative features of polyps, result-ing in much less accurate classification by other imaging means. Therefore, a well-designed WL-based CADx system is in urgent need for better colorectal polyp recognition using only WL images. In recent studies, a deep learning model proposed by Yang et al. [22] presented a promising performance in clas-sifying colorectal lesions only on WL colonoscopy images. To further narrow the gap between conventional WL images and enhanced images, Wang et al. [17] enhanced low representation WL features with NBI images through domain alignment and contrastive learning which improved the WL-based classification results.

In this paper, we propose a simple yet effective method to improve the dis-criminative representation of WL images by conducting structured cross-modal representation consistency, realizing accurate WL colonoscopy image classifica-tion. During the training, the hierarchical feature representations of the paired WL and NBI images are extracted by a shared Transformer. A carefully designed novel Spatial Attention Modules (SAM) is further exploited to align the output class tokens of the paired images, while their spatial attention maps are further constrained to be consistent at different levels, enhancing the feature represen-tation of WL images from the global and local perspectives simultaneously.

The main contributions are three-fold. (1) An effective scheme based on global and local consistent learning is proposed to extract more discriminative representations of WL images for colorectal polyps classification (CPC). (2) We introduce a general framework of multi-modal learning for medical image anal-ysis based on Transformer where the unilateral advantages can be propagated across domains through introducing external attention modules but dropped out these external auxiliary modules in inference, efficiently reducing the model com-plexity in applications. (3) The proposed method outperforms state-of-the-arts of CPC by a margin, promoting the development of related methods in clinically assisted colorectal polyp recognition.

2   Related Work

Domain Alignment. Domain Alignment mainly focuses on reducing the dis-crepancy between the associated distributions of the projected source and target features. Two different schemes are usually considered for aligning feature representations based on recent deep domain alignment methods: (i) extracted features are learned in a share subspace, aiming to minimize the distance between the source and the target distributions. (ii) Another scheme concentrates on reducing the Maximum Mean Discrepancy and is commonly adopted in the case of missing target domain labels. In [14], a nonlinear transformation is learned to align correlations of layer activations in deep neural networks (Deep CORAL). Moreover, by utilizing the intra-class variation in the target domain, Chen et al. [4] proposed the Progressive Feature Alignment Network to align the discriminative feature across different domains.

Attention and Vision Transformer. To gain a more vivid representation of related different positions from a single sequence or sentence, the attention mechanism was redefined as the self-attention and later adopted by the Transformer architecture [16] that based solely on it. Soon after that, extensive works [18,19,24] have been improved by leveraging the attention mechanism. Currently, Dosovitskiy et al. [5] applied Transformer architecture from NLP to computer vision as Vision Transformer (ViT), and showed that the sequences of image patches could perform very well with a pure transformer on image classification, while the convolutional networks usually suffer from difficulty in capturing and storing long-distance dependent information due to the limited receptive field. Following the ViT, many other vision transformer variants are proposed [3,15], and some of them have achieved great performance on various medical tasks [2,8,20,21,25] with the strong representation capabilities of transformer.

3 Method

3.1 Problem Formulation and Framework Overview

Given a WL image and its corresponding NBI image $\{I_w, I_n\}$ with their shared label G, the proposed cross-modal learning framework aims to explore structured semantic information by utilizing a shared class proxy embedding that bridges the modality gap after the training process. Finally, this framework produces a unified accurate CPC model for WL-only predictions, outperforming ones trained with single-modal data with a margin.

As illustrated in Fig. 1, we introduce a Transformer-based framework with cross-modal global alignment (CGA) and spatial attention module (SAM). We first feed a dual-modal image pair as input $\{I_w, I_n\} \in \mathbb{R}^{H \times W \times 3}$ to the framework. The image-pair is then divided into $P \times P$ image patches. After a linear projection layer, each patch is embedded into a patch token with embedded dimension $d = \frac{3P^2}{2}$. Note that here we reduce the embedded dimension to $1/2$ as standard so as to release the computational overhead and drop model parameters. Patch embedding is implemented by a convolutional layer of a $P \times P$ kernel. In this way, we could reassign the dimension of the patch token inputs as $\{X_w, X_n\} \in \mathbb{R}^{N \times d}$, where $N = \frac{HW}{P^2}$. Then, $\{X_w, X_n\}$ will be successively

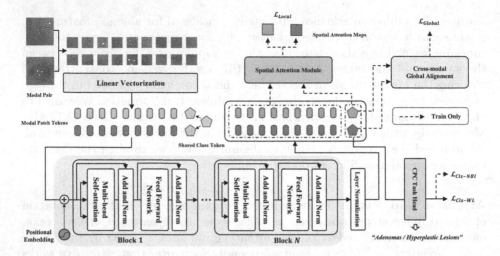

Fig. 1. Overview of our proposed cross-modal shared colorectal polyp recognition framework via single Transformer architecture and the proposed CGA and SAM. Note that all parts linked by dotted lines can be removed during the inference phase.

concatenated with a shared learnable class token $\mathbf{c} \in \mathbb{R}^d$. To compensate for the missing 2-D structural information, positional embedding $E \in \mathbb{R}^{(N+1) \times d}$ is supplemented to $\{X_w, X_n\}$ and $\mathbf{c}$ by element-wise addition. As a result, information about the relative or absolute position is provided in patch tokens.

Next, dual-modal patch tokens are separately passed through a series of shared Transformer blocks for deep feature extraction and comprehensive context modeling to generate modality-specific global representations on two class tokens $\{\mathbf{c}_w, \mathbf{c}_n\} \in \mathbb{R}^d$. During training, $\{\mathbf{c}_w, \mathbf{c}_n\}$ are then fed into CGA to align dual-modal image pair's global representation. Meanwhile, SAM furthers cross-modal local alignment by comparing two modalities' response maps between their global representation and local instance-level information.

The inference stage only relies on the single Transformer architecture and discards any modality-specific or cross-modal modules for CPC prediction.

3.2 Shared Transformer Block

The shared Transformer blocks are employed to learn pixel-level contextual dependencies adapted to both modalities. As shown in Fig. 1, the share Transformer blocks consist of B layers, and each is composed of two components, multi-head self-attention (MSA) and a feed-forward network (FFN). Layer normalization (LN) is applied before each component while skipping connection after each component. The FFN is a multi-layer perceptron (MLP) which includes two linear layers with a medium GeLU activation. In addition, considering a pair of arbitrary dual-modal token sequences $\{X_w, X_n\} \in \mathbb{R}^{\hat{N} \times d}$ as input, single-head self-attention (SA) formulated as:

$$\text{SA}(X_m^i) = \text{Softmax}(\frac{X_m^i \mathbf{W}_Q^i (X_m^i \mathbf{W}_K^i)^\text{T}}{\sqrt{d'}})(X_m^i \mathbf{W}_V^i) \tag{1}$$

where $m \in \{w, n\}$. $\{\mathbf{W}_Q^i, \mathbf{W}_K^i, \mathbf{W}_V^i\} \in \mathbb{R}^{d \times d'}$ donate the triple parameter matrices of ith layer and d' is the dimension of each head. Here, we omit LN layers for simplicity. MSA is a concatenation of h parallel SA modules with linear projection to rearrange their outputs. In our experiments, we employ $h = 6$, $d = 384$, and $d' = d/h = 64$. As depicted in Fig. 1, the whole calculation can be formulated as:

$$X_m^i = \text{MSA}(X_m^{i-1}) + \text{FFN}(X_m^{i-1} + \text{MSA}(X_m^{i-1})) \tag{2}$$

3.3   Cross-Modal Global Alignment

Diminishing the discrepancy between WL and NBI images allows a model to leverage cross-modal knowledge from either of modalities. To this end, we propose Cross-modal Global Alignment (CGA), an auxiliary module only used during the training stage. CGA maintains the cross-modal consistency in the shared model, or in other words, drives the model to learn cross-domain knowledge while only WL images are given. In practice, to begin with, CGA computes the cosine difference between two modal-specific class tokens of two paired images. Then, imposing a loss function on their cosine similarity reduces the average cosine distance between image-pairs of two modalities. This way, feature representations from paired images may match up more closely, and the model may learn to capture hard features from NBI images (e.g. textures clear in NBI images but unclear in WL images).

As figured in Fig. 1, after being passed through an external LN, the average distance between two modalities' paired samples is narrowed based on cosine similarity. Given two modality-specific class tokens $\{\mathbf{c}_w, \mathbf{c}_n\}$, their alignment loss is computed by:

$$\mathcal{L}_{global} = 1 - \text{dist}_{\cos}(\mathbf{c}_\text{w}, \mathbf{c}_\text{n}) = 1 - \frac{\mathbf{c}_\text{w} \cdot \mathbf{c}_\text{n}}{||\mathbf{c}_\text{w}||_2 ||\mathbf{c}_\text{n}||_2} \tag{3}$$

3.4   Spatial Attention Module

Despite the global alignment of modal-specific class tokens, we also propose the Spatial Attention Module (SAM) to pursue the multi-level structured semantic consistency between two modalities. First, we obtain through SAM the globally guided affinity, i.e., the response map between each image's global representation and local regions. Subsequently, we align two modalities' local semantics by limiting the distance between two modalities' response maps.

Concretely, as illustrated in Fig. 1, SAM takes as input the output patch-token features from the external LN layer $\{F_w, F_n\} \in \mathbb{R}^{N \times d}$ as well as the modal-specific class tokens $\{\mathbf{c}_w, \mathbf{c}_n\} \in \mathbb{R}^d$. In practice, we utilize $\{F_w, F_n\}$ to generate the key of Spatial Attention (SpA) operation through linear projection, while the query of SpA is obtained based on $\{c_w, c_n\}$, described as follows:

$$R_m = \text{SpA}(F_m, c_m) = \text{Softmax}(c_m \mathbf{W}_Q (F_m \mathbf{W}_K)^T) \tag{4}$$

where $m \in \{w, n\}$. $\{R_w, R_n\} \in \mathbb{R}^N$ are two response maps for WL and NBI images respectively, representing global-to-local affinity of each modality. Similar to Eq. 3, we compute cosine similarity between two response maps by:

$$\mathcal{L}_{local} = \lambda(1 - \text{dist}_{\cos}(R_w, R_n)) \tag{5}$$

where λ is a ratio coefficient to normalize the magnitude of $\mathcal{L}_{local}$ to balance it with $\mathcal{L}_{global}$ (here we set $\lambda = 0.3$). Taking the global representation as standard, the distribution of semantics over local regions is obtained by Eq. 4. In Eq. 5, $\mathcal{L}_{local}$ computes the distance between two global-to-local affinity distributions. Here, as two modalities' local semantics are extracted by a shared model, they are in shared representation space. Therefore, by minimizing $\mathcal{L}_{local}$, SAM could align two modalities in terms of their local semantics. Further, with local semantics bridged between WL and NBI image pairs, the model is learning to mine hidden local features from WL images, thus bridging the accuracy gap between them. Arguably, in our experiments, SAM further brings a 0.7% accuracy improvement. In this way, the overall training loss contains four parts:

$$\mathcal{L} = \mathcal{L}_{cls-WL} + \mathcal{L}_{cls-NBI} + \mathcal{L}_{Global} + \mathcal{L}_{Local} \tag{6}$$

in which $\mathcal{L}_{cls-WL} = CrossEntropy(H(c_w))$, $\mathcal{L}_{cls-NBI} = CrossEntropy(H(c_n))$ to supervise the CPC binary classification task (*i.e.*, adenomatous or hyperplastic) given the ground truth label G. Note that at the inference stage, the model performs binary classification for WL images only.

4   Experiments and Results

4.1   Dataset

The dataset we used is named CPC-Paired Dataset [17]. The dataset has two paired image modalities (WL-NBI) and consists of two parts due to the insufficient amount of each part. Every WL image and corresponding NBI image share the same label of a colorectal polyp. There are 307 image pairs labeled as adenomas and 116 images labeled as hyperplastic lesions in total. One part of the dataset was extracted from ISIT-UMR Colonoscopy Dataset [12] and another part was collected from the hospital. Specifically, the ISIT-UMR part contains 102 adenomatous and 63 hyperplastic paired frames of two modals, collated by 40 adenomas and hyperplastic lesions sequences from a total of 76 short categorized video data. And the clinical part was composed of 258 WL-NBI pairs, 205 adenoma images, and 53 hyperplastic polyp images in detail from 123 patients. What's more, the dataset has annotated bounding box data for subsequently cropping lesion area corresponding to all images of two modalities.

Table 1. Comparisons of our full model with previous best studies on the test sets with respect to accuracy metric. ‡ refers to that the result is directly cited from the original paper based on their own divisions of training-validation sets.

Methods	Backbone	Params (M)	Fold 1	Fold 2	Fold 3	Fold 4	Fold 5	Mean
	VGG	138.36	78.2%	79.5%	77.3%	78.0%	77.9%	78.2%
Yang [22]‡	InceptionV3	24.73	81.1%	82.1%	80.5%	82.0%	81.3%	81.5%
	ResNet50	22.56	79.5%	80.5%	78.0%	80.3%	78.8%	79.7%
Wang [17]‡	ResNet50	22.56	85.9%	86.1%	84.2%	85.8%	85.2%	85.3%
Ours	ViT-Small	**21.67**	87.2%	88.5%	94.3%	92.7%	75.8%	**87.7%**

4.2 Implementation Details

In our work, we conducted 5-fold cross-validation for all experiments. The training and validation set is 8:2, and the partition in each fold was generated randomly based on every subject. We implemented our model with the PyTorch toolkit on a single NVIDIA V100 GPU. The input size of the image is 224 × 224, and the data augmentation was adopted by random resized cropping and horizontal flipping. Our model's embedding dimension is 384 with 6 attention heads and 12 block layers based on the setting of the ViT small version. We choose SGD optimizer and cosine annealing learning rate scheduler for network optimization, Following [17], the batch size is 16, and learning rate is $1e-3$ with a maximum of 500 epochs. We also use the momentum of 0.9 and the weight decay of $5e-5$.

4.3 Results and Analysis

Through massive experiments, we verify the effectiveness of our proposed method. To be specific, as shown by a 5-fold cross-validation result in Table 1, despite the fewer parameters, we can observe that our approach outperforms existing state-of-the-art methods. This also proves the stronger representation power and higher parameter efficiency of our model.

The ablation study further examines the effectiveness of each component, which is shown in Table 2. "Trans with WL-only" means vanilla Transformer-

Table 2. The performance of the baseline and our full model on the test sets with respect to accuracy metric. "Δ M" indicates the gain of current average result compared with the former row, and the first row is recent state-of-the-art result.

Methods	Fold 1	Fold 2	Fold 3	Fold 4	Fold 5	Mean	Δ M
Wang [17]	85.9%	86.1%	84.2%	85.8%	85.2%	85.3%	–
Trans with WL-only	84.6%	88.5%	93.1%	82.9%	76.5%	85.1%	−0.2%
Trans + CGA	87.2%	88.5%	93.1%	87.8%	77.4%	86.8%	**+1.7%**
Trans + CGA + SAM	87.2%	88.5%	94.3%	92.7%	75.8%	**87.7%**	**+0.9%**

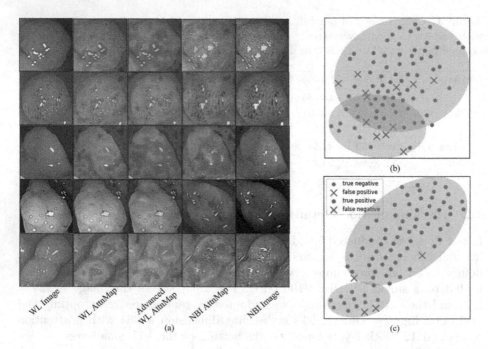

(a)   (c)

Fig. 2. (a) is the comparative study of different attention maps. "WL/NBI AttnMap" is produced by the WL/NBI vanilla model and "Advanced WL AttnMap" by the proposed model. (b) displays the predictive distribution of WL images given by WL-only Transformer over the 2-dimensional representation space offered by t-SNE, while (c) shows the corresponding distribution obtained from our proposed shared Transformer.

based baseline of WL images. "Trans + X" indicates the proposed framework combined with only CGA or both CGA and SAM, which improve the baseline by 1.7% or 2.6%. This ablation study demonstrates that the accuracy gain is more heavily from our proposed modules other than the Transformer architecture.

The qualitative analysis is illustrated in Fig. 2. Precisely, we extract the attention maps of the last Transformer layer of our model and vanilla models for each modality and remap them on the original images [7]. As is shown in Fig. 2(a), the advanced WL attention maps from our model are more similar to the NBI attention maps, covering nearly the entire lesion region. In contrast, the WL vanilla one underperforms the former, focusing on fractional attentive areas, irrelevant corners, or backgrounds. There is an apparent visual gap between WL attention maps and our advances, proving our model's semantic consistency between the two modalities.

Figure 2(b)(c) respectively shows the WL-only model and our proposed model's discrimination between adenomatous or hyperplastic lesion through t-SNE visualization [11]. In Fig. 2(b), there is an obvious overlapping region between two predictive distributions in which the WL-only model is dramatically less discriminative, resulting in a higher number of false samples (marked by forks). However, in Fig. 2(c), our proposed model separates two distributions with a clear margin, with only two outliers for each class. This result shows that our proposed shared Transformer can achieve higher performance on white-light colonoscopy image classification by untangling two predictive distributions.

5 Conclusion

In this paper, a novel Transformer-based framework is introduced to tackle WL-only CPC, which proposed the Cross-modal Global Alignment (CGA) and a newly designed Spatial Attention Module (SAM) to pursue the structured semantic consistency, *i.e.* modality-aware global representations and instance-aware local correlations. In the inference phase, all of the modules used for modality alignment can be removed, guaranteeing WL-only simple but accurate prediction for CPC task. Extensive experimental results and visualizations demonstrate the effectiveness of our proposed multi-level consistency learning method.

Acknowledgement. The work is supported in part by the Young Scientists Fund of the National Natural Science Foundation of China under grant No. 62106154, by Natural Science Foundation of Guangdong Province, China (General Program) under grant No. 2022A1515011524 and by the Guangdong Provincial Key Laboratory of Big Data Computing, The Chinese University of Hong Kong, Shenzhen.

References

1. Bisschops, R., et al.: BASIC (BLI adenoma serrated international classification) classification for colorectal polyp characterization with blue light imaging. Endoscopy **50**(03), 211–220 (2018)
2. Cao, H., et al.: Swin-UNet: UNet-like pure transformer for medical image segmentation. arXiv preprint arXiv:2105.05537 (2021)
3. Carion, N., Massa, F., Synnaeve, G., Usunier, N., Kirillov, A., Zagoruyko, S.: End-to-end object detection with transformers. In: Vedaldi, A., Bischof, H., Brox, T., Frahm, J.-M. (eds.) ECCV 2020. LNCS, vol. 12346, pp. 213–229. Springer, Cham (2020). https://doi.org/10.1007/978-3-030-58452-8_13
4. Chen, C., et al.: Progressive feature alignment for unsupervised domain adaptation. In: Proceedings of the IEEE/CVF Conference on Computer Vision and Pattern Recognition, pp. 627–636 (2019)
5. Dosovitskiy, A., et al.: An image is worth 16 × 16 words: transformers for image recognition at scale. In: International Conference on Learning Representations (2021)
6. Fonollà, R., et al.: A CNN CADx system for multimodal classification of colorectal polyps combining WL, BLI, and LCI modalities. Appl. Sci. **10**(15), 5040 (2020)
7. Gao, W., et al.: TS-CAM: token semantic coupled attention map for weakly supervised object localization. In: Proceedings of the IEEE/CVF International Conference on Computer Vision, pp. 2886–2895 (2021)

8. Ji, Y., et al.: Multi-compound transformer for accurate biomedical image segmentation. In: de Bruijne, M., et al. (eds.) MICCAI 2021. LNCS, vol. 12901, pp. 326–336. Springer, Cham (2021). https://doi.org/10.1007/978-3-030-87193-2_31

9. Keum, N., Giovannucci, E.: Global burden of colorectal cancer: emerging trends, risk factors and prevention strategies. Nat. Rev. Gastroenterol. Hepatol. **16**(12), 713–732 (2019)

10. Komeda, Y., et al.: Magnifying narrow band imaging (NBI) for the diagnosis of localized colorectal lesions using the Japan NBI expert team (JNET) classification. Oncology **93**(Suppl. 1), 49–54 (2017)

11. Van der Maaten, L., Hinton, G.: Visualizing data using t-SNE. J. Mach. Learn. Res. **9**(11) (2008)

12. Mesejo, P., et al.: Computer-aided classification of gastrointestinal lesions in regular colonoscopy. IEEE Trans. Med. Imaging **35**(9), 2051–2063 (2016)

13. Sierra-Jerez, F., Martínez, F.: A deep representation to fully characterize hyperplastic, adenoma, and serrated polyps on narrow band imaging sequences. Heal. Technol. **12**(2), 401–413 (2022). https://doi.org/10.1007/s12553-021-00633-8

14. Sun, B., Saenko, K.: Deep CORAL: correlation alignment for deep domain adaptation. In: Hua, G., Jégou, H. (eds.) ECCV 2016. LNCS, vol. 9915, pp. 443–450. Springer, Cham (2016). https://doi.org/10.1007/978-3-319-49409-8_35

15. Touvron, H., Cord, M., Douze, M., Massa, F., Sablayrolles, A., Jégou, H.: Training data-efficient image transformers & distillation through attention. In: International Conference on Machine Learning, pp. 10347–10357. PMLR (2021)

16. Vaswani, A., et al.: Attention is all you need. In: Advances in Neural Information Processing Systems, vol. 30 (2017)

17. Wang, Q., et al.: Colorectal polyp classification from white-light colonoscopy images via domain alignment. In: de Bruijne, M., et al. (eds.) MICCAI 2021. LNCS, vol. 12907, pp. 24–32. Springer, Cham (2021). https://doi.org/10.1007/978-3-030-87234-2_3

18. Wang, T., Zhang, R., Lu, Z., Zheng, F., Cheng, R., Luo, P.: End-to-end dense video captioning with parallel decoding. In: Proceedings of the IEEE/CVF International Conference on Computer Vision, pp. 6847–6857 (2021)

19. Wei, J., Hu, Y., Zhang, R., Li, Z., Zhou, S.K., Cui, S.: Shallow attention network for polyp segmentation. In: de Bruijne, M., et al. (eds.) MICCAI 2021. LNCS, vol. 12901, pp. 699–708. Springer, Cham (2021). https://doi.org/10.1007/978-3-030-87193-2_66

20. Xie, Y., Zhang, J., Shen, C., Xia, Y.: CoTr: efficiently bridging CNN and transformer for 3D medical image segmentation. In: de Bruijne, M., et al. (eds.) MICCAI 2021. LNCS, vol. 12903, pp. 171–180. Springer, Cham (2021). https://doi.org/10.1007/978-3-030-87199-4_16

21. Yang, J., Zhang, R., Wang, C., Li, Z., Wan, X., Zhang, L.: Toward unpaired multimodal medical image segmentation via learning structured semantic consistency. arXiv preprint arXiv:2206.10571 (2022)

22. Yang, Y.J., et al.: Automated classification of colorectal neoplasms in white-light colonoscopy images via deep learning. J. Clin. Med. **9**(5), 1593 (2020)

23. Zauber, A.G., et al.: Colonoscopic polypectomy and long-term prevention of colorectal-cancer deaths. N. Engl. J. Med. **366**, 687–696 (2012)

24. Zhang, R., et al.: Scan: self-and-collaborative attention network for video person re-identification. IEEE Trans. Image Process. **28**(10), 4870–4882 (2019)

25. Zhou, H.Y., Guo, J., Zhang, Y., Yu, L., Wang, L., Yu, Y.: nnFormer: interleaved transformer for volumetric segmentation. arXiv preprint arXiv:2109.03201 (2021)

TGANet: Text-Guided Attention for Improved Polyp Segmentation

Nikhil Kumar Tomar[1,2], Debesh Jha[3], Ulas Bagci[3],
and Sharib Ali[4,5,6(✉)]

[1] NepAl Applied Mathematics and Informatics Institute for Research (NAAMII),
Kathmandu, Nepal
[2] School of Computer and Information Sciences, Indira Gandhi National Open
University, New Delhi, India
[3] Machine and Hybrid Intelligence Lab, Department of Radiology,
Northwestern University, Chicago, USA
[4] School of Computing, University of Leeds, Leeds, UK
s.s.ali@leeds.ac.uk
[5] Institute of Biomedical Engineering, University of Oxford,
Oxford OX3 7DQ, UK
[6] NIHR Oxford Biomedical Research Centre, Oxford, UK

Abstract. Colonoscopy is a gold standard procedure but is highly operator-dependent. Automated polyp segmentation, a precancerous precursor, can minimize missed rates and timely treatment of colon cancer at an early stage. Even though there are deep learning methods developed for this task, variability in polyp size can impact model training, thereby limiting it to the size attribute of the majority of samples in the training dataset that may provide sub-optimal results to differently sized polyps. In this work, we exploit *size-related* and *polyp number-related* features in the form of text attention during training. We introduce an auxiliary classification task to weight the text-based embedding that allows network to learn additional feature representations that can distinctly adapt to differently sized polyps and can adapt to cases with multiple polyps. Our experimental results demonstrate that these added text embeddings improve the overall performance of the model compared to state-of-the-art segmentation methods. We explore four different datasets and provide insights for size-specific improvements. Our proposed *text-guided attention network* (TGANet) can generalize well to variable-sized polyps in different datasets. Codes are available at https://github.com/nikhilroxtomar/TGANet.

Keywords: Label embedding · Polyp · Multi-scale features · Attention

1 Introduction

Colorectal cancer (CRC) is one of the leading causes of cancer-related deaths [16] worldwide. However, high operator-dependence and subjectivity during gold

Supplementary Information The online version contains supplementary material available at https://doi.org/10.1007/978-3-031-16437-8_15.

© The Author(s), under exclusive license to Springer Nature Switzerland AG 2022
L. Wang et al. (Eds.): MICCAI 2022, LNCS 13433, pp. 151–160, 2022.
https://doi.org/10.1007/978-3-031-16437-8_15

standard colonoscopic procedures remain high. This is also due to the complex topology of organ, severe artefacts, constant deformation of organ, debris and stool etc. Even though cleansing of the bowel is done to improve detection rates of cancer and cancer precursor lesions such as polyps, the missed rate is still high that accounts for 26.8% for polyps located on the right colon and 21.4% to polyps on the left colon [10,12]. In addition, the missed rate for flat or sessile polyps (diminutive polyps) is grim (nearly 32.7%). An automated system is thus clearly needed to minimize the operator subjectivity and missed rate. Semantic segmentation can classify each pixel into a class category, allowing the opportunity to learn meaningful semantic representations of polyps and their complex surroundings. Several methods do exist in the literature [3,7,14] but most them focus on exploiting only localized spatial context. However, the nature and occurrence of polyps in colonic mucosa can be confused with colonic folds. Exploiting associated attributes such as size and occurrence (one or a few) could be used to infer and improve segmentation for hard samples.

Encoder-decoder networks has been widely used for polyp segmentation using various modifications to boost network performance [3,7,14]. PraNet [3] applied area and boundary cues in reverse attention to focus on the polyp boundary regions. The high-level feature aggregation and boundary attention blocks in the network help to calibrate some of the misaligned predictions and improve the segmentation accuracy. Similarly, HRENet [14] designed an informative context enhancement technique and adaptive feature aggregation module and trained the model on their edge and structure consistency aware loss, and obtained superior performance. Other works such as PolypSeg [18] and MSRFNet [15] uses modules incorporating multiple-scale information. An adaptive scale context module (ASCM) and semantic global context module (SGCM) was used in PolypSeg [18]. The ASCM tackles the size variations among the polyp and improves the better feature representation capability, while SGCM enhances the feature fusion between the high-level and low-level features and remove noise in the low-level features to improve the segmentation accuracy. Similarly, MSRFNet [15] integrated cross-scale fusion modules to propagate both high resolution and low-resolution features and an added shape stream network to prune polyp boundaries.

Most of these works in the literature [3,14,15] focuses on size variation, boundary curves, background regions, dense skip connections and dense residual scale fusions that can boost performance. However, these adjustments are made using additional layers and explicit extensions of networks and their connections. This adds to the complexity of the model that can adversely affect the generalization of test samples coming from a similar distribution and require a large dataset. In addition, it can also affect images with under represented polyp sizes. In this work, we propose incorporating a text guided attention mechanism using a simple byte-pair encoding for the attributes comprising polyp number and its size. In addition, we use the same encoder layer of the network to provide weights for each of these attributes.

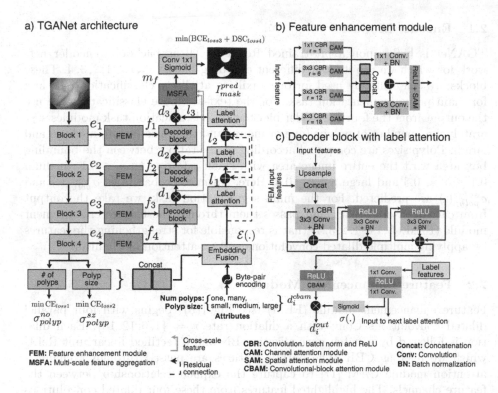

Fig. 1. Block diagram of the proposed TGANet

The main contributions of the presented work include - 1) *text guided attention* (TGA) to learn different features in the context of the number of polyps presence (one or many) and size (small, medium and large), 2) *feature enhancement module* to strengthen the features of the encoder and pass them to the decoder, and 3) *multi-scale feature aggregation* to capture features learned by different decoder blocks. TGA using attributes such as size variability and occurrence of polyps has not been explored before. In addition, steering the label attention with feature attributes from classification heads fused together with attribute embeddings allow to establish refined label attention to the network. We have evaluated our TGANet on four publicly available polyp datasets and compared it with five SOTA medical image segmentation methods.

2   Method

The proposed TGANet is a polyp segmentation architecture with text guided attention that enables to enhance feature representations such that the polyps present in images are segmented optimally independent of their size variability and occurrence. Our TGANet architecture consists of various components that are shown in Fig. 1 and elaborated below.

2.1   Encoder Module

TGANet is built upon a pre-trained ResNet50 [4] as backbone encoder network for which we use its four different encoding blocks, $e_i, i \in 1, 2, 3, 4$. These blocks are consecutively used for our auxiliary attribute classification task and for main polyp segmentation task. For the text-attribute classification, we use the output from the fourth encoder block as two classification task modules separately, i.e., number of polyps (one or many) and their size (small, medium and large). Polyp sizes are computed according to the ratio r between the bounding box area w.r.t the entire image area with small referring to $r < 0.1$, medium $0.1 \leq r < 0.3$ and large as $r \geq 0.3$. Here, softmax probabilities $\sigma_{polyp}^{no}(.)$ and $\sigma_{polyp}^{sz}(.)$ are predicted. For the main segmentation task, we take the output from each ResNet50 block and passes them through the feature enhancement module (FEM, $f_i, i \in 1, 2, 3, 4$) that is responsible for strengthening the features by applying multiple dilated convolutions and an attention mechanism.

2.2   Feature Enhancement Module

Feature enhancement module (FEM) (see Fig. 1(b)) begins with four parallel dilated convolutions Conv with a dilation rate $r = \{1, 6, 12, 18\}$. Each dilation is followed by a batch normalization BN and a rectified linear unit ReLU which we refer as CBR. The output features are passed through a channel-attention module CAM [17] to capture the explicit relationship between the feature channels. The highlighted features from these four dilated convolutions are then concatenated and passed through a $Conv_{3\times3}$ followed by BN layer and added with the original input features through a $Conv_{1\times1}$. The resulting features are then followed by a ReLU activation function, and a spatial attention mechanism SAM [17] is applied to suppress the irrelevant regions.

2.3   Label Attention

Label attention module is designed to provide learned text-based attention to the output features of the decoder blocks in our TGANet. Here, we use three label attention modules, $l_i, i \in 1, 2, 3$, as soft channel-wise attention to the three decoder outputs that enables larger weights to the representative features and suppress the redundant ones. The first label attention module uses the output of the embedding fusion $\mathcal{E}(.)$ obtained by element-wise dot product between the softmax probability concatenation $\{\sigma_{one}, \sigma_{many}, \sigma_{small}, \sigma_{medium}, \sigma_{large}\}$ with the encoded text embedding. Say, $\mathcal{A} = \{one, many, small, medium, large\}$ be the attributes that are encoded using byte-pair encoding (BPE, a simple form of data compression) [5] and denoted by $\mathcal{A}_{encode}$ with $\{a_j^k\}$ as vector embedding for each attribute j of length $|k|$, then $\mathcal{E}(.)$ that is given by:

$$\mathcal{E} = \sigma_j \odot a_j^k, \quad \forall k. \tag{1}$$

The output of the label attention module is referred to as label features l_f. Here, $\mathcal{A}_{encode}$ is thus a collection of pre-trained sub-word embeddings as BPE.

2.4   Decoder

The decoder in the proposed TGANet is comprised of three different decoder blocks $d_i, i \in 1, 2, 3$, of which each takes the input features to upsample it and pass it through some convolutional layers to produce the output. This output is refined using the label attention module l_i and passed to the subsequent decoder blocks d_i (see Fig. 1(c)). The first decoder block takes the output of the fourth FEM f_4 to upsample it using bilinear interpolation by a factor of two, and then it is concatenated with the output features from the third FEM f_3. The resulting concatenated feature is passed through a $\text{Conv}_{1 \times 1}$-BN-ReLU referred as CBR followed by a sequence of three $\text{Conv}_{3 \times 3}$-BN, further accompanied by their multiple residual connections and a ReLU activation function with subsequent convolutional block attention module represented as d_i^{cbam}. An element-wise multiplication is done to allow additional soft-attention from the computed label features l_f using a sigmoid function for each decoder block output $d_i^{out}, i \in 1, 2, 3$ given by:

$$d_i^{out} = d_i^{cbam} \odot \sigma([\text{Conv} - \text{ReLU} - \text{Conv}]l_f), \quad \forall i \in 1, 2, 3 \qquad (2)$$

2.5   Multi-scale Feature Aggregation

Multi-scale feature aggregation (MSFA) module (see supplementary Fig. 1) is used to fuse multi-scale feature representations at various decoder outputs $d_i^{out}, i \in 1, 2, 3$ that allows to capture learned features. We take the first two features $\{d_1^{out}, d_2^{out}\}$ and pass them through a bilinear upsampling to ensure that all three features have the exact spatial dimensions followed by linear 1×1 convolution layers, BN and ReLU activation before concatenation. To boost the capture of non-linear features we further apply a series of convolutional layers, BN and ReLU together with multiple residual connections for improved flow of information. The output is represented as m_f which is responsible for our predicted segmentation map I_{mask}^{pred} given by: $I_{mask}^{pred} = \sigma(\text{Conv}_{1 \times 1}(m_f))$.

2.6   Joint Loss Optimization

We jointly minimize loss for both the auxiliary classification tasks (cross-entropy losses, CE_{loss1}, CE_{loss2}) and the segmentation task (binary cross entropy, BCE_{loss3} and dice loss, DSC_{loss4}) with equal weights.

3   Experiments and Results

3.1   Datasets

To evaluate the performance of our TGANet, we have used four publicly available polyp segmentation benchmark datasets including Kvasir-SEG [9], CVC-ClinicDB [1], BKAI [11], and Kvasir-Sessile [8] (details are presented in supplementary Table 1). Relevant to our experiment, Kvasir-Sessile [8] contains 196 small, diminutive, sessile and flat polyps that are less than 10 mm in size.

Table 1. Quantitative results on the experimented polyp datasets.

Method	Backbone	mIoU	mDSC	Recall	Precision	F2
Dataset: Kvasir-SEG [9]						
U-Net [13]	-	0.7472	0.8264	0.8504	0.8703	0.8353
HarDNet-MSEG [6]	HardNet68	0.7459	0.8260	0.8485	0.8652	0.8358
ColonSegNet [7]	-	0.6980	0.7920	0.8193	0.8432	0.7999
DeepLabV3+ [2]	ResNet50	0.8172	0.8837	0.9014	0.9028	0.8904
PraNet [3]	Res2Net	0.8296	0.8942	0.9060	**0.9126**	0.8976
TGANet (Ours)	ResNet50	**0.8330**	**0.8982**	**0.9132**	0.9123	**0.9029**
Dataset: CVC-ClinicDB [1]						
U-Net [13]	-	0.8428	0.8978	0.9001	0.9209	0.8981
HarDNet-MSEG [6]	HardNet68	0.8388	0.8967	0.8929	0.9216	0.8938
ColonSegNet [7]	-	0.8248	0.8862	0.8828	0.9017	0.8826
DeepLabV3+ [6]	ResNet50	0.8973	0.9391	**0.9441**	0.9442	0.9389
PraNet [3]	Res2Net	0.8866	0.9318	0.9347	0.9479	0.9333
TGANet (Ours)	ResNet50	**0.8990**	**0.9457**	0.9437	**0.9519**	**0.9439**
Dataset: BKAI [11]						
U-Net [13]	-	0.7599	0.8286	0.8295	0.8999	0.8264
HarDNet-MSEG	HardNet68	0.6734	0.7627	0.7532	0.8344	0.7528
ColonSegNet [7]	-	0.6881	0.7748	0.7852	0.8711	0.7746
DeepLabV3+ [2]	ResNet50	0.8314	0.8937	0.8870	**0.9333**	0.8882
PraNet [3]	Res2Net	0.8264	0.8904	0.8901	0.9247	0.8885
TGANet (Ours)	ResNet50	**0.8409**	**0.9023**	**0.9026**	0.9208	**0.9002**
Dataset: Kvasir-Sessile [8]						
U-Net [13]	-	0.2472	0.3688	0.7237	0.3264	0.4635
HarDNet-MSEG	HardNet68	0.1565	0.2558	0.5403	0.2236	0.3298
ColonSegNet [7]	-	0.2113	0.3278	0.5234	0.3336	0.3868
DeepLabV3+ [2]	ResNet50	0.5927	0.7078	0.7085	0.8225	0.7009
PraNet [3]	Res2Net	0.6671	0.7736	0.8069	0.8244	0.7871
TGANet (Ours)	ResNet50	**0.6910**	**0.7980**	**0.7925**	**0.8588**	**0.7879**
Training dataset: Kvasir-SEG − Test dataset: CVC-ClinicDB						
U-Net [13]	-	0.5433	0.6336	0.6982	0.7891	0.6563
HarDNet-MSEG [6]	HardNet68	0.6058	0.6960	0.7173	0.8528	0.7010
ColonSegNet [7]	-	0.5090	0.6126	0.6564	0.7521	0.6246
DeepLabV3+ [2]	ResNet50	0.7388	0.8142	**0.8331**	0.8735	0.8198
PraNet [3]	Res2Net	0.7286	0.8046	0.8188	0.8968	0.8077
TGANet (Ours)	ResNet50	**0.7444**	**0.8196**	0.8290	**0.8879**	**0.8207**

Table 2. mDSC for different *sizes* and *polyp counts* on Kvasir-SEG [9]

Method	Small	Medium	Large	One	Many
DeepLabV3+ [2]	0.8776	0.9003	0.8633	0.8922	0.8289
PraNet [3]	0.8826	0.9079	**0.8900**	0.9071	0.8106
TGANet	**0.8869**	**0.9203**	0.8769	**0.9075**	**0.8378**

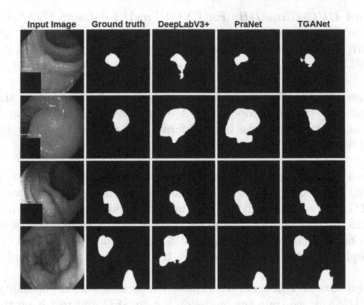

Fig. 2. Qualitative results comparison on the Kvasir-SEG dataset.

3.2 Implementation Details

All models are trained on NVIDIA GeForce RTX 3090 GPU, and images are resized to 256×256 pixels with 80:10:10 training, validation, and testing splits except for Kvasir-SEG, where we adopted the official split of 880/120 for training and testing. Simple data augmentation strategy including random rotation, vertical flipping, horizontal flipping, and coarse dropout are used. All models are trained on a similar hyperparameters configuration with a learning rate of $1e^{-4}$, batch size of 16, and optimized with Adam optimizer. An early stopping mechanism and ReduceLROnPlateau is used to prevent models from overfitting. Standard medical image segmentation metrics such as mean intersection over union (mIoU), mean Sørensen-dice coefficient (mDSC), recall, precision, F2-score and frame per second (FPS) are used.

3.3 Results

We have compared our results with five SOTA methods that include UNet [13], HarDNet-MSEG [6], ColonSegNet [7], DeepLabv3+ [2], and PraNet [3]. These algorithms are widely used baselines in both polyp segmentation and general medical image segmentation. The quantitative results are presented in Table 1.

Results on Kvasir-SEG: Table 1 shows that TGANet outperforms all the SOTA methods with a mIoU of 0.8330 and mDSC of 0.8982. Our TGANet outperforms most competitive PraNet [3] by 1.58% in mIoU and 1.45% mDSC.

Results on CVC-ClinicDB: For CVC-ClinicDB dataset, TGANet outperforms all SOTA methods reporting the highest mIoU and mDSC of 0.8990 and 0.9457, respectively. Our method outperformed the most competitive DeepLabV3+ [2] with a mIoU of 0.17% and mDSC of 0.66%.

Results on BKAI: Table 1 shows the comparison of the result on the BKAI dataset that show that our proposed TGANet obtains mIoU of 0.8409 and mDSC of 0.9023 and outperforms the best performing DeepLabV3+ [2] by 0.95% on mIoU and 0.86% on mDSC.

Results on Kvasir-Sessile: Kvasir-Sessile dataset is clinically most relevant as it has flat and sessile polyps. On this dataset, it can be observed (see Table 1) that our TGANet surpasses all the other methods in all the evaluation metrics. It outperforms the best performing PraNet [3] by a large margin of 2.39% on mIoU and 2.44% on mDSC. Similarly, almost 10% improvement is observed compared to the DeepLabV3+ [2] in this case which is a significant improvement.

Results on Cross Dataset: To explore the generalization capability of our proposed TGSNet, we train the model on Kvasir-SEG and test it on the CVC-ClinicDB. The cross-dataset test (Table 1) also suggested improvements compared to all SOTA methods and obtained an increment of 0.56% on mIoU and 0.54% on mDSC compared to the SOTA DeepLabv3+ [2].

Results on Size and Number-Based Sampled Polyps: To show the effectiveness of our proposed TGANet, we evaluated test samples of Kvasir-SEG-based on the attributes used in training. It can be observed in Table 2 that our model outperforms the best SOTA methods for almost all cases. For the 'small', 'medium' and 'many cases', the improvement ranges from nearly 1–2%. Our qualitative results (see Fig. 2) demonstrate a clear improvement of our text-based attention method for different sizes and number polyp samples. It is evident that both PraNet [3] and DeepLabV3+ [2] failed to capture sample with two polyps (4th row) and also provided over segmentation for the small (1st row) and medium polyps (2nd row). Additionally, we have provided the total number of parameters, flops and FPS in supplementary material Table 2.

3.4   Ablation Study

To validate the effectiveness and importance of the core components used in the network, we compare TGANet with its five variants, which is presented in

Table 3. Ablation study of TGANet on Kvasir-SEG

No	Method	mIoU	mDSC	Recall	Precision	F2
#1	TGANet w/o label and classifier	0.8104	0.8786	0.8987	0.8970	0.8850
#2	TGANet w/o MSFA	0.8151	0.8832	0.9061	0.8999	0.8907
#3	TGANet w/o FEM	0.8084	0.8766	0.8968	0.9010	0.8838
#4	TGANet w/o (MSFA + FEM)	0.8055	0.8739	0.8982	0.8962	0.8827
#5	TGANet w/o (label+classifier+ MSFA+FEM)	0.8063	0.8747	0.8963	0.8971	0.8798
#6	**TGANet (Ours)**	**0.8330**	**0.8982**	**0.9132**	**0.9123**	**0.9029**

Table 3. The results suggest that the introduction of the text guided attention along with the label boosts the performance of the network. The results show that TGANet improves the baseline without the label and classifier (#1) by 2.26% on mIoU and 1.96% on mDSC.

4   Conclusion

We proposed a text-guided attention architecture (TGANet) to tackle polyps' variable size and number for robust polyp segmentation. We have used multiple feature enhancement modules connected with different encoder blocks to achieve this. An auxiliary task is learned together with the main task to compliment both the size-based and number-based feature representations and used as label attentions in the decoder blocks. Additionally, the multi-scale fusion of the features at the decoder enabled our network to deal with these attribute changes. Our experimental results demonstrated the effectiveness of our TGANet outperformed and provided higher segmentation performance on flat and sessile polyps that are clinically important.

Acknowledgement. This project is partially supported by the NIH funding: R01-CA246704 and R01-CA240639.

References

1. Bernal, J., Sánchez, F.J., Fernández-Esparrach, G., Gil, D., Rodríguez, C., Vilariño, F.: WM-DOVA maps for accurate polyp highlighting in colonoscopy: validation vs. saliency maps from physicians. Comput. Med. Imaging Graph. **43**, 99–111 (2015)
2. Chen, L.C., Zhu, Y., Papandreou, G., Schroff, F., Adam, H.: Encoder-decoder with atrous separable convolution for semantic image segmentation. In: Proceedings of the European Conference on Computer Vision (ECCV), pp. 801–818 (2018)
3. Fan, D.-P., et al.: PraNet: parallel reverse attention network for polyp segmentation. In: Martel, A.L., et al. (eds.) MICCAI 2020. LNCS, vol. 12266, pp. 263–273. Springer, Cham (2020). https://doi.org/10.1007/978-3-030-59725-2_26

4. He, K., Zhang, X., Ren, S., Sun, J.: Deep residual learning for image recognition. In: Proceedings of the IEEE Conference on Computer Vision and Pattern Recognition (CVPR), pp. 770–778 (2016)
5. Heinzerling, B., Strube, M.: BPEmb: tokenization-free pre-trained subword embeddings in 275 languages. In: Proceedings of the International Conference on Language Resources and Evaluation (LREC 2018) (2018)
6. Huang, C.H., Wu, H.Y., Lin, Y.L.: HarDNet-MSEG a simple encoder-decoder polyp segmentation neural network that achieves over 0.9 mean dice and 86 FPS. arXiv preprint arXiv:2101.07172 (2021)
7. Jha, D., et al.: Real-time polyp detection, localization and segmentation in colonoscopy using deep learning. IEEE Access **9**, 40496–40510 (2021)
8. Jha, D., et al.: A comprehensive study on colorectal polyp segmentation with ResUNet++, conditional random field and test-time augmentation. IEEE J. Biomed. Health Inform. **25**(6), 2029–2040 (2021)
9. Jha, D., et al.: Kvasir-SEG: a segmented polyp dataset. In: Ro, Y.M., et al. (eds.) MMM 2020. LNCS, vol. 11962, pp. 451–462. Springer, Cham (2020). https://doi.org/10.1007/978-3-030-37734-2_37
10. Kim, N.H., et al.: Miss rate of colorectal neoplastic polyps and risk factors for missed polyps in consecutive colonoscopies. Intestinal Res. **15**(3), 411 (2017)
11. Lan, P.N., et al.: NeoUNet: towards accurate colon polyp segmentation and neoplasm detection. arXiv preprint arXiv:2107.05023 (2021)
12. Rex, D.K., et al.: Colonoscopic miss rates of adenomas determined by back-to-back colonoscopies. Gastroenterology **112**(1), 24–28 (1997)
13. Ronneberger, O., Fischer, P., Brox, T.: U-Net: convolutional networks for biomedical image segmentation. In: Navab, N., Hornegger, J., Wells, W.M., Frangi, A.F. (eds.) MICCAI 2015. LNCS, vol. 9351, pp. 234–241. Springer, Cham (2015). https://doi.org/10.1007/978-3-319-24574-4_28
14. Shen, Y., Jia, X., Meng, M.Q.-H.: HRENet: a hard region enhancement network for polyp segmentation. In: de Bruijne, M., et al. (eds.) MICCAI 2021. LNCS, vol. 12901, pp. 559–568. Springer, Cham (2021). https://doi.org/10.1007/978-3-030-87193-2_53
15. Srivastava, A., et al.: MSRF-Net: a multi-scale residual fusion network for biomedical image segmentation. IEEE J. Biomed. Imaging Health Inform. **26**, 2252–2263 (2021)
16. Sung, H., et al.: Global cancer statistics 2020: GLOBOCAN estimates of incidence and mortality worldwide for 36 cancers in 185 countries. CA: Cancer J. Clin. **71**(3), 209–249 (2021)
17. Woo, S., Park, J., Lee, J.Y., Kweon, I.S.: CBAM: convolutional block attention module. In: Proceedings of the European Conference on Computer Vision (ECCV), pp. 3–19 (2018)
18. Zhong, J., Wang, W., Wu, H., Wen, Z., Qin, J.: PolypSeg: an efficient context-aware network for polyp segmentation from colonoscopy videos. In: Martel, A.L., et al. (eds.) MICCAI 2020. LNCS, vol. 12266, pp. 285–294. Springer, Cham (2020). https://doi.org/10.1007/978-3-030-59725-2_28

Computer Aided Diagnosis

SATr: Slice Attention with Transformer for Universal Lesion Detection

Han Li[1,2], Long Chen[2,3], Hu Han[2(✉)], and S. Kevin Zhou[1,2(✉)]

[1] School of Biomedical Engineering and Suzhou Institute for Advanced Research
Center for Medical Imaging, Robotics, and Analytic Computing and Learning
(MIRACLE), University of Science and Technology of China, Suzhou 215123, China
han.li@miracle.ict.ac.cn, skevinzhou@ustc.edu.cn
[2] Key Lab of Intelligent Information Processing of Chinese Academy of Sciences
(CAS), Institute of Computing Technology, CAS, Beijing 100190, China
long.chen@miracle.ict.ac.cn, hanhu@ict.ac.cn
[3] School of Computer Science and Technology,
University of the Chinese Academy of Science, Beijing, China

Abstract. Universal Lesion Detection (ULD) in computed tomography plays an essential role in computer-aided diagnosis. Promising ULD results have been reported by multi-slice-input detection approaches which model 3D context from multiple adjacent CT slices, but such methods still experience difficulty in obtaining a global representation among different slices and within each individual slice since they only use convolution-based fusion operations. In this paper, we propose a novel Slice Attention Transformer (SATr) block which can be easily plugged into convolution-based ULD backbones to form hybrid network structures. Such newly formed hybrid backbones can better model long-distance feature dependency via the cascaded self-attention modules in the Transformer block while still holding a strong power of modeling local features with the convolutional operations in the original backbone. Experiments with five state-of-the-art methods show that the proposed SATr block can provide an almost free boost to lesion detection accuracy without extra hyperparameters or unique network designs. Code: https://github.com/MIRACLE-Center/A3D_SATr.

Keywords: Universal lesion detection · Slice attention · Transformer

1 Introduction

Universal Lesion Detection (ULD) in computed tomography (CT) [1–18], aiming to localize different types of lesions instead of identifying lesion types [19–28],

This research was supported in part by the Natural Science Foundation of China (grants 61732004 and 62176249).

Supplementary Information The online version contains supplementary material available at https://doi.org/10.1007/978-3-031-16437-8_16.

© The Author(s), under exclusive license to Springer Nature Switzerland AG 2022
L. Wang et al. (Eds.): MICCAI 2022, LNCS 13433, pp. 163–174, 2022.
https://doi.org/10.1007/978-3-031-16437-8_16

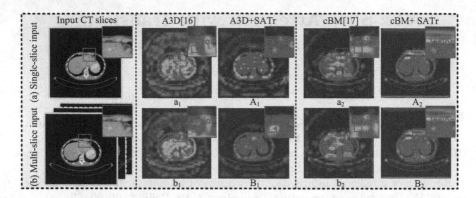

Fig. 1. Visualization of the CAMs of two state-of-the-art ULD methods with and without using our SATr block under (a) single-slice-input and (b) multi-slice-input scenarios.

plays an essential role in computer-aided diagnosis (CAD) [29,30]. ULD is a challenging task because different lesions have diverse shapes and sizes, easily leading to false positive and false negative detections. Mainly inspired by the clinical fact that radiologists need several adjacent slices for locating and diagnosing lesions on one CT slice, most existing ULD methods take several adjacent 2D CT slices as the inputs to a 2D network architecture [3,4,6–10,12,15–18] or directly adopt 3D network designs [10] that take 3D volume as input to extract more 3D context information. While both 2D and 3D methods have yielded great ULD performances, the multi-slice-input based 2D detection methods are much more popular than pure 3D fashion because 2D networks benefit from robust 2D models pretrained from large-scale data whereas publicly available 3D medical datasets are not large enough for robust 3D pretraining.

While achieving success in ULD, the multi-slice-input based 2D approaches have inherent limitations: (i) *Weak global context modeling within each slice.* Most existing ULD methods are fueled by the successful Convolutional Neural Networks (CNN) pretrained backbones (e.g., DenseNet [31]) which learn local features via convolution operations in a hierarchical fashion (e.g., FPN [32]) as powerful image representations. Despite the strong power of local feature extraction, CNNs show their weakness in dealing with global (long-distance spatial) contexts [33]. As shown in Fig. 1 (a), the single-slice CAMs of two CNN-based ULD backbones contain flocks of redundant and inaccurate activation regions which is due to the limitation of CNN backbones in handling global contexts within each slice for ULD tasks. (ii) *Weak inter-slice context modeling.* When dealing with the multi-slice input, most ULD methods [4,6–8,10,12,15,18] adopt independent 2D convolutional operations for each slice to extract independent features, and the independent features further go through a 3D convolution for feature fusion. Such convolution-based fusion methods are good at dealing with local features, but they unfortunately deteriorate in handling global features among different slices. To tackle this, some ULD approaches [9,16,17] propose to reshuffle feature channels among different slices which relieve this issue to

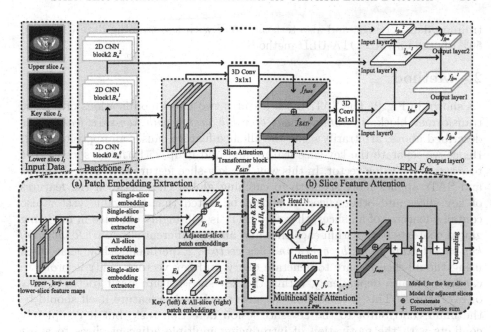

Fig. 2. The network architecture of ULD with the proposed SATr block.

some degree. But their ability in capturing rich global representations is still weak due to the use of pure convolutional operations. As shown in the CAMs in Fig. 1, with the help of convolution based multi-slice fusion, the number of redundant and inaccurate activation regions can be decreased than single-slice-input based methods; however, these methods still cannot capture rich global contextual information due to the limit of CNN receptive field.

Recently, CNN-transformer hybrid architectures have been introduced to several visual tasks [33–35]. Thanks to the Multihead Self-Attention (MSA) mechanism, Multilayer Perceptron (MLP) structure and CNN-transformer hybrid design can reflect the complex spatial transforms for both global and local feature dependency. Unfortunately, such hybrid architectures are designed for single-image-input (or single-slice-input) cases, and thus may not work well in the multi-slice-input scenarios. The main reason is that they may unavoidably introduce a lot of redundant information from adjacent slices (e.g., non-lesion slices or too-far slices), which may be harmful for ULD. Equally using features from every slice and feeding them to the naive transformer block will largely diminish the representation of the key CT slice.

To address the above issues, we hereby propose a novel Slice Attention Transformer (SATr) block, which works properly with multi-slice-input and assists can effectively extract the global representations from both each individual slice and multiple adjacent slices. The SATr block can be easily plugged into the popular convolution-based ULD detection backbones to form hybrid network structures, thus reaping the advantages from both Transformer and CNN worlds.

To validate the effectiveness of our method, we conduct extensive experiments on the DeepLesion dataset [36] based on five SOTA ULD methods under all training data settings, and also test the SATr with less training data (25% and 50%) based on two SOTA ULD methods.

2   Method

As shown in Fig. 2, The SATr block features two novel modifications of a naive transformer block: *i) Enhancement of value vector with key-slice feature.* As described above, arbitrarily using the all-slice feature causes an opposite effect. Hence we separate the key-slice feature and add it up with the all-slice feature to serve as the value vector. In this way, the key-slice feature is largely preserved and SATr applies more strength to catching global cues between the features of key slices and other slices. It should be noted that the value vector also contains features of all slices; therefore SATr is still highly effective in modeling global contexts both within each slice and among different slices. *ii) Removal of key-slice feature from the query and key vectors.* Experiments show that when using the all-slice feature to generate query and key vectors, SATr is prone to catching feature dependency within the key-slice feature itself (network lazy or overfitting). This is reasonable because the key-slice feature itself should be the main contributor to the key-slice lesion detection task, but it completely conflicts with the motivation of introducing multiple adjacent slices to assist ULD. Admittedly, this removal sacrifices the feature dependency learned within the key slice, but it forces the SATr to learn more dependency to better make up the weakness of CNN backbone.

We hereby take the three-slice-input ULD method as a working example to illustrate our method while most SOTA results are reached under 7-slice or 9-slice settings. The SATr blocks are inserted between the feature extractor block (i.e., blocks in backbones) and each feature collector (e.g., FPN) without extra modifications to the original network. Section 2.1 details the common backbone of multi-slice-input ULD methods; Sect. 2.2 explains the newly introduced SATr block, and Sect. 2.3 presents the hybrid network with SATr blocks.

2.1   Multi-slice-Input Backbone

In the multi-slice-input fashion, the ULD method is trained to localize the lesion in the key slice I_k, while the adjacent slices, including the same number of upper slices, $I_u = [I_u^1, ..., I_u^N]$, and lower slices, $I_u = [I_l^1, ..., I_l^N]$, are used to assist the lesion detection for the key slice. Without loss of generality, we set $N = 1$. Hence, the input data can be formulated as $I = [I_u, I_k, I_l] \in \mathcal{R}^{3 \times 1 \times W \times H}$, where W and H are the width and height of the input CT slices, respectively.

Similar to most backbones, the multi-slice-input backbone F_b consists of several continuous CNN blocks $B = [B^0, ..., B^M]$ (e.g., Dense blocks in DenseNet), and further each block utilizes several separate sub CNN blocks $B^m = [B_u^m, B_k^m, B_l^m]$ to deal with the multi-slice input; thus the features from different slices $f = [f_u, f_k, f_l]$ are independent after these blocks. Specifically,

$$f^m = [f_u^m, f_k^m, f_l^m], \ f^{m+1} = [f_u^{m+1}, f_k^{m+1}, f_l^{m+1}] = [B_u^m(f_u^m), B_k^m(f_k^m), B_l^m(f_l^m)],$$
(1)

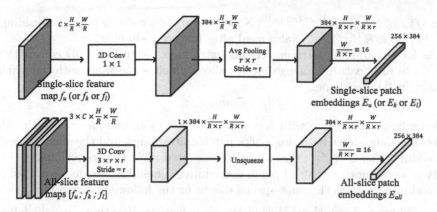

Fig. 3. The architecture of the proposed single-slice patch embedding extractor (upper) and all-slice patch embedding extractor (lower).

where B_k^m is the sub-CNN block for the key-slice feature in the m-th CNN block B^m, and $f^m \in \mathcal{R}^{3 \times C^m \times \frac{W}{R^m} \times \frac{H}{R^m}}$ (or f^{m+1}) is the input (or the output) of B^m. R^m and C^m are the downsampling ratio and channel number of CNN block B^m.

Afterward, the output feature $f^m \in \mathcal{R}^{3 \times C^m \times \frac{W}{R^m} \times \frac{H}{R^m}}$ is fed into a 3D convolution $Conv_{3d}$ with $3 \times 1 \times 1$ Kernal to fuse the feature among different slices:

$$f_{fuse}^m = Unsqueeze(Conv_{3d}(f^m), 0), \quad f_{fuse} = [f_{fuse}^0, ..., f_{fuse}^M], \qquad (2)$$

where $f_{fuse}^m \in \mathcal{R}^{C^m \times \frac{W}{R^m} \times \frac{H}{R^m}}$ acts as the m-th input layer l_{fpn}^m in FPN F_{fpn}:

$$l_{fpn}^m = f_{fuse}^m, \quad f_{fpn} = F_{fpn}(l_{fpn}^0, ..., l_{fpn}^M). \qquad (3)$$

The f_{fpn} is the final output of the FPN F_{fpn}.

2.2 Slice Attention Transformer

As shown in Fig. 2, the SATr block contains two stages, i.e., patch embedding extraction and slice feature attention.

Patch Embedding Extraction. Motivated by [33,37,38], as shown in Fig. 3, we use the 'Convolutional + pooling' manner for single-slice patch embedding extraction. Considering that the extraction is the same among different slices and different CNN blocks B^m, we only formulate the extraction processes of key-slice $f_k \in \mathcal{R}^{C \times \frac{W}{R} \times \frac{H}{R}}$ and remove the CNN block index m in this section:

$$f_k^E = AP(Conv_{2d}(f_k), r), \quad r = \frac{W}{16R} = \frac{H}{16R}, \qquad (4)$$

where $Conv_{2d}$ denotes a 2D convolution with $Kernel = (1 \times 1)$ and 384 output channels, the AP is the average pooling operation with pooling size r and stride r. We change the pooling size accordingly to get the fixed size output feature

$f^e = [f_u^E, f_k^E, f_l^E] \in \mathcal{R}^{3 \times 384 \times 16 \times 16}$. Now the size of each slice's patch embeddings $[E_u, E_k, E_l] \in \mathcal{R}^{3 \times 256 \times 384}$ is also fixed after feature reshaping.

As for the all-slice patch embedding extraction, we utilize a 3D convolution $Conv_{3d}$ to fuse features among different slices which is the same with the fusion manner in the backbone.

$$f_{all}^E = Conv_{3d}(f). \tag{5}$$

Differently, the $Conv_{3d}$ is with $Kernal = (3 \times r \times r)$ and $stride = r$, so the f_{all}^E shares the same shape with single-slice patch embeddings, and the all-slice patch embeddings $E_{all} \in \mathcal{R}^{256 \times 384}$ can be also generated after feature reshaping.

It is worth noting that all patch embeddings, including single- and all-slice embeddings, are with the same spacial size to fit the followed transformer block.

Slice Feature Attention. Within the slice feature attention, we still follow the '$<q, k, v>$ head-MSA-MLP' design as in the naive transformer block, but we modify the input of $q\&k\&v$ according to the ULD multi-slice-input scene.

As for the query head H_q and key head H_k, we concatenate the patch embedding of all adjacent slices ($E_u\&E_l$) to act as the input:

$$f_q = H_q([E_u; E_l]), f_k = H_k([E_u; E_l]), \tag{6}$$

where f_q and f_k are the q value and k value, respectively. As for the value head H_v, the sum of key-slice and all-slice patch embedding is the input:

$$f_v = H_v(E_{all} + E_k), \tag{7}$$

where f_v is the v value and '+' denotes the element-wise sum.

The afterward design follows the naive Transformer block including MSA F_{msa} and MLP F_{mlp}:

$$f_{msa} = F_{msa}(f_q, f_k, f_v), \quad f_{mlp} = F_{mlp}(f_{msa} + f_v), \quad f_{tr} = f_{mlp} + f_{msa}. \tag{8}$$

The generated output feature $f_{tr} \in \mathcal{R}^{256 \times 384}$ will be interpolated and reshaped to the original size to get the final output of our proposed SATr,

$$f_{SATr} = ReShape(Interpolate(f_{tr})), \tag{9}$$

where $f_{SATr} \in \mathcal{R}^{C^m \times \frac{W}{R^m} \times \frac{H}{R^m}}$ is the output of the SATr.

2.3   Hybrid Network

The proposed SATr F_{SATr} can be easily implemented into convolution-based detection backbones to form hybrid network structures:

$$f_{SATr}^m = F_{SATr}^m(f^m), \quad \hat{f}_{fuse} = Conv_{3d}([f_{SATr}^m; f_{fuse}^m]), \tag{10}$$

where $Conv_{3d}$ denotes a 3D convolution with $Kernal = (2 \times 1 \times 1)$, the outputs of 'SATr' f_{SATr}^m and the original backbone f_{fuse}^m will be concatenated and further fused by $Conv_{3d}$ to serve as the m-th input layer l_{fpn}^m of FPN:

$$l_{fpn}^m = \hat{f}_{fuse}^m, \quad f_{fpn} = F_{fpn}(l_{fpn}^0, ..., l_{fpn}^M). \tag{11}$$

Table 1. Sensitivity (%) at various FPPI under full training dataset settings on the testing dataset of DeepLesion [36].

Methods	Data	Slices	@0.5	@1	@2	@4	Avg. [0.5, 1, 2, 4]
Faster R-CNN [39]	100%	3	57.17	68.82	74.97	82.43	70.85
Faster R-CNN+SATr	100%	3	62.52 (**5.35**↑)	73.91 (**5.09**↑)	79.40 (**4.43**↑)	86.03 (**3.60**↑)	75.47(**4.62**↑)
Faster R-CNN+cBM [17]	100%	3	65.37	76.31	81.03	87.98	77.67
Faster R-CNN+cBM+SATr	100%	3	67.41 (2.04↑)	78.02 (1.71↑)	82.43 (1.40↑)	88.90 (0.92↑)	79.19(1.52↑)
3DCE [6]	100%	9	59.32	70.68	79.09	84.34	73.36
3DCE+SATr	100%	9	64.38 (5.06↑)	75.55 (4.87↑)	82.74 (3.65↑)	87.78 (3.44↑)	77.61(4.25↑)
3DCE+cBM [17]	100%	9	66.98	77.25	83.64	88.41	79.07
3DCE+cBM+SATr	100%	9	68.12 (1.14↑)	78.33 (1.08↑)	84.57 (0.93↑)	89.21 (0.80↑)	80.06(0.99↑)
MVP-Net [7]	100%	9	70.07	78.77	84.91	87.33	80.27
MVP-Net+SATr	100%	9	72.34 (2.27↑)	80.27 (1.50↑)	86.11 (1.20↑)	88.21 (0.88↑)	81.73(1.46↑)
MVP-Net+cBM [17]	100%	9	73.05	81.41	87.22	89.37	82.76
MVP-Net+cBM+SATr	100%	9	74.11 (1.06↑)	82.31 (0.90↑)	88.14 (0.92↑)	90.13 (0.76↑)	83.67(0.91↑)
AlignShift [9]	100%	7	77.20	84.38	89.03	92.31	85.73
AlignShift+SATr	100%	7	78.98 (1.78↑)	85.82 (1.44↑)	90.21 (1.18↑)	93.27 (0.96↑)	87.07(1.34↑)
AlignShift+cBM [17]	100%	7	79.17	85.71	89.80	92.65	86.83
AlignShift+cBM+SATr	100%	7	79.98 (0.81↑)	86.36 (0.65↑)	90.22 (0.42↑)	92.99 (0.34↑)	87.39(0.56↑)
A3D w/o Fusion [16]	100%	7	72.47	81.35	86.68	90.41	82.73
A3D w/o Fusion +SATr	100%	7	74.68(2.21↑)	83.17(1.82↑)	88.24(1.56↑)	91.58(1.17↑)	84.42(1.69↑)
A3D [16]	100%	7	79.24	85.04	89.15	92.71	86.54
A3D+SATr	100%	7	**81.03** (1.79↑)	**86.64** (1.60↑)	**90.70** (1.55↑)	**93.30**(0.59↑)	**87.92**(1.38↑)

3 Experiments

3.1 Dataset and Setting

Our experiments are conducted on the ULD dataset DeepLesion [36]. The dataset contains 32,735 lesions on 32,120 axial slices from 10,594 CT studies of 4,427 unique patients. Most existing datasets typically focus on one type of lesion, while DeepLesion contains a variety of lesions with large diameter range (from 0.21 to 342.5 mm). The 12-bit intensity CT is rescaled to [0,255] with different window range settings used in different frameworks. Also, every CT slice is resized and interpolated according to the detection frameworks' setting. We follow the official split, i.e., 70% for training, 15% for validation and 15% for testing. To further test our method's performance on a small dataset, we also conduct experiments based on 25% and 50% training data. The number of false positives per image (FPPI) is used as the evaluation metric. For training, we use the original network architecture and settings (Table 1).

3.2 Lesion Detection Performance

Five state-of-the-art ULD approaches [6,7,9,16,17] and one natural image [39] detection method are compared to evaluate SATr's effectiveness.

Full Training Dataset Results. As shown in Table 2, our method brings promising detection performance improvements for all baselines with full training dataset. The improvements of Faster R-CNN [39], 3DCE, 3DCE w/ cBM and

Table 2. Sensitivity (%) at various FPPI on the testing dataset of DeepLesion [36] under 25% and 50% training data settings.

Methods	Data	Slices	@0.5	@1	@2	@4	Avg. [0.5, 1, 2, 4]
AlignShift [9]	25%	7	52.17	62.17	69.50	75.24	64.77
AlignShift+SATr	25%	7	56.68(4.51↑)	65.04(2.87↑)	72.03(2.53↑)	77.83(**2.59**↑)	67.90(3.13↑)
AlignShift+cBM [17]	25%	7	56.84	64.96	71.86	81.08	68.69
AlignShift+cBM+SATr	25%	7	**62.31 (5.47**↑)	**70.13 (5.17**↑)	**76.79 (4.93**↑)	**81.17** (0.09↑)	**72.60(3.91**↑)
A3D [16]	25%	7	55.67	65.39	73.35	79.31	68.43
A3D+SATr	25%	7	59.99 (4.32↑)	68.05 (2.66↑)	74.67 (1.32↑)	79.09 (0.22↓)	70.45(2.02↑)
AlignShift [9]	50%	7	68.69	76.73	82.25	86.54	78.55
AlignShift+SATr	50%	7	71.10(2.41↑)	78.32 (1.59↑)	83.51 (1.26↑)	87.99 (**1.45**↑)	80.23(1.68↑)
AlignShift+cBM [17]	50%	7	70.30	78.16	83.35	87.99	79.95
AlignShift+cBM+SATr	50%	7	73.53 (**3.23**↑)	79.91 (1.75↑)	84.89 (**1.54**↑)	88.50 (0.51↑)	81.71(**1.76**↑)
A3D [16]	50%	7	72.52	80.27	86.14	90.15	82.27
A3D+SATr	50%	7	**75.24** (2.72↑)	**82.19** (1.92↑)	**86.99** (0.85↑)	**90.96** (0.81↑)	**83.85**(1.58↑)

MVP-Net are more pronounced than those of AlignShift [9] and A3D [16]. This is because AlignShift and A3D introduce channel-fusion mechanism among different slices in backbone, thus the v value enhancement design in SATr brings less advances. Anyway, SATr still endows A3D and AlignShift with strong global dependency catching power to provide notable improvements. Meantime, the A3D w/ SATr reached the SOTA result.

Partial Training Dataset Results. As shown in Table 2, under the 25% and 50% training data settings, our method brings more improvements than that in the full training data scene. Although the bigger performance improving space plays a huge role, we can also attribute this to our value removal design. When the training data is limited, SATr forces the network to learn more dependency to avoid overfitting. Besides, our experiments also show some results contradicted with the full-training dataset experiment, e.g., AlignShift outperformed A3D.

Classification Activation Maps Results. We showcase CAMs (based on [40] [41]) of two SOTA ULD methods to evaluate the global context modeling power of SATr. As shown in Fig. 1, SATr helps the model catch useful dependencies (e.g., lesion area) and depress redundant dependencies (e.g., air area) within a single slice (a_1v.s.A_1 and a_2v.s.A_2) and among different slices (b_1v.s.B_1 and b_2v.s.B_2). It is worth noting that some anatomy-aware objects and anatomies, e.g., the treatment couch of CT scanner and the outline of the patient's body, should also be the contributor to the task. More cases are included in Suppl. Material.

3.3 Ablation Study

An ablation study is provided to evaluate the importance of the two key designs: (i) Enhancement of value vector with key-slice feature and (ii) Removal of key-slice feature from the query and key vectors. Adding naive transformer block for baseline, the performance is increased by 0.44%, enhancing the value vector with the key-slice feature. We obtain a 0.94% (0.82%+0.04%+0.08%) improvement over the naive transformer block. Further removing the key-slice feature from the query and key vectors accounts for another 0.83% improvement and gives the best performance. We also report two different q-k-v head designs to support the superiority of using key-slice feature in v head. Ablation study for different SATr blocks, which is included in Suppl. Material, shows that all SATr blocks are important (Table 3).

Table 3. Ablation study of our method at various FPs per image (FPPI).

A3D [16] w/o fusion	Naive transformer block	Enhancement of value vector	$q\&k$ head	v head	$FPPI = 0.5$	$FPPI = 1$
✓					72.47	81.35
✓	✓				72.91(0.44↑)	81.68(0.33↑)
✓	✓	✓	Key	All	72.99(0.08↑)	81.81(0.13↑)
✓	✓	✓	Key	Adjacent	73.03(0.04↑)	82.11(0.30↑)
✓	✓	✓	All	Key	73.85(0.82↑)	82.48(0.37↑)
✓	✓	✓		Adjacent	Key	**74.68(0.83 ↑)**	**83.17(0.69↑)**

4   Conclusion

Multi-slice-input based ULD methods using CNN backbones have inherent limitations in capturing the global contextual information within each individual slice and among multiple adjacent slices. To address this issue, we propose a slice attention transformer (SATr) block that can be integrated with conventional CNN backbones to obtain better global representation. Extensive experiments using several SOTA ULD methods as baselines show that the proposed can be easily integrated with many existing methods to boost their performance without extra hyperparameters or unique network designs.

References

1. Zlocha, M., Dou, Q., Glocker, B.: Improving RetinaNet for CT lesion detection with dense masks from weak RECIST labels. In: Shen, D., et al. (eds.) MICCAI 2019. LNCS, vol. 11769, pp. 402–410. Springer, Cham (2019). https://doi.org/10.1007/978-3-030-32226-7_45
2. Tao, Q., Ge, Z., Cai, J., Yin, J., See, S.: Improving deep lesion detection using 3D contextual and spatial attention. In: Shen, D., et al. (eds.) MICCAI 2019. LNCS, vol. 11769, pp. 185–193. Springer, Cham (2019). https://doi.org/10.1007/978-3-030-32226-7_21
3. Zhang, N., et al.: 3D anchor-free lesion detector on computed tomography scans. arXiv:1908.11324 (2019)
4. Zhang, N., et al.: 3D aggregated faster R-CNN for general lesion detection. arXiv:2001.11071 (2020)

5. Tang, Y., et al.: Uldor: a universal lesion detector for CT scans with pseudo masks and hard negative example mining. In: IEEE ISBI, pp. 833–836 (2019)

6. Yan, K., Bagheri, M., Summers, R.M.: 3D context enhanced region-based convolutional neural network for end-to-end lesion detection. In: Frangi, A.F., Schnabel, J.A., Davatzikos, C., Alberola-López, C., Fichtinger, G. (eds.) MICCAI 2018. LNCS, vol. 11070, pp. 511–519. Springer, Cham (2018). https://doi.org/10.1007/978-3-030-00928-1_58

7. Li, Z., Zhang, S., Zhang, J., Huang, K., Wang, Y., Yu, Y.: MVP-Net: multi-view FPN with position-aware attention for deep universal lesion detection. In: Shen, D., et al. (eds.) MICCAI 2019. LNCS, vol. 11769, pp. 13–21. Springer, Cham (2019). https://doi.org/10.1007/978-3-030-32226-7_2

8. Yan, K., et al.: MULAN: multitask universal lesion analysis network for joint lesion detection, tagging, and segmentation. In: Shen, D., et al. (eds.) MICCAI 2019. LNCS, vol. 11769, pp. 194–202. Springer, Cham (2019). https://doi.org/10.1007/978-3-030-32226-7_22

9. Yang, J., et al.: *AlignShift*: bridging the gap of imaging thickness in 3D anisotropic volumes. In: Martel, A.L., et al. (eds.) MICCAI 2020. LNCS, vol. 12264, pp. 562–572. Springer, Cham (2020). https://doi.org/10.1007/978-3-030-59719-1_55

10. Cai, J., et al.: Deep volumetric universal lesion detection using light-weight pseudo 3D convolution and surface point regression. In: Martel, A.L., et al. (eds.) MICCAI 2020. LNCS, vol. 12264, pp. 3–13. Springer, Cham (2020). https://doi.org/10.1007/978-3-030-59719-1_1

11. Li, H., Han, H., Zhou, S.K.: Bounding maps for universal lesion detection. In: Martel, A.L., et al. (eds.) MICCAI 2020. LNCS, vol. 12264, pp. 417–428. Springer, Cham (2020). https://doi.org/10.1007/978-3-030-59719-1_41

12. Zhang, S., et al.: Revisiting 3D context modeling with supervised pre-training for universal lesion detection in CT slices. In: Martel, A.L., et al. (eds.) MICCAI 2020. LNCS, vol. 12264, pp. 542–551. Springer, Cham (2020). https://doi.org/10.1007/978-3-030-59719-1_53

13. Yan, K., et al.: Learning from multiple datasets with heterogeneous and partial labels for universal lesion detection in CT. IEEE Trans. Med. Imaging **40**, 2759–2770 (2020)

14. Cai, J., et al.: Deep lesion tracker: monitoring lesions in 4D longitudinal imaging studies. In: IEEE CVPR, pp. 15159–15169 (2021)

15. Tang, Y., et al.: Weakly-supervised universal lesion segmentation with regional level set loss. In: de Bruijne, M., et al. (eds.) MICCAI 2021. LNCS, vol. 12902, pp. 515–525. Springer, Cham (2021). https://doi.org/10.1007/978-3-030-87196-3_48

16. Yang, J., He, Y., Kuang, K., Lin, Z., Pfister, H., Ni, B.: Asymmetric 3D context fusion for universal lesion detection. In: de Bruijne, M., et al. (eds.) MICCAI 2021. LNCS, vol. 12905, pp. 571–580. Springer, Cham (2021). https://doi.org/10.1007/978-3-030-87240-3_55

17. Li, H., Chen, L., Han, H., Chi, Y., Zhou, S.K.: Conditional training with bounding map for universal lesion detection. In: de Bruijne, M., et al. (eds.) MICCAI 2021. LNCS, vol. 12905, pp. 141–152. Springer, Cham (2021). https://doi.org/10.1007/978-3-030-87240-3_14

18. Lyu, F., Yang, B., Ma, A.J., Yuen, P.C.: A segmentation-assisted model for universal lesion detection with partial labels. In: de Bruijne, M., et al. (eds.) MICCAI 2021. LNCS, vol. 12905, pp. 117–127. Springer, Cham (2021). https://doi.org/10.1007/978-3-030-87240-3_12

19. Boot, T., Irshad, H.: Diagnostic assessment of deep learning algorithms for detection and segmentation of lesion in mammographic images. In: Martel, A.L., et al. (eds.) MICCAI 2020. LNCS, vol. 12264, pp. 56–65. Springer, Cham (2020). https://doi.org/10.1007/978-3-030-59719-1_6

20. Yu, X., et al.: Deep attentive panoptic model for prostate cancer detection using biparametric MRI scans. In: Martel, A.L., et al. (eds.) MICCAI 2020. LNCS, vol. 12264, pp. 594–604. Springer, Cham (2020). https://doi.org/10.1007/978-3-030-59719-1_58

21. Ren, Y., et al.: Retina-match: ipsilateral mammography lesion matching in a single shot detection pipeline. In: de Bruijne, M., et al. (eds.) MICCAI 2021. LNCS, vol. 12905, pp. 345–354. Springer, Cham (2021). https://doi.org/10.1007/978-3-030-87240-3_33

22. Baumgartner, M., Jäger, P.F., Isensee, F., Maier-Hein, K.H.: nnDetection: a self-configuring method for medical object detection. In: de Bruijne, M., et al. (eds.) MICCAI 2021. LNCS, vol. 12905, pp. 530–539. Springer, Cham (2021). https://doi.org/10.1007/978-3-030-87240-3_51

23. Shahroudnejad, A., et al.: TUN-Det: a novel network for thyroid ultrasound nodule detection. In: de Bruijne, M., et al. (eds.) MICCAI 2021. LNCS, vol. 12901, pp. 656–667. Springer, Cham (2021). https://doi.org/10.1007/978-3-030-87193-2_62

24. Luo, L., Chen, H., Zhou, Y., Lin, H., Heng, P.-A.: OXnet: deep omni-supervised thoracic disease detection from chest X-rays. In: de Bruijne, M., et al. (eds.) MICCAI 2021. LNCS, vol. 12902, pp. 537–548. Springer, Cham (2021). https://doi.org/10.1007/978-3-030-87196-3_50

25. Chen, J., Zhang, Y., Wang, J., Zhou, X., He, Y., Zhang, T.: EllipseNet: anchor-free ellipse detection for automatic cardiac biometrics in fetal echocardiography. In: de Bruijne, M., et al. (eds.) MICCAI 2021. LNCS, vol. 12907, pp. 218–227. Springer, Cham (2021). https://doi.org/10.1007/978-3-030-87234-2_21

26. Yang, H.-H., et al.: Leveraging auxiliary information from EMR for weakly supervised pulmonary nodule detection. In: de Bruijne, M., et al. (eds.) MICCAI 2021. LNCS, vol. 12907, pp. 251–261. Springer, Cham (2021). https://doi.org/10.1007/978-3-030-87234-2_24

27. Lin, C., Wu, H., Wen, Z., Qin, J.: Automated Malaria cells detection from blood smears under severe class imbalance via importance-aware balanced group softmax. In: de Bruijne, M., et al. (eds.) MICCAI 2021. LNCS, vol. 12908, pp. 455–465. Springer, Cham (2021). https://doi.org/10.1007/978-3-030-87237-3_44

28. Zhao, Z., Pang, F., Liu, Z., Ye, C.: Positive-unlabeled learning for cell detection in histopathology images with incomplete annotations. In: de Bruijne, M., et al. (eds.) MICCAI 2021. LNCS, vol. 12908, pp. 509–518. Springer, Cham (2021). https://doi.org/10.1007/978-3-030-87237-3_49

29. Kevin Zhou, S., et al.: A review of deep learning in medical imaging: imaging traits, technology trends, case studies with progress highlights, and future promises (2021)

30. Kevin Zhou, S., Rueckert, D., Fichtinger, G.: Handbook of Medical Image Computing and Computer Assisted Intervention. Academic Press (2019)

31. Huang, G., et al.: Densely connected convolutional networks. In: IEEE CVPR, pp. 4700–4708 (2017)

32. Lin, T., et al.: Feature pyramid networks for object detection. In: IEEE CVPR, pp. 2117–2125 (2017)

33. Peng, Z., et al.: Conformer: local features coupling global representations for visual recognition. In: IEEE ICCV, pp. 367–376 (2021)

34. Xu, Y., et al.: ViTAE: vision transformer advanced by exploring intrinsic inductive bias. In: NeurIPS, vol. 34 (2021)

35. Mao, M., et al.: Dual-stream network for visual recognition. In: NeurIPS, vol. 34 (2021)
36. Yan, K., et al.: Deep lesion graphs in the wild: relationship learning and organization of significant radiology image findings in a diverse large-scale lesion database. In: IEEE CVPR, pp. 9261–9270 (2018)
37. Zhu, X., et al.: Deformable detr: deformable transformers for end-to-end object detection. arXiv preprint arXiv:2010.04159 (2020)
38. Carion, N., Massa, F., Synnaeve, G., Usunier, N., Kirillov, A., Zagoruyko, S.: End-to-end object detection with transformers. In: Vedaldi, A., Bischof, H., Brox, T., Frahm, J.-M. (eds.) ECCV 2020. LNCS, vol. 12346, pp. 213–229. Springer, Cham (2020). https://doi.org/10.1007/978-3-030-58452-8_13
39. Ren, S., He, K., Girshick, R., Sun, J.: Faster R-CNN: towards real-time object detection with region proposal networks. In: NIPS, pp. 91–99 (2015)
40. Gildenblat, J., et al.: Pytorch library for cam methods (2021). https://github.com/jacobgil/pytorch-grad-cam
41. Muhammad, M.B., et al. Eigen-CAM: class activation map using principal components. In: IEEE IJCNN, pp. 1–7 (2020)

MAL: Multi-modal Attention Learning for Tumor Diagnosis Based on Bipartite Graph and Multiple Branches

Menglei Jiao[1,6], Hong Liu[1]($\boxtimes$), Jianfang Liu[2], Hanqiang Ouyang[3,4,5], Xiangdong Wang[1], Liang Jiang[3,4,5], Huishu Yuan[2], and Yueliang Qian[1]

[1] Beijing Key Laboratory of Mobile Computing and Pervasive Device, Institute of Computing Technology, Chinese Academy of Sciences, Beijing 100190, China
hliu@ict.ac.cn
[2] Department of Radiology, Peking University Third Hospital, Beijing 100191, China
[3] Department of Orthopaedics, Peking University Third Hospital, Beijing 100191, China
[4] Engineering Research Center of Bone and Joint Precision Medicine, Beijing 100191, China
[5] Beijing Key Laboratory of Spinal Disease Research, Beijing 100191, China
[6] University of Chinese Academy of Sciences, Beijing 100086, China

Abstract. The multi-modal fusion of medical images has been widely used in recent years. Most methods focus on images with a single plane, such as the axial plane with different sequences (T1, T2) or different modalities (CT, MRI), rather than multiple planes with or without cross modalities. Further, most methods focus on segmentation or classification at the image or sequence level rather than the patient level. This paper proposes a general and scalable framework named MAL for the classification of benign and malignant tumors at the patient level based on multi-modal attention learning. A bipartite graph is used to model the correlations between different modalities, and then modal fusion is carried out in feature space by attention learning and multi-branch networks. Thereafter, multi-instance learning is adopted to obtain patient-level diagnostic results by considering different modal pairs of patient images to be bags and the edges in the bipartite graph to be instances. The modal and intra-type similarity losses at the patient level are calculated using the feature similarity matrix to encourage the model to extract high-level semantic features with high correlation. The experimental results confirm the effectiveness of MAL on three datasets with respect to different multi-modal fusion tasks, including axial and sagittal MRI, axial CT and sagittal MRI, and T1 and T2 MRI sequences. And the application of MAL can also significantly improve the diagnostic accuracy and efficiency of doctors. Code is available at https://github.com/research-med/MAL.

Keywords: Multi-modal fusion · Attention learning · Tumor diagnosis

Supplementary Information The online version contains supplementary material available at https://doi.org/10.1007/978-3-031-16437-8_17.

© The Author(s), under exclusive license to Springer Nature Switzerland AG 2022
L. Wang et al. (Eds.): MICCAI 2022, LNCS 13433, pp. 175–185, 2022.
https://doi.org/10.1007/978-3-031-16437-8_17

1 Introduction

The diagnosis of benign and malignant tumors based on medical images is critical for early treatment. If not diagnosed in time, benign tumors may cause local damage and even exhibit invasive growth into surrounding tissues, whereas malignant tumors may induce systemic multisystem metastasis and threaten the lives of patients. The use of computer technology based on medical image data, such as computed tomography (CT), or magnetic resonance imaging (MRI), to diagnose benign and malignant tumors is of considerable clinical significance to ensure early therapy.

In a clinical environment, the comprehensive diagnosis of patients usually requires multiple modalities, such as different sequences (e.g., T1-weighted image: T1, T2-weighted image: T2); modalities (e.g., CT, MRI); planes (e.g., axial, sagittal); and directions of the 3D reconstructed data. Artificial intelligence-based methods for medical image processing can be classified as single [1–4] and multi-modality methods [5–10]. However, in the case of some complex medial tasks, such as tumor detection or diagnosis, single-modality data may be insufficient. In such cases, multi-modality methods can improve diagnostic accuracy based on multi-modal data analysis, e.g., fusion of axial CT and axial MRI for detecting the thermal ablation of liver tumors [5]; fusion of axial MRI and ultrasound for volume registration [6]; fusion of axial MRI and positron emission computed tomography (PET) for automated dementia diagnosis [7]; and fusion of different sequences of axial CT and axial MRI (T1, T2, Flair, etc.) for tumor segmentation [8–10]. In addition, Liu et al. [11] fused multiple sequences and patient age information, and achieved good results. The performance of these multi-modal methods is better than that of single-modality methods.

Recently, the attention mechanism [12] has attracted considerable attention in the field of natural language processing. Several studies have demonstrated that the attention mechanism can also be effectively applied in computer vision based on large-scale image training [13–15]. In medical imaging, Dai et al. [16] spliced two modalities from the T1 and T2 sequences of axial MRI in the input stage and then fused them using Transformer [12] based on an attention mechanism to classify parotid gland tumors and knee joint injuries. Wang et al. [17] spliced different modalities from the T1, T1c, T2, and Flair sequences of axial brain MRI in the input stage and then performed feature fusion using Transformer to segment brain tumors.

However, most of the aforementioned multi-modal fusion methods focus on specific tasks on a single plane, such as the axial plane with different sequences (T1, T2) or cross modalities (CT, MRI), rather than multiple planes with or without cross modalities, such as the combination of axial and sagittal planes. Further, most methods focus on detection, segmentation, and classification at the image or sequence level rather than the patient level. To address these limitations, this study aimed to introduce a general, scalable framework for multi-modal fusion by constructing and exploring the internal correlations between different modality data for patient-level tumor diagnosis.

Data corresponding to one modality often corresponds to series data corresponding to another modality, such as multiple-plane data. Inspired by this clinical observation, herein, we propose a graph to represent this correlation. A general and scalable framework, MAL, is proposed based on multi-modal attention learning for patient-level benign and malignant tumor diagnosis. The proposed framework is capable of comprehensively considering multiple planes with or without cross modalities. The primary contributions of this paper are as follows. (1) We construct a bipartite graph to model the correlations between data with varying modalities, with the vertices represented by the data of each modality and edges represented by the connections between different modalities. (2) We propose an attention learning and multi-branch network to efficiently predict the edges of the bipartite graph and explore the correlations between the data of different modalities. (3) We present PMSLoss and PiTSLoss with modal and intra-type similarities at the patient level to encourage the extraction of high-level semantic features with high correlation by the model corresponding to the same modality or type at the patient level. (4) We adopt multi-instance learning considering image pairs of different modalities as bags and edges as instances for patient-level diagnosis. Experimental results confirm the effectiveness of the MAL framework on three datasets with respect to different multi-modal fusion tasks.

2 Proposed Framework

The proposed multi-modal attention learning framework, MAL, is depicted in Fig. 1. First, a bipartite graph is used to model the correlations between data with different modalities. Thereafter, the attention learning and multi-branch network is used to predict all edges of the bipartite graph. Finally, multi-instance learning is implemented to obtain patient-level diagnosis results by considering the different modal pairs of patient images as bags and the edges in the bipartite graph as instances. MAL performs multi-modal fusion and classifies tumors based on manually pre-detected tumor regions (represented by green boxes in the patient images). The training stage includes a multi-modal matching module based on the bipartite graph, a multi-modal fusion module based on attention learning and multiple branches, PMSLoss based on modal similarity at the patient level, and PiTSLoss based on intra-type similarity at the patient level. During the inference stage, the trained multi-modal fusion module is used to predict each edge of the bipartite graph and the corresponding probability. Then the discriminative edges are selected for patient-level diagnosis. The proposed modules are described using an example of the multi-modal fusion of axial CT and sagittal MRI.

Multi-modal Matching Module Based on the Bipartite Graph: A bipartite graph is a special structure in graph theory. Let $G = (V, E)$ be an undirected graph. If the vertex set, V, can be divided into two disjoint subsets, A and B, such that the two vertices, i and j, associated with each edge, $e(i, j)$, in the graph belong to different vertex sets, i.e., $(i \in A, j \in B)$, then the graph, G, is called a bipartite graph. In the aforementioned example, the multi-modal data of patients are divided into different sets. First, all images of axial CT are considered to set A, and those of sagittal MRI are considered to set B. Thereafter, vertices belonging to A and B are connected to create a complete bipartite graph, as depicted in Fig. 1. As in the case of multi-instance learning, the ground truth of each edge in the bipartite graph is considered to be the overall ground truth of the patient. During the training stage, the edges in the bipartite graph corresponding to each patient are randomly activated and feature fusion is performed during each iteration. The two vertices of each activated edge correspond to different modalities. During the inference stage, all edges in the bipartite graph are activated, and each edge is considered to be an instance and is predicted using the multi-modal fusion module. Finally, the overall diagnostic result of each patient is obtained from the patient-level fusion module depicted in Fig. 1. To make the model more general, MAL performs post fusion on data of different modalities at the feature level instead of fusion at the input level.

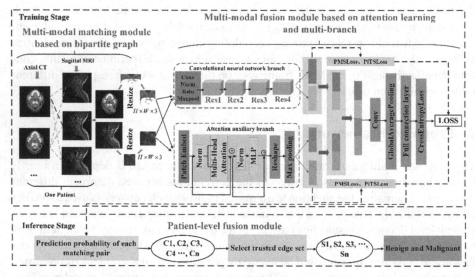

Fig. 1. Proposed multi-modal fusion framework, MAL. (Color figure online)

Multi-modal Fusion Module Based on Attention Learning and Multiple Branches: Each edge in the aforementioned bipartite graph connects two images of different modalities. The multi-modal fusion module is utilized to first extract the features of different modalities and then combine them. First, a shared convolutional neural network (CNN) branch is used to extract the features of tumor areas with different modalities (M_1, M_2), and the features, F_1 and F_2, corresponding to the modalities, M_1 and M_2, are obtained. A general ResNet model capable of extracting local features of images but not global information can be considered as the CNN branch. Thus, an auxiliary branch based on attention learning is proposed. Via the shared auxiliary branch, the features, L_1 and L_2, of the modalities, M_1 and M_2, are obtained. Unlike models completely based on Transformer [13, 14], the attention auxiliary branch proposed in this study is a simple structure. First, the input tumor areas are divided into patches of 8×8 pixels, and each patch is embedded. Thereafter, a multi-head attention block is used to correlate global features. Finally, a max-pooling operation is implemented to reserve the area with the largest response value as the feature, L.

For the modalities, M_1 and M_2, the features, F_1, F_2, L_1, L_2, are obtained via the two aforementioned branches F_1 and F_2 are obtained using the CNN branch, and L_1 and L_2 are obtained using the attention auxiliary branch. Thereafter, feature fusion is performed using $H = (F_1 \oplus F_2) \oplus (L_1 \oplus L_2)$, where $\oplus$ denotes the concatenating operation. The fused feature, H, is subjected to conv, global average pooling, and full connection operations to calculate the final output and CrossEntropy loss.

PMSLoss and PiTSLoss: CrossEntropy loss focuses on optimizing the model based on a single instance without considering the correlations between different instances corresponding to the same patient, which may lead to suboptimal solutions. Hence, we propose the modal similarity loss function PMSLoss and intra-type similarity loss function PiTSLoss at the patient level. These losses are not directly calculated based on the ground truth but by constructing a similarity matrix to encourage the extraction of similar high-level semantic features by the model corresponding to different modalities or of identical type at the patient level. Let us consider PMSLoss as an example. Each batch may contain multiple samples or instances corresponding to the same patient. It is assumed that the patients' identification (ID) set in the current batch is $\{i^1, i^2, \ldots, i^m\} \in ID$, and that the features extracted by any branch of the multi-modal fusion module for the modalities, M_1 and M_2, are T_1 and T_2, respectively. The features extracted by the model corresponding to the respective samples or instances of each patient in this batch based on the patients' ID set, and multiple features corresponding to each patient in each modality are column-wise averaged to integrate features corresponding to different instances and the same patient, as given by Eq. (1).

$$\tilde{T}_1^{m \times d} = \left(\tilde{T}_1^{m1} \ldots \tilde{T}_1^{md}\right), \tilde{T}_2^{m \times d} = \left(\tilde{T}_2^{m1} \ldots \tilde{T}_2^{md}\right) \tag{1}$$

Here, $\tilde{T}_1^{m \times d}$ and $\tilde{T}_2^{m \times d}$ denotes the patient-level modal feature matrix of M_1 and M_2. The feature similarity matrix of M_1 and M_2 is calculated using $\tilde{T} = \frac{\tilde{T}_1 \cdot (\tilde{T}_2)^T}{\left\|\tilde{T}_1\right\| \left\|(\tilde{T}_2)^T\right\|}$. The size of the matrix, $\tilde{T}$, is $m \times m$, where m denotes the number of patients in this batch. The determinant of $\tilde{T}$ represents the similarity between the features of M_1 and M_2 extracted by the model corresponding to the same patient. We attempt to maximize the determinant and make its value as close to 1 as possible while simultaneously minimizing the values in other positions. The loss is obtained by calculating the mean square error (MSE) between $\tilde{T}$ and the identity matrix. The calculation of PiTSLoss is almost identical to that of PMSLoss. However, for the latter, after $\tilde{T}_1$ and $\tilde{T}_2$ are obtained, the two matrices are averaged by bit to obtain matrix $\tilde{T}_{mean}$, which is used to calculate the self-similar matrix and MSE. The final loss of the model is taken as the average of CrossEntropy loss, PMSLoss, and PiTSLoss.

Patient-Level Fusion Module: As in the case of multi-instance learning [18], this study considers the bipartite graph comprising different modal image pairs as a bag and its edges as instances. The ground truths of all instances are identical to the patients' pathological reports. Patient-level fusion is implemented to aggregate the predictions of the multi-modal fusion module corresponding to each instance and obtain the overall diagnostic results of each patient. It is assumed that the multi-modal fusion module creates a set C of the predictions corresponding to each instance in the bag. The Top-K elements in C constitute the discriminative edge set, S. An element in S is considered a malignant result if its value is greater than T and a benign result otherwise. The number of benign results in set S is N_b, and the number of malignant results is N_m, with $p_b = N_b/(N_b + N_m)$, $p_m = N_m/(N_b + N_m)$. If $p_b > p_m$, the patient is diagnosed with a benign tumor; otherwise, the patient is diagnosed with a malignant tumor.

3 Experiment and Results

Dataset: The tumor classification performance of the proposed method was evaluated on three different spine tumor datasets including MRI-Axi&Sag(133), CT-Axi&MRI-Sag(84), and MRI-T1&T2(134). The results obtained on the third dataset are presented in the Supplementary Material. The collection of datasets on primary spine tumors between January 2006 and December 2019 from The Third Hospital of Peking University was approved beforehand by the Institutional Review Board. The labels, benign and malignant, were obtained from pathological reports. The locations of tumors were manually marked by doctors, and the results were verified with each other. The MRI-Axi&Sag(133) dataset contains the data of 422 patients or cases with both axial and sagittal MRI image data. Among these, the data of 289 patients were included in the training set (197 benign, 92 malignant), and those of 133 patients were included in the test set (84 benign, 49 malignant), with a total of 13,681 labeled images in the training set and 6,161 images in the test set. The CT-Axi&MRI-Sag(84) dataset contains the data of 315 patients with both axial CT and sagittal MRI image data. Among these, the data of 231 patients were included in the training set (157 benign, 74 malignant), and those of 84 patients were included in the test set (52 benign, 32 malignant), with a total of 12,481 labeled images in the training set and 4,372 images in the test set.

Metrics: The samples of malignant tumors were considered to be positive. The area under the curve (AUC), accuracy (ACC), sensitivity (SE), and specificity (SP) at the patient level were used as evaluation metrics to compare the performances of different AI-based methods. The performances of AI-based models and doctors were compared in terms of ACC, SE, and SP, with respect to a classification threshold of 0.5 to ensure fair comparison, which can be adjusted according to SE or SP.

Experimental Settings: Three experiments were conducted to evaluate the proposed method. (1) Patient-level benign and malignant diagnosis was performed based on single-modality. (2) The two modalities were intermixed during patient-level diagnosis. (3) The multi-modal fusion method based on the bipartite graph was implemented, with or without the attention auxiliary branch, PMSLoss, and PiTSLoss.

All methods used ResNet18 as the backbone (CNN branch). During the training stage, all methods used identical values of hyperparameters, with a learning rate of 0.0002, and SGD was used as the optimizer. The number of epochs was considered to be 20, tumor area was resized to 224 × 224, and batch size was assumed to be 32. The patch size in the attention auxiliary branch was 8, embedding dimension was 768, and number of attention heads was 4. All methods used the same patient-level fusion module to obtain the patient-level diagnostic results.

We also compared the performance of the proposed AI-based model with those of three doctors including an orthopedic doctor (D1: 8 y of experience) and two radiologists (D2: 3 y of experience; D3: 11 y of experience). The doctors independently predicted benign and malignant tumors in each patient in the testing sets of the first two datasets. In addition, we evaluated the results of doctors assisted by AI-based models.

Performances of Different Methods: In Table 1, Axi represents the experiments that rely solely on axial images, Sag represents the experiments that rely solely on sagittal images, and Axi-Sag represents mixed training methods depending on both axial and sagittal images. Mg-Axi-Sag represents methods of multi-modal fusion based on the bipartite graph, including the results of ablation experiments. As summarized in Table 1 the results obtained by simply mixing the axial and sagittal planes during training (Axi-Sag) on the two datasets were inferior to those obtained via training on single-modality Axi or Sag. This may be attributed to the fact that simply mixing multiple-modality data during training merely accomplishes data amplification without exploring the correlations between the data of different modalities. When only bipartite graph-based multi-modal fusion (Mg-Axi-Sag) was used, without an attention branch, PMSLoss, and PiTSLoss, its AUCs on the two datasets were higher (7.7% and 12.4%) than those of single-modal (Axi, Sag) or multi-modal mixed training (Axi-Sag). This indicates the effectiveness of the proposed bipartite graph-matching strategy. On the multiple-plane MRI dataset, MRI-Axi&Sag(133), the AUC of the multi-modal fusion method with the attention branch was 1.7% higher than that of the original version. When PMSLoss and PiTSLoss were adopted, the AUC was further improved and attained a maximum value of 87.4%. On cross-modality data drawn from the multiple-plane dataset, CT-Axi&MRI-Sag(84), the AUCs of the proposed method with and without the attention branch, PMSLoss, and PiTSLoss were similar. However, the ACC of the proposed method with the attention branch (83.3%) was 5.9% higher than that of the version without the attention branch on this dataset, and the ACC was further improved by 9.5% after adding PMSLoss and PiTSLoss. Thus, the experimental results confirmed the effectiveness of the proposed MAL method. In addition, the visualization results of the attention auxiliary branch are shown in Fig. 2. Revealing regions that the attention auxiliary branch focused on corresponding to data of varying modalities. The features of these regions can be fused to improve the classification performance.

Although the total numbers of cases in the two datasets were different, some interesting clinical conclusions could be drawn. The AUC and ACC of the proposed method based on the bipartite graph on CT-Axi&MRI-Sag(84) were significantly higher than those on MRI-Axi&Sag(133), which indicates that cross-modal data (CT, MRI) corresponding to multiple planes (axial, sagittal) may contain a greater number of distinguishable features. The correlated information be further explored using the proposed multi-modal attention learning-based method. This indicates that doctors should comprehensively refer to CT, MRI, and multiple-plane data for accurate diagnosis. The fusion experiment on different sequences (T1, T2) under the same modality is presented in the Supplementary Materials to illustrate the generality of MAL framework.

Table 1. Results of tumor diagnosis by proposed method on two datasets (%, 95% CI)

Method	AB	MS	TS	MRI-Axi&Sag(133)				CT-Axi&MRI-Sag(84)			
				AUC	ACC	SE	SP	AUC	ACC	SE	SP
Axi	-	-	-	79.1 ± 6.9	75.2 ± 7.3	55.1 ± 8.5	86.9 ± 5.7	85.0 ± 7.6	78.6 ± 8.8	84.4 ± 7.8	75.0 ± 9.3
Sag	-	-	-	82.5 ± 6.5	74.4 ± 7.4	49.0 ± 8.5	89.3 ± 5.3	79.2 ± 8.7	76.2 ± 9.1	71.9 ± 9.6	78.9 ± 8.7
Axi-Sag	-	-	-	82.9 ± 6.4	75.2 ± 7.3	49.0 ± 8.5	90.5 ± 5.0	82.9 ± 8.1	73.8 ± 9.4	65.6 ± 10.2	78.9 ± 8.7
Mg-Axi-Sag (MAL)				86.8 ± 5.8	77.4 ± 7.1	83.7 ± 6.3	73.8 ± 7.5	91.6 ± 5.9	77.4 ± 8.9	90.6 ± 6.2	69.2 ± 9.9
		√		86.9 ± 5.7	**79.0 ± 6.9**	83.7 ± 6.3	**76.2 ± 7.2**	91.1 ± 6.1	79.8 ± 8.6	90.6 ± 6.2	73.1 ± 9.5
			√	85.9 ± 5.9	76.7 ± 7.2	85.7 ± 5.9	71.4 ± 7.7	91.0 ± 6.1	81.0 ± 8.4	90.6 ± 6.2	75.0 ± 9.3
		√	√	85.7 ± 5.9	76.7 ± 7.2	87.8 ± 5.6	70.2 ± 7.8	**92.0 ± 5.8**	81.0 ± 8.4	84.4 ± 7.8	78.9 ± 8.7
	√			86.2 ± 5.9	74.4 ± 7.4	87.8 ± 5.6	66.7 ± 8.0	91.7 ± 5.9	83.3 ± 8.0	87.5 ± 7.1	80.8 ± 8.4
	√	√		86.1 ± 5.9	78.2 ± 7.0	83.7 ± 6.3	75.0 ± 7.4	90.3 ± 6.3	83.3 ± 8.0	90.6 ± 6.2	78.9 ± 8.7
	√		√	86.8 ± 5.8	78.2 ± 7.0	83.7 ± 6.3	75.0 ± 7.4	90.7 ± 6.2	83.3 ± 8.0	87.5 ± 7.1	80.8 ± 8.4
	√	√	√	**87.4 ± 5.6**	78.2 ± 7.0	**89.8 ± 5.1**	71.4 ± 7.7	91.3 ± 6.0	**86.9 ± 7.2**	**90.6 ± 6.2**	**84.6 ± 7.7**

*AB: Attention Branch, MS: PMSLoss, TS: PiTSLoss

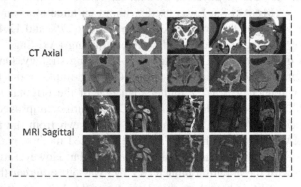

Fig. 2. Heat maps of the attention auxiliary branch.

In addition, the proposed method is scalable, i.e., it can be used to combine more than two modalities by extending the bipartite graph to a graph comprising multiple vertex sets. Multi-class tasks can be performed using such graphs by appropriately modifying the multi-modal fusion module and the patient-level fusion module.

Comparison of Performances of the Proposed Method and Human Doctors: We also compared the performance of the proposed method, MAL, with those of three doctors in terms of ACC, SE, SP, P and average classification duration on the two datasets presented in Table 2. On the two datasets, the ACC of MAL was observed to be higher than those of D1 and D2, and similar to that of D3. As evident from Table 2, D3 exhibited similar accuracies compared to the proposed model on both datasets. Further, the average durations required to obtain prediction corresponding to a single patient by D1 and D2 were 74.9 and 108.1 s, respectively, while that required by MAL was only 0.7 s. D3 required the highest average duration, which also improved its accuracy.

Performance of Doctors with the Help of the AI-Based Model: To evaluate whether the AI-based model can improve the diagnosis accuracy of doctors, we recorded the

Table 2. Comparison of tumor diagnoses between doctors and AI-based model (%, 95% CI)

	MRI-Axi&Sag(133)					CT-Axi&MRI-Sag(84)				
	ACC	SE	SP	P value	Avg. time (s)	ACC	SE	SP	P value	Avg. time (s)
D1	54.1 ± 8.5	95.9 ± 3.4	29.8 ± 7.8	<0.005	18.8	71.4 ± 9.7	84.4 ± 7.8	63.5 ± 10.3	0.167	18.8
D2	70.7 ± 7.7	89.8 ± 5.1	59.5 ± 8.3	0.219	29.5	77.4 ± 8.9	84.4 ± 7.8	73.1 ± 9.5	0.539	19.7
D3	79.7 ± 6.8	69.4 ± 7.8	85.7 ± 5.9	0.006	74.9	86.9 ± 8.2	84.4 ± 7.8	88.5 ± 6.8	0.534	108.1
MAL	**78.2 ± 7.0**	**89.8 ± 5.1**	**71.4 ± 7.7**	-	**0.7**	**86.9 ± 7.2**	**90.6 ± 6.2**	**84.6 ± 7.7**	-	**0.7**

Table 3. Comparison of tumor diagnoses by doctors with the help of AI-based model (%)

	MRI-Axi&Sag(133)				CT-Axi&MRI-Sag(84)			
	ACC	SE	SP	Avg. time (s)	ACC	SE	SP	Avg. time (s)
D1	78.2(+24.1)	91.8(−4.1)	70.2(+40.4)	15.4(−3.4)	83.3(+11.9)	84.4(+0.0)	82.7(+19.2)	8.4(−10.4)
D2	79.7(+9.0)	89.8(+0.0)	73.8(+14.3)	11.5(−18.0)	82.1(+4.7)	81.3(−3.1)	82.7(+9.6)	11.1(−8.6)
D3	**84.2(+4.5)**	**81.6(+12.2)**	**85.7(+0.0)**	**24.6(−50.3)**	**89.3(+2.4)**	**81.3(−3.1)**	**94.2(−5.7)**	**35.5(−72.6)**

diagnoses of three doctors on test cases in the two aforementioned datasets based on a combination of patients' medical images and additional malignant probabilities obtained using the MAL model. The diagnostic accuracy and efficiency of doctors were observed to considerably improve with the application of the proposed model, as summarized in Table 3. The integration of D3 with manual diagnosis further increased its high ACCs by 4.5% and 2.4%, even surpassing the performance of the proposed AI-based model. The diagnosis time also decreased significantly, the main reason is that the model greatly reduces the analysis time of doctors.

4 Conclusion

This paper proposed a multi-modal attention learning framework named MAL for the diagnosis of benign and malignant tumors at the patient level that is capable of estimating the correlations between data of varying modalities. A bipartite graph was incorporated within a multi-modal matching model, enabling the modeling of correlations between data of various modalities. Attention learning and multi-branch strategies in MIL were implemented in the multi-modal fusion module to explore and combine multi-modal relationship features and achieve patient-level diagnosis. The experimental results on three different datasets and comparison of its performance with those of doctors confirmed the effectiveness of the MAL framework in tumor diagnosis. Further, the proposed method was demonstrated to be generalizable and scalable which can be extended to combine more than two modalities and perform multi-class tasks. We intend to conduct further research on the pathological subtype diagnosis of tumors using multi-modal medical images in the future.

Acknowledgements. This work was supported by the Beijing Natural Science Foundation (Z190020), National Natural Science Foundation of China (82171927, 81971578), Capital's Funds for Health Improvement and Research (2020-4-40916).

References

1. Ben-Cohen, A., Diamant, I., Klang, E., Amitai, M., Greenspan, H.: Fully convolutional network for liver segmentation and lesions detection. In: Carneiro, G., et al. DLMIA LABELS 2016. LNCS, vol. 10008, pp. 77–85. Springer, Cham (2016). https://doi.org/10.1007/978-3-319-46976-8_9
2. Christ, P.F., et al.: Automatic liver and tumor segmentation of CT and MRI volumes using cascaded fully convolutional neural networks. arXiv preprint arXiv:1702.05970 (2017)
3. Zhang, J., Xie, Y., Wu, Q., Xia, Y.: Medical image classification using synergic deep learning. Med. Image Anal. **54**, 10–19 (2019)
4. Isensee, F., Jäger, P.F., Kohl, S.A., Petersen, J., Maier-Hein, K.H.: Automated design of deep learning methods for biomedical image segmentation, vol. 1, pp. 1–8. arXiv preprint arXiv: 1904.08128 (2019)
5. Wei, D., et al.: Synthesis and inpainting-based MR-CT registration for image-guided thermal ablation of liver tumors. In: Shen, D., et al. MICCAI 2019. LNCS, vol. 11768, pp. 512–520. Springer, Cham (2019). https://doi.org/10.1007/978-3-030-32254-0_57
6. Song, X., et al.: Cross-modal attention for mri and ultrasound volume registration. In: de Bruijne, M., et al. MICCAI 2021. LNCS, vol. 12904, pp. 66–75. Springer, Cham (2021). https://doi.org/10.1007/978-3-030-87202-1_7
7. Zhou, T., et al.: Deep multi-modal latent representation learning for automated dementia diagnosis. In: Shen, D., et al. MICCAI 2019. LNCS, vol. 11767, pp. 629–638. Springer, Cham (2019). https://doi.org/10.1007/978-3-030-32251-9_69
8. Zhang, Y., et al.: Modality-aware mutual learning for multi-modal medical image segmentation. In: de Bruijne, M., et al. MICCAI 2021. LNCS, vol. 12901, pp. 589–599. Springer, Cham (2021). https://doi.org/10.1007/978-3-030-87193-2_56
9. Zhang, Y., et al.: Multi-phase liver tumor segmentation with spatial aggregation and uncertain region inpainting. In: de Bruijne, M., et al. MICCAI 2021. LNCS, vol. 12901, pp. 68–77. Springer, Cham (2021). https://doi.org/10.1007/978-3-030-87193-2_7
10. Syazwany, N.S., Nam, J.-H., Lee, S.-C.: MM-BiFPN: multi-modality fusion network with Bi-FPN for MRI brain tumor segmentation. IEEE Access **9**, 160708–160720 (2021)
11. Liu, H., et al.: Benign and malignant diagnosis of spinal tumors based on deep learning and weighted fusion framework on MRI. Insights Imaging **13**(1), 1–11 (2022)
12. Vaswani, A., et al.: Attention is all you need. In: Advances in neural information processing systems 30 (2017)
13. Dosovitskiy, A., et al.: An image is worth 16x16 words: transformers for image recognition at scale. arXiv preprint arXiv:2010.11929 (2020)
14. Liu, Z., et al.: Swin transformer: hierarchical vision transformer using shifted windows. In: Proceedings of the IEEE/CVF International Conference on Computer Vision, pp. 10012–10022 (2021)
15. Carion, N., Massa, F., Synnaeve, G., Usunier, N., Kirillov, A., Zagoruyko, S.: End-to-end object detection with transformers. In: Vedaldi, A., Bischof, H., Brox, T., Frahm, JM. (eds.) ECCV 2020. LNCS, vol. 12346, pp. 213–229. Springer, Cham (2020). https://doi.org/10. 1007/978-3-030-58452-8_13
16. Dai, Y., Gao, Y., Liu, F.: Transmed: transformers advance multi-modal medical image classification. Diagnostics **11**(8), 1384 (2021)

17. Wang, W., Chen, C., Ding, M., Yu, H., Zha, S., Li, J.: TransBTS: multimodal brain tumor segmentation using transformer. In: de Bruijne, M., et al. MICCAI 2021. LNCS, vol. 12901, pp. 109–119. Springer, Cham (2021). https://doi.org/10.1007/978-3-030-87193-2_11

18. Li, B., Li, Y., Eliceiri, K.W.: Dual-stream multiple instance learning network for whole slide image classification with self-supervised contrastive learning. In: Proceedings of the IEEE/CVF Conference on Computer Vision and Pattern Recognition, pp. 14318–14328 (2021)

Optimal Transport Based Ordinal Pattern Tree Kernel for Brain Disease Diagnosis

Kai Ma, Xuyun Wen, Qi Zhu, and Daoqiang Zhang[✉]

College of Computer Science and Technology,
MIIT Key Laboratory of Pattern Analysis and Machine Intelligence,
Nanjing University of Aeronautics and Astronautics, Nanjing 211106, China
dqzhang@nuaa.edu.cn

Abstract. Finding a faithful connection pattern of brain network is a challenging task for most of the existing methods in brain network analysis. To process this problem, we propose a novel method called ordinal pattern tree (OPT) for representing the connection pattern of network by using the ordinal pattern relationships of edge weights in brain network. On OPT, nodes are connected by ordinal edges which make nodes have hierarchical structures. The changes of edge weights in brain network will affect ordinal edges and result in the differences of OPTs. We further leverage optimal transport distances to measure the transport costs between the nodes on the paired of OPTs. Based on these optimal transport distances, we develop a new graph kernel called optimal transport based ordinal pattern tree kernel to measure the similarity between the paired brain networks. To evaluate the effectiveness of the proposed method, we perform classification and regression experiments in functional magnetic resonance imaging data of brain diseases. The experimental results demonstrate that our proposed method can achieve significant improvement compared with the state-of-the-art graph kernel methods on classification and regression tasks.

Keywords: Graph kernel · Ordinal pattern tree · Brain network · Classification · Optimal transport

1   Introduction

Brain networks characterize functional or structural interconnection of human brain, where brain regions correspond to nodes and functional or anatomical associations between nodes are considered as edges. Brain connectivity network is widely applied to the classification tasks of brain diseases, including Alzheimer's disease [13], attention deficit hyperactivity disorder (ADHD) [7], major depressive disorder [4] and schizophrenia [25]. In these studies, various network measurements, e.g., degree, clustering coefficient [15,19], are first extracted from connectivity networks as features for classification. However, these network measurements are regarded as feature vectors in brain disease classification, which ignore topological information of brain network. Therefore, it is a challenge for

© The Author(s), under exclusive license to Springer Nature Switzerland AG 2022
L. Wang et al. (Eds.): MICCAI 2022, LNCS 13433, pp. 186–195, 2022.
https://doi.org/10.1007/978-3-031-16437-8_18

the previous methods to measure the topological similarity between a pair of brain networks.

The graph kernels, which can measure the topological similarity of brain networks, have shown promising performance on many kinds of classification problems [16,21]. There are a variety of graph kernels having been proposed. Most of the existing graph kernels can be divided into three categories, they are respectively path kernel, walk kernel, and subgraph kernel. For example, shortest-path kernel [2] and truncated tree based graph kernels (Tree++) [24] belong to path kernels which use the paths between nodes to measure the similarity of the paired graphs. These graph kernels ignore the hierarchical structure information of nodes in the graphs. Random walk graph kernels [6,23] and return probability graph kernel [27] are walk kernels which measure the similarities of the paired graphs by matching random walks. Subgraph matching kernel [11] is a kind of subgraph kernel which measures the similarity between a pair of graphs by summing the similarity of subgraphs. Walk and subgraph kernels are usually constructed on unweighted graphs with edges present or not, and thus neglect the valuable weight information of edges in the graphs. In brain network, edge weights convey the strengths of temporal correlation or fiber connection between brain regions, and are very important for brain network analysis.

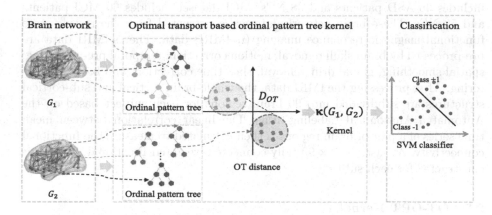

Fig. 1. Illustration of the optimal transport based ordinal pattern tree kernel.

To tackle these problems, we develop an optimal transport based ordinal pattern tree (OT-OPT) kernel for measuring the similarity between the paired brain networks. In brain network, we use the ordinal pattern relationship of edge weights to define an ordinal pattern tree (OPT) for each node. The OPT reflects the hierarchical structure relationship of nodes. After obtaining OPTs, the optimal transport (OT) distances are used to measure the transport costs between the nodes on the paired of OPTs. Finally, the OT-OPT kernel is calculated based on these OT distances. We apply the proposed OT-OPT kernel to support vector machine (SVM) classifier and support vector regression (SVR) for brain disease diagnosis and the prediction of clinical scores. Figure 1 presents the schematic diagram of the proposed framework with each network representing a specific subject. Specifically, our work has following advantages:

- The proposed ordinal pattern tree can be used to explore the hierarchical structure relationship of nodes in brain network.
- The proposed OT-OPT kernel can be applied to kernel two-sample test for distinguishing the discriminative tree structures from patients with brain disease.
- The proposed OT-OPT kernel can be used to diagnose brain diseases and predict clinical scores related to brain diseases.

2   Methods

2.1   Data and Preprocessing

The brain network data used in the experiments are based on three datasets of brain diseases, they are ADHD dataset[1], autistic spectrum disorder (ASD) dataset[2], and mild cognitive impairment (MCI) dataset[3]. In ADHD dataset, there are 110 ADHD patients and 100 normal controls (NCs). ASD dataset includes 36 ASD patients and 38 NCs. MCI dataset includes 60 MCI patients and 58 NCs. These brain network data are constructed from the resting state functional magnetic resonance imaging (rs-fMRI) data. The rs-fMRI data are pre-processed by brain skull removal, motion correction, temporal pre-whitening, spatial smoothing, global drift removal, slice time correction, and band pass filtering. After processing the fMRI data, the whole-brain cortical and sub-cortical structures are subdivided into 90 brain regions for each subject based on the Automated Anatomical Labeling atlas. The linear correlation between mean time series of a pair of brain regions is then calculated to measure the functional connectivity. At last, a 90×90 fully-connected weighted functional network is constructed for each subject.

2.2   OT-OPT Kernel

Most of the existing network methods (e.g., ordinal pattern descriptor) [8,9,26] neglect the hierarchical structure information of nodes in brain network. Tree is an efficient way that can capture the hierarchical structure of the data [18]. In our work, we generalize the ordinal pattern to the tree structure called OPT. Then, we define OT-OPT kernel by using OT distance to calculate the transport cost between the nodes on the paired of OPTs.

OPT: Given a weighted network or graph G, there are a set of nodes V, edges E and weight vectors W in G, $G = (V, E, W)$. W is the weight vector for those

[1] http://www.nitrc.org/plugins/mwiki/index.php/neurobureau:AthenaPipeline.
[2] http://fcon_1000.projects.nitrc.org/indi/abide/.
[3] http://adni.loni.usc.edu/.

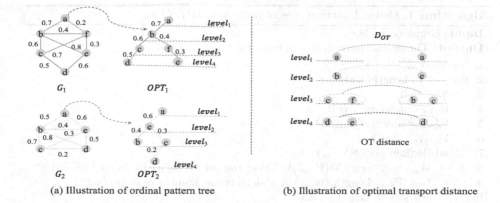

(a) Illustration of ordinal pattern tree   (b) Illustration of optimal transport distance

Fig. 2. Illustration of (a) ordinal pattern tree, and (b) optimal transport distance. $level_i$ represents the i-th level of ordinal pattern tree.

edges with the i-th element $W(e_i)$ representing the connection strength of the edge e_i, $e_i \in E$. The ordinal pattern (OP) defined in graph G is a set including ordinal nodes and ordinal edges $OP = (V_{op}, E_{op})$. E_{op} is an ordinal edge set, $E_{op} = \{e_1, e_2, \cdots, e_i, e_j, \cdots, e_M\} \subseteq E$, all $0 < i < j \leq M$, $W(e_i) > W(e_j)$, e_i and e_j are called ordinal edges. V_{op} is a vertex set where vertexes are connected by ordinal edges included in E_{op}. We generalize the OP to the tree structure which is called OPT. $\forall u \in V$, the neighborhood vertex set of a vertex u is represented as $\delta(u) = \{v : (u,v) \in E, v \in V\}$, the edge weight set between vertex u and its neighborhood vertices is represented as $\mathbb{W}(u) = \{W(u,v) : v \in \delta(u), (u,v) \in E\}$. In graph G, given a node v_0, we construct an OPT for node v_0 by using Algorithm 1. The OPT is shown in Fig. 2(a). From Fig. 2(a), we can find that the nodes in OPT have hierarchical structure relationships.

OT Distance: OT was originally presented to investigate the translocation of masses [10]. OT is utilized to determine the difference between two probability distributions or histograms based on some given ground distances [12,28]. The distance induced by OT is also called Earth Mover's distance or Wasserstein distance [28]. Recently, OT is becoming extremely popular in image processing, computer vision and graph similarity measurement, such as image reconstruction [29] and graph kernels [22].

The OT distances are very powerful in measuring two probability distributions or histograms, especially when the probability space has geometrical structures. For r and c are two probability distributions from the simplex Σ_n and Σ_m ($\Sigma_m := \{x \in \mathbb{R}^m_+ : x^T \mathbf{1}_m = 1\}$), we regard $\Gamma(r,c)$ as the set of all transportation plans of r and c, Γ is a $n \times m$ matrices, $\Gamma(r,c) := \{P \in \mathbb{R}^{n \times m}_+ | P\mathbf{1}_m = r, P^T \mathbf{1}_n = c\}$.

The OT distance based on $\Gamma(r,c)$ is not isometric. The graph kernel based on this distance is not positive definite [22], which leads non-convex optimization problem. To acquire the positive definite graph kernel, we add the entropic constraints to $\Gamma(r,c)$, which is defined as:

Algorithm 1. Ordinal pattern tree of node v_0: $OPT(G, v_0)$

Input: Graph G, node v_0
Output: The ordinal pattern tree of node v_0: V_{set}
1: $v = v_0$; $V_{set}[v_0] = 1$;
2: **for** i=1 to Length($\delta(v)$) **do**
3:     $Sum = 0$
4:     $W_{des} = descend(\mathbb{W}(v))$
5:     $V_{next} = index(W_{des}[i])$
6:     $V_{set}[V_{next}] = 1$
7:     **while** !isempty $(\delta(V_{next}))$ **do**
8:         $W_{des} = descend(\mathbb{W}(V_{next}))$ % Sort the set $\mathbb{W}(V_{next})$ in descending order
9:         Len-W_{des}=Length (W_{des}) % Calculate the number of edges in W_{des}
10:         **for** j=1 to Len-W_{des} **do**
11:             **if** $W_{des}[j] < W(V_{next}, v)$ **then**
12:                 $v = V_{next}$
13:                 $V_{next} = index(W_{des}[j])$
14:                 $V_{set}[V_{next}] = 1$
15:                 break
16:             **else**
17:                 Sum=Sum+1
18:             **if** Sum==Len-W_{des} **then**
19:                 break
20: **return** V_{set}

$$\Gamma'(r, c) := \{P \in \Gamma(r, c) | KL(P||rc^T) = 0\} \tag{1}$$

where $KL(P||rc^T)$ is Kullback-Leibler divergence of $P, rc^T \in \Gamma(r, c)$.

Given a $n \times m$ cost matrix M defined on the ground distance $d(x, y)$, $x \in r, y \in c$, the total cost of mapping r to c with a transportation matrix P is represented as $\langle P, M \rangle$. The OT distance is defined as:

$$d_{OT}(r, c) := min_{P \in \Gamma'(r,c)} \langle P, M \rangle \tag{2}$$

OT Distance on OPT: Given two graphs G_1 and G_2, we construct the OPTs for them using algorithm 1, the OPT of G_1 and G_2 is respectively OPT_1 and OPT_2. For nodes at each level of OPT_1 and OPT_2, we use the OT distance defined in Eq.(2) to measure their transport costs, as shown in Fig. 2(b). Then, we use the average OT distance of all levels on OPT_1 and OPT_2 to represent the transport cost between OPT_1 and OPT_2. The OT distance between OPT_1 and OPT_2 is defined as follows:

$$D_{OT}(OPT_1, OPT_2) := \frac{1}{level} \sum_{l=1}^{level} d_{OT}(f_l(OPT_1), f_l(OPT_2)) \tag{3}$$

where $f_l(OPT_1)$ represents the embeddings of the nodes at l-th level of the OPT_1. The $level$ is the hierarchical number of the OPT_1.

OT-OPT Kernel: Given two graphs G_1 and G_2, OPT_1 and OPT_2 are their OPT structures. An OPT structure can be used to represent a graph and we calculate the OT-OPT kernel between the graphs G_1 and G_2 as follows:

$$K_{OT-OPT}(G_1, G_2) := e^{-\lambda D_{OT}(OPT_1, OPT_2)} \tag{4}$$

2.3 OT-OPT Kernel Based Learning

We utilize the image processing method introduced in data and preprocessing to process the rs-fMRI data of all subjects and construct the brain functional network for each subject. In the brain functional network, nodes represent brain regions, and edges represent the functional connections between the paired brain regions. Given the brain functional networks of all subjects, we compute the proposed OT-OPT kernels on them with Eq. (4) and apply SVM and SVR for disease classification and regression, respectively.

3 Experiments

Table 1. Comparison of different methods on three classification tasks

Method	ADHD vs. NCs		ASD vs. NCs		MCI vs. NCs	
	ACC(%)	AUC	ACC(%)	AUC	ACC(%)	AUC
SP kernel [2]	65.43	0.623	66.77	0.643	69.27	0.672
WL-ST kernel [20]	70.13	0.682	72.35	0.697	75.31	0.723
WL-SP kernel [20]	68.25	0.664	67.55	0.643	68.34	0.631
RW kernel [23]	69.78	0.643	70.11	0.691	72.36	0.705
PM kernel [17]	74.04	0.721	73.44	0.719	76.24	0.731
WWL kernel [22]	73.63	0.688	72.26	0.715	80.15	0.762
GH kernel [5]	67.56	0.639	71.36	0.674	79.24	0.751
Tree++ kernel [24]	71.49	0.692	71.54	0.703	82.66	0.803
OT-OPT kernel (Proposed)	**76.13**	**0.741**	**78.38**	**0.778**	**89.36**	**0.862**

3.1 Experimental Setup

In the experiments, we compare our kernel with the state-of-the-art graph kernels including shortest path kernel (SP) [2], Weisfeiler-Lehman subtree kernel (WL-ST) [20], Weisfeiler-Lehman shortest path kernel (WL-SP) [20], random walk kernel (RW) [23], pyramid match kernel (PM) [17], Wasserstein Weisfeiler-Lehman graph kernel (WWL) [22], GraphHopper kernel (GH) [5] , Tree++ [24].

SVM [3] as the final classifier is exploited to conduct the classification experiment. The eigen-decomposition strategy is applied to the brain network of each subject and then each subject has a bag-of-vectors where each row represents the embedding feature of the node. We perform the leave-one-out cross-validation for all the classification experiments. In the experiments, uniform weight λ is chosen from $\{10^{-2}, 10^{-1}, \cdots, 10^{2}\}$ and the tradeoff parameter C in the SVM is selected from $\{10^{-3}, 10^{-2}, \cdots, 10^{3}\}$.

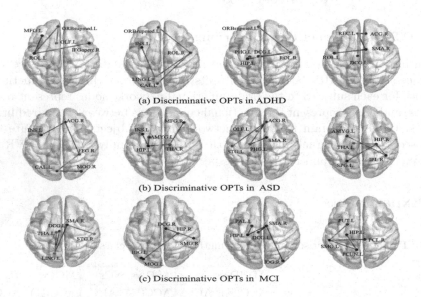

(a) Discriminative OPTs in ADHD

(b) Discriminative OPTs in ASD

(c) Discriminative OPTs in MCI

Fig. 3. The discriminative OPTs in ADHD, ASD and MCI.

3.2   Classification Results

We compare the proposed OT-OPT kernel with the state-of-the-art graph kernels on three classification tasks, i.e., ADHD vs. NCs, ASD vs. NCs and MCI vs. NCs classification. Classification performance is evaluated by accuracy (ACC) and area under receiver operating characterisitc curve (AUC). The classification results are shown in Table 1. From Table 1, we can find that our proposed method achieves the best performance on three tasks. For instance, the accuracy achieved by our method is respectively 76.13%, 78.38%, and 89.36% in ADHD vs. NCs, ASD vs. NCs and MCI vs. NCs classification, which is better than the second best result obtained by PM and WWL kernel. This demonstrates that the proposed OT-OPT kernel is good at distinguishing the patients with brain diseases (i.e., ADHD, ASD, and MCI) from NCs, compared with the state-of-the-art graph kernels.

3.3   Discriminative OPTs

From the construction of OPTs in the graphs, we can find that different initialized nodes will generate different OPT structures. In brain network, the different

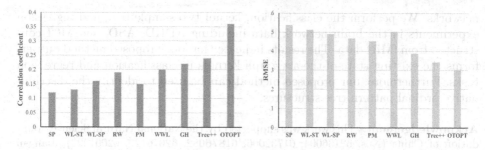

Fig. 4. Regression results of MMSE scores on different methods.

nodes represent the different brain regions which have functional interaction. Hence each brain region has its own OPT structure. In order to investigate the discriminate OPTs between the patients with brain diseases (e.g., ADHD, ASD, and MCI) and NCs, We calculate the OT-OPT kernel for each brain region and apply the obtained OT-OPT kernels to kernel two-sample test (KTST) [1,14] which is a statistical analysis method based on kernel function. We plot the discriminative OPTs for four brain regions in ADHD, ASD, and MCI in Fig. 3. For convenience of display, these OPTs only involve five brain regions. The four brain regions in the discriminative OPTs in ADHD is respectively IFGoperc.R, INS.L, ORBsupmed.L, and ROL.L, as shown in Fig. 3(a). The four brain regions in the discriminative OPTs in ASD is respectively FFG.R, INS.L, STG.L, and THA.L, as shown in Fig. 3(b). The four brain regions in the discriminative OPTs in MCI is respectively DCG.L, HIP.R, IOG.R, and PCUN.L, as shown in Fig. 3(c).

3.4 Regression Results

We conduct the regression task for mini-mental state examination (MMSE) scores of the subjects in the MCI dataset. We compare the proposed OT-OPT kernel with SP WL-ST, WL-SP, RW, PM, WWL, GH, Tree++ kernel methods and apply them to SVR. We use correlation coefficient and root mean square error (RMSE) to evaluate the regression results of all methods, as shown in Fig. 4. The correlation coefficient and RMSE values acquired by our proposed method is respectively 0.36 and 3.01. From Fig. 4, we can find that our method achieves higher correlation coefficient and lower RMSE values, compared with the competing methods. These results further imply that the proposed kernel based on OPT structure can help facilitate the predicting performance of clinical scores.

4   Conclusion

In this paper, we propose an optimal transport based ordinal pattern tree kernel, which can make full use of the weight information of edges and node hierarchical structure relationships to measure the similarities between a pair of brain networks. We perform the classification, kernel two-sample test, and regression experiments in the brain network data including ADHD, ASD, and MCI constructed from fMRI data. The results indicate that our proposed method outperforms the existing state-of-the-art graph kernels in classification and regression tasks. Furthermore, our proposed method can be used to identify the discriminative ordinal pattern tree structures.

Acknowledgement. This work was supported by the National Natural Science Foundation of China (Nos. 62136004, 61732006, 61876082, 62076129, 62001222), Jiangsu Funding Program for Excellent Postdoctoral Talent, and also by the National Key R&D Program of China (Grant Nos. 2018YFC2001600, 2018YFC2001602).

References

1. Arthur, G., Karsten, M., Malte, J., Bernhard, S., Alexander, S.: A kernel two-sample test. J. Mach. Learn. Res. **13**(1), 723–773 (2012)
2. Borgwardt, K.M., Kriegel, H.: Shortest-path kernels on graphs. In: Fifth IEEE International Conference on Data Mining, pp. 74–81 (2005)
3. Chang, C.C., Lin, C.J.: Libsvm: a library for support vector machines. ACM Trans. Intell. Syst. Technol. **2**(3), 1–27 (2011)
4. Fee, C., Banasr, M., Sibille, E.: Somatostatin-positive gamma-aminobutyric acid interneuron deficits in depression: Cortical microcircuit and therapeutic perspectives. Biol. Psychiat. **82**(8), 549–559 (2017)
5. Feragen, A., Kasenburg, N., Petersen, J., De Bruijne, M., Borgwardt, K.M.: Scalable kernels for graphs with continuous attributes. In: Advances in Neural Information Processing Systems, pp. 216–224 (2013)
6. Gärtner, T., Flach, P., Wrobel, S.: On graph kernels: hardness results and efficient alternatives. In: Schölkopf, B., Warmuth, M.K. (eds.) COLT-Kernel 2003. LNCS (LNAI), vol. 2777, pp. 129–143. Springer, Heidelberg (2003). https://doi.org/10.1007/978-3-540-45167-9_11
7. Hartmut, H., Thomas, H., Moll, G.H., Oliver, K.: A bimodal neurophysiological study of motor control in attention-deficit hyperactivity disorder: a step towards core mechanisms? Brain **4**, 1156–1166 (2014)
8. Ho, M.C., Shen, H.A., Chang, Y.P.E., Weng, J.C.: A CNN-based autoencoder and machine learning model for identifying betel-quid chewers using functional MRI features. Brain Sci. **11**(6), 809 (2021)
9. Jie, B., Liu, M., Lian, C., Shi, F., Shen, D.: Designing weighted correlation kernels in convolutional neural networks for functional connectivity based brain disease diagnosis. Med. Image Anal. **63**, 101709 (2020)
10. Kantorovitch, L.: On the translocation of masses. Manag. Sci. **5**(1), 1–4 (1958)
11. Kriege, N.M., Mutzel, P.: Subgraph matching kernels for attributed graphs. In: International Conference on Machine Learning, pp. 291–298 (2012)
12. Le, T., Yamada, M., Fukumizu, K., Cuturi, M.: Tree-sliced variants of Wasserstein distances. In: Advances in Neural Information Processing Systems, pp. 12283–12294 (2019)
13. Liu, M., Zhang, J., Adeli, E., Shen, D.: Joint classification and regression via deep multi-task multi-channel learning for Alzheimer's disease diagnosis. IEEE Trans. Biomed. Eng. **66**(5), 1195–1206 (2019)

14. Ma, K., Shao, W., Zhu, Q., Zhang, D.: Kernel based statistic: identifying topological differences in brain networks. Intell. Med. **2**(1), 30–40 (2022)
15. Ma, K., Yu, J., Shao, W., Xu, X., Zhang, Z., Zhang, D.: Functional overlaps exist in neurological and psychiatric disorders: a proof from brain network analysis. Neuroscience **425**, 39–48 (2020)
16. Morris, C., Kriege, N.M., Kersting, K., Mutzel, P.: Faster kernels for graphs with continuous attributes via hashing. In: 16th IEEE International Conference on Data Mining, pp. 1095–1100 (2016)
17. Nikolentzos, G., Meladianos, P., Vazirgiannis, M.: Matching node embeddings for graph similarity. In: Association for the Advance of Artificial Intelligence, pp. 2429–2435 (2017)
18. Rishi, S., Anna, C.G.: Tree! i am no tree! i am a low dimensional hyperbolic embedding. In: Advances in Neural Information Processing Systems, vol. 33, pp. 845–856 (2020)
19. Rubinov, M., Sporns, O.: Complex network measures of brain connectivity: uses and interpretations. Neuroimage **52**(3), 1059–1069 (2010)
20. Shervashidze, N., Schweitzer, P., Jan, E., Leeuwen, V., Borgwardt, K.M.: Weisfeiler-lehman graph kernels. J. Mach. Learn. Res. **12**(3), 2539–2561 (2011)
21. Tian, Y., Zhao, L., Peng, X., Metaxas, D.N.: Rethinking kernel methods for node representation learning on graphs. In: Advances in Neural Information Processing Systems, pp. 11686–11697 (2019)
22. Togninalli, M., Ghisu, E., Llinares-Lpez, F., Rieck, B., Borgwardt, K.: Wasserstein weisfeiler-lehman graph kernels. In: Advances in Neural Information Processing Systems, pp. 6439–6449 (2019)
23. Vishwanathan, S.V.N., Schraudolph, N.N., Kondor, R., Borgwardt, K.M.: Graph kernels. J. Mach. Learn. Res. **11**(2), 1201–1242 (2010)
24. Ye, W., Wang, Z., Redberg, R., Singh, A.: Tree++: truncated tree based graph kernels. IEEE Trans. Knowl. Data Eng. **33**(4), 1778–1789 (2021)
25. Yu, Q., Sui, J., Kiehl, K.A., Pearlson, G.D., Calhoun, V.D.: State-related functional integration and functional segregation brain networks in schizophrenia. Schizophr. Res. **150**(2), 450–458 (2013)
26. Zhang, D., Huang, J., Jie, B., Du, J., Tu, L., Liu, M.: Ordinal pattern: a new descriptor for brain connectivity networks. IEEE Trans. Med. Imaging **37**(7), 1711–1722 (2018)
27. Zhang, Z., Wang, M., Xiang, Y., Huang, Y., Nehorai, A.: Retgk: graph kernels based on return probabilities of random walks. In: Advances in Neural Information Processing Systems, pp. 3964–3974 (2018)
28. Zhao, P., Zhou, Z.: Label distribution learning by optimal transport. In: Association for the Advancement of Artificial Intelligence, pp. 4506–4513 (2018)
29. Zheng, W., Yan, L., Zhang, W., Gou, C., Wang, F.: Guided cyclegan via semi-dual optimal transport for photo-realistic face super-resolution. In: International Conference on Image Processing, pp. 2851–2855 (2019)

Dynamic Bank Learning for Semi-supervised Federated Image Diagnosis with Class Imbalance

Meirui Jiang[1], Hongzheng Yang[2], Xiaoxiao Li[3], Quande Liu[1],
Pheng-Ann Heng[1], and Qi Dou[1(✉)]

[1] Department of Computer Science and Engineering,
The Chinese University of Hong Kong, Sha Tin, Hong Kong
qidou@cuhk.edu.hk
[2] Department of Artificial Intelligence, Beihang University, Beijing, China
[3] Department of Electrical and Computer Engineering,
The University of British Columbia, Vancouver, Canada

Abstract. Despite recent progress on semi-supervised federated learning (FL) for medical image diagnosis, the problem of imbalanced class distributions among unlabeled clients is still unsolved for real-world use. In this paper, we study a practical yet challenging problem of class imbalanced semi-supervised FL (imFed-Semi), which allows all clients to have only unlabeled data while the server just has a small amount of labeled data. This imFed-Semi problem is addressed by a novel dynamic bank learning scheme, which improves client training by exploiting class proportion information. This scheme consists of two parts, i.e., the dynamic bank construction to distill various class proportions for each local client, and the sub-bank classification to impose the local model to learn different class proportions. We evaluate our approach on two public real-world medical datasets, including the intracranial hemorrhage diagnosis with 25,000 CT slices and skin lesion diagnosis with 10,015 dermoscopy images. The effectiveness of our method has been validated with significant performance improvements (7.61% and 4.69%) compared with the second-best on the accuracy, as well as comprehensive analytical studies. Code is available at https://github.com/med-air/imFedSemi.

Keywords: Semi-supervised FL · Class imbalance · Image diagnosis

1 Introduction

Federated learning (FL), i.e., collaboratively training a model with decentralized datasets, is increasingly important for medical image diagnosis [4,5,13,20,21, 28]. To significantly enlarge cohorts, developing semi-supervised FL is crucial to include massive unlabeled data from different training clients [10,29,31]. Though some progress on semi-FL has been made for image diagnosis tasks [2,14], there are still key challenges remaining to be solved. First, existing methods typically

© The Author(s), under exclusive license to Springer Nature Switzerland AG 2022
L. Wang et al. (Eds.): MICCAI 2022, LNCS 13433, pp. 196–206, 2022.
https://doi.org/10.1007/978-3-031-16437-8_19

assume one or several client(s) to be fully labeled, without allowing the extreme yet practically favorable situation that *all* clients are unlabeled. Second, *class imbalance* issue among clients (due to different patient demographics and disease incidence rate [19,21]) hinders model accuracy, yet not been carefully considered.

In this regard, we address a new problem setting of class imbalanced semi-supervised FL (named as *imFed-Semi*), in which we assume that only the server holds a small amount of labeled data, while all clients only provide unlabeled data with the presence of class imbalance. This is a difficult problem, but closer to real-world use compared with previous ideal ones. Existing semi-supervised methods typically apply the consistency regularization [8,14,23,24] or use the pseudo-labeling [1,2,29,31] on unlabeled samples. For instance, FedPerl [2] obtains pseudo labels for unlabeled data by ensembling models for a cluster of similar clients, and FedIRM [14] imposes the consistency regularization by adding a class relation constraint between labeled and unlabeled clients. Unfortunately, these methods require some clients to be partially or fully labeled. Moreover, they suffer from performance drop upon the class imbalance problem, i.e., unlabeled clients have different proportions of samples from each disease category. Such a limitation is because that, the sample-level supervision on unlabeled data leads to each client model training to be locally dominated by its own majority class(es), therefore affecting the model aggregation at the server in FL.

The key point to address the imFed-Semi lies in how to design class proportion-aware supervision to enhance training for unlabeled clients. Our insight is to rely on labeled data at the server to help unlabeled clients for estimating class-specific proportions. Inspired by the learning from label proportion works [6,15,17], we can arrange data into several subsets and leverages the information of label proportion of subsets to weakly supervise the unlabeled client model training. In other words, we can split data inside the client to distill various label proportions from pseudo labels, because it makes more sense to use the soft global information instead of unreliable sample-level pseudo labels to train the model, especially when all samples in the client are unlabeled. Then, how to obtain an accurate proportion estimation becomes an important step, directly using pseudo labels is unreliable due to misclassification issues on hard samples.

In this paper, we propose a novel dynamic bank learning method for the imFed-Semi problem. Our method consists of two parts, the dynamic bank construction to extract various class proportion information within each client, and the sub-bank classification to enforce the local model to learn different class proportions. Specifically, the dynamic bank iteratively collects highly-confident samples during the training to estimate the client class distribution, and splits samples into sub-banks with the presence of different pseudo label proportions. Furthermore, a prior transition function is designed to transform the original classification task into the sub-bank classification task, which explicitly utilizes different class proportions to train the local model. Owing to such label-proportion-awared supervision, the local client training is enhanced to learn different distributions of imbalanced classes to avoid being dominated by the local majority class(es). We validate our method on two large-scale real-world medical datasets, including the intracranial hemorrhage diagnosis with 25,000 CT slices

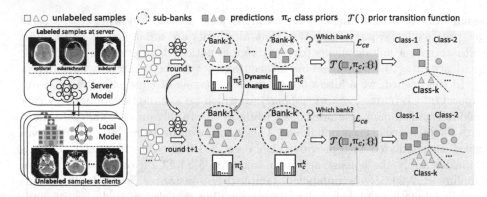

Fig. 1. Method overview. Our approach constructs sub-banks to collect confident samples and estimate class priors (π_c^k) for each client c. Then a sub-bank classification task is designed via learning different class proportions using the prior transition function. Finally, local model is promoted to learn discriminative decision boundaries.

and the skin lesion diagnosis with 10,015 dermoscopy images. The effectiveness of our method has been validated with significant performance improvements on both tasks compared with a number of state-of-the-art semi-supervised learning and FL methods, as well as comprehensive analytical studies.

2 Method

2.1 Overview of the imFed-Semi Framework

Hereby, we formulate the semi-supervised federated classification task with class imbalance. Figure 1 is an overview of our proposed imFed-Semi solution. We consider a K-class classification problem with input space $\mathcal{X} \subset \mathbb{R}^d$ and label space $\mathcal{Y}$, where d is the input dimension. Let $\boldsymbol{x} \in \mathcal{X}$ and $y \in \mathcal{Y}$ be the input and output following an underlying joint distribution with density $p(\boldsymbol{x}, y)$, which is identified by the class priors $\{\pi^k = p(y = k)\}_{k=1}^K$ and the class-conditional densities $\{p(\boldsymbol{x} \mid y = k)\}_{k=1}^K$. We have n_s labeled samples $\mathcal{S} = \{(\boldsymbol{x}_i^s, y_i^s)\}_{i=1}^{n_s}$ at server, and for C clients, each client $c \in [C]$ has n_c ($n_c \gg n_s$) unlabeled samples $\mathcal{U}_c = \{(\boldsymbol{x}_i^c)\}_{i=1}^{n_c}$. The goal of the classification task is to obtain a global FL model $f_g : \mathcal{X} \to \mathcal{Y}$ by using several labeled samples at the server and large-scale unlabeled data with the presence of such class-imbalanced clients.

The FL training paradigm of our framework adopts the popular FedAvg [16] algorithm, where the server collects client models f_c and aggregates them to obtain the global model $f_g = \sum_{c=1}^C \frac{n_c}{N} f_c$, where $N = \sum_{c=1}^C n_c$. Then, the global model is broadcast back to clients for local training, and such a process repeats until convergence. The special practice in our label-at-server setting is, before broadcasting, the global model is updated for an extra round of gradient descents with labeled samples at the server. Overall, the FL loss function is written as:

$$\mathcal{L}_{overall} = \mathcal{L}_s(f_g(\boldsymbol{x}^s), y^s) + \frac{1}{C} \sum_{c=1}^C \mathcal{L}_u(f_g(\boldsymbol{x}^c)), \tag{1}$$

where the global model f_g performs supervised learning on labeled sample pairs $\mathcal{S} = \{(\boldsymbol{x}_i^s, y_i^s)\}_{i=1}^{n_s}$ by minimizing the cross-entropy loss $\mathcal{L}_s$ at server. The local clients perform unsupervised learning which minimizes $\mathcal{L}_u$ (cf. Eq. (6)) on unlabeled inputs $\mathcal{U}_c = \{(\boldsymbol{x}_i^c)\}_{i=1}^{n_c}$ with our designed new scheme in the following.

2.2 Dynamic Bank Construction for Unlabeled Clients

Considering current semi-supervised methods are prone to make the local training dominated by majority class(es) at each client, we instead encourage the FL model to learn discriminative decision boundaries by taking advantage of different class proportions from a global perspective. Inspired by learning from class proportions of unlabeled samples [15], we design a *dynamic bank learning* scheme, which estimates the class proportions for client training. Specifically, we first build a dynamic bank to collect confident samples based on the thresholding of high prediction probabilities. Then, we split the dynamic bank into K sub-banks to present different class proportions locally. By further designing a class prior transition function, we convert the K-class classification problem into K-bank classification via learning different class proportions.

Dynamic Bank Construction. We build the dynamic bank by progressively storing confident samples as training goes on. Given an unlabeled sample $\boldsymbol{x}_i^c$ at each client, we denote p_i (c is omitted for ease of notation) as the highest prediction probability across all classes by the local model, i.e., $p_i = \max f_c(\boldsymbol{x}_i^c)$. We use a threshold τ_α to select confident samples exceeding a high probability. In addition, due to class imbalance, the minority class(es) tend to be underrepresented [22], leading to lower prediction probability by the local model. We further use another threshold τ_β to rescue underrepresented unlabeled samples, which helps maintain the class diversity. With these considerations, we design the following threshold scheme to dynamically collect samples:

$$\mathcal{B}_{c,t} = \mathcal{B}_{c,t-1} \setminus \{\mathbb{1}_{(p_i^{t-1} < \tau_\alpha)} \cdot \boldsymbol{x}_i^c\}_{i=1}^{n_c} \cup \{\mathbb{1}_{(p_i^t > \tau_\beta)} \cdot \boldsymbol{x}_i^c\}_{i=1}^{n_c}. \tag{2}$$

Specifically, $\mathcal{B}_{c,t}$ is the bank for client c at the t-th communication round and p_i^t is predicted probability at the round t. The bank is initialized as an empty set at the beginning, i.e., $\mathcal{B}_{c,0} = \phi$. Before each round, the bank collects samples with probability exceeding the threshold τ_β for training. Then it is adjusted by only preserving confident samples with prediction probability exceeding the τ_α. By building the bank in such a dynamic way, we can gradually gather and make use of more samples at each round to estimate class distributions. In particular, almost all samples are promised to be included in the bank as training goes on, which helps the local model to get rid of neglecting minor classes.

Class Prior Estimation. We further use the proportion of classes in the constructed dynamic bank to provide supervision for the training of unlabeled clients. The idea is to emphasize the class prior knowledge by proposing an auxiliary task of sub-bank classification, which explicitly connects the original classification task with class priors. The ground-truth for the sub-bank classification task can be easily obtained during training, thus providing more reliable supervision than conventional pseudo-labeling. In practice, for each bank $\mathcal{B}_c$ of client c, we randomly split it to K non-overlapping sub-banks, i.e., $\mathcal{B}_c = \{\mathcal{B}_c^m\}_{m=1}^K$, where $\mathcal{B}_c^m = \{\boldsymbol{x}_i^{c,m}\}_{i=1}^{n_{c,m}}$ is the m-th sub-bank with sample size $n_{c,m}$. Notably, we impose that not all sub-banks have exactly the same class distribution, considering diverse training cases for real-world applications. Denoting $\bar{y}$ as the index of these K sub-banks, we utilize the index as proxy labels to transform unlabeled samples to labeled pairs $\mathcal{B}_c = \{(\boldsymbol{x}_i^c, \bar{y}_i)\}_{i=1}^{n_b} \sim \bar{p}_c(\boldsymbol{x}, \bar{y})$, where n_b is the bank size and $\bar{p}_c(\boldsymbol{x}, \bar{y})$ is the underlying joint distribution for the random variables $\boldsymbol{x} \in \mathcal{X}$ and $\bar{y} \in [K]$. Since we assume the original input and output follow the distribution $p(\boldsymbol{x}, y)$, which can be identified by class priors and class-conditional densities, the data samples of each sub-bank can be seen as drawn from a mixture of original class-conditional densities as follows:

$$\mathcal{B}_c^m \sim p_c^m(\boldsymbol{x}) = \sum_{k=1}^K \pi_c^{m,k} p(\boldsymbol{x} \mid y = k), \tag{3}$$

where $\pi_c^{m,k}$ denotes the class prior of class k at the m-th sub-bank. We can calculate the class prior via pseudo labels from samples in the sub-bank:

$$\pi_c^{m,k} = \sum_{i=1}^{n_{c,m}} \mathbb{1}_{[\arg\max(p_i)=k]} / n_{c,m}. \tag{4}$$

For the prior regarding proxy label m, we have $\bar{\pi}_c^m = \bar{p}_c(\bar{y} = m)$, which can be estimated as $\bar{\pi}_c^m = n_{c,m}/n_b$. Comparing Eq. (3) with $\mathcal{B}_c \sim \bar{p}_c(\boldsymbol{x}, \bar{y})$, we have the class-conditional density $\bar{p}_c(\boldsymbol{x} \mid \bar{y} = m)$ corresponds to the original density $p_c^m(\boldsymbol{x})$, thereby connecting the proxy labels with class labels. In the following, we will design a prior transition function to convert the original classification task into a new sub-bank classification for the local client training.

2.3 Model Training on Unlabeled Client

To rely on class priors to supervise client training, we introduce the prior transition function to explicitly connect sub-bank priors and class priors. Denote g_c as the model to perform the sub-bank classification for client c. Let $g_c(\boldsymbol{x})_m = \bar{p}_c(\bar{y} = m \mid \boldsymbol{x})$ and $f_c(\boldsymbol{x})_k = p(y = k \mid \boldsymbol{x})$, by the Bayes' rule, for $\forall m \in [K]$:

$$\begin{aligned}
g_c(\boldsymbol{x})_m &= \frac{\bar{p}_c(\boldsymbol{x}, \bar{y} = m)}{\bar{p}_c(\boldsymbol{x})} = \frac{\bar{p}_c(\boldsymbol{x} \mid \bar{y} = m)\bar{p}_c(\bar{y} = m)}{\sum_{m=1}^K \bar{p}_c(\boldsymbol{x} \mid \bar{y} = m)\bar{p}_c(\bar{y} = m)} \\
&= \frac{p_c^m(\boldsymbol{x})\bar{\pi}_c^m}{\sum_{m=1}^K p_c^m(\boldsymbol{x})\bar{\pi}_c^m} = \frac{\bar{\pi}_c^m \sum_{k=1}^K \pi_c^{m,k} f_c(\boldsymbol{x})_k / \pi^k}{\sum_{m=1}^K \bar{\pi}_c^m \sum_{k=1}^K \pi_c^{m,k} f_c(\boldsymbol{x})_k / \pi^k},
\end{aligned} \tag{5}$$

where the class prior π^k can be estimated by adding up all class priors of sub-banks. By transforming the priors in Eq. (5) into matrix form, we define the prior transition function as $\mathcal{T}(f_c(\boldsymbol{x}^c); \Pi_c, \bar{\boldsymbol{\pi}}_c, \boldsymbol{\pi})$, where $\Pi_c \in \mathbb{R}^{K \times K}$ is the matrix of all $\pi_c^{m,k}$, and $\bar{\boldsymbol{\pi}}_c = [\bar{\pi}_c^1, \ldots, \bar{\pi}_c^K]$, $\boldsymbol{\pi} = [\pi^1, \ldots, \pi^K]$. The prior transition function naturally bridges the class probability and the proxy labels. We could learn the class discrimination knowledge via the sub-bank classification. Given unlabeled client data x^c and proxy labels $\bar{y}^c$, we use the prior transition function to perform the sub-bank classification with the cross-entropy loss:

$$\mathcal{L}_u = \mathcal{L}_{ce}(\mathcal{T}(f_c(x^c); \Pi_c, \bar{\pi}_c, \pi), \bar{y}^c). \tag{6}$$

Therefore, the overall training loss in Eq. (1) is completed with the new form of $\mathcal{L}_u$, which is class distribution-aware and subject to each unlabeled client.

3 Experiments

3.1 Dataset and Experiment Setup

Datasets. We evaluate our method on two medical image classification tasks: 1) intracranial hemorrhage (ICH) diagnosis for brain CT slices, 2) skin lesion diagnosis for dermoscopy images. For ICH diagnosis, we use the RSNA ICH dataset [7] and follow the FedIRM [14] to randomly sample 25,000 slices which contains 5 subtypes of ICH. For skin lesion diagnosis, we use the HAM10000 dataset [25], which contains 10,015 dermoscopy images with 7 skin lesion subtypes. For both datasets, we use 70% for training, 10% for validation, and 20% for test. For data pre-processing, we apply random transformations of rotation, translation, and flipping on 2D images, resize the images into 224×224 and normalize them before feeding into an ImageNet [3] pre-trained model.

Experiment Setup. We set the number of clients C to 10. For each class of server labeled samples S_k, we set $S_k = 15$ for ICH diagnosis and $S_k = 10$ for skin lesion diagnosis, making the number of labeled data as 75 and 70 respectively. Following previous studies [11,26,30], we use Dirichlet distribution to simulate the imbalanced classes i.e., sampling $p_k \sim Dir_N(\gamma)$ with $\gamma = 1.5$, and allocating a $p_{k,c}$ proportion of instances of class k to client c. This yields the proportion of each class in client c in the range $[0.05, 0.5]$. We select the best global model from the validation and report the test performance regarding five metrics including AUC, Accuracy, Specificity, Sensitivity, and F1 score over 3 independent runs.

Implementation Details. Following the top-rank solution [27], we use the DenseNet121 [9]. To make a fair comparison, all compared methods are trained on both server and client data, and follow the FedIRM [14] to apply the warming up for 30 epochs. The thresholds τ_β and τ_α are set to 0.5 and 0.9. We use the Adam optimizer with momentum terms of 0.9 and 0.99, and the batch size is 24. The total communication rounds is 200 with the local training epoch set as 1.

3.2 Comparison with State-of-the-Art Methods

We compare our method with recent state-of-the-art semi-supervised FL methods, including the **FSSL** (MIA'21) [29] which introduces consistency loss and pseudo labeling into FL, the **FedIRM** (MICCAI'21) [14] which enhances the consistency regularization with an inter-client relation matching, and the **Fed-Match** (ICLR'21) [10] which applies inter-client consistency and decomposes

Table 1. Results comparison with state-of-the-art methods on two diagnosis tasks.

Methods	AUC	Accuracy	Specificity	Sensitivity	F1
Intracranial hemorrhage diagnosis					
FedAvg-SL	88.59 ± 0.85	91.22 ± 0.17	93.39 ± 0.09	63.64 ± 0.80	59.54 ± 0.78
FedIRM [14]	60.41 ± 0.93	72.27 ± 0.40	82.93 ± 0.29	30.13 ± 0.55	22.93 ± 0.31
FedMatch [10]	64.15 ± 1.76	73.09 ± 0.55	84.01 ± 0.19	33.44 ± 2.60	24.84 ± 1.36
FSSL [29]	60.63 ± 1.92	72.61 ± 1.47	83.57 ± 0.79	24.88 ± 0.39	21.98 ± 0.93
FedAvg-FM	61.54 ± 1.63	74.50 ± 1.31	86.12 ± 1.79	25.01 ± 0.55	22.55 ± 0.07
FedProx-FM [12]	62.49 ± 3.41	74.95 ± 3.03	85.91 ± 3.05	26.18 ± 5.24	22.84 ± 1.55
FedAdam-FM [18]	62.42 ± 3.93	73.85 ± 0.66	85.63 ± 1.15	25.33 ± 3.62	22.08 ± 1.02
imFed-Semi (Ours)	$\mathbf{82.96 \pm 1.26}$	$\mathbf{82.56 \pm 0.58}$	$\mathbf{90.66 \pm 0.21}$	$\mathbf{54.58 \pm 3.75}$	$\mathbf{47.71 \pm 3.59}$
Skin lesion diagnosis					
FedAvg-SL	87.58 ± 0.28	93.32 ± 0.27	89.64 ± 0.57	59.09 ± 1.34	57.09 ± 0.68
FedIRM [14]	65.29 ± 2.26	79.05 ± 1.33	90.39 ± 1.08	28.72 ± 2.22	23.52 ± 1.21
FedMatch [10]	70.90 ± 1.25	84.25 ± 2.34	$\mathbf{93.33 \pm 1.74}$	29.56 ± 1.90	29.13 ± 2.69
FSSL [29]	70.86 ± 1.26	83.20 ± 1.65	93.39 ± 0.21	28.32 ± 0.89	27.90 ± 1.43
FedAvg-FM	70.61 ± 1.79	82.67 ± 0.91	91.92 ± 1.67	30.65 ± 0.91	29.09 ± 0.94
FedProx-FM [12]	69.86 ± 1.48	82.01 ± 1.66	91.45 ± 2.86	27.87 ± 2.69	25.21 ± 1.14
FedAdam-FM [18]	70.58 ± 1.86	83.22 ± 2.25	92.92 ± 1.98	28.97 ± 1.87	27.85 ± 1.68
imFed-Semi (Ours)	$\mathbf{77.47 \pm 1.81}$	$\mathbf{88.94 \pm 1.50}$	89.81 ± 2.64	$\mathbf{37.48 \pm 2.71}$	$\mathbf{33.79 \pm 1.75}$

the model parameter for server and client training. Moreover, we incorporate the widely-used semi-supervised method (FixMatch [24], abbreviated as FM) into the baseline FedAvg [16] and existing FL algorithms for the class imbalance problem, including the **FedProx** [12] (MLSys'20) and the **FedAdam** (ICLR'21) [18].

Table 1 lists the results, where FedAvg-SL denotes the supervised learning with full labels on all clients (upper bound). It can be observed that all semi-supervised methods present a great performance drop without sufficient supervision. FedMatch, with the model decomposition on top of the consistency, outperforms other consistency-based semi-supervised FL methods on all metrics. The combination of FM and FL methods shows improvements on the accuracy. Compared with all these methods, our approach achieves the best performance on almost all metrics, with 7.61% and 4.69% accuracy boost on both tasks over the second-best. The significant improvement is benefited from our dynamic bank learning scheme, which effectively provides the class distribution-specific supervision to alleviate the model being dominated by majority classes.

3.3   Analytical Studies on Key Components of the Approach

Learning Procedure with the Dynamic Bank. We study the learning behavior of our proposed method regarding the dynamic bank construction and class prior estimation on the ICH dataset. As shown in Fig. 2(a), two curves are the test accuracy and the estimation error (we use the Frobenius norm between

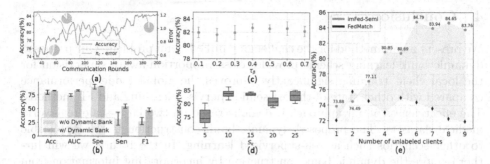

Fig. 2. Analysis of our approach: (a) learning behaviors of the dynamic bank, (b) ablation on the dynamic bank, (c) effect of different threshold τ_β, (d) model performance as increasing server's labeled samples, (e) effect of number of unlabeled clients.

the estimated and real class priors), and the pie chart denotes the percentage of data samples collected in the bank. In the beginning, the dynamic bank $\mathcal{B}_c$ gradually collects more confident samples, making the class prior estimation close to the real one and the accuracy is quickly improved (from 74% to 82%). With more samples collected, the model takes time to gradually fit different class proportions and the bank finally includes almost all samples (98%). The local training becomes stable with a high accuracy and low estimation error.

Effectiveness of the Dynamic Bank Construction. We further investigate the dynamic selection by comparing with a fixed threshold (i.e., 0.9, which is widely adopted). As shown in Fig. 2(b), the dynamic banks improve the performance on all metrics with smaller standard deviations. Especially, it has 20% boost on both sensitivity and F1-score. Such improvements benefit from the dynamic way of gathering samples, which helps the model train on diverse classes, thus avoiding fitting to majority classes only. Besides, we study the choice of different thresholds τ_β, which can affect the confidence for selecting minority classes. As shown in Fig. 2(c), the change of τ_β does not affect the test accuracy significantly (less than 1%), indicating that the way of dynamic construction is the key factor to improve the performance.

Studies on Scalability with Labeled/Unlabeled Samples. We finally analyze the scalability of our method with different numbers of labeled samples at the server and different unlabeled client numbers. The results are shown in Fig. 2(d) and (e), respectively. For the change of labeled samples, the global model suffers a great performance drop with very few labeled samples, i.e., $S_k = 5$. With more samples available, the test accuracy remains stable in a certain range, demonstrating that the performance bottleneck lies in client-side learning. For the scalability of unlabeled clients, the class imbalance issue may be enlarged with more clients involved, thereby making client training more difficult. Notably, we find that our method benefits more from the increased unlabeled clients, while the FedMatch might endure a performance drop after introducing more clients.

4   Conclusion

We present a new method for the challenging imFed-Semi problem. Our proposed dynamic bank learning scheme provides class proportion-aware supervision for the local client training, significantly improves the global model performance compared with other state-of-the-art semi-supervised learning and FL methods. The effectiveness of our approach is demonstrated on two large-scale real-world medical datasets. The proposed dynamic bank construction is also applicable to other scenarios such as self-supervised learning. In the future, we will further explore the dynamic bank construction by incorporating information from other clients to solve a potential limitation for handling the more severe class imbalance, where each local client may not cover all possible disease categories.

Acknowledgement. This work was supported in part by the Hong Kong Innovation and Technology Fund (Projects No. ITS/238/21 and No. GHP/110/19SZ), in part by the CUHK Shun Hing Institute of Advanced Engineering (project MMT-p5-20), in part by the Shenzhen-HK Collaborative Development Zone, and in part by NSERC Discovery Grant (DGECR-2022-00430).

References

1. Bai, W., et al.: Semi-supervised learning for network-based cardiac MR image segmentation. In: Descoteaux, M., Maier-Hein, L., Franz, A., Jannin, P., Collins, D.L., Duchesne, S. (eds.) MICCAI 2017. LNCS, vol. 10434, pp. 253–260. Springer, Cham (2017). https://doi.org/10.1007/978-3-319-66185-8_29
2. Bdair, T., Navab, N., Albarqouni, S.: FedPerl: semi-supervised peer learning for skin lesion classification. In: de Bruijne, M., et al. (eds.) MICCAI 2021. LNCS, vol. 12903, pp. 336–346. Springer, Cham (2021). https://doi.org/10.1007/978-3-030-87199-4_32
3. Deng, J., Dong, W., Socher, R., Li, L.J., Li, K., Fei-Fei, L.: ImageNet: a large-scale hierarchical image database. In: CVPR, pp. 248–255. IEEE (2009)
4. Dong, N., Voiculescu, I.: Federated contrastive learning for decentralized unlabeled medical images. In: de Bruijne, M., et al. (eds.) MICCAI 2021. LNCS, vol. 12903, pp. 378–387. Springer, Cham (2021). https://doi.org/10.1007/978-3-030-87199-4_36
5. Dou, Q., et al.: Federated deep learning for detecting COVID-19 lung abnormalities in CT: a privacy-preserving multinational validation study. NPJ Digit. Med. **4**(1), 1–11 (2021)
6. Dulac-Arnold, G., Zeghidour, N., Cuturi, M., Beyer, L., Vert, J.P.: Deep multi-class learning from label proportions. arXiv preprint arXiv:1905.12909 (2019)
7. Flanders, A.E., et al.: Construction of a machine learning dataset through collaboration: the RSNA 2019 brain CT hemorrhage challenge. Radiol. Artif. Intell. **2**(3), e190211 (2020)
8. Gyawali, P.K., Ghimire, S., Bajracharya, P., Li, Z., Wang, L.: Semi-supervised medical image classification with global latent mixing. In: Martel, A.L., et al. (eds.) MICCAI 2020. LNCS, vol. 12261, pp. 604–613. Springer, Cham (2020). https://doi.org/10.1007/978-3-030-59710-8_59

9. Huang, G., Liu, Z., Van Der Maaten, L., Weinberger, K.Q.: Densely connected convolutional networks. In: CVPR, pp. 4700–4708 (2017)
10. Jeong, W., Yoon, J., Yang, E., Hwang, S.J.: Federated semi-supervised learning with inter-client consistency & disjoint learning. In: ICLR (2021). https://openreview.net/forum?id=ce6CFXBh30h
11. Li, Q., He, B., Song, D.: Model-contrastive federated learning. In: CVPR, pp. 10713–10722 (2021)
12. Li, T., Sahu, A.K., Zaheer, M., Sanjabi, M., Talwalkar, A., Smith, V.: Federated optimization in heterogeneous networks. Proc. Mach. Learn. Syst. **2**, 429–450 (2020)
13. Li, X., Jiang, M., Zhang, X., Kamp, M., Dou, Q.: FedBN: federated learning on non-IID features via local batch normalization. In: ICLR (2021). https://openreview.net/forum?id=6YEQUn0QICG
14. Liu, Q., Yang, H., Dou, Q., Heng, P.-A.: Federated semi-supervised medical image classification via inter-client relation matching. In: de Bruijne, M., et al. (eds.) MICCAI 2021. LNCS, vol. 12903, pp. 325–335. Springer, Cham (2021). https://doi.org/10.1007/978-3-030-87199-4_31
15. Lu, N., Wang, Z., Li, X., Niu, G., Dou, Q., Sugiyama, M.: Unsupervised federated learning is possible. In: ICLR (2022). https://openreview.net/forum?id=WHA8009laxu
16. McMahan, B., Moore, E., Ramage, D., Hampson, S., Arcas, B.A.: Communication-efficient learning of deep networks from decentralized data. In: Artificial Intelligence and Statistics, pp. 1273–1282 (2017)
17. Quadrianto, N., Smola, A.J., Caetano, T.S., Le, Q.V.: Estimating labels from label proportions. JMLR **10**(10), 2349–2374 (2009)
18. Reddi, S.J., et al.: Adaptive federated optimization. In: ICLR (2021). https://openreview.net/forum?id=LkFG3lB13U5
19. Rieke, N., et al.: The future of digital health with federated learning. NPJ Digit. Med. **3**(1), 1–7 (2020)
20. Roth, H.R., et al.: Federated learning for breast density classification: a real-world implementation. In: Albarqouni, S., et al. (eds.) DART/DCL -2020. LNCS, vol. 12444, pp. 181–191. Springer, Cham (2020). https://doi.org/10.1007/978-3-030-60548-3_18
21. Sheller, M.J., et al.: Federated learning in medicine: facilitating multi-institutional collaborations without sharing patient data. Sci. Rep. **10**(1), 1–12 (2020)
22. Shen, Z., Cervino, J., Hassani, H., Ribeiro, A.: An agnostic approach to federated learning with class imbalance. In: ICLR (2022). https://openreview.net/forum?id=Xo0lbDt975
23. Shi, X., Su, H., Xing, F., Liang, Y., Qu, G., Yang, L.: Graph temporal ensembling based semi-supervised convolutional neural network with noisy labels for histopathology image analysis. MIA **60**, 101624 (2020)
24. Sohn, K., et al.: Fixmatch: simplifying semi-supervised learning with consistency and confidence. NeurIPS **33**, 596–608 (2020)
25. Tschandl, P., Rosendahl, C., Kittler, H.: The HAM10000 dataset, a large collection of multi-source dermatoscopic images of common pigmented skin lesions. Sci. Data **5**(1), 1–9 (2018)
26. Wang, H., Yurochkin, M., Sun, Y., Papailiopoulos, D., Khazaeni, Y.: Federated learning with matched averaging. In: ICLR (2020). https://openreview.net/forum?id=BkluqlSFDS

27. Wang, X., et al.: A deep learning algorithm for automatic detection and classification of acute intracranial hemorrhages in head CT scans. NeuroImage Clin. **32**, 102785 (2021)
28. Wu, Y., Zeng, D., Wang, Z., Shi, Y., Hu, J.: Federated contrastive learning for volumetric medical image segmentation. In: de Bruijne, M., et al. (eds.) MICCAI 2021. LNCS, vol. 12903, pp. 367–377. Springer, Cham (2021). https://doi.org/10.1007/978-3-030-87199-4_35
29. Yang, D., et al.: Federated semi-supervised learning for Covid region segmentation in chest CT using multi-national data from China, Italy, Japan. MIA **70**, 101992 (2021)
30. Yurochkin, M., Agarwal, M., Ghosh, S., Greenewald, K., Hoang, N., Khazaeni, Y.: Bayesian nonparametric federated learning of neural networks. In: ICML, pp. 7252–7261. PMLR (2019)
31. Zhang, Z., Yao, Z., Yang, Y., Yan, Y., Gonzalez, J.E., Mahoney, M.W.: Benchmarking semi-supervised federated learning. arXiv preprint arXiv:2008.11364 17 (2020)

Coronary R-CNN: Vessel-Wise Method for Coronary Artery Lesion Detection and Analysis in Coronary CT Angiography

Yu Zhang, Jun Ma, and Jing Li[✉]

Raysight Intelligent Medical Technology Co., Ltd., Shenzhen, China
jing.li@raysightmed.com

Abstract. In recent decades, coronary artery disease (CAD) is the leading cause of death worldwide. Therefore, automatic diagnostic methods are strongly necessary with the progressively increasing number of CAD patients. However, it is difficult for physicians to recognize the lesion from Coronary CT Angiography (CCTA) scans as the coronary plaques have complicated appearance and patterns. Previous studies are mostly based on the single image patch around a lesion, which are often limited by the field of view of the local sample patch. To address this problem, in this paper we propose a novel vessel-wise object detection method. Different with previous approaches, we directly input the whole curved planar reformation (CPR) volume along the coronary artery centerline into our deep learning network, and then predict the plaque type and stenosis degree simultaneously. This enables the network to learn the dependencies between distant locations. In addition, two cascade modules are used to decompose the challenging problem into two simpler tasks and this also yields better interpretability. We evaluated our method on a dataset of 1031 CCTA images. The experimental results demonstrated the efficacy of our presented approach.

Keywords: Automatic lesion detection · Stenosis analysis · Plaque classification · CCTA

1 Introduction

In recent decades, coronary stenosis generated by atherosclerosis plaque is the leading cause of myocardial ischemia, and the patient would suffer the acute coronary syndromes when the plaque rapture occurs [1]. Therefore, detection of coronary artery plaque and stenosis as early as possible is the key to lower the risk of CAD. CCTA as a non-invasive way of imaging, is widely used in early screening of CAD. Besides, CCTA has been substantiated to be an effective way for detecting coronary lesions [2]. However, it is tedious and time-consuming for physicians to analyze them, especially when a large volume of patients' CCTA images need to be diagnosed in some hospitals.

Supplementary Information The online version contains supplementary material available at https://doi.org/10.1007/978-3-031-16437-8_20.

© The Author(s), under exclusive license to Springer Nature Switzerland AG 2022
L. Wang et al. (Eds.): MICCAI 2022, LNCS 13433, pp. 207–216, 2022.
https://doi.org/10.1007/978-3-031-16437-8_20

Semi- or fully-automatic studies have been presented to tackle these challenges. According to the best of our knowledge, published methods can be roughly divided into three categories: intensity-based, lumen-based and point-based method.

Intensity-based algorithms were proposed to mainly analyze the intensity of plaques. Damini et al. [7] proposed a method using the scan-specific attenuation thresholds to classify the lumen, non-calcified plaque and calcified plaque. However, intensity-based methods are easily affected by the image artifacts like motion, blooming, and stair-step artifacts.

Some lumen-based methods firstly segmented the lumen region and then detected significant stenoses or plaques [3–6]. Those methods usually searched for the contours of the lumen and plaques using geometric models, e.g., level-set [8]. These lumen-based methods highly depend on the accuracy of contours and generality of the geometric models. It is worth noting that due to the CCTA resolution limitation and the complex plaque representation, it is a challenging task to identify the arterial wall precisely.

Point-based methods usually take classification or regression approaches to process the patches or slices extracted from coronary centerline points. A texture-based method [10] was introduced to detect the significant stenosis, in which multi-view 2D images of different angles (i.e., sagittal and coronal views) were input to extract texture information. In [11], radiomic features were firstly extracted and then a CNN network was applied for classifying significant stenosis, and the results were directly compared with the clinical decision of revascularization. Recently, [12] was the first to employ transformer in stenosis classification. It utilized 3D CNN to extract rich local semantic information and then the transformer blocks were used to associate these features to acquire global contextual information. More comprehensively, a multi-task recurrent CNN architecture was proposed to detect significant stenosis and characterize the plaque type simultaneously [13]. A 3D CNN was deployed to extract features along the coronary centerline and a RNN was performed to obtain the multi-task results subsequently. However, due to the limitation of patch size, these methods often suffer from the limited field of view.

In this paper, in order to tackle the pervious limitations, we propose a novel vessel-wise multi-task detection architecture. We are enlightened by the practice of how physicians analyze plaques. They first quickly view the entire vessel and find out the coarse locations of plaques, and then they will focus on these candidate regions as well as analyze stenosis degree and plaque type carefully. Unlike point-based methods which analyze the patches or slices extracted from coronary centerline points, our method directly feeds the entire vessel into the Coronary R-CNN network and exports the plaque region proposals, just like Faster R-CNN [14]. Our method consists of two cascading modules. Firstly, detection module searches the abnormal regions that may contain plaques. Secondly, the multi-head analysis module judges whether the proposal includes lesion, and if yes, then predicts its plaque type and stenosis degree.

This paper's contributions to automated diagnosis of CAD are summarized as follows:

1. We proposed a novel vessel-wise detection architecture inspired by Faster R-CNN [14]. To the best of our knowledge, it is the first application of this type of method on coronary artery lesion detection.

2. We proposed a multi-head analysis module that can predict not only plaque types but also the exact stenosis degree, rather than just significant stenosis classification.
3. Our method achieved an outstanding performance on a dataset consisting of 1031 CCTA images with 7961 vessels.

2   Method

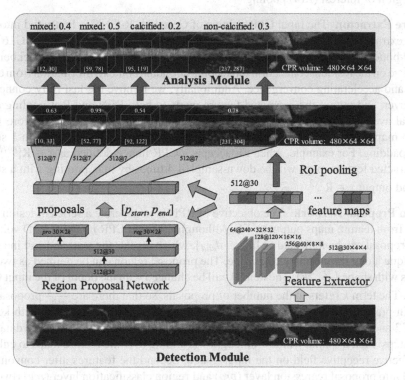

Fig. 1. A schematic illustration of two main modules of Coronary R-CNN whose input is a CPR volume with size 480 × 64 × 64.

As shown in Fig. 1, the proposed Coronary R-CNN can be divided into two main components: detection module and analysis module. The detection module is firstly used to obtain the plaque location and size. Instead of feeding a sequence of point-wise 2D slices or 3D volume cubes to the network, we directly input the entire curved planar reformation (CPR) image (vessel-wise) stacked from extracted coronary centerline, and arbitrary CPR length is acceptable. Besides, the vessel region may only occupy a small region of the cross-section view along the centerline, so we transformed the CPR image from Cartesian to polar coordinates that can enrich the context from every angle of the vessel. Then the analysis module is exploited to predict the stenosis degree and plaque type simultaneously. The proposed two modules decouple the difficult problem into several easier tasks, and meanwhile they can be employed end-to-end. Details will be explained as below.

2.1 Detection Module for Proposals Localization

Inspired by Faster R-CNN [14], we separate the challenging detection task into two crucial procedures. In most cases, the input image may contain far more normal or background information than regions with our critical concerns. Hence the first and the most essential module is to localize the abnormal proposals. The detection module consists of three primary parts: Feature Extractor (FE), Region Proposal Network (RPN) and Region of Interest (RoI) pooling.

Feature Extractor. The input CPR volume of Coronary R-CNN is firstly fed into this unit to extract features with convolutional neural network. Instead of using VGG16 [15] as backbone like the original paper [14], ResNet [16] is adopted as the FE backbone to capitalize on the residual connection for keeping abundant context information from early layers and accelerating convergence. Additionally, we remove the last fully-connected (FC) layer, and output the generated feature maps to both RPN and RoI pooling after a global average pooling on the last two channels. In Fig. 1, the number before @ is feature map channels, the conv-layers' (if not specified) kernel size is 3 with 1 stride and 1 padding. For example, when FE accepts a CPR image defined as $x \in \mathbf{R}^{D \times H \times W}$, the extracted feature maps will be down-sampled 4 times by max pooling with a stride of 2 and output $y \in \mathbf{R}^{512 \times \frac{D}{16}}$.

Region Proposal Network. The objective of RPN is to generate a series of lesion proposals from feature maps output by FE. Although the input CPR image is a 3D volume, each proposal can be indicated as $[d_{start}, d_{end}]$, which reveals the start and end index of the plaque along the coronary centerline. The proposal region can be defined as two 2D vectors with shape $\left(\frac{D}{16}, 2k\right)$. The item D will be changed with the depth of the input CPR image. The term k refers to the number of proposals, so that there are $\frac{D \times k}{16}$ proposals in total. In practice, RPN is a small network established by a 1D conv-layer with kernel size of 3 and two sibling FC layers to integrate local information globally. The dataflow has 2 steps: first, a spatial window of length 3 is sliding over the feature maps to enlarge the effective receptive field on the input image. Second the features after convolution are fed into proposal regression layer (*pro*) and region classification layer (*reg*) (two FC layers mentioned above) respectively. *Pro* represents the position of proposals and *reg* is a classifier to classify the proposals belonging to lesion or background.

RoI Pooling. This part solves the problem of fixed image size requirement for analysis module. After receiving the proposals from RPN and the extracted feature maps from FE, we firstly map the position index of lesion proposals generated by *pro* to the same scale ratio as 16x down-sampled features. Next, the features with different length in proposal region will be reshaped to fixed length by RoI pooling. Finally, the detection module outputs the filtered proposals and corresponding captured features to the next analysis module.

2.2 Multi-head Analysis Module for Plaque Analyzing

We develop a multi-head analysis module for analyzing the plaque type, the stenosis degree and the precise lesion region. Figure 2 illustrates the schematic of three crucial heads including plaque classification (*pla*), stenosis regression (*ste*) and RoI regression (*roi*). The proposals containing positions and features generated by the previous detection module are fed into the three heads simultaneously.

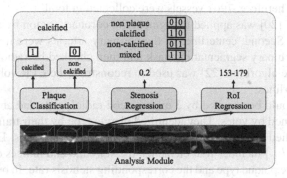

Fig. 2. Multi-head analysis module.

Plaque Classification. This head is utilized to classify a proposal as one of the four types: calcified, non-calcified, mixed or background. In contrast to most classification methods that use a FC layer to classify categories, we use two independent single FC layers to recognize calcified and non-calcified plaque, respectively. As shown in Fig. 2, we propose a novel two-digits binary code method to solve this classification problem. The first digit code indicates the existence of calcified plaque, and the second refers to non-calcified plaque. For each of them, "1" stands for presence of this plaque while "0" means absence. For example, "10" denotes calcified plaque, and "11" denotes mixed plaque. By this way, we simplify a difficult four type classification task to two straight-forward binary classification tasks. The ablation experiment results demonstrate the efficiency of this method.

Stenosis Regression. In previous publications [10–13], the stenosis analysis methods are mostly defined as a simple binary classification problem of whether significant stenosis (stenosis rate > 50%) existing or not. However, this may not fully meet the clinical needs in the real-world clinical practice. Generally, radiologists need to recognize the patients with intermediate stenosis (stenosis rate 30%–70%) to make further assessments [19], like invasive coronary angiography or CCTA–derived fractional flow reserve (CT-FFR). Therefore, we output the precise stenosis degree between [0, 1] by a FC layer in our *ste* part.

RoI Regression. This head is exploited to refine the coarse localization of lesion proposals from the detection module via a FC layer. The final output of roi are the refined start and end indexes of plaques along the CPR image centerline.

3 Data

We totally collected 1031 CCTA scans from patients with suspected CAD at 6 clinical centers in China. According to [17], we only extracted the centerlines of main coronary artery branches. Therefore, 7961 vessels were collected in total.

First, UNet++ [20] was applied to segment the coronary region by a trained model with annotations. Second, centerlines of the coronary arteries were initially extracted [21] based on the binary segmentation mask and then checked by radiologists. Finally, the marching cube algorithm [22] was used to reconstruct the CPR volumes along the direction of centerlines.

Every case would be checked by 3 experienced radiologists and any disagreement would be determined by votes. They identified plaque type by their trainable skills and calculated quantified stenosis rate by measurement on CCTA scans. Each annotation $([l_{start}, l_{end}], l_{pla}, l_{ste})$ includes four elements: the start and end indices of the CPR volume centerline, the plaque type and the corresponding stenosis rate. In our dataset, there were 3366 lesions including 788 calcified plaques, 1285 non-calcified plaques and 1293 mixed plaques. The length of plaque was between 16 to 349 points and mostly shorter than 100 points. Data augmentations were applied during training, including randomly rotation around the axial central axis, translation with a uniform random offset within ±3 points along the CPR centerline and vertical/horizontal flip.

4 Experiments and Results

We conducted a serious of experiments to evaluate the efficacy of our approach. First, we evaluated the effect of three different settings on the performance of Coronary R-CNN. Next, we compared our proposed method with previous works. We performed 5-fold cross-validation on patient-level for all experiments. Our models were trained with 400 epochs and were validated every 5 epochs, where 10% of training dataset was used for validation. The weights of models with best performance were saved for testing phase.

As for stenosis degree, we calculated the mean absolute error (MAE_{ste}) between regression results and ground truth. Regarding plaque type, we evaluated the classification accuracy (ACC_{pla}), F1-score of calcified ($F1_{calc}$), non-calcified ($F1_{non-calc}$), mixed ($F1_{mix}$) plaque and the average of them with background ($F1_{background}$). All the metrics mentioned above were measured on segment-level, which meant a coronary tree would be divided into 18 segments by SCCT 2014 [17] and these segments were the minimum units. For one segment, if more than half of the points were predicted correctly, the result of this segment was considered to be correct. Specifically, MAE_{ste} is defined as

$$MAE_{ste} = \frac{1}{N} \sum_i \left| p_{ste}^i - l_{ste}^i \right| \tag{1}$$

and ACC_{pla} can be represented as

$$ACC_{pla} = \frac{1}{N} \sum_i p_{pla}^i = = l_{pla}^i \tag{2}$$

where N indicates the total number of segments, $\{p^i_{ste}, p^i_{pla}\}$ and $\{l^i_{ste}, l^i_{pla}\}$ are the predicted and labeled {stenosis rate, plaque type} of the i_{th} segment.

$F1_{type} \in \{F1_{calc}, F1_{non-calc}, F1_{mix}, F1_{background}\}$ is calculated as

$$F1_{type} = \frac{2 \times \mathbf{p}_{pla} \times \mathbf{l}_{pla}}{\mathbf{p}_{pla} + \mathbf{l}_{pla}} \tag{3}$$

where $\mathbf{l}_{pla} = \{l^i_{pla} \in \{0, 1\} | 0 < i < K\}$ is a vector indicating the K segments are labeled as *type* plaque or not, and $\mathbf{p}_{pla}$ is the predicted vector with the same format of $\mathbf{l}_{pla}$. $F1_{all}$ equals to the average of F1-scores across all types.

4.1 Ablation Experiments

The investigated three settings combined with our baseline model: pretrained model, polar coordinates and two-digits binary code. Firstly, to make good use of plentiful CCTA images without annotations, we implemented a 3D U-Net to restore the transformed CPR volumes to their original appearances by a self-supervised training scheme like models genesis [18]. This approach has been proved beneficial for learning common representations from medical image and improving downstream tasks [18]. We initialized the FE by the weights of the 3D U-Net encoder part that was pretrained on 5130 unlabeled CCTA images (39530 vessels). Secondly, before the CPR volumes were input into our model, each cross-section view along the centerline was transformed to polar coordinates. Thirdly, we compared the proposed two-digits binary code method with regular four-type classification.

Table 1. Ablation experiment results on segment-level. (A: Pre-train, B: Polar, C: Two-digits binary code)

Method	MAE_{ste}	ACC_{pla}	$F1_{all}$	$F1_{calc}$	$F1_{non-calc}$	$F1_{mix}$
Baseline	0.152	0.863	0.812	0.761	0.874	0.754
Baseline + A	0.140	0.893	0.844	0.801	0.886	0.787
Baseline + B	0.143	0.876	0.829	0.792	0.879	0.802
Baseline + C	0.146	0.879	0.833	0.783	0.889	0.793
Baseline + A + B + C	**0.138**	**0.908**	**0.856**	**0.802**	**0.891**	**0.839**

As shown in Table 1, combination with each one of the three settings can improve the Coronary R-CNN performance. Moreover, the method integrated with all three achieved the best results. Besides, they have verified the efficacy of simultaneous plaque and stenosis analysis compared with being done separately in [13], so we did not repeat the similar experiments.

4.2 Comparison with Other Methods

We compared our method with several state-of-the-art methods [10–13]. As shown in Table 2, our method achieved best performance in terms of both MAE_{ste}, ACC_{pla}, $F1_{non\text{-}calc}$ and $F1_{mix}$, and third best performance in both $F1_{all}$ and $F1_{calc}$. The following number in brackets after each metric is the p-value between the method and our method. The stenosis result of our method is significant lower than most of other methods. The plaque result is either significant better than some others or close to the best of them.

Table 2. Quantitative results of experiments compared with other methods on segment-level.

Method	MAE_{ste}	ACC_{pla}	$F1_{all}$	$F1_{calc}$	$F1_{non\text{-}calc}$	$F1_{mix}$
2D CNN [10]	0.26 ($<$0.05)	0.834 ($<$0.05)	0.799 ($<$0.05)	0.793 (0.67)	0.823 ($<$0.05)	0.767 ($<$0.05)
2D RCNN [11]	0.245 ($<$0.05)	0.855 ($<$0.05)	0.814 ($<$0.05)	0.782 (0.15)	0.87 (0.22)	0.755 ($<$0.05)
3D CNN	0.216 ($<$0.05)	0.882 (0.47)	0.833 ($<$0.05)	0.798 (0.73)	0.862 ($<$0.05)	0.817 ($<$0.05)
3D RCNN [13]	0.186 ($<$0.05)	0.885 (0.46)	**0.858** (0.96)	0.806 (0.85)	0.822 ($<$0.05)	0.809 ($<$0.05)
Transformer [12]	0.147 (0.17)	0.894 (0.63)	0.857 (0.99)	**0.812** (0.44)	0.883 (0.76)	0.826 (0.27)
Our method	**0.138**	**0.908**	0.856	0.802	**0.891**	**0.839**

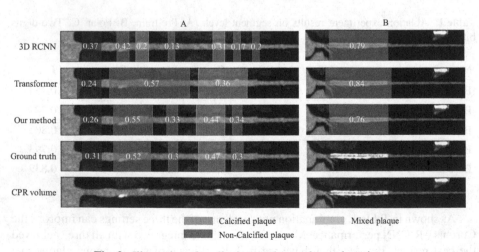

Fig. 3. Illustration of qualitative results on plaque detection.

Figure 3 displays the qualitative comparison among [12, 13] and our method. In column A, we can see that the 3D RCNN method [13] captured plenty of shorter plaques.

It obtained results point by point but lacked global context. Transformer method [12] sampled input cubes every 5 points along the vessel centerline, so the detected plaques were longer and may leak to normal region between two adjacent plaques. Benefited from multi-head analysis module, our Coronary R-CNN yields results closer to annotations. Column B reveals a normal vessel but all methods recognize the stent as a calcified plaque because of its high CT value. This common limitation that should be considered to improve the adaptation on vessels with stents.

5 Conclusion

In this paper, we proposed a novel vessel-wise method called Coronary R-CNN, and applied it to detecting and analyzing lesions of coronary arteries in CCTA images. It is an object detection architecture that is the first trial in this challenging task. Our multi-task model can simultaneously predict not only the location and type of plaque, but also an exact stenosis rate. Evaluation on 1031 cases demonstrated that the presented method achieved equivalent or better results compared to the state-of-the-art methods. Investigating other object detection models is an interesting direction for future research, for that the proposed method provides an important basis.

References

1. Mendis, S., Davis, S., Norrving, B.: Organizational update: the World Health Organization global status report on noncommunicable diseases 2014. Stroke **46**(5), e121–e122 (2015)
2. Dewey, M., Rutsch, W., Schnapauff, D., Teige, F., Hamm, B.: Coronary artery stenosis quantification using multislice computed tomography. Invest. Radiol. **42**(2), 78–84 (2007)
3. Kirişli, H., et al.: Standardized evaluation framework for evaluating coronary artery stenosis detection, stenosis quantification and lumen segmentation algorithms in computed tomography angiography. Med. Image Anal. **17**(8), 859–876 (2013)
4. Kim, Y.J., et al.: Quantification of coronary artery plaque using 64-slice dual-source CT: comparison of semi-automatic and automatic computer-aided analysis based on intravascular ultrasonography as the gold standard. Int. J. Cardiovasc. Imaging **29**(2), 93–100 (2013)
5. Sankaran, S., Schaap, M., Hunley, S.C., Min, J.K., Taylor, C.A., Grady, L.: HALE: healthy area of lumen estimation for vessel stenosis quantification. In: Ourselin, S., Joskowicz, L., Sabuncu, M.R., Unal, G., Wells, W. (eds.) MICCAI 2016. LNCS, vol. 9902, pp. 380–387. Springer, Cham (2016). https://doi.org/10.1007/978-3-319-46726-9_44
6. Shahzad, R., et al.: Automatic segmentation, detection and quantification of coronary artery stenoses on CTA. Int. J. Cardiovasc. Imaging **29**(8), 1847–1859 (2013). https://doi.org/10.1007/s10554-013-0271-1
7. Dey, D., et al.: Automated 3-dimensional quantification of noncalcified and calcified coronary plaque from coronary CT angiography. J. Cardiovasc. Comput. Tomogr. **3**(6), 372–382 (2009)
8. Osher, S., Sethian, J.A.: Fronts propagating with curvature-dependent speed: algorithms based on Hamilton-Jacobi formulations. J. Comput. Phys. **79**(1), 12–49 (1988)
9. Ronneberger, O., Fischer, P., Brox, T.: U-Net: convolutional networks for biomedical image segmentation. In: Navab, N., Hornegger, J., Wells, W., Frangi, A. (eds.) MICCAI 2015. LNCS, vol. 9351, pp. 234–241. Springer, Cham. https://doi.org/10.1007/978-3-319-24574-4_28
10. Tejero-de-Pablos, A., et al.: Texture-based classification of significant stenosis in CCTA multi-view images of coronary arteries. In: Shen, D., et al. MICCAI 2019. LNCS, vol. 11765, pp. 732–740. Springer, Cham (2019). https://doi.org/10.1007/978-3-030-32245-8_81

11. Denzinger, F., et al.: Coronary artery plaque characterization from CCTA scans using deep learning and radiomics. In: Shen, D., et al. MICCAI 2019. LNCS, vol. 11767, pp. 593–601. Springer, Cham (2019). https://doi.org/10.1007/978-3-030-32251-9_65

12. Ma, X., Luo, G., Wang, W., Wang, K.: Transformer network for significant stenosis detection in CCTA of coronary arteries. In: de Bruijne, M., et al. (eds.) MICCAI 2021. LNCS, vol. 12906, pp. 516–525. Springer, Cham (2021). https://doi.org/10.1007/978-3-030-87231-1_50

13. Zreik, M., Van Hamersvelt, R.W., Wolterink, J.M., Leiner, T., Viergever, M.A., Išgum, I.: A recurrent CNN for automatic detection and classification of coronary artery plaque and stenosis in coronary CT angiography. IEEE Trans. Med. Imaging 38(7), 1588–1598 (2018)

14. Ren, S., He, K., Girshick, R., Sun, J.: Faster R-CNN: towards real-time object detection with region proposal networks. Adv. Neural. Inf. Process. Syst. 28, 91–99 (2015)

15. Simonyan, K., Zisserman, A.: Very deep convolutional networks for large-scale image recognition. arXiv preprint arXiv:1409.1556 (2014)

16. He, K., Zhang, X., Ren, S., Sun, J.: Deep residual learning for image recognition. In: Proceedings of the IEEE Conference on Computer Vision and Pattern Recognition, 770–778 (2016)

17. Leipsic, J., et al.: SCCT guidelines for the interpretation and reporting of coronary CT angiography: a report of the Society of Cardiovascular Computed Tomography Guidelines Committee. J. Cardiovasc. Comput. Tomogr. 8(5), 342–358 (2014)

18. Zhou, Z., Sodha, V., Pang, J., Gotway, M.B., Liang, J.: Models genesis. Med. Image Anal. 67, 101840 (2021)

19. Nørgaard, B.L., et al.: Myocardial perfusion imaging versus computed tomography angiography–derived fractional flow reserve testing in stable patients with intermediate-range coronary lesions: influence on downstream diagnostic workflows and invasive angiography findings. J. Am. Heart Assoc. 6(8), e005587 (2017)

20. Zhou, Z., Rahman Siddiquee, M.M., Tajbakhsh, N., Liang, J.: UNet++: a nested U-Net architecture for medical image segmentation. In: Stoyanov, D., et al. DLMIA ML-CDS 2018. LNCS, vol. 11045, pp. 3–11. Springer, Cham (2018). https://doi.org/10.1007/978-3-030-008 89-5_1

21. Guo, Z., et al.: DeepCenterline: a multi-task fully convolutional network for centerline extraction. In: Chung, A., Gee, J., Yushkevich, P., Bao, S. (eds.) IPMI 2019. LNCS, vol. 11492, pp. 441–453 (2019). Springer, Cham. https://doi.org/10.1007/978-3-030-20351-1_34

22. Rajon, D.A., Bolch, W.E.: Marching cube algorithm: review and trilinear interpolation adaptation for image-based dosimetric models. Comput. Med. Imaging Graph. 27(5), 411–435 (2003)

Flat-Aware Cross-Stage Distilled Framework for Imbalanced Medical Image Classification

Jinpeng Li[1], Guangyong Chen[2(✉)], Hangyu Mao[3], Danruo Deng[1], Dong Li[3], Jianye Hao[3,4], Qi Dou[1], and Pheng-Ann Heng[1]

[1] Department of Computer Science and Engineering, The Chinese University of Hong Kong, Hong Kong, China
[2] Zhejiang Lab, Zhejiang, China
gychen@zhejianglab.com
[3] Noah's Ark Lab, Huawei, Shenzhen, China
[4] College of Intelligence and Computing, Tianjin University, Tianjin, China

Abstract. Medical data often follow imbalanced distributions, which poses a long-standing challenge for computer-aided diagnosis systems built upon medical image classification. Most existing efforts are conducted by applying re-balancing methods for the collected training samples, which improves the predictive performance for the minority class but at the cost of decreasing the performance for the majority. To address this paradox, we adopt a flat-aware cross-stage distilled framework (FCD), where we first search for flat local minima of the base training objective function on the original imbalanced dataset, and then continuously finetune this classifier within the flat region on the re-balanced one. To further prevent the performance decreasing for the majority, we propose a cross-stage distillation regularizing term to promote the optimized features to remain in the common optimal subspace. Extensive experiments on two imbalanced medical image datasets demonstrate the effectiveness of our proposed framework and its generality in improving the performance of existing imbalanced methods. The code of this work will be released publicly.

Keywords: Imbalanced image classification · Flat-aware optimization · Cross-stage distillation

1 Introduction

Deep learning based medical image classification plays an increasingly important role in intelligent diagnosing systems [19,27]. It is well known that deep learning based methods heavily rely on the quality of training datasets [6], thus their

Supplementary Information The online version contains supplementary material available at https://doi.org/10.1007/978-3-031-16437-8_21.

© The Author(s), under exclusive license to Springer Nature Switzerland AG 2022
L. Wang et al. (Eds.): MICCAI 2022, LNCS 13433, pp. 217–226, 2022.
https://doi.org/10.1007/978-3-031-16437-8_21

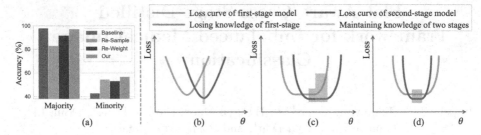

Fig. 1. (a) Accuracies of different methods on the majority and minority classes of Eyepacs testing set which occupy 73.5% and 4.5% data, respectively. (b)–(d) Loss curves of two-stage learning strategies. Gray region denotes only the second-stage model achieves local optima while the first-stage model has a large loss. Brown region indicates two models both obtain local optima. (b) Vanilla two-stage training. (c) Two-stage training with flat local minima. (d) Two-stage training with flat local minima and cross-stage distillation. (Color figure online)

usage would be limited in clinical scenarios since medical images often follow imbalanced distribution in their nature. For example, there are far more healthy people than patients with proliferative diabetic retinopathy (DR) in the Eyepacs dataset for DR grading [1].

Most previous efforts solve this challenge by re-balancing strategies, categorized into re-sampling data and re-weighting loss [5,13,24]. Re-sampling methods over-sample the minority classes or under-sample the majority classes to re-balance data [3,11,18]. Re-weighting methods adjust weights of classes in loss functions, and show promising performance in different imbalance tasks [5,16]. Due to the high flexibility of re-balancing methods, they are explored to combine with other techniques, such as Mixup [8] and contrastive learning [23]. However, re-balance based methods manually disturb the original data distributions, thus are prone to learn sub-optimal feature representations [11]. Moreover, they increase the performance on the minority classes at the cost of decreasing the performance on the majority classes, as shown in Fig. 1(a). In contrast, our method can maintain the performance on the majority classes while effectively increasing the performance on minority classes.

Recent advances have been achieved by adopting two-stage training [4,5,11], which first pretrains on the original imbalanced data to obtain good feature representation and then finetunes with re-balancing techniques to mitigate the problem of low performance on minority classes lied in the pretraining stage.

However, the data distributions of two stages are different, directly finetuning may cause catastrophic forgetting [15]. Decoupling method [11] tries to solve it by reformulating the second stage as only learning the classifier while freezing the feature extractor. Although the frozen feature extractor can retain all features learned in the first stage, only training the classifier by re-balance methods limits the power of two-stage training and lowers the performance on majority classes.

Given the achievements of two-stage methods, we are motivated to fully leverage the power of two-stage training by finetuning the whole network without freezing the feature extractor. However, as shown in Fig. 1(b), neural networks

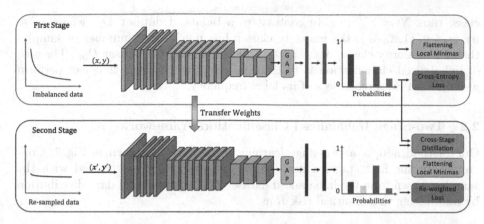

Fig. 2. Illustration of proposed FCD framework which follows a two-stage training pipeline. The first-stage model learns from the imbalanced data distribution and searches for flat local minima. We transfer the first-stage model weights to the second-stage model and finetune it with cross-stage distillation and re-balancing strategies.

are prone to find sharp local minima [7,12], thus directly finetuning is hard to find a solution where both two models obtain small loss, and the knowledge learned in the first stage may be lost. To address this catastrophic forgetting issue, we propose a novel two-stage framework with flat local minima [7,20] and cross-stage distillation for imbalanced medical image classification. Specifically, we regularize the geometric of loss landscape to find flat minima in which the solutions have similar optimal values. The flat minima of two stages form a common optimal subspace (brown region in Fig. 1(c)) where two models can obtain local optimal values thus maintaining knowledge learned from both stages. However, the second-stage training cannot sense the states of the first-stage model, the model weights may still converge to a region (gray region in Fig. 1(c)) corresponding to a large loss value for the first-stage model. To explicitly promote the learning procedure within this common optimal subspace, we design a cross-stage distillation that limits the divergence of two models' outputs. Thus, the second-stage training is designed to find the solution that is optimal for both models as shown in Fig. 1(d), which maintains knowledge learned in both stages. Experimental results show that our framework outperforms state-of-the-art imbalanced classification methods on various evaluation metrics.

2 Method

2.1 Problem Formulation

In imbalanced medical image classification, both training and testing sets follow imbalanced data distributions over the categories. Let $D = \{(x_i, y_i)\}_{i=1}^{N}$ be a dataset with N samples, where x_i is the ith image and y_i is its label. Let M denote the number of classes and n_i denote the number of samples in the ith class, thus $N = \sum_{i=1}^{M} n_i$. In contrast to a balanced dataset D_{ba} where $n_i \approx n_j, \forall i, j \in \{1, ..., M\}$, the majority class p has much larger number of samples than the minority class q, i.e., $n_p \gg n_q$, in an imbalanced dataset D_{im}. The goal of imbalanced classification is to learn an effective classifier that can perform well on each class regardless of its label frequency.

2.2 Two-Stage Imbalanced Classification Framework

Our method adopts a two-stage learning framework as shown in Fig. 2. Concretely, in the first stage, a deep neural network $f(x; \theta)$ is trained with the samples drawn from the imbalanced dataset D_{im} with original data distribution by minimizing the empirical risk R as,

$$R(\theta_1) = \frac{1}{N} \sum_{i \in D_{im}} L(f(x_i; \theta_1), y_i), \tag{1}$$

where θ is the parameters of f, and L is the loss function. This learning strategy ensures that each sample has the same occurrence frequency in the training phase, thus the network can make full use of all samples to learn the most knowledge and feature representation. However, as indicated by previous works [11, 24], good feature representation is not enough to directly achieve promising classification accuracy in the imbalanced tasks. In the second stage, we initialize the network with the model weights from the first stage, and finetune the whole model by re-balancing strategies which can be implemented by data re-sampling and/or loss re-weighting as follow:

$$R(\theta_2) = \frac{1}{N} \sum_{i \in D_{ba}} w(i) L(f(x_i; \theta_2), y_i), \tag{2}$$

where $w(i)$ is the re-weighting factor of ith sample in the re-balanced dataset D_{ba}. Note that the network architectures of these two stages are the same, and the difference only lies in the strategies of data sampling and loss function.

2.3 Flattening Local Minima

In the first stage, our framework learns the parameters on the original data distribution to find a local optimum that provides representative features. However, this local optima does not guarantee that it is a good starting point for the second stage. Ideally, the learning procedure of our two-stage framework, as a pretraining-to-finetuning paradigm, should have strong stability to make sure that the second-stage model does not easily escape from the local optima of pretraining. Otherwise, the network will lose the old information when learning new knowledge in the second stage.

In this work, we achieve this strong stability by finding the flat minima on the loss landscape [7, 20], which means that the points within the neighboring area of minima have similar optimal values. This neighboring area provides a parameter subspace where the finetuning procedure can learn new knowledge by updating parameters while preserving similar optimal value on the initial data distribution as the first-stage model. Instead of searching flat minima by adding random noise [20], we adopt the sharpness-aware minimization [7] to measure the flatness of current parameters θ by the largest increment of loss value within their neighboring area:

$$\max_{||\epsilon||_2 \leq \rho} L(\theta + \epsilon) - L(\theta), \tag{3}$$

where $\theta + \epsilon$ denotes a neighbouring point and $||\cdot||_2$ is ℓ^2 norm. $||\epsilon||_2 \leq \rho$ limits the neighbouring area into a hypersphere with the radius of ρ. ρ is experimentally selected based on the performance on the validation set of Eyepacs [1] and set to 0.05 in all experiments. Adding this flatness term into the loss function, the model optimization is formulated as:

$$\arg\min_{\theta} \max_{||\epsilon||_2 \leq \rho} L(\theta + \epsilon) - L(\theta) + L(\theta) = \arg\min_{\theta} \max_{||\epsilon||_2 \leq \rho} L(\theta + \epsilon). \tag{4}$$

Compared to the minimization problem in the original optimization procedure for neural networks, this formulation involves an extra maximization step, thus becoming a problem of min-max optimization. We approximately solve the maximization step by Taylor expansion and dual norm [7] to get the value of ϵ^* as

$$\epsilon^* = \rho \nabla_\theta L(\theta) / (||\nabla_\theta L(\theta)||_2). \tag{5}$$

Thus, our method includes two updating steps in each training iteration, which first updates parameters to $\theta^* = \theta + \epsilon^*$ along the direction of gradient ascent, and then performs gradient descent at θ^*.

2.4 Cross-Stage Distillation

The flat areas provide common optimal subspaces for two-stage models to explore optima. However, since the network is repeatedly updated during finetuning, the final parameters may still escape the common optimal subspaces and suffer from the knowledge forgetting, which degenerates to the one-stage re-balancing method and decreases the performance on the imbalanced distribution.

Inspired by the knowledge distillation method [10] which is initially used to compress large models into small ones, we propose a cross-stage distillation loss which explicitly promotes the consistency of the outputting probabilities between the models in two stages by minimizing their divergence to address the above issue of large weights shift. Compared to the hard parameter clamping [20], our distillation method is end-to-end differentiable and efficient in learning new knowledge. Specifically, for each training image x, we feed it to the fixed first-stage network $f_1(x; \theta_1)$ and the second-stage network $f_2(x; \theta_2)$ to generate the probability distributions of $\bar{p}$ and $\hat{p}$, respectively, from the last fully-connected layer with the Softmax activation. The dimension of the probability distribution is the same as the number of classes. Our cross-stage distillation loss encourages $\hat{p}$ to approximate $\bar{p}$ for preserving the representative features across stages by

$$L_D(f_1, f_2) = -\sum_{i=1}^{B} \sum_{j=1}^{M} \bar{p}_i^{(j)} \log \hat{p}_i^{(j)}, \tag{6}$$

where B is the batch size, and $p_i^{(j)}$ denotes the probability of the jth class of the ith sample. Note that we only compute the gradient with respect to the weights θ_2 of the second-stage network, and the first-stage network is frozen during the second-stage training. The network architectures in our two-stage framework are the same instead of compressing a big model into a small one as [10].

Similar to the normal classification task, cross-entropy loss L_C is employed to directly supervise $\hat{y}$ towards groundtruth y. Thus, the overall loss function of the second-stage training is composed of a cross-entropy loss and a cross-stage distillation loss as:

$$L = L_C + \lambda L_D, \tag{7}$$

where λ is a weight factor to control the magnitude of preserving the first-stage features, and we simply set it to 0.1 for our models.

3   Experiments

3.1   Dataset and Evaluation Metrics

We evaluate our methods on two public imbalanced medical image datasets including Eyepacs [1] and Hyper-Kvasir [2]. Eyepacs is the largest fundus image dataset for DR grading, which includes 31,613 training images, 3,513 validation images, and 53,577 testing images. All images are labeled into 5 DR gradings which represent increasing severity from no DR to proliferative DR. Hyper-Kvasir is a public dataset for gastrointestinal images (GI) classification, which samples are acquired from endoscopy videos. Endoscopy is currently the standard way to examine the abnormalities and diseases of digestive system. Hyper-Kvasir contains 10,662 images with 23 classes following a long-tailed distribution. Imbalanced ratio β is used to measure the degree of imbalance and is defined as the number of the most frequent class divided by the least frequent one. The imbalanced ratios of Eyepacs and Hyper-Kvasir are 36 and 171, respectively. Following the evaluation protocol as [8], we test our methods by the metrics of quadratic weighted kappa (Quad-k), Kendall-π and Matthews correlation coefficient (MCC) on the testing set of Eyepacs, and by MCC, Macro-F1, and balanced accuracy (B-ACC) on the Hyper-Kvasir dataset under 5 folds cross-validation.

3.2   Implementation Details

We employ MobileNet-V2 [17] as the backbone for both stages. The model weights pretrained on ImageNet [6] are only used in the first stage. Batch size is set to 8. Stochastic gradient descent is used as the optimizer. In the first stage, the model is trained for 160 epochs with the initial learning rate set as 0.01. In the second stage, we reset the learning rate to 0.001 and finetune the models for 10, and 30 epochs on the tasks of DR grading and GI classification, respectively. The input image sizes for DR grading and GI classification are 512×512 and 512×640, respectively. For fair comparisons, we follow the data augmentation methods used in the previous state-of-the-art method [8]. All models are implemented with the PyTorch library, and trained on a GTX 3090 GPU.

Table 1. Quantitative comparisons with state-of-the-art imbalanced classification methods on Eyepacs and Hyper-Kvasir datasets.

Method	Year	Eyepacs			Hyper-Kvasir		
		Quad-k	MCC	Kendall-π	MCC	B-ACC	Macro-F1
Focal Loss [13]	2017	78.69	61.58	73.36	90.52	61.37	61.32
Sqrt-RS [14]	2018	79.12	59.69	72.14	90.42	62.77	63.10
CB-Focal [5]	2019	76.38	55.07	68.53	86.85	64.31	61.69
LDAM loss [4]	2019	77.97	60.55	72.86	90.64	61.46	60.89
BBN-RS [26]	2020	77.18	56.67	70.62	90.23	61.57	61.41
PG-RS [11]	2020	78.51	59.14	71.93	90.40	62.44	62.26
cRT [11]	2020	77.55	52.78	67.38	88.78	62.36	61.78
EQL [22]	2020	79.23	57.71	73.16	89.30	62.03	61.22
EQL V2 [21]	2021	79.59	62.30	74.01	90.78	61.59	61.38
Bal-Mxp [8]	2021	79.25	62.44	73.84	90.63	63.22	62.58
LAS [25]	2021	78.95	60.35	72.63	90.32	61.23	60.28
DiVE [9]	2021	78.66	54.71	69.61	88.80	63.85	62.15
RS Loss [16]	2021	78.47	61.31	72.97	89.36	60.29	59.84
Ours (RS)		81.34	**63.37**	75.38	91.13	**64.76**	63.84
Ours (RW)		**81.67**	63.16	**75.42**	**91.14**	64.64	**64.09**

3.3 Comparison to the State-of-the-Arts

We compare the performance of our models with recent state-of-the-art (SOTA) imbalanced classification approaches from different spectra including data re-sampling (RS) methods (such as Sqrt-RS [14] and PG-RS [11]), loss re-weighting (RW) methods (such as Focal loss [13] and EQL [22]), and two-stage methods (such cRT [11] and DiVE [9]). We re-train all these methods with the same backbone and data augmentations as our models for fair comparisons. We implement two versions of our models, i.e., Ours (RS) and Ours (RW), which use the re-balancing strategies of RS and RW in the second stage, respectively. Table 1 shows the imbalanced classification results on the Eyepacs and Hyper-Kvasir datasets. We observe that our models outperform all the other methods in all metrics on both datasets. Compared with Sqrt-RS which is a one-stage RS method, Ours (RS) uses the same RS strategy while surpassing it by a large margin of +2.22% Quad-k, 3.68% MCC, and 3.24% Kendall-π on the Eyepacs dataset, demonstrating the importance of sequentially learning the representative features and discriminating features in our framework. Our methods outperform cRT [11] which also employs the two-stage pipeline. The performance gap may be due to that cRT only learns the classifier and freezes the feature extractor in the second stage, while our method finetunes the whole network and effectively maintains the knowledge learned in two stages. The two variants

Table 2. Results of ablation study on the testing set of Eyepacs. TS: Two-stage framework; FLM: Flattening local minima; CSD: Cross-stage distillation.

Modules			Eyepacs		
TS	FLM	CSD	Quad-k	MCC	Kendall-π
			78.67	61.65	73.42
√			79.72	60.34	72.38
√	√		**81.40**	62.70	75.03
√		√	79.51	61.84	73.66
√	√	√	81.34	**63.37**	**75.38**

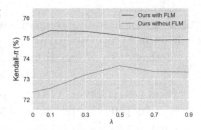

Fig. 3. Comparative results on the testing set of Eyepacs with different weight factors λ, using our framework with and without FLM.

of our methods achieve similar performance on both datasets, which shows that our framework is suitable for different re-balancing methods.

3.4   Ablation Studies

We report the results of ablation studies in Table 2 to show the effectiveness of proposed components, i.e., two-stage training (TS), flattening local minima (FLM), and cross-stage distillation (CSD). Our baseline is the first stage of our framework. TS improves Quad-k while decreasing the other two metrics, indicating that the catastrophic forgetting issue happens and directly finetuning is not enough to fully maintain the knowledge learned from two stages. Adding FLM to TS further improves Quad-k from 79.72% to 81.40%, and also achieves better performance than baseline in the other metrics. These improvements demonstrate that optimizing models into flat local minima enables two-stage training to effectively transfer the knowledge across stages. We further investigate the importance of CSD by testing the performance of TS+CSD and TS+FLM+CSD. The results show that incorporating CSD achieves better and comparable results than the counterparts without it, due to CSD explicitly maintains the knowledge of models trained in two stages by limiting their outputs' divergence. To further investigate how CSD works, Fig. 3 shows the performance of CSD with different values of λ which controls the relative magnitude of CSD loss to cross-entropy loss. We observe that increasing the value of λ first increases the performance of Kendall-π and then decreases it, demonstrating that maintaining the first-stage knowledge and learning the new knowledge in the second stage are both valuable. Besides, we find that TS+CSD needs a large λ than TS+FLM+CSD to obtain the peak performance, since it has sharper local minima and needs a stronger constraint to prevent knowledge forgetting. We further compare the performance of baseline and our method on the majority and minority classes of Eyepacs and Hyper-Kvasir datasets. The baseline achieves 97.86%, 42.89%, 96.48%, and 44.26% accuracies on the majority and minority classes of Eyepacs, and majority and minority classes of Hyper-Kvasir, while our method achieves

97.26%, 56.78%, 96.21%, and 50.82% on the same classes, respectively. These results show that our method can maintain the performance on the majority class while effectively increasing the performance on minority classes.

4 Conclusion

We propose a new two-stage framework, FCD, for imbalanced medical image classification. FCD sequentially learns the representative and discriminating features from the imbalanced data distribution and re-balancing strategies, respectively. To address the catastrophic forgetting difficulty in two-stage learning, our method flattens the local minima to form a common optimal region and further employs the cross-stage distillation to facilitate the network optimization within this region. Extensive experiments on two imbalanced datasets demonstrate the superiority of our framework over existing imbalanced classification methods.

Acknowledgement. This work was supported by National Key R&D Program of China (2022YFE0200700), National Natural Science Foundation of China (Project No. 62006219), Natural Science Foundation of Guangdong Province (2022A1515011579), Hong Kong Innovation and Technology Fund Project No. GHP/110/19SZ and ITS/170/20.

References

1. Diabetic retinopathy detection. In: Kaggle. https://www.kaggle.com/c/diabetic-retinopathy-detection
2. Endotect challenge, ICPR 2020. https://endotect.com/
3. Buda, M., Maki, A., Mazurowski, M.A.: A systematic study of the class imbalance problem in convolutional neural networks. Neural Netw. **106**, 249–259 (2018)
4. Cao, K., Wei, C., Gaidon, A., Aréchiga, N., Ma, T.: Learning imbalanced datasets with label-distribution-aware margin loss. In: Advances in Neural Information Processing Systems 32: Annual Conference on Neural Information Processing Systems, NeurIPS, pp. 1565–1576 (2019)
5. Cui, Y., Jia, M., Lin, T., Song, Y., Belongie, S.J.: Class-balanced loss based on effective number of samples. In: IEEE Conference on Computer Vision and Pattern Recognition, CVPR, pp. 9268–9277 (2019)
6. Deng, J., Dong, W., Socher, R., Li, L., Li, K., Fei-Fei, L.: ImageNet: a large-scale hierarchical image database. In: 2009 IEEE Computer Society Conference on Computer Vision and Pattern Recognition, CVPR, pp. 248–255 (2009)
7. Foret, P., Kleiner, A., Mobahi, H., Neyshabur, B.: Sharpness-aware minimization for efficiently improving generalization. In: 9th International Conference on Learning Representations, ICLR (2021)
8. Galdran, A., Carneiro, G., González Ballester, M.A.: Balanced-MixUp for highly imbalanced medical image classification. In: de Bruijne, M., et al. (eds.) MICCAI 2021. LNCS, vol. 12905, pp. 323–333. Springer, Cham (2021). https://doi.org/10.1007/978-3-030-87240-3_31
9. He, Y., Wu, J., Wei, X.: Distilling virtual examples for long-tailed recognition. In: IEEE International Conference on Computer Vision, ICCV (2021)

10. Hinton, G., Vinyals, O., Dean, J., et al.: Distilling the knowledge in a neural network. arXiv preprint arXiv:1503.02531 (2015)
11. Kang, B. et al.: Decoupling representation and classifier for long-tailed recognition. In: 8th International Conference on Learning Representations, ICLR (2020)
12. Li, H., Xu, Z., Taylor, G., Studer, C., Goldstein, T.: Visualizing the loss landscape of neural nets. In: Advances in Neural Information Processing Systems 31: Annual Conference on Neural Information Processing Systems, NeurIPS, pp. 6391–6401 (2018)
13. Lin, T., Goyal, P., Girshick, R.B., He, K., Dollár, P.: Focal loss for dense object detection. In: IEEE International Conference on Computer Vision, ICCV, pp. 2999–3007 (2017)
14. Mahajan, D. et al.: Exploring the limits of weakly supervised pretraining. In: Computer Vision - ECCV - 15th European Conference, pp. 185–201 (2018)
15. Mai, Z., Li, R., Jeong, J., Quispe, D., Kim, H., Sanner, S.: Online continual learning in image classification: an empirical survey. Neurocomputing **469**, 28–51 (2022)
16. Oksuz, K., Cam, B.C., Akbas, E., Kalkan, S.: Rank & sort loss for object detection and instance segmentation. In: IEEE International Conference on Computer Vision, ICCV (2021)
17. Sandler, M., Howard, A., Zhu, M., Zhmoginov, A., Chen, L.C.: MobileNetV 2: inverted residuals and linear bottlenecks. In: IEEE Conference on Computer Vision and Pattern Recognition, CVPR, pp. 4510–4520 (2018)
18. Sarafianos, N., Xu, X., Kakadiaris, I.A.: Deep imbalanced attribute classification using visual attention aggregation. In: Computer Vision - ECCV - 15th European Conference, pp. 708–725 (2018)
19. Shen, D., Wu, G., Suk, H.I.: Deep learning in medical image analysis. Ann. Rev. Biomed. Eng. **19**(1), 221–248 (2017)
20. Shi, G., Chen, J., Zhang, W., Zhan, L., Wu, X.: Overcoming catastrophic forgetting in incremental few-shot learning by finding flat minima. In: Advances in Neural Information Processing Systems 34: Annual Conference on Neural Information Processing System, NeurIPS, pp. 6747–6761 (2021)
21. Tan, J., Lu, X., Zhang, G., Yin, C., Li, Q.: Equalization loss v2: a new gradient balance approach for long-tailed object detection. In: IEEE Conference on Computer Vision and Pattern Recognition, CVPR, pp. 1685–1694 (2021)
22. Tan, J. et al.: Equalization loss for long-tailed object recognition. In: 2020 IEEE/CVF Conference on Computer Vision and Pattern Recognition, CVPR, pp. 11659–11668 (2020)
23. Wang, P., Han, K., Wei, X., Zhang, L., Wang, L.: Contrastive learning based hybrid networks for long-tailed image classification. In: IEEE Conference on Computer Vision and Pattern Recognition, CVPR, pp. 943–952 (2021)
24. Zhang, Y., Wei, X., Zhou, B., Wu, J.: Bag of tricks for long-tailed visual recognition with deep convolutional neural networks. In: Thirty-Fifth AAAI Conference on Artificial Intelligence, AAAI, pp. 3447–3455 (2021)
25. Zhong, Z., Cui, J., Liu, S., Jia, J.: Improving calibration for long-tailed recognition. In: IEEE Conference on Computer Vision and Pattern Recognition, CVPR, pp. 16489–16498 (2021)
26. Zhou, B., Cui, Q., Wei, X., Chen, Z.: BBN: bilateral-branch network with cumulative learning for long-tailed visual recognition. In: 2020 IEEE/CVF Conference on Computer Vision and Pattern Recognition, CVPR, pp. 9716–9725 (2020)
27. Zhou, S.K., et al.: A review of deep learning in medical imaging: imaging traits, technology trends, case studies with progress highlights, and future promises. Proc. IEEE **109**, 820–838 (2021)

CephalFormer: Incorporating Global Structure Constraint into Visual Features for General Cephalometric Landmark Detection

Yankai Jiang[1], Yiming Li[1], Xinyue Wang[1], Yubo Tao[1(✉)], Jun Lin[2], and Hai Lin[1(✉)]

[1] State Key Laboratory of CAD&CG, College of Computer Science and Technology, Zhejiang University, Hangzhou, China
{taoyubo,lin}@cad.zju.edu.cn
[2] Department of Stomatology, The First Affiliated Hospital, College of Medicine, Zhejiang University, Hangzhou, China

Abstract. Accurate cephalometric landmark detection is a crucial step in orthodontic diagnosis and therapy planning. However, existing deep learning-based methods lack the ability to explicitly model the complex dependencies among visual features and landmarks. Therefore, they fail to adaptively encode the landmark's global structure constraint into the representation of visual concepts and suffer from large biases in landmark localization. In this work, we propose CephalFormer, which exploits the correlations between visual concepts and landmarks to provide meaningful guidance for accurate 2D and 3D cephalometric landmark detection. CephalFormer explores local-global anatomical contents in a coarse-to-fine fashion and consists of two stages: (1) a new efficient Transformer-based architecture for coarse landmark localization; (2) a novel paradigm based on self-attention to represent visual clues and landmarks in one coherent feature space for fine-scale landmark detection. We evaluated CephalFormer on two public cephalometric landmark detection benchmarks and a real-patient dataset consisting of 150 skull CBCT volumes. Experiments show that CephalFormer significantly outperforms the state-of-the-art methods, demonstrating its generalization capability and stability to naturally handle both 2D and 3D scenarios under a unified framework.

Keywords: Cephalometric landmark detection · Transformer

1 Introduction

Accurate anatomical landmark detection is a fundamental procedure in cephalometric analysis, which is widely adopted in orthodontic diagnosis, treatment

Y. Jiang and Y. Li—Equally contributed.

Supplementary Information The online version contains supplementary material available at https://doi.org/10.1007/978-3-031-16437-8_22.

© The Author(s), under exclusive license to Springer Nature Switzerland AG 2022
L. Wang et al. (Eds.): MICCAI 2022, LNCS 13433, pp. 227–237, 2022.
https://doi.org/10.1007/978-3-031-16437-8_22

planning and evaluation [7]. In clinical practice, manual annotation is typically tedious and subjective [19]. Therefore, automatic and accurate landmark detection has attracted great interest from both academia and industry.

Conventionally, cephalometric analysis is performed on 2D cephalograms. In recent years, convolutional neural networks (CNNs) have achieved significant success in 2D cephalometric landmark detection [1,12–14,25,27,29]. Most existing methods design complex network architectures or use powerful mechanisms such as self-attention to improve the detection accuracy. For example, Chen *et al.* [3] proposed an attentive feature pyramid fusion module (AFPF) to shape semantically enhanced features for landmark localization. Zhou *et al.* [27] developed a learning-to-learn framework to search objective metrics for regressing multiple landmark heatmaps dynamically. Zeng *et al.* [25] proposed a cascaded three-stage CNNs for cephalometric landmark detection.

Variations on individual structures, appearance ambiguity, and image complexity are the main factors that hinder accurate landmark detection. A potential solution, as described in [12,13], is to incorporate prior knowledge on spatial structure configurations of landmarks into the learned heatmaps, which could enable the network to simplify the modeling of the underlying distribution of anatomical features. More recently, Li *et al.* [10] proposed a deep adaptive graph learning approach to model landmarks as a graph to exploit prior structural knowledge. Despite popularity and success, these methods only explore the relationships among landmarks via additional modules or GCNs [8] and fail to represent input image features and landmarks in one coherent graph. As a result, they can not explicitly model the dependencies between all visual features and anatomical landmarks. This major drawback makes them susceptible to large anatomical variations and imperfect overlaps of anatomical structures. In practice, The landmarks are usually associated with anatomical features [22]. Building the correlations between visual concepts and landmarks can help the network adaptively learn the global structure constraint on spatial configurations of landmarks. So far, learning explicit dependencies between all features and anatomical landmarks via representing them jointly in one coherent feature space has not yet been investigated.

Another challenge in this field is the lack of an efficient and accurate 3D cephalometric landmark detection methodology. 2D cephalograms only provide a projected 2D view of real 3D anatomical structures, resulting in problems such as geometric distortions and superimpositions [24]. However, the expensive computational cost and complex search space make it impractical to transfer the state-of-the-art methods from 2D to 3D directly.

Several research efforts have been conducted to explore 3D cephalometric landmark detection. Lang *et al.* [9] proposed a two-stage coarse-to-fine method, utilizing 3D mask R-CNN [5] and GCNs to detect landmarks. However, it takes minutes to infer a CBCT volume and lacks the consideration of the global structure constraint. Chen *et al.* [4] also proposed a two-stage coarse-to-fine framework based on LSTM [6] for 3D landmark detection. However, it still relys on an additional graph attention module to encode the landmark's global struc-

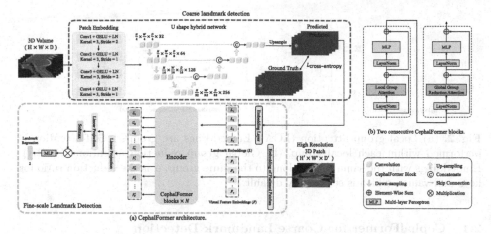

Fig. 1. Overview of CephalFormer. It first locates the coarse landmarks via heatmap regression and then performs fine-scale landmark detection. These two stages are shown in the purple and blue dashed box respectively. (Color figure online)

ture constraint and fails to input the image features and landmark embeddings jointly. Moreover, the recursive prediction process of LSTM is slow and suffers from the vanishing gradient problem.

Motivated by the aforementioned limitations, we propose a general end-to-end deep learning method, CephalFormer, for automatic 2D and 3D cephalometric landmark detection. The contribution of our work is threefold. (1) We propose the first general Transformer-based framework that naturally handles both 2D and 3D scenarios. (2) We are the first to study how to represent visual features and landmarks into a coherent feature space to explicitly incorporate the global structure constraint for accurate cephalometric landmark detection. (3) Our method significantly outperforms the state-of-the-art methods on two public cephalometric landmark detection benchmarks and a real-patient dataset.

2   Methods

As shown in Fig. 1(a), CephalFormer adopts a coarse-to-fine two-stage strategy. In the first stage, a new U shape hybrid structure is built based on the combination of interleaved convolution and CephalFormer blocks (see Fig. 1(b)) for coarse landmark localization. In the second stage, we input the visual features of high-resolution cropped patches and landmark embeddings jointly into a Transformer encoder, which allows rich information exchange and adaptively incorporates the global structural constraint for accurate coordinate refinement. The two-stage models are trained end-to-end. In the following, we take the 3D scene as an example to present details of these two stages and give more insights into our method. 2D scenes can be easily handled by replacing configurations such as 3D convolution operations with 2D ones.

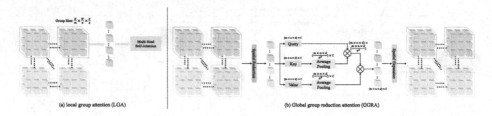

(a) local group attention (LGA)   (b) Global group reduction attention (GGRA)

Fig. 2. (a) Local group attention (LGA). Embeddings are grouped. Self-attention is performed within each local group. (b) Global group reduction attention (GGRA). Embeddings with the same color belong to the same group. α is the reduction ratio for keys and values, which is set to 7 by default.

2.1   CephalFormer for Coarse Landmark Detection

We first separate each input as non-overlapping patches ($size = 4 \times 4 \times 4$). Then, all the patches are flattened and projected to a higher dimension. This patch embedding process is fulfilled through four successive convolutional layers, each followed by a LayerNorm (LN) layer and a GELU.

CNN-based models inherently lack the ability to explicitly capture long-range feature dependencies between visual clues. However, pure transformer-based models lack the ability to capture details within each patch [21]. To overcome these limitations, we choose to build a U shape hybrid network for coarse landmark detection. Different from approaches that simply treated transformers as assisted modules [2, 20], we follow nnFormer [28] to interleavely take advantages of convolution and self-attention. Concretely, each stage in the encoder and decoder involves one convolutional layer and two CephalFormer blocks. Different from nnFormer, our CephalFormer can process the whole volumes instead of 3D local patches thanks to the proposed local group attention (LGA) and global group reduction attention (GGRA).

Self-attention [17] incurs huge memory and computation costs, which motivates us to reduce its computational complexity and enable its application for the whole 3D volumes. As shown in Fig. 1(b), LGA and GGRA appear alternately in different CephalFormer blocks.

For LGA (see Fig. 2(a)), given patch embeddings $X \in \mathbb{R}^{H_i \times W_i \times D_i \times C_i}$, we split X into multiple $m_i \times n_i \times d_i$ sub-groups in the spatial dimension. Considering that the visual dependencies between patches nearby are usually stronger than those far away, we perform the fine-grained self-attention only in local groups. Each group contains $\frac{H_i W_i D_i}{m_i n_i d_i}$ elements, and thus the computation cost of the self-attention in each group is $O(\frac{H_i^2 W_i^2 D_i^2}{m_i^2 n_i^2 d_i^2} C_i)$. The total cost of all groups is $O(\frac{H_i^2 W_i^2 D_i^2}{m_i n_i d_i} C_i)$, which is significantly more efficient and grows linearly with $H_i W_i D_i$ when n_i, m_i, d_i is close to H_i, W_i, D_i. In comparison, the cost of original self-attention $O(H_i^2 W_i^2 D_i^2 C_i)$ would exceed the limitation of GPU memory.

In LGA, the image is divided into non-overlapping groups. This makes the receptive field small and would significantly degrade the performance. Thus, we propose GGRA to enable cross-group information exchange. As shown in Fig. 2(b), we first downsample each group to a single representative to summarize the important information by a spatial reduction module, which consists of two convolutional layers and a global average pooling layer. Then we feed the representative of each group to the multi-head self-attention layer to compute queries, keys and values. To further reduce computation cost, we use subsampled keys and values to communicate with other groups. Finally, the patch tokens are upsampled by a spatial expansion module, which consists of two deconvolutional layers. GGRA reduces the cost of self-attention to $O(\frac{m_i^2 n_i^2 d_i^2}{\alpha^3} C_i)$. The coarse landmark detection is accomplished via heatmap regression on the output feature maps of the U shape hybrid network.

2.2 Fine-Scale Coordinate Refinement

Following most two-stage coarse-to-fine methods [4,9], we sample high resolution 3D patches $P \in \mathbb{R}^{H' \times W' \times D' \times C}$ in the vicinity of the landmark locations estimated in the first stage. There is no mechanism to encode interactions between visual information and prior knowledge on landmark topology in one joint feature space in traditional models. To address this drawback, we propose a technique to easily incorporate landmark configurations into visual features.

First, we split P into $m' \times n' \times d'$ fixed-size patches and linearly embed each of them to obtain a set of visual feature embeddings $F = \{F_1, F_2, ...F_v\}$, $F_j \in \mathbb{R}^g$, $v = m' \times n' \times d'$. Then we consider each vector $L_k \in \mathbb{R}^g$ from L, with k ranging from 1 to t (where t is the number of landmarks), to be representative of the kth possible landmark. Landmark embeddings are learned from a linear embedding layer which takes features extracted from the coarse predicted heatmaps as inputs to encode the information of each landmark. We concatenate F and L, add position embeddings and feed the resulting sequence of vectors to a Transformer encoder built with N CephalFormer blocks. Compared with previous methods [10] that only model landmarks as a graph, our Transformer encoder can be viewed as a fully connected graph, which is able to learn any relationships between visual features and landmarks. In this way, landmarks are associated with anatomical features explicitly and the network can adaptively learn the global structure constraint on the spatial configuration of landmarks. Such an advantage is very important for cephalometric landmark detection, because the feasible location of a landmark is not distributed uniformly in image space, but is constrained by the locations of other anatomical landmarks [13].

At last, an attention module, which consists of two linear layers and a softmax layer, is utilized to extract the weight matrix for the output token embeddings (F' and L'). This attention module helps to fully utilize the contributions of all output tokens while distinguishing their differences. Finally, we assign the attention weights to the learned token embeddings and feed them to a multilayer perceptron to accomplish the final landmark regression. If the model in the first stage misses some landmarks in the prediction, the second-stage model still can get them back since the second-stage model has direct access to the input high resolution image patches.

3 Experiments

3.1 Settings

Dataset: In the 2D scenario, we evaluate our method on the public benchmark from the IEEE ISBI 2015 cephalometric landmark detection Challenge [18,19]. It contains 400 cephalograms with resolution of 1935×2400. 150 images are used as training set, the rest 150 images and 100 images are used as test set 1 and test set 2. Each cephalometric image contains 19 landmarks. The size of each pixel is 0.1 by 0.1 mm. We follow [3] to resize the Cephalometric X-rays to 800×640.

For 3D tasks, we make experiments on two datasets. One is the Public Domain Database for Computational Anatomy (PDDCA) dataset form MICCAI Head-Neck Challenge 2015 [15]. It contains 48 patient CT images, of which 33 CT images are labeled with 5 bony landmarks. The dataset is equally distributed into three folds to perform three-fold cross-validation, and the average accuracy of three folds is reported. Following [4], we standardize the voxel spacing to 1 mm. The average resolution is $518 \times 518 \times 384$ after standardization. Besides, an in-house skull dataset is also used for evaluation, which consists of 150 skull CBCT in DICOM format with resolution of $512 \times 512 \times 512$. The isotropic voxel spacing is 0.3 mm. Each data has 33 manually annotated landmarks on the skull. We split the dataset into three folds with 50, 50 and 50 samples to perform three-fold cross-validation. We report the average accuracy of three folds.

Metrics: We follow the original evaluation protocol designated by the official challenge. The evaluation metrics are the mean radial error (MRE), standard deviation (SD), and the successful detection rate (SDR) under 2 mm, 2.5 mm, 3mm and 4 mm [19]. Besides, we also report the inference time, model paramerters and floating point operations (FLOPs) for efficiency comparison.

Implementation Details: We build two architecture variants for 2D and 3D tasks on Pytorch, respectively. For 2D tasks, the input image size in the first and second stage are both 800×640. In the encoder of the U shape hybrid architecture, the group size of LGA is gradually decreased from 32×24, 16×12, 8×6 to 4×3. The encoder and the decoder are symmetric. In the fine-scale detection stage, the number of CephalFormer blocks N is set as 12 while the embedding dimension g is set as 96. For 3D tasks, the input volume size in the first and second stage are both $518 \times 518 \times 384$ for PDDCA dataset ($512 \times 512 \times 512$ for in-house skull dataset). The group size of LGA is gradually decreased from $32 \times 32 \times 24$, $16 \times 16 \times 12$, $8 \times 8 \times 6$ to $4 \times 4 \times 3$. In the fine-scale detection stage, N is set as 8 while g is set as 64. We adopt cross-entropy loss for heatmap regression and L_1 loss for coordinate regression. The total loss is a weighted sum of these two losses. All experiments are performed on 8 V100 GPUs. We employ an AdamW optimizer for 1000 epochs using a cosine decay learning rate scheduler and 20 epochs of linear warm-up. A mini-batch size of 8, an initial learning rate of 0.001, and a weight decay of 0.05 are used. We use gradient clipping with a max norm of 1.0 to stabilize the training

Table 1. Evaluation metrics of different models on ISBI dataset. * represents the performances copied from the original paper. - represents that no experimental results can be found in the original paper. The best results are in bold.

Model	Time (S)	FLOPs (G)	#param. (M)	Test 1 dataset					Test 2 dataset				
				MRE (SD)	2 mm	2.5 mm	3 mm	4 mm	MRE (SD)	2 mm	2.5 mm	3 mm	4 mm
Arik et al. [1]*	–	–	–	-(-)	75.37	80.91	84.32	88.25	-(-)	67.68	74.16	79.11	84.63
Qian et al. [14]*	1.04	10.5	42.5	1.28(1.17)	82.51	86.25	89.30	90.62	1.54(1.05)	72.40	76.15	79.65	85.90
Chen et al. [3]*	**0.07**	–	–	1.17(-)	86.67	92.67	95.54	98.53	1.48(-)	75.05	82.84	88.53	95.05
DRM [26]	–	–	–	1.12(0.88)	86.91	91.82	94.88	97.90	1.42(0.84)	76.00	82.90	88.74	94.32
Cascaded [25]*	2.03	0.6	69.1	1.34(0.92)	81.37	89.09	93.79	97.86	1.64(0.91)	70.58	79.53	86.05	93.32
Zhou et al. [27]*	–	–	–	-(-)	–	–	–	–	1.39(1.32)	76.11	84.21	88.79	94.84
Swin-CE [23]*	1.74	56.7	88.3	1.13(-)	87.31	92.93	95.87	98.74	1.44(-)	76.23	83.36	89.15	95.24
GU2Net [29]	0.22	**0.5**	**4.4**	1.12(1.27)	87.21	92.55	95.60	97.70	1.37(1.16)	74.11	85.05	89.39	94.04
SCN [13]	0.29	9.2	29.6	1.34(1.22)	81.47	89.36	93.15	97.01	1.65(1.12)	69.94	78.84	85.74	93.89
DACFL [12]*	0.15	–	–	1.18(1.01)	86.20	91.20	94.41	97.68	1.47(0.82)	75.92	83.44	89.27	94.73
DAG [10]	0.67	0.8	23.9	1.04(0.95)	88.49	93.12	95.72	98.42	1.43(0.80)	76.57	83.68	88.21	94.31
CephalFormer-2D	0.10	2.9	25.1	**0.92(0.84)**	**89.79**	**94.11**	**96.33**	**98.94**	**1.26(0.75)**	**80.44**	**87.63**	**91.40**	**96.29**

process. Augmentations such as rotation, scaling, gaussian blur, and mirroring are utilized during the training process.

3.2 Comparisons with State-of-the-Art Methods

The results on ISBI datset are shown in Table 1. Our method significantly outperforms the state-of-the-art methods. Especially in the precision of 2 mm, the improvements of SDR on two ISBI test datasets are 1.3% and 3.9%, respectively. Moreover, our CephalFormer-2D behaves much better than SCN, DACFL and DAG. These three methods also incorporate prior global structural constraint but use additional modules or GCNs. CephalFormer differs from these approaches in how it models global structural constraint. Above all, previous methods using graphs or additional CNN modules to incorporate structural prior knowledge do not allow for interactions between visual clues and landmarks while CephalFormer is capable of learning any relationship between visual features and landmarks. Compared with Swin-CE [23], which combines Swin Transformer [11] with a CNN encoder for 2D cephalometric landmark detection, our CephalFormer-2D outperforms it with 3x fewer parameters and 18x lower FLOPs. Swin Transformer also models dependencies within local sub-windows, but it cannot model interactions between different windows. While in our GGRA, sub-groups are connected in an efficient way, which significantly improves the detection performance.

The results on PDDCA datset and in-house skull dataset are shown in Table 2. PDDCA dataset has much fewer samples (33 samples vs. 150 samples) than our in-house dataset. However, our CephalFormer-3D is robust to such a small dataset and significantly outperforms other methods. The SDRs is increased by 14% in the precision of 2 mm. As for our in-house dataset, the average error of three-fold cross-validation is 1.33 mm, and 82.11% of landmarks are within the 2 mm clinically acceptable errors. The inference time of our method on these two datasets is 0.14 and 0.33 s, respectively. CephalFormer-3D is significantly faster and more efficient than the other two-stage methods and has much

Table 2. Comparison with state-of-the-art methods on 3D tasks.

Model	#param. (M)	PDDCA dataset							In-house skull dataset						
		Time (S)	FLOPs (G)	MRE (SD)	2 mm	2.5 mm	3 mm	4 mm	Time (S)	FLOPs (G)	MRE	2 mm	2.5 mm	3 mm	4 mm
LA-GCN [9]	288.1	93.21	107.4	3.23 (2.52)	35.68	46.76	58.19	69.48	287.38	222.9	2.13(1.72)	67.22	70.01	79.33	85.17
SA-LSTM [4]	75.9	0.27	26.5	2.37 (1.60)	56.36	71.60	80.00	89.99	1.29	68.1	1.81(1.27)	72.57	80.46	84.69	88.62
CephalFormer-3D	35.3	0.14	8.3	1.85(1.44)	70.31	80.29	88.16	92.34	0.33	26.4	1.33 (1.07)	82.11	89.53	93.45	96.77

Table 3. Ablation study of network components.

Model	ISBI test 2 dataset					PDDCA dataset				
	MRE (SD)	2 mm	2.5 mm	3 mm	4 mm	MRE (SD)	2 mm	2.5 mm	3 mm	4 mm
U-Net [16]	1.62 (1.46)	70.11	75.61	80.20	85.39	3.86 (2.55)	30.14	39.26	47.92	60.88
nnFormer [28]	1.63 (1.55)	69.13	79.66	86.19	90.51	3.39 (2.64)	32.91	44.73	56.98	67.53
ϕ_c	1.45 (1.38)	76.29	84.33	88.80	94.82	2.33 (1.68)	55.48	70.87	79.16	89.10
$\phi_c + GCNs$	1.39 (0.77)	77.81	85.05	89.10	95.14	2.11 (1.57)	62.29	74.33	83.75	90.08
CephalFormer	1.26 (0.75)	80.44	87.63	91.40	96.29	1.85 (1.44)	70.31	80.29	88.16	92.34

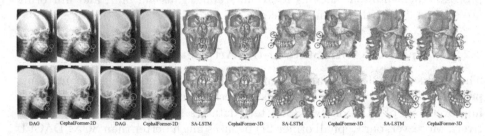

Fig. 3. Qualitative comparison of different models on ISBI dataset and our in-house skull dataset. The red points are the learned landmarks while the green points are the ground truth labels. We mark significant improvements with circles. It is clear that our model's results have more overlap regions of red and green points than other methods. (Color figure online)

fewer paramerters and lower FLOPs. The qualitative comparison of Cephal-Former and suboptimal methods is shown in Fig. 3. Note that directly transferring 2D methods to 3D leads to deficient performance, so their results are not reported.

3.3   Ablation Study

We perform ablation study on the two public benchmarks. First, in order to demonstrate the effectiveness of our LGA and GGRA, we replace the U shape hybrid structure in the first stage with several other close alternatives, i.e., U-Net [16] and nnFormer [28]. Second, we evaluate the performance on ϕ_c (with coarse landmark detection network only), $\phi_c + GCNs$ (with GCNs to incorporate the global structural constraint for fine-scale coordinate refinement), and CephalFormer to demonstrate the effectiveness of modeling interactions between visual features and landmarks via the Transformer encoder in the second stage.

As shown in Table 3, when comparing CephalFormer with U-Net and nnFormer, it is evident that our U shape hybrid structure consisting of LGA and GGRA improves the performance by a large margin. This is attributed to the efficiency and effectiveness of our new self-attention mechanism, which enables whole-volume inference, whereas U-Net lacks the ability to model long-range dependencies and patch-based nnFormer has limited receptive fields. By comparing CephalFormer with ϕ_c and $\phi_c + GCNs$, we observe much improvement of detection accuracy, which demonstrates the effectiveness of embeddings learned with the dependencies between visual features and landmarks.

4 Conclusion

In this paper, we propose a novel end-to-end Transformer-based framework called CephalFormer for 2D and 3D cephalometric landmark detection. It first locates the coarse landmarks and then refines them via exploring complex dependencies between visual features and landmarks. Our method adaptively learns the global structure constraint on the spatial configuration of landmarks and thus significantly improves the cephalometric landmark detection accuracy in the clinically accepted precision range of 2.0 mm.

Acknowledgements. This research was partially supported by the National Natural Science Foundation of China under Grant 61972343, the National Major Scientific Research Instrument Development Project under Grant 81827804, and the Key Research and Development Program of Zhejiang Province under Grant 2021C03032.

References

1. Arik, S.Ö., Ibragimov, B., Xing, L.: Fully automated quantitative cephalometry using convolutional neural networks. J. Med. Imaging **4**(1), 014501 (2017)
2. Chen, J., et al.: TransUNet: transformers make strong encoders for medical image segmentation. arXiv preprint arXiv:2102.04306 (2021)
3. Chen, R., Ma, Y., Chen, N., Lee, D., Wang, W.: Cephalometric landmark detection by attentive feature pyramid fusion and regression-voting. In: Shen, D., et al. (eds.) MICCAI 2019. LNCS, vol. 11766, pp. 873–881. Springer, Cham (2019). https://doi.org/10.1007/978-3-030-32248-9_97
4. Chen, R., et al.: Structure-aware long short-term memory network for 3d cephalometric landmark detection. IEEE Trans. Med. Imaging (2022)
5. He, K., Gkioxari, G., Dollár, P., Girshick, R.: Mask R-CNN. In: Proceedings of the IEEE International Conference on Computer Vision, pp. 2961–2969 (2017)
6. Hochreiter, S., Schmidhuber, J.: Long short-term memory. Neural Comput. **9**(8), 1735–1780 (1997)
7. Hurst, C.A., Eppley, B.L., Havlik, R.J., Sadove, A.M.: Surgical cephalometrics: applications and developments. Plast. Reconstr. Surg. **120**(6), 92e–104e (2007)
8. Kipf, T.N., Welling, M.: Semi-supervised classification with graph convolutional networks. arXiv preprint arXiv:1609.02907 (2016)

9. Lang, Y., et al.: Automatic localization of landmarks in craniomaxillofacial CBCT images using a local attention-based graph convolution network. In: Martel, A.L., et al. (eds.) MICCAI 2020. LNCS, vol. 12264, pp. 817–826. Springer, Cham (2020). https://doi.org/10.1007/978-3-030-59719-1_79

10. Li, W., et al.: Structured landmark detection via topology-adapting deep graph learning. In: Vedaldi, A., Bischof, H., Brox, T., Frahm, J.-M. (eds.) ECCV 2020. LNCS, vol. 12354, pp. 266–283. Springer, Cham (2020). https://doi.org/10.1007/978-3-030-58545-7_16

11. Liu, Z., et al.: Swin transformer: hierarchical vision transformer using shifted windows. In: Proceedings of the IEEE/CVF International Conference on Computer Vision, pp. 10012–10022 (2021)

12. Oh, K., Oh, I.S., Lee, D.W., et al.: Deep anatomical context feature learning for cephalometric landmark detection. IEEE J. Biomed. Health Inform. **25**(3), 806–817 (2020)

13. Payer, C., Štern, D., Bischof, H., Urschler, M.: Integrating spatial configuration into heatmap regression based CNNs for landmark localization. Med. Image Anal. **54**, 207–219 (2019)

14. Qian, J., Cheng, M., Tao, Y., Lin, J., Lin, H.: CephaNet: an improved faster R-CNN for cephalometric landmark detection. In: 2019 IEEE 16th International Symposium on Biomedical Imaging (ISBI 2019), pp. 868–871. IEEE (2019)

15. Raudaschl, P.F., et al.: Evaluation of segmentation methods on head and neck CT: auto-segmentation challenge 2015. Med. Phys. **44**(5), 2020–2036 (2017)

16. Ronneberger, O., Fischer, P., Brox, T.: U-net: convolutional networks for biomedical image segmentation. In: Navab, N., Hornegger, J., Wells, W.M., Frangi, A.F. (eds.) MICCAI 2015. LNCS, vol. 9351, pp. 234–241. Springer, Cham (2015). https://doi.org/10.1007/978-3-319-24574-4_28

17. Vaswani, A., et al.: Attention is all you need. Advances in Neural Inf. Process. Syst. **30** (2017)

18. Wang, C.W., et al.: Evaluation and comparison of anatomical landmark detection methods for cephalometric x-ray images: a grand challenge. IEEE Trans. Med. Imaging **34**(9), 1890–1900 (2015)

19. Wang, C.W., et al.: A benchmark for comparison of dental radiography analysis algorithms. Med. Image Anal. **31**, 63–76 (2016)

20. Wang, W., Chen, C., Ding, M., Yu, H., Zha, S., Li, J.: TransBTS: multimodal brain tumor segmentation using transformer. In: de Bruijne, M., et al. (eds.) MICCAI 2021. LNCS, vol. 12901, pp. 109–119. Springer, Cham (2021). https://doi.org/10.1007/978-3-030-87193-2_11

21. Xu, W., Xu, Y., Chang, T., Tu, Z.: Co-scale conv-attentional image transformers. In: Proceedings of the IEEE/CVF International Conference on Computer Vision, pp. 9981–9990 (2021)

22. Yao, Q., Quan, Q., Xiao, L., Kevin Zhou, S.: One-shot medical landmark detection. In: de Bruijne, M., et al. (eds.) MICCAI 2021. LNCS, vol. 12902, pp. 177–188. Springer, Cham (2021). https://doi.org/10.1007/978-3-030-87196-3_17

23. Yueyuan, A., Hong, W.: Swin transformer combined with convolutional encoder for cephalometric landmarks detection. In: 2021 18th International Computer Conference on Wavelet Active Media Technology and Information Processing (ICCWAMTIP), pp. 184–187. IEEE (2021)

24. Yun, H.S., Hyun, C.M., Baek, S.H., Lee, S.H., Seo, J.K.: Automated 3D cephalometric landmark identification using computerized tomography. arXiv preprint arXiv:2101.05205 (2020)

25. Zeng, M., Yan, Z., Liu, S., Zhou, Y., Qiu, L.: Cascaded convolutional networks for automatic cephalometric landmark detection. Med. Image Anal. **68**, 101904 (2021)
26. Zhong, Z., Li, J., Zhang, Z., Jiao, Z., Gao, X.: An attention-guided deep regression model for landmark detection in cephalograms. In: Shen, D., et al. (eds.) MICCAI 2019. LNCS, vol. 11769, pp. 540–548. Springer, Cham (2019). https://doi.org/10.1007/978-3-030-32226-7_60
27. Zhou, G.Q., et al.: Learn fine-grained adaptive loss for multiple anatomical landmark detection in medical images. IEEE J. Biomed. Health Inform. **25**(10), 3854–3864 (2021)
28. Zhou, H.Y., Guo, J., Zhang, Y., Yu, L., Wang, L., Yu, Y.: nnFormer: interleaved transformer for volumetric segmentation. arXiv preprint arXiv:2109.03201 (2021)
29. Zhu, H., Yao, Q., Xiao, L., Zhou, S.K.: You only learn once: universal anatomical landmark detection. In: de Bruijne, M., et al. (eds.) MICCAI 2021. LNCS, vol. 12905, pp. 85–95. Springer, Cham (2021). https://doi.org/10.1007/978-3-030-87240-3_9

ORF-Net: Deep Omni-Supervised Rib Fracture Detection from Chest CT Scans

Zhizhong Chai[1]([✉]), Huangjing Lin[1], Luyang Luo[2], Pheng-Ann Heng[2,3], and Hao Chen[4]

[1] Imsight AI Research Lab, Shenzhen, China
chaizhizhong@imsightmed.com
[2] Department of Computer Science and Engineering,
The Chinese University of Hong Kong, Hong Kong, China
[3] Guangdong Provincial Key Laboratory of Computer Vision and Virtual Reality
Technology, Shenzhen Institutes of Advanced Technology,
Chinese Academy of Sciences, Shenzhen, China
[4] Department of Computer Science and Engineering,
The Hong Kong University of Science and Technology, Hong Kong, China

Abstract. Most of the existing object detection works are based on the bounding box annotation: each object has a precise annotated box. However, for rib fractures, the bounding box annotation is very labor-intensive and time-consuming because radiologists need to investigate and annotate the rib fractures on a slice-by-slice basis. Although a few studies have proposed weakly-supervised methods or semi-supervised methods, they could not handle different forms of supervision simultaneously. In this paper, we proposed a novel omni-supervised object detection network, which can exploit multiple different forms of annotated data to further improve the detection performance. Specifically, the proposed network contains an omni-supervised detection head, in which each form of annotation data corresponds to a unique classification branch. Furthermore, we proposed a dynamic label assignment strategy for different annotated forms of data to facilitate better learning for each branch. Moreover, we also design a confidence-aware classification loss to emphasize the samples with high confidence and further improve the model's performance. Extensive experiments conducted on the testing dataset show our proposed method outperforms other state-of-the-art approaches consistently, demonstrating the efficacy of deep omni-supervised learning on improving rib fracture detection performance.

Keywords: Omni-supervised learning · Rib fracture · Object detection

1 Introduction

Rib fractures are the most common disease in thoracic injuries [20]. Although most fractures only require conservative treatment, the complications caused by

© The Author(s), under exclusive license to Springer Nature Switzerland AG 2022
L. Wang et al. (Eds.): MICCAI 2022, LNCS 13433, pp. 238–248, 2022.
https://doi.org/10.1007/978-3-031-16437-8_23

rib fractures can be excruciating to patients [18]. In addition, the mortality rate will rise with the increasing number of rib fractures [20]. Therefore, the accurate recognition and location of rib fracture have important clinical value for patients with thoracic trauma. However, the identification of rib fractures in Computed Tomography (CT) images is tedious and labor-intensive, especially for subtle rib fractures and buckle rib fractures, resulting in missed and misdiagnosed rib fractures in clinical practice [2,17]. Recently, deep learning (DL) has achieved comparable performance with the experienced radiologists [24,25,29], based on enormous annotated data. Although detailed annotations are essential to DL-based disease detection [10], they are time-consuming and expertise-demanding, which is a considerable challenge for annotating rib fractures.

To ease the dependence on expensive annotated data, some studies have proposed methods [5,9,19,22] based on weakly supervised learning or semi-supervised learning. For example, Wang et al. [23] proposed an adaptive asymmetric label sharping algorithm to address the class imbalance problem for weakly supervised object detection. Zhou et al. [28] developed an adaptive consistency cost function to regularize the predictions from different components. However, practical applications are usually faced with a variety of different supervision labels. To address the above issue, omni-supervised learning [13] was proposed to leverage different types of available labels to jointly train the model and boost the performance. For example, Ren et al. [15] proposed a unified object detection framework that can handle several different forms of supervision simultaneously. Luo et al. [11] presented a unified framework that can simultaneously utilize strong-annotated, weakly-annotated, and unlabeled data for chest X-ray disease detection. However, both of the above methods are based on the anchor-based detection networks, which have many hyper-parameters that require careful design and tuning.

A potential challenge of omni-supervised learning is that it requires designing reliable label assignment strategies for different forms of annotated data, so as to enable the model to be effectively trained with different supervision. The label assigning strategies commonly used in exiting object detection works can be separated into two categories: fixed label assignment and dynamic label assignment. The fixed label assignment-based methods [14,21] are heavily based on hand-crafted rules. For example, Faster-RCNN [14] uses the IoU thresholds to define the positive and negative proposals generated from the region proposal network (RPN). FCOS [21] regards the pixels in the center region of the object bounding box as positives with a centerness score to refne the confidence. The dynamic label assignment-based methods [27,30] propose adaptive mechanisms to determine the positives and negatives. ATSS [27] adaptively selects positive and negative samples according to the statistical characteristics of objects. AutoAssign [30] proposed an appearance-aware and fully differentiable weighting mechanism for label assignment in both spatial and scale dimensions.

Inspired by the above dynamic label assignment-based methods for object detection, we design an omni-supervised framework that can utilize different annotation forms of data in a label assignment manner. Specifically, the pro-

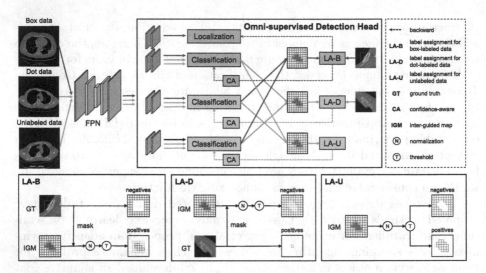

Fig. 1. Overview of our proposed framework. The network consists of a Feature Pyramid Network (FPN [7]) as the backbone, and an omni-supervised detection head to predict the classification score and localization information. For each form of annotated data, there is a corresponding classification branch that is trained using a dynamic label assignment strategy. The proposed confidence-aware classification loss is adopted on the uncertain regions of different annotated forms of data.

posed omni-supervised network contains an omni-supervised detection head to handle the training of different annotated forms of data. Furthermore, to dynamically conduct the label assignment for each annotation form of data, we design a label assignment strategy that uses the prediction maps from the multiple classification branches as guidance. Moreover, a confidence-aware classification loss is also introduced to emphasize the samples with high confidence and suppress the noise candidates. Extensively experimental results show our proposed method outperforms other state-of-the-art approaches consistently, demonstrating the efficacy of deep omni-supervised learning on the task of rib fracture detection.

2   Method

2.1   Problem Statement and Formulation

Formally, let us define the training dataset as $D = [\mathcal{D}_b, \mathcal{D}_d, \mathcal{D}_u]$, where $\mathcal{D}_b$ is a bounding box annotated dataset, $\mathcal{D}_d$ is a dot annotated dataset with a single point for each object, and $\mathcal{D}_u$ is an unlabeled dataset. Intuitively, there are two types of the region for each form of labeled data: certain and uncertain. For box-labeled data, the region outside the labeled boxes can be regarded as certain negative samples to supervise the model. However, for the regions inside the labeled boxes, we cannot directly define their positive and negative attributes. If we simply treat all the samples in the box as positive samples, it will undoubt-edly bring learning difficulties and performance degradation to the model [30]. Similarly, for dot-labeled data, the labeled points can be regarded as certain positive samples, but the locations in other regions have no explicit positive and negative attributes. In this paper, we aim to design a label assignment strategy that can dynamically select high-quality samples from these uncertain regions during training for omni-supervised object detection.

Specifically, we build our model based on the one-stage detection network FCOS [21], which predicts the objects in an anchor-free manner. Figure 1 illustrates the overview of our proposed method. The proposed omni-supervised detection framework consists of a Feature Pyramid Network (FPN [7]) and an omni-supervised detection head. The omni-supervised detection head contains a localization head to regress the localization offsets, and multiple classification branches to predict the probability score for different annotated forms of data. Since dot-labeled data and unlabeled data do not have specific labeling boxes, we only train the localization branch with the box-labeled data.

2.2 Omni-Supervision with Dynamic Label Assignment

To enable effectively learning from different labeled data and unlabeled data, we design a dynamic label assignment strategy that combines the prediction scores from the multiple classification branches and its own annotations to guide the model's training. Inspired by the success of co-training [1,26], which minimizes the divergence by assigning pseudo labels between each view on unlabeled data, we use the prediction maps from the other branches to generate the confidence map, which we called the inter-guided map (IGM). By this mutual supervision mechanism, we could prompt different branches to maximize their agreement on different forms of annotated data for better performance. Suppose P_b, P_d, P_u represent the predictions of the box-labeled branch, the dot-labeled branch, and the unlabeled branch, respectively, we calculate the inter-guided map with the prediction maps from the other two branches as follows:

$$W_b = N((P_d * P_u)^{\frac{1}{2}}) , W_d = N((P_b * P_u)^{\frac{1}{2}}) , \text{ and } W_u = N((P_b * P_d)^{\frac{1}{2}}) \quad (1)$$

where N denotes the normalization function which linearly rescales the values to the range of $[0, 1]$ and ensures effective learning of hard instances which usually have a small value for the corresponding positive locations. Note that we only use the uncertain region to generate the corresponding IGM map. For box-labeled data, the region in each annotated box is normalized separately. We then discuss the label assignment strategy for each annotated form of data separately.

Box Supervision: Different from FCOS [21] which uses a fixed fraction of the center area as positive samples for each object, we use a prediction-guided map to dynamically filter the locations of the bounding box as the reliable positive samples to train the model. Specifically, given an annotated bounding box j, we first calculate its iter-guided map W_b^j, and then we assign the location whose probability is greater or equal to the threshold t in W_b^j as a positive sample to train the box-labeled classification branch. For the negative regions S_n outside the labeled boxes of the box-labeled data, the focal loss [8] is adopted to handle the severe imbalance problem. The positive loss and negative loss for box annotated data are computed as follows:

$$\mathcal{L}_b^p = -\sum_j^m \sum_{W_b^{ij}>=t}^{S_p^j} (1 - P_b^{ij})^\gamma \log P_b^{ij} \tag{2}$$

$$\mathcal{L}_b^n = -\sum_i^{S_n} (P_b^i)^\gamma \log (1 - P_b^i) \tag{3}$$

where m denotes the number of annotated boxes in the slice, i denotes the i-th location from the FPN feature maps, and S_p^j denotes the number of locations in the bounding box j. For the localization branch, we use the generalized IoU (GIoU) [16] loss (L_{GIoU}) as the objective for bounding box localization:

$$\mathcal{L}_b^{reg} = \sum_j^m \sum_{W_b^i>=t}^{S_p^j} L_{\text{GIoU}}(k_i, \hat{k}_i) \tag{4}$$

where k and $\hat{k}$ denote the predicted bounding boxes and the corresponding ground-truth boxes.

Dot Supervision: For dot-annotated data, we assume the annotated locations are certain positive samples, and the other locations are uncertain regions where we should conduct the label assignment strategy. Specifically, we use the focal loss [8] to train the model on the annotated points. For the locations beyond the labeled points, We first calculate the inter-guided map according to Eq. (1) and then treat the locations whose probability is smaller than the threshold t as negative samples to train the dot classification branch. The positive loss and negative loss for dot-labeled data are computed as follows:

$$\mathcal{L}_d^p = -\sum_i^l (1 - P_d^i)^\gamma \log P_d^i \tag{5}$$

$$\mathcal{L}_d^n = -\sum_{W_d^i<t}^{R_n} (P_d^i)^\gamma \log (1 - P_d^i) \tag{6}$$

where l denotes the number of annotated dots in the slice, and R_n denotes the number of unlabeled locations on the dot-labeled feature maps.

Unlabeled Supervision: Since there is no certain region in the unlabeled data, we directly conduct the label assignment strategy on all locations on the feature maps. Specifically, we compute the inter-guided map of the locations and assign the locations whose probability is greater or equal to the threshold t as positive samples, the others as negative samples. The loss function is as follows:

$$\mathcal{L}_u^p = - \sum_{W_u^i >= t}^{T_p} (1 - P_u^i)^\gamma \log P_u^i \tag{7}$$

$$\mathcal{L}_u^n = - \sum_{W_u^i < t}^{T_n} (P_u^i)^\gamma \log (1 - P_u^i) \tag{8}$$

where T_p and T_n denote the number of assigned positive samples and negative samples, respectively.

Confidence-Aware Classification Loss: Although the above label assignment strategy can enable the model to be trained on data with different annotation types, directly dividing the positive and negative samples with a threshold may limit the detection performance of the model. To address this problem, we propose a confidence-aware (CA) classification loss for the uncertain regions of each annotated form of data to further emphasize the locations with high confidence. The classification losses are computed as follows:

$$\mathcal{L}^p = - \sum_{W^i >= t}^{S} (1 - P^i)^\gamma \log (P^i(1 - W^i)) \tag{9}$$

$$\mathcal{L}^n = - \sum_{W^i < t}^{S} (P^i)^\gamma \log (1 - P^i(1 - W^i)) \tag{10}$$

where W and P denote the corresponding inter-guided map and prediction map, and S denotes the number of locations on the FPN feature maps.

2.3   Optimization and Implementation Details

The final training loss is defined as follows:

$$\mathcal{L}_{total} = \mathcal{L}_b^{reg} + \mathcal{L}_b^p + \mathcal{L}_b^n + \lambda(\mathcal{L}_d^p + \mathcal{L}_d^n) + \beta(\mathcal{L}_u^p + \mathcal{L}_u^n) \tag{11}$$

where λ and β are empirically set to 1. We use FCOS [21] with ResNet-50 [4] backbone pre-trained on ImageNet [3] and FPN as our base model. All experiments are conducted based on Pytorch [12]. During training, we adopt 1 TITAN Xp GPU with a batch size of 3, where the three different annotation types of data are equally sampled. During the testing phase, we first ensemble the prediction results of different classification branches, and then combine the regression result from the localization branch to generate the final detection result. Random flip is applied for data augmentation. We use Stochastic Gradient Descent (SGD) with a momentum of 0.9 to update the weight of the model, and the initial learning rate is 0.001 and multiplied by 0.1 every 30000 iterations. We set threshold t as 0.5 for all the annotated forms of data to filter the training samples from the uncertain regions. We adopt non-maximum suppression (NMS) with IoU threshold of 0.6 for post-processing in all experiments.

3   Experiments and Results

3.1   Dataset and Evaluation Metrics

Dataset: We collect a total of 2239 CT images from patients with rib fractures, of which 685 were box-annotated data, 450 were dot-annotated data, and 1104 were unlabeled data. Then we divide the 685 cases of box annotation data into training dataset (224), validation dataset (151), and testing dataset (310). **Evaluation Metrics:** The mean Average Precision (mAP) from AP40 to AP75 with an interval of 5^1 and AP50[2] are adopted as the evaluation metrics.

3.2   Comparison with State-of-the-Art Methods

To the best of our knowledge, few studies have been proposed to simultaneously leverage the box-labeled data, dot-labeled data, and unlabeled data for object detection. Hence, we first train the supervised object detection model FCOS [21] as the baseline model. Then we implement several semi-supervised methods including STAC [19], Unbiased Teacher [9], Π Model [6] and AALS [23]. Note that, we enable the learning from the dot labeled data by training the supervised or semi-supervised model with only the annotated positive points.

Table 1. Quantitative comparisons with different methods on the testing dataset.

Method	#Number of CT scans			Metrics	
	Box-labeled	Dot-labeled	Unlabeled	mAP	AP50
FCOS [21]	224	0	0	39.9	53.7
FCOS [21]	224	450	0	41.3	54.4
ORF-Net	224	450	0	**42.3**	**56.3**
STAC [19]	224	450	1104	40.0	56.1
UT [9]	224	450	1104	42.6	56.3
Π Model [6]	224	450	1104	42.9	56.3
AALS [23]	224	450	1104	43.4	57.2
ORF-Net	224	450	1104	**44.3**	**59.1**

---

[1] https://www.kaggle.com/c/rsna-pneumonia-detection-challenge.
[2] https://cocodataset.org/#detection-eval.

Table 2. Ablation studies on the validation dataset.

Method	#Number of CT scans			Metrics	
	Box-labeled	Dot-labeled	Unlabeled	mAP	AP50
FCOS [21]	224	0	0	36.3	51.8
ORF-Net (SGM)	224	450	0	38.8	53.7
ORF-Net (IGM)	224	450	0	**39.9**	55.2
ORF-Net (IGM+CA)	224	450	0	**39.9**	**56.6**
ORF-Net (SGM)	224	450	1104	40.2	56.6
ORF-Net (IGM)	224	450	1104	41.6	57.8
ORF-Net (IGM+CA)	224	450	1104	**42.8**	**57.9**

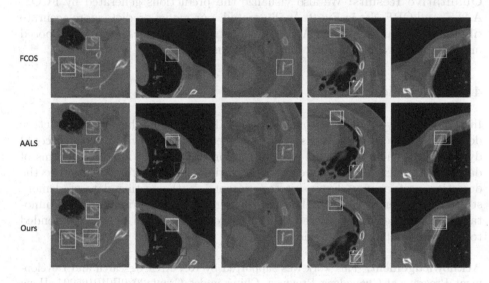

Fig. 2. Qualitative comparisons of the FCOS [21], AALS [23], and Our proposed method on the testing set. Red rectangles stand for ground truths, green rectangles stand for true positives, and blue rectangles stand for false positives. (Color figure online)

Quantitative Results: As shown in Table 1, by simply training the annotated positive samples from the dot labeled data, the fully supervised method FCOS [21] can also achieve an improvement of 1.4% and 0.7% on mAP and AP50. Our method achieved a large improvement on both mAP and AP50 (2.4%, 2.6%), which demonstrated the effectiveness of the proposed ORF-Net on leveraging the dot-labeled data. By training with all the different forms of annotation, all the semi-supervised methods improved the fully supervised baseline, demonstrating effectiveness in utilizing the unlabeled data. The proposed method outperformed all other methods with at least 0.9% in mAP, and 1.9% in AP50, demonstrating the efficacy of omni-supervised learning on the task of rib fracture detection.

Ablation Study: For a better comparison, we also implement a self-guided map (SGM) based label assignment strategy which uses its own classification scores and annotations to conduct the label assignment. As shown in Table 2, compared with the SGM based label assignment strategy, the IGM based label assignment strategy has an improvement of 1.% and 1.5% on mAP and AP50 with the dot labeled data, and an improvement of 1.4% and 1.5% on the mAP and AP50 with the dot labeled data and unlabeled data, demonstrating the effectiveness of the proposed IGM-based label assignment strategy. For the confidence-aware (CW) loss, we observe an improvement of 1.4% on AP50 with the dot labeled data, and an improvement of 1.2% on the mAP with both the dot labeled data and unlabeled data, which validate the efficiency of the proposed soft regularization.

Qualitative Results: We also visualize the predictions generated by FCOS, AALS, and ORF-Net in Fig. 2. As illustrated, our model predicts more accurate rib fractures than FCOS and AALS, demonstrating the efficacy of the proposed method on leveraging different annotated forms of data.

4   Conclusion

In this paper, we present an omni-supervised learning method for rib fracture detection from chest CT scans. The proposed omni-supervised network could dynamically conduct the label assignment for the different annotated forms of data and thus exploit the supervision of various levels to further improve the detection performance. Extensive experiments on the testing dataset demonstrate the efficiency of our method in utilizing the various granularities of annotations. Moreover, the proposed method is general and can be easily extended to other tasks of object detection.

Acknowledgement. This work was supported by Key-Area Research and Development Program of Guangdong Province, China under Grant 2020B010165004, Hong Kong RGC TRS Project No. T42-409/18-R, and Shenzhen Science and Technology Innovation Committee Funding (Project No. SGDX20210823103201011).

References

1. Blum, A., Mitchell, T.: Combining labeled and unlabeled data with co-training. In: Proceedings of the Eleventh Annual Conference on Computational Learning Theory, pp. 92–100 (1998)
2. Cho, S., Sung, Y., Kim, M.: Missed rib fractures on evaluation of initial chest CT for trauma patients: pattern analysis and diagnostic value of coronal multiplanar reconstruction images with multidetector row CT. Br. J. Radiol. **85**(1018), e845–e850 (2012)
3. Deng, J., Dong, W., Socher, R., Li, L.J., Li, K., Fei-Fei, L.: ImageNet: a large-scale hierarchical image database. In: 2009 IEEE Conference on Computer Vision and Pattern Recognition, pp. 248–255. IEEE (2009)

4. He, K., Zhang, X., Ren, S., Sun, J.: Deep residual learning for image recognition. In: Proceedings of the IEEE Conference on Computer Vision and Pattern Recognition, pp. 770–778 (2016)
5. Jeong, J., Lee, S., Kim, J., Kwak, N.: Consistency-based semi-supervised learning for object detection. Adv. Neural Inf. Process. Syst. **32** (2019)
6. Laine, S., Aila, T.: Temporal ensembling for semi-supervised learning. arXiv preprint arXiv:1610.02242 (2016)
7. Lin, T.Y., Dollár, P., Girshick, R., He, K., Hariharan, B., Belongie, S.: Feature pyramid networks for object detection. In: Proceedings of the IEEE Conference on Computer Vision and Pattern Recognition, pp. 2117–2125 (2017)
8. Lin, T.Y., Goyal, P., Girshick, R., He, K., Dollár, P.: Focal loss for dense object detection. In: Proceedings of the IEEE International Conference on Computer Vision, pp. 2980–2988 (2017)
9. Liu, Y.C., et al.: Unbiased teacher for semi-supervised object detection. arXiv preprint arXiv:2102.09480 (2021)
10. Luo, L., et al.: Rethinking annotation granularity for overcoming deep short-cut learning: a retrospective study on chest radiographs. arXiv preprint arXiv:2104.10553 (2021)
11. Luo, L., Chen, H., Zhou, Y., Lin, H., Heng, P.-A.: OXnet: deep omni-supervised thoracic disease detection from chest X-rays. In: de Bruijne, M., et al. (eds.) MICCAI 2021. LNCS, vol. 12902, pp. 537–548. Springer, Cham (2021). https://doi.org/10.1007/978-3-030-87196-3_50
12. Paszke, A., et al.: PyTorch: an imperative style, high-performance deep learning library. Adv. Neural Inf. Process. Syst. **32** (2019)
13. Radosavovic, I., Dollár, P., Girshick, R., Gkioxari, G., He, K.: Data distillation: towards omni-supervised learning. In: Proceedings of the IEEE Conference on Computer Vision and Pattern Recognition, pp. 4119–4128 (2018)
14. Ren, S., He, K., Girshick, R., Sun, J.: Faster r-cnn: Towards real-time object detection with region proposal networks. Adv. Neural Inf. Process. Syst. **28** (2015)
15. Ren, Z., Yu, Z., Yang, X., Liu, M.-Y., Schwing, A.G., Kautz, J.: UFO^2: a unified framework towards omni-supervised object detection. In: Vedaldi, A., Bischof, H., Brox, T., Frahm, J.-M. (eds.) ECCV 2020. LNCS, vol. 12364, pp. 288–313. Springer, Cham (2020). https://doi.org/10.1007/978-3-030-58529-7_18
16. Rezatofighi, H., Tsoi, N., Gwak, J., Sadeghian, A., Reid, I., Savarese, S.: Generalized intersection over union: a metric and a loss for bounding box regression. In: Proceedings of the IEEE/CVF Conference on Computer Vision and Pattern Recognition, pp. 658–666 (2019)
17. Ringl, H., et al.: The ribs unfolded-a CT visualization algorithm for fast detection of rib fractures: effect on sensitivity and specificity in trauma patients. Eur. Radiol. **25**(7), 1865–1874 (2015)
18. Sirmali, M., et al.: A comprehensive analysis of traumatic rib fractures: morbidity, mortality and management. Eur. J. Cardiothorac. Surg. **24**(1), 133–138 (2003)
19. Sohn, K., Zhang, Z., Li, C.L., Zhang, H., Lee, C.Y., Pfister, T.: A simple semi-supervised learning framework for object detection. arXiv preprint arXiv:2005.04757 (2020)
20. Talbot, B.S., Gange, C.P., Jr., Chaturvedi, A., Klionsky, N., Hobbs, S.K., Chaturvedi, A.: Traumatic rib injury: patterns, imaging pitfalls, complications, and treatment. Radiographics **37**(2), 628–651 (2017)
21. Tian, Z., Shen, C., Chen, H., He, T.: Fcos: A simple and strong anchor-free object detector. IEEE Trans. Pattern Anal. Mach. Intell. (2020)

22. Wang, D., Zhang, Y., Zhang, K., Wang, L.: FocalMix: semi-supervised learning for 3D medical image detection. In: Proceedings of the IEEE/CVF Conference on Computer Vision and Pattern Recognition, pp. 3951–3960 (2020)
23. Wang, Y., et al.: Knowledge distillation with adaptive asymmetric label sharpening for semi-supervised fracture detection in chest X-rays. In: Feragen, A., Sommer, S., Schnabel, J., Nielsen, M. (eds.) IPMI 2021. LNCS, vol. 12729, pp. 599–610. Springer, Cham (2021). https://doi.org/10.1007/978-3-030-78191-0_46
24. Weikert, T., et al.: Assessment of a deep learning algorithm for the detection of rib fractures on whole-body trauma computed tomography. Korean J. Radiol. **21**(7), 891 (2020)
25. Wu, M., et al.: Development and evaluation of a deep learning algorithm for rib segmentation and fracture detection from multicenter chest CT images. Radiol.: Artif. Intell. **3**(5), e200248 (2021)
26. Xia, Y., et al.: Uncertainty-aware multi-view co-training for semi-supervised medical image segmentation and domain adaptation. Med. Image Anal. **65**, 101766 (2020)
27. Zhang, S., Chi, C., Yao, Y., Lei, Z., Li, S.Z.: Bridging the gap between anchor-based and anchor-free detection via adaptive training sample selection. In: 2020 IEEE/CVF Conference on Computer Vision and Pattern Recognition (CVPR) (2020)
28. Zhou, H.Y., et al.: SSMD: semi-supervised medical image detection with adaptive consistency and heterogeneous perturbation. Med. Image Anal. **72**, 102117 (2021)
29. Zhou, Q.Q., et al.: Automatic detection and classification of rib fractures on thoracic CT using convolutional neural network: accuracy and feasibility. Korean J. Radiol. **21**(7), 869 (2020)
30. Zhu, B., et al.: AutoAssign: differentiable label assignment for dense object detection. arXiv preprint arXiv:2007.03496 (2020)

Point Beyond Class: A Benchmark for Weakly Semi-supervised Abnormality Localization in Chest X-Rays

Haoqin Ji[1,2,3], Haozhe Liu[1,2,3], Yuexiang Li[3], Jinheng Xie[1,2], Nanjun He[3(✉)], Yawen Huang[3], Dong Wei[3], Xinrong Chen[4], Linlin Shen[1,2(✉)], and Yefeng Zheng[3]

[1] Computer Vision Institute, College of Computer Science and Software Engineering, Shenzhen University, Shenzhen, China
llshen@szu.edu.com
[2] AI Research Center for Medical Image Analysis and Diagnosis, Shenzhen University, Shenzhen, China
[3] Jarvis Lab, Tencent, Shenzhen, China
nanjunhe91@163.com
[4] Academy for Engineering and Technology, Fudan University, Shanghai, China

Abstract. Accurate abnormality localization in chest X-rays (CXR) can benefit the clinical diagnosis of various thoracic diseases. However, the lesion-level annotation can only be performed by experienced radiologists, and it is tedious and time-consuming, thus difficult to acquire. Such a situation results in a difficulty to develop a fully-supervised abnormality localization system for CXR. In this regard, we propose to train the CXR abnormality localization framework via a weakly semi-supervised strategy, termed Point Beyond Class (PBC), which utilizes a small number of fully annotated CXRs with lesion-level bounding boxes and extensive weakly annotated samples by points. Such a point annotation setting can provide weakly instance-level information for abnormality localization with a marginal annotation cost. Particularly, the core idea behind our PBC is to learn a robust and accurate mapping from the point annotations to the bounding boxes against the variance of annotated points. To achieve that, a regularization term, namely multi-point consistency, is proposed, which drives the model to generate the consistent bounding box from different point annotations inside the same abnormality. Furthermore, a self-supervision, termed symmetric consistency, is also proposed to deeply exploit the useful information from the weakly annotated data for abnormality localization. Experimental results on RSNA and VinDr-CXR datasets justify the effectiveness of the proposed method. When ≤20% box-level labels are used for training, an improvement of ~5% in mAP can be achieved by our PBC, compared to the current state-of-the-art method (*i.e.*, Point DETR). Code is available at https://github.com/HaozheLiu-ST/Point-Beyond-Class.

Keywords: Weakly supervised learning · Semi-supervised learning · Regularization consistency

H. Ji, H. Liu and Y. Li—Equal Contribution.

© The Author(s), under exclusive license to Springer Nature Switzerland AG 2022
L. Wang et al. (Eds.): MICCAI 2022, LNCS 13433, pp. 249–260, 2022.
https://doi.org/10.1007/978-3-031-16437-8_24

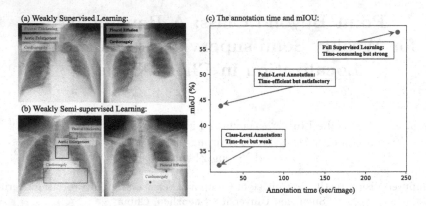

Fig. 1. Our solution to reduce the cost of annotation for abnormality localization in chest X-Rays. (a) is a general solution, i.e., weakly supervised learning, which only utilizes class-wise annotation for lesion detection. Compared to weakly supervised learning (a), the proposed solution (b) adopts limited bounding boxes and extensive point-level labels to train the detector. Based on the analysis of annotation time and performance (c) carried on PASCAL VOC [1], point-level annotation do not significantly increase the time cost but improve performance effectively [1].

1   Introduction

As a noninvasive diagnostic imaging examination, chest X-rays are widely used for screening various thoracic diseases [12,22]. Radiologists routinely need to screen hundreds CXRs per day, which are extremely laborious. To alleviate the workload of radiologists, an accurate automated abnormality localization system for CXRs is worthwhile to develop. With the recent advances in deep neural networks [6,7,11,20], numerous modern object detectors [8–10], such as FCOS [19] and Faster R-CNN [15,16], have been proposed, which can be adopted for abnormality localization. However, training these detectors often requires extensive data annotated with lesion-level bounding boxes. Such lesion-level annotations are difficult to acquire, since the annotation process yields an over-heavy workload for radiologists. Therefore, reducing the annotation cost gradually becomes the core challenge for the development of automated CXR abnormality localization frameworks.

To address such a problem, various weakly supervised learning methods [2,21,23–25] have been proposed. Concretely, the weakly supervised object detection methods adopt the data with weak annotations, instead of lesion-level bounding boxes, for network training. As shown in Fig. 1(a), a typical solution for weakly supervised object detection is using the image-level annotations. These image-level-annotation-based methods localize the objects through region proposals [2], which generally depend on the boundary of the objects. However, the boundaries of lesions in chest X-ray images are commonly not clear; therefore, massive invalid proposals may be generated by these image-level-annotation-based methods. Due to this reason, current weakly supervised object detection methods could not achieve competitive performance against the fully-supervised counterparts. As shown in Fig. 1(c), the empirical study carried by Bearman *et al.* [1] is a solid evidence. The model can only achieve a mean IoU of 32.2% under the weakly-supervised setting with the image-level annotations, which is boosted to 58.3% by switching to the fully-supervised strategy. Recent study [5] proposed a new setting, called weakly semi-supervised object detection (WSSOD), which may be a potential solution for the problem of invalid proposals occurring in current weakly supervised approaches based on image-level annotation. Concretely, as shown in Fig. 1(b), a novel annotation (*i.e.*, a point inside the object) was adopted to train the object detector together with a small number of fully-labeled samples. According to the reported result [1] shown in Fig. 1(c), such a weakly semi-supervised setting can significantly improve the performance of object detection with a marginal extra annotation cost, since the point-level annotation can provide weakly instance-level information.

In this paper, we evaluate the effectiveness of weakly semi-supervised learning strategy for abnormality localization task with chest X-rays, and accordingly establish a publicly-available benchmark by including various baselines, *e.g.,* image-level-annotation-only weakly supervised, semi-supervised and fully-supervised approaches. While transferring the existing WSSOD framework (Point DETR [5]) to CXR abnormality localization task, we notice that the framework is very sensitive to the positions of point annotations. This is because the semantic information contained in the center and boundary of lesion area is different—the points closer to the center provide more useful information for the generation of pseudo bounding boxes. To this end, we propose a regularization term, namely multi-point consistency, to enforce the detector to yield a consistent pseudo bounding box for different points locating in the same lesion area. Furthermore, a self-supervised constraint, termed symmetric consistency, is also proposed to more reasonably explore the weakly annotated data and accordingly boost the abnormality localization performance. Integrating the proposed multi-point consistency and symmetric consistency into the Point DETR [5], we form a new framework, namely Point Beyond Class (PBC), for abnormality localization with CXRs. Experimental results on publicly available CXR datasets show that the proposed PBC method can significantly outperform all the baselines.

2   Revisit of Point DETR

In this section, we firstly review the pipeline and the network architecture of Point DETR [5], and then analyze the challenges unsolved in this scheme.

Pipeline. Referring to the Point DETR, the conventional WSSOD for abnormality localization in chest X-rays can be represented as a multi-stage scheme: 1) Train a teacher model with a small number of CXRs labeled with both points (randomly selected inside the boxes) and lesion-level bounding boxes; 2) Generate pseudo bounding boxes for the CXRs with point-level annotations only using the well-trained teacher model; and 3) Train a student detector by utilizing the samples with ground truth box labels and the ones with pseudo labels.

Network Structure. The Point DETR adopts a point encoder $\mathcal{F}_p(\cdot)$ to embed the point annotation $\mathbb{X}_p$ into a latent space for object query. Specifically, $\mathcal{F}_p(\cdot)$ decomposes $\mathbb{X}_p$ into position $(x, y) \in [0, 1]^2$ and class-wise annotation c. By utilizing a fixed positional encoding method [3, 14, 20], (x, y) can be transferred to a q-dimensional code vector $\mathbf{V}_p \in \mathbb{R}^q$, while we predefine a learnable embedding code vector with the same dimension ($\mathbf{V}_c \in \mathbb{R}^q$) for the category information c. Hence, the object query $\mathbf{V}_f$ can be formulated as: $\mathbf{V}_f = \mathbf{V}_p + \mathbf{V}_c$. Apart from the point encoder, the Point DETR has an image encoder, consisting of a convolutional neural network (CNN) and a Transformer encoder, to encode the CXR images. Concretely, the image encoder firstly embeds the image to feature maps via the CNN. Then, the feature maps are flattened with the positional encoding and fed to the Transformer encoder. The object query $\mathbf{V}_f$ is attended to the image features extracted by image encoder via a Transformer decoder. After that, the output of Transformer decoder is sent to a shared feed forward network (FFN) [3] for bounding box prediction.

Although Point DETR achieved a satisfactory performance on object detection in natural images, there are still some challenges unsolved for adapting the framework for lesion localization in chest X-rays:

In **step 1**, since there is less contextual information contained in gray-scale CXRs, compared to the natural color images, the performance of teacher model might be affected by the positions of point-annotations. Specifically, the points closer to the center of lesion area provides more useful information for pseudo label generation than the ones locating around the lesion boundary.

In **step 2**, the images with point-level annotations are only employed to generate pseudo labels in the pipeline of Point DETR. The rich information contained in the massive weakly-annotated data is not fully exploited.

3   Method: Point Beyond Class

To address the aforementioned challenges, we propose two regularization terms, namely multi-point consistency and symmetric consistency, and form a novel framework, *i.e.,* Point Beyond Class (PBC), by integrating the two terms into Point DETR. The overview of our PBC is shown in Fig. 2, where the proposed multi-point consistency and symmetric consistency are implemented to the step 1 and step 2 of original Point DETR, respectively.

Step 1: Multi-point Consistency for Box-Level Annotation

As shown in Fig. 2, the first step of our PBC is to train a teacher model with fully labeled data, where the model is trained to generate the bounding boxes from the point annotations. The Point DETR [5] is adopted as backbone to process the input CXRs together with their point annotations. Denoted Point DETR as $\mathcal{F}_d(\cdot, \cdot)$, the process of bounding box prediction can be written as:

$$\mathcal{F}_d(\mathbb{X}_p, \mathbb{X}_i) = \hat{\mathbb{Y}}, \tag{1}$$

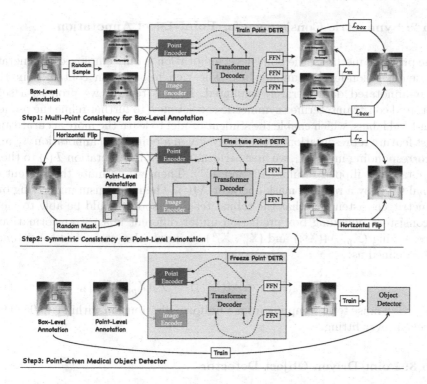

Fig. 2. The pipeline of our proposed method, denoted as PBC, where annotations of different levels are processed separately by following a multi-step strategy.

where $\mathbb{X}_p = (x, y, c)$ is the point annotation with (x, y) for the positional information and c for the abnormality category; $\mathbb{X}_i$ refers to the corresponding CXR and $\hat{\mathbb{Y}}$ is the predicted bounding box. The full objective $\mathcal{L}$ of this step is:

$$\mathcal{L} = \mathcal{L}_{box} + \mathcal{L}_m, \tag{2}$$

where $\mathcal{L}_{box}$ and $\mathcal{L}_m$ are object detection loss and multi-point consistency loss, respectively. Here, we used the same object detection loss defined by DETR [3,5] as $\mathcal{L}_{box}$. Due to the limited amount of fully-labeled data using in this step, the performance of Point DETR is sensitive to the position variation of point annotations (locating around center $vs.$ boundary of bounding box). To mitigate this problem, we propose an auxiliary regularization term, $i.e.$, multi-point consistency. As shown in Fig. 2, given $\mathbb{X}_i$ as an input image, two point annotations can be generated by randomly sampling inside the lesion-level bounding boxes (denoted as $\mathbb{X}_p^1$ and $\mathbb{X}_p^2$). Our multi-point consistency $\mathcal{L}_m$ aims to narrow down the distance between network predictions using $\mathbb{X}_p^1$ and $\mathbb{X}_p^2$, which can be formulated as:

$$\mathcal{L}_m = ||\mathcal{F}_d(\mathbb{X}_p^1, \mathbb{X}_i) - \mathcal{F}_d(\mathbb{X}_p^2, \mathbb{X}_i)||_2. \tag{3}$$

Step 2: Symmetric Consistency for Point-Level Annotation

In the previous methods, the point-level annotation is only employed to generate the pseudo lesion-level bounding box. The rich information contained in the weakly-annotated data is rarely exploited. In this regard, we propose a self-supervised constraint, termed symmetric consistency, to further refine the pseudo lesion-level labels, which enable the student model to learn the more accurate and robust feature representation. Particularly, given a point-level annotation $\mathbb{X}_p$ and the corresponding image $\mathbb{X}_i$, we first perform the flipping operation $\mathcal{T}(\cdot)$ to them and obtain the flipped results ($\mathbb{X}_p^m$ and $\mathbb{X}_i^m$). Then, we permute the content of original CXR by a random mask operator $\mathcal{M}(\cdot)$. The mechanism underlying our symmetric consistency is that the robust teacher model should be able to yield the consistent bounding box prediction under different image transformations. Hence, taking ($\mathbb{X}_p$, $\mathcal{M}(\mathbb{X}^i)$) and ($\mathbb{X}_p^m$, $\mathbb{X}_i^m$) as input, the symmetric consistency can be defined as:

$$\mathcal{L}_c = ||\mathcal{T}(\mathcal{F}_d(\mathbb{X}_p + \sigma, \mathcal{M}(\mathbb{X}_i^m))) - \mathcal{F}_d(\mathbb{X}_p^m, \mathbb{X}_i^m)||_2, \qquad (4)$$

where σ is a noise term sampling from a uniform distribution within $[-0.05, 0.05]$ to prevent over fitting.

Step 3: Point-Driven Object Detector

After the above two steps, we get a well-trained Point DETR ($F_d(\cdot, \cdot)$), which is regarded as the teacher model to generate pseudo box labels for point-level weakly-annotated data. The generated pseudo labels can be employed to train the student model $\mathcal{F}_s(\cdot)$ using any modern detector, *e.g.*, FCOS [19] and Faster R-CNN [15,16], as backbone. The training process can be written as:

$$\min_{\mathcal{F}_s} \underbrace{\mathcal{L}_o(\mathcal{F}_s(\mathbb{X}_i), \mathcal{F}_d(\mathbb{X}_p, \mathbb{X}_i))}_{\text{Point-Level Annotation}} + \underbrace{\mathcal{L}_o(\mathcal{F}_s(\mathbb{X}_i), \mathbb{Y})}_{\text{Box-Level Annotation}}, \qquad (5)$$

where $\mathbb{Y}$ is the box-level annotation of $\mathbb{X}_i$, and $\mathcal{L}_o$ follows the loss function of $\mathcal{F}_s(\cdot)$ adopted by existing studies [15,16,19].

4 Experiments

To validate the effectiveness of our PBC, extensive experiments are conducted on publicly-available datasets, including RSNA [22] and VinDr-CXR [13]. In this section, we first introduce the information of datasets and implementation details, and then construct the benchmarking for weakly semi-supervised abnormality localization in CXRs by comparing our PBC with the state-of-the-art weakly-supervised, semi-supervised and fully-supervised object detectors. Please note that, we have also tested our method on medical segmentation. More details can be found at our ArXiv Version.

Datasets. The RSNA dataset came from a pneumonia detection challenge.[1] The dataset consists of 26,684 CXRs, which can be categorized to negative or pneumonia. The lesion areas in pneumonia CXRs were identified and localized by experienced radiologists. VinDr-CXR dataset,[2] consists of 15,000 CXR images. The dataset providers invited experienced radiologist to annotate lesion areas of 14 thoracic diseases, *e.g.,* aortic enlargement and cardiomegaly.

In this study, we separate each dataset to training and test sets according to the ratio of 80:20. Referring to the setting of weakly semi-supervised learning, we randomly sample 5%, 10%, 20%, 30%, 40%, 50% of training images as fully-labeled samples, while the rest only has the point-level annotation.[3] Since the VinDr-CXR dataset has a long-tailed distribution, *i.e.,* some abnormal categories only contain less than ten samples, we group eight categories with the least numbers of CXRs into one class (denoted as 'Others') to stabilize the network training.

Implementation Details. For a fair comparison, the network architecture of teacher model (*i.e.,* our PBC) is consistent to [5]. The Adam optimizer is adopted for network optimization with an initial learning rate of 1×10^{-4}. Our PBC is observed to converge after 108 epochs of training. For the student detectors, we involve the widely-used FCOS and Faster R-CNN for evaluation. The student detectors are trained with stochastic gradient descent (SGD) optimizer. The model converges after 12 epochs of training. The setting of all hyper-parameters in the student model, including learning rate, weight decay and momentum, follows MMDetection [4].

Baselines and Evaluation Criterion. For the competing methods, we set the fully-supervised detector as the upper bound and point DETR [5] without any regularization as the lower bound. In order to give a more comprehensive analysis of the proposed method, we also include a semi-supervised object detector [17] for comparison. The mean average precision (mAP) is adopted as the evaluation metric. Note that we also evaluate several image-level-annotation-based weakly supervised approaches [18] on the two datasets. However, due to the unclear boundaries of lesion areas, the region proposals are totally inaccurate, which results in an mAP $\leq 5\%$. Hence, we do not include the results in the benchmark.

4.1   Ablation Study

To quantify the contribution of each regularization term in the proposed PBC, we evaluate the performance of the variants with/without each constraint. The evaluation results are presented in Table 1. As shown, the proposed method outperforms the baseline (raw Point DETR) significantly with different numbers of box-level annotations. Using the multi-point consistency constraint, improvements of +4.0% and +9.6% can be achieved with 50% fully labeled data on RSNA and VinDr-CXR datasets, respectively. The performance can be further

---

[1] https://www.kaggle.com/c/rsna-pneumonia-detection-challenge/overview.

[2] https://vindr.ai/datasets/cxr.

[3] The strategy of point-level annotation generation is the same to [5].

Table 1. The ablation study carried on RSNA [22]/VinDr-CXR [13] based on Point
DETR (teacher model) [5] in the terms of mAP (%).

Baseline	Multi-Point	Symmetric Consistency	5%	10%	20%	30%	40%	50%
√	×	×	2.1/8.9	2.8/9.5	9.0/10.1	18.8/10.6	23.7/10.9	25.1/12.3
√	√	×	3.4/9.4	9.5/9.3	18.3/13.4	23.8/15.8	25.6/20.1	29.1/21.9
√	×	√	4.1/8.0	8.4/9.6	14.8/9.9	24.6/10.5	30.6/12.4	32.4/11.8
√	√	√	**6.8/9.5**	**17.8/13.0**	**28.1/15.4**	**34.4/21.9**	**36.9/27.5**	**39.2/28.5**

Table 2. mAP (%) on RSNA [22]/VinDr-CXR [13] based on different student detectors
trained with various methods.

Detector	Method	5%	10%	20%	30%	40%	50%	100%
FCOS [19]	Only Box	14.1/0.9	22.5/19.2	29.7/15.9	33.1/19.7	33.4/24.1	34.4/29.9	43.1/39.2
	Weak-Sup.*	≤5%						
	Semi-Sup.	17.4/1.4	24.2/14.9	32.0/25.0	36.5/24.1	37.8/31.7	38.9/33.5	
	Point DETR	21.8/19.4	28.3/24.9	34.4/27.1	37.3/27.7	38.5/32.3	39.8/34.7	
	PBC (Ours)	**25.8/21.8**	**32.4/25.1**	**37.1/29.9**	**40.5/31.3**	**40.2/34.4**	**42.6/36.4**	
Faster R-CNN	Only Box	14.0/2.0	17.0/13.3	24.9/27.4	29.9/31.2	33.4/34.4	35.6/35.7	43.0/39.5
	Weak-Sup.	≤5%						
	Semi-Sup.	14.6/13.5	19.5/21.5	30.4/29.9	33.7/32.0	37.3/34.6	39.5/36.5	
	Point DETR	17.4/21.8	19.4/25.2	31.7/29.8	38.4/31.9	39.2/34.7	40.8/36.2	
	PBC (Ours)	**23.3/23.3**	**30.4/26.7**	**37.7/31.9**	**39.3/33.7**	**40.7/36.2**	**43.1/37.2**	

* Since there is less contextual information contained in CXRs, the re-implemented
WSOD method (PCL [18]) cannot obtain competitive results. More details can be
found in Appendix A.

boosted by using our symmetric consistency: the mAP reaches 39.2% on RSNA
and 28.5% on VinDr-CXR, respectively, which surpasses the baseline by a large
margin of ∼15%.

4.2   Performance Benchmark

To construct the benchmark for weakly semi-supervised abnormality localization
with CXRs, the performance of our PBC method is compared with approaches
under different training strategies on RSNA and VinDr-CXR datasets. As shown
in Table 2, our PBC method consistently outperforms other methods on both
datasets. Concretely, with 30% full labeled data, the FCOS trained with the
proposed PBC method achieves an mAP of 40.5%, which is comparable to the
one with 100% fully-labeled data (43.1%). The experimental results indicate
that the proposed PBC method can balance the trade-off between annotation
cost and model accuracy. Furthermore, compared to the listed semi-supervised
and weakly semi-supervised methods, our PBC yields more significant improvements to box-only models, especially with extremely limited fully-labeled data.
Specifically, the proposed method boosts the mAP of Faster R-CNN to 23.3%
on both datasets with only 5% fully-labeled samples, i.e., +5.9% and +1.5%
higher than the runner-up (Point DETR), which demonstrates the effectiveness

of our regularization terms on refining the pseudo bounding boxes for the student model.

5   Conclusion

In this paper, we constructed a benchmark for weakly semi-supervised abnormality localization with CXRs. A novel framework, namely point beyond class (PBC), was formed, which consists of two novel regularization terms (multi-point consistency and symmetric consistency). In particular, our multi-point consistency drives the model to localize the consistent bounding boxes from different points inside the same lesion area. While, the proposed symmetric consistency enforces the network to yield consistent predictions for the same CXR permuted by different transformations. These two regularization terms thereby improve the robustness of learned features against the variety of point annotations. The effectiveness of the proposed PBC method has been validated on two publicly available datasets, *i.e.*, RSNA and VinDr-CXR.

Acknowledgment. This work was supported in part by the National Natural Science Foundation of China (Grant No. 91959108), Key-Area Research and Development Program of Guangdong Province, China (No. 2018B010111001), National Key R&D Program of China (2018YFC2000702) and the Scientific and Technical Innovation 2030-"New Generation Artificial Intelligence" Project (No. 2020AAA0104100).

A   The Appendix of Point Beyond Class

(See Figs. 3, 4, 5 and 6)

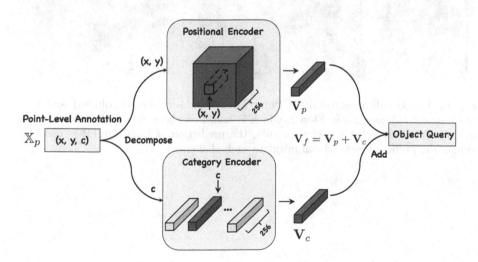

Fig. 3. The pipeline of point encoder

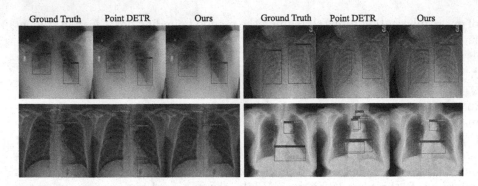

Fig. 4. The visualization result of Point DETR and ours based on 50% point-level annotations and 50% box-level annotations. The bounding boxes are carried by Faster R-CNN with 0.5 threshold. First row refers to the result on RSNA and the second row stands for VinDr-CXR.

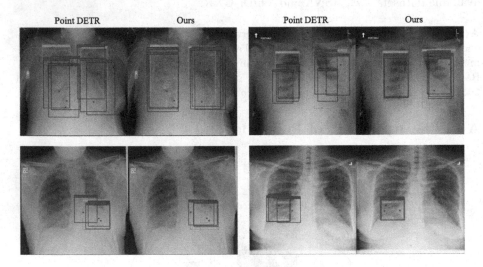

Fig. 5. The visualization result of Point DETR (odd-numbered column) and Ours (even-numbered column) in Step 1, which is based on the points with different localization. When changing the given points, the prediction of Point DETR is unstable. While, the proposed method can mitigate such challenge.

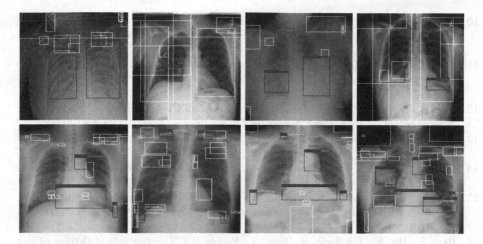

Fig. 6. The region proposals of weakly supervised object detection. The boundaries of lesions in chest X-rays images are commonly not clear, therefore, massive invalid proposals are generated by these image-level-annotation-based methods.

References

1. Bearman, A., Russakovsky, O., Ferrari, V., Fei-Fei, L.: What's the point: semantic segmentation with point supervision. In: Leibe, B., Matas, J., Sebe, N., Welling, M. (eds.) ECCV 2016. LNCS, vol. 9911, pp. 549–565. Springer, Cham (2016). https://doi.org/10.1007/978-3-319-46478-7_34
2. Bilen, H., Pedersoli, M., Tuytelaars, T.: Weakly supervised object detection with convex clustering. In: Proceedings of the IEEE Conference on Computer Vision and Pattern Recognition, pp. 1081–1089 (2015)
3. Carion, N., Massa, F., Synnaeve, G., Usunier, N., Kirillov, A., Zagoruyko, S.: End-to-end object detection with transformers. In: Vedaldi, A., Bischof, H., Brox, T., Frahm, J.-M. (eds.) ECCV 2020. LNCS, vol. 12346, pp. 213–229. Springer, Cham (2020). https://doi.org/10.1007/978-3-030-58452-8_13
4. Chen, K., et al.: MMdetection: open MMLab detection toolbox and benchmark. arXiv preprint arXiv:1906.07155 (2019)
5. Chen, L., Yang, T., Zhang, X., Zhang, W., Sun, J.: Points as queries: weakly semi-supervised object detection by points. In: Proceedings of the IEEE/CVF Conference on Computer Vision and Pattern Recognition, pp. 8823–8832 (2021)
6. Goodfellow, I., Bengio, Y., Courville, A.: Deep Learning. MIT Press, Cambridge (2016)
7. He, K., Zhang, X., Ren, S., Sun, J.: Deep residual learning for image recognition. In: Proceedings of the IEEE Conference on Computer Vision and Pattern Recognition, pp. 770–778 (2016)
8. Law, H., Deng, J.: CornerNet: detecting objects as paired keypoints. In: Proceedings of the European Conference on Computer Vision, pp. 734–750 (2018)
9. Lin, T.Y., Dollár, P., Girshick, R., He, K., Hariharan, B., Belongie, S.: Feature pyramid networks for object detection. In: Proceedings of the IEEE Conference on Computer Vision and Pattern Recognition, pp. 2117–2125 (2017)

10. Lin, T.Y., Goyal, P., Girshick, R., He, K., Dollár, P.: Focal loss for dense object detection. In: Proceedings of the IEEE International Conference on Computer Vision, pp. 2980–2988 (2017)
11. Liu, H., Wu, H., Xie, W., Liu, F., Shen, L.: Group-wise inhibition based feature regularization for robust classification. In: Proceedings of the IEEE/CVF International Conference on Computer Vision, pp. 478–486 (2021)
12. Luo, L., Chen, H., Zhou, Y., Lin, H., Pheng, P.A.: OXNet: omni-supervised thoracic disease detection from chest X-rays. arXiv preprint arXiv:2104.03218 (2021)
13. Nguyen, H.Q., et al.: VinDr-CXR: an open dataset of chest X-rays with radiologist's annotations. arXiv preprint arXiv:2012.15029 (2020)
14. Parmar, N., et al.: Image transformer. In: International Conference on Machine Learning, pp. 4055–4064. PMLR (2018)
15. Ren, S., He, K., Girshick, R., Sun, J.: Faster R-CNN: towards real-time object detection with region proposal networks. Adv. Neural. Inf. Process. Syst. **28**, 91–99 (2015)
16. Ren, S., He, K., Girshick, R., Sun, J.: Faster R-CNN: towards real-time object detection with region proposal networks. IEEE Trans. Pattern Anal. Mach. Intell. **39**(6), 1137–1149 (2016)
17. Sohn, K., Zhang, Z., Li, C.L., Zhang, H., Lee, C.Y., Pfister, T.: A simple semi-supervised learning framework for object detection. arXiv preprint arXiv:2005.04757 (2020)
18. Tang, P., et al.: PCL: proposal cluster learning for weakly supervised object detection. IEEE Trans. Pattern Anal. Mach. Intell. 1 (2018)
19. Tian, Z., Shen, C., Chen, H., He, T.: FCOS: fully convolutional one-stage object detection. In: Proceedings of the IEEE/CVF international Conference on Computer Vision, pp. 9627–9636 (2019)
20. Vaswani, A., et al.: Attention is all you need. In: Advances in Neural Information Processing Systems, pp. 5998–6008 (2017)
21. Wang, S., et al.: CPNet: cycle prototype network for weakly-supervised 3D renal compartments segmentation on CT images. In: de Bruijne, M., et al. (eds.) MICCAI 2021. LNCS, vol. 12902, pp. 592–602. Springer, Cham (2021). https://doi.org/10.1007/978-3-030-87196-3_55
22. Wang, X., Peng, Y., Lu, L., Lu, Z., Bagheri, M., Summers, R.M.: ChestX-ray8: hospital-scale chest X-ray database and benchmarks on weakly-supervised classification and localization of common thorax diseases. In: Proceedings of the IEEE Conference on Computer Vision and Pattern Recognition, pp. 2097–2106 (2017)
23. Xie, J., Hou, X., Ye, K., Shen, L.: CLIMS: cross language image matching for weakly supervised semantic segmentation. In: Proceedings of the IEEE/CVF Conference on Computer Vision and Pattern Recognition, pp. 4483–4492 (2022)
24. Xie, J., Luo, C., Zhu, X., Jin, Z., Lu, W., Shen, L.: Online refinement of low-level feature based activation map for weakly supervised object localization. In: Proceedings of the IEEE/CVF International Conference on Computer Vision (ICCV), pp. 132–141 (2021)
25. Xie, J., Xiang, J., Chen, J., Hou, X., Zhao, X., Shen, L.: C2AM: contrastive learning of class-agnostic activation map for weakly supervised object localization and semantic segmentation. In: Proceedings of the IEEE/CVF Conference on Computer Vision and Pattern Recognition, pp. 989–998 (2022)

Did You Get What You Paid For? Rethinking Annotation Cost of Deep Learning Based Computer Aided Detection in Chest Radiographs

Tae Soo Kim, Geonwoon Jang, Sanghyup Lee, and Thijs Kooi[✉]

Lunit Inc., Seoul, South Korea
tkooi@lunit.io

Abstract. As deep networks require large amounts of accurately labeled training data, a strategy to collect sufficiently large and accurate annotations is as important as innovations in recognition methods. This is especially true for building Computer Aided Detection (CAD) systems for chest X-rays where domain expertise of radiologists is required to annotate the presence and location of abnormalities on X-ray images. However, there lacks concrete evidence that provides guidance on how much resource to allocate for data annotation such that the resulting CAD system reaches desired performance. Without this knowledge, practitioners often fall back to the strategy of collecting as much detail as possible on as much data as possible which is cost inefficient. In this work, we investigate how the cost of data annotation ultimately impacts the CAD model performance on classification and segmentation of chest abnormalities in frontal-view X-ray images. We define the cost of annotation with respect to the following three dimensions: quantity, quality and granularity of labels. Throughout this study, we isolate the impact of each dimension on the resulting CAD model performance on detecting 10 chest abnormalities in X-rays. On a large scale training data with over 120K X-ray images with gold-standard annotations, we find that cost-efficient annotations provide great value when collected in large amounts and lead to competitive performance when compared to models trained with only gold-standard annotations. We also find that combining large amounts of cost efficient annotations with only small amounts of expensive labels leads to competitive CAD models at a much lower cost.

Keywords: Annotation cost · Classification · Segmentation · Chest X-ray · Chest abnormality

---

T. S. Kim and G. Jang—*Authors equally contributed.*

Supplementary Information The online version contains supplementary material available at https://doi.org/10.1007/978-3-031-16437-8_25.

© The Author(s), under exclusive license to Springer Nature Switzerland AG 2022
L. Wang et al. (Eds.): MICCAI 2022, LNCS 13433, pp. 261–270, 2022.
https://doi.org/10.1007/978-3-031-16437-8_25

1   Introduction

Deep neural network driven Computer Aided Detection (CAD) systems for chest X-ray images have become powerful aids to radiologists and have been shown to surpass human performance [5]. One of the biggest driving forces behind the innovation is the collection of large scale chest X-ray datasets with accurate annotations provided by expert radiologists [4,14,15]. However, curation of accurately annotated large scale datasets for CAD systems in X-rays is challenging because the expertise of radiologists is required to annotate abnormalities.

Without access to radiologists, an alternative is to generate labels by parsing radiology reports and training a language model to predict the existence of findings from associated text reports [4,12,14]. Although this can minimize the cost of annotation, the performance of the resulting CAD system inevitably hinges on the accuracy of the generated labels [9], as we also later verify in this paper. There has been attempts to leverage unlabeled data [7] and use mixed set of supervision [6] which can also contribute towards saving annotation cost. However, it is still unclear how different dimensions of annotation cost quantitatively impact the resulting CAD system performance.

In this work, we investigate how annotation cost impacts the performance of CAD systems for classifying and localizing ten chest abnormalities in frontal view (PA) chest X-ray. We define the total cost of annotation (i.e., the effort it takes to annotate a scan) as a function of three factors: the *quantity*, *quality* and *granularity* of annotations. We highlight some of the key motivations and insights provided by this study with respect to each dimension of annotation cost.

1. **Granularity:** A radiologist can curate a given chest radiograph at an image-level and pixel-level. We further decompose the pixel-level granularity by the amount of effort and thus time required by a radiologist to annotate a given image. For example, a radiologist drawing accurate contours around chest abnormalities will require more time to annotate the same amount of chest radiographs than the radiologist defining a rough bounding box around the lesion. The quickest and the lowest level of granularity is to collect only image-level annotations (whether the lesion exists or not). We quantify the trade-off between CAD model performance and the added cost of collecting granular labels.

2. **Quality:** A more cost efficient alternative to the radiologist-provided gold standard is to get image-level annotations automatically by extracting labels using natural language processing (NLP) algorithms [4,12]. The caveat of such algorithms is that they are vulnerable to multiple factors [10] that lead to inaccurate labeling of X-ray images [4,6,9]. How do annotation quality and the amount of label noise impact the CAD model performance?

3. **Quantity:** Lastly, the cost of annotation is determined by the number of scans that are annotated. Although several papers investigate how the number of annotated samples affect the model's performance under label noise [6] and limited/mixed supervision [11], there has not been a sufficiently large enough set of X-ray images with gold standard annotations to enable analysis at

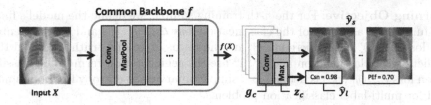

Fig. 1. Overview of the model architecture. We use common backbone architectures for f and append per-class classification and segmentation heads.

the scale presented in this paper. Our study includes more than four times as many annotations from radiologists than the currently available largest dataset with annotations from experts [1]. Furthermore, all images used in this study contain pixel-level labeling of radiologist's opinion on the location of the findings.

To the best of our knowledge, this is the first study performed at this scale with over 120K gold standard annotations where we can quantitatively measure how CAD model performance is affected by the quantity, quality, and granularity of annotations.

2   Methods

To investigate the effect of different types of annotation on multi-label classification and segmentation performance, we make use of a general, flexible, and model agnostic framework. We first describe our simple neural network approach that we use for both tasks in Sect. 2.1. Then, we describe in detail the different annotation types used in this study in Sect. 2.2.

2.1   Common Architecture

We use the same architecture for all experiments for both classification and segmentation tasks in this study. The architecture is similar to FCN [8] with a common Residual Network (ResNet-34) backbone [3] but outputs a smaller spatial resolution segmentation results.

Let $X \in \mathbb{R}^{1 \times H \times W}$ be a CXR image input to our model. The model is a neural network with a feature extractor $f : \mathbb{R}^{1 \times H \times W} \mapsto \mathbb{R}^{d \times h \times w}$ which maps an input CXR image to a feature space defined by d channels and down-sampled h, w spatial dimensions. Given the feature $f(X)$, the segmentation head $g : \mathbb{R}^{d \times h \times w} \mapsto \mathbb{R}^{C \times h \times w}$ produces a pixel-wise prediction of whether a pixel belongs to one of C classes. The classification head $z : \mathbb{R}^{C \times h \times w} \mapsto \mathbb{R}^{C}$ maps $g(f(X))$ to a multi-label classification of existence of chest abnormalities. Figure 1 depicts the architecture used to evaluate the cost of annotation for both tasks in this study.

Learning Objective: For the n-th training sample, we define the model's final loss function $\mathcal{L}$ as a sum of the classification loss $\mathcal{L}_{cls}(y_l^n, \hat{y}_l^n)$ and the segmentation loss $\mathcal{L}_{seg}(y_s^n, \hat{y}_s^n)$, where $y_l^n, \hat{y}_l^n$ are the terms of ground truth label and the prediction for classification and $y_s^n, \hat{y}_s^n$ are the segmentation terms. The classification loss $\mathcal{L}_{cls}(y_l^n, \hat{y}_l^n)$ is defined as a simple binary cross-entropy loss commonly used for multi-label classification problems.

When pixel-wise ground truth for the n-th sample $y_s^n(h, w)$ is available for class c such that $y_s^n(h, w) = \{0, 1\}$, the model can receive additional supervision using the segmentation prediction head. We define the total segmentation loss as:

$$\mathcal{L}_{seg}(y_s^n, \hat{y}_s^n) = \mathcal{L}_{pbce} + \mathcal{L}_{dice} \tag{1}$$

where $\mathcal{L}_{pbce}$ is a pixel-wise binary cross entropy term and $\mathcal{L}_{dice}$ is a generalized dice loss [13] commonly used to train segmentation models.

The final loss term $\mathcal{L}$ is defined as a combination of the classification and segmentation loss terms summed over all N cases in the training set such that:

$$\mathcal{L} = \sum_{n=1}^{N} \mathcal{L}_{cls}(y_l^n, \hat{y}_l^n) + \mathbb{1}_{y_s^n} * \mathcal{L}_{seg}(y_s^n, \hat{y}_s^n) \tag{2}$$

where $\mathbb{1}_{y_s^n}$ is an indicator variable such that $\mathbb{1}_{y_s^n} = 1$ if pixel-wise labels exists for the n-th training case X^n.

2.2   Annotation Granularity Definitions

We define four levels of annotation granularity: image-level labels extracted from text reports using NLP algorithms such as [4], image-level labels from radiologists, pixel-level annotations defined using bounding boxes around findings, and annotations drawn as polygons (contours). We note that we do not actually use models such as CheXPert [4] to generate labels but manually control the amount of label noise that we introduce to the set of gold-standard annotations to mimic the NLP algorithm's behavior. Throughout this paper, we use the notation NLP@F1$_X$ to represent a model trained using a set of labels that are generated with an NLP algorithm with an F1 score of X. By controlling the F1 score of the NLP based label generator, we measure how the cheaper yet potentially inaccurate labels affect the CAD model performance and report the results in Sect. 3.2.

Compared to image-level annotations, pixel-level annotations potentially provide more information to the model by localizing the lesions in X-ray images. We define different granularities in providing this localization information. Our intuition is that drawing coarse outlines such as bounding boxes around the lesion takes less effort and thus less time than drawing accurate contours around the lesions. To investigate this, we simulate bounding box annotations by generating them from our gold-standard contour labels. The generated bounding boxes provide pixel-level supervision to the model via $\mathcal{L}_{seg}$ where $y_s(h, w) = 1$ for all pixels (h, w) within the bounding box.

3 Experiments

3.1 Data and Settings

The dataset used in this study contains a total of $121,356$ training samples where each image is a frontal view chest X-ray image with a set of gold-standard annotations provided by expert radiologists. Table 1 summarizes the number of positive and negative training/test samples for the ten chest abnormalities defined in this study. The test set consists of a total of $3,546$ independent cases also with gold-standard annotations. For more details on the data and experimental settings used in this paper, please refer to the supplementary material. We report the average of three independent runs and the associated 95% confidence intervals for all experiments reported in this paper.

Table 1. Number of positive and negative samples for each chest abnormality used in this study. Atl: Atelectasis, Calc: Calcification, Cm: Cardiomegaly, Csn: Consolidation, Fib: Fibrosis, MW: Mediastinal widening, Ndl: Nodule, PEf: Pleural Effusion, Ppm: Pneumoperitoneum, Ptx: Pneumothorax

Train	Atl	Calc	Cm	Csn	Fib	MW	Ndl	PEf	Ppm	Ptx
Pos	16939	9308	10602	43095	14704	1607	10500	21290	3108	6018
Neg	104417	112048	110754	78261	106652	119749	110856	100066	118248	115338
Test										
Pos	272	267	187	664	272	42	694	510	152	437
Neg	1842	1881	1880	1785	1793	2071	2351	2775	1961	3109

3.2 Classification

In this section, we present how the quantity, quality and granularity of annotation ultimately impacts the CAD model's classification performance.

Impact of Granularity: The impact of annotation granularity on the resulting classification models is summarized in Table 2. We use macro averaged AUCROC over ten lesions to measure the classification performance of the models.

Our experiments show that on relatively small scale datasets (less than 12K training samples in total), the models performed similarly even when additional supervision in the form of bounding boxes or contours were provided during training. However, when more training data points were available, additional pixel-wise supervision from bounding boxes or contours improves over the image-level models by over 2 points when using all 121K training cases. For providing pixel-level supervision to the classification model, we did not observe meaningful differences between models trained with bounding boxes and contours. Our results suggest that collecting bounding box level labels is a more cost effective alternative to collecting contours for classification problems.

Models trained with noisy labels consistently performed worse than the models supervised with ground truth provided by radiologists. However, the performance gap between the NLP@F1$_{0.80}$ and the image-level models was much

Table 2. How annotation granularity impacts CAD model performance across various dataset scales. Performance reported in AUCROC ($\times 100$) with 95% confidence intervals.

	Dataset size					
	1.2K (1%)	6K (5%)	12K (10%)	30K (25%)	60K (50%)	121K
NLP @ $F1_{0.80}$	55.7 ± 5.0	66.9 ± 9.2	73.2 ± 3.0	84.5 ± 1.6	88.4 ± 1.8	92.9 ± 0.7
Image level	65.3 ± 5.1	84.0 ± 3.2	90.4 ± 0.1	92.4 ± 0.2	92.9 ± 0.1	93.7 ± 0.7
Bbox	67.5 ± 1.9	81.2 ± 1.2	88.5 ± 0.7	94.3 ± 0.1	94.9 ± 0.3	95.8 ± 0.01
Contours	68.9 ± 1.3	82.1 ± 0.5	89.8 ± 1.3	94.6 ± 0.08	95.3 ± 0.03	95.9 ± 0.03

Table 3. How annotation noise commonly introduced by NLP derived labels impacts CAD model performance across various dataset scales. Performance reported in AUCROC with 95% confidence intervals.

NLP F1 score	Dataset size					
	1.2K (1%)	6K (5%)	12K (10%)	30K (25%)	60K (50%)	121K
0.75	52.9 ± 3.8	54.6 ± 2.9	71.6 ± 9.3	77.2 ± 4.2	86.2 ± 2.3	90.0 ± 0.1
0.80	55.7 ± 5.0	66.9 ± 9.2	73.2 ± 3.0	84.5 ± 1.6	88.4 ± 1.8	92.9 ± 0.7
0.85	57.7 ± 6.0	72.4 ± 3.1	83.2 ± 9.3	86.1 ± 1.1	92.2 ± 0.5	94.1 ± 0.1
0.90	61.8 ± 5.2	69.6 ± 7.7	82.9 ± 3.8	90.8 ± 2.8	93.2 ± 0.4	94.0 ± 0.3
0.95	60.1 ± 2.6	82.9 ± 0.6	87.5 ± 3.9	92.75 ± 0.3	93.3 ± 0.2	93.9 ± 0.8

smaller (less than 1 point) when using 121K training samples. The performance gap between the two widens as less training data is available. The patterns are consistent with the literature on learning with noisy labels for medical image analysis [6]. This suggests that using cheap labels that can be automatically extracted from text reports have the potential to yield accurate classification models given large enough training datasets.

Impact of Quality: Given that the annotations extracted using NLP algorithms is potentially a very cost effective approach for training classification models for chest radiographs, we scrutinize the CAD model classification performance as the accuracy of NLP models improves. The impact of this label noise is summarized in Table 3.

As the automatic NLP based label extracting algorithms get more accurate, the resulting CAD classification models improve as well. Our results show that a relatively small gain of 0.05 in F1 score leads to considerable CAD model performance gains (92.9 of NLP@$F1_{0.80}$ vs 94.1 of NLP@$F1_{0.85}$). Using more accurate label generators improves data efficiency. Our results show that we can use only half the amount of data (60K) to achieve 92.2 AUCROC when using NLP@$F1_{0.85}$ whereas all 121K cases are needed to get to 92.9 AUCROC when using NLP@$F1_{0.80}$ algorithm. When using generated labels, it is important to invest in improving the NLP algorithm itself to maximize CAD model performance with the same amount of annotation budget.

Table 4. How annotation granularity impacts CAD model segmentation performance across various dataset scales. Performance measured in mean over per-lesion Dice Coefficients with 95% confidence intervals.

	Dataset size					
	1.2K (1%)	6K (5%)	12K (10%)	30K (25%)	60K (50%)	121K
Bbox	.046 ± .02	.229 ± .02	.341 ± .01	.404 ± .01	.421 ± .02	.427 ± .02
Contours	.111 ± .07	.327 ± .03	.422 ± .005	.469 ± .009	.486 ± .009	.500 ± .007

3.3   Segmentation

In this section, we present our findings on how factors that determine annotation cost ultimately impact the segmentation performance of CAD models. We measure the segmentation performance using an aggregated Sørenson-Dice coefficient [2] macro averaged over all lesions.

Impact of Granularity and Quality: In Table 4, we report the segmentation performance difference between a model trained using pixel-level supervision extracted from bounding boxes and a model trained using gold-standard contours. The segmentation performances of both bounding-box and contour based models improve as more data is available. However, in contrast to the classification setting where the performance gap gradually decreased across different models as more data was available, the segmentation performance gap between the two models was kept consistent throughout all dataset scales. In fact, the results trend similarly with the experiments in Table 3 where we measure performance across different noise levels. We observe that a bounding box is actually a polygon with much fewer degrees of freedom than a contour. We can view bounding boxes as lower quality (noisier) alternatives to contour annotations. The visualizations of model predictions in Fig. 1 of the supplementary material highlight that the model trained with contours yields much cleaner segmentation results.

Improving Performance by Mixing Levels of Granularity: We investigate whether we can train segmentation models in a more economic manner by adding large amounts of cost-efficient annotations to a small set of contour annotations. For example, as shown in Table 4, we can improve Dice coefficient of the model trained with 6K training images by about 0.14 points by adding 24K extra training samples with contour annotations. In our experiments, we observe that we can improve the same model (Contour-6K) by almost a similar amount (0.12) by adding the same 24K extra training samples but with only *image-level* annotations. In Fig. 2, we show how adding various amounts of training images with only image-level annotations benefits the original segmentation model trained with small amounts (6K) of training data. Interestingly, we observe that combining image-level annotations with contours is a better strategy than combining bounding boxes with contours. At the largest dataset scale, adding bounding box annotations leads to a slightly worse segmentation model than adding the noisy NLP@F1$_{0.75}$ (Yellow in Fig. 2) label. For a more quali-

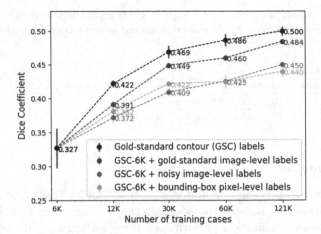

Fig. 2. Effects of adding additional gold-standard contours, pixel-level information from bounding boxes, gold-standard image-level and noisy (NLP@F1$_{0.75}$) annotations on the segmentation model performance. (Color figure online)

tative comparison of segmentation results of the models, please refer to Fig. 1 from the supplementary material. When there is not enough budget available for collecting large amounts of pixel-level annotations, the results suggest that collecting large amounts of image-level annotations and combining them with available pixel-level data is a cost efficient strategy for training segmentation models.

4   Discussion and Conclusions

We presented a large-scale analysis of how annotation cost impacts CAD systems for classification and segmentation of chest abnormalities in frontal view chest X-rays. We defined quantity, quality and granularity as the factors that determine annotation cost. We found that the more cost efficient bounding boxes are as useful as the accurate contours when provided as additional supervision to the classification model. We also showed that relatively small improvements to the label extracting algorithms lead to gains in classification performance. Lastly, we reported that we can achieve strong segmentation performance by mixing image-level labels with only small amounts of pixel-level contour labels. A limitation of the study is how we modeled the errors introduced by an NLP label extractor. We generated errors randomly at a given F1 operating point but the errors produced by the actual NLP model will be much more nuanced. As part of future work, we wish to design a study with labels generated with actual NLP models. As for the next follow-up study, we hope to analyze how the cost actually paid to the annotator changes depending on the method and amount of annotation. Moreover, we hope to investigate whether our findings generalize to other modalities such as CT, breast mammography and MRIs. We hope the results presented in this paper provide practical guidance for practitioners building automated CAD systems for chest radiographs to ultimately improve patient care by assisting healthcare providers and medical experts.

References

1. Bustos, A., Pertusa, A., Salinas, J.M., de la Iglesia-Vayá, M.: PadChest: a large chest x-ray image dataset with multi-label annotated reports. Med. Image Anal. **66**, 101797 (2020)
2. Dice, L.R.: Measures of the amount of ecologic association between species. Ecology **26**(3), 297–302 (1945)
3. He, K., Zhang, X., Ren, S., Sun, J.: Deep residual learning for image recognition. In: 2016 IEEE Conference on Computer Vision and Pattern Recognition (CVPR), pp. 770–778 (2016)
4. Irvin, J., et al.: CheXpert: a large chest radiograph dataset with uncertainty labels and expert comparison. In: Proceedings of the AAAI Conference on Artificial Intelligence, vol. 33, pp. 590–597 (2019)
5. Jin, K.N., et al.: Diagnostic effect of artificial intelligence solution for referable thoracic abnormalities on chest radiography: a multicenter respiratory outpatient diagnostic cohort study. Eur. Radiol. **32**(5), 3469–3479 (2022)
6. Karimi, D., Dou, H., Warfield, S.K., Gholipour, A.: Deep learning with noisy labels: exploring techniques and remedies in medical image analysis. Med. Image Anal. **65**, 101759 (2020). https://doi.org/10.1016/j.media.2020.101759
7. Li, Y., Chen, J., Xie, X., Ma, K., Zheng, Y.: Self-loop uncertainty: a novel pseudo-label for semi-supervised medical image segmentation. In: Martel, A.L., et al. (eds.) MICCAI 2020. LNCS, vol. 12261, pp. 614–623. Springer, Cham (2020). https://doi.org/10.1007/978-3-030-59710-8_60
8. Long, J., Shelhamer, E., Darrell, T.: Fully convolutional networks for semantic segmentation. In: 2015 IEEE Conference on Computer Vision and Pattern Recognition (CVPR), pp. 3431–3440 (2015)
9. Oakden-Rayner, L.: Exploring large-scale public medical image datasets. Acad. Radiol. **27** (2019). https://doi.org/10.1016/j.acra.2019.10.006
10. Olatunji, T., Yao, L., Covington, B., Upton, A.: Caveats in generating medical imaging labels from radiology reports with natural language processing. In: International Conference on Medical Imaging with Deep Learning - Extended Abstract Track, London, UK, 08–10 July 2019
11. Peng, J., Wang, Y.: Medical image segmentation with limited supervision: a review of deep network models. IEEE Access **9**, 36827–36851 (2021)
12. Smit, A., Jain, S., Rajpurkar, P., Pareek, A., Ng, A., Lungren, M.: Combining automatic labelers and expert annotations for accurate radiology report labeling using BERT. In: Proceedings of the 2020 Conference on Empirical Methods in Natural Language Processing (EMNLP), pp. 1500–1519. Association for Computational Linguistics (2020). https://doi.org/10.18653/v1/2020.emnlp-main.117, https://aclanthology.org/2020.emnlp-main.117
13. Sudre, C.H., Li, W., Vercauteren, T., Ourselin, S., Jorge Cardoso, M.: Generalised dice overlap as a deep learning loss function for highly unbalanced segmentations. In: Cardoso, M.J., et al. (eds.) DLMIA/ML-CDS -2017. LNCS, vol. 10553, pp. 240–248. Springer, Cham (2017). https://doi.org/10.1007/978-3-319-67558-9_28

14. Wang, X., Peng, Y., Lu, L., Lu, Z., Bagheri, M., Summers, R.M.: ChestX-ray8: hospital-scale chest X-ray database and benchmarks on weakly-supervised classification and localization of common thorax diseases. In: CVPR, pp. 3462–3471. IEEE Computer Society (2017)
15. Çallı, E., Sogancioglu, E., van Ginneken, B., van Leeuwen, K.G., Murphy, K.: Deep learning for chest X-ray analysis: a survey. Med. Image Anal. **72**, 102125 (2021). https://doi.org/10.1016/j.media.2021.102125, https://www.sciencedirect.com/science/article/pii/S1361841521001717

Context-Aware Transformers for Spinal Cancer Detection and Radiological Grading

Rhydian Windsor[1]([✉]), Amir Jamaludin[1], Timor Kadir[1,2],
and Andrew Zisserman[1]

[1] Visual Geometry Group, Department of Engineering Science, University of Oxford, Oxford, UK
rhydian@robots.ox.ac.uk
[2] Plexalis Ltd., Oxford, UK

Abstract. This paper proposes a novel transformer-based model architecture for medical imaging problems involving analysis of vertebrae. It considers two applications of such models in MR images: (a) detection of spinal metastases and the related conditions of vertebral fractures and metastatic cord compression, (b) radiological grading of common degenerative changes in intervertebral discs. Our contributions are as follows: (i) We propose a Spinal Context Transformer (SCT), a deep-learning architecture suited for the analysis of repeated anatomical structures in medical imaging such as vertebral bodies (VBs). Unlike previous related methods, SCT considers all VBs as viewed in all available image modalities together, making predictions for each based on context from the rest of the spinal column and all available imaging modalities. (ii) We apply the architecture to a novel and important task – detecting spinal metastases and the related conditions of cord compression and vertebral fractures/collapse from multi-series spinal MR scans. This is done using annotations extracted from free-text radiological reports as opposed to bespoke annotation. However, the resulting model shows strong agreement with vertebral-level bespoke radiologist annotations on the test set. (iii) We also apply SCT to an existing problem – radiological grading of inter-vertebral discs (IVDs) in lumbar MR scans for common degenerative changes. We show that by considering the context of vertebral bodies in the image, SCT improves the accuracy for several gradings compared to previously published models.

Keywords: Metastasis · Vertebral fracture · Metastatic cord compression · Radiological reports · Radiological grading · Transformers

Supplementary Information The online version contains supplementary material available at https://doi.org/10.1007/978-3-031-16437-8_26.

© The Author(s), under exclusive license to Springer Nature Switzerland AG 2022
L. Wang et al. (Eds.): MICCAI 2022, LNCS 13433, pp. 271–281, 2022.
https://doi.org/10.1007/978-3-031-16437-8_26

1   Introduction

When a radiologist is reporting on a patient's imaging study, they will typically make assessments based on multiple scans of the same patient. For example, in a MRI study they will often consider multiple pulse sequences showing the same region and note differences between them. Related conditions may appear similar in one modality and only be distinguishable when more context is given via an additional imaging modality. Similarly, when reporting on spinal images, radiologists can learn a lot about a specific vertebra based on its appearance *relative to the vertebrae neighbouring it*. For example, a region of hyperintensity may represent a lesion or may just be an artifact of the scanning protocol used; this often can be elucidated by looking at the rest of the spine.

Accordingly, automated models for spinal imaging tasks should also be able to make vertebra-level predictions with context from multiple sequences *and from neighbouring vertebrae*. In this work, we propose the *Spinal Context Transformer* (SCT), a model which aims to do exactly this. Crucially, SCT leverages lightweight transformer-based models which allow for variable numbers of inputs (both during training and at test-time) in terms of both: (a) the number of vertebral levels, and (b) the number of input modalities depicting each vertebra.

As well as introducing SCT, this paper also explores a new application of deep learning in medical imaging: automated diagnosis of spinal metastases and related conditions. This is an important task for several reasons. Firstly, metastases are very common; overall 2 in 3 cancer cases will metastasise [25] (rising to nearly 100% incidence in patients who die of cancer [20]). The spine is one of the most common places for these metastases to occur [23]. Secondly, early detection is vital; metastases generally indicate advanced cancer and, if left untreated, cause conditions such as vertebral fractures and metastatic cord compression, both of which can lead to significant pain and disability for the patient and are difficult and expensive to treat [28,29]. Current clinical practice relies on heuristic-based scoring systems to quantify the development of metastases, such as SINS [7] and the Tokuhashi score [27]. We aim to develop models to aid in the rapid detection and consistent quantification of this serious condition. To further validate our model, we also test it on a previously published dataset for grading common intervertebral disc (IVD) degenerative changes in lumbar MR scans.

Another theme explored in this work is the ability of free-text radiological reports to generate supervisory signals for medical imaging tasks. The common praxis when training models on radiological images is to extract a relevant dataset from an imaging centre, pseudo-anonymise it and then annotate the dataset for a condition of interest. While this method is simple, effective and very popular, the additional step of annotating the dataset can be a significant bottleneck in the size of datasets used. This is a more serious problem than in conventional computer vision datasets as medical image annotation generally requires a specialist, whose time is limited and expensive. A much more scalable method of curating annotated datasets is to use existing hospital records for generating supervisory signals. However, in radiology, these records are usually in the form of free-text reports. Due to variation and lack of structure in reporting, extracting useful information from these reports has been considered a difficult task [5], and, as such, most studies annotate their own data retrospectively. In this paper, we explore how useful the information in free-text reports can be for training models for our specific task of detecting spinal cancer, and propose methods for dealing with ambiguities and omissions from the text.

2 Related Work

The SCT, and its training and application, builds on three related areas of work: deep learning for analysis of spinal scans, obtaining supervision from free-text radiology reports and transformers for aggregating multiple sources of information. Automated detection and labelling of vertebrae is a long standing task in spinal MRI [18] and CT [10] analysis, with deep learning now the standard approach [3,8,26,31,32]. One study particularly relevant to this work is that of Tao *et al.* [26] who use a transformer architecture for this task. Beyond vertebra detection and labelling, deep learning has also been used to assess spinal MRIs for degenerative changes. For example, SpineNet of Jamaludin *et al.* [15] and Windsor *et al.* [33] is a multi-task classification model acting on intervertebral discs which grades for multiple common conditions. We compare to this approach here. DeepSpine of Lu *et al.* [19] and also Lewandrowski *et al.* [17] use ground truth derived from radiological reports to train models for grading stenosis and disc herniation respectively, a theme we also explore in this work. In terms of spinal cancer, [34] and [30] both train models to detect metastases based on 2D images extracted from MRI studies; and [22] proposes a model to automatically detect metastatic cord compression, although this is restricted to axial T2 scans or the cervical region. Several other works have considered vertebra and metastases detection in other modalities such as CT (e.g. [2,10–12]), although MR remains the clinical gold-standard in early spinal cancer detection. Finally, our approach also has parallels with works in video sequence analysis that combine a 2D CNN backbone with a temporal transformer for tasks such as representation learning, tracking, object detection and segmentation [1,4,9,21]; though in our case, visual features are extracted from multiple vertebrae shown in multiple MR sequences rather than consecutive frames.

3 Spinal Context Transformer

Like other models used to assess vertebral disorders such as SpineNet [15,33] and DeepSpine [19], the Spinal Context Transformer (SCT) exploits the repetitive nature of the spine by using the same CNN model to extract visual features from each vertebra given as input. However, there are two key conceptual differences between how SCT and these other models operate which are described in this section.

Firstly, instead of using 3D convolutions and pooling layers to extract features from multislice MR images, SCT instead extracts features from each slice independently using a standard 2D ResNet18 [13] architecture adapted for single-channel images. The extracted feature vectors for each slice are then aggregated using an attention mechanism. This is shown in Fig. 1b. This design choice is made for several reasons: (i) 3D convolutions assume all images have the same slice thickness. In real-world clinical practice a variety of scanning protocols are used. This means slice thickness can vary significantly from sample to sample. Of course, resampling can be employed in pre-processing to ensure a consistent slice thickness for input images, however this introduces visual artifacts which may degrade performance. (ii) Trained encoder models can operate on 2D or 3D scans without adaptation. This also means image CNNs trained on large 2D datasets such as ImageNet [6] can be used to initialize the encoder backbone. (iii) The attention mechanism forces the model to explicitly determine the importance of each slice for its prediction. This allows peripheral slices with partial-volume effects to be ignored and can be useful when it comes to interpreting the output of the model. An example of this effect is shown in the supplementary material.

Secondly, as well as using attention to aggregate feature vectors from each slice, SCT also uses attention to collate information from each MR sequence and vertebra in the spinal column. This is done by feeding the vertebra's visual embedding vectors from each MR sequence to a lightweight 2-layer transformer encoding model, with additional embedding vectors describing the level of each input vertebra and the imaging modality used. Both these additional embeddings are calculated by linear layers operating on a one-hot encoding of the vertebra's name and sequence. The output feature vectors for the same vertebra shown in different sequences are then pooled together using a final attention mechanism identical to that used to pool features across slices. This creates a single output vector for each vertebra. Linear classification layers convert this output vector into predictions for each classification task. The full process is shown in Fig. 1a. Annotated PyTorch-style pseudo-code is given in the supplementary material.

4   Detecting Spine Cancer Using Radiological Reports

This section describes applying SCT to a novel task: detecting the presence of spinal metastases in whole spine clinical MR scans, as well as the related conditions of vertebral fractures/collapse and spinal cord compression, employing information from free-text radiological reports for supervision. We use a dataset of anonymised clinical MRI scans and associated reports extracted from a local hospital trust PACS (Oxford University Hospitals Trust) with appropriate ethical clearance. Inclusion criteria were that the patient was over 18 years old, had at least one whole spine study, and was referred from a cancer-related specialty. In order to ensure a mix of positive and negative cases, all patients that matched this criteria from April 2015 until April 2021 were included in the dataset, regardless of whether or not they had metastases.

To avoid having to annotate each scan independently we instead rely on the radiological reports written at the time the scan was taken to provide supervision for training our models. A major advantage of this approach is that it is much

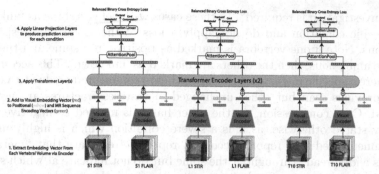

(a) The SCT architecture illustrated for the spine cancer prediction task operating on multiple MRI sequences of three vertebrae. In the actual cancer detection task, the whole spinal column is used, and SCT is able to handle arbitrary variations in the number and type of input sequences and vertebrae without adaptation.

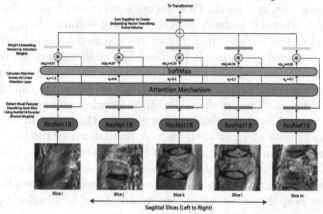

(b) The visual encoder network used to extract features from each vertebral volume, This model can be trained by itself or in conjunction with the transformer architecture as shown in (a). The same model weights are used for all MR sequences.

Fig. 1. Architecture diagrams for SCT: (a) the full SCT architecture; (b) the encoder used to extract visual features from each vertebra. STIR and FLAIR sequences are shown here, however the radiological grading task uses strictly T1 and T2 sequences, whereas the spinal cancer task uses a range of common sequences (T1, T2, STIR, FLAIR etc.).

faster and thus readily scalable to larger datasets – since no radiologist is required to review each image independently, an annotation set can be obtained much quicker than would otherwise be possible. Furthermore, significantly less clinical expertise is required to generate these annotations – only a basic understanding of vocabulary related to the conditions of interest is needed.

A particular challenge of using annotations derived from reports for training is dealing with ambiguities in the text. Firstly, reports are often inconclusive; e.g. "The nature of this lesion is indeterminate" or "Metastases are a possibility but further investigation is required". In these cases we label vertebrae as 'unknown' for a specific condition and do not apply a loss to the model's corresponding predictions. Note if one vertebra is marked as positive in a spine and the other levels are ambiguous, then the others are marked as 'unknown'. This accounts for cases where a specific metastasis/fracture is being commented on due to a change from a previous scan but other unchanged metastases/fractures are not fully described. Cord compression, on the other hand, is marked as negative unless explicitly stated otherwise, as it is a severe condition which is highly unlikely to go unmentioned in a report. Secondly, reports often state that "metastatic disease is widespread" throughout the spine but do not indicate at which specific levels this disease occurs. To derive annotations from these ambiguous reports, we introduce an additional, global label for each study, indicating if the given condition is present in the scan at any vertebral level or not. We then aggregate each vertebra-level prediction to produce a spinal-column level prediction of whether the condition is present. This setting is an example of multiple-instance learning (MIL), whereby instead of each training instance being individually labelled, sets of instances (known as 'bags') are assigned a single label, indicating if at least one instance in the bag is positive. The challenge is then to determine which instances in the bag are positives. In this case each bag represents an entire spinal column and each instance is a single vertebra. We can then train by a hybrid single-instance and multiple-instance learning approach; vertebra-level annotations are used for supervision where given by the report in addition to the scan-level annotations given in each case. A breakdown of the labels extracted by this method is given in Table 1. Examples of vertebral bodies from each class are shown in Fig. 2. Further explanation and example annotations are given in the supplementary material. To ensure our model produces results similar to those given by bespoke annotations, our test dataset is labelled by an expert spinal surgeon.

5   Radiological Grading of Spinal Degenerative Changes

To validate SCT on an existing vertebra analysis problem, we also measure its performance at grading common degenerative changes around the intervertebral discs. Specifically, we grade the following common conditions: Pfirrmann Grading, Disc Narrowing, Endplate Defects (Upper/Lower), Marrow Changes (Upper/Lower), Spondylolisthesis and Central Canal Stenosis. We do this on Genodisc, a dataset of clinical lumbar MRIs of 2295 patients from 6 different clinical imaging sites. This problem setting closely follows that discussed in [15] and [33], which give the existing state-of-the-art methods for many of these grading tasks. Each vertebral disc from L5/S1 to T12/L1 in each scan is graded by an expert radiologist for the aforementioned gradings. In this setting, SCT takes as input volumes surrounding intervertebral discs (IVD) and outputs predictions for each grading. The study protocol for scans vary, though the vast majority of scans have at least a T1-weighted and T2-weighted sagittal scan. In this work we only consider sagittally sliced scans. However, there is no reason why SCT

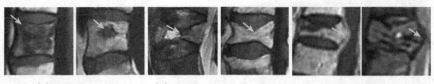

(a) Metastases   (b) Collapsed Vertebrae   (c) Cord Compression

Fig. 2. Example vertebrae with the three spinal-cancer related conditions of interest. Note that in many cases vertebrae have a combination of these conditions.

Table 1. The dataset used to train SCT to detect metastases, fractures and compression. The first-three rows indicate the number of labels extracted from free-text reports for the training, validation and test splits. The bottom row indicates the same test dataset with each vertebra independently annotated by an expert. Vertebra are labelled as positive (+), negative (−), or unknown (?). Note the expert annotator was required to make their best guess in cases of uncertainty.

Split	Patients	Studies	Total	Vertebral bodies								
				Metastases			Fractures			Compression		
				+	−	?	+	−	?	+	−	?
Training	258	558	12459	829	6099	5531	216	12124	119	90	12369	0
Validation	33	63	1391	81	829	482	25	1360	7	15	1377	0
Test (reports)	32	66	1430	92	646	692	22	1408	0	11	1419	0
Test (expert)	32	66	1430	374	1056	0	46	1384	0	31	1399	0

could not also operate on axial scans in conjunction. A more complete breakdown of the Genodisc dataset is given in [14]. We compare to SpineNet V1 [15] and SpineNet V2 [33], existing models trained on the same dataset.

6   Experimental Results

This section describes experiments to evaluate the performance of SCT at the tasks of: (a) detecting spinal cancer; (b) grading common degenerative changes.

Preprocessing, Implementation & Training Details: For both tasks, vertebrae are detected and labelled in the scans using the automated method described in [33]. This is then used to extract volumes surrounding VBs (for the cancer task) and IVDs (for the grading task). Volumes are resampled to $S \times 112 \times 112$ and $S \times 112 \times 224$ respectively, where S is the number of sagittal slices. Additional training details are given in the supplementary material.

Baselines: For each task, we compare SCT to baseline models operating on a single vertebra/IVD at a time. For the cancer task we compare to the visual encoder shown in Fig. 1b operating alone on a single sequence (since this cannot be trivially extended to multiple sequences). For the radiological grading task we compare with SpineNet V1 [15], SpineNet V2 [33] and the visual encoder alone. We also train SCT on T1 and T2 scans independently.

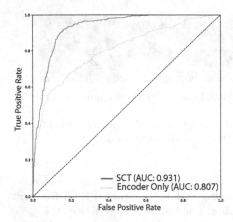

Fig. 3. ROC curves for the metastases task for SCT and the encoder baseline. Curves for the other tasks are given in the supplementary.

Table 2. AUC scores for the three tasks with various models. We compare to the report-extracted annotations and also expert annotations of each image.

Expert-Labelled Test Set Annotations

	ROC AUC		
	Mets	Frac.	Cmprs.
Baseline	0.80	0.975	**0.930**
SCT	**0.931**	**0.980**	0.868

Report-Extracted Test Set Annotations

	ROC AUC		
	Mets	Frac.	Cmprs.
Baseline	0.934	0.902	**0.955**
SCT	**0.944**	**0.901**	0.918

Results: Table 2 shows the performance of all models at the cancer task. The SCT model performs well across all tasks. In particular, it performs much better at detecting subtle metastases (AUC: **0.80** → **0.931** for the expert labels). This effect can be seen clearly in Fig. 3. This makes sense as patients will often present with multiple metastases and thus obvious metastases in one area of the spine will inform predictions on marginal cases elsewhere in the spine. We note slightly depreciated performance at the compression task. We believe this is due to overfitting as there are relatively few compression cases in the training set.

Table 3 shows the performance of all models at the radiological grading tasks, and compares to the state-of-the-art SpineNet model [14]. SCT outperforms SpineNet V1 & V2 on the same dataset for all tasks except central canal stenosis where the difference in performance is minimal. In total average performance increases from **85.9%** → **87.4%**. As expected, the multiple sequence model exceeds the performance of single sequence models in most tasks. The largest improvements can be seen in the endplate defect task (**82.9/87.8%** → **87.2/90.7%**), disc narrowing (**76.1%** → **77.4%**) and Pfirrmann grading (**71.0%** → **73.0%**). The Pfirrmann grading score of the T1-only model is far worse than the models with T2 sequences (64.5%). This matches expectations as one of the criteria of Pfirrmann grading is the intensity of the intervertebral discs in T2 sequences [24]. Overall, the improved performance is a clear indication of the benefit of considering context from multiple sequences and IVDs together.

Table 3. Results of the IVD grading task. The balanced accuracy for each sub-task is shown. CCS represents central canal stenosis. [†]For SpineNetV2, CCS is originally graded with 4 degrees of severity. We combine the severity classes 2–4 (mild, moderete & severe CCS) into a single class compared to class 1 (no CCS).

Model # Classes:	Pfirrmann	Disc narrowing	C.C.S.	Spondylolisthesis
	5	4	2	2
SpineNet V1 (T2) [16]	71.0	76.1	**95.8**	95.4
SpineNet V2 (T2) [33]	70.9	76.3	93.2[†]	95.0
Baseline: SCT Encoder (T2)	70.8	74.6	94.3	96.2
SCT (T1)	64.5	73.9	93.6	**96.6**
SCT (T2)	71.7	76.9	93.6	96.2
SCT (T1, T2)	**73.0**	**77.4**	94.9	95.5

Model # Classes:	Endplate defect		Marrow change		Average
	Upper	Lower	Upper	Lower	
	2	2	2	2	
SpineNet V1 (T2) [16]	82.9	87.8	89.2	88.4	85.8
SpineNet V2 (T2) [33]	84.9	89.8	88.9	88.2	85.9
Baseline: SCT Encoder (T2)	87.2	88.2	90.0	89.2	86.3
SCT (T1)	84.7	87.0	88.4	88.7	84.7
SCT (T2)	85.3	89.3	**91.0**	**90.2**	86.8
SCT (T1, T2)	**87.2**	**90.7**	90.1	89.9	**87.4**

7   Conclusion

In this paper we present SpinalContextTransformer (SCT) for the analysis of multiple vertebrae and multiple MRI sequences. We demonstrate SCT being applied to a new task, detecting conditions related to spinal cancer including metastases, vertebral collapses and metastatic cord compression. We also show that the model improves on existing models for radiological grading of common spinal conditions and can be used flexibly with varying input imaging modalities. Finally, it is worth noting that the SCT architecture is applicable to the analysis of other modalities (e.g. CT, X-Ray), fields-of-view (e.g. coronal, axial) and to other repeated anatomical sequential structures, such as teeth or ribs.

Acknowledgements and Ethics:. Ethics for the spinal cancer dataset extraction are provided by OSCLMRIC (IRAS project ID: 207857). We are grateful to Dr. Sarim Ather, Dr. Jill Urban and Prof. Jeremy Fairbank for insightful conversations on the clinical aspect of this work as well as Prof. Ian McCall for annotating the data. Finally, we thank to our funders: Cancer Research UK via the EPSRC AIMS CDT and EPSRC Programme Grant Visual AI (EP/T025872/1).

References

1. Bain, M., Nagrani, A., Varol, G., Zisserman, A.: Frozen in time: a joint video and image encoder for end-to-end retrieval. In: IEEE International Conference on Computer Vision (2021)
2. Burns, J.E., Yao, J., Wiese, T.S., Muñoz, H.E., Jones, E.C., Summers, R.M.: Automated detection of sclerotic metastases in the thoracolumbar spine at CT. Radiology **268**(1), 69–78 (2013)
3. Cai, Y., Osman, S., Sharma, M., Landis, M., Li, S.: Multi-modality vertebra recognition in arbitrary views using 3D deformable hierarchical model. IEEE Trans. Med. Imaging **34**(8), 1676–1693 (2015)
4. Carion, N., Massa, F., Synnaeve, G., Usunier, N., Kirillov, A., Zagoruyko, S.: End-to-end object detection with transformers. In: Vedaldi, A., Bischof, H., Brox, T., Frahm, J.-M. (eds.) ECCV 2020. LNCS, vol. 12346, pp. 213–229. Springer, Cham (2020). https://doi.org/10.1007/978-3-030-58452-8_13
5. Casey, A., et al.: A systematic review of natural language processing applied to radiology reports. BMC Med. Inf. Decis. Making **21**(1), 1–18 (2021)
6. Deng, J., Dong, W., Socher, R., Li, L.J., Li, K., Fei-Fei, L.: ImageNet: a large-scale hierarchical image database. In: Proceedings of CVPR (2009)
7. Fisher, C.G., et al.: A novel classification system for spinal instability in neoplastic disease: an evidence-based approach and expert consensus from the spine oncology study group. Spine **35**(22), E1221–E1229 (2010)
8. Forsberg, D., Sjöblom, E., Sunshine, J.L.: Detection and labeling of vertebrae in MR images using deep learning with clinical annotations as training data. J. Digit. Imaging **30**(4), 406–412 (2017)
9. Gabeur, V., Sun, C., Alahari, K., Schmid, C.: Multi-modal transformer for video retrieval. In: Vedaldi, A., Bischof, H., Brox, T., Frahm, J.-M. (eds.) ECCV 2020. LNCS, vol. 12349, pp. 214–229. Springer, Cham (2020). https://doi.org/10.1007/978-3-030-58548-8_13
10. Glocker, B., Feulner, J., Criminisi, A., Haynor, D.R., Konukoglu, E.: Automatic localization and identification of vertebrae in arbitrary field-of-view CT scans. In: Ayache, N., Delingette, H., Golland, P., Mori, K. (eds.) MICCAI 2012. LNCS, vol. 7512, pp. 590–598. Springer, Heidelberg (2012). https://doi.org/10.1007/978-3-642-33454-2_73
11. Glocker, B., Zikic, D., Konukoglu, E., Haynor, D.R., Criminisi, A.: Vertebrae localization in pathological spine CT via dense classification from sparse annotations. In: Mori, K., Sakuma, I., Sato, Y., Barillot, C., Navab, N. (eds.) MICCAI 2013. LNCS, vol. 8150, pp. 262–270. Springer, Heidelberg (2013). https://doi.org/10.1007/978-3-642-40763-5_33
12. Hammon, M., et al.: Automatic detection of lytic and blastic thoracolumbar spine metastases on computed tomography. Eur. Radiol. **23**(7), 1862–1870 (2013)
13. He, K., Zhang, X., Ren, S., Sun, J.: Deep residual learning for image recognition. In: Proceedings of CVPR (2016)
14. Jamaludin, A., Kadir, T., Zisserman, A.: Self-supervised learning for spinal MRIs. In: Cardoso, M.J., et al. (eds.) DLMIA/ML-CDS -2017. LNCS, vol. 10553, pp. 294–302. Springer, Cham (2017). https://doi.org/10.1007/978-3-319-67558-9_34
15. Jamaludin, A., Kadir, T., Zisserman, A.: SpineNet: automated classification and evidence visualization in spinal MRIs. Med. Image Anal. **41**, 63–73 (2017)
16. Jamaludin, A., et al.: Automation of reading of radiological features from magnetic resonance images (MRIs) of the lumbar spine without human intervention is comparable with an expert radiologist. Eur. Spine J. **26**, 1374–1383 (2017)

17. Lewandrowskl, K.U., et al.: Feasibility of deep learning algorithms for reporting in routine spine magnetic resonance imaging. Int. J. Spine Surg. **14**, S86–S97 (2022)
18. Lootus, M., Kadir, T., Zisserman, A.: Vertebrae detection and labelling in lumbar MR images. In: Yao, J., Klinder, T., Li, S. (eds.) Computational Methods and Clinical Applications for Spine Imaging. LNCVB, vol. 17, pp. 219–230. Springer, Cham (2014). https://doi.org/10.1007/978-3-319-07269-2_19
19. Lu, J.T., et al.: Deep spine: automated lumbar vertebral segmentation, disc-level designation, and spinal stenosis grading using deep learning. In: Machine Learning for Healthcare (2018)
20. Maccauro, G., Spinelli, M.S., Mauro, S., Perisano, C., Graci, C., Rosa, M.A.: Physiopathology of spine metastasis. Int. J. Surg. Oncol. **2011** (2011)
21. Meinhardt, T., Kirillov, A., Leal-Taixe, L., Feichtenhofer, C.: TrackFormer: multi-object tracking with transformers. In: Proceedings of CVPR (2022)
22. Merali, Z., Wang, J.Z., Badhiwala, J.H., Witiw, C.D., Wilson, J.R., Fehlings, M.G.: A deep learning model for detection of cervical spinal cord compression in MRI scans. Sci. Rep. **11**(1), 1–11 (2021)
23. Ortiz Gomez, J.: The incidence of vertebral body metastases. Int. Orthop. **19**(5), 309–311 (1995)
24. Pfirrmann, C.W.A., Metzdorf, A., Zanetti, M., Hodler, J., Boos, N.: Magnetic resonance classification of lumbar intervertebral disc degeneration. Spine **26**(17), 1873–1878 (2001)
25. Shaw, B., Mansfield, F.L., Borges, L.: One-stage posterolateral decompression and stabilization for primary and metastatic vertebral tumors in the thoracic and lumbar spine. J. Neurosurg. **70**(3), 405–410 (1989)
26. Tao, R., Zheng, G.: Spine-transformers: vertebra detection and localization in arbitrary field-of-view spine CT with transformers. In: de Bruijne, M., et al. (eds.) MICCAI 2021. LNCS, vol. 12903, pp. 93–103. Springer, Cham (2021). https://doi.org/10.1007/978-3-030-87199-4_9
27. Tokuhashi, Y., Uei, H., Oshima, M., Ajiro, Y.: Scoring system for prediction of metastatic spine tumor prognosis. World J. Orthop. **5**(3), 262–271 (2014)
28. van Tol, F.R., Massier, J.R.A., Frederix, G.W.J., Öner, F.C., Verkooijen, H.M., Verlaan, J.J.: Costs associated with timely and delayed surgical treatment of spinal metastases. Glob. Spine J. (2021)
29. van Tol, F.R., Versteeg, A.L., Verkooijen, H.M., Öner, F.C., Verlaan, J.J.: Time to surgical treatment for metastatic spinal disease: identification of delay intervals. Glob. Spine J. (2021)
30. Wang, J., Fang, Z., Lang, N., Yuan, H., Su, M.Y., Baldi, P.: A multi-resolution approach for spinal metastasis detection using deep Siamese neural networks. Comput. Biol. Med. **84**, 137–146 (2017)
31. Windsor, R., Jamaludin, A.: The ladder algorithm: finding repetitive structures in medical images by induction. In: IEEE ISBI (2020)
32. Windsor, R., Jamaludin, A., Kadir, T., Zisserman, A.: A convolutional approach to vertebrae detection and labelling in whole spine MRI. In: Martel, A.L., et al. (eds.) MICCAI 2020. LNCS, vol. 12266, pp. 712–722. Springer, Cham (2020). https://doi.org/10.1007/978-3-030-59725-2_69
33. Windsor, R., Jamaludin, A., Kadir, T., Zisserman, A.: SpineNetV2: automated detection, labelling and radiological grading of clinical MR scans. In: Technical report arXiv:2205.01683 (2022)
34. Zhao, S., Chen, B., Chang, H., Wu, X., Li, S.: Discriminative dictionary-embedded network for comprehensive vertebrae tumor diagnosis. In: Martel, A.L., et al. (eds.) MICCAI 2020. LNCS, vol. 12266, pp. 691–701. Springer, Cham (2020). https://doi.org/10.1007/978-3-030-59725-2_67

End-to-End Evidential-Efficient Net for Radiomics Analysis of Brain MRI to Predict Oncogene Expression and Overall Survival

Yingjie Feng[1], Jun Wang[1], Dongsheng An[2], Xianfeng Gu[2], Xiaoyin Xu[3], and Min Zhang[4(✉)]

[1] School of Software Technology, Zhejiang University, Hangzhou, China
[2] Department of Computer Science, Stony Brook University, Stony Brook, NY, USA
[3] Department of Radiology, Brigham and Women's Hospital, Harvard Medical School, Boston, MA, USA
[4] College of Computer Science and Technology, Zhejiang University, Hangzhou, China
min_zhang@zju.edu.cn

Abstract. We presented a novel radiomics approach using multimodality MRI to predict the expression of an oncogene (O6-Methylguanine-DNA methyltransferase, MGMT) and overall survival (OS) of glioblastoma (GBM) patients. Specifically, we employed an EffNetV2-T, which was down scaled and modified from EfficientNetV2, as the feature extractor. Besides, we used evidential layers based to control the distribution of prediction outputs. The evidential layers help to classify the high-dimensional radiomics features to predict the methylation status of MGMT and OS. Tests showed that our model achieved an accuracy of 0.844, making it possible to use as a clinic-enabling technique in the diagnosing and management of GBM. Comparison results indicated that our method performed better than existing work.

Keywords: Evidential deep learning · EfficientNet-V2 · Radiomics · MGMT promoter methylation prediction · Brain tumor

1 Introduction

GBM is the most lethal type of brain tumor, constituting 60% of malignant adult brain tumors [24]. Diverse MRI modalities are sensitive to different tissue and thus can provide rich information about GBM, including shedding new insight into the oncogenetic status of GBM. With the rapid development of deep learning techniques, good performance in classification [8,16] and regression tasks [4,7,10,13,22] on MRI to address clinical questions has been achieved. In this study, we designed an end-to-end deep learning model for predicting OS and the status of MGMT methylation using multimodality MRI (Fig. 1).

X. Gu and X. Xu were partially supported by NIH R01LM012434. X. Gu was partially supported by NSF 2115095, NSF 1762287, NIH 92025.

© The Author(s), under exclusive license to Springer Nature Switzerland AG 2022
L. Wang et al. (Eds.): MICCAI 2022, LNCS 13433, pp. 282–291, 2022.
https://doi.org/10.1007/978-3-031-16437-8_27

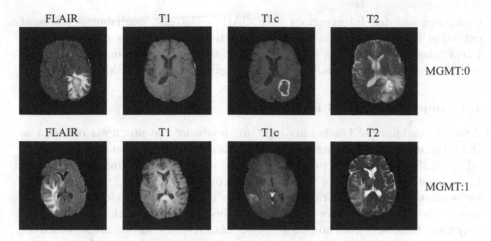

Fig. 1. Four MRI modalities of patients with different MGMT promoter methylation status.

1.1   Importance of MGMT Prediction

One important oncogenetic characteristic of GBM is the expression of MGMT. The methylation status of MGMT can be used not only as a diagnostic basis but also for prognostic evaluation, predicting the sensitivity of radiotherapy and chemotherapy, and providing effective information for precise treatment plans [11]. Though the status of MGMT can be assessed by biopsy, it is often beneficial to predict whether a GBM expresses MGMT at the earliest time point for better design of treatment and predicting the progress of the tumor. There has been great interest in using radiomics methods to infer the methylation status of MGMT in GBM [14,17,21]. Recently, several studies have shown appealing results in identifying the methylation status of MGMT promoters using deep learning methods [8,16,27]. Chang et al. designed a CNN based on ResNet to classify the methylation status of MGMT on multimodality MRI scans (T1c, T2, FLAIR) and reached a mean accuracy score of 83% on 5-fold cross-validation [8]. In a work by Yogananda et al., researchers proposed a MGMT-net based on 3D-Dense-UNets to use T2 MRI only for predicting the status of MGMT [27] and achieved a mean accuracy of 94.73% with an AUC of 0.93 on 3-fold cross-validation. Using T2 MRI, Korfiatis et al. found that ResNet50, with an accuracy of 94.90%, gave the most accurate results in this task.

While most methods predict MGMT status by focusing on features extracted from the tumor region, studies using whole brain MRI are not widely reported. On the whole brain MRI, Han and Kamdar designed a new model to predict the status of MGMT without tumor segmentation [11]. Their results showed an accuracy of 67% on the validation data and 62% on the test data [11], indicating the usefulness of whole brain analysis in such applications. In 2021, the Brain Tumor Radiogenomic Classification challenge [3], organized by the Radiological Society of North America (RSNA) and the Medical Image Computing and

Computer-Assisted Interventions (MICCAI) conference, contributed the most extensive image dataset for MGMT prediction without tumor segmentation. Participants were not able to achieve more than 0.62 AUC on the validation set [19], pointing to the challenge in accurately predicting MGMT status.

1.2 Importance of OS Prediction

Accurate estimate of OS is important for assessing the prognosis of GBM as the estimate is used to design appropriate treatment [18]. Most previous methods [1,9,25,26,28] for OS prediction employ a two-step pipeline that includes: 1) segmenting whole tumor region into necrotic, edema, and enhancing tumor regions; and 2) extracting radiomics features to train a prediction model. However, these approaches have some obvious shortcomings. On the one hand, the majority of datasets from hospitals do not contain segmentation maps, which are time-consuming and labor-intensive to acquire. On the other hand, annotation disagreements among experts can also cause inconsistency in tumor segmentation. Therefore, existing methods that predict OS are limited by the requirement for segmentation maps.

1.3 Challenges

In predicting MGMT status and OS, challenges experienced by existing works are: 1) the predicted values of the output are concentrated near the mean of ground truth, and the models do not have a good discriminating ability; and 2) given an input, a model must and can only create a single predicted value, and it is not clear that the uncertainty in prediction is sufficiently incorporated into the model. In other words, a model does not know its own limitation in dealing with uncertainty.

1.4 Contribution of This Work

A novel contribution of this work is that we proposed an EDL-based approach as a classifier for predicting the methylation status of MGMT. We used EfficientNet as a feature extractor and reached the best performance on the task of predicting the methylation status of MGMT. Our model also achieved similar performance on predicting patients' OS, demonstrating its broad applicability. Novelties of this work are as follows.

1. In general, when applying radiomics analysis with deep-learning methods, the learning models do not know the uncertainty of the prediction. To address this problem, we used Evidential-Regression to implement the final prediction, with uncertainty information attached to the prediction.
2. In Evidential-Regression, we used *a priori* NIG (Normal Inverse-Gamma) of a Gaussian distribution to fit the data. And we implemented a network structure similar to a multilayer perceptron to enhance the performance of Evidential-Regression.

2   Methods

2.1   Overview

Evidential Deep Learning. As an important branch of prediction uncertainty modeling, EDL builds on previous works of uncertainty estimation and modeling probability distributions using neural networks. Unlike BNNs, which indirectly infer prediction uncertainty through weight uncertainties, EDL employs the theory of subjective logic to explicitly estimate the uncertainty [20]. EDL treats the prediction as subjective opinions and uses a deterministic neural network to accumulate evidence that leads to these opinions. Though EDL is usually employed to address the "know unknown" flaws, for its powerfulness, as shown in [20], EDL can be used for handling uncertainties in classification. In this work, we used EDL to produce evidential distributions that better separate features arisen from binary or multi-nary populations. Then starting with the EDL-generated distributions, the learner can achieve higher performance in classification. In our case, we consider MGMT prediction as a binary classification task such that, given a GBM case, a regression model assigns probabilities to whether the case is MGMT mutant or not. So we use Evidential-Regression in EDL to perform classification for both MGMT and OS tasks.

Workflow of Our Algorithm. Our algorithm consists of several major steps, namely, feature extraction, EDL for generating evidential distributions, and classifiers. After being pre-processed, data are input into feature extractor to generate high dimensional feature maps. Evidential predictor outputs the evidential distribution based on the feature maps. Last, the algorithm computes the result from the distribution parameters. For MGMT classification, the result is a binary assignment. For OS prediction, the result is a probability on whether the GBM patient will have a short-, medium-, or long-term survival (Fig. 2).

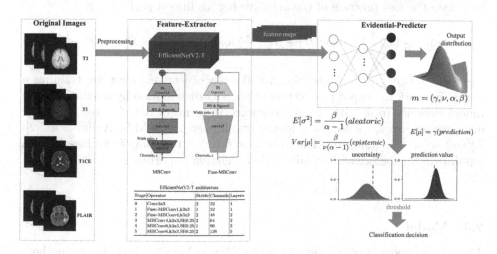

Fig. 2. Architecture of our proposed Evidential-Efficient-Net.

2.2 Evidential Regression

For regression problems, generally, loss function $L_i(w) = \frac{1}{2}\|y_i - f(x_i; w)\|^2$ is used for optimization, but such a loss function can only characterize how close the predicted value is to the data. In other words, it can only be used to represent uncertainty in the data, also known as the aleatoric uncertainty [15]. However, using evidential regression with uncertainty in the prediction results, it is possible to explicitly calculate the part of uncertainty caused by the model's predictions.

Although the ground-truth labels only have binary value of 0 or 1, combined with the prediction of the NormalCNN shown in Fig. 3 and other related knowledge, it is expected that the classification probability values should be close to the normal distribution. To reach the approximation for the true posterior which is close to the normal distribution, the evidential distribution takes the form of the Gaussian conjugate prior – NIG distribution – such that

$$P(\mu, \sigma^2 | \gamma, \nu, \alpha, \beta) = \frac{\beta^\alpha \sqrt{\nu}}{\Gamma(\alpha) \sqrt{2\pi\sigma^2}} \left(\frac{1}{\sigma^2}\right)^{\alpha+1} \exp\left\{-\frac{2\beta + \nu(\gamma - \mu)^2}{2\sigma^2}\right\}.$$

With the NIG distribution, we can calculate the prediction, aleatoric uncertainty, also known as statistical or data uncertainty, and epistemic uncertainty, which presents the estimated uncertainty in the prediction, as follows:

$$E[\mu] = \gamma \text{ (prediction)}, E[\sigma^2] = \frac{\beta}{\alpha - 1} \text{ (aleatoric)}, Var[\mu] = \frac{\beta}{\nu(\alpha - 1)} \text{ (epistemic)}.$$

After obtaining an evidential distribution expression that captures both uncertainties at the same time, model training becomes a process of accumulating evidence on the model that supports our observations or maximizing the ability of the model to fit and reduce the impact of erroneous evidence on the model. In terms of accumulating evidence, we use the Student-t distribution derivation to obtain the loss function of the negative log likelihood part:

$$L_i^{NLL}(w) = \frac{1}{2}\log\left(\frac{\pi}{\nu}\right) - \alpha\log(\Omega) + \left(\alpha + \frac{1}{2}\right)\log((y_i - \gamma)^2\nu + \Omega) + \log\left(\frac{\Gamma(\alpha)}{\Gamma(\alpha + \frac{1}{2})}\right)$$

where $\Omega = 2\beta(1 + \nu)$. This loss provides an objective target for training a neural network to output parameters of an NIG distribution by maximizing the model evidence to fit with the observations. In terms of reducing the impact of evidence on errors, we use the evidence regularizer proposed by Amini et al. [2]: $L_i^R(w) = |y_i - \gamma| \cdot (2\nu + \alpha)$. The total loss $L_i(w)$ is the sum of two losses and a regularization coefficient λ to adjust their relative importance:

$$L_i(w) = L_i^{NLL}(w) + \lambda L_i^R(w)$$

2.3 Model

The core component of our method is the EfficientNetV2, which, in our method, was scaled down to a tiny net that we named EfficientNetV2-T. It preserved the

MBConv and Fused-MBConv block in the original architecture searching space of EfficientNet. We adjusted the channels and the SE (squeeze and excitation) layer inside the blocks to accommodate four MRI modalities. As the original EfficientNetV2-S proves to be too deep for our task since it caused overfitting and reached a low bottleneck, we introduced dropout layers and adjusted regularization in our design to address this problem. With the extracted feature map, the evidential predictor exports the parameters $m = (\gamma, \nu, \alpha, \beta)$ of the evidential distribution. In the end, we used output distribution to calculate predicted values, classification results, aleatoric and epistemic uncertainties.

3   Experiments and Results

3.1   Datasets and Implementation

Dataset and Evaluation Metrics for MGMT Predication. This study used the dataset from the Brain Tumor Radiogenomic Classification challenge [3]. Multimodality MRI scans (T1, T1c, T2, FLAIR) of 585 GBM patients were provided in DICOM format. Pre-processing included skull stripping, isotropic resolution uniformization, and co-registration to the same anatomical template (SRI24). The dataset consists of 307 methylated cases and 278 unmethylated cases. Three cases were removed due to data quality issue. We then randomly separated the cases into a training group of 466 patients and a test group of 116 patients. For each modality, we converted and resampled DICOM to 3D $16 \times 256 \times 256$ NIFITI data. Four modalities data constitute the four input channels to our model. To evaluate the performance of our model, we implemented 5-fold cross-validation on the dataset and calculated several performance metrics for each patients. In this study, six widely used performance metrics are used to compare our model with several state-of-the-art models for predicting MGMT methylation, including overall accuracy (OA), sensitivity (SN), specificity (SP), positive predictive value (PPV), negative predictive value (NPV), the area under the receiver operating characteristic (ROC) curve (AUC).

Dataset and Evaluation Metrics for OS Prediction. We trained and tested the model for predicting OS of GBM patients with BraTS2019 datasets [5,6,18], which contain four MRI modality scans and survival labels of 210 patients. Like with BraTS2021, we resampled scans of each modality to $128 \times 128 \times 128$ and sent them to four channels of the feature-extractor. The ages of patients were appended with feature maps and sent to the evidential predictor. For this task, we calculated MSE (mean squared error) and classification accuracy for each patient. Classification is based on an official evaluation setup that categorizes lengths of OS into three groups: 1) short-term survivors (less than 300 days), 2) mid-term survivors (between 300 and 450 days), and 3) long-term survivors (more than 450 days).

Table 1. Results on MGMT prediction.

Methods	AUC	OA	SN	SP	PPV	NPV
Kaggle winner	0.605	0.610	0.702	0.509	0.615	0.618
ResNet [12]	0.611	0.64	–	–	–	–
Saeed et al. [19]	0.630	–	–	–	–	–
EfficientNetV2-S [23]	0.637	0.67	–	–	–	–
Proposed methods	**0.809**	**0.844**	**0.819**	**0.886**	**0.921**	**0.750**
Our method (without EDL)	0.613	0.62	–	–	–	–
EfficientNetV2-S (with EDL)	0.729	0.73	–	–	–	–

Implementation. Our model was implemented in a PyTorch 16.1 environment. The training and testing process was performed on a PC equipped with eight NVIDIA GTX 2080Ti GPUs. In training, we chose the Adam optimizer. The learning rate was 0.01, the batch-size was set to 16, and the model parameters were randomly initialized.

3.2   Results of Predicting MGMT

For MGMT prediction, the test result of our model is shown in Table 1. Compared with two state-of-the-art methods, it is seen that our method had significantly better performance across all metrics. Also, we designed ablation experiments to compare and verify the effectiveness of our components. The EfficentNetV2-S with EDL gave an increase of 0.092 on AUC and 0.06 on accuracy, respectively, as compared to the original network. Together with the comparison between proposed methods and our method without EDL, we observed that the effect of EDL is significant. Besides, the better results of proposed methods over EfficientNetV2-S with EDL proves that our modification of EfficientNet leads to improved performance.

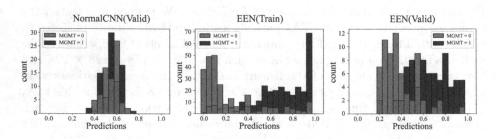

Fig. 3. Histograms of the predicted classification probabilities made by our EEN and ordinary CNN on training and validation data in predicting MGMT.

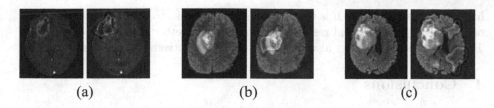

(a)   (b)   (c)

Fig. 4. Visualization of heat-maps generated by gradCAM in MGMT prediction.

On the predicted value distribution of the final output, as shown in Fig. 3, we compared the results of our method with the predicted output value of our network without EDL. Although the distribution of the predicted values after using EDL still resembles a normal distribution, it had better differentiation ability than networks that do not use EDL. In Fig. 4, we present heatmaps generated by the gradCAM algorithm to show that our model learns more information from tumor areas. Specifically, tumor areas in most cases contribute mainly to classification while other regions provide supplementary support to the classification process like the case a and b. However, in a few cases (e.g., Fig. 4(c)), the non-tumor area also contributes a lot to the final classification results, which may reveal some limitations of the model.

Table 2. Comparison and ablation results on OS prediction.

Method	Accuracy	MSE
Kaggle first winner [1]	0.586	105,062
Kaggle second winner [25]	0.488	100,000
Post-hoc [13]	0.517	105,746
EfficientNetV2-S [23]	0.421	129,547
Proposed methods	**0.513**	**79,265**
Regression (using age only)	0.387	152,619
Ours (without EDL)	0.409	136,475

3.3   Results of Predicting OS

Our model achieves the results of Table 2 on the test dataset. Compared with two state-of-the-art methods, it is seen that our method had better performance. Currently, on this dataset, most high-performance models use segmented tumor regions for analysis, such as [1,25]. While our method does not require tumor segmentation and directly processes MRI of the whole brain still achieves good performance close to the champion [1] in terms of accuracy. Compared with other models that do not require segmentation [13,23], our performances also have certain advantages. It is worth noting that, thanks to the good fitting ability of EDL, our experimental results are particularly better as measured by MSE. The ablation experiments also demonstrate the effectiveness of our method.

4   Conclusions

We presented a novel End-to-End Evidential-Efficient Net for radiomics analysis that incorporates EDL layers as a classifier to predict MGMT expression and OS of brain tumor patients. We compared the proposed method to the state-of-the-art method and the results showed our model obtains better results. Our model also requires a short training time and demonstrates high stability.

References

1. Agravat, R.R., Raval, M.S.: Brain tumor segmentation and survival prediction. In: Crimi, A., Bakas, S. (eds.) BrainLes 2019. LNCS, vol. 11992, pp. 338–348. Springer, Cham (2020). https://doi.org/10.1007/978-3-030-46640-4_32
2. Amini, A., Schwarting, W., Soleimany, A., Rus, D.: Deep evidential regression. Adv. Neural. Inf. Process. Syst. **33**, 14927–14937 (2020)
3. Baid, U., et al.: The RSNA-ASNR-MICCAI BraTS 2021 benchmark on brain tumor segmentation and radiogenomic classification. arXiv preprint arXiv:2107.02314 (2021)
4. Baid, U., et al.: Deep learning radiomics algorithm for gliomas (DRAG) model: a novel approach using 3D UNET based deep convolutional neural network for predicting survival in gliomas. In: Crimi, A., Bakas, S., Kuijf, H., Keyvan, F., Reyes, M., van Walsum, T. (eds.) BrainLes 2018. LNCS, vol. 11384, pp. 369–379. Springer, Cham (2019). https://doi.org/10.1007/978-3-030-11726-9_33
5. Bakas, S., et al.: Advancing the cancer genome atlas glioma MRI collections with expert segmentation labels and radiomic features. Sci. Data **4**(1), 1–13 (2017)
6. Bakas, S., et al.: Identifying the best machine learning algorithms for brain tumor segmentation, progression assessment, and overall survival prediction in the brats challenge. arXiv preprint arXiv:1811.02629 (2018)
7. Carver, E., et al.: Automatic brain tumor segmentation and overall survival prediction using machine learning algorithms. In: Crimi, A., Bakas, S., Kuijf, H., Keyvan, F., Reyes, M., van Walsum, T. (eds.) BrainLes 2018. LNCS, vol. 11384, pp. 406–418. Springer, Cham (2019). https://doi.org/10.1007/978-3-030-11726-9_36
8. Chang, P., et al.: Deep-learning convolutional neural networks accurately classify genetic mutations in gliomas. Am. J. Neuroradiol. **39**(7), 1201–1207 (2018)
9. Feng, X., Dou, Q., Tustison, N., Meyer, C.: Brain tumor segmentation with uncertainty estimation and overall survival prediction. In: Crimi, A., Bakas, S. (eds.) BrainLes 2019. LNCS, vol. 11992, pp. 304–314. Springer, Cham (2020). https://doi.org/10.1007/978-3-030-46640-4_29
10. Feng, X., Tustison, N.J., Patel, S.H., Meyer, C.H.: Brain tumor segmentation using an ensemble of 3D U-nets and overall survival prediction using radiomic features. Front. Comput. Neurosci. **14**, 25 (2020)
11. Han, L., Kamdar, M.R.: MRI to MGMT: predicting methylation status in glioblastoma patients using convolutional recurrent neural networks. In: Pacific Symposium on Biocomputing 2018: Proceedings of the Pacific Symposium, pp. 331–342. World Scientific (2018)

12. He, K., Zhang, X., Ren, S., Sun, J.: Deep residual learning for image recognition. In: Proceedings of the IEEE Conference on Computer Vision and Pattern Recognition, pp. 770–778 (2016)
13. Hermoza, R., Maicas, G., Nascimento, J.C., Carneiro, G.: Post-hoc overall survival time prediction from brain MRI. In: 2021 IEEE 18th International Symposium on Biomedical Imaging (ISBI), pp. 1476–1480. IEEE (2021)
14. Hu, L.S., et al.: Radiogenomics to characterize regional genetic heterogeneity in glioblastoma. Neuro-Oncol. **19**(1), 128–137 (2016). https://doi.org/10.1093/neuonc/now135
15. Kendall, A., Gal, Y.: What uncertainties do we need in Bayesian deep learning for computer vision? Adv. Neural Inf. Process. Syst. **30**, 5580–5590 (2017)
16. Korfiatis, P., Kline, T.L., Lachance, D.H., Parney, I.F., Buckner, J.C., Erickson, B.J.: Residual deep convolutional neural network predicts MGMT methylation status. J. Digit. Imaging **30**(5), 622–628 (2017)
17. Levner, I., Drabycz, S., Roldan, G., De Robles, P., Cairncross, J.G., Mitchell, R.: Predicting MGMT methylation status of glioblastomas from MRI texture. In: Yang, G.-Z., Hawkes, D., Rueckert, D., Noble, A., Taylor, C. (eds.) MICCAI 2009. LNCS, vol. 5762, pp. 522–530. Springer, Heidelberg (2009). https://doi.org/10.1007/978-3-642-04271-3_64
18. Menze, B.H., et al.: The multimodal brain tumor image segmentation benchmark (BRATS). IEEE Trans. Med. Imaging **34**(10), 1993–2024 (2014)
19. Saeed, N., Hardan, S., Abutalip, K., Yaqub, M.: Is it possible to predict MGMT promoter methylation from brain tumor MRI scans using deep learning models? Proc. Mach. Learn. Res. **1**, 16 (2022)
20. Sensoy, M., Kaplan, L., Kandemir, M.: Evidential deep learning to quantify classification uncertainty. Adv. Neural Inf. Process. Syst. **31**, 3183–3193 (2018)
21. Stupp, R., et al.: Radiotherapy plus concomitant and adjuvant temozolomide for glioblastoma. N. Engl. J. Med. **352**(10), 987–996 (2005)
22. Suter, Y., et al.: Deep learning versus classical regression for brain tumor patient survival prediction. In: Crimi, A., Bakas, S., Kuijf, H., Keyvan, F., Reyes, M., van Walsum, T. (eds.) BrainLes 2018. LNCS, vol. 11384, pp. 429–440. Springer, Cham (2019). https://doi.org/10.1007/978-3-030-11726-9_38
23. Tan, M., Le, Q.: EfficientNetV2: smaller models and faster training. In: International Conference on Machine Learning, pp. 10096–10106. PMLR (2021)
24. Taylor, O.G., Brzozowski, J.S., Skelding, K.A.: Glioblastoma multiforme: an overview of emerging therapeutic targets. Front. Oncol. **9**, 963 (2019)
25. Wang, F., Jiang, R., Zheng, L., Meng, C., Biswal, B.: 3D U-net based brain tumor segmentation and survival days prediction. In: Crimi, A., Bakas, S. (eds.) BrainLes 2019. LNCS, vol. 11992, pp. 131–141. Springer, Cham (2020). https://doi.org/10.1007/978-3-030-46640-4_13
26. Wang, S., Dai, C., Mo, Y., Angelini, E., Guo, Y., Bai, W.: Automatic brain tumour segmentation and biophysics-guided survival prediction. In: Crimi, A., Bakas, S. (eds.) BrainLes 2019. LNCS, vol. 11993, pp. 61–72. Springer, Cham (2020). https://doi.org/10.1007/978-3-030-46643-5_6
27. Yogananda, C.G.B., et al.: MRI-based deep learning method for determining methylation status of the o6-methylguanine-DNA methyltransferase promoter outperforms tissue based methods in brain gliomas. bioRxiv (2020)
28. Zhou, T., et al.: M^{2}Net: multi-modal multi-channel network for overall survival time prediction of brain tumor patients. In: Martel, A.L., et al. (eds.) MICCAI 2020. LNCS, vol. 12262, pp. 221–231. Springer, Cham (2020). https://doi.org/10.1007/978-3-030-59713-9_22

Denoising of 3D MR Images Using a Voxel-Wise Hybrid Residual MLP-CNN Model to Improve Small Lesion Diagnostic Confidence

Haibo Yang[1,2], Shengjie Zhang[1,2], Xiaoyang Han[1,2], Botao Zhao[1,2], Yan Ren[3], Yaru Sheng[3], and Xiao-Yong Zhang[1,2(✉)]

[1] Institute of Science and Technology for Brain-Inspired Intelligence,
Fudan University, Shanghai, China
[2] Key Laboratory of Computational Neuroscience and Brain-Inspired Intelligence,
Fudan University, Ministry of Education, Shanghai, China
xiaoyong_zhang@fudan.edu.cn
[3] Department of Radiology, Huashan Hospital, Fudan University, Shanghai, China

Abstract. Small lesions in magnetic resonance imaging (MRI) images are crucial for clinical diagnosis of many kinds of diseases. However, the MRI quality can be easily degraded by various noise, which can greatly affect the accuracy of diagnosis of small lesion. Although some methods for denoising MR images have been proposed, task-specific denoising methods for improving the diagnosis confidence of small lesions are lacking. In this work, we propose a voxel-wise hybrid residual MLP-CNN model to denoise three-dimensional (3D) MR images with small lesions. We combine basic deep learning architecture, MLP and CNN, to obtain an appropriate inherent bias for the image denoising and integrate each output layers in MLP and CNN by adding residual connections to leverage long-range information. We evaluate the proposed method on 720 T2-FLAIR brain images with small lesions at different noise levels. The results show the superiority of our method in both quantitative and visual evaluations on testing dataset compared to state-of-the-art methods. Moreover, two experienced radiologists agreed that at moderate and high noise levels, our method outperforms other methods in terms of recovery of small lesions and overall image denoising quality. The implementation of our method is available at https://github.com/laowangbobo/Residual_MLP_CNN_Mixer.

Keywords: MRI · Denoising · Small lesion · Deep learning

1 Introduction

Magnetic resonance imaging (MRI) is a high-resolution and non-invasive medical imaging technique that plays an important role in clinical diagnosis [1–3]. In particular, the diagnosis of small lesions in vivo, such as white matter hyperintensities (WMH) lesions, cerebral small vessel disease (CSVD), stroke, etc.,

© The Author(s), under exclusive license to Springer Nature Switzerland AG 2022
L. Wang et al. (Eds.): MICCAI 2022, LNCS 13433, pp. 292–302, 2022.
https://doi.org/10.1007/978-3-031-16437-8_28

depends largely on image quality [4–6]. However, the MRI quality can be often degraded by the noise generated during image acquisition [7], thereby affecting detection, segmentation [8,9] and diagnosis accuracy of small lesions [10,11].

Previous non-learning methods for MRI denoising can be divided into filtering-based and transform-based methods. Filtering-based methods apply linear or non-linear filters to remove noise in MR images, such as non-local means (NLM) filter [12], Perona-Malik (PM) model [13], and anisotropic diffusion filter (ADF) [14]. Transform-based methods reduce noise with a new image representation by transforming images from spatial domain into other domains, including wavelet-based methods [15] and discrete cosine transform -based methods [16]. These traditional non-learning methods suffer from several limitations, such as relying on complex algorithms based on image priors, heavy computational burden, etc.

Currently, deep learning (DL) methods have been introduced into MRI denoising and achieved better performance than traditional methods [17,18]. For example, several pioneering studies have been reported. Jiang et al. [19] and You et al. [20] applied a plain convolutional neural network (CNN) to learn the noise in MR images. Ran et al. [21] proposed a residual encoder-decoder Wasserstein generative adversarial network (RED-WGAN) to directly generate the noise-free MR images. Although the results produced by these DL methods are promising, their performance are not satisfied for denoising MR images. On the other hand, these methods have not been evaluated in clinical diagnosis. Thus, novel models are still needed to improve MRI denoising quality, especially for small lesion diagnosis that is easily contaminated by image noise.

In this paper, we propose a new DL method based on a voxel-wise hybrid residual MLP-CNN model to denoise three-dimensional (3D) MR images. We combine multilayer perceptrons (MLPs) and convolutional neural networks (CNNs) to obtain appropriate intrinsic bias for image denoising, thereby providing an appropriate number of parameters to avoid overfitting and thus improve model performance. Additionally, we integrate each output layers in MLP and CNN by adding residual connections to leverage long-range information. The structure of residual MLP is inspired by masked autoencoders (MAE) [22], which is used to reconstruct the random missing pixels. MAE uses vision transformer (ViT) [23] as encoder to reconstruct random masked patches in a image. Due to computational limitation, instead of using ViT as encoder, a simple MLP with residual connection was used to encode each noising voxel in MR images in our model. In short, the main contributions of this paper are: (a) our method can significantly denoise 3D MR images; (b) our method shows a superior performance for the recovery of small lesions in MR images compared to several state-of-the-art (SOTA) methods; (c) the diagnosis confidence of small lesion is confirmed by experienced radiologists.

2   Methods

The main purpose of MRI denoising is to recover the clean image $X \in R^{H \times W}$ from noisy image $Y \in R^{H \times W}$. Assuming that $\hat{Y} = f(Y)$ is the denoised image by f, MR image denoising can be simplified to find the optimal approximation of f:

$$\arg\min_f \|X - \hat{Y}\|_2^2 \tag{1}$$

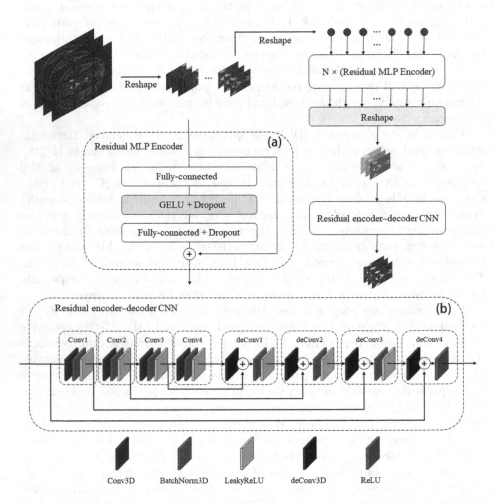

Fig. 1. Overall architecture of our proposed model. It consists of several residual MLP encoders and a residual encoder-decoder CNN. The details of residual MLP encoders and residual encoder-decoder CNN are shown in two black dotted boxes (a) and (b) respectively.

An overview of our architecture is illustrated in Fig. 1. It consists of several residual MLP encoders and a residual encoder-decoder CNN. To process the 3D images, we reshape the noisy image $\mathbf{X} \in \mathbb{R}^{H \times W \times C}$ into a sequence of flatten 3D patches, $\mathbf{X}_p \in \mathbb{R}^{K \times (P^2 \cdot C)}$, where (H, W) is the resolution of the original image,

C is the number of slices, (P, P) is the resolution of each image patch, K is the resulting number of patches.

Each MLP block contains two fully-connected layers and a nonlinearity applied independently to each flatten patches. Residual MLP encoders can be written as follows:

$$\mathbf{z}_l = \mathbf{z}_{l-1} + \mathbf{W}_2 \delta \left(\mathbf{W}_1 \, \mathrm{LN} \, (\mathbf{z}_{l-1}) \right), \quad \text{for } l = 1 \ldots L \qquad (2)$$

Here δ is an element-wise nonlinearity (GELU). $\mathbf{z}_l$ is result of the lth MLP encoder, $\mathbf{z}_0$ serves as one flatten 3D patches, the size of $\mathbf{z}_0$ is $(P^2 \cdot C \times 1 \times 1)$. $\mathbf{W}_1$ and $\mathbf{W}_2$ is the first and the second fully-connected layer weights. L is the number of Residual MLP encoders. Layernorm (LN) is applied before every block, and residual connections after every block.

Table 1. Qualitative image evaluation criteria and scoring definitions.

Score	Small lesion conspicuity	Overall image quality
1~2	No visualization	No noise reduction
3~4	Poor visualization	Noise was reduced slightly, and only a few tissue contours were discernible
5~6	Visualized with location but have blurred margins	Some noise is reduced and some details of the tissue could be identified, but not clear
7~8	Most of lesions are clearly visible, some are blurred	Noise is almost reduced, tissue details can be clearly identified, some noise remains but does not affect the diagnosis
9~10	Almost the same as the noise-free image	Remove the noise completely

Residual encoder-decoder CNN has an encoder-decoder structure composed of J convolutional and J deconvolutional layers. Skip residual connections link the corresponding convolution-deconvolutional layer pairs. Except for the last layer, the rest layers contain a 3D convolution, a batch-normalization and a LeakyReLU operation in order, while the last layer only contains a 3D convolution and a ReLU operation. Residual encoder-decoder CNN can be written as follows:

$$\mathbf{x}_j = \sigma \left(\mathrm{BN} \left(\mathbf{H}_j \mathbf{x}_{j-1} \right) \right), \qquad \text{for } j = 1 \ldots J \qquad (3)$$
$$\mathbf{y}_j = \sigma \left(\mathrm{BN} \left(\mathbf{U}_j \mathbf{y}_{j-1} + \mathbf{x}_{J-l} \right) \right), \qquad \text{for } j = 1 \ldots J \qquad (4)$$
$$\mathbf{y}_J = \theta (\mathbf{U}_J \mathbf{y}_{J-1} + \mathbf{x}_0), \qquad (5)$$
$$\mathbf{y}_0 = \mathbf{x}_J, \qquad (6)$$

Here σ is a LeakyReLU operation, θ is a LeakyReLU operation. $\mathbf{x}_j$ is result of the jth convolution layer, $\mathbf{y}_j$ is result of the jth deconvolution layer. $\mathbf{H}_j$ is the convolutional weight, $\mathbf{U}_j$ is the deconvolutional weight. $\mathbf{x}_0$ is reshaped from $\mathbf{z}_L$, The size of $\mathbf{x}_0$ is changed from $(P^2 \cdot C \times 1 \times 1)$ to $(P \times P \times C)$. BN represents batch-normalization operation.

Table 2. Ablation study of models with different block combinations. The average PSNR and SSIM are measured on validation set with 15% noise level. MLP stands for the residual MLP encoders block and CNN stands for the residual encoder-decoder CNN block.

Model	PSNR	SSIM
MLP+MLP	29.8472	0.8118
CNN+CNN	29.8285	0.8131
MLP+CNN	**32.2679**	**0.8690**

Mean squared error (MSE) loss was used to train the propose model. It can be calculated as follows:

$$L_{MSE} = \frac{1}{HWC}\|G(X) - Y\|^2 \tag{7}$$

where H, W, and C represent the dimensions of the image, X is the noisy image, Y is the noise-free image, G represents the function of the model. Note that the concise DL-based architecture and MSE loss make our model easy to train.

Table 3. The average PSNR and SSIM measures of different methods on testing images with different noise levels.

Method	3%		9%		15%	
	PSNR	SSIM	PSNR	SSIM	PSNR	SSIM
BM4D	31.2357	0.5184	22.5407	0.3590	18.1125	0.2915
PRINLM	37.0051	0.9167	29.7914	0.6699	24.6919	0.3994
RED	35.7440	0.8845	31.4146	0.8363	28.8303	0.7721
RED-WGAN	35.9356	0.9006	32.0610	0.8572	29.5153	0.8011
Ours	**38.8119**	**0.9378**	**34.8806**	**0.9000**	**32.4347**	**0.8536**

3 Experiments and Discussion

3.1 Dataset

The MRI data from UK Biobank (http://www.ukbiobank.ac.uk) were used to validate the performance of the proposed method. Details of the image acquisi-

tion and processing are freely-available on the website (http://biobank.ctsu.ox.
ac.uk/crystal/refer.cgi?id=2367).

120 T2-FLAIR brain MRI volumes (720 images) with white matter hyperintensities (WMH) lesions were randomly selected. 100 brain volumes (600 images)
were used as the training set, 10 volumes (60 images) used as the validation set,
the rest 10 volumes (60 images) used as testing set. For each volume, the central
slice and its neighboring slices were extracted to form one input 3D MRI volume.
The number of slices that we use to denoise 3D image volume is 6. All the image
volumes were reshaped into size of 176 × 256 × 6. The training MRI volumes
were cropped into 16 × 16 × 6 patches with the stride 10, which means a total
of 42,500 training patches with size of 16 × 16 × 6 were acquired.

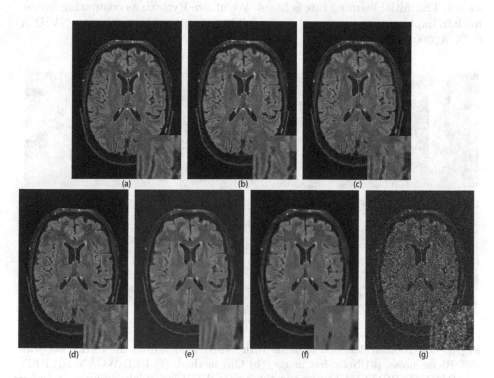

Fig. 2. A visualization of different denoising methods on T2-FLAIR images with 15%
Rician noise. (a) Noise-free image, (b) Our method, (c) RED-WGAN, (d) RED, (e)
BM4D, (f) PRINLM, (g) Noisy image. The small picture in the lower right corner of
each image is an amplified version of the area inside the red box in (a). (Color figure
online)

Previous studies have shown that the noise in MRI is governed by a Rician
distribution, both real and imaginary parts are corrupted by Gaussian noise with
equal variance [24]. We simulated noisy images by manually adding Rician noise
to the images, as follows:

$$A = \sqrt{(I + \alpha L n_r)^2 + \alpha L n_i^2}\qquad(8)$$

Here I is the original 3D image volume, α is the maximum value of I, L is the noise level, n_r and n_i are independent and identical $\mathcal{N}(0,1)$ Gaussian distributed noise, and A is the noisy image. Three levels of noise, 3%, 9% and 15% were added into original T2-FLAIR volumes.

3.2 Training Details

In our experiments, the kernel size of 3D convolution and deconvolution operation is $3 \times 3 \times 3$. The Adam [25] algorithm was used to optimize the proposed model. The initial learning rate is $5e-4$. We utilize Pytorch as computing framework to implement the proposed method. The network is trained on two NVIDIA RTX A6000 GPUs with 48 GB memory.

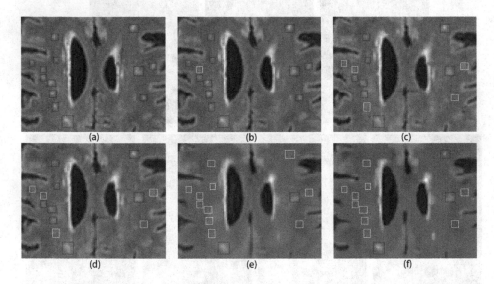

Fig. 3. One zoomed denoised example with WMH lesions from the testing set with 15% Rician noise. (a) Noise-free image, (b) Our method, (c) RED-WGAN, (d) RED, (e) BM4D, (f) PRINLM. The area in the box is WMH lesion labeled by experienced reader, the red box indicates that WMH lesions can be identified from the image, while the yellow box indicates that WMH lesion can not be identified from the image. (Color figure online)

3.3 Evaluation Methods

To validate the performance of the proposed method, two quantitative metrics were employed, including peak signal-to-noise ratio (PSNR) and structural similarity index measure (SSIM) [26]. PSNR calculates the distortion between stored images and ground truth images. The metrics of SSIM is calculated based on three comparative measurements, including luminance, contrast, and structure.

To further demonstrate the diagnostic confidence of our method, we invited two experienced radiologists as readers to score the overall image quality and the small lesion conspicuity in the restored testing images. The scoring criteria is shown in Table 1.

3.4 Results and Discussion

Ablation experiments conducted at validation set with 15% noise level show the effectiveness of combination of residual MLP encoders and residual encoder-decoder CNN. Table 2 shows the performance improvements when we combine the MLP block and the CNN block.

Table 4. Summary of image evaluation.

Reader	Method	3%		9%		15%	
		Lesions	Overall	Lesions	Overall	Lesions	Overall
Reader1	BM4D	7.4	8.3	5.8	6.3	4.7	5.2
	PRINLM	8.2	8.7	6.7	6.8	5.8	6.3
	RED	9.1	9.0	8.0	8.2	7.2	7.5
	RED-WGAN	8.8	8.8	8.0	8.2	7.7	7.5
	Ours	**9.6**	**9.6**	**8.5**	**8.5**	**8.4**	**8.2**
Reader2	BM4D	8.8	8.9	6.2	6.2	5.6	5.6
	PRINLM	**9.7**	**9.7**	6.8	6.8	6.2	6.2
	RED	9.3	9.3	7.2	7.3	6.8	6.8
	RED-WGAN	9.1	9.1	7.8	7.8	6.9	7.1
	Ours	9.2	9.2	**8.1**	**8.1**	**7.6**	**7.7**

To demonstrate the performance of the proposed method, four different SOTA methods are used for the comparison, including BM4D [27], PRINLM [28], RED [29], RED-WGAN [21].

The average quantitative results from different methods at various noise levels on testing set are summarized in Table 3. The proposed method significantly outperforms other methods on all metrics under various noise levels. At 3% noise level, the traditional non-learning method, PRINLM, yields a better performance than methods based deep learning, RED and RED-WGAN. While as the noise level increases, at 9% and 15% noise level, the deep learning based methods outperform traditional non-learning methods, especially on the SSIM metrics.

Figure 2 provides a visual evaluations of different methods for a 15% noise level task. Compared with other methods, the proposed method is most similar to the noise-free image in terms of the clarity of brain wrinkles and folds. Note that some brain structural details are missing from the RED and RED-WGAN results, while most brain structural details are missing from the BM4D and PRINTLN results.

Figure 3 shows representative small lesions recovery results by different methods. Our method restores almost all labeled WMH lesions, which is the best of all methods. In contrast, the traditional non-learning methods, such as BM4D and PRINLM, miss most of labeled WMH lesions. Table 4 shows the scores of the two readers on the testing dataset. Reader1 believes that the proposed method outperforms other methods in both small lesion recovery scores and overall image quality scores at all noise levels, while Reader2 believes that PRINLM outperforms other methods at 3% noise level. Both readers agree that our proposed method outperforms the other methods at 9% and 15% noise levels.

3.5 Conclusions

We propose a voxel-wise hybrid residual MLP-CNN model to denoise 3D MR images and improve diagnostic confidence for small lesions. We compared the proposed method with four SOTA methods on 720 T2-FLAIR brain images with WMH lesions by adding 3%, 9% and 15% noise. The results demonstrate that our method outperforms other methods in quantitative and visual evaluations at testing dataset. In addition, two experienced readers agreed that our method outperforms other methods in terms of small lesions recovery and overall image denoising quality at moderate and high noise levels.

Acknowledgments. This study was supported in part by the National Natural Science Foundation of China (81873893, 82171903), Science and Technology Commission of Shanghai Municipality (20ZR1407800), and Shanghai Municipal Science and Technology Major Project (No.2018SHZDZX01).

References

1. Jiang, Y., et al.: A novel distributed multitask fuzzy clustering algorithm for automatic MR brain image segmentation. J. Med. Syst. **43**(5), 1–9 (2019)
2. Mohan, J., Krishnaveni, V., Guo, Y.: A survey on the magnetic resonance image denoising methods. Biomed. Signal Process. Control, **9**, 56–69 (2014)
3. Zhang, X., et al.: Denoising of 3D magnetic resonance images by using higher-order singular value decomposition. Med. Image Anal. **19**(1), 75–86 (2015)
4. González, R.G.: Clinical MRI of acute ischemic stroke. J. Magn. Reson. Imaging **36**(2), 259–271 (2012)
5. Ovbiagele, B., Saver, J.L.: Cerebral white matter hyperintensities on MRI: current concepts and therapeutic implications. Cerebrovasc. Dis. **22**(2–3), 83–90 (2006)
6. Zwanenburg, J.J.M., van Osch, M.J.P.: Targeting cerebral small vessel disease with MRI. Stroke **48**(11), 3175–3182 (2017)
7. Gudbjartsson, H., Patz, S.: The Rician distribution of noisy MRI data. Magn. Reson. Med. **34**(6), 910–914 (1995)

8. Caligiuri, M.E., Perrotta, P., Augimeri, A., Rocca, F., Quattrone, A., Cherubini, A.: Automatic detection of white matter hyperintensities in healthy aging and pathology using magnetic resonance imaging: a review. Neuroinformatics **13**(3), 261–276 (2015). https://doi.org/10.1007/s12021-015-9260-y

9. Jiong, W., Zhang, Y., Wang, K., Tang, X.: Skip connection U-Net for white matter hyperintensities segmentation from MRI. IEEE Access **7**, 155194–155202 (2019)

10. Jiang, Y., et al.: Seizure classification from EEG signals using transfer learning, semi-supervised learning and TSK fuzzy system. IEEE Trans. Neural Syst. Rehabil. Eng. **25**(12), 2270–2284 (2017)

11. Khademi, A., Venetsanopoulos, A., Moody, A.R.: Robust white matter lesion segmentation in FLAIR MRI. IEEE Trans. Biomed. Eng. **59**(3), 860–871 (2011)

12. Coupé, P., Yger, P., Barillot, C.: Fast non local means denoising for 3D MR images. In: Larsen, R., Nielsen, M., Sporring, J. (eds.) MICCAI 2006. LNCS, vol. 4191, pp. 33–40. Springer, Heidelberg (2006). https://doi.org/10.1007/11866763_5

13. Perona, P., Malik, J.: Scale-space and edge detection using anisotropic diffusion. IEEE Trans. Pattern Anal. Mach. Intell. **12**(7), 629–639 (1990)

14. Sijbers, J., den Dekker, A.J., Van der Linden, A., Verhoye, M., Van Dyck, D.: Adaptive anisotropic noise filtering for magnitude MR data. Magn. Reson. Imaging **17**(10), 1533–1539 (1999)

15. Anand, C.S., Sahambi, J.S.: MRI denoising using bilateral filter in redundant wavelet domain. In: TENCON 2008–2008 IEEE Region 10 Conference, pp. 1–6. IEEE (2008)

16. Hu, J., Pu, Y., Wu, X., Zhang, Y., Zhou, J.: Improved DCT-based nonlocal means filter for MR images denoising. Comput. Math. Methods Med. 2012 (2012)

17. Ledig, C., et al.: Photo-realistic single image super-resolution using a generative adversarial network. In: Proceedings of the IEEE Conference on Computer Vision and Pattern Recognition, pp. 4681–4690 (2017)

18. Zhang, K., Zuo, W., Chen, Y., Meng, D., Zhang, L.: Beyond a gaussian denoiser: residual learning of deep CNN for image denoising. IEEE Trans. Image Process. **26**(7), 3142–3155 (2017)

19. Jiang, D., Dou, W., Vosters, L., Xiayu, X., Sun, Y., Tan, T.: Denoising of 3D magnetic resonance images with multi-channel residual learning of convolutional neural network. Jpn. J. Radiol. **36**(9), 566–574 (2018). https://doi.org/10.1007/s11604-018-0758-8

20. You, X., Cao, N., Hao, L., Mao, M., Wanga, W.: Denoising of MR images with Rician noise using a wider neural network and noise range division. Magn. Reson. Imaging **64**, 154–159 (2019)

21. Ran, M., Jinrong, H., Yang Chen, H., Chen, H.S., Zhou, J., Zhang, Y.: Denoising of 3D magnetic resonance images using a residual encoder-decoder Wasserstein generative adversarial network. Med. Image Anal. **55**, 165–180 (2019)

22. He, K., Chen, X., Xie, S., Li, Y., Dollár, P., Girshick, R.: Masked autoencoders are scalable vision learners. arXiv preprint arXiv:2111.06377 (2021)

23. Dosovitskiy, A., et al.: An image is worth 16x16 words: transformers for image recognition at scale. arXiv preprint arXiv:2010.11929 (2020)

24. Li, S., Zhou, J., Liang, D., Liu, Q.: MRI denoising using progressively distribution-based neural network. Magn. Reson. Imaging **71**, 55–68 (2020)

25. Kingma, D.P., Ba, J.: Adam: a method for stochastic optimization. arXiv preprint arXiv:1412.6980 (2014)

26. Kala, R., Deepa, P.: Adaptive hexagonal fuzzy hybrid filter for Rician noise removal in MRI images. Neural Comput. Appl. **29**(8), 237–249 (2018)

27. Maggioni, M., Katkovnik, V., Egiazarian, K., Foi, A.: Nonlocal transform-domain filter for volumetric data denoising and reconstruction. IEEE Trans. Image Process. **22**(1), 119–133 (2012)
28. Manjón, J.V., Coupé, P., Martí-Bonmatí, L., Collins, D.L., Robles, M.: Adaptive non-local means denoising of MR images with spatially varying noise levels. J. Magn. Reson. Imaging **31**(1), 192–203 (2010)
29. Mao, X., Shen, C., Yang, Y. B.: Image restoration using very deep convolutional encoder-decoder networks with symmetric skip connections. In: Advances in neural information processing systems, vol. 29 (2016)

ULTRA: Uncertainty-Aware Label Distribution Learning for Breast Tumor Cellularity Assessment

Xiangyu Li[1], Xinjie Liang[1], Gongning Luo[1(✉)], Wei Wang[1],
Kuanquan Wang[1(✉)], and Shuo Li[2]

[1] Harbin Institute of Technology, Harbin, China
{luogongning,wangkq}@hit.edu.cn
[2] Western University, London, ON, Canada

Abstract. Neoadjuvant therapy (NAT) for breast cancer is a common treatment option in clinical practice. Tumor cellularity (TC), which represents the percentage of invasive tumors in the tumor bed, has been widely used to quantify the response of breast cancer to NAT. Therefore, automatic TC estimation is significant in clinical practice. However, existing state-of-the-art methods usually take it as a TC score regression problem, which ignores the ambiguity of TC labels caused by subjective assessment or multiple raters. In this paper, to efficiently leverage the label ambiguities, we proposed an **U**ncertainty-aware **L**abel dis**TR**ibution le**A**rning (**ULTRA**) framework for automatic TC estimation. The proposed ULTRA first converted the single-value TC labels to discrete label distributions, which effectively models the ambiguity among all possible TC labels. Furthermore, the network learned TC label distributions by minimizing the Kullback-Leibler (KL) divergence between the predicted and ground-truth TC label distributions, which better supervised the model to leverage the ambiguity of TC labels. Moreover, the ULTRA mimicked the multi-rater fusion process in clinical practice with a multi-branch feature fusion module to further explore the uncertainties of TC labels. We evaluated the ULTRA on the public BreastPathQ dataset. The experimental results demonstrate that the ULTRA outperformed the regression-based methods for a large margin and achieved state-of-the-art results. The code will be available from https://github.com/PerceptionComputingLab/ULTRA.

Keywords: Breast cancer · Neoadjuvant therapy · Tumor cellularity · Label distribution learning · Label ambiguity

1 Introduction

Breast cancer is one of the most common cancers for women worldwide [1]. For breast cancer treatment, neoadjuvant therapy (NAT) [2], which aims to reduce the tumor size and avoid mastectomy for patients, is a common treatment option in clinical practice [3,4]. Tumor cellularity (TC), which represents

© The Author(s), under exclusive license to Springer Nature Switzerland AG 2022
L. Wang et al. (Eds.): MICCAI 2022, LNCS 13433, pp. 303–312, 2022.
https://doi.org/10.1007/978-3-031-16437-8_29

the percentage of invasive tumors in the tumor bed, has been widely used to quantify the response of breast cancer to NAT [5–7]. Besides, TC is also a significant indicator of the residual cancer burden (RCB), which predicts cancer recurrence and patients' survival [8]. In the current clinical practice, TC is estimated by the pathologists on hematoxylin and eosin (H&E)-stained slides, which is rather time-consuming and suffers from inter-rater and intra-rater variability [9]. Therefore, it is highly desirable to develop automatic and reliable methods for TC estimation. However, it is still challenging for automatic TC estimation methods because of the textural variations of different tissue types and the tissue color variations induced by differences in the slide generation process [10,11].

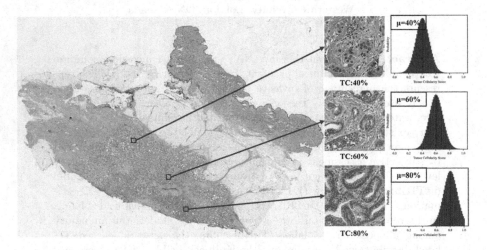

Fig. 1. The illustration of the tumor cellularity (TC) assessment task and the core idea of the proposed ULTRA which transfers the traditional TC regression problem to a label distribution learning [13] problem.

Still, there are lots of methods proposed trying to achieve accurate TC estimation in a fully automatic manner. Peikari et al. [12] utilized the traditional machine learning method to estimate tumor cellularity. They first conducted nuclei segmentation and malignant epithelial figures classification, then performed cellularity estimation based on the malignant images. Likewise, Akbar et al. [14] proposed to conduct tumor cellularity assessment by measuring the proportion of malignant cells in the region of interest (ROI). Although the above methods have made some progress on TC estimation, they need nuclei segmentation labels which are expensive to obtain, and the performance is relatively poor. To address the above problems, Rakhlin et al. [15] performed TC estimation with a cellularity score regression method, which skipped the intermediate segmentation and achieved superior results. Similarly, Akbar et al. [16] proposed to regress tumor cellularity with ResNet [17] directly. They trained a series of ResNet architectures to conduct regression tasks, and achieved better results than traditional machine learning methods. Although those techniques achieved superior performance compared to the segmentation-based methods, they took TC estimation as a simple regression problem and ignored the intrinsic ambiguity of the TC labels caused by subjective assessment or multiple raters, which further restricted the performance improvements.

Inspired by [13,18,19], we proposed an **U**ncertainty-aware **L**abel dis**TR**ibution le**A**rning (**ULTRA**) framework that fully leverages the label ambiguity (uncertainty) for TC estimation. The core idea is illustrated in Fig. 1. The ULTRA first converted the single-value TC labels to discrete label distributions, which effectively models the ambiguity among all possible TC labels. Furthermore, the network learned the TC label distributions by minimizing the Kullback-Leibler (KL) divergence between the predicted and ground-truth TC label distributions, which better supervised the model to leverage the ambiguity of TC labels caused by subjective assessment. Moreover, the ULTRA mimicked the multi-rater fusion process in clinical routines with a multi-branch feature fusion module to further explore the uncertainties of TC labels from multiple raters. The main contributions of our paper are three-fold:

- We are the first to model label ambiguity (uncertainty) of TC labels by transferring the TC score regression to a label distribution learning problem.
- We proposed a multi-branch feature fusion module by mimicking the multi-rater fusion process in clinical routines, which effectively leveraged the label uncertainty and significantly improved the TC estimation performance.
- Our ULTRA outperforms both segmentation-based and regression-based methods on the TC estimation task and achieved state-of-the-art results on the SPIE-AAPM-NCI BreastPathQ dataset.

2   Methods

The ULTRA aims to effectively exploit the uncertainty of the labels in breast tumor cellularity tasks by modeling the label ambiguity with a label distribution rather than a single value. As is illustrated in Fig. 2, it mainly consists of three parts: label distribution generation, multi-branch feature fusion, and label distribution learning. The details of each module are as follows:

2.1   Label Distribution Generation

Following [18,19], we first converted the single-value TC labels to discrete label distributions, which effectively models the ambiguity among all possible TC labels. The TC distribution comprises a group of probability values which represent the description degree of each TC value for the input image. Specifically, to generate label distributions based on the TC scores, we first quantize the TC score ranges $[0,1]$ into an ordered label set $t \in \{t_1, t_2, t_i, ...t_n\}$, where n is the number of discretized labels, $t_i \in [0,1]$ are possible TC scores. We set $n = 100$

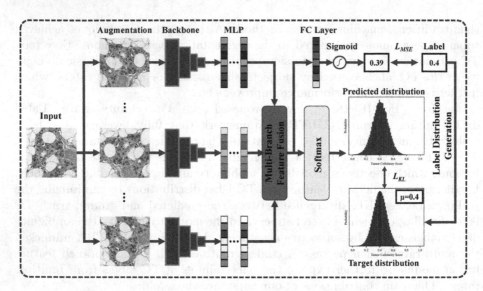

Fig. 2. The overview of the proposed ULTRA. The input image patches are first augmented and fed into the backbone network to extract features. Furthermore, the feature maps are processed by MLPs in different branches and fused with multi-branch feature fusion. Finally, the network is optimized by jointly performing TC distribution learning and score regression.

according to the precision of TC labels. Given a TC value s of a specific image, we then constructed a Gaussian distribution with label set t:

$$G\left(t_i \mid s, \sigma\right) = \frac{1}{\sqrt{2\pi}\sigma} \exp\left(-\frac{(t_i - s)^2}{2\sigma^2}\right) \tag{1}$$

where s is the TC label and also the mean value of the Gaussian distribution. σ is the hyper-parameter that defines the sharpness of the Gaussian distribution and can be also taken as the uncertainty of TC score. We set $\sigma = 0.04$ in our implementation since it achieved better results than other settings. Considering that a label distribution $y_i, i \in \{1, 2, ...n\}$ should satisfy two constrains, i.e., $y_i \in [0, 1]$, and $\sum_i y_i = 1$, we finally generate the TC label distribution by normalizing the Gaussian distribution:

$$y_i = \frac{G\left(t_i \mid s, \sigma\right)}{\sum_k G\left(t_k \mid s, \sigma\right)} \tag{2}$$

2.2 Multi-branch Feature Fusion

To further explore the TC label uncertainties caused by multiple raters, we mimicked the multi-rater fusion process with a Multi-Branch Feature Fusion module (MBFF). The proposed MBFF significantly improved the TC estimation performance by effectively modeling the estimation process of multiple raters and leveraging the ambiguity among them. Specifically, given an input image I, we designed multiple branches to process it, and each branch represents the annotation process of a single rater. In a specific branch, we first applied augmentation methods to increase the variabilities of the input and further improve the generalization performance of the network; $\hat{I}_k$ denotes the processed image with augmentations at branch k. Then, the augmented image was further processed by a backbone network and an MLP in each branch, which modeled the decision process for a single rater. Finally, each branch generated various rich hierarchical features, which encoded valuable information for the annotation of the input image:

$$f^k = \mathcal{F}_{MLP}\left(\mathcal{F}_{Backbone}\left(\hat{I}_k\right)\right) \tag{3}$$

where $\mathcal{F}_{MLP}$ and $\mathcal{F}_{Backbone}$ represent the MLP and the backbone network. f^k are the feature maps at branch k. We used ResNet34 as the backbone network and three fully connected layers as the MLP in our implementation.

Furthermore, to better exploit the uncertainties among different raters, we introduced MBFF which calculated the weighted average among features from different branches. Finally, we achieved the enhanced feature maps as follows:

$$f_{enhanced} = \frac{1}{N}\sum_{k=1}^{N} W_k f^k \tag{4}$$

where W_k is the relative importance of different branch, in our implementation, we set $W_k = 1$ in all branches, empirically.

2.3   Label Distribution Learning

The proposed ULTRA transfers the traditional TC regression problem to a label distribution learning problem (LDL), which better supervises the model to leverage the ambiguity of TC labels. Based on the hierarchical features generated from the MBFF, the predicted TC distribution can be achieved by applying a softmax activation function:

$$p_i = \frac{\exp\left(f_{enhanced}^i\right)}{\sum_i \exp\left(f_{enhanced}^i\right)}, \quad i = 0, 1, 2, \cdots, n \tag{5}$$

where $f_{enhanced}^i$ denotes the ith channel of enhanced feature from MBFF. Finally, we minimize the Kullback-Leibler (KL) divergence between the estimated TC distribution p_i and the normalized Gaussian distribution y_i:

$$\mathcal{L}_{KL} = KL\left\{\boldsymbol{p_i}\|\boldsymbol{y_i}\right\} = \sum_i p_i \log \frac{p_i}{y_i} \tag{6}$$

Moreover, existing label distribution learning suffers from severe object mismatch problem [20,21]. Therefore, different from existing methods in [18,19], which simply supervise the model with the KL divergence loss, we added an extra regression branch with mean square error loss function to mitigate the object mismatch problem, and finally jointly performed TC distribution learning and TC score regression in a multi-task learning manner:

$$\mathcal{L} = \mathcal{L}_{KL} + \alpha\mathcal{L}_{MSE} \tag{7}$$

where α represents the relative weight to balance the importance of label distribution learning and TC score regression. Specifically, we empirically set $\alpha = 1$ in our experiments. $\mathcal{L}_{MSE}$ denotes the mean square error between predicted TC score and labels.

3   Experiments and Results

3.1   Materials and Evaluation Metrics

To demonstrate the effectiveness of the proposed method, we utilized the public SPIE-AAPM-NCI BreastPathQ[1] dataset [22] in our study. The dataset provides 2394 and 185 patches with TC labels ranging from 0% to 100% in the training and validation set, and 1119 patches for testing whose TC labels are unavailable.

Following [15], we adopted intra-class correlation (ICC) [23], Cohen's kappa (Kappa) [24] and mean square error (MSE) as the evaluation metrics.

3.2   Implementation Details

Our model was trained on the NVIDIA RTX 2080Ti GPU for 150 epochs. In the training phase, we first normalized the input data by subtracting the mean and dividing it by the standard deviation. In addition, we randomly performed different augmentations in MBFF including horizontal, vertical flips, elastic transforms, and etc. Besides, we utilized the Adam optimizer and set the initial learning rate to 1e−4, and the learning rate decayed by multiplying 0.1 every 100 epochs. We set the batch size to 8 in all our experiments empirically. We performed a two-stage training scheme: first stage for backbone training and the second stage for the whole framework training. In the testing phase, the input patches are also randomly augmented to perform multi-branch feature fusion, and then the network predicts the TC distributions for input patches. After obtaining the predicted TC distribution by performing the softmax function, the TC assessment is obtained by selecting the TC value with the largest probability among all possible TC scores. Finally, the predicted TC score is achieved by averaging the predictions from regression and label distribution branches.

3.3   Experimental Results and Discussion

Effectiveness of the MBFF: Table 1 shows the ablation study on the MBFF module. We tested three ULTRA variations with different numbers of branches, $N = \{1, 2, 3\}$. Experimental results demonstrate that the ULTRA with MBFF

Table 1. Ablations of the MBFF, N: number of branches. ↑: The larger, the better; ↓: The smaller, the better.

Model	ICC↑	Kappa↑	MSE↓
ULTRA ($N=1$)	0.919	0.688	0.013
ULTRA ($N=2$)	0.921	0.693	0.014
ULTRA ($N=3$)	**0.941**	**0.703**	**0.011**

Table 2. Ablations of the LDL and regression for the ULTRA. ↑: The larger, the better; ↓: The smaller, the better.

Model	ICC↑	Kappa↑	MSE↓
ULTRA w/o KL	0.918	0.600	0.012
ULTRA w/o MSE	0.926	0.650	0.014
ULTRA	**0.941**	**0.703**	**0.011**

module outperforms its variants without fusion (i.e., N = 1). This further proves the effectiveness of the MBFF.

Effectiveness of the LDL: Table 2 illustrates the experimental results in different settings. It demonstrates that the ULTRA with only LDL branch outperforms that of regression branch in ICC and Kappa, while slightly inferior in MSE. Moreover, the above two variants are inferior to the normal ULTRA in all metrics. This proves that the LDL effectively leveraged the label uncertainty and improved the TC estimation results. More importantly, the proposed ULTRA further enhanced the performance by combining the LDL and TC regression.

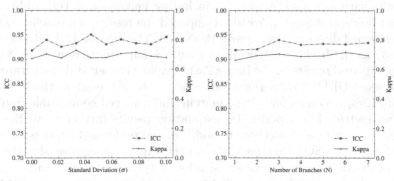

(a) Results on different Standard Deviation of Gaussian distribution

(b) Results on different number of branches

Fig. 3. The experimental results on different hyper-parameter settings

Different Hyper-Parameter Settings: We performed experiments to investigate the impact of two significant hyper-parameter settings: (1) Different σ for Gaussian distribution. Following [18], we uniformly sampled 10 values for σ in range $[0, 0.1]$. Figure 3(a) illustrates the results for different σ. It shows that σ should be set neither too large nor too small. This is reasonable since σ controls

---

1 https://breastpathq.grand-challenge.org/.

Table 3. Quantitative results (Mean with 95% confidence intervals) of state-of-the-art methods on BreastPathQ validation set. Bold text denotes the best result for that column. "–" means the authors didn't report that metric.

Methods	Metrics		
	ICC↑	Kappa↑	MSE↓
Baseline	0.901[0.870,0.930]	0.688[0.602,0.774]	0.015[0.011,0.019]
Peikari et al. [12]	0.750[0.710,0.790]	0.380–0.420	–
Akbar et al. [14]	0.830[0.790,0.860]	–	–
Rakhlin et al. [15]	0.883[0.858,0,905]	0.689[0.642,0,734]	**0.010[0.009,0.012]**
Ours(ULTRA)	**0.941[0.920,0.950]**	**0.703[0.620,0.787]**	0.011[0.007,0.014]

the uncertainty of the TC label, large σ would introduce extra label noise, while small σ would not be enough to represent the ambiguity. (2) Different number of branches. Except for $N = \{1, 2, 3\}$, we tested more settings on the number of branches. Figure 3(b) shows the results for different branches. The experimental results first improve and then decrease, finally tend to stable with the improvement of branches. $N = 3$ achieves the best results.

Comparing with State-of-the-Art Methods: We compared the experimental results with some SOTA methods including Peikari et al. [12], Akbar et al. [14], and Rakhlin et al. [15]. We also compared the results of a baseline network that performed direct regression with ResNet34. The proposed ULTRA takes the ResNet34 as the backbone network to ensure a fair comparison. Table 3 shows the experimental results on the BreastPathQ validation set. It demonstrates that the proposed ULTRA outperforms all other state-of-the-art methods on both ICC and Kappa metrics for a large margin and achieved comparable results on the MSE metric. The superior TC estimation results further prove the effectiveness of the proposed method. In addition, we performed t-test between the ULTRA and other SOTA methods, the experimental results prove that the superiority of the ULTRA is statistically significant (p-value < 0.001, p-value < 0.005, p-value < 0.01 for ICC, Kappa and MSE, respectively). Moreover, we also compared the corresponding TC scores generated by each method on WSIs of the BreastPathQ validation set, which is illustrated in Fig. 4. It demonstrates that the TC scores generated from the proposed methods are closer to the ground-truth labels than other techniques, proving that the proposed method can effectively assess breast tumor cellularity.

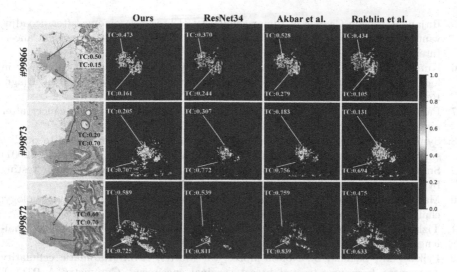

Fig. 4. TC scores generated on WSIs of the BreastPathQ validation set. The blue color denotes healthy tissue (TC = 0%) and red denotes malignant (TC = 100%). (Color figure online)

4   Conclusion

In this paper, we proposed ULTRA to address the problem of ignoring label ambiguity for the tumor cellularity assessment task. The proposed ULTRA transformed TC regression to a label distribution learning problem, which significantly exploited the label ambiguity of the TC scores. Moreover, by mimicking the multi-rater fusion process in clinical routines, the framework further leveraged the label uncertainty and improved the TC estimation performance. Experimental results prove that the proposed ULTRA achieved superior performance compared to many state-of-the-art methods.

Acknowledgments. This work was supported by the National Natural Science Foundation of China under Grant 62001144 and 62001141, and by Science and Technology Innovation Committee of Shenzhen Municipality under Grant JCYJ20210324131800002 and RCBS20210609103820029.

References

1. Key, T.J., Verkasalo, P.K., Banks, E.: Epidemiology of breast cancer. Lancet Oncol. **2**(3), 133–140 (2001)
2. Thompson, A., Moulder-Thompson, S.: Neoadjuvant treatment of breast cancer. Ann. Oncol. **23**, x231–x236 (2012)
3. Loibl, S., Denkert, C., von Minckwitz, G.: Neoadjuvant treatment of breast cancer-clinical and research perspective. Breast **24**, S73–S77 (2015)
4. Rubovszky, G., Horváth, Z.: Recent advances in the neoadjuvant treatment of breast cancer. J. Breast Cancer **20**(2), 119–131 (2017)

5. Rajan, R., et al.: Change in tumor cellularity of breast carcinoma after neoadjuvant chemotherapy as a variable in the pathologic assessment of response. Cancer: Interdisc. Int. J. Am. Cancer Soc. **100**(7), 1365–1373 (2004)
6. Kumar, S., Badhe, B.A., Krishnan, K., Sagili, H.: Study of tumour cellularity in locally advanced breast carcinoma on neo-adjuvant chemotherapy. J. Clin. Diagn. Res.: JCDR **8**(4), FC09 (2014)
7. Park, C.K., Jung, W.H., Koo, J.S.: Pathologic evaluation of breast cancer after neoadjuvant therapy. J. Pathol. Transl. Med. **50**(3), 173 (2016)
8. Symmans, W.F., et al.: Measurement of residual breast cancer burden to predict survival after neoadjuvant chemotherapy. J. Clin. Oncol. **25**(28), 4414–4422 (2007)
9. Smits, A.J., et al.: The estimation of tumor cell percentage for molecular testing by pathologists is not accurate. Mod. Pathol. **27**(2), 168–174 (2014)
10. Madabhushi, A., Lee, G.: Image analysis and machine learning in digital pathology: challenges and opportunities. Med. Image Anal. **33**, 170–175 (2016)
11. Tizhoosh, H.R., Pantanowitz, L.: Artificial intelligence and digital pathology: challenges and opportunities. J. Pathol. Inf. **9** (2018)
12. Peikari, M., Salama, S., Nofech-Mozes, S., Martel, A.L.: Automatic cellularity assessment from post-treated breast surgical specimens. Cytometry A **91**(11), 1078–1087 (2017)
13. Geng, X.: Label distribution learning. IEEE Trans. Knowl. Data Eng. **28**(7), 1734–1748 (2016)
14. Akbar, S., Peikari, M., Salama, S., Panah, A.Y., Nofech-Mozes, S., Martel, A.L.: Automated and manual quantification of tumour cellularity in digital slides for tumour burden assessment. Sci. Rep. **9**(1), 1–9 (2019)
15. Rakhlin, A., Tiulpin, A., Shvets, A.A., Kalinin, A.A., Iglovikov, V.I., Nikolenko, S.: Breast tumor cellularity assessment using deep neural networks. In: Proceedings of the IEEE/CVF International Conference on Computer Vision Workshops (2019)
16. Akbar, S., Peikari, M., Salama, S., Nofech-Mozes, S., Martel, A.L.: Determining tumor cellularity in digital slides using ResNet. In: Medical Imaging 2018: Digital Pathology, vol. 10581, pp. 233–239. International Society for Optics and Photonics (2018)
17. He, K., Zhang, X., Ren, S., Sun, J.: Deep residual learning for image recognition. In: Proceedings of the IEEE Conference on Computer Vision and Pattern Recognition, pp. 770–778 (2016)
18. Gao, B.B., Xing, C., Xie, C.W., Wu, J., Geng, X.: Deep label distribution learning with label ambiguity. IEEE Trans. Image Process. **26**(6), 2825–2838 (2017)
19. Tang, Y., et al.: Uncertainty-aware score distribution learning for action quality assessment. In: Proceedings of the IEEE/CVF Conference on Computer Vision and Pattern Recognition, pp. 9839–9848 (2020)
20. Wang, J., Geng, X.: Label distribution learning machine. In: International Conference on Machine Learning, pp. 10749–10759. PMLR (2021)
21. Wang, J., Geng, X., Xue, H.: Re-weighting large margin label distribution learning for classification. IEEE Trans. Pattern Anal. Mach. Intell. (2021)
22. Petrick, N., et al.: SPIE-AAPM-NCI BreastPathQ challenge: an image analysis challenge for quantitative tumor cellularity assessment in breast cancer histology images following neoadjuvant treatment. J. Med. Imaging **8**(3), 034501 (2021)
23. Shrout, P.E., Fleiss, J.L.: Intraclass correlations: uses in assessing rater reliability. Psychol. Bull. **86**(2), 420 (1979)
24. McHugh, M.L.: Interrater reliability: the kappa statistic. Biochemia Med. **22**(3), 276–282 (2012)

Test-Time Adaptation with Calibration of Medical Image Classification Nets for Label Distribution Shift

Wenao Ma[1], Cheng Chen[1], Shuang Zheng[2,3], Jing Qin[4], Huimao Zhang[2,3]($\boxtimes$), and Qi Dou[1]($\boxtimes$)

[1] Department of Computer Science and Engineering,
The Chinese University of Hong Kong, Shatin, Hong Kong
qdou@cse.cuhk.edu.hk

[2] Department of Radiology, The First Hospital of Jilin University, Changchun, China
huimao@jlu.edu.cn

[3] Jilin Provincial Key Laboratory of Medical Imaging & Big Data,
Changchun, China

[4] Centre for Smart Health, The Hong Kong Polytechnic University,
Kowloon, Hong Kong

Abstract. Class distribution plays an important role in learning deep classifiers. When the proportion of each class in the test set differs from the training set, the performance of classification nets usually degrades. Such a label distribution shift problem is common in medical diagnosis since the prevalence of disease vary over location and time. In this paper, we propose the first method to tackle label shift for medical image classification, which effectively adapt the model learned from a single training label distribution to arbitrary unknown test label distribution. Our approach innovates distribution calibration to learn multiple representative classifiers, which are capable of handling different one-dominating-class distributions. When given a test image, the diverse classifiers are dynamically aggregated via the consistency-driven test-time adaptation, to deal with the unknown test label distribution. We validate our method on two important medical image classification tasks including liver fibrosis staging and COVID-19 severity prediction. Our experiments clearly show the decreased model performance under label shift. With our method, model performance significantly improves on all the test datasets with different label shifts for both medical image diagnosis tasks. Code is available at https://github.com/med-air/TTADC.

Keywords: Test-time adaptation · Label distribution shift · Medical image classification

Supplementary Information The online version contains supplementary material available at https://doi.org/10.1007/978-3-031-16437-8_30.

© The Author(s), under exclusive license to Springer Nature Switzerland AG 2022
L. Wang et al. (Eds.): MICCAI 2022, LNCS 13433, pp. 313–323, 2022.
https://doi.org/10.1007/978-3-031-16437-8_30

1   Introduction

Intelligent medical image diagnosis has witnessed great success on accurate predictions for various tasks such as disease staging [14,23], lesion diagnosis [9,15], and severity prediction [11,25]. However, real-world use of classification models is challenged by the inevitable shift in class distributions on test data at deployment [1,27,32,33]. Usually, the proportion of samples belonging to each class is associated with patient demographics and region-related prevalence of disease, which differs from one hospital to another. This issue is called *label distribution shift*, which means that the label distribution can change across training and test datasets. As label distribution plays a vital role in classification tasks [7,16], such shift can make the learned classifier become suboptimal on unseen datasets, thus suffering from performance degradation in testing.

Label distribution shifts are very common in medical diagnosis as the disease distributions vary across location and time. For example, the prevalence of liver diseases significantly differs among regions due to the difference in vaccination coverage [31]. Such label shifts often degrade the performance of a learned classifier on test data, leading to erroneous predictions as observed in prior works [3,4,6]. For example, Davis et al. find that the prediction accuracy of their machine learning models decreases due to the declining incidence of acute kidney injury over time [6]. Since the proportion of normal and disease cases differs between the screening and diagnostic scenarios, an accurate model for screening purpose could perform poorly for diagnosis purpose, even for the same disease [3]. Park et al. [20] show in three disease classification models that dataset shifts including the label shift can lead to unreasonable predictions. Despite being observed in many real applications, the problem of label distribution shift has not yet been tackled for medical image diagnosis, severely hindering the large-scale deployment of deep models in clinical practice.

To generalize model under label shift, if the label distribution of test data can be known, such as the uniform distribution assumption made in [24,30], the label shift can be alleviated by re-sampling training data or adjusting the prediction probability in the softmax loss [22,24] accordingly. In practical scenarios, however, it is unlikely to anticipate the label distribution of test data, which is usually unknown and arbitrary, and may even continuously change. In this regard, we aim to mitigate label shift in a highly practical yet challenging setting, where the test label distribution is unknown and the trained model itself must accommodate label shift by utilizing the test data only. To tackle this problem, we consider two key ingredients. Firstly, since the test label distribution can be arbitrary, it is important to enlarge the capacity of models for an extensive label distribution space. The difficulty lies in how to establish such a representative space during model learning from the training set with a fixed label distribution. Secondly, motivated by the recent test-time learning works [28,29], although the knowledge of test dataset is unknown during model training, it can be explored from the test data at inference time.

In this paper, to our best knowledge, we present the first work to effectively tackle the label distribution shift in medical image classification. Our method

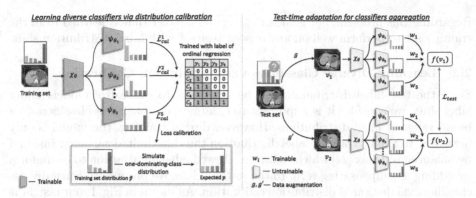

Fig. 1. Overview of our proposed method for test-time adaptation by calibration of medical image classification networks for label distribution shift.

learns representative classifiers with distribution calibration, by extending the concept of balanced softmax loss [24,34] to simulate multiple distributions that one class dominates other classes. Compared with [34], our method can be more flexible and be more targeted for ordinal classification, as our one-dominating-class distributions can represent more diverse label distributions and we use ordinal encoding instead of one-hot encoding to train the model. Then, at model deployment to new test data, we dynamically combine the representative classifiers by adapting their outputs to the label distribution of test data. The test-time adaptation is driven by a consistency regularization loss to adjust the weights of different classifier. We evaluate our method on two important medical applications of liver fibrosis staging and COVID-19 severity prediction. With our proposed method, the label shift can be largely mitigated with consistent performance improvement.

2   Method

2.1   Problem Formulation of Label Distribution Shift

For disease diagnosis, consider a classification task that aims to train a model to predict the disease class y correctly given an input image x. Let $\hat{p}(x, y)$ and $\tilde{p}(x, y)$ denote the training and test set distributions respectively. In practice, a deployed model often suffers from label distribution shift, which means the label distribution of training set $\hat{p}(y)$ is different from that of test set $\tilde{p}(y)$, i.e., $\hat{p}(y) \neq \tilde{p}(y)$, but the conditional distributions are consistent, i.e., $\hat{p}(x|y) = \tilde{p}(x|y)$. This phenomenon is especially common in medical image classification, where the disease label y is often the causal variable and the image data x can be regarded as the manifestations of a disease [26,32]. According to the Bayesian inference $\hat{p}(y|x) = \frac{\hat{p}(x|y)\hat{p}(y)}{\hat{p}(x)}$, the model prediction $\hat{p}(y|x)$ is strongly coupled with the label distribution $\hat{p}(y)$, thus the shift in $\hat{p}(y)$ can cause erroneous prediction of $\hat{p}(y|x)$.

Regarding this problem, our goal is to adapt a classifier that is learned from the training set to perform well on any unseen test set with label distribution shift.

2.2 Learning Diverse Classifiers via Distribution Calibration

Since the test label distribution can be arbitrary, to generalize models under label shift, we consider it is important to enlarge the capacity of classifiers to a broad range of label distributions. However, during training, the model is only presented to a fixed training label distribution thus has limited capacity. Inspired by balanced softmax [24] which calibrates skewed label distribution to be uniform by adding a compensating term to the softmax loss, we propose to learn diverse classifiers via dedicated distribution calibration. As shown in Fig. 1, our insight is to simulate representative one-dominating-class distributions so that the proper combination of learned classifiers can handle arbitrary test label distribution.

Before introducing how to achieve distribution calibration, we first clarify the ordinal encoding in our classification task. To encourage classification network to learn the commonness of all classes and the distinctions between different classes, we use ordinal encoding [18] instead of one-hot encoding for the ordinal classes in our liver fibrosis staging and COVID-19 severity prediction tasks. This ordinal encoding performs multiple binary classifications with sigmoid function and combines the multiple binary outputs by taking the highest class that is predicted as 1 as the final prediction. Furthermore, for distribution calibration in our ordinal regression, we extend the balanced softmax to the sigmoid function and derive the corresponding compensating term. Let $p(y_i = 1|x)$ be the desired conditional probability for the expected label distribution, and $\hat{p}(y_i = 1|x)$ be the desired conditional probability of the training set, and assume $p(y_i = 1|x)$ is expressed by the standard sigmoid function of the network output ϕ_i in i-th ordinal vector: $p(y_i = 1|x) = \frac{e^{\phi_i}}{1+e^{\phi_i}}$, then the $\hat{p}(y_i = 1|x)$ with the same output ϕ_i can be expressed as:

$$\hat{p}(y_i = 1|x) = \frac{e^{\phi_i - \log\left(\frac{r_i'}{1-r_i'} \cdot \frac{1-r_i}{r_i}\right)}}{1 + e^{\phi_i - \log\left(\frac{r_i'}{1-r_i'} \cdot \frac{1-r_i}{r_i}\right)}}, \tag{1}$$

where r_i' and r_i are the positive label proportion in the i-th ordinal vector for the expected label distribution p and factual label distribution respectively $\hat{p}$, and the term $\log\left(\frac{r_i'}{1-r_i'} \cdot \frac{1-r_i}{r_i}\right)$ is the compensating term. The proof of Eq. (1) is provided in the supplementary material.

In this way, the calibrated loss function is:

$$\bar{\mathcal{L}}_{cal} = -\sum_{i=1}^{K-1} \left(y_i \log \hat{p}(y_i = 1|x) + (1 - y_i) \log\left(1 - \hat{p}(y_i = 1|x)\right)\right), \tag{2}$$

where K denotes the total number of classes. This calibrated loss function enables the model learned on the training label distribution to generate the prediction for the expected label distribution.

Moreover, we aim to properly construct different r'_i to simulate K one-dominating-class distributions for K classifiers. Assume the proportion of dominating class j is λ times other classes, then the value of r'_i can be calculated as:

$$r'_i = 1 - \frac{i - 1_{i \geq j} \cdot (1 - \lambda)}{\lambda + K - 1}, \tag{3}$$

where $1_{i \geq j}$ is the indicator function. The derivation of Eq. (3) can be found in the supplementary material. Notably, our distribution-calibrated networks use independent parameters only at the last stages and fully-connected layer of networks, while share the parameters at other layers (see the shared network χ_θ and the independent networks ψ_θ in Fig. 1). This is motivated by the observation that decoupling the representation learning and classification gives more generalizable representations [10]. In this way, we obtain diverse classifiers to handle different label distributions, but adding only minimal computational cost.

2.3   Test-Time Adaptation for Dynamic Classifier Aggregation

After obtaining diverse distribution-calibrated classifiers during training phase, then at test time, the key is how to aggregate these classifiers to handle the unknown test label distribution with the given inference samples. To build the connection between the obtained classifiers and the test data, we aggregate the outputs of all classifiers with learnable weights, which are dynamically adapted using information implicitly provided by the test data. It's worth to mention that a set of test data, which can reflect the label distribution in the test center, should be accessible simultaneously during this phase.

Specifically, the aggregated output is defined as $\hat{p}_{\text{agg}} = \sum_{k=1}^{K} w_k \hat{p}_k$, where $\sum_{k=1}^{K} w_k = 1$ and $\hat{p}_k$ is the output of k-th classifiers with the form of ordinal vector. As different combination of $\{w_1, w_2, ..., w_K\}$ can enable the model to deal with different test label distributions, the aim of our test-time adaptation is to find the optimal combination for a given test set. Our assumption is that if the aggregated model has adapted to a particular test label distribution, for the test images generated from such a label distribution, the model should give similar predictions to perturbed versions of the same image. Based on this assumption, we design a consistency regularization mechanism to drive the test-time learning. Given an input x, we generate two augmented views $g(x) = v_1$ and $g'(x) = v_2$ using the data augmentation approaches, including rotating, flipping, and shifting the images, and adding Gaussian noise to the images. The two views are then forwarded to the trained model $f(\cdot)$ respectively, yielding the ordinal encoded output $f(v_1) = \hat{p}_{\text{agg}} = w_1 \cdot \hat{p}_1 + w_2 \cdot \hat{p}_2 + \cdots + w_K \cdot \hat{p}_K$ and $f(v_2) = \hat{p}'_{\text{agg}} = w_1 \cdot \hat{p}'_1 + w_2 \cdot \hat{p}'_2 + \cdots + w_K \cdot \hat{p}'_K$. The consistency regularization for the outputs of the two views is imposed with a cosine similarity loss:

$$\mathcal{L}_{\text{test}} = -cos(f(v_1), f(v_2)) = -\frac{f(v_1) \cdot f(v_2)}{\|f(v_1)\|_2 \times \|f(v_2)\|_2}, \tag{4}$$

The loss $\mathcal{L}_{\text{test}}$ drives the updates of the weights set $\{w_1, w_2, ..., w_K\}$ with the implicit knowledge of label distribution on the test set, while the other network parameters of $f(\cdot)$ are frozen. This implicit knowledge is reflected by the consistency that measures whether the aggregated model has adapted to the test label distribution successfully. Each weight of $\{w_1, w_2, ..., w_K\}$ is initialized to $\frac{1}{K}$ and we use softmax function to maintain the sum of them equals to one after each iteration. As a result, the test results can be obtained after the test-time adaptation given the optimized weights set.

3   Experiment

3.1   Dataset and Experimental Setup

Datasets. We have validated our proposed method on two tasks: 1) liver fibrosis staging with an in-house abdominal CT dataset, and 2) COVID-19 severity prediction with a public chest CT dataset (iCTCF [17]). The liver CT dataset consists of three centers with different label distributions, including 823 cases from our center, 99 cases from external center A and 50 cases from external center B. The ground truths of the liver fibrosis staging come from the pathology results of liver biopsy. The liver fibrosis disease is divided into 5 stages, including no fibrosis (F0), portal fibrosis without septa (F1), portal fibrosis with few septa (F2), numerous septa without cirrhosis (F3) and cirrhosis (F4). Segmentation of the liver is pre-computed with an out-of-the-box tool in a related clinical study, so we adopt it in our paper as the region of interest for classification. The slice thickness of the CT images is 5 mm and the in-plane resolution is 512 × 512. For the COVID-19 dataset, it contains 969 cases from HUST-Union Hospital for training and 370 cases from HUST-Liyuan Hospital for test. The severity of COVID-19 is divided to 6 levels: control (S0), suspected (S1), mild (S2), regular (S3), severe (S4) and critically (S5). The preprocessing and automatic lung segmentation process are the same as a recent work [2] on this dataset.

Experimental Setting. For liver fibrosis staging, we take 630 cases from our center as the training set, 193 cases from our center as evaluation set and the data from two external centers as two different test sets. For COVID-19 severity prediction, we use the data from HUST-Union Hospital for training and data from HUST-Liyuan Hospital for test. Label distribution statistics of different centers for both datasets are provided in supplementary.

Evaluation Metrics. For both tasks, the diagnosis performance is evaluated with accuracy, area under the receiver operating characteristic curve (AUC) and Obuchowski index (OI) [19], as reported in related works [2,5,21]. Considering the AUC is defined for binary classification while ours are multi-class classification tasks, we combine the classes and convert the multi-class classification to several binary classifications. Specifically, we calculate the AUC of F0 vs F1-4, F0-1 vs F2-4, F0-2 vs F3-4 and F0-3 vs F4 for the liver fibrosis staging, and the AUC of S0 vs S1-5, S0-1 vs S2-5, S0-2 vs S3-5, S0-3 vs S4-5 and S0-4 vs S5 for COVID-19 severity prediction. We report the average of all the AUC values as overall performance. The Obuchowski index (OI) is a metric which is proved to

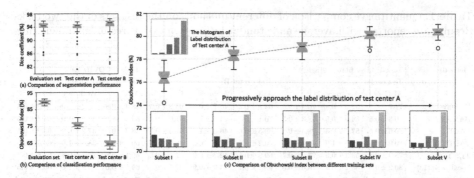

Fig. 2. Analysis of model performance with label distribution shift.

have no bias when label distributions are different between training and test sets [12].

Implementation Details. Considering model efficiency while still capturing 3D information in CT scans, we use ResNet-50 to get a vector of spatial features and then forward the features of adjacent slices to a LSTM module and a fully connected layer for classification. We train the models using Adam with an initial learning rate of $1e-5$, a weight decay of $1e-4$ and batch size of 4. Our models are implemented using a workstation with four NVIDIA TITAN Xp GPUs.

3.2 Experimental Results

Observation of Label Distribution Shift. Label distribution shift and data distribution shift are two types of dataset shift, as introduced in previous work [27]. We first clearly show in the multi-center liver CT datasets that under label distribution shift, the performance of classification model would degrade. Figure 2(a) and (b) present that the segmentation performance for region-of-interest liver extraction is consistent between the evaluation set and test sets, while the final classification performance of Obuchowski index largely decreases by 12.9% and 27.5% at test set A and B. It worth to mention that the label distribution of evaluation set is consistent with the training set while the test sets are not. In Fig. 2(c), we progressively adjust the class distribution of training set to approach the label distribution of test set A, by random sampling a certain proportion of images belonging to each class. We can see that the classification performance increases when the class distribution of the training set becomes closer to the test set. These experiments clearly demonstrate it is indeed the label shift causes the performance drop of classification model in our datasets.

Comparison with State-of-the-Art Methods. We here compare our method with state-of-the-art approaches for label shift in natural images as strong competitors, including **BALMS** [24], which calibrates the training label distribution to be uniform, **LADE** [8], which disentangles the training label

Table 1. Quantitative comparison of different methods on the test sets of the two tasks. Results are reported with average and standard deviation over three independent runs.

| Methods | Task 1: Liver fibrosis staging | | | | | | Task 2: COVID-19 | | |
| | Test center A | | | Test center B | | | Severity prediction | | |
	AUC	Accuracy	OI	AUC	Accuracy	OI	AUC	Accuracy	OI
Baseline	77.7±0.7	52.5±0.8	76.3±0.5	68.8±0.6	40.7±0.9	66.3±0.5	68.4±0.8	36.2±1.0	65.2±0.6
BALMS [24]	80.3±0.5	54.9±0.5	78.3±0.5	70.1±0.7	44.0±1.6	67.0±0.4	69.5±0.6	36.2±1.0	66.4±0.5
LADE [8]	80.6±0.5	57.6±0.8	78.5±0.5	69.3±0.6	46.0±1.6	67.9±0.5	68.3±0.6	37.2±0.9	66.2±0.5
TADE [34]	80.9±0.6	59.9±1.9	79.2±0.5	70.2±0.8	47.3±0.9	68.5±0.7	69.6±0.8	38.3±1.6	68.4±0.6
TENT [29]	78.9±0.8	53.2±0.5	77.0±0.7	69.9±0.5	42.7±0.9	67.1±0.5	69.1±0.8	36.4±0.8	65.6±0.7
Focal Loss [13]	80.2±0.6	53.2±0.5	78.0±0.5	69.1±0.7	43.3±0.9	67.7±0.6	69.5±0.6	36.5±1.0	66.5±0.5
TTADC (ours)	**82.3±0.4**	**61.0±1.0**	**80.2±0.4**	**72.4±0.6**	**50.7±0.9**	**69.6±0.4**	**71.1±0.6**	**40.2±1.1**	**69.8±0.5**

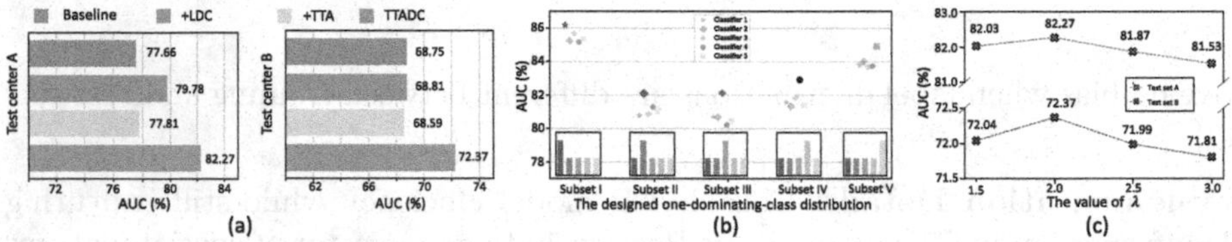

Fig. 3. Ablation analysis of our method on liver CT dataset. (a) Contribution of LDC and TTA in our method; (b) Performance of our learned diverse classifiers on different one-dominating-class distributions; (c) Effect of the value of λ on model performance.

distribution from the model prediction, and **TADE** [34], which also proposes to train multiple networks with different expertise but their networks are less representative than ours. Note that BALMS, LADE, and TADE need to use our derived compensating term in Eq. 1 to be applied in our classification tasks with ordinal regression. We also compare our method with **TENT** [29], which is a general test-time adaptation approach for domain shift problem, and **Focal Loss** [13], which can alleviate class imbalance by increasing the focus on hard samples.

Table 1 presents the comparison results on the test centers of both liver fibrosis staging and COVID-19 severity prediction. Our TTADC significantly improves the model performance over baseline on all test sets, with 4.6%, 3.6%, 2.7% increase in AUC, 8.5%, 10.0%, 4.0% increase in Accuracy, and 3.9%, 3.3%, 4.6% increase in OI respectively, outperforming all the comparison methods. The results validate the effectiveness of our distribution calibration and test-time adaptation on addressing arbitrary label shift. Our method clearly outperforms the domain adaptation method TENT, showing the necessity of designing app-roach specifically for label shift. Although not significant, Focal loss can also generally improve over baseline, indicating the alleviation of class imbalance may help reduce the effect of label shift. The other methods on tackling label shift generally outperform TENT and Focal loss. Our method and TADE which learn multiple classifiers obtain better performance than BALMS and LADE which use uniform distribution assumption, showing the importance of enlarging the model capacity for proper test-time adaptation. Our method also clearly outperforms TADE, demonstrating the combination of our one-dominating-class distributions can represent more diverse test label distributions.

Ablation Analysis. Comprehensive ablation studies have been conducted with the liver CT dataset to analyze the key ingredients regarding our TTADC. As shown in Fig. 3(a), adding only the learning distribution-calibrated classifier (LDC) or test-time adaptation (TTA) over baseline is not able to improve over baseline. This is as expected since the two key components are strongly coupled, i.e., the diverse classifiers need to be properly aggregated at test time for the unknown label distribution. In Fig. 3(b), we manually sample a few images from the training center to construct the test subsets with different one-dominating-class distributions. We can see that given the test subset with k-th class dominating other classes, the best performance comes from the k-th classifier, demonstrating that our proposed distribution calibration successfully generate classifiers that have expertise on different one-dominating-class distributions. Moreover, Fig. 3(c) compares the model performance trained with different λ in Eq. 3. The results show that the optimal choice of λ is 2.

4   Conclusion

We present, to our best knowledge, the first method to generalize deep classifiers to unknown test label distributions for medical image classification. Our methods innovates distribution calibration to learn multiple representative classifiers during training, which are then dynamically aggregated via test-time adaptation to deal with arbitrary label shift. Our method is general and experiments on two important medical diagnosis tasks demonstrate the effectiveness of our method.

Acknowledgement. This work was supported in part by the Hong Kong Innovation and Technology Fund (Project No. ITS/238/21), in part by the CUHK Shun Hing Institute of Advanced Engineering (project MMT-p5-20), in part by the Shenzhen-HK Collaborative Development Zone, in part by Jilin Provincial Key Laboratory of Medical Imaging & Big Data (20200601003JC), Radiology, and in part by Technology Innovation Center of Jilin Province (20190902016TC).

References

1. Azizzadenesheli, K., Liu, A., Yang, F., Anandkumar, A.: Regularized learning for domain adaptation under label shifts. In: International Conference on Learning Representations (2019)
2. Bao, G., et al.: COVID-MTL: multitask learning with Shift3D and random-weighted loss for COVID-19 diagnosis and severity assessment. Pattern Recogn. **124**, 108499 (2022)
3. Challen, R., Denny, J., Pitt, M., Gompels, L., Edwards, T., Tsaneva-Atanasova, K.: Artificial intelligence, bias and clinical safety. BMJ Qual. Saf. **28**(3), 231–237 (2019)

4. Chen, I.Y., Joshi, S., Ghassemi, M., Ranganath, R.: Probabilistic machine learning for healthcare. Annu. Rev. Biomed. Data Sci. **4**, 393–415 (2021)
5. Choi, K.J., et al.: Development and validation of a deep learning system for staging liver fibrosis by using contrast agent-enhanced CT images in the liver. Radiology **289**(3), 688–697 (2018)
6. Davis, S.E., Lasko, T.A., Chen, G., Siew, E.D., Matheny, M.E.: Calibration drift in regression and machine learning models for acute kidney injury. J. Am. Med. Inform. Assoc. **24**(6), 1052–1061 (2017)
7. Galar, M., Fernandez, A., Barrenechea, E., Bustince, H., Herrera, F.: A review on ensembles for the class imbalance problem: bagging-, boosting-, and hybrid-based approaches. IEEE Trans. Syst. Man Cybern. Part C (Appl. Rev.) **42**(4), 463–484 (2011)
8. Hong, Y., Han, S., Choi, K., Seo, S., Kim, B., Chang, B.: Disentangling label distribution for long-tailed visual recognition. In: Proceedings of the IEEE/CVF Conference on Computer Vision and Pattern Recognition, pp. 6626–6636 (2021)
9. Hussein, S., Kandel, P., Bolan, C.W., Wallace, M.B., Bagci, U.: Lung and pancreatic tumor characterization in the deep learning era: novel supervised and unsupervised learning approaches. IEEE Trans. Med. Imaging **38**(8), 1777–1787 (2019)
10. Kang, B., et al.: Decoupling representation and classifier for long-tailed recognition. In: International Conference on Learning Representations (2020)
11. Konwer, A., et al.: Attention-based multi-scale gated recurrent encoder with novel correlation loss for COVID-19 progression prediction. In: de Bruijne, M., et al. (eds.) MICCAI 2021. LNCS, vol. 12905, pp. 824–833. Springer, Cham (2021). https://doi.org/10.1007/978-3-030-87240-3_79
12. Lambert, J., Halfon, P., Penaranda, G., Bedossa, P., Cacoub, P., Carrat, F.: How to measure the diagnostic accuracy of noninvasive liver fibrosis indices: the area under the ROC curve revisited. Clin. Chem. **54**(8), 1372–1378 (2008)
13. Lin, T.Y., Goyal, P., Girshick, R., He, K., Dollár, P.: Focal loss for dense object detection. In: Proceedings of the IEEE International Conference on Computer Vision, pp. 2980–2988 (2017)
14. Liu, M., Zhang, D., Shen, D.: Relationship induced multi-template learning for diagnosis of Alzheimer's disease and mild cognitive impairment. IEEE Trans. Med. Imaging **35**(6), 1463–1474 (2016)
15. Mesejo, P., et al.: Computer-aided classification of gastrointestinal lesions in regular colonoscopy. IEEE Trans. Med. Imaging **35**(9), 2051–2063 (2016)
16. Moreno-Torres, J.G., Raeder, T., Alaiz-Rodríguez, R., et al.: A unifying view on dataset shift in classification. Pattern Recogn. **45**(1), 521–530 (2012)
17. Ning, W., et al.: Open resource of clinical data from patients with pneumonia for the prediction of COVID-19 outcomes via deep learning. Nat. Biomed. Eng. **4**(12), 1197–1207 (2020)
18. Niu, Z., Zhou, M., Wang, L., Gao, X., Hua, G.: Ordinal regression with multiple output CNN for age estimation. In: CVPR, pp. 4920–4928 (2016)
19. Obuchowski, N.A., Goske, M.J., Applegate, K.E.: Assessing physicians' accuracy in diagnosing paediatric patients with acute abdominal pain: measuring accuracy for multiple diseases. Stat. Med. **20**(21), 3261–3278 (2001)
20. Park, C., Awadalla, A., Kohno, T., Patel, S.: Reliable and trustworthy machine learning for health using dataset shift detection. In: NeurIPS, vol. 34 (2021)
21. Park, H.J., et al.: Radiomics analysis of gadoxetic acid-enhanced MRI for staging liver fibrosis. Radiology **290**(2), 380–387 (2019)

22. Peng, J., Bu, X., Sun, M., Zhang, Z., Tan, T., Yan, J.: Large-scale object detection in the wild from imbalanced multi-labels. In: Proceedings of the IEEE/CVF Conference on Computer Vision and Pattern Recognition, pp. 9709–9718 (2020)

23. Ren, J., Hacihaliloglu, I., Singer, E.A., Foran, D.J., Qi, X.: Adversarial domain adaptation for classification of prostate histopathology whole-slide images. In: Frangi, A.F., Schnabel, J.A., Davatzikos, C., Alberola-López, C., Fichtinger, G. (eds.) MICCAI 2018. LNCS, vol. 11071, pp. 201–209. Springer, Cham (2018). https://doi.org/10.1007/978-3-030-00934-2_23

24. Ren, J., Yu, C., Ma, X., Zhao, H., Yi, S., et al.: Balanced meta-softmax for long-tailed visual recognition. Adv. Neural. Inf. Process. Syst. **33**, 4175–4186 (2020)

25. Roy, S., et al.: Deep learning for classification and localization of COVID-19 markers in point-of-care lung ultrasound. IEEE Trans. Med. Imaging **39**(8), 2676–2687 (2020)

26. Schölkopf, B., Janzing, D., Peters, J., Sgouritsa, E., Zhang, K., Mooij, J.: On causal and anticausal learning. In: ICML (2012)

27. Subbaswamy, A., Saria, S.: From development to deployment: dataset shift, causality, and shift-stable models in health AI. Biostatistics **21**(2), 345–352 (2020)

28. Sun, Y., Wang, X., Liu, Z., Miller, J., Efros, A., Hardt, M.: Test-time training with self-supervision for generalization under distribution shifts. In: International Conference on Machine Learning, pp. 9229–9248. PMLR (2020)

29. Wang, D., Shelhamer, E., Liu, S., Olshausen, B., Darrell, T.: Tent: fully test-time adaptation by entropy minimization. In: International Conference on Learning Representations ICLR (2021)

30. Wang, X., Lian, L., Miao, Z., Liu, Z., Yu, S.X.: Long-tailed recognition by routing diverse distribution-aware experts. In: International Conference on Learning Representations (2021)

31. Williams, R.: Global challenges in liver disease. Hepatology **44**(3), 521–526 (2006)

32. Wu, R., Guo, C., Su, Y., Weinberger, K.Q.: Online adaptation to label distribution shift. In: Advances in Neural Information Processing Systems, vol. 34 (2021)

33. Zhang, K., Schölkopf, B., Muandet, K., Wang, Z.: Domain adaptation under target and conditional shift. In: ICML, pp. 819–827. PMLR (2013)

34. Zhang, Y., Hooi, B., Hong, L., Feng, J.: Test-agnostic long-tailed recognition by test-time aggregating diverse experts with self-supervision. arXiv preprint arXiv:2107.09249 (2021)

Interaction-Oriented Feature Decomposition for Medical Image Lesion Detection

Junyong Shen[1], Yan Hu[1(✉)], Xiaoqing Zhang[1], Zhongxi Qiu[1], Tingming Deng[3], Yanwu Xu[2], and Jiang Liu[1,4(✉)]

[1] Research Institute of Trustworthy Autonomous Systems and Department of Computer Science and Engineering, Southern University of Science and Technology, Shenzhen 518055, China
{huy3,liuj}@sustech.edu.cn
[2] Intelligent Healthcare Unit, Baidu, Beijing 100000, China
[3] Department of Ophthalmology, Shenzhen People's Hospital, Shenzhen 518020, Guangdong, China
[4] Guangdong Provincial Key Laboratory of Brain-inspired Intelligent Computation, Department of Computer Science and Engineering, Southern University of Science and Technology, Shenzhen 518055, China

Abstract. Common lesion detection networks typically use lesion features for classification and localization. However, many lesions are classified only by lesion features without considering the relation with global context features, which raises the misclassification problem. In this paper, we propose an Interaction-Oriented Feature Decomposition (IOFD) network to improve the detection performance on context-dependent lesions. Specifically, we decompose features output from a backbone into global context features and lesion features that are optimized independently. Then, we design two novel modules to improve the lesion classification accuracy. A Global Context Embedding (GCE) module is designed to extract global context features. A Global Context Cross Attention (GCCA) module without additional parameters is designed to model the interaction between global context features and lesion features. Besides, considering the different features required by classification and localization tasks, we further adopt a task decoupling strategy. IOFD is easy to train and end-to-end in terms of training and inference. The experimental results for datasets in two modalities outperform state-of-the-art algorithms, which demonstrates the effectiveness and generality of IOFD. The source code is available at https://github.com/mklz-sjy/IOFD

Keywords: Lesion detection · Context embedding · Cross attention · Medical image

1 Introduction

Lesion detection is to find and report all possible abnormal regions from images, which is vital for disease diagnosis. With the increase in imaging modalities [22]

© The Author(s), under exclusive license to Springer Nature Switzerland AG 2022
L. Wang et al. (Eds.): MICCAI 2022, LNCS 13433, pp. 324–333, 2022.
https://doi.org/10.1007/978-3-031-16437-8_31

and the widespread use of 3D imaging [18], it is time-consuming and labor-intensive for doctors to detect all abnormalities, which leads to patients' long-time waiting and possible wrong report [5]. Automatic lesion detection reduces doctors' workload and provides accurate detection results on a consistent bias [20].

Most of the existing works on lesion detection are based on region-based networks, such as Faster RCNN [1,16], Cascade RCNN [2], Dynamic R-CNN [24] and SABL [21]. They have achieved great success in many medical image detection tasks. They consist of a shared backbone, a Region Proposal Network (RPN) that generates high-quality region proposals, and two task heads for classification and localization, respectively. However, they all encounter a misclassification problem. Many types of lesions are distinguished by the position and the relative proportion of the lesion to the tissue, but Region Of Interest (ROI) features only contain the lesion itself. The misclassification problem is more serious than mislocalization, which leads to misdiagnosis, delayed treatment, and deterioration of diseases.

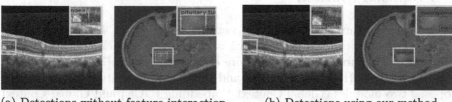

(a) Detections without feature interaction   (b) Detections using our method

Fig. 1. Two context-dependent samples with feature interaction or not. The red, green and cyan bounding box are the ground truth, the right prediction and the wrong prediction. (Color figure online)

Recently, researchers proposed to integrate context information to solve the above problems. Context information is applied to seek out the relation with lesions severing as auxiliary features for classifying lesions [15]. Some works [14,19] focus on fusing the neighbor context information. But neighbor context information is limited and cannot contain the global tissue information of the lesion. Other works focus on fusing the global context information. For example, a cascade structure [25] concatenates the entire image features to fuse global context features, HCE [3] concatenates object-level contexts with region features for both classification and localization. However, as shown in Fig. 1(a), a wet Age-related Macular Degeneration (wAMD) sample and a meningioma sample are still wrongly classified. The misclassification problem still exists because most of these methods simply aggregate features implicitly and cannot effectively emphasize the relationship between lesion region and global features.

Therefore, a novel lesion detection network is proposed to reduce the misclassification rate, which simulates the way a physician diagnoses a disease based on the interaction between a lesion and other tissues. We design two novel modules, including one for extracting global context features and the other for modeling the interaction between global context features and lesion features. Besides, considering the features misalignment between tasks [23], we propose to adopt decoupled task heads to further improve classification accuracy. As shown in Fig. 1(b), our network successfully corrects these mentioned misclassification samples. Therefore, the main contributions of the paper are summarized as follows: (1) A novel Interaction-Oriented Feature Decomposition (IOFD) network is proposed for lesion detection to solve the misclassification problem. (2) Two novel modules are proposed: Global Context Embedding (GCE) module for extracting the global context features and Global Context Cross Attention (GCCA) module for modeling the interaction between lesion features and global context features without additional parameters. (3) The proposed IOFD network superiors the state-of-the-arts algorithms for lesion detection based on two modal datasets, including a private Optical Coherence Tomography (OCT) dataset and a publicly available Magnetic Resonance Imaging (MRI) dataset.

2   Proposed Method

The overview of our proposed IOFD network is shown in Fig. 2. Our IOFD is made of two task branches, including a localization branch and a classification branch. The former contains Feature Pyramid Network (FPN) [12], Region Proposal Network (RPN), ROI align, and box head. The latter consists of the proposed two modules (Global Context Embedding (GCE) and Global Context Cross Attention (GCCA) modules) and class head. The network architecture and modules are described as follows.

2.1   IOFD Architecture

An image ($X \in C, H, W$, where C, H, W are the channel, height, and width of the image.) is first input into the backbone and then processed in two branches for different tasks, respectively. For the localization task, features out from the backbone are enhanced by FPN. Then, N lesion proposals (B) are generated by RPN. Two examples are displayed in Fig. 2, an orange box (B_1) and a green box (B_N). Lesion features are obtained by ROI align ($L = \{L_1, \cdots, L_N\}$) and finally mapped to the bounding boxes ($y_b = \{y_b^1, \cdots, y_b^N\}$) by box head. For example, B_1 is processed to L_1 and finally mapped to the localization (y_b^1). For the classification tasks, GCE processes features output from the backbone to obtain global context features (G). GCCA models the interaction between each set of lesion features and the set of shared global context features to generate fused features. Fused features ($F = \{F_1, \cdots, F_N\}$) are finally mapped by class head to obtain classification results ($y_c = \{y_c^1, \cdots, y_c^N\}$). For example, GCCA models the interaction between L_1 and G and generates F_1, which finally is mapped to the classification result y_c^1.

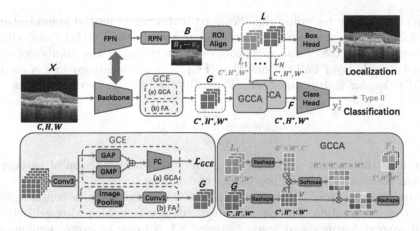

Fig. 2. Overview of proposed Interaction-Oriented Feature Decomposition (IOFD) network, where X, B, G, L, F, N denote input image, proposals sets, global context features, lesion features sets, fused features sets, and the total number of proposals. We propose two modules to solve the misclassification problem: Global Context Embedding (GCE), consisting of Global Context Auxiliary (GCA) and Feature Adaptation (FA), and Global Context Cross Attention (GCCA). For brevity, we take L_1 as an example and get results (y_b^1 and y_c^1). In fact, N results are generated and then are post-processed for filtering out redundant results. In particular, the dotted bounding box is the original proposal, and the solid box is the final regression result.

The final detection results are post-processed by Non-Maximum Suppression (NMS) [17]. The whole network is trained end-to-end and constrained by four parts of losses: GCE loss, RPN loss, box loss, and class loss. Formally, the losses of IOFD can be described as:

$$\mathcal{L} = \mathcal{L}_{GCE} + \mathcal{L}_{RPN} + \mathcal{L}_{box} + \mathcal{L}_{class}, \tag{1}$$

where $\mathcal{L}_{RPN}$, $\mathcal{L}_{box}$ and $\mathcal{L}_{class}$ all are the corresponding losses to Faster RCNN [16]. $\mathcal{L}_{GCE}$ is described in detail in the GCE module below. Particularly, four loss terms are equally important and the training strategy has no trick.

2.2 Global Context Embedding (GCE)

Features extracted by the backbone are high-level features with low resolution and contain more global semantic information including the information of the surrounding structure of the lesion. To extract and optimize global context features, as shown in Fig. 2, GCE is designed. Firstly, we adopt a 3×3 convolutional layer (Conv3) on input features of the module to avoid the optimization effect of the localization branch. Then, features are processed concurrently by Global Context Auxiliary (GCA) and Feature Adaptation (FA).

Global Context Auxiliary: To extract effective global context features, GCA employs an auxiliary task for image-level classification like current lesion classification methods [13]. The auxiliary task can preserve the beneficial features for lesion classification including global context features and partial lesion features. In detail, features are aggregated by Global Max-Pooling (GMP) and Global Average-Pooling (GAP) and mapped to the dimension of the number of categories by single layer Fully Connection (FC). The cross-entropy loss is used as the GCE loss for feature constraint. Formally, the loss can be described as:

$$\mathcal{L}_{GCE} = - \sum_{i=0}^{C-1} y_i \log{(p_i)} \tag{2}$$

where C is the number of categories, y_i denotes whether lesion of category i exists in the image, p_i is the probability of category i.

Feature Adaptation: With the help of GCA, input features contain global contextual information, but the size mismatch needs to be addressed between global context features and lesion features. FA is made of image pooling and 1×1 convolutional layer (Conv1). Image pooling is similar to ROI align and the input proposal is replaced by the whole image, which is to solve the width and height mismatch and reduce the impact of different pooling. The convolutional layer is to solve the channel mismatch. Finally, output features of GCE have rich global contextual information and can match lesion features.

2.3   Global Context Cross Attention (GCCA)

Most of the current methods[3, 25] use the concatenation or addition to fuse context features. Their methods do not extract explicit global contextual features and cannot effectively reflect the interaction between lesion features and global contextual features. With the help of GCE, we obtain global context features, which have a larger theoretical receptive field than lesion features and contain partial features about lesion [8]. Therefore, we take global context features as the base and enhance partial lesion features by lesion features to model the interaction. Inspired by the way self-attention mechanisms construct a set of queries, keys, and values to model the interaction [6], we design Global Context Cross Attention (GCCA) to model the interaction of lesion features and global context features, which employ non-parameter operation to generate interaction weight matrix adaptively.

As shown in Fig. 2, there are N sets of lesion features $(L_1, L_2 \cdots L_N)$ and a set of global context features $(G \in C^*, H^*, W^*$, where C^*, H^*, W^* are the channel, height, and width of features.) for an image. The global context features set is shared by all sets of lesion features. The input of GCCA is a set of global context features and a set of lesion features. In detail, G stands for both key (K) and value (V). It is reshaped into $\mathbb{R}^{H^* \times W^*, C^*}$. For example, a set of lesion features (L_1) stands for query (Q) and is reshaped into $\mathbb{R}^{C^*, H^* \times W^*}$. To find lesion features from global context features, we perform the matrix multiplication to calculate the similarity of Q and K. Next, the weight matrix can be obtained by the softmax function and the shape is $\mathbb{R}^{H^* \times W^*, H^* \times W^*}$. The fused features (F_1) are formed by matrix multiplication of weight matrix and V. Finally, the fused features are reshaped into $\mathbb{R}^{C^*, H^*, W^*}$. This process of generating weights imitates the operation that doctors focus on observing the lesions after browsing the whole picture. In the end, each set of lesion features corresponds to a set of fused features, which is used for classification.

3   Experiments

In this section, we prove the efficacy of our IOFD based on extensive experiments. We illustrate two modal datasets, metrics for evaluation, and implementation details. Then the ablation study and comparison experiments based on these datasets are listed.

3.1   Datasets, Evaluations and Implementation Details

Datasets: Two modal datasets are adopted to evaluate our proposed IOFD network, including a private wAMD (OCT) dataset and a publicly available Brain_tumor (MRI) dataset [4], which are suitable to construct the problem that lesion classification focuses more on the contextual information of lesion.

(1) **wAMD (OCT):** Wet AMD is characterized by Choroidal NeoVascularization (CNV) and can be classified by the relative position of the lesion and the Retinal Pigment Epithelium (RPE) into type I CNV and type II CNV [9]. The dataset is collected in an outpatient clinic, aging range from 53 to 80. It includes 23 cases that 9 cases as type I and 14 cases as type II. Each case has an OCT volume containing 384 OCT B-scans and lesion images are selected and labeled by professional ophthalmologists. In this experiment, there are 5063 OCT images for training and testing. 18 cases (7 cases of type I and 11 cases of type II) are randomly selected as the train set. The rest are selected as the test set.

(2) **Brain_tumor (MRI):** A brain tumor can be divided into glioma, meningioma, and pituitary tumor based on occurring different positions of the brain. The public dataset contains 3064 MRI T1-weighted contrast-enhanced images from 233 patients with three kinds of brain tumor, namely, meningioma (708 slices), glioma (1426 slices), and pituitary tumor (930 slices). We randomly split it into train, validation, and test set based on indices provided by the public dataset.

Evaluation Metrics: We adopt four metrics: mean Average Precision (mAP), Accuracy (Acc), Recall, and Precision. In detail, Acc is the proportion of correct classification boxes in all ground truths, which directly reflect the lesion classification performance. Recall and Precision are the proportion of correct detections in all ground truths and all detections.

Implementation Details: We implemented our method with the Pytorch framework and ran it on a server with an NVIDIA GeForce RTX 2080 Ti. We took ResNet50 [10] as the backbone and initialized it from the model pre-trained on ImageNet [11]. We set N as 2000 during training and 1000 during inference. The optimizer, weight decay, momentum, epoch, and batch size are set as SGD, $5e-4$, 0.9, 50, and 1, respectively. The initial learning rate is 0.0005 and decreased to 0.1 for every 20 epochs.

Table 1. Performance of ablation study on wAMD dataset

Baseline	GCE	GCCA	mAP	Acc	Recall	Precision
√			0.6304	0.8049	0.7030	0.7126
√	√		0.6508	0.8392	0.7165	0.7165
√		√	0.7191	0.8736	0.7754	0.7792
√	√	√	**0.7548**	**0.9030**	**0.7963**	**0.7963**

3.2   Ablation Study

We construct a classification and localization decoupling network as the baseline. Classification features are the concatenation of features out from the backbone and lesion features. Localization features are lesion features. We separately add GCE and GCCA to compare model performances equipped with different modules on the wAMD dataset, as shown in Table 1. Compared with the baseline, both modules significantly increase four metrics and obtain better performance. The baseline with GCE is better than the baseline, which indicates GCE obtains effective global context information for classification. In particular, the recall is equal to the precision, which means the number of detections is equal to the number of ground truths. Therefore, GCE can help to detect the lesion as many as possible and can avoid lesion missing detection. Similarly, the baseline with GCCA indicates modeling features interaction is more effective than the simple feature concatenation. The final results with GCE and GCCA are the best, which indicates that the final model combines the advantage of two modules and these modules can complement each other.

3.3   Comparison Experiments

This paper uses nine state-of-the-art methods to evaluate the effectiveness of our IOFD. Faster RCNN [1,16], Dynamic R-CNN [24], SABL[21] and Cascade RCNN[2] are region-based networks. Grid R-CNN [14], MSB[19] and HCE [3] are classic networks of classification and regression fusing context information. Double Heads [23] and TOOD [7] all adopt classification and regression decoupling structures. In terms of data, to demonstrate the generality of our method, comparison experiments are also performed on the public MRI brain tumor dataset, in addition to our own OCT image dataset.

The quantitative results are shown in Table 2. Compared with region-based methods, our method improves Acc by about 6% and mAP by about 3%, which indicates global context features extracted by our GCE are beneficial for lesion classification. Compared with other context fusion methods, we improve Acc by about 5% on the wAMD dataset. Meanwhile, recall and precision are improved by about 1% on the brain tumor dataset. Our IOFD successfully reduces the misclassification rate and therefore achieves better mAP. The results show that modeling specific feature interaction by GCCA is more effective than simple feature aggregation. Compared with the decoupling structure, our IOFD considers

Table 2. The comparison results among proposed method and other methods

	wAMD(**OCT**)				Brain_tumor(**MRI**)			
	mAP	Acc	Recall	Precision	mAP	Acc	Recall	Precision
Faster RCNN [1,16]	0.7195	0.8392	0.7509	0.7640	0.7658	0.9253	0.7916	0.8015
Cascade RCNN [2]	0.7081	0.8147	0.7423	0.7487	0.7216	0.9082	0.7651	0.7772
Dynamic RCNN [24]	0.7222	0.8417	0.7558	0.7595	0.7523	0.9206	0.7776	0.7923
SABL [21]	0.7271	0.8453	0.7595	0.7613	0.7295	0.9129	0.7760	0.7833
MSB [19]	0.7112	0.8233	0.7546	0.7765	0.6954	0.8942	0.7589	0.7774
HCE [3]	0.7344	0.8515	0.6969	0.7082	0.7800	0.9533	0.8164	0.8164
Grid RCNN [14]	0.6458	0.7766	0.7055	0.7063	0.6783	0.8989	0.7216	0.7216
Double Heads [23]	0.6637	0.7730	0.6920	0.6945	0.7348	0.9035	0.7729	0.7964
TOOD [7]	0.6950	0.7963	0.7472	0.7472	0.7591	0.9393	0.7822	0.7822
IOFD(Ours)	**0.7548**	**0.9030**	**0.7963**	**0.7963**	**0.7930**	**0.9548**	**0.8242**	**0.8242**

the difference of input features for classification and localization, which further improves the metrics. In summary, the proposed method obtains better overall performance in lesion detection with different modalities in different metrics.

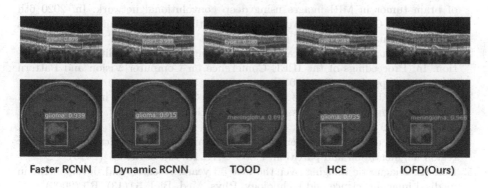

Faster RCNN   Dynamic RCNN   TOOD   HCE   IOFD(Ours)

Fig. 3. Lesion detection results. The red bounding boxes represent ground truth annotations, the cyan bounding boxes represent misclassification results, the green bounding boxes represent right detections and the numbers above bounding boxes represent the score of predicted classification. (Color figure online)

We choose Faster RCNN, Dynamic RCNN, TOOD, and HCE as representatives since their results are better than others. Figure 3 illustrates some context-dependent samples on wAMD (OCT) and Brain_tumor (MRI) by the aforementioned methods. The classification of the lesion is the results we focus on. Except for TOOD and our IOFD, other methods suffer from misclassification problems. We achieve better brain tumor classification scores (0.966) than TOOD (0.692), demonstrating the effectiveness of our fused features.

4   Conclusion

In this paper, we present an Interaction-Oriented Feature Decomposition (IOFD) network to model the interaction between global context features and lesion features to solve the misclassification problem. The experimental results indicate that modeling the interaction of features is better than features concatenation. Compared with other detection methods, our method effectively reduces the misclassification rate in lesion detection with different modalities in different metrics.

Acknowledgement. This work was supported in part by The National Natural Science Foundation of China (8210072776), Guangdong Provincial Department of Education (2020ZDZX3043), Guangdong Basic and Applied Basic Research Foundation(2021A1515012195), Guangdong Provincial Key Laboratory (2020B121201001), Shenzhen Natural Science Fund (JCYJ20200109140820699) and the Stable Support Plan Program (20200925174052004).

References

1. Bhanothu, Y., Kamalakannan, A., Rajamanickam, G.: Detection and classification of brain tumor in MRI images using deep convolutional network. In: 2020 6th International Conference on Advanced Computing and Communication Systems (ICACCS), pp. 248–252. IEEE (2020)
2. Cai, Z., Vasconcelos, N.: Cascade r-cnn: delving into high quality object detection. In: Proceedings of the IEEE Conference on Computer Vision and Pattern Recognition, pp. 6154–6162 (2018)
3. Chen, Z.M., Jin, X., Zhao, B.R., Zhang, X., Guo, Y.: HCE: hierarchical context embedding for region-based object detection. IEEE Trans. Image Process. **30**, 6917–6929 (2021)
4. Cheng, J., et al.: Enhanced performance of brain tumor classification via tumor region augmentation and partition. PLoS ONE **10**(10), e0140381 (2015)
5. Doi, K.: Diagnostic imaging over the last 50 years: research and development in medical imaging science and technology. Phys. Med. Biol. **51**(13), R5 (2006)
6. Dosovitskiy, A., et al.: An image is worth 16×16 words: transformers for image recognition at scale. arXiv preprint arXiv:2010.11929 (2020)
7. Feng, C., Zhong, Y., Gao, Y., Scott, M.R., Huang, W.: Tood: task-aligned one-stage object detection. In: Proceedings of the IEEE/CVF International Conference on Computer Vision, pp. 3510–3519 (2021)
8. Girshick, R.: Fast r-cnn. In: Proceedings of the IEEE International Conference on Computer Vision, pp. 1440–1448 (2015)
9. Grossniklaus, H., Gass, J.D.: Clinicopathologic correlations of surgically excised type 1 and type 2 submacular choroidal neovascular membranes. Am. J. Ophthalmol. **126**(1), 59–69 (1998)
10. He, K., Zhang, X., Ren, S., Sun, J.: Deep residual learning for image recognition. IEEE (2016)
11. Jia, D., Wei, D., Socher, R., Li, L.J., Kai, L., Li, F.F.: Imagenet: a large-scale hierarchical image database, pp. 248–255 (2009)

12. Lin, T.Y., Dollár, P., Girshick, R., He, K., Hariharan, B., Belongie, S.: Feature pyramid networks for object detection. In: Proceedings of the IEEE Conference on Computer Vision and Pattern Recognition, pp. 2117–2125 (2017)

13. Lopez, A.R., Giro-i Nieto, X., Burdick, J., Marques, O.: Skin lesion classification from dermoscopic images using deep learning techniques. In: 2017 13th IASTED International Conference on Biomedical Engineering (BioMed), pp. 49–54. IEEE (2017)

14. Lu, X., Li, B., Yue, Y., Li, Q., Yan, J.: Grid r-cnn. In: Proceedings of the IEEE/CVF Conference on Computer Vision and Pattern Recognition, pp. 7363–7372 (2019)

15. McRobert, A.P., Causer, J., Vassiliadis, J., Watterson, L., Kwan, J., Williams, M.A.: Contextual information influences diagnosis accuracy and decision making in simulated emergency medicine emergencies. BMJ Qual. Saf. 22(6), 478–484 (2013)

16. Ren, S., He, K., Girshick, R., Sun, J.: Faster r-cnn: towards real-time object detection with region proposal networks. Adv. Neural. Inf. Process. Syst. 28, 91–99 (2015)

17. Rosenfeld, A., Thurston, M.: Edge and curve detection for visual scene analysis. IEEE Trans. Comput. 100(5), 562–569 (1971)

18. Sansoni, G., Trebeschi, M., Docchio, F.: State-of-the-art and applications of 3d imaging sensors in industry, cultural heritage, medicine, and criminal investigation. Sensors 9(1), 568–601 (2009)

19. Shao, Q., Gong, L., Ma, K., Liu, H., Zheng, Y.: Attentive CT lesion detection using deep pyramid inference with multi-scale booster. In: Shen, D., et al. (eds.) MICCAI 2019. LNCS, vol. 11769, pp. 301–309. Springer, Cham (2019). https://doi.org/10.1007/978-3-030-32226-7_34

20. Ting, D.S., Liu, Y., Burlina, P., Xu, X., Bressler, N.M., Wong, T.Y.: Ai for medical imaging goes deep. Nat. Med. 24(5), 539–540 (2018)

21. Wang, J.: Side-aware boundary localization for more precise object detection. In: Vedaldi, A., Bischof, H., Brox, T., Frahm, J.-M. (eds.) ECCV 2020. LNCS, vol. 12349, pp. 403–419. Springer, Cham (2020). https://doi.org/10.1007/978-3-030-58548-8_24

22. White, S.C., Pharoah, M.J.: The evolution and application of dental maxillofacial imaging modalities. Dent. Clin. North Am. 52(4), 689–705 (2008)

23. Wu, Y., et al.: Rethinking classification and localization for object detection. In: Proceedings of the IEEE/CVF Conference on Computer Vision and Pattern Recognition, pp. 10186–10195 (2020)

24. Zhang, H., Chang, H., Ma, B., Wang, N., Chen, X.: Dynamic R-CNN: towards high quality object detection via dynamic training. In: Vedaldi, A., Bischof, H., Brox, T., Frahm, J.-M. (eds.) ECCV 2020. LNCS, vol. 12360, pp. 260–275. Springer, Cham (2020). https://doi.org/10.1007/978-3-030-58555-6_16

25. Zhong, Q., Li, C., Zhang, Y., Xie, D., Yang, S., Pu, S.: Cascade region proposal and global context for deep object detection. Neurocomputing 395, 170–177 (2020)

Prototype Learning of Inter-network Connectivity for ASD Diagnosis and Personalized Analysis

Eunsong Kang[1], Da-Woon Heo[2], and Heung-Il Suk[1,2(✉)]

[1] Department of Brain and Cognitive Engineering, Korea University,
Seoul, Republic of Korea
{eunsong1210,hisuk}@korea.ac.kr
[2] Department of Artificial Intelligence, Korea University, Seoul, Republic of Korea
daheo@korea.ac.kr

Abstract. In recent studies, deep learning has shown great potential to explore topological properties of functional connectivity (FC), *e.g.*, graph neural networks (GNNs), for brain disease diagnosis, *e.g.*, Autism spectrum disorder (ASD). However, many of the existing methods integrate the information locally, *e.g.*, among neighboring nodes in a graph, which hinders from learning complex patterns of FC globally. In addition, their analysis for discovering imaging biomarkers is confined to providing the most discriminating regions without considering individual variations over the average FC patterns of groups, *i.e.*, patients and normal controls. To address these issues, we propose a unified framework that globally captures properties of inter-network connectivity for classification and provides individual-specific group characteristics for interpretation via prototype learning. In our experiments using the ABIDE dataset, we validated the effectiveness of the proposed framework by comparing with competing topological deep learning methods in the literature. Furthermore, we individually analyzed functional mechanisms of ASD for neurological interpretation.

Keywords: Prototype learning · Transformer · Inter-network connectivity · Resting-State functional magnetic resonance imaging · Autism spectrum disorder

1 Introduction

Autism spectrum disorder (ASD) is a neurodevelopmental disorder that mainly shows difficulty with communications, social interactions and behaviors. Therefore, early diagnosis of ASD is important to get proper intervention for patient's quality of life. Since ASD has a wide spectrum of symptoms, a patient needs

Supplementary Information The online version contains supplementary material available at https://doi.org/10.1007/978-3-031-16437-8_32.

© The Author(s), under exclusive license to Springer Nature Switzerland AG 2022
L. Wang et al. (Eds.): MICCAI 2022, LNCS 13433, pp. 334–343, 2022.
https://doi.org/10.1007/978-3-031-16437-8_32

personalized treatment depending on their symptoms. To identify pathological mechanisms that vary from one person to another, resting-state functional magnetic resonance imaging (rs-fMRI) has been widely used in the literature [6,7,12]. Specifically, functional connectivity (FC), temporal correlations of rs-fMRI signals among spatially distant brain regions, has been investigated to detect the brain's normal or abnormal functional characteristics for diagnosis.

Recently, many studies have proposed deep learning methods to discover or learn the topological features in FC [8,9,11] by regarding FC as a graph structure. For example, BrainNetCNN [8], a convolutional neural network (CNN)-based model, encodes FC with 1D convolution operations that leverage the topological locality of brain networks. Although it showed promising results in predicting cognitive and motor development outcomes in their experiments, row-/column-wise 1D convolution operations over FC are limited to cover the global context of FC. Meanwhile, graph neural networks (GNNs) have received attention and been successfully applied to analyze an FC-based graph for brain disorder identification or task prediction [11,17]. For example, [11] developed BrainGNN that uses an ROI-aware graph convolutional operation to consider topological and functional information of fMRI. However, these GNN models may overlook ROIs' interactions with weak correlations while mainly focusing on neighborhood nodes with high correlations when propagating and aggregating features.

To address the issues of local dependency, we devise a novel approach of learning global inter-network relations by leveraging Transformer [4,13], which has been making remarkable successes in natural language processing and computer vision. In particular, we regard FC as a set of seed-based functional networks (*i.e.*, rows in FC) and learn functional-network-level representations using multi-head self-attention mechanisms that can be understood as a fully-connected graph but adaptively changing the edge connections [16].

In the meantime, it is essential to discover the most discriminating regions for neuroimaging-based diagnosis. The previous studies have used a post-hoc analysis of trained deep models, such as saliency map and gradient-weighted class activation mapping [10,19]. On the other hand, the prototype-based model predicts a class label based on the similarity with the learned representative prototypes [2], thus naturally providing interpretability for a decision. In the same line of approach, we propose a novel architecture that learns class-representative prototypes and uses them to estimate how much an individual's FC deviates from the group average. Notably, our method can identify the distinctive characteristics of an individual, marked by subject-specific damaged regions and the extent of severity.

Contributions. Three main contributions of this work[1] can thus be summarized: (1) We propose a topological relational learning method that summarizes high-order inter-network functional connectivity; (2) Combined with prototype

---

[1] The code of our proposed model is available at https://github.com/ku-milab/PL-FC.

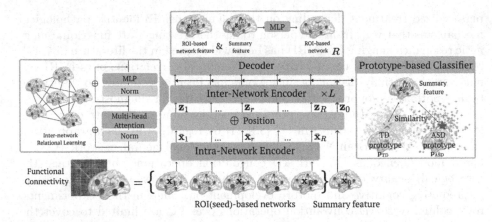

Fig. 1. An overview of the proposed model that consists of three main components; intra-network and inter-network encoder, decoder, and prototype-based classifier.

learning, our model can capture representative characteristics, so-called proto-types, of a patient group and a normal group, along with individual character-istics interpretation. (3) With the group prototypes, our model can also provide two types of neuroscientific explanation: inter-class variations of FC, *i.e.*, the changes if the normal is diagnosed with ASD, and intra-class variations of FC, *i.e.*, the functional differences within ASD. We validated our method on Autism Brain Imaging Data Exchange I (ABIDE I) dataset.

2   Dataset and Preprocessing

We used resting-state fMRI (rs-fMRI) data from the Autism Imaging Data Exchange I (ABIDE I[2]). ABIDE dataset provides image data from 17 inter-national organizations, a total of 1,112 subjects. In our study, the data is pre-processed with Configurable Pipeline for the Analysis of Connectomes (CPAC), band-pass filtering (0.01–0.1Hz), and without global signal regression. The brain was parcellated into 110 ROIs using Harvard-Oxford (HO) atlas [3]. We excluded subjects of missing files and those with incomplete ROIs of FC, hence, we used 985 subjects; 479 ASD patients and 506 subjects of typical development (TD).

3   Proposed Method

Our framework is composed of three main components as depicted in Fig. 1: (1) Intra-network and inter-network encoder, (2) Prototype-based classifier, and (3) Decoder. The inter-netwok encoder is devised to learn a representation of global

---

[2] Scan procedure and protocols can be found at http://fcon_1000.projects.nitrc.org/indi/abide/.

inter-network relations by using a Transformer network [4,13]. The prototype-based classifier calculates the similarity between group-wise (*i.e.*, ASD and TD) prototypes and individual features for diagnosis. At the same time, these prototypes are learned by individual features, which have high similarities with each of them. Meanwhile, the decoder not only encourages to learn expressive features by reconstructing the original FC, but also visualizes individual or group representative features in terms of FC.

3.1 Intra-network and Inter-network Encoder

By regarding the FC of a given input rs-fMRI as a sequence of ROI-based (seed-based) networks, $\mathbf{x} = \{\mathbf{x}_1, \ldots, \mathbf{x}_r, \ldots, \mathbf{x}_R\} \in \mathbb{R}^{R \times R}$, *i.e.*, $\mathbf{x}_r$ denotes the temporal correlations of an ROI r and the other ROIs, we first learn intra-network representations by embedding $\mathbf{x}_r$ to $\bar{\mathbf{x}}_r \in \mathbb{R}^{1 \times D}$ with a non-linear projection. Then, we prepend a learnable vector $\mathbf{x}_{\text{summary}} = \mathbf{x}_0 \in \mathbb{R}^{1 \times D}$ to represent the global features of the FC at a functional-network level. Also, we add a sinusoidal positional encoding matrix $\mathbf{e} \in \mathbb{R}^{(R+1) \times D}$ to give relative regional information of ROIs and the summary feature,

$$\mathbf{z} = [\mathbf{x}_0; \bar{\mathbf{x}}] + \mathbf{e}. \tag{1}$$

Therefore, inter-network encoder receives $\mathbf{z}$ as input, which consists of R number of ROI-based network features and the summary vector.

The inter-network encoder block is composed of multi-head self-attention (MSA) and multi-layer perceptron (MLP) layers. Layer normalization (LN) is applied before MSA and MLP layers, while residual connections are adopted after MSA and MLP layers. A self-attention (SA) mechanism plays a major role in learning global inter-network relations:

$$\text{SA} = \text{Self-Attention}(Q, K, V) = \text{Softmax}(\frac{QK^\top}{\sqrt{D}})V, \tag{2}$$

where $Q, K, V \in \mathbb{R}^{(N+1) \times D}$ represent the linear transformation of input $\mathbf{z}$, $Q = \mathbf{z}W_Q, K = \mathbf{z}W_K$, and $V = \mathbf{z}W_V$, respectively. W_Q, W_K, and W_V are learnable parameters. Furthermore, we use multi-head SA to jointly encode multiple representations of those relations, rather than a single-head SA that averages multiple relations:

$$\text{MSA} = \text{Multi-head-SA}(Q, K, V) = \text{Concat}(SA_1, SA_2, ..., SA_H)W_{\text{MSA}}, \tag{3}$$

where H is the number of heads, and each single-head SA is calculated by their respective $Q_h, K_h, V_h \in \mathbb{R}^{(N+1) \times (D/H)}$. The outputs of single-head SA are concatenated, then linearly transformed by $W_{\text{MSA}} \in \mathbb{R}^{D \times D}$. Stacking L encoder blocks, we finally obtain an output $\mathbf{z}^{(L)} \in \mathbb{R}^{(N+1) \times D}$ of the same size with input as follows:

$$\mathbf{z}' = \text{MSA}(\text{LN}(\mathbf{z}^{(l-1)})) + \mathbf{z}^{(l-1)} \tag{4}$$

$$\mathbf{z}^{(l)} = \text{MLP}(\text{LN}(\mathbf{z}')) + \mathbf{z}', \tag{5}$$

where $l = \{1, ..., L\}$. MLP consists of two feed-forward layers shared by all ROI-based functional networks.

Note that the output representation of the summary vector at the last layer $\mathbf{z}_0^{(L)}$ is obtained using the ROIs' representations over the multiple layers. Hence, we can interpret this as a global functional network-level representation and utilize it for classification as well as reconstruction.

3.2   Prototype-Based Classifier

For classification, the summary feature vector at the last layer is compared to each class prototype $\mathbf{p}_c \in \mathbb{R}^{1 \times D}$ by calculating the similarity. These prototypes are learnable parameters randomly initialized and trained via prototype learning [15]. The probability of $p(c|\mathbf{x})$ is denoted as follows:

$$p(c|\mathbf{x}) = \frac{e^{s(\mathbf{z}_0^{(L)}, \mathbf{p}_c)}}{\sum_{c=1}^{C} e^{s(\mathbf{z}_0^{(L)}, \mathbf{p}_c)}}, \tag{6}$$

where c represents a class index, and s is a function to measure cosine similarity. Then, we classify the individual summary feature vector as follows:

$$\mathcal{L}_{\text{class}} = -\log p(\mathbf{y}|\mathbf{x}) = -\log p(c|\mathbf{x}). \tag{7}$$

By comparing similarities between prototypes and individual features, the individual summary feature vector is classified as the prototype(s) label with the highest similarity. Concurrently, the trainable prototypes become class-representative by increasing the similarity with the assigned summary feature vector. Moreover, a regularization loss [18] is added to distinguish between class prototypes with adding margin m,

$$\mathcal{L}_{\text{proto}}(\mathbf{P}_{\text{ASD}}, \mathbf{P}_{\text{TD}}, m) = \frac{f(\mathbf{P}_{\text{ASD}}) - f(\mathbf{P}_{\text{TD}}) + m}{\|\mathbf{P}_{\text{ASD}} - \mathbf{P}_{\text{TD}}\| + \epsilon}, \tag{8}$$

where f is a linear layer and ϵ is a small constant. As prototypes are trainable parameters, this margin-based loss helps prevent prototypes from having insignificant differences.

3.3   Decoder

Decoder for Reconstruction. To improve the reconstruction quality, the summary feature vector is concatenated to the ROI-based network features as follows:

$$\hat{\mathbf{x}}_r = g([\mathbf{z}_r^{(L)} \| \mathbf{z}_0^{(L)}]), \tag{9}$$

where $r = \{1, ..., R\}$, and g represents our decoder, defined by a fully connected layer with a nonlinear activation function. We use L_1 loss for reconstruction:

$$\mathcal{L}_{\text{recon}} = \|\hat{\mathbf{x}} - \mathbf{x}\|, \tag{10}$$

where $\hat{\mathbf{x}}$ is the reconstructed FC.

Decoder for Personalized Analysis. Not only can the decoder reproduce an individual's FC, but the proposed decoder can also produce an individual's FC characterized by group prototypes. As visualized in Fig. 2, we can substitute an individual's summary feature vector to one of the prototypes and decode an individual's ROI-based network features concatenated with the group prototype to produce prototype-guided FC, $\hat{\mathbf{x}}^c$, as follows:

$$\hat{\mathbf{x}}_r^c = g([\mathbf{z}_r^{(L)} \| \mathbf{p}_c]), \tag{11}$$

where c represents the class index of prototypes. Consequently, the prototype-guided FC still maintains individual characteristics, while individual class-related characteristics are replaced with the group-representative characteristics. To be specific, if a summary feature vector of a TD subject is replaced with the ASD prototype ($\mathbf{p}_c = \mathbf{p}_{\text{ASD}}$), we can predict the functional degradation of FC as if the subject were suffering from ASD. Meanwhile, for the ASD subjects, we can infer the distinguishing variations of FC from the general FC patterns of ASD, which is exceptional to each individual. It should be noted that our proposed method can generate *counterfactual* FC patterns for a subject, which is greatly beneficial to obtain deeper insights into the functional characteristics of a brain in regard to ASD.

4   Results

4.1   Experimental Settings

Learning Strategy Transformer-based models have less inductive bias, so they need a bigger dataset for stable training and robust results. We truncated the ROI-mean time series fMRI signals using the sliding window technique, then calculated Pearson's correlation coefficient. Since we are aware that the results can be affected to window and stride sizes, we gathered both static FC and all FCs estimated from a size set of sliding windows, (window size, stride size) $\in$ $\{(30, 15), (50, 25), (70, 35), (100, 50)\}$, for data augmentation purpose. We took subject-wise 5-fold cross-validation for evaluation. At test time, we used static FC to make a decision.

Our framework was trained in two steps. First, we focused on the reconstruction part, where the inter-network encoder and decoder are involved in. After finishing the pre-training of the encoder and decoder, we added the prototype-based classifier. While training the classifier, the inter-network encoder and decoder were fine-tuned jointly[3]. Large-scale pre-training before a downstream task is a typical learning strategy to make the Transformer encoder more competitive.

---

[3] With a total loss of $\mathcal{L}_{\text{all}} = \lambda_1 \mathcal{L}_{\text{recon}} + \lambda_2 \mathcal{L}_{\text{class}} + \lambda_3 \mathcal{L}_{\text{proto}}$, where $\lambda_{1/2/3}$ denote weight parameters.

Table 1. Performance on the classification between ASD and TD. P: Prototypes, AUC: Area under the receiver operating characteristic

Model	AUC	Accuracy (%)	Sensitivity (%)	Specificity (%)
BrainNetCNN	0.7140 ± 0.0206	65.15 ± 1.33	53.71 ± 9.58	75.95 ± 8.65
BrainNetCNN+P	0.7102 ± 0.0441	66.81 ± 3.02	61.40 ± 6.44	71.98 ± 8.97
GAT	0.6265 ± 0.0381	62.75 ± 4.19	63.74 ± 5.51	61.56 ± 3.56
GraphSage	0.6586 ± 0.0325	66.32 ± 3.20	70.51 ± 4.34	61.20 ± 2.97
BrainGNN	0.6432 ± 0.0342	63.20 ± 1.85	62.68 ± 7.34	63.70 ± 9.02
Ours	$\mathbf{0.7226} \pm 0.0194$	$\mathbf{69.47} \pm 3.48$	67.09 ± 3.12	71.73 ± 3.75

4.2 Experimental Results

To verify the effectiveness of our method, we compared our model with the following topological deep learning methods: BrainNetCNN [8], BrainNetCNN with prototypes (BrainNetCNN+P) [19], graph attention model (GAT) [14], GraphSage [5], and BrainGNN [11]. The parameters of the proposed method and competitive methods can be found in Supplementary Materials.

Table 1 shows the averaged classification performance of the proposed and the comparative methods. It is noteworthy that our method achieved the best AUC of 0.7226, the best accuracy of 69.47%, and a more balanced performance between sensitivity and specificity. Regarding the results of BrainNetCNN and BrainNetCNN+P that cover the local context of FC, we observed that our approach of learning global inter-network relations helped to increase the diagnostic performance. In comparison with the results of GNN models that predefine ROIs' relations, we demonstrated that our multi-head self-attention mechanism effectively learns those relations for ASD identification. Moreover, we conducted a Wilcoxon signed-rank test ($p < 0.05$) with accuracy to demonstrate the statistical significance. Our proposed method is statistically significant compared with all competing methods ($p = 0.0313$).

5 Personalized FC Analysis

We randomly selected four subjects for personalized FC analysis: two ASD subjects and two TD subjects. To compare the FC connection strengths between reconstructed FC and ASD prototype-guided FC, we calculated the differences by subtracting the ASD prototype-guided FC from the reconstructed FC, then averaged them over ROI-based networks to offer ROI-level interpretation. The averaged connections of ROIs from individual reconstructed-FC are represented as gray, while those from ASD prototype-guided FC are colored red. In Fig. 2, the bar graphs on the upper side represent inter-class variations for TD subjects, whereas the graphs on the lower side show intra-class variations for ASD subjects. We extracted the top 5 ROIs with the largest differences.

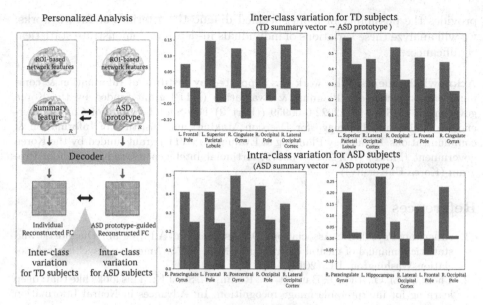

Fig. 2. Comparison of averaged connection strengths between individual reconstructed FC (gray) and ASD prototype-guided FC (red) for randomly selected four subjects. L: left, R: right (Color figure online)

The top 5 ROIs of inter-class variations are identical for two TD subjects (*i.e.*, L.frontal pole, L.superior parietal lobule, R.cingulate gyrus, R.occipital pole, and R.lateral occipital cortex). However, the degree of degradation varies among ROIs and individuals, which indicates the varying severity for subjects. When considering the top 5 ROIs of intra-class variations, four ROIs are the same across subjects, but the other ROIs are different (*i.e.*, R.postcentral gyrus, L.hippocampus). Especially, the right ASD subject has noticeably lower connections in the L.hippocampus than the average. It should be noted that functional deficits in specific brain regions can provide insights into understanding individuals' symptoms.

From those results, we observed that each person has a different pathological mechanism of ASD. We believed that our personalized FC analysis could help to categorize subtypes of ASD and the severity levels for ASD, which are diagnostic criteria for ASD, presented in the Diagnostic and Statistical Manual of Mental Disorders (DSM)-5 [1].

6   Conclusion

In this paper, we proposed a novel framework that learns global functional-network-level representations and provides explainability of FC variations based on group prototypes. In our experiments on the ABIDE dataset, the proposed relational learning showed superiority over comparative methods in ASD classification. Also, we investigated individually specified characteristics of ASD, which provides the possibility of more specified diagnostic criteria. For future works, we will analyze these variations of individuals in-depth to demonstrate statistical significance.

Acknowledgements. This work was supported by Institute of Information & communications Technology Planning & Evaluation (IITP) grant funded by the Korea government (MSIT) No. 2022-0-00959 ((Part 2) Few-Shot Learning of Causal Inference in Vision and Language for Decision Making) and Institute of Information & communications Technology Planning & Evaluation (IITP) grant funded by the Korea government (MSIT) (No. 2019-0-00079, Artificial Intelligence Graduate School Program (Korea University))

References

1. American Psychiatric Association, D., Association, A.P., et al.: Diagnostic and statistical manual of mental disorders: DSM-5, vol. 5. American Psychiatric Association Washington, DC (2013)
2. Chen, C., Li, O., Tao, D., Barnett, A., Rudin, C., Su, J.K.: This looks like that: deep learning for interpretable image recognition. In: Advances in Neural Information Processing Systems, vol. 32 (2019)
3. Craddock, R.C., James, G.A., Holtzheimer, P.E., III., Hu, X.P., Mayberg, H.S.: A whole brain fMRI atlas generated via spatially constrained spectral clustering. Hum. Brain Mapp. **33**(8), 1914–1928 (2012)
4. Dosovitskiy, A., et al.: An image is worth 16x16 words: transformers for image recognition at scale. In: International Conference on Learning Representations (2021)
5. Hamilton, W., Ying, Z., Leskovec, J.: Inductive representation learning on large graphs. In: Advances in Neural Information Processing Systems, vol. 30 (2017)
6. Jun, E., Kang, E., Choi, J., Suk, H.I.: Modeling regional dynamics in low-frequency fluctuation and its application to autism spectrum disorder diagnosis. Neuroimage **184**, 669–686 (2019)
7. Kam, T.E., Suk, H.I., Lee, S.W.: Multiple functional networks modeling for autism spectrum disorder diagnosis. Hum. Brain Mapp. **38**(11), 5804–5821 (2017)
8. Kawahara, J., et al.: BrainNetCNN: convolutional neural networks for brain networks; towards predicting neurodevelopment. Neuroimage **146**, 1038–1049 (2017)
9. Kazeminejad, A., Sotero, R.C.: Topological properties of resting-state fMRI functional networks improve machine learning-based autism classification. Front. Neurosci. **12**, 1018 (2019)
10. Kim, B.H., Ye, J.C.: Understanding graph isomorphism network for rs-fMRI functional connectivity analysis. Front. Neurosci. **14** (2020)
11. Li, X., et al.: BrainGNN: interpretable brain graph neural network for fMRI analysis. Med. Image Anal. **74**, 102233 (2021)
12. Suk, H.I., Wee, C.Y., Lee, S.W., Shen, D.: Supervised discriminative group sparse representation for mild cognitive impairment diagnosis. Neuroinformatics **13**(3), 277–295 (2015)
13. Vaswani, A., et al.: Attention is all you need. In: Advances in Neural Information Processing Systems, vol. 30 (2017)
14. Veličković, P., Cucurull, G., Casanova, A., Romero, A., Liò, P., Bengio, Y.: Graph attention networks. In: International Conference on Learning Representations (2018)

15. Yang, H.M., Zhang, X.Y., Yin, F., Liu, C.L.: Robust classification with convolutional prototype learning. In: Proceedings of the IEEE Conference on Computer Vision and Pattern Recognition, pp. 3474–3482 (2018)
16. Ying, C., et al.: Do transformers really perform badly for graph representation? In: Advances in Neural Information Processing Systems, vol. 34 (2021)
17. Zhao, K., Duka, B., Xie, H., Oathes, D.J., Calhoun, V., Zhang, Y.: A dynamic graph convolutional neural network framework reveals new insights into connectome dysfunctions in ADHD. Neuroimage **246**, 118774 (2022)
18. Zhao, Q., Liu, Z., Adeli, E., Pohl, K.M.: Longitudinal self-supervised learning. Med. Image Anal. **71**, 102051 (2021)
19. Zhi, D., et al.: BNCPL: Brain-network-based convolutional prototype learning for discriminating depressive disorders. In: 2021 43rd Annual International Conference of the IEEE Engineering in Medicine & Biology Society, pp. 1622–1626 (2021)

Effective Opportunistic Esophageal Cancer Screening Using Noncontrast CT Imaging

Jiawen Yao[1][(✉)], Xianghua Ye[2], Yingda Xia[3], Jian Zhou[4], Yu Shi[5], Ke Yan[1],
Fang Wang[2], Lili Lin[2], Haogang Yu[2], Xian-Sheng Hua[1], Le Lu[3], Dakai Jin[3],
and Ling Zhang[3]

[1] DAMO Academy, Alibaba Group, Beijing, China
yaojiawen.yjw@alibaba-inc.com
[2] The First Affiliated Hospital of Zhejiang University, Hangzhou, China
[3] DAMO Academy, Alibaba Group, New York, USA
[4] Sun Yat-sen University Cancer Center, Guangzhou, China
[5] Shengjing Hospital of China Medical University, Shenyang, China

Abstract. Esophageal cancer is the second most deadly cancer. Early detection of resectable/curable esophageal cancers has a great potential to reduce mortality, but no guideline-recommended screening test is available. Although some screening methods have been developed, they are expensive, might be difficult to apply to the general population, and often fail to achieve satisfactory sensitivity for identifying early-stage cancers. In this work, we investigate the feasibility of esophageal tumor detection and classification (cancer or benign) on the noncontrast CT scan, which could potentially be used for opportunistic cancer screening. To capture the global context, a novel position-sensitive self-attention is proposed to augment nnUNet with non-local interactions. Our model achieves a sensitivity of 93.0% and specificity of 97.5% for the detection of esophageal tumors on a holdout testing set with 180 patients. In comparison, the mean sensitivity and specificity of four doctors are 75.0% and 83.8%, respectively. For the classification task, our model outperforms the mean doctors by absolute margins of 17%, 31%, and 14% for cancer, benign tumor, and normal, respectively. Compared with established state-of-the-art esophageal cancer screening methods, e.g., blood testing and endoscopy AI system, our method has comparable performance and is even more sensitive for early-stage cancer and benign tumor. Our proposed method is a novel, non-invasive, low-cost, and highly accurate tool for opportunistic screening of esophageal cancer.

Keywords: Esophageal cancer · Cancer screening · Self-attention · Noncontrast CT

1 Introduction

Esophageal cancer (EC) is the second most deadly cancer, with a 5-year survival rate of only 20% [18], and even less than 5% in many developing countries [2].

© The Author(s), under exclusive license to Springer Nature Switzerland AG 2022
L. Wang et al. (Eds.): MICCAI 2022, LNCS 13433, pp. 344–354, 2022.
https://doi.org/10.1007/978-3-031-16437-8_33

This poor survival is mainly due to patients are usually diagnosed at advanced stages with unresectable tumors [2, 5, 21], because their signs and symptoms tend to be latent and non-specific [13]. Early-stage disease, however, is associated with a substantially higher 5-year survival rate of 80%–90% [16, 21]. Therefore, the early detection of resectable/curable esophageal cancers (ideally, before symptoms) has a great potential to reduce mortality [21]. Unfortunately, EC has no guideline-recommended screening tests available [1]. Several tools have been developed, and some are implemented in high-risk areas, such as endoscopic techniques [13], Cytosponge procedure [6], and blood-based biomarkers [11, 15]. However, they are difficult to apply to the general population due to moderate sensitivity and high-cost [2, 15]. Novel screening methods that are noninvasive with low-cost, ready to distribute, and highly accurate are eagerly needed.

Routine CT imaging performed for other clinical indications offers an opportunity for opportunistic screening of diseases at no additional cost or radiation exposure to patients. Previous studies show that abdominal and chest CT with or without contrast enhancement provide values for incidental osteoporosis screening [9] and cardiovascular event prediction [14]. For cancer detection, researchers have found that pancreatic cancer could be detected with high accuracy by deep learning from noncontrast CT [22], which has long been thought to be impossible (only detectable from contrast-enhanced CT). However, detection of esophageal cancer on noncontrast CT can be extremely challenging. The early-stage esophageal carcinoma tumors can be very small, invading only lamina propria (stage I) and muscle layer (stage II) [17]. Given the extremely poor contrast between the tumor and normal esophageal tissues in the noncontrast CT (e.g., chest CT), the early-stage tumor detection task is highly challenging. Actually, even on the contrast-enhanced CT, human experts often require substantial effort and expertise to detect early-stage esophageal tumors by referring to several other clinical information, such as endoscopy, endoscopic ultrasound, and FDG-PET; even so, some tiny tumors are still hard to be detected on CT.

So far, studies on deep learning-based esophageal cancer image analysis all focus on the tumor segmentation task by improving the local image feature extraction/modeling [26, 28] or fusion of multi-modal imaging [10, 25] to improve the segmentation accuracy. In this study, we propose the first deep learning-based tool for opportunistic esophageal cancer screening using noncontrast CT – specifically, detecting the esophageal tumor if there exists and then classifying the detection as cancer or benign. As discussed above, the local image texture could be insufficient to detect esophageal tumors in noncontrast CT. In clinical practice, global context features of the esophagus, such as "asymmetric esophageal wall thickening" and "squeezed esophageal wall", are key signs to diagnose esophageal cancer, especially early-stage ones. On the other hand, for deep learning, each convolutional kernel could only attend a local-subset of voxels or local patterns rather than the global context. Therefore, we incorporate global attention layers with positional embedding to enhance the ability to model global context as well as long-range dependencies in 3D medical image segmentation. This design could improve the ability of tumor distinction especially for early stage tumors. We collect a multi-center dataset including two main esophageal tumor types (ESCC and leiomyoma) and normal esophagus from 741 patients. On the holdout test set, our model achieves an AUC of 0.990, sensitivity of 93.0%, and specificity of 97.5% for tumor detection, surpassing the average sensitivity of 75.0% and specificity of 83.8% of four doctors. The main contributions of this paper can be summarized as follows:

- We present a deep learning method to detect and classify esophageal tumors from noncontrast CT, a novel, non-invasive, low-cost, ready-to-distribute, and highly accurate tool, for screening esophageal cancer.
- The position-sensitive full-attention layer shows its better use of positional information and long-range dependencies in 3D noncontrast CT, therefore, could improve the performance over a strong baseline nnUNet model [8].
- Compared with doctors' reading of noncontrast CT, our automated method shows substantially higher accuracy in both detection and classification. Compared with established state-of-the-art esophageal cancer screening methods, e.g., blood testing [11] and endoscopy AI system [13], our screening tool has comparable performance and is even more sensitive for early-stage cancer and benign tumor.

2   Methods

We aim at a three-class classification problem in noncontrast CT scans. We denote the whole dataset as $S = \{(\mathbf{X_i}, \mathbf{Y_i}, \mathbf{P_i}) | i = 1, 2, ..N\}$, where $\mathbf{X_i} \in \mathbb{R}^{H_i \times W_i \times D_i}$ is a 3D CT volume of the i-th patient. $\mathbf{Y_i} \in \mathcal{L}^{H_i \times W_i \times D_i}$ is a voxel-wise annotated label with the same (H_i, W_i, D_i) three dimensional size as $\mathbf{X_i}$ and represents our segmentation targets, *i.e.*, background, esophagus, esophageal cancer, and benign tumor. To obtain tumor annotations in $\mathbf{Y}$ (upper panel in Fig. 1), for those patients who have radiotherapy CT, we directly use their gross tumor volume (GTV) masks. For others, manual tumor annotation is first performed on the contrast-enhanced CT phase, referring to clinical and panendoscopy reports when necessary. Then, a robust image registration method, DEEDS [7], is used to register the annotated mask from the contrast-enhanced CT to the noncontrast CT, followed by a manual correction during quality check. $\mathbf{P_i} \in \mathcal{L}$ is the patient-level label (esophageal cancer, benign, and normal), confirmed by either pathology or radiology reports with follow-up.

Segmentation for Classification with Self-attention. Segmentation for classification is the most straightforward method and has been successfully adopted for the task of tumor detection [3,22,24,29]. In those methods, a localization UNet [4] is first trained to locate the ROI. However, vanilla UNet-based segmentation networks have limited receptive field and heavily rely on local textual patterns rather than the global context. It will definitely affect the later classification task if the first segmentation model fails to segment the target. Therefore, it is important to build a more robust segmentation model that is sensitive to tumors especially early stage tumors. In this paper, we propose

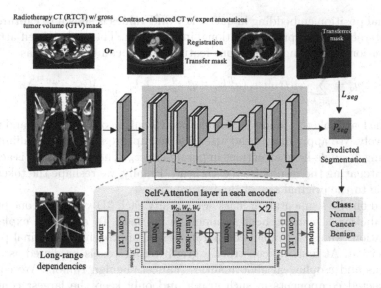

Fig. 1. The main architecture diagram of the proposed model. GTV masks and transferred masks from registration are used as labels to train the model. To capture long-rang dependencies and global context, the position-sensitive self-attention is added to augment convolutional layers with non-local interactions in the encoder block.

an architectural improvement by integrating the global self-attention layer to enhance the model's ability on modeling global context. As shown in Fig. 1, we add self-attention layers after each convolutional layers in the encoder. Let us consider an input feature map $x \in \mathbb{R}^{C_{in} \times H \times W \times D}$. The output at position $o = (i, j, k)$, $y_o \in \mathbb{R}^{C_{out} \times H \times W \times D}$ of a self-attention layer is computed by pooling over the projected input as the following:

$$y_o = \sum_{p \in \mathcal{N}} softmax_p(q_o^T k_p) v_p \tag{1}$$

where $\mathcal{N}$ is the whole location lattice, and queries $q_o = W_Q x_o$, keys $k = W_K x_o$ and values $v_o = W_V x_o$ are all projections of the input x_o. W_Q, W_K, W_V are all learnable matrices. The $softmax_p$ denotes a softmax function applied to all possible positions. When applying self-attention to vision problems, one of the main obstacles is that the computational complexity of attention mechanism scales and this is even more of a problem when dealing with 3D data in medical segmentation. Another issue is that self-attention layer does not utilize any positional information when computing the non-local context. Motivated by recent advances applying attention to multi-dimensional data [19,20], we add local constraints to serve as a memory bank for computing the output y_o which could significantly reduces the original computation of Eq. 1. For each location o, a local region $\mathcal{M} = \mathcal{N}_{m_h \times m_w \times m_d}(o)$ is extracted in each self-attention layer. Additionally, a learned relative positional encoding term is introduced and the additional position embedding in query, key and values is shown to capture long-range interaction with precise positional information. The updated self-attention with positional encoding given input feature map x can be written as

$$y_o = \sum_{p \in \mathcal{M}} softmax_p(q_o^T k_p + q_o^T r_{p-o}^q + k_p^T r_{p-o}^k)(v_p + r_{p-o}^v) \qquad (2)$$

where the learnable r_{p-o}^k and r_{p-o}^v are the positional encoding for keys and values, respectively. We apply multi-head attention to capture a mixture of affinities by computing N single-head attentions in parallel on x_o. The final output is achieved by concatenating the results from each head. Finally, we reshape the tokens and upsample to the original size of the feature map.

Based on the segmentation mask, we classify each 3D volume as one of three target labels, i.e., cancer, benign tumor, or normal. To achieve a explainable classification, we use a simple, non-parametrized approach to give final patient-level decision. At first, we construct a graph on all voxels predicted as normal esophagus and esophageal abnormalities (cancer+benign tumor). We compute all connected components in such graph and only keep the largest connected comment to filter out the outliers. Then, a 3D volume is considered as normal if less than K mm³ of voxels are predicted as abnormal. K is tuned to achieve a specificity of 99% for the model on the validation set. To further classify the abnormal case as cancer or benign, the label with more segmented voxels is predicted.

3   Experiments

Datasets and Annotation: We build a dataset of CT scans of 741 patients and among them, 481 have esophageal tumors, either cancer or benign (leiomyomas), and the other 260 are normal cases. The dataset is randomly split into a training and a testing set but according to the distribution of cancer stages and tumor types. The resulting training set includes 324 cancer, 57 benign, and 180 normal. The testing set has 80 cancer, 20 benign, and 80 normal. Both cancer and benign cases are confirmed by pathology reports. Normal cases are confirmed by radiology reports and 2 years of follow-up. An experienced radiation oncologist (10 yrs esophageal specialist) manually annotates tumors in the training set (on either radiotherapy CT or contrast-enhanced CT). Esophagus masks are automatically generated by a segmentation model which is trained on a public dataset [12] with a self-learning process [27] on our training set.

Implementation Details: Each CT volume is resampled into $0.8 \times 0.8 \times 3.0$ mm spacing and then normalized into zero mean and unit variance. In self-attention layer, (m_h, m_w, m_d) is set as $(12, 12, 6)$ and N=4 heads is used. During training, extensive data augmentation [8] was applied on the fly to improve the generalization, including random rotation and scaling, elastic deformation, additive brightness, and gamma scaling. The objective for optimization is the sum of binary cross entropy and the dice loss. The networks were optimized with

Table 1. Results on two-class classification: abnormal (esophageal cancer + benign) vs. normal, and three-class classification: esophageal cancer vs. benign vs. normal. WOTC: without time constraint. Sens.: Sensitivity. Spec.: Specificity. *for $p < 0.05$, **for $p < 0.01$.

Method	Two-class			Three-class(%)		
	AUC	Sens. (%)	Spec. (%)	Cancer	Benign	Normal
Ours	0.990	93.1	97.5	91.3	60.0	97.5
	(0.986–0.993)	(91.8–94.4)	(96.6–98.4)			
nnUNet-S4C [8,29]	0.961*	85.1	100.0	90.0	45.0	100.0
	(0.954–0.967)	(83.3–86.8)	(100.0–100.0)			
LENS (detection) [23]	0.954**	88.1	96.3	91.3	15.0	96.3
	(0.947–0.961)	(86.5–89.6)	(95.3–97.3)			
Mean doctors WOTC	-	75.0	83.8	74.4	28.8	83.8

RAdam with the initial learning rate as 0.001 and we set the maximum epoch as 1000 following the nnUNet.

Evaluation Metrics and Reader Study. Evaluation metrics include AUC, sensitivity, and specificity in the 2-class classification task (abnormal vs. normal), and class accuracy in the 3-class classification task. To measure the performance of tumor localization, we compare the segmentation mask with the ground truth and define the localization is successful if the intersection (Dice score) over the human annotation is larger than 0.1. Four readers, including two radiation oncologists (5 and 14 yr [12 yr esophageal specialist], respectively) and one radiologist (8 yr) from the 1st affiliated hospital of Zhejiang University and one radiologist (13 yr [6 yr esophageal specialist]) from Sun Yat-sen University Cancer Center (SYSUCC), are invited for the reader study. They read the 180 noncontrast CTs in the testing set without time constraints and give a forced three-class decision for each CT: esophageal cancer, leiomyomas, or normal. Patient information and records are not provided. Readers are informed that the dataset might contain more tumor cases than the standard prevalence observed in screening, but the proportion of case types is not informed. The first three readers view and interpret these CTs in the radiation planning viewer used in their daily work environment. The reader from SYSUCC uses ITK-SNAP, with three years of user experience for research purposes.

Comparison with Other Algorithms and Readers. We compared our method with two approaches representing strong segmentation-based and detection-based methods. The segmentation-based one uses the "Segmentation for classification (S4C)" [29] paradigm but its standard U-Net is updated with the more powerful nnUNet [8]. The detection-based one has proven its performances as a competitive universal lesion detector [23]. Standard deviation of the AUCs, Sens. and Spec. values are obtained from 1000 bootstrap replicas of the test dataset. DeLong test is performed for statistical analysis between two AUCs (Ours vs. comparison method). Results can be seen in Table 1 and two illustrative

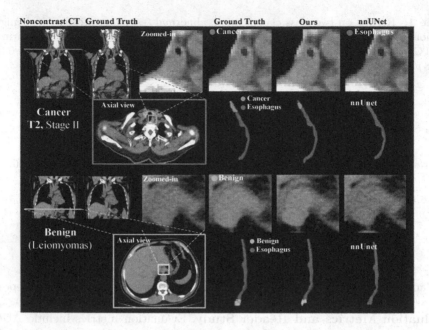

Fig. 2. Two cases (a cancer and a benign) miss-detected by all readers in the test set. Our methods can successfully locate the tumors while nnUNet fails in these cases.

examples are shown in Fig. 2. Most medical segmentation and detection models focus on local texture and structure and thus lack the ability to model the global context. From the table and figure, we could see improved results of our method in finding abnormal patients and detecting benign tumors. The performance of all four doctors is below our model's predictions (ROC curve, Fig. 3 (A)). Our sensitivity in finding tumor (93%) outperforms the best sensitive doctor (79%) by a large margin and also performs better in predicting normal patients than the best specific doctor (Specificity 98% vs. 96%). For the 3-class task, our model perform much better than the mean doctors by absolute margins of 17%, 31%, and 14% for cancer, benign, and normal, respectively (Table 1; confusion matrix, Fig. 3 (B)).

Subgroup Analysis Stratified by Tumor Stage. More specificity, we evaluate performance of our model in benign and malignant tumor, as shown in Table 2. We first report patient-level detection rate values across benign and each T stage cancer. Then we compare predictions with ground-truth annotations to see if predicted tumors are detected correctly. Tumor-level localization evaluates how segmented masks overlap with the ground-truth cancer or leiomyoma regions (Dice > 0.1 is used for correct detection). Radiologist 2 (the best sensitive reader) in finding abnormalities is also compared. A detection is considered successful if the intersection (between the ground truth and segmentation mask) over the ground truth is > 0.1; otherwise, it is considered a misdetection. Our model performs better in detecting benign and T2 stage tumor and providing more accurate tumor location.

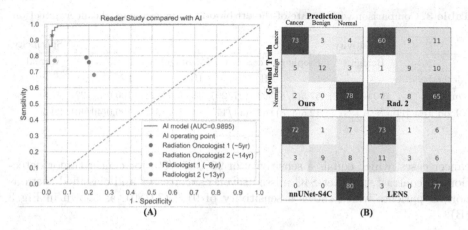

Fig. 3. (A) ROC curve for our model results versus all participated experts' referrals on the test set of $n = 180$ patients for 2-class classification (abnormal vs. normal). (B) Confusion matrices with patient numbers of predictions for our model, the two baseline methods, and Radiologist 2 (the best 3-class accuracy reader).

Table 2. Patient-level detection and tumor-level localization results over the two types of esophageal abnormalities (benign and cancer).

Method	Criteria	Benign	Esophageal cancer			
		–	T1	T2	T3	T4
Ours	Patient-level	60.0(12/20)	44.4(4/9)	94.1(16/17)	97.1(33/34)	100(20/20)
	Tumor-level	40.0(8/20)	44.4(4/9)	88.2(15/17)	97.1(33/34)	100(20/20)
nnUNet-S4C	Patient-level	45.0(9/20)	55.6(5/9)	82.4(14/17)	97.1(33/34)	100(20/20)
	Tumor-level	35.0(7/20)	44.4(4/9)	70.6(12/17)	97.1(33/34)	100(20/20)
Radiologist 2	Patient-level	45.0(9/20)	55.6(5/9)	64.7(11/17)	76.5(26/34)	90.0(18/20)

Comparison with Established Screening Tools. Compared with a state-of-the-art blood test screening [11], at a similar specificity level, our solution achieves much better results in detecting early-stage esophageal cancer (stage I-II) using the similar size of testing patients (Table 3). Moreover, our method can detect benign tumors (leiomyomas, which needs surgery) while the blood test cannot. We also compare our model with a state-of-the-art endoscopy AI system (trained on 15,000 patients data with fully supervision) [13] for upper gastrointestinal cancer detection. Note that the definition of sensitivity and specificity in the endoscopy screening scenario is slightly different from our radiological scenario. To facilitate the comparison, we report the results of cancer vs. (benign + normal) by following [13]. We could see a (slightly) lower (91.3% vs. 94.2%) sensitivity with a slightly higher specificity (92.9% vs 92.3%). In fact, for the 80 cancer cases in our test set, our method detects 76 of them, with three being misclassified as benign, which are leiomyomas, and surgery will be recommended in the noncontrast CT screening scenario. As such, our method only misses four

Table 3. Comparison with a state-of-the-art blood test on esophageal cancer detection.

	Sensitivity					Specificity
	Benign	Esophageal cancer (TNM Stage)				Normal
Method	-	I	II	III	IV	-
Ours	60.0(12/20)	37.5(3/8)	93.1(27/29)	100.0(21/21)	100.0(22/22)	97.5
Blood Test [11]	N/A	12.5(1/8)	64.7(11/17)	94.1(32/34)	100.0(40/40)	99.5

cancer cases, which equals a sensitivity of 95% (76/80) for cancer tumor detection. In contrast, nnUNet-S4C missed seven cancer cases and classified them as normal patients which shows a sensitivity of 91% (73/80), as shown in Fig. 3 (B).

4   Conclusion

In this paper, we investigate a relatively convenient, simple opportunistic screening solution of esophageal cancer with noncontrast CT scans. To better capture global context and detect early-stage tumors, we propose a position-sensitive self-attention to augment convolutional layers with non-local interactions in the encoder block. We achieve high sensitivity and specificity on a large-scale dataset and outperform the mean doctors by large margins. Compare with other tools like blood test and endoscopy, our work suggests the good feasibility of using noncontrast CT scans as a promising clinical tool for large-scale esophageal cancer opportunistic screening with no extra costs.

References

1. U.S. Preventive Services Task Force (USPSTF), "Recommendations,". https://www.uspreventiveservicestaskforce.org/uspstf/topic_search_results?topic_status=P
2. Arnal, M.J.D., Arenas, Á.F., Arbeloa, Á.L.: Esophageal cancer: risk factors, screening and endoscopic treatment in western and eastern countries. World J. Gastroenterol. **21**(26), 7933 (2015)
3. Cheng, N.M., et al.: Deep learning for fully automated prediction of overall survival in patients with oropharyngeal cancer using FDG-PET imaging. Clin. Cancer Res. **27**(14), 3948–3959 (2021)
4. Çiçek, Ö., Abdulkadir, A., Lienkamp, S.S., Brox, T., Ronneberger, O.: 3D U-Net: learning dense volumetric segmentation from sparse annotation. In: Ourselin, S., Joskowicz, L., Sabuncu, M.R., Unal, G., Wells, W. (eds.) MICCAI 2016. LNCS, vol. 9901, pp. 424–432. Springer, Cham (2016). https://doi.org/10.1007/978-3-319-46723-8_49
5. Doki, Y., et al.: Nivolumab combination therapy in advanced esophageal squamous-cell carcinoma. N. Engl. J. Med. **386**(5), 449–462 (2022)

6. Gehrung, M., Crispin-Ortuzar, M., Berman, A.G., O'Donovan, M., Fitzgerald, R.C., Markowetz, F.: Triage-driven diagnosis of Barrett's esophagus for early detection of esophageal adenocarcinoma using deep learning. Nat. Med. **27**(5), 833–841 (2021)

7. Heinrich, M.P., Jenkinson, M., Brady, M., Schnabel, J.A.: MRF-based deformable registration and ventilation estimation of lung CT. IEEE Trans. Med. Imaging **32**(7), 1239–1248 (2013)

8. Isensee, F., Jaeger, P.F., Kohl, S.A., Petersen, J., Maier-Hein, K.H.: nnU-Net: a self-configuring method for deep learning-based biomedical image segmentation. Nat. Methods **18**(2), 203–211 (2021)

9. Jang, S., Graffy, P.M., Ziemlewicz, T.J., Lee, S.J., Summers, R.M., Pickhardt, P.J.: Opportunistic osteoporosis screening at routine abdominal and thoracic CT: normative L1 trabecular attenuation values in more than 20 000 adults. Radiology **291**(2), 360–367 (2019)

10. Jin, D., et al.: Deeptarget: gross tumor and clinical target volume segmentation in esophageal cancer radiotherapy. Med. Image Anal. **68**, 101909 (2021)

11. Klein, E., et al.: Clinical validation of a targeted methylation-based multi-cancer early detection test using an independent validation set. Ann. Oncol. **32**(9), 1167–1177 (2021)

12. Lambert, Z., Petitjean, C., Dubray, B., Kuan, S.: Segthor: segmentation of thoracic organs at risk in CT images. In: 2020 Tenth International Conference on Image Processing Theory, Tools and Applications (IPTA), pp. 1–6. IEEE (2020)

13. Luo, H., et al.: Real-time artificial intelligence for detection of upper gastrointestinal cancer by endoscopy: a multicentre, case-control, diagnostic study. Lancet Oncol. **20**(12), 1645–1654 (2019)

14. Pickhardt, P.J., et al.: Automated CT biomarkers for opportunistic prediction of future cardiovascular events and mortality in an asymptomatic screening population: a retrospective cohort study. Lancet Dig. Health **2**(4), e192–e200 (2020)

15. Qin, Y., Wu, C.W., Taylor, W.R., Sawas, T., Burger, K.N., Mahoney, D.W., Sun, Z., Yab, T.C., Lidgard, G.P., Allawi, H.T., et al.: Discovery, validation, and application of novel methylated DNA markers for detection of esophageal cancer in plasma. Clin. Cancer Res. **25**(24), 7396–7404 (2019)

16. Rice, T., Ishwaran, H., Hofstetter, W., Kelsen, D., Apperson-Hansen, C., Blackstone, E.: Recommendations for pathologic staging (PTNM) of cancer of the esophagus and esophagogastric junction for the 8th edition AJCC/UICC staging manuals. Dis. Esophagus **29**(8), 897–905 (2016)

17. Thompson, W.M.: Esophageal carcinoma. Abdom. Imaging **22**(2), 138–142 (1997). https://doi.org/10.1007/s002619900158

18. Siegel, R.L., Miller, K.D., Jemal, A.: Cancer statistics, 2021. CA: A Cancer Journal for Clinicians **71**(1), 7–333 (2021)

19. Valanarasu, J.M.J., Oza, P., Hacihaliloglu, I., Patel, V.M.: Medical transformer: gated axial-attention for medical image segmentation. In: de Bruijne, M., et al. (eds.) MICCAI 2021. LNCS, vol. 12901, pp. 36–46. Springer, Cham (2021). https://doi.org/10.1007/978-3-030-87193-2_4

20. Wang, H., Zhu, Y., Green, B., Adam, H., Yuille, A., Chen, L.-C.: Axial-DeepLab: stand-alone axial-attention for panoptic segmentation. In: Vedaldi, A., Bischof, H., Brox, T., Frahm, J.-M. (eds.) ECCV 2020. LNCS, vol. 12349, pp. 108–126. Springer, Cham (2020). https://doi.org/10.1007/978-3-030-58548-8_7

21. Wei, W.Q., Chen, Z.F., He, Y.T., Feng, H., Hou, J., Lin, D.M., Li, X.Q., Guo, C.L., Li, S.S., Wang, G.Q., et al.: Long-term follow-up of a community assignment, one-

time endoscopic screening study of esophageal cancer in china. J. Clin. Oncol. **33**(17), 1951 (2015)

22. Xia, Y., et al.: Effective pancreatic cancer screening on non-contrast CT scans via anatomy-aware transformers. In: de Bruijne, M., et al. (eds.) MICCAI 2021. LNCS, vol. 12905, pp. 259–269. Springer, Cham (2021). https://doi.org/10.1007/978-3-030-87240-3_25

23. Yan, K., et al.: Learning from multiple datasets with heterogeneous and partial labels for universal lesion detection in CT. IEEE Trans. Med. Imaging **40**, 2759–2770 (2021)

24. Yao, J.: Deepprognosis: preoperative prediction of pancreatic cancer survival and surgical margin via comprehensive understanding of dynamic contrast-enhanced CT imaging and tumor-vascular contact parsing. Med. Image Anal. **73**, 102150 (2021)

25. Ye, X., et al.: Multi-institutional validation of two-streamed deep learning method for automated delineation of esophageal gross tumor volume using planning-ct and fdg-petct. arXiv preprint arXiv:2110.05280 (2021)

26. Yousefi, S., et al.: Esophageal gross tumor volume segmentation using a 3D convolutional neural network. In: Frangi, A.F., et al. (eds.) MICCAI 2018. LNCS, vol. 11073, pp. 343–351. Springer, Cham (2018). https://doi.org/10.1007/978-3-030-00937-3_40

27. Zhang, L., Gopalakrishnan, V., Lu, L., Summers, R.M., Moss, J., Yao, J.: Self-learning to detect and segment cysts in lung CT images without manual annotation. In: 2018 IEEE 15th International Symposium on Biomedical Imaging (ISBI 2018), pp. 1100–1103. IEEE (2018)

28. Zhou, D., et al.: Eso-net: a novel 2.5 d segmentation network with the multi-structure response filter for the cancerous esophagus. IEEE Access **8**, 155548–155562 (2020)

29. Zhu, Z., et al.: Multi-scale coarse-to-fine segmentation for screening pancreatic ductal adenocarcinoma. In: Shen, D. (ed.) MICCAI 2019. LNCS, vol. 11769, pp. 3–12. Springer, Cham (2019). https://doi.org/10.1007/978-3-030-32226-7_1

Joint Prediction of Meningioma Grade and Brain Invasion via Task-Aware Contrastive Learning

Tianling Liu[1], Wennan Liu[2], Lequan Yu[3], Liang Wan[1(✉)], Tong Han[4], and Lei Zhu[5,6]

[1] College of Intelligence and Computing, Tianjin University, Tianjin, China
{liu_dling,lwan}@tju.edu.cn
[2] Medical College of Tianjin University, Tianjin, China
[3] The University of Hong Kong, Hong Kong, China
[4] Brain Medical Center of Tianjin University, Huanhu Hospital, Tianjin, China
[5] The Hong Kong University of Science and Technology (Guangzhou), Guangzhou, China
[6] The Hong Kong University of Science and Technology, Hong Kong, China

Abstract. Preoperative and noninvasive prediction of the meningioma grade is important in clinical practice, as it directly influences the clinical decision making. What's more, brain invasion in meningioma (*i.e.*, the presence of tumor tissue within the adjacent brain tissue) is an independent criterion for the grading of meningioma and influences the treatment strategy. Although efforts have been reported to address these two tasks, most of them rely on hand-crafted features and there is no attempt to exploit the two prediction tasks simultaneously. In this paper, we propose a novel task-aware contrastive learning algorithm to jointly predict meningioma grade and brain invasion from multi-modal MRIs. Based on the basic multi-task learning framework, our key idea is to adopt contrastive learning strategy to disentangle the image features into task-specific features and task-common features, and explicitly leverage their inherent connections to improve feature representation for the two prediction tasks. In this retrospective study, an MRI dataset was collected, for which 800 patients (containing 148 high-grade, 62 invasion) were diagnosed with meningioma by pathological analysis. Experimental results show that the proposed algorithm outperforms alternative multi-task learning methods, achieving AUCs of 0.8870 and 0.9787 for the prediction of meningioma grade and brain invasion, respectively. The code is available at https://github.com/IsDling/predictTCL.

Keywords: Meningioma grading · Brain invasion · Preoperative prediction · Contrastive learning · Feature disentanglement

Supplementary Information The online version contains supplementary material available at https://doi.org/10.1007/978-3-031-16437-8_34.

© The Author(s), under exclusive license to Springer Nature Switzerland AG 2022
L. Wang et al. (Eds.): MICCAI 2022, LNCS 13433, pp. 355–365, 2022.
https://doi.org/10.1007/978-3-031-16437-8_34

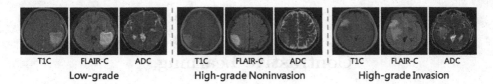

Fig. 1. MRI examples for low-grade meningiomas, high-grade meningiomas without/with brain invasion.

1   Introduction

Meningiomas are the most common primary intracranial tumors in adults, comprising 38.3% of central nervous system tumors [22]. According to the World Health Organization (WHO), meningiomas can be subdivided into three grades, *i.e.*, grade I (80%), grade II (18%) and grade III (2%). Grade I represents low-grade meningiomas, grade II and grade III are high-grade meningiomas. The WHO grading is an important factor in determining treatment options and overall prognosis for meningiomas. Specifically, low-grade meningiomas can be treated with surgery or beam radiation, and rarely recur after resection, while high-grade meningiomas should be treated with both means, and subjected to universal recurrence for grade III and 20–75% recurrence rates for grade II [5,13]. In another aspect, brain invasion is taken as a stand-alone pathological criterion for distinguishing between grade I and grade II in 2016 WHO classification [19]. Recent researches also uncover a link between brain invasion and increased risks of tumor progression, disease recurrence and poor prognosis [4,17,21].

In real clinical diagnosis, pathological analysis provides the gold standard for the determination of brain invasion and meningioma grading [18,19]. However, to conduct pathological analysis, clinicians are required to sample tissues from the core and surrounding areas of the tumor during *invasive* resection or biopsy, while some important treatment decision has been made without the knowledge of brain invasion and meningioma grading. Moreover, the accuracy of brain invasion determination heavily depends on the clinician's experience [29]. If the brain tissue samples do not fall in the invasion area, there is a risk that the patient will be misdiagnosed and his prognosis can be affected. Given these practical concerns, accurate *preoperative* and *non-invasive* assessment of meningioma grade and brain invasion is clinically essential to facilitate treatment decisions.

Many recent studies in the clinical field are devoted to predicting the grade of meningioma or identification of brain invasion by analyzing brain MRIs [10,11,15,23,28,29,31]. Hale *et al.* extracted radiographic features by traditional statistical methods and verified that machine learning-based methods can help to predict meningioma grade [10]. Later on, several followed-up studies are reported to extract different hand-crafted features or CNN features and then use machine learning-based classifiers, such as SVM or random forest, to predict the meningioma grade [11,23,31], or for the determination of brain invasion [15,29]. Moreover, Zhang *et al.* applied the widely-used deep-learning net-

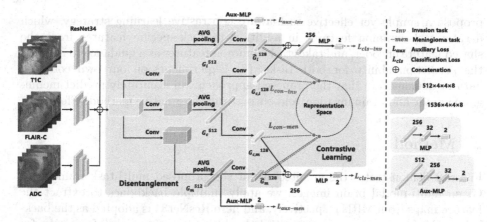

Fig. 2. The overview of the proposed multi-task learning framework. We disentangle the fused feature maps into two task-specific features and one task-common feature. A task-aware contrastive learning strategy is devised to ensure that task-common features can better guide both tasks.

work, ResNet, to predict the meningioma grade [28]. Overall, most of the previous studies extracted hand-crafted features and conducted the two tasks, without exploring the powerful feature representation learning of neural networks. Furthermore, to the best of our knowledge, there is no previous study performing both tasks simultaneously, although there exist clinical connections between them.

In this paper, as a preliminary exploration, we develop a novel multi-task learning algorithm to make joint prediction of brain invasion and meningioma grade from multi-modal MRIs, including post-contrast T1-weighted (T1C), constrat T2 fluid attenuation inversion recovery (FLAIR-C), and apparent diffusion coefficient (ADC) images, following clinical practice (Fig. 1). Note that the common multi-task learning strategy adopts a shared backbone and then learns task-specific features with multiple heads to perform separate tasks [9]. Existing methods are usually focused to tune the loss function to balance contributions between different tasks [8,16], which is still weak to ensure the feature representation ability. We notice that multi-task methods assume that the tasks are related to each other, and considering tasks simultaneously can help to improve the feature representation ability. Hence, in our work, we propose a novel task-aware contrastive learning strategy, by respecting both the coherence between tasks and the distinctness of each individual task. Our approach disentangles the multi-modal MRI features into *task-specific* features, which are sensitive to a certain classification task, as well as *task-common* features that are helpful to both prediction tasks. Then we align the task-common features to each task, and enforce the feature emeddings contributing to the same task to be more similar than the embeddings contributing to different tasks, via contrastive losses.

In summary, our contributions are three folds: (1) This is the first study to simultaneously predict meningioma grade and identify brain invasion. (2) We propose a simple yet effective task-aware contrastive learning strategy, which derives task-common features in addition to task-specific features from the shared feature encoder, and takes task-common features as a guidance to improve the prediction ability for both tasks; (3) Experiments on our own collected dataset demonstrate that the proposed approach can accurately predict meningioma grade and brain invasion, and outperforms alternative methods effectively.

2   Method

Figure 2 demonstrates the architecture of the proposed multi-task framework. Given multi-modal brain images, we apply multiple backbones to extract the feature maps from MRIs respectively, and here ResNet34 is adopted as the backbone. The fused features from multiple backbones contain rich information for performing multiple tasks. We then disentangle the fused feature maps to task-specific features and task-common features, and a new task-aware contrastive learning strategy is leveraged to exploit the "comparative" relation between task-specific features and task-common features.

2.1   Feature Disentanglement

As demonstrated by [6,7], the disentangled representation, which decouples the entangled features into task-specific features that are easier to predict, is beneficial to the realization of multi-task learning. Like most multi-task applications, our goal is to find the most discriminative features for each prediction task. Although the general practice is to disentangle the fused features into multiple task-specific features, we consider that there are features that are simultaneously effective for multiple tasks, since multi-task learning assumes that the tasks are related to each other. In our work, we disentangle the fused features into the task-specific feature for invasion identification task (denoted as G_i), the task-specific feature for meningioma grading task (G_m), and the task-common feature (G_c). Here, we adopt a convolution layer and an average pooling to realize feature disentanglement, which is formulated as follows,

$$G_k = AP(Conv(Concat(F_T, F_F, F_A))), k = i, m, c, \qquad (1)$$

where $\{F_T, F_F, F_A \in \mathbb{R}^{C \times h \times w \times d}\}$ denote the feature maps extracted from T1C, Flair-C and ADC MRIs respectively, and their sizes are all $512 \times 4 \times 4 \times 8$. $Concat(\cdot)$ means channel concatenation. $Conv(\cdot)$ means 2×2 convolution operation, $AP(\cdot)$ is average pooling. After flattening, the resultant disentangled features $\{G_i, G_m, G_c\}$ are feature vectors with size of 512. Note that to realize feature disentanglement, we will rely on the task-aware contrastive learning as well as auxiliary classification branches, discussed in the following subsections.

2.2   Task-Aware Contrastive Learning

The proposed task-aware contrastive learning strategy has two functions, one is to help feature disentanglement, and the other is to improve the predictive ability of each task-specific features for the corresponding task by leveraging the relationship between task-common features and task-specific features. Contrastive learning (CL) has shown great potential in the natural image field, and has been applied in the medical image filed in recent years. Its core idea is "learn to compare": given an anchor point, distinguish a similar (or positive) sample from a set of dissimilar (or negative) samples, in a projected embedding space [25]. Most current CL researches are image-level and pixel-level. Image-level CL [12,26] takes multiple views of the same image as positive samples and different images as negative samples. Pixel-level CL [25,27] take pixels from the same class as positive samples and pixels from different classes as negative samples. Different from CL methods mentioned above, we propose a new task-aware contrastive learning strategy. As the task-common feature is supposed to be helpful for both tasks, we align it to each task and enforce the feature emeddings contributing to the same task to be more similar than the embeddings contributing to different tasks, via contrastive losses. This simple strategy respects the coherence between tasks and strengthens the distinctness of each individual task, thus improving the feature representation ability of the network.

Taking the invasion identification task as an example, we align the task-common feature G_c to this task to get $G_{c,i}$ via a fully connected layer, yielding $G_{c,i} \in \mathbb{R}^{128}$. Then $G_{c,i}$ is expected to be similar to the task-specific feature for invasion identification G_i, and not like the task-specific feature for meningioma grading G_m. To this end, we also transform two task-specific features into $\hat{G}_i$ and $\hat{G}_m$ via fully connected layers, respectively, where $\hat{G}_i, \hat{G}_m \in \mathbb{R}^{128}$. We define the task-aware contrastive loss for the invasion identification task as

$$\mathcal{L}_{con-inv} = -log \frac{exp(sim(G_{c,i}, \hat{G}_i)/\tau)}{exp(sim(G_{c,i}, \hat{G}_i)/\tau) + exp(sim(G_{c,i}, \hat{G}_m)/\tau)}, \quad (2)$$

where $sim(\cdot)$ denotes cosine similarity between two feature vectors; τ is the temperature parameter and set as 0.07 empirically. Similarly, the task-aware contrastive loss for the meningioma grading task is defined as follows,

$$\mathcal{L}_{con-men} = -log \frac{exp(sim(G_{c,m}, \hat{G}_m)/\tau)}{exp(sim(G_{c,m}, \hat{G}_m)/\tau) + exp(sim(G_{c,m}, \hat{G}_i)/\tau)}, \quad (3)$$

where $G_{c,m} \in \mathbb{R}^{128}$ denotes the manipulated feature vector via aligning the task-common feature G_c to meningioma grading.

2.3   Overall Loss Functions and Training Strategy

Besides to estimate the task-aware contrastive losses, the features for the same task are further concatenated and undergo a three-layer MLP (with dimensions of 256, 32 and 2), yielding the prediction loss for this task, i.e. $\mathcal{L}_{cls-inv}$ and

Table 1. Details of the dataset.

Characteristics	Low grade	High grade	
		Invasion	Noninvasion
Number	652	62	86
Age(years±SD)	56.75±10.7	55.77±11.97	55.52±12.46
Male	139	26	39
Female	513	36	47

$\mathcal{L}_{cls-men}$. In order to promote the guiding ability of task-common features, we introduce auxiliary classification branches to the disentangled task-specific features; see Fig. 2, which offers two auxiliary classification losses, i.e. $\mathcal{L}_{aux-inv}$ and $\mathcal{L}_{aux-men}$. Each auxiliary branch is a four-layer MLP with dimensions of 512, 256, 32 and 2. In summary, our multi-task learning loss is defined as

$$\mathcal{L} = \mathcal{L}_{cls-inv} + \mathcal{L}_{cls-men} + \alpha(\mathcal{L}_{con-inv} + \mathcal{L}_{con-men}) \\ + \beta(\mathcal{L}_{aux-inv} + \mathcal{L}_{aux-men}). \tag{4}$$

We use cross entropy loss as the classification loss; α and β are weights to balance the contribution of different losses, which are set as 1 and 0.7 empirically. Besides, to ensure the contrastive learning strategy work well, we add the contrastive learning loss after a period of training, empirically set as 30 epoches.

3   Experiments

Dataset and Preprocessing. We collected an MRI dataset of meningiomas for patients with tumor resection between March 2016 and March 2021 in Brain Medical Center of Tianjin University, Tianjin Huanhu Hospital. MRI scans were performed with four 3.0T MRI scanners (i.e., Skyra, Trio, Avanto, Prisma from Siemens). Table 1 presents details of the dataset, which contains 800 MRI volumes with a size of $256 \times 256 \times 24$ and $1mm$ spacing. Every MRI volume contains three modals, *i.e.*, T1C (contrast-enhanced T1), FLAIR-C (contrast-enhanced T2 FLAIR) and ADC, and has two labels (*i.e.*, grading and invasion classification labels) for each patient.

During experiments, due to the small amount of invasion samples, we use randomly drawn data division to alleviate overfitting. Specifically, we randomly draw training and testing sets in three runs to ensure training and testing sets have similar distribution of low/high, invasion yes/no. Then there are 214 MRIs for training and the remaining MRIs as the testing dataset in each run. Specifically, the training dataset contains 44 invasion MRIs and 170 noninvasion MRIs, 69 high grade MRIs and 145 low grade MRIs. For preprocessing, radiologists are asked to crop the tumor ROIs following previous works [1,2,14]. In order to maintain the shape of tumor and edema area, we zero pad the cropped image into a square and resize them into $128 \times 128 \times 24$ as the network input.

Table 2. Quantitative comparison. The best and second best results for each metric are highlighted in red and blue, respectively.

Methods		EFMT	MFMT	MMoE	MAML	Proposed
Invasion	Sensitivity	0.7593	0.6852	0.7963	0.7407	0.7593
	Specificity	0.9707	0.9771	0.9630	0.9824	0.9789
	Accuracy	0.9642	0.9681	0.9579	0.9750	0.9721
	G-Means	0.8555	0.8181	0.8756	0.8529	0.8619
	Balanced Accuracy	0.8650	0.8312	0.8797	0.8616	0.8691
	MCC	0.5787	0.5691	0.5529	0.6486	0.6252
	AUPRC	0.3607	0.3558	0.3329	0.4490	0.4157
	AUC	0.9574	0.9527	0.9425	0.9668	0.9787
Meningioma	Sensitivity	0.5950	0.6582	0.6498	0.6920	0.6878
	Specificity	0.8475	0.9132	0.8126	0.8586	0.9277
	Accuracy	0.8134	0.8788	0.7907	0.8362	0.8953
	G-Means	0.7099	0.7748	0.7262	0.7690	0.7981
	Balanced Accuracy	0.6554	0.7493	0.6460	0.6998	0.7806
	MCC	0.3706	0.5327	0.3671	0.4675	0.5860
	AUPRC	0.2814	0.4088	0.2773	0.3509	0.4599
	AUC	0.8137	0.8707	0.8163	0.8618	0.8870

Evaluation Metrics and Implementation Details. The metrics used to evaluate the network performance are Sensitivity, Specificity, Accuracy, G-Means, Balanced Accuracy [3], MCC, AUPRC, and AUC. Also note that MCC and AUPRC are two metrics which can be used with imbalanced datasets. The proposed algorithm is built with PyTorch on a NVIDIA RTX 3090 GPU. We use the Adam optimizer and set the initial learning rate to $1e-3$. To prevent overfitting, we add dropout of 0.5 and $L2$ regularization with regularization parameter as $1e-3$. Flip, Gaussian noise and random crop are employed in data augmentation. Our model has a parameter size of 200M, trained with 100 epoches, and takes an average inference time of 0.098s for one image.

Comparison with Other Methods. As there are no existing methods directly available for our joint classification tasks, we built four alternative methods for comparison. All the compared methods use ResNet34 as the backbone, and are trained on the collected dataset to get best results. (1) EFMT (Early Fusion with Multi-task). We built a model that concatenates multi-modal MRIs at the input and the extracted features are connected to two classifiers that perform the corresponding tasks. (2) MFMT (Middle Fusion with Multi-task). The MFMT model contains three ResNet34 to extract the features of multi-modal MRIs respectively, and the concatenated features are fed to two classifiers to finish both tasks. (3) MMoE [20]. We adapt this method to our multi-task topic, which uses three expert networks and two gating networks to generate task-specific features. We adopt ResNet34 as the expert network and ResNet18 as the gating

Table 3. The results of ablation experiments. TC, Aux and L_{con} mean Task-common branch, auxiliary branch and contrastive loss respectively. The best and second best results for each metric are highlighted in red and blue, respectively.

Baseline		Baseline1	Baseline2	Baseline3	Baseline4	Proposed
	TC	×	✓	✓	✓	✓
Ablation	L_{con}	×	×	✓	×	✓
	Aux	×	×	×	✓	✓
Invasion	Sensitivity	0.6111	0.7778	0.7407	0.7037	0.7593
	Specificity	0.9730	0.9736	0.9783	0.9794	0.9789
	Accuracy	0.9619	0.9676	0.9710	0.9710	0.9721
	G-Means	0.7629	0.8699	0.8504	0.8291	0.8619
	Balanced Accuracy	0.7921	0.8757	0.8595	0.8416	0.8691
	MCC	0.5379	0.5990	0.6097	0.5974	0.6252
	AUPRC	0.3216	0.3859	0.4026	0.3882	0.4157
	AUC	0.9542	0.9695	0.9727	0.9717	0.9787
Meningioma	Sensitivity	0.5781	0.6878	0.7131	0.6962	0.6878
	Specificity	0.9211	0.8941	0.8692	0.8988	0.9277
	Accuracy	0.8749	0.8663	0.8481	0.8714	0.8953
	G-Means	0.7240	0.7841	0.7859	0.7909	0.7981
	Balanced Accuracy	0.7486	0.7289	0.7070	0.7336	0.7806
	MCC	0.4934	0.5157	0.4903	0.5272	0.5860
	AUPRC	0.3728	0.3935	0.3676	0.4016	0.4599
	AUC	0.8851	0.8753	0.8853	0.8815	0.8870

network. (4) MAML [30]. This method extracts features for each modality via multiple encoders, and estimates an modality-aware attention map to obtain boosted features. We change the output of this method to two classifiers. It is noted that the classifier adopted in the above compared methods is the same as the auxiliary branch in our proposed framework; see Fig. 2.

Table 2 summarizes the comparison results. Among the compared methods, for the brain invasion identification task, MMoE gets the highest sensitivity, g-means, and balanced accuracy (0.7963, 0.8756, 0.8797). MAML achieves the best specificity, accuracy, MCC, AUPRC, and AUC (0.9824, 0.9750, 0.6486, 0.4490, 0.9668). For the meningioma grade prediction task, MAML gets the best sensitivity (0.6920) and MFMT achieves the best specificity, accuracy, g-means, balanced accuracy, MCC, AUPRC, and AUC (0.9132, 0.8788, 0.7748, 0.7493, 0.5327, 0.4088, 0.8701). In comparison, our proposed algorithm achieves the best AUC (0.9787, 0.0119 better than the MAML) for the brain invasion identification task; and the best specificity, accuracy, g-means, balanced accuracy, MCC, AUPRC, and AUC (0.9277, 0.8953, 0.7981, 0.7806, 0.5860, 0.4599, 0.8870) for the meningioma grade prediction task. It is noted that for MCC and AUPRC metrics, our method reports the best results for meningioma grade prediction and the second best results for brain invasion identfication. The standard deviations of metrics results can be found in the supplementary material. We also calculate AUC differences between compared methods and ours using ROC-kit [24], while p-values are less than 0.05.

Ablation Analysis. To demonstrate the effectiveness of the TC (Task-common branch), Aux (auxiliary branch) and L_{con} (contrastive learning loss), we conduct ablation studies; see Table 3. Compared with the proposed algorithm, Baseline1 removes the task-common branch, auxiliary branch and contrastive learning loss. Baseline2 adds the task-common branch, while Baseline3 and Baseline4 add auxiliary branch and contrastive learning loss, respectively.

It can be seen from the first two columns that MCC and AUPRC increase from (0.5379, 0.3216) to (0.5590, 0.3859) for invasion task and from (0.4934, 0.3728) to (0.5157, 0.3935) for meningioma task by adding TC, which proves the rationality of the task-common branch. On this basis, L_{con} is added and improves MCC and AUPRC to (0.6097,0.4026) for invasion task. What's more, respecting the Baseline3, the proposed method adds L_{aux} to further improve MCC and AUPRC to (0.6252,0.4157) for invasion task and (0.5860, 0.4599) for meningioma task, which verifies that the auxiliary branch can help the contrastive learning strategy to play a better role. In comparison, the sensitivity of meningioma grade is somehow lower, for which we would like to further explore in the future work.

4   Conclusion

Joint prediction of brain invasion and meningioma grade is a novel topic and an urgent clinical need. As far as we know, there is no study so far that solves both prediction tasks based on brain MRIs simultaneously. In this paper, we propose a novel contrastive learning-based multi-task algorithm, which respects the coherence between tasks and also enhances the distinctness of each invididual task. We first use a middle-fusion strategy to fuse the feature maps from the multi-modal MRIs. Then, we disentangle the fused image features into task-specific features, which focus on a separate task, and task-common features that can perform both tasks. A new contrastive loss is leveraged to use task-common features as guidance to improve the prediction ability of two tasks. In the future work, we would like to explore the influence of each modality MRI on the two tasks to further improving the results of both tasks.

Acknowledgements. This work was supported by the grant from Tianjin Natural Science Foundation (Grant No. 20JCYBJC00960) and HKU Seed Fund for Basic Research (Project No. 202111159073).

References

1. Adeli, A., et al.: Prediction of brain invasion in patients with meningiomas using preoperative magnetic resonance imaging. Oncotarget **9**(89), 35974 (2018)
2. Behling, F., Hempel, J.M., Schittenhelm, J.: Brain invasion in meningioma-a prognostic potential worth exploring. Cancers **13**(13), 3259 (2021)

3. Brodersen, K.H., Ong, C.S., Stephan, K.E., Buhmann, J.M.: The balanced accuracy and its posterior distribution. In: 2010 20th International Conference on Pattern Recognition, pp. 3121–3124. IEEE (2010)

4. Brokinkel, B., Hess, K., Mawrin, C.: Brain invasion in meningiomas-clinical considerations and impact of neuropathological evaluation: a systematic review. Neuro Oncol. **19**(10), 1298–1307 (2017)

5. Champeaux, C., Houston, D., Dunn, L., meningioma, A.: A study on recurrence and disease-specific survival. Neurochirurgie **63**, 272–281 (2017)

6. Chartsias, A., et al.: Disentangled representation learning in cardiac image analysis. Med. Image Anal. **58**, 101535 (2019)

7. Chartsias, A., et al.: Disentangle, align and fuse for multimodal and semi-supervised image segmentation. IEEE Trans. Med. Imaging **40**(3), 781–792 (2020)

8. Chen, Z., Badrinarayanan, V., Lee, C.Y., Rabinovich, A.: Gradnorm: gradient normalization for adaptive loss balancing in deep multitask networks. In: International Conference on Machine Learning, pp. 794–803. PMLR (2018)

9. Doersch, C., Zisserman, A.: Multi-task supervised visual learning. In: Proceedings of the IEEE/CVF International Conference on Computer Vision (2017)

10. Hale, A.T., Stonko, D.P., Wang, L., Strother, M.K., Chambless, L.B.: Machine learning analyses can differentiate meningioma grade by features on magnetic resonance imaging. Neurosurg. Focus **45**(5), E4 (2018)

11. Han, Y., Wang, T., Wu, P., Zhang, H., Chen, H., Yang, C.: Meningiomas: preoperative predictive histopathological grading based on radiomics of MRI. Magn. Reson. Imaging **77**, 36–43 (2021)

12. Henaff, O.: Data-efficient image recognition with contrastive predictive coding. In: International Conference on Machine Learning, pp. 4182–4192. PMLR (2020)

13. Huang, R.Y., et al.: Imaging and diagnostic advances for intracranial meningiomas. Neuro-oncology **21**(Supplement_1), i44–i61 (2019)

14. Joo, L., et al.: Extensive peritumoral edema and brain-to-tumor interface MRI features enable prediction of brain invasion in meningioma: development and validation. Neuro Oncol. **23**(2), 324–333 (2021)

15. Kandemirli, S.G., et al.: Presurgical detection of brain invasion status in meningiomas based on first-order histogram based texture analysis of contrast enhanced imaging. Clin. Neurol. Neurosurg. **198**, 106205 (2020)

16. Kendall, A., Gal, Y., Cipolla, R.: Multi-task learning using uncertainty to weigh losses for scene geometry and semantics. In: Proceedings of the IEEE Conference on Computer Vision and Pattern Recognition, pp. 7482–7491 (2018)

17. Li, N., et al.: A clinical semantic and radiomics nomogram for predicting brain invasion in who grade ii meningioma based on tumor and tumor-to-brain interface features. Front. Oncol. 4362 (2021)

18. Louis, D.N., et al.: The 2007 who classification of Tumours of the central nervous system. Acta Neuropathol. **114**(2), 97–109 (2007)

19. Louis, D.N., et al.: The 2016 world health organization classification of tumors of the central nervous system: a summary. Acta Neuropathol. **131**(6), 803–820 (2016)

20. Ma, J., Zhao, Z., Yi, X., Chen, J., Hong, L., Chi, E.H.: Modeling task relationships in multi-task learning with multi-gate mixture-of-experts. In: Proceedings of the 24th ACM SIGKDD International Conference on Knowledge Discovery & Data Mining, pp. 1930–1939 (2018)

21. Nowosielski, M., et al.: Diagnostic challenges in meningioma. Neuro Oncol. **19**(12), 1588–1598 (2017)

22. Ostrom, Q.T., Patil, N., Cioffi, G., Waite, K., Kruchko, C., Barnholtz-Sloan, J.S.: Cbtrus statistical report: Primary brain and other central nervous system tumors diagnosed in the united states in 2013–2017. Neuro Oncol. **22**, iv1-iv96 (2020)
23. Park, Y.W., et al.: Radiomics and machine learning may accurately predict the grade and histological subtype in meningiomas using conventional and diffusion tensor imaging. Eur. Radiol. **29**(8), 4068–4076 (2018). https://doi.org/10.1007/s00330-018-5830-3
24. Pesce LL, Papaioannu J, M.C.: Roc-kit software (2004). http://radiology.uchicago.edu/?q=MetzROCsoftware
25. Wang, W., Zhou, T., Yu, F., Dai, J., Konukoglu, E., Van Gool, L.: Exploring cross-image pixel contrast for semantic segmentation. In: Proceedings of the IEEE/CVF International Conference on Computer Vision, pp. 7303–7313 (2021)
26. Xie, J., Zhan, X., Liu, Z., Ong, Y.S., Loy, C.C.: Delving into inter-image invariance for unsupervised visual representations. arXiv preprint arXiv:2008.11702 (2020)
27. Xie, Z., Lin, Y., Zhang, Z., Cao, Y., Lin, S., Hu, H.: Propagate yourself: exploring pixel-level consistency for unsupervised visual representation learning. In: Proceedings of the IEEE/CVF Conference on Computer Vision and Pattern Recognition, pp. 16684–16693 (2021)
28. Zhang, H., et al.: Deep learning model for the automated detection and histopathological prediction of meningioma. Neuroinformatics **19**(3), 393–402 (2021)
29. Zhang, J., et al.: A radiomics model for preoperative prediction of brain invasion in meningioma non-invasively based on MRI: a multicentre study. EBioMedicine **58**, 102933 (2020)
30. Zhang, Y., et al.: Modality-aware mutual learning for multi-modal medical image segmentation. In: de Bruijne, M., et al. (eds.) MICCAI 2021. LNCS, vol. 12901, pp. 589–599. Springer, Cham (2021). https://doi.org/10.1007/978-3-030-87193-2_56
31. Zhu, Y., et al.: A deep learning radiomics model for preoperative grading in meningioma. Eur. J. Radiol. **116**, 128–134 (2019)

Residual Wavelon Convolutional Networks for Characterization of Disease Response on MRI

Amir Reza Sadri, Thomas DeSilvio, Prathyush Chirra, Sneha Singh, and Satish E. Viswanath[✉]

Case Western Reserve University, Cleveland, OH 44106, USA
{sev21,satish.viswanath}@case.edu

Abstract. Wavelets have shown significant promise for medical image decomposition and artifact pre-processing by representing inputs via shifted and scaled components of a specified mother wavelet function. However, wavelets could also be leveraged within deep neural networks as activation functions for neurons (called *wavelons*) in the hidden layer. Integrating wavelons into a convolutional neural network architecture (termed a "wavelon network" (WN)) offers additional flexibility and stability during optimization, but the resulting model complexity has caused it to be limited to low-dimensional applications. Towards addressing these issues, we present the Residual Wavelon Convolutional Network (RWCN), a novel integrated WN architecture that employs weighted skip connections (to enable residual learning) together with image convolutions and wavelet activation functions to more efficiently capture high-dimensional disease response-specific patterns from medical imaging data. In addition to developing the analytical basis for wavelet activation functions as used in this work, we implemented RWCNs by adapting the popular VGG and ResNet architectures. Evaluation was conducted within three different challenging clinical problems: (a) predicting pathologic complete response (pCR) to neoadjuvant chemoradiation via 153 pre-treatment T2-weighted (T2w) MRI scans in rectal cancers, (b) evaluating pCR after chemoradiation via 100 post-treatment T2w MRIs in rectal cancers, as well as (c) risk stratifying patients who will or will not require surgery after aggressive medication in Crohn's disease using 73 baseline MRI scans. In comparison to 4 state-of-the-art alternative models (VGG-16, VGG-19, ResNet-18, ResNet-50), RWCN architectures yielded significantly improved and more efficient classifier performance on unseen data in multi-institutional validation cohorts (hold-out accuracies of 0.82, 0.85, and 0.88, respectively).

Keywords: Wavelet network · Residual learning · Convolutional neural network · Deep learning · Disease response

Supplementary Information The online version contains supplementary material available at https://doi.org/10.1007/978-3-031-16437-8_35.

© The Author(s), under exclusive license to Springer Nature Switzerland AG 2022
L. Wang et al. (Eds.): MICCAI 2022, LNCS 13433, pp. 366–375, 2022.
https://doi.org/10.1007/978-3-031-16437-8_35

1 Introduction

Wavelets have been widely used for capturing multi-scale, multi-oriented responses [1] from medical imaging data towards denoising [2], image super resolution [3], as well as within radiomics models [4]. More recently, using wavelets as activation units for neurons in the hidden layer of deep neural networks [5] have offered the ability for comprehensive image representation via dilation and orientation of a mother wavelet function. However, issues related to random wavelet initialization within "wavelet neuron" or *wavelon* networks (which could result in zero-valued responses) worsen the complexity of such models, potentially limiting their usage to low-dimensional problems [1]. This limitation gets magnified when wavelets are integrated with convolutional neural networks (CNNs) resulting in vanishing/exploding gradients [6] or saturated performance [7].

While batch normalization or regularization [6] are often invoked to account for these issues, the recent incorporation of residual learning [6] (i.e. "skip connections" that bypass activations from one layer to the next) has enabled construction of more powerful and efficiently optimizable CNN architectures, such as ResNet [8], which have seen significant use in medical imaging applications [9]. Integrating residual learning within wavelon building blocks could allow for WNs to be better integrated with CNNs; the hypothesis being that skip connections will allow convolutional responses to be learned more effectively and efficiently during backpropagation while wavelet activation functions with trainable parameters could be used to better optimize network training.

2 Previous Work and Novel Contributions

Initial efforts to incorporate wavelets within CNNs include ScatNet [10], where a set of robust features were extracted by cascading wavelet transform convolutions and average-pooling, as well as its extensions that leveraged Rectified Linear Unit (ReLu) activation [11]. Similarly, WaveCNet [7] utilized wavelets as down-sampling or up-sampling operators within CNNs. Here wavelets were utilized as pre-defined "action units" alone (i.e. feature extraction or image decomposition) without learnable parameters. Alternately, leveraging wavelet functions as activation units within CNNs [5,12] have demonstrated better flexibility and stability over linear activation functions (e.g. ReLu) [13]. Unfortunately, CNNs with wavelet activation functions also tend to be significantly more complex, making them difficult to train and optimize [5]. While residual learning can help alleviate model complexity (popularly leveraged in ResNet [8]), WNs have only been implemented with direct internal connections from the input to the output thus far [14]. To our knowledge, no previous work has specifically examined the properties of residual skip connections within WNs in conjunction with CNNs.

In this work, we introduce a novel residual wavelon convolutional network (RWCN) that combines internal skip connections within WN blocks to enable effective use of wavelet functions as activation units for convolutional responses. We also outline the theoretical underpinnings and practical advantages of wavelet activation functions within to our RWCN formulation. In total, five different types of trainable parameters make up the final RWCN which makes the network uniquely capable of extracting more complicated patterns from medical images that may be relevant to distinguishing challenging disease response phenotypes. We implemented RWCNs as adaptations of the popular VGG [15] and ResNet [8] architectures (termed RWCN-VGG-6 and RWCN-ResNet-15) and evaluated them in the context of three clinical problems with multi-institutional validation ($n > 320$ scans total): (i) predicting pathologic complete responders in rectal cancers via pre-treatment MRI, (ii) evaluating pathologic response after chemoradiation via post-treatment MRI, and (iii) identifying high-risk patients who need earlier surgery after aggressive medication for Crohn's disease via MRI.

3   Methodology

Wavelon Network: A WN usually has the form of a three-layer network. In the input layer, the explanatory variable $\mathbf{x} = [x_1, \ldots, x_n]^T \in \mathbb{R}^n$ is inserted into the network. In the hidden layer the input variable is transformed to a dilated and translated version of a mother wavelet ψ, resulting in $\Omega = \left\{ \det(\mathbf{D}_i^{\frac{1}{2}})\psi(\mathbf{D}_i(\mathbf{x} - \mathbf{t}_i)) : \mathbf{t}_i \in \mathbb{R}^n, \mathbf{D}_i = \mathrm{diag}(\mathbf{d}_i), \mathbf{d}_i \in \mathbb{R}_+^n, i \in \mathbb{Z} \right\}$, where $\mathbf{t}_i$ are translation vectors, and $\mathbf{d}_i$ are dilation vectors specifying the diagonal dilation matrices $\mathbf{D}_i$. The hidden layer consists of *wavelons*, (analogous to neurons), which are combined to produce the output of the WN. Figure 1(a) illustrates the topological structure of a $(n \times q \times c)$ WN with n inputs, q wavelons as hidden layers, and c outputs. The pth output of the WN can be obtained using a scaled and shifted version of input $\mathbf{x}$ within: $y_p = \sum_{i=1}^q w_{ip}\psi(\mathbf{D}_i(\mathbf{x} - \mathbf{t}_i)) + \mathbf{v}_p^T\mathbf{x} + b_p$, where w_{ip} are the weights between the hidden layer and output layer and b_p are the bias of the wavelons. $\mathbf{v}_p = [v_{1p}, \ldots, v_{np}]^T$ corresponds to weights for the skip connections. After selecting an appropriate mother wavelet, the network has five different parameters (weights $\mathbf{w}_p = [w_{1p}, \ldots, w_{qp}]^T$ and $\mathbf{v}_p$, dilations $\mathbf{d}_i$, translations $\mathbf{t}_i$, and biases b_p) which can be optimized during learning.

Advantages of Wavelet Activation Functions: Nonlinearity of activation functions [16] is a key aspect to model performance, where the latter can be classified as *homeomorphism* or *non-homeomorphism* transforms [17] (see Definition 1 in Supplementary Materials). As activation functions derived from non-homeomorphic maps induce rapid topology changes, associated models often generalize better across a variety of datasets, in addition to converging faster with improved classification performance [17]. Lemma 1 in the Supplementary Materials shows that wavelet activation functions create a non-homeomorphism map, thus offering wider modulation and non-monotonicity. As the wavelet's energy is bounded (shown in Definition 2 of the Supplementary Materials), model stability is improved when wavelet activation functions are implemented in conjunction with high-dimensional CNNs [1]. Wavelets also allow the activation unit to be more flexible by including trainable shift and scale parameters.

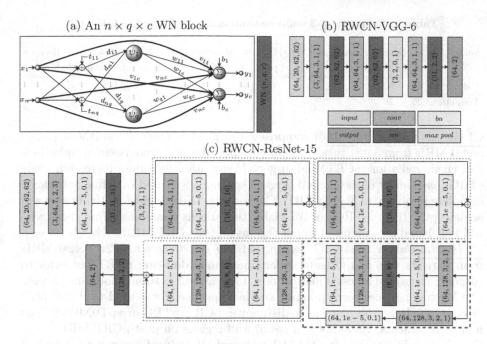

Fig. 1. (a) Three layered WN structure with n inputs, q wavelons, and c outputs. Learnable parameters including wavelets' shift (t_{ij}), scale (d_{ij}), wavelon weights (w_{ik}), bias (b_k) and weighting of skip connections (v_{jk}) (where $i = \{1, \ldots, q\}$, $j = \{1, \ldots, n\}$, and $k = \{1, \ldots, c\}$) are highlighted with different colors. (b) RWCN-VGG-6 with 3 convolution layers, 1 max pooling, and 3 WNs. (c) RWCN-ResNet-15 with 10 convolution layers, 10 batch normalization, 1 max pooling, and 6 WNs. Note WN blocks in RWCN-VGG-6 and RWCN-ResNet-15 include internal skip connections, shown in (a). (Color figure online)

Residual Wavelon Convolutional Network (RWCN) leverages a CNN architecture together with wavelet activation functions as well as skip connections within the wavelon blocks. As stated in [6,18], the use of skip connections helps preserve the gradient norm and leads to stable back-propagation via residual learning. Weighting skip connections within the WN structure results in an uninterrupted gradient flow from the network input layer to its output, thus tackling the vanishing gradient problem. By building a deep wavelon network with internal skip connections in a sequential fashion, later layers can learn lesser semantic information derived from initial input layers [19]. This also mitigates information loss in the size reduction blocks (e.g. pooling layers).

4   Experimental Design

4.1   Data Description

All experiments were conducted using three different multi-institutional retrospectively accrued cohorts that were segregated into independent discovery (70% of each class) and validation (the remaining 30%) sets; summarized in Table 1.

Table 1. Summary of 3 multi-institutional cohorts used in this work

Split	C1 (Rectal cancer, pre-CRT)		C2 (Rectal cancer, post-CRT)		C3 (Crohn's disease)	
	pCR	Non-pCR	pCR	Non-pCR	Low-risk	High-risk
Discovery	45	62	24	41	30	22
Validation	20	26	11	24	13	8

C1 (Rectal Cancer, pre-CRT) comprised 153 *pre-treatment* T2-weighted (T2w) rectal MRIs from 3 institutions, from patients who later underwent standard-of-care chemoradiation (nCRT) and surgery. Histopathologic tumor regression grade (TRG) assessment of the excised surgical specimen was used to define pathologic complete response (pCR) to nCRT. The goal was to distinguish patients who will achieve pCR (i.e., ypTRG0 or 0% viable tumor cells remaining) from those who will not, based on annotated tumor regions on pre-nCRT MRI.

C2 (Rectal Cancer, post-CRT) comprised 100 *post-treatment* T2w rectal MRIs from patients who had already undergone standard-of-care nCRT but prior to surgery; as curated across 3 institutions. TRG as well as tumor-nodal-metastasis (TNM) staging from corresponding surgical specimens was used to define pCR after nCRT. Here, the goal was to distinguish pCR (ypTRG or ypT0N0M0) from non-pCR patients, via annotated rectal wall regions on post-nCRT MRI.

C3 (Crohn's Disease) comprised 73 T2w bowel MRI scans from patients who had been endoscopically confirmed with Crohn's disease. The goal was to distinguish high-risk patients who needed surgery within one year of MRI and initiation of agggressive immunosuppressive therapy, from low-risk patients (stable for up to 5 years in follow-up); using annotated terminal ileum regions on baseline MRIs.

Pre-processing and Annotation of MRIs: All MRI scans underwent trilinear resampling to ensure isotropic $(1 \times 1 \times 1\text{mm})$ voxel resolution. Bounding boxes were then estimated to span the annotated region on each MRI scan (after connected component analysis to omit regions smaller than 20 pixels), and resized to $62 \times 62 \times 20$ voxels based on average annotation size across all cohorts.

4.2 Model Implementation and Evaluation

Two different RWCN architectures were developed, as adaptations of the widely-used VGG [15] and ResNet [8] architectures, as follows:

- **RWCN-VGG-6** [Fig. 1(b)]: Comprised 6 blocks including learnable parameters, with two blocks performing a regular convolution (kernel size 3, stride 1, padding 1), followed by a $(62 \times 62 \times 62)$ WN, and a max pool layer (kernel size 2, stride 2, padding 0, and dilation 1). Two blocks containing the same convolution plus a $(31, 2, 2)$ WN were used in the last layer.
- **RWCN-ResNet-15** [Fig. 1(c)]: Comprised 15 blocks including learnable parameters. The first four blocks consisted of a regular convolution (kernel size 7, stride 2, padding 3), followed by a batch normalization (epsilon $1e-5$, momentum 0.1), a $(31 \times 31 \times 31)$ WN, and a max pool layer (kernel size 3, stride 2, padding 1, and dilation 1). The gain was fed into four sequential residual blocks where three are categorized as non-transition blocks (preserve dimen-

sionality, gray dotted outline) and one is a transition block (change dimensionality, magenta dashed outline). Each of these residual blocks encloses a sequence of a convolution, a batch normalization, a WN, a convolution, and a batch normalization, respectively. The residual blocks were integrated with an output $(128 \times 2 \times 2)$ WN for the final classification layer.

As alternative strategies, VGG-16, VGG-19, ResNet-18, and ResNet-50 were also implemented. For RWCNs, wavelet activations used the Mexican hat function. All model were implemented in 3D, with parameters optimized in the discovery cohort using ten-fold cross validation, with training over 150 epochs, a batch size of 64, a learning rate of $1e - 4$, a cross-entropy loss function, and the SGD optimization function (weight decay $= 1e - 2$). Implementation was done in Python 3.9 and PyTorch 1.0.0 in Google Colab on a desktop computer with Intel(R) Core(TM) i7 CPU 930 (3.60 GHz), 32 GB RAM, 64 bit Windows 10, and NVIDIA GeForce GT610 GPU with 4.0 GB memory.

Model Evaluation: The optimal weighting scheme for skip connections within the RWCN architectures was evaluated by implementing four variant RWCN-VGG-6 and four variant RWCN-ResNet-15 models; based on changing $\mathbf{v}_p$. Performance for each of the 6 models considered was evaluated in terms of accuracy as well as area under the ROC curve (AUC) in the discovery cohorts, with the mean and standard deviation computed over all cross-validation runs. The model with performance closest to the average was then evaluated in hold-out validation. Pairwise Wilcoxon testing with multiple comparison correction was used to assess significant differences in model performance. Shapley Additive Explanations (SHAP) [20] was implemented via the shap [21] library to assign a class relevance value to each image pixel based on a game-theoretic approach used to explain machine learning model outputs [22]. SHAP maps were visualized using a red-blue heatmap such that any regions shaded red indicate increased confidence of the model in predicting the target class, blue indicates relevance to the non-target class, and white implies no importance.

5   Experimental Results and Discussion

5.1   Experiment 1: Optimizing Skip Connection Weights in RWCN

Figure 2 (top panel) visualizes the loss functions for eight different RWCN models based on changing the skip connection weighting, for the discovery cohort in each of C1, C2, and C3. Here, cells with overall lighter shading correspond to faster convergence by the model to a lower loss function (which would be ideal). RWCN-VGG-6 and RWCN-ResNet-15 with identity skip connections ($\mathbf{v}_p = 1$) appear to exhibit the fastest learning phase, suggesting that enabling the skip connection allows faster and more accurate model convergence. By comparison, disabling the skip connection ($\mathbf{v}_p = 0$) has markedly poorer convergence while equal weighting ($\mathbf{w}_p = \mathbf{v}_p$) closely trails the performance of identity skip connections.

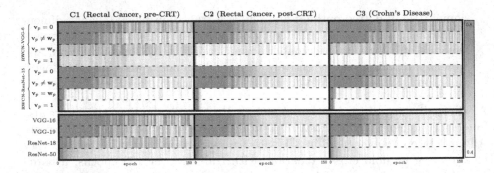

Fig. 2. Classification loss cellmap visualization for all models in all 3 cohorts. Top panel: 4 variants for each of RWCN-VGG-6 and RWCN-ResNet-15 models, based on with different weightings for skip connections. Within each cell, as the number of epochs increases (X-axis), cells are iteratively shaded such that the cell color changes from darker (learning phase) to lighter (tuning phase). Bottom panel: corresponding cellmap visualizations for four alternative CNN models.

5.2   Experiment 2: Comparison of RWCNs Against Alternatives

Table 2 summarizes classification accuracy in both discovery and hold-out validation for all six models considered, while Table S-1 in Supplementary Materials summarizes corresponding AUC performance. RWCN-ResNet-15 with identity skip connections was found to yield significantly higher model performance in both discovery and hold-out validation compared to any alternative RWCN-based model as well as all of VGG-16, VGG-19, ResNet-18, and ResNet-50. Notably the latter CNN models are seen to perform equivalently to each other, suggesting that any improvements in performance may be a result of adapting them to incorporate WN blocks with skip connections. Further, RWCNs using WN blocks without skip connections ($\mathbf{v}_p = \mathbf{0}$) perform significantly worse than CNN-based strategies, suggesting the importance of residual learning in these tasks. Notably, all RWCN models exhibit markedly lower runtimes and faster convergence than any CNN model, in addition to requiring significantly fewer parameters than ResNet and VGG architectures (see Table 2 and S-1).

Table 2. Model accuracy and number of parameters for distinguishing between patient groupings in each of C1, C2, and C3, with best model in bold. * indicates $p < 0.005$ in pairwise Wilcoxon ranksum testing between top-ranked RWCN model and alternatives.

Approach	Parameters	C1 (Rectal Cancer, pre-CRT)		C2 (Rectal Cancer, post-CRT)		C3 (Crohn's Disease)	
		Accuracy		Accuracy		Accuracy	
		Discovery	Validation	Discovery	Validation	Discovery	Validation
RWCN-VGG-6 ($\mathbf{v}_p = 0$)	842,562	0.74 ± 0.15	0.72	0.76 ± 0.12	0.75	0.71 ± 0.12	0.71
RWCN-VGG-6 ($\mathbf{v}_p = 1$)	**781,388**	0.88 ± 0.08	0.78	0.89 ± 0.06	0.80	0.86 ± 0.08	0.81
RWCN-ResNet-15 ($\mathbf{v}_p = 0$)	967,314	0.75 ± 0.06	0.72	0.77 ± 0.08	0.77	0.74 ± 0.09	0.73
RWCN-ResNet-15 ($\mathbf{v}_p = 1$)	1,848,966	$\mathbf{0.94 \pm 0.02^*}$	**0.82**	$\mathbf{0.95 \pm 0.02^*}$	**0.85**	$\mathbf{0.92 \pm 0.03^*}$	**0.88**
VGG-16	138,357,544	0.86 ± 0.06	0.73	0.87 ± 0.06	0.79	0.85 ± 0.06	0.83
VGG-19	143,667,240	0.87 ± 0.08	0.75	0.90 ± 0.06	0.80	0.86 ± 0.07	0.84
ResNet-18	11,178,580	0.84 ± 0.09	0.75	0.90 ± 0.03	0.81	0.87 ± 0.05	0.85
ResNet-50	23,512,130	0.88 ± 0.04	0.78	0.91 ± 0.04	0.81	0.89 ± 0.04	0.86

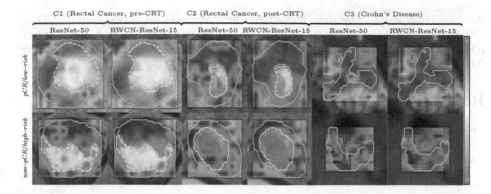

Fig. 3. SHAP importance maps for best RWCN model (RWCN-ResNet-15) vs best CNN model (ResNet-50), for representative pCR/low-risk (top row) and non-pCR/high-risk (bottom row) patients, respectively. Input regions outlined in green in each image: (a) tumor for C1, (b) rectal wall for C2, (c) terminal ileum for C3. In-plane rectal wall (pink) and lumen (yellow) were excluded from model input in C1 and C2. (Color figure online)

Figure 3 depicts SHAP maps overlaid on original MR images for the best-performing RWCN-ResNet-15 model as well as the best-performing CNN model (ResNet-50) for representative patients from each of C1, C2, and C3. RWCN-ResNet-15 appears to have correctly and more homogeneously assigned target-class relevance to annotated input regions in non-pCR/high-risk patients (more deep red shading within green outlines in the bottom row) in comparison to pCR/low-risk (top row shows more consistent blue shading within green outlines). However, ResNet-50 assigns mixed importance to annotated regions (lighter shading, mixture of red and blue within all outlines); suggesting lower model confidence as well as suboptimal learning of class-specific responses.

6   Concluding Remarks

In this work, we introduced a new residual wavelon network (RWCN) which is the first attempt to integrate wavelon networks with internal skip connections into a CNN architecture. RWCNs yielded significantly higher performance across independent discovery and hold-out validation cohorts within three challenging clinical problems involving characterization of disease response in rectal cancers and Crohn's disease via MRI. Parameter sensitivity, analytical evaluation, and interpretability mapping revealed the importance of both residual skip connections as well as WN blocks within RWCNs for capturing disease response-specific patterns. Future work will involve validation of our novel architecture in other disease domains, examining other mother wavelet functions, as well as more comprehensively evaluating model generalizability.

Acknowledgments. Research supported by NCI (1U01CA248226-01), DOD/ CDMRP (W81XWH-21-1-0345), and NIDDK (1F31DK130587-01A1). Content solely responsibility of the authors and does not necessarily represent the official views of the NIH, DOD, or the United States Government.

References

1. Billings, S.A.: Nonlinear System Identification: NARMAX Methods in the Time, Frequency, and Spatio-Temporal Domains. Wiley, Hoboken (2013)
2. Gu, J., Yang, T.S., Ye, J.C., Yang, D.H.: CycleGAN denoising of extreme low-dose cardiac CT using wavelet-assisted noise disentanglement. Med. Image Anal. **74**, 102209 (2021)
3. Chen, L., et al.: Super-resolved enhancing and edge deghosting (SEED) for spatiotemporally encoded single-shot MRI. Med. Image Anal. **23**(1), 1–14 (2015)
4. Lao, J., et al.: A deep learning-based radiomics model for prediction of survival in glioblastoma multiforme. Sci. Rep. **7**(1), 1–8 (2017)
5. Liu, J., Li, P., Tang, X., Li, J., Chen, J.: Research on improved convolutional wavelet neural network. Sci. Rep. **11**(1), 1–14 (2021)
6. Zaeemzadeh, A., Rahnavard, N., Shah, M.: Norm-preservation: why residual networks can become extremely deep? IEEE Trans. Pattern Anal. Mach. Intell. **43**(11), 3980–3990 (2020)
7. Li, Q., Shen, L., Guo, S., Lai, Z.: Wavecnet: wavelet integrated CNNs to suppress aliasing effect for noise-robust image classification. IEEE Trans. Image Process. **30**, 7074–7089 (2021)
8. He, K., Zhang, X., Ren, S., Sun, J.: Deep residual learning for image recognition. In: Proceedings of the IEEE Conference on Computer Vision and Pattern Recognition, pp. 770–778 (2016)
9. Xie, H., et al.: Cross-attention multi-branch network for fundus diseases classification using SLO images. Med. Image Anal. **71**, 102031 (2021)
10. Bruna, J., Mallat, S.: Invariant scattering convolution networks. IEEE Trans. Pattern Anal. Mach. Intell. **35**(8), 1872–1886 (2013)
11. Wiatowski, T., Bölcskei, H.: A mathematical theory of deep convolutional neural networks for feature extraction. IEEE Trans. Inf. Theor. **64**(3), 1845–1866 (2017)
12. Rodriguez, M.X.B., et al.: Deep adaptive wavelet network. In: Proceedings of the IEEE/CVF Winter Conference on Applications of Computer Vision, pp. 3111–3119 (2020)
13. Gu, J., et al.: Recent advances in convolutional neural networks. Pattern Recogn. **77**, 354–377 (2018)
14. Alexandridis, A.K., Zapranis, A.D.: Wavelet Neural Networks: with Applications in Financial Engineering, Chaos, and Classification. Wiley, Hoboken (2014)
15. Simonyan, K., Zisserman, A.: Very deep convolutional networks for large-scale image recognition. arXiv preprint arXiv:1409.1556 (2014)
16. Biswas, K., Kumar, S., Banerjee, S., Pandey, A.K.: Tanhsoft-dynamic trainable activation functions for faster learning and better performance. IEEE Access **9**, 120613–120623 (2021)
17. Naitzat, G., Zhitnikov, A., Lim, L.H.: Topology of deep neural networks. J. Mach. Learn. Res. **21**(184), 1–40 (2020)
18. Oyedotun, O.K., Al Ismaeil, K., Aouada, D.: Why is everyone training very deep neural network with skip connections? IEEE Trans. Neural Netw. Learn. Syst. (2022)

19. Furusho, Y., Ikeda, K.: Theoretical analysis of skip connections and batch normalization from generalization and optimization perspectives. APSIPA Trans. Sig. Inf. Process. **9** (2020)
20. Lundberg, S.M., et al.: From local explanations to global understanding with explainable AI for trees. Nature Mach. Intell. **2**(1), 56–67 (2020)
21. Lundberg, S.M., et al.: Explainable machine-learning predictions for the prevention of hypoxaemia during surgery. Nature Biomed. Eng. **2**(10), 749 (2018)
22. Ancona, M., Oztireli, C., Gross, M.: Explaining deep neural networks with a polynomial time algorithm for Shapley value approximation. In: International Conference on Machine Learning, pp. 272–281. PMLR (2019)

Self-Ensembling Vision Transformer (SEViT) for Robust Medical Image Classification

Faris Almalik⑩, Mohammad Yaqub⑩, and Karthik Nandakumar$^{(\boxtimes)}$⑩

Mohamed Bin Zayed University of Artificial Intelligence, Abu Dhabi, UAE
{faris.almalik,mohammad.yaqub,karthik.nandakumar}@mbzuai.ac.ae

Abstract. Vision Transformers (ViT) are competing to replace Convolutional Neural Networks (CNN) for various computer vision tasks in medical imaging such as classification and segmentation. While the vulnerability of CNNs to adversarial attacks is a well-known problem, recent works have shown that ViTs are also susceptible to such attacks and suffer significant performance degradation under attack. The vulnerability of ViTs to carefully engineered adversarial samples raises serious concerns about their safety in clinical settings. In this paper, we propose a novel self-ensembling method to enhance the robustness of ViT in the presence of adversarial attacks. The proposed Self-Ensembling Vision Transformer (SEViT) leverages the fact that feature representations learned by initial blocks of a ViT are relatively unaffected by adversarial perturbations. Learning multiple classifiers based on these intermediate feature representations and combining these predictions with that of the final ViT classifier can provide robustness against adversarial attacks. Measuring the consistency between the various predictions can also help detect adversarial samples. Experiments on two modalities (chest X-ray and fundoscopy) demonstrate the efficacy of SEViT architecture to defend against various adversarial attacks in the gray-box (attacker has full knowledge of the target model, but not the defense mechanism) setting. Code: https://github.com/faresmalik/SEViT

Keywords: Adversarial attack · Vision transformer · Self-ensemble

1 Introduction

Convolutional Neural Networks (CNNs) have been the de facto models for medical image analysis tasks such as segmentation [30], landmark localization [24], and classification [15]. However, it is well-known that CNNs trained on natural or medical images are vulnerable to adversarial attacks [10,13], which add imperceptible perturbations to input images to deliberately mislead a target model. The work in [8] has identified two weak links in the healthcare economy that are susceptible to adversarial attacks. Firstly, automated systems deployed by insurance companies to process reimbursement claims can be fooled using

© The Author(s), under exclusive license to Springer Nature Switzerland AG 2022
L. Wang et al. (Eds.): MICCAI 2022, LNCS 13433, pp. 376–386, 2022.
https://doi.org/10.1007/978-3-031-16437-8_36

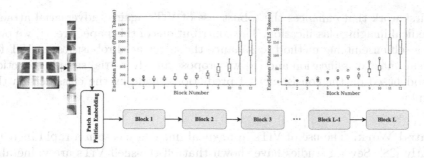

Fig. 1. Euclidean distance between class and patch tokens of adversarial images and the corresponding clean version. The distance increases moving towards the final blocks.

adversarial samples to trigger specific diagnostic codes and obtain higher payouts. Secondly, automated systems deployed by regulators to confirm results of clinical trials can be circumvented by malicious manufacturers, who can employ adversarial test samples to pass clinical trials successfully without being noticed. In addition, the rapid growth of telemedicine during the COVID-19 pandemic and the emergence of "as-a-service" business models based on cloud computing (e.g., radiology-as-a-service) have created an environment where medical images will be increasingly processed remotely. Often, automated machine learning algorithms will perform the diagnosis, which human experts may optionally verify. However, such medical imaging scenarios will be highly vulnerable to adversarial attacks. Thus, a robust defensive strategy must be devised before automated medical imaging systems can be securely deployed.

Recently, self-attention-based neural network architectures such as Vision Transformers (ViT) [6] have proven to be successful in image classification. Consequently, ViTs are competing to replace CNNs in various tasks related to medical images [3,32,33]. Despite the superior performance of ViTs, recent works have confirmed that they are also susceptible to malicious perturbations [1,18]. ViTs typically partition the input image into multiple patches and learn feature representations (called *patch tokens*) for each patch as they pass through multiple Transformer blocks. In addition, each block distills information from the patch tokens into a global representation (called *class token*). The class token output by the final Transformer block is often used for classification.

Adversarial attacks can be expected to push the feature representations of the perturbed samples away from that of the clean samples. However, a careful analysis of ViTs indicates that the impact of perturbations is more pronounced in the final set of blocks and the patch tokens learned by the initial blocks remain relatively unaffected (see Fig. 1). This phenomenon raises two main questions: (i) are the features learned by initial blocks useful for classification? and (ii) can we use these intermediate features to enhance the robustness of ViTs?

We propose a novel method that utilizes the patch tokens learned by initial blocks in ViTs along with the final class token to enhance the classification robustness and detect adversarial samples. To the best of our knowledge, this is the first work that enhances the robustness of ViTs against adversarial attacks in medical imaging classification. The contributions of this paper are: (i) we propose a self-ensembling method to enhance the adversarial robustness of ViT for medical image classification and (ii) we propose an adversarial sample detection method based on the consistency of predictions made by the classifiers in the ensemble.

Related Work. The use of ViTs in medical imaging has seen a rapid increase recently [28]. Several studies have shown that off-the-shelf ViTs are vulnerable to adversarial perturbations [10,14,17]. However, they were found to be more robust against adversarial attacks compared to CNNs [1,21,29]. Low transferability of adversarial samples between CNNs and ViTs has also been observed [18,22]. While these initial results were promising, recent works have demonstrated that it is indeed possible to fool ViTs using adversarial samples. For example, the Patch-Fool attack proposed in [9] targets the self-attention mechanisms in ViTs making them weaker learners compared to CNNs. Similarly, an ensemble (of CNN and ViT) defense approach was found to be ineffective against white-box adversaries [18]. Furthermore, Nasser et al. [22] show that it is possible to generate more powerful adversarial attacks against ViTs by targeting an ensemble of intermediate representations in addition to the final class token.

Many defense mechanisms have been proposed to improve the robustness of machine learning models against adversarial attacks. These include adversarial training [10,17,31], defensive distillation [23] and adversarial purification [27]. Algorithms have also been proposed to detect and filter out adversarial samples [7,16,19]. However, almost all these defense mechanisms have been designed for CNNs and there is limited work on enhancing the adversarial robustness of ViTs. The most notable approaches are PatchVeto [11] and Robust Self-Attention [20], both of which are designed to defend ViTs against adversarial patch attacks.

2   Proposed Method

Let $\mathbf{x}$ be a medical image with true label y provided as input to a ViT-based classifier $\mathbf{f}(\cdot)$. The goal of an adversarial attack is to craft a perturbed image $\mathbf{x}'$ such that $\mathbf{f}(\mathbf{x}') \neq y$ with high probability under the constraint that $\mathbf{x}'$ is *close* to $\mathbf{x}$. In this work, we use L_∞ norm as the distance metric and ensure that $||\mathbf{x} - \mathbf{x}'||_\infty \leq \epsilon$. The objective of our defense mechanism is to obtain a robust classifier $\tilde{\mathbf{f}}$ from $\mathbf{f}$ such that both the clean accuracy ($P(\tilde{\mathbf{f}}(\mathbf{x}) = y)$) and robust accuracy ($P(\tilde{\mathbf{f}}(\mathbf{x}') = y)$) are high. The aim of the detection mechanism is to discriminate between $\mathbf{x}$ and $\mathbf{x}'$ with high accuracy, especially when the attack is successful ($\mathbf{f}(\mathbf{x}') \neq y$).

Vanilla ViTs divide the input image $\mathbf{x}$ into N patches of size $P \times P$ each. Patches are then flattened into an one-dimensional (1D) vector and embedded into D dimensions via a linear layer. Since patch embeddings do not preserve the positional information, a 1D learnable position embedding is added to create N patch tokens ($\{pt_0^j\}_{j=1}^N$). Furthermore, a separate class token (ct_0) is added to

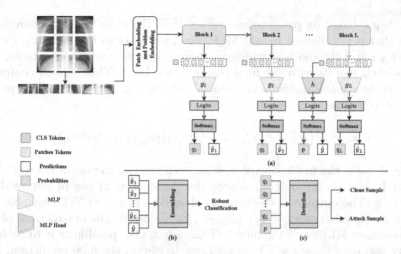

Fig. 2. The proposed SEViT framework extracts the patch tokens from the initial blocks and trains separate MLPs as shown in (a). (b) A self-ensemble of these MLPs with the final ViT classifier enhances the robustness of ViT. (c) Consistency between the predictions in the ensemble can be used to detect adversarial samples.

consolidate information from all the patch tokens. These patch and class tokens get refined as they pass through L Transformer blocks. Let $\{pt_i^j\}_{j=1}^N$ and ct_j be the patch and class tokens (respectively) output by the ith block, $i = 1, 2, \cdots, L$. The final class token ct_L output by the Lth block is passed through a multi-layer perceptron (MLP) classification head ($\mathbf{h}_\theta$) to obtain the final classification result, i.e., $\mathbf{f}(\mathbf{x}) = \mathbf{h}_\theta(ct_L)$. The MLP typically includes a final softmax layer to produce a probability distribution $\mathbf{p}$ over the possible classes.

2.1 Self-Ensembling for Robust Classification

Vanilla ViTs rely only on the final class token ct_L for image classification and ignores the rest of the learned features. As illustrated in Fig. 1, adversarial attacks have a significant impact on the patch and class tokens learned by the later blocks, while the corresponding tokens produced by the initial blocks remain relatively robust. Our main hypothesis is that the intermediate feature representations output by the initial blocks are useful for classification and harder to attack. Hence, we propose to add a MLP classifier at the end of each block. Each intermediate MLP classifier (denoted as $\mathbf{g}_{\beta_i}$, $i = 1, 2, \cdots, (L-1)$) utilizes the patch tokens of the corresponding block to produce a probability distribution $\mathbf{q}_i$ over the class labels. This results in a self-ensemble of L classifiers that can be fused to obtain the final classification result. We refer to this architecture as Self-Ensembling Vision Transformer (SEViT), which is illustrated in Fig. 2.

Unlike the final class token ct_L, we observed that the intermediate class tokens are not discriminative enough. In contrast, the intermediate patch tokens contained useful information. This is the reason for constructing the intermediate MLPs $\mathbf{g}_{\beta_i}$ based on patch tokens rather than the class token. Moreover, we observe that it is sufficient to add MLP classifiers only to the first m $(m < L)$ blocks (see Fig. 3 (b) and (c)), which are more robust to adversarial attacks, and combine their results with the final classification head. This reduces computational complexity and increases adversarial robustness. Thus, the SEViT model can be expressed as:

$$\tilde{\mathbf{f}}(\mathbf{x}) = \mathcal{F}\left(\mathbf{g}_{\beta_1}(\{pt_1^j\}_{j=1}^N), \cdots, \mathbf{g}_{\beta_m}(\{pt_m^j\}_{j=1}^N), \mathbf{h}_\theta(ct_L)\right) \tag{1}$$

where $\mathcal{F}$ denotes the fusion operator, θ denotes the parameters of the final ViT classifier $\mathbf{h}$, and β_1 through β_m denote the parameters of the m intermediate classifiers $\mathbf{g}$. The outputs of the $(m+1)$ classifiers in the SEViT ensemble can be fused in a number of ways. Let $\{\hat{y}_1, \cdots, \hat{y}_m, \hat{y}\}$ be the class predictions of the m intermediate MLPs and the final ViT classifier. One possibility is to perform majority voting of these $(m+1)$ predictions to obtain the final prediction. The SEViT ensemble could also be formed by randomly choosing only c out of the initial m intermediate classifiers, $c \in \{1, \ldots, m-1\}$ and decisions can be made based on the $(c+1)$ classifier ensemble (including the final ViT classifier). This random selection approach can be expected to be more robust against white-box adversaries, who have full knowledge of the ViT model and intermediate MLPs.

2.2 Adversarial Sample Detection

As noted earlier, adversarial attacks are expected to adversely affect the predictions of the final ViT classifier, whereas the intermediate classifiers at the initial blocks are relatively unaffected. Hence, the predictions of the intermediate classifiers can be expected to be different from that of the final ViT classifier, when an adversarial sample is presented. In contrast, the predictions of all classifiers in SEViT are expected to be in agreement for a clean sample. This phenomenon can be leveraged to detect an adversarial sample in the following way. Let $\mathbf{A}$ be a $((m+1) \times (m+1))$ matrix containing the Kullback-Leibler divergence (D_{KL}) between the probability distributions output by the $(m+1)$ classifiers in SEViT. Clearly, the diagonal elements of this matrix will always be zero. For a clean sample, all the non-diagonal entries in this matrix are also expected to be close to zero. On the other hand, at least some of the non-diagonal entries in this matrix (especially elements involving the final ViT classifier) are expected to be large for the adversarial sample. Hence, we compute Frobenius norm of the KL-matrix as follows:

$$\|\mathbf{A}\|_F = \sqrt{\sum_{i=1}^{m+1}\sum_{j=1}^{m+1}|a_{i,j}|^2}, \tag{2}$$

$a_{i,j} = D_{KL}(\mathbf{q}_i, \mathbf{q}_j)$, $\forall$ $i, j = 1, 2, \cdots, m$ and $a_{i,m+1} = D_{KL}(\mathbf{q}_i, \mathbf{p})$, $\forall$ $i = 1, 2, \cdots, m$, and $a_{m+1,m+1} = 0$. The Frobenius norm of the KL-matrix is compared to a threshold τ and the input is detected as adversarial if $\|\mathbf{A}\|_F > \tau$.

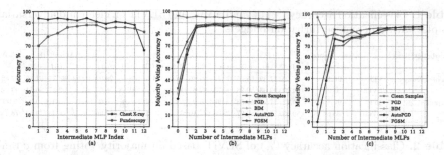

Fig. 3. Accuracy (%) of (a) intermediate MLPs on clean test set for both datasets. (b) majority voting from different number of MLPs for chest X-ray dataset. (c) majority voting from different number of MLPs for Fundoscopy dataset

3 Result and Discussion

Datasets. We conduct our experiments on two medical imaging datasets. The first dataset [25] consists of 7,000 chest X-ray images and the classification task is binary, Normal or Tuberculosis. We randomly split 80% of the dataset for training, 10% for validation, and 10% for testing. The second dataset is APTOS2019 [12] for diabetic retinopathy (DR). It has 5 classes and 3,662 retina images. We again convert it to a binary classification task by labeling the images as DR or Normal and randomly select 80% of the images for training and the rest for testing. We refer to this dataset as Fundoscopy.

Vision Transformer and MLPs. For ViT, we choose the ViT-B/16 model pretrained on ImageNet [5]. For the intermediate classifiers, we design a 4-layer MLP, which takes the patch tokens as input. We train 12 separate MLPs, one after each block. Adam optimizer was used with 10^{-3} as the initial learning rate and a decay of 0.1 every 10 epochs. Different augmentations including random rotation, color jitter, random horizontal and vertical translation are applied during ViT fine-tuning. ViT achieved 96.38% and 97.64% accuracy on the original clean test set for chest X-ray and Fundoscopy, respectively. The accuracy of intermediate classifiers on the original clean test set is depicted in Fig. 3 (a). We conduct our experiments using one Nvidia RTX 6000 GPU with 24 GB memory.

Adversarial Attacks. We use the Foolbox library [26] to generate various L_∞ attacks such as FGSM [10], BIM [14] and PGD [17], and L_2 C&W [2] attacks. The attacks are generated on the original test set in a gray-box setting (adversary has full access to the fine-tuned ViT, but is not aware of the intermediate classifiers). For FGSM, PGD, BIM and AutoPGD [4] attacks, we generate samples with $\epsilon = 0.003, 0.01, 0.03$, whereas for C&W attacks we set the Lagrange multiplier to 2 and limit the number of steps to 4000. All other parameters are set to default values.

Table 1. Classification accuracy (%) of vanilla ViT and SEViT on clean and adversarial samples with different perturbation budgets.

Perturbation size ϵ		Clean Accuracy	FGSM			PGD			BIM			AutoPGD			C&W
		–	0.003	0.01	0.03	0.003	0.01	0.03	0.003	0.01	0.03	0.003	0.01	0.03	–
Chest X-ray	ViT	**96.38**	91.59	77.39	55.65	92.17	69.86	32.32	91.01	66.38	28.99	92.64	63.77	21.30	47.83
	SEViT	94.64	94.20	92.03	89.28	94.35	92.61	88.41	94.06	91.88	86.67	94.20	92.17	86.51	93.62
Fundoscopy	ViT	**97.64**	35.42	17.92	16.53	1.40	0.0	0.0	0.30	0.0	0.0	0.0	0.0	0.0	9.00
	SEViT	85.62	80.69	80.69	80.97	78.06	77.40	77.64	77.50	77.78	77.06	77.36	78.33	79.03	78.33

Table 2. Classification accuracy (%) of SEViT based on majority voting from c randomly selected intermediate classifiers along with the final ViT classifier. All reported results are obtained by averaging over 5 trials.

Number of MLPs		Chest X-ray				Fundoscopy			
		1	2	3	4	1	2	3	4
Clean Samples	–	**94.25**	**95.26**	**94.72**	**95.30**	**84.30**	**87.87**	**85.50**	**85.40**
FGSM	$\epsilon = 0.003$	91.62	94.29	93.62	94.84	69.89	78.42	78.33	81.31
	$\epsilon = 0.01$	83.54	91.36	91.40	92.32	57.14	81.11	80.75	87.97
	$\epsilon = 0.03$	72.84	88.06	88.10	89.28	55.06	83.81	83.67	81.92
PGD	$\epsilon = 0.003$	92.32	94.26	93.77	94.87	44.06	71.36	72.67	78.94
	$\epsilon = 0.01$	83.39	90.93	91.07	92.35	42.08	74.75	73.28	79.42
	$\epsilon = 0.03$	66.20	85.94	87.51	88.26	42.06	78.64	78.01	79.50
BIM	$\epsilon = 0.003$	91.51	93.97	93.48	94.58	42.83	70.94	72.25	78.78
	$\epsilon = 0.01$	81.07	90.46	90.87	91.86	41.94	75.72	75.11	79.31
	$\epsilon = 0.03$	62.35	84.49	85.16	86.75	41.72	79.63	79.01	79.50
AutoPGD	$\epsilon = 0.003$	92.55	94.46	93.83	94.89	42.43	70.83	72.11	78.64
	$\epsilon = 0.01$	81.10	90.49	90.75	92.14	42.06	76.50	76.31	80.11
	$\epsilon = 0.03$	60.1	84.06	85.28	87.30	42.06	79.89	76.20	79.70
C&W	–	69.33	92.26	93.70	93.68	48.53	74.89	76.69	80.97

3.1 Results on Adversarial Robustness of SEViT

We compare the clean and robust accuracy of vanilla ViT and SEViT with majority voting based on $m = 5$ intermediate classifiers. Quantitative results in Table 1 show that vanilla ViTs are highly susceptible to various adversarial attacks, especially for higher values of perturbation budget ϵ as can be observed from the drastic fall in robust accuracy (Rows 1 and 3 of Table 1). On the other hand, SEViT reduced the clean accuracy by 2% and 12% for chest X-ray and Fundoscopy datasets, respectively. However, the robust accuracy of SEViT is significantly higher compared to that of vanilla ViTs for both datasets. In fact, SEViT boosts the adversarial robustness and attains very high robust accuracy across all attacks and for various perturbation budgets (Rows 2 and 4 of Table 1).

For SEViT models with a random selection of c classifiers instead of choosing all the m intermediate classifiers, the accuracy for different values of c are reported in Table 2. The results show that SEViT has reasonable robust accu-

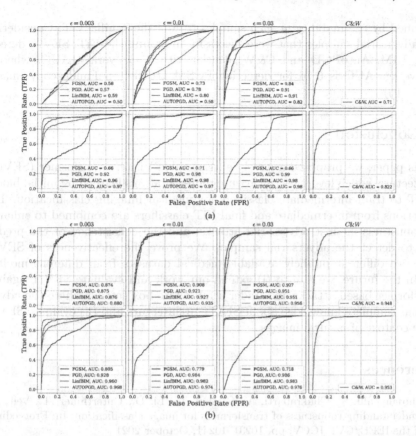

Fig. 4. ROC curves along with AUC values for (a) all adversarial samples (b) adversarial samples that succeeded in fooling vanilla ViT. First row in each sub graph corresponds to Chest X-ray dataset, the second row corresponds to Fundoscopy dataset.

racy even with $c = 1$. However, as c increases to 4, the robust accuracy is close to that of the full SEViT model with $m = 5$. These results validate our hypothesis that patch tokens from initial ViT blocks can be utilized along with the class token of the last ViT block to enhance the robustness of the model.

3.2 Results on Adversarial Sample Detection

We evaluate the ability of SEViT to detect adversarial samples, regardless of the success of these attacks in fooling the vanilla ViT. Figure 4 (a) show the ROC curves along with the Area Under ROC (AUC). SEViT detects attacks with larger perturbation budget more effectively for both datasets. However, the detection performance is low for smaller perturbation budget and for FGSM attacks on Fundoscopy images. This is mainly because attacks with smaller perturbation budget introduce only negligible changes to the input images. This is confirmed by the inability of such samples to fool the vanilla ViT. Considering only adversarial samples that were successful in fooling the ViT, SEViT detects PGD, BIM, AutoPGD and C&W attacks with 0.93 average AUC, whereas the average AUC is 0.76 for detecting FGSM attacks on Fundoscopy images (Fig. 4 (b)).

4   Conclusion

In this paper, we proposed a novel Self-Ensemble Vision Transformer (SEViT) architecture, which leverages the feature representations learned by initial blocks of ViT to train intermediate classifiers for medical imaging classification. The predictions from intermediate and final ViT classifiers are combined to enhance the robustness against adversarial attacks, and the consistency of the predictions to detect the adversarial samples. We prove the effectiveness of SEViT using two different publicly available medical datasets from different modalities. In the future, we aim to (i) extend our work and evaluate SEViT against Transformer-based attacks and under a full white-box setting where the adversary has complete knowledge about SEViT architecture and (ii) evaluate SEViT in the context of natural images.

References

1. Bhojanapalli, S., Chakrabarti, A., Glasner, D., Li, D., Unterthiner, T., Veit, A.: Understanding robustness of transformers for image classification. In: Proceedings of the IEEE/CVF ICCV, pp. 10231–10241, October 2021
2. Carlini, N., Wagner, D.: Towards evaluating the robustness of neural networks. In: 2017 IEEE Symposium on Security and Privacy (SP), pp. 39–57 (2017). https://doi.org/10.1109/SP.2017.49
3. Chen, J., et al.: Transunet: transformers make strong encoders for medical image segmentation (2021). https://doi.org/10.48550/ARXIV.2102.04306
4. Croce, F., Hein, M.: Reliable evaluation of adversarial robustness with an ensemble of diverse parameter-free attacks. In: ICML (2020)
5. Deng, J., Dong, W., Socher, R., Li, L.J., Li, K., Fei-Fei, L.: Imagenet: a large-scale hierarchical image database. In: CVPR, pp. 248–255 (2009)
6. Dosovitskiy, A., et al.: An image is worth 16 × 16 words: transformers for image recognition at scale. In: ICLR (2021). https://openreview.net/forum?id=YicbFdNTTy
7. Feinman, R., Curtin, R.R., Shintre, S., Gardner, A.B.: Detecting adversarial samples from artifacts (2017). https://doi.org/10.48550/ARXIV.1703.00410
8. Finlayson, S.G., Bowers, J.D., Ito, J., Zittrain, J.L., Beam, A.L., Kohane, I.S.: Adversarial attacks on medical machine learning. Science **363**(6433), 1287–1289 (2019). https://doi.org/10.1126/science.aaw4399
9. Fu, Y., Zhang, S., Wu, S., Wan, C., Lin, Y.: Patch-fool: are vision transformers always robust against adversarial perturbations? In: ICLR (2022)
10. Goodfellow, I., Shlens, J., Szegedy, C.: Explaining and harnessing adversarial examples. In: ICLR (2015)

11. Huang, Y., Li, Y.: Zero-shot certified defense against adversarial patches with vision transformers (2021). https://doi.org/10.48550/ARXIV.2111.10481
12. Kaggle: Aptos 2019 blindness detection. Kaggle (2019). https://www.kaggle.com/c/aptos2019-blindness-detection/data
13. Kotia, J., Kotwal, A., Bharti, R.: Risk susceptibility of brain tumor classification to adversarial attacks. In: Gruca, A., Czachórski, T., Deorowicz, S., Harezlak, K., Piotrowska, A. (eds.) Man-Machine Interactions 6, pp. 181–187. Springer International Publishing, Cham (2020). https://doi.org/10.1007/978-3-030-31964-9_17
14. Kurakin, A., Goodfellow, I.J., Bengio, S.: Adversarial machine learning at scale. In: ICLR (2017). https://openreview.net/forum?id=BJm4T4Kgx
15. Liu, S., Liu, S., Cai, W., Pujol, S., Kikinis, R., Feng, D.: Early diagnosis of Alzheimer's disease with deep learning. In: ISBI, pp. 1015–1018. IEEE (2014)
16. Ma, X., et al.: Characterizing adversarial subspaces using local intrinsic dimensionality. In: ICLR (2018)
17. Madry, A., Makelov, A., Schmidt, L., Tsipras, D., Vladu, A.: Towards deep learning models resistant to adversarial attacks. In: ICLR (2018)
18. Mahmood, K., Mahmood, R., van Dijk, M.: On the robustness of vision transformers to adversarial examples. In: Proceedings of the IEEE/CVF International Conference on Computer Vision (ICCV), pp. 7838–7847, October 2021
19. Meng, D., Chen, H.: Magnet: a two-pronged defense against adversarial examples. In: Proceedings of the 2017 ACM SIGSAC Conference on Computer and Communications Security, pp. 135–147. CCS 2017, Association for Computing Machinery, New York (2017). https://doi.org/10.1145/3133956.3134057
20. Mu, N., Wagner, D.: Defending against adversarial patches with robust self-attention. In: ICML 2021 Workshop on Uncertainty and Robustness in Deep Learning (2021)
21. Naseer, M., Ranasinghe, K., Khan, S., Hayat, M., Khan, F., Yang, M.H.: Intriguing properties of vision transformers. In: Beygelzimer, A., Dauphin, Y., Liang, P., Vaughan, J.W. (eds.) NeurIPS (2021)
22. Naseer, M., Ranasinghe, K., Khan, S., Khan, F., Porikli, F.: On improving adversarial transferability of vision transformers. In: ICLR (2022)
23. Papernot, N., McDaniel, P., Wu, X., Jha, S., Swami, A.: Distillation as a defense to adversarial perturbations against deep neural networks. In: 2016 IEEE Symposium on Security and Privacy (SP), pp. 582–597. IEEE Computer Society, Los Alamitos, May 2016. https://doi.org/10.1109/SP.2016.41
24. Payer, C., Štern, D., Bischof, H., Urschler, M.: Regressing heatmaps for multiple landmark localization using CNNs. In: Ourselin, S., Joskowicz, L., Sabuncu, M.R., Unal, G., Wells, W. (eds.) MICCAI 2016. LNCS, vol. 9901, pp. 230–238. Springer, Cham (2016). https://doi.org/10.1007/978-3-319-46723-8_27
25. Rahman, T., et al.: Reliable tuberculosis detection using chest X-ray with deep learning, segmentation and visualization. IEEE Access 8, 191586–191601 (2020)
26. Rauber, J., Zimmermann, R., Bethge, M., Brendel, W.: Foolbox native: fast adversarial attacks to benchmark the robustness of machine learning models in Pytorch, Tensorflow, and Jax. Journal of Open Source Software 5(53), 2607 (2020)
27. Samangouei, P., Kabkab, M., Chellappa, R.: Defense-GAN: protecting classifiers against adversarial attacks using generative models. In: ICLR (2018)
28. Shamshad, F., et al.: Transformers in medical imaging: a survey (2022)
29. Shao, R., Shi, Z., Yi, J., Chen, P.Y., Hsieh, C.J.: On the adversarial robustness of vision transformers (2022). https://openreview.net/forum?id=O0g6uPDLW7

30. Tang, Y., Tang, Y., Zhu, Y., Xiao, J., Summers, R.M.: E^{2}Net: an edge enhanced network for accurate liver and tumor segmentation on CT scans. In: Martel, A.L., et al. (eds.) MICCAI 2020. LNCS, vol. 12264, pp. 512–522. Springer, Cham (2020). https://doi.org/10.1007/978-3-030-59719-1_50

31. Tramèr, F., Kurakin, A., Papernot, N., Goodfellow, I., Boneh, D., McDaniel, P.: Ensemble adversarial training: attacks and defenses. In: ICLR (2018)

32. Yu, S., et al.: MIL-VT: multiple instance learning enhanced vision transformer for fundus image classification. In: de Bruijne, M., et al. (eds.) MICCAI 2021. LNCS, vol. 12908, pp. 45–54. Springer, Cham (2021). https://doi.org/10.1007/978-3-030-87237-3_5

33. Zhu, X., Su, W., Lu, L., Li, B., Wang, X., Dai, J.: Deformable {detr}: deformable transformers for end-to-end object detection. In: ICLR (2021)

A Penalty Approach for Normalizing Feature Distributions to Build Confounder-Free Models

Anthony Vento[1], Qingyu Zhao[1], Robert Paul[2], Kilian M. Pohl[1,3], and Ehsan Adeli[1(✉)]

[1] Stanford University, Stanford, CA 94305, USA
eadeli@stanford.edu
[2] Missouri Institute of Mental Health, St. Louis, MO 63121, USA
[3] SRI International, Menlo Park, CA 94025, USA

Abstract. Translating machine learning algorithms into clinical applications requires addressing challenges related to interpretability, such as accounting for the effect of confounding variables (or metadata). Confounding variables affect the relationship between input training data and target outputs. When we train a model on such data, confounding variables will bias the distribution of the learned features. A recent promising solution, MetaData Normalization (MDN), estimates the linear relationship between the metadata and each feature based on a non-trainable closed-form solution. However, this estimation is confined by the sample size of a mini-batch and thereby may cause the approach to be unstable during training. In this paper, we extend the MDN method by applying a Penalty approach (referred to as PDMN). We cast the problem into a bi-level nested optimization problem. We then approximate this optimization problem using a penalty method so that the linear parameters within the MDN layer are trainable and learned on all samples. This enables PMDN to be plugged into any architectures, even those unfit to run batch-level operations, such as transformers and recurrent models. We show improvement in model accuracy and greater independence from confounders using PMDN over MDN in a synthetic experiment and a multi-label, multi-site dataset of magnetic resonance images (MRIs).

Keywords: Confounders · Neuroscience · Fairness · Deep learning

1 Introduction

Modern machine learning approaches rely on automatically learning features from data [28] using approaches such as convolutional neural networks (CNNs) [2,8] and attention-based transformer models [6,9]. Although these methods solve challenging problems, they are known to capture spurious associations and biases introduced by confounding or protected variables [27]. These limitations confine the neuroscientific impact of these algorithms, in which controlling for

© The Author(s), under exclusive license to Springer Nature Switzerland AG 2022
L. Wang et al. (Eds.): MICCAI 2022, LNCS 13433, pp. 387–397, 2022.
https://doi.org/10.1007/978-3-031-16437-8_37

(and explaining the effects of) confounding variables is crucial. To remedy this, several approaches have been proposed, such as based on adversarial training [15,27], counterfactual generative models [13,19], disentanglement [16,22], and correlation fair inference [5]. They learn features that are invariant or conditionally independent to the confounding variables.

These training methods reduce the error from confounders with minimum compromise to model accuracy. However, adversarial models or those based on disentanglement and correlation are inefficient when accounting for multiple confounders (or metadata) and only partially remove the effects from feature maps of a single layer in the network [27]. Methods based on counterfactual require reliable generative models with respect to arbitrary variables, which is added complexity. To remove confounding effects at different feature layers, MetaData Normalization (MDN) [17] can be plugged into a CNN and remove the effects of multiple confounders (or metadata) from the features while training the network. MDN aims to fix the distribution shift [3] caused by the confounding variables using a closed-form solution to linear regression capturing the relationship between confounders and each feature.

The closed-form solution in MDN requires building a linear model (relationship between metadata and each feature) as a batch-level operation. It requires large batch sizes to obtain accurate approximations of the linear model. However, batch-level statistics in MDN (similar to batch normalization) face several challenges, including (1) instability when using small batch sizes, (2) increased training time due to the calculation of closed-form solutions for each feature at each iteration, (3) inconsistent results from training and inference since there are no batches during inference, (4) inability to use MDN for online training, in which the model is trained incrementally by feeding the samples in a sequential manner, and finally (5) inability to apply MDN to Recurrent Neural Networks [4] and selected transformer models [23,24]. To overcome these limitations, we now introduce a new penalty method that turns MDN to a layer with parameters that can be optimized with other components of the network during training.

Referred to as a Penalty approach for MetaData Normalizing (PMDN), our method improves upon the batch-level MDN operation. Specifically, PMDN can be applied to all architectures and any number of confounding variables. We show that PMDN is not dependent on the batch size. We apply PDMN to a synthetic dataset to analyze and validate the method within a controlled setting. We then examine applicability of PDMN compared to MDN in classifying multi-site MRIs into 4 diagnostic groups with image acquisition site, sex, and age as confounders.

2   Methodology

Given a dataset of N training samples, we define the metadata matrix as $\mathbf{M} = [m_1, m_2, \ldots, m_N]^\top \in \mathbb{R}^{N \times K}$. Each row of $\mathbf{M}$, $m_i \in \mathbb{R}^K$, defines the metadata for sample i. Also, let $\mathbf{f} = [f_1, f_2, \ldots, f_N]^\top \in \mathbb{R}^N$ be the features for all training samples extracted at a particular layer. The goal of MDN is to remove confounding information from the features and use the residual component, $\mathbf{r}$, as

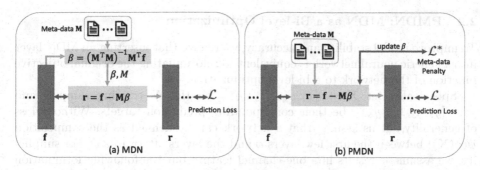

Fig. 1. (a) MDN calculates residuals **r** based on the parameter β being determined in closed-form, while PMDN (b) turns β into network trainable parameters and simultaneously optimizes both features f and β by penalizing the prediction loss.

input to the next layer of the network. The next subsection reviews how MDN performs this task via batch-level operations (Sect. 2.1), while Sect. 2.2 reformulates MDN so that it can be parameterized with respect to all training samples (Fig. 1).

2.1 MDN Review

Lu et al. [17] implemented the MDN layer as a general linear model (GLM), i.e.,

$$\mathbf{f} = \mathbf{M}\beta + \mathbf{r}, \tag{1}$$

where $\beta \in \mathbb{R}^K$ is an unknown set of parameters, $\mathbf{M}\beta$ describes the component in **f** that is relevant to the metadata, and **r** is the residual component that is irrelevant to the metadata. Then, the MDN operation is defined as

$$\mathbf{r} = \mathrm{MDN}(\mathbf{f}; \mathbf{M}) = \mathbf{f} - \mathbf{M}\beta. \tag{2}$$

The MDN layer is not trained but instead β is determined by the closed-form solution of least squares, i.e.,

$$\beta_{ls} = (\mathbf{M}^\top \mathbf{M})^{-1} \mathbf{M}^\top \mathbf{f}. \tag{3}$$

The underlying model assumes that the computation is performed across the features of all N training samples. However, training of deep learning is generally confined to batches of b samples producing $\mathbf{f} \in \mathbb{R}^b$. Therefore, [17] approximates β_{ls} as

$$\beta_{ls} = N(\mathbf{M}^\top \mathbf{M})^{-1} \mathbb{E}[m^\top f] \approx \frac{N}{b}(\mathbf{M}^\top \mathbf{M})^{-1} \sum_{i=1}^{b} m_i f_i. \tag{4}$$

As Eq. (4) approximates the expectation $\mathbb{E}[\cdot]$ only using data from a batch, the β_{ls} estimates are generally inaccurate for a small batch size and are likely to vary from batch to batch resulting in model instability, similar to Batch Norm [25].

2.2   PMDN: MDN as a Bi-level Optimization

To improve model stability and accuracy, we realize that inserting an MDN layer into a generic neural network is equivalent to reformulating the original objective function of the network to a bi-level optimization.

Specifically, let $\mathbf{X} = [x_1, x_2, \ldots, x_N]^\top$ be the N training samples and $\mathbf{y} = [y_1, y_2, \ldots, y_N]^\top$ be their corresponding prediction targets. Without loss of generality, let us assume that a network can be defined as the composition $\psi(\phi(\mathbf{X}))$ between the first few layers ϕ and the layers afterwards ψ. For simplicity, we assume ϕ results in a one-channel feature but the following formulation generalizes to multi-channel features. Let $\mathbf{W}$ be the network parameters of ψ and ϕ, then training of the network often reduces to solving the minimization problem

$$\min_{\mathbf{W}} \mathcal{L}(\psi(\phi(\mathbf{X})), \mathbf{y}). \tag{5}$$

Adding an MDN layer after ϕ changes the minimization problem to

$$\min_{\mathbf{W}} \mathcal{L}(\psi(\phi(\mathbf{X}) - \mathbf{M}\beta_{ls}), \mathbf{y}) \tag{6}$$

$$s.t. \ \beta_{ls} = \arg\min_{\beta} \mathcal{L}^*(\phi(\mathbf{X}); \mathbf{M}) = \arg\min_{\beta} ||\phi(\mathbf{X}) - \mathbf{M}\beta||^2. \tag{7}$$

In other words, the constraint itself is a nested optimization, which aims to maximally remove the metadata effect from the feature learned by ϕ.

To solve this bi-level optimization problem, PMDN (a Penalty approach for MDN) determines the minimum to a proxy objective function that combines the two minimization problems:

$$\min_{\beta, \mathbf{W}} \mathcal{L}(\psi(\phi(\mathbf{X}) - \mathbf{M}\beta), \mathbf{y}) + \lambda \mathcal{L}^*(\phi(\mathbf{X}); \mathbf{M}). \tag{8}$$

Now, Eq. (8) is a well-defined, differentiable function that can be optimized by any gradient descent algorithm. Unlike MDN that sets β to different values according to the batch construction, the β estimates in PMDN can converge to a local optimum defined with respect to all training data. Here, we use an alternating optimization schema for removing metadata effects (Algorithm. 1). As can be seen in lines 6 and 9, each of the two objectives have their own learning rates which are then consolidated into the optimizer (e.g., Adam [12]), making the implementation independent from the hyperparameter λ.

Although Algorithm 1 only is based on one PMDN layer, multiple PMDN layers can be added without loss of generality to further remove any remaining residual confounding effects. If we perform the metadata normalization after each of the C features (from different layers or channels), $\mathcal{L}^*$ in Eq. (8) is the sum of all PMDN losses $\lambda \frac{1}{C} \sum_{i=1}^{C} ||\mathbf{f}^i - \mathbf{M}\beta^i||^2$, where $\mathbf{f}^i$ and β^i are the feature vector and parameters of the ith PMDN, respectively. Furthermore, Algorithm 1 uses Stochastic Gradient Descent (SGD) [21] to update $\mathbf{W}$ and β. However, SGD can be replaced with other optimizers such as Adam [12].

Algorithm 1. Optimizing a network with PMDN

1: **procedure** PMDN
2:    **Initialize:** network parameters $\mathbf{W}$, PMDN parameters β, learning rates η_1, η_2
3:    **for** t **in** $(0,1,\cdots,T)$:
4:        Freeze $\mathbf{W}^{(t)}$, Unfreeze $\beta^{(t)}$
5:        $\hat{\mathbf{y}} = \psi(\phi(\mathbf{X}) - \mathbf{M}\beta^{(t)})$ ▷ Forward pass
6:        $\beta^{(t+1)} = \beta^{(t)} - \eta_1 \nabla_{\beta^{(t)}} \mathcal{L}^*(\phi(\mathbf{X}); \mathbf{M})$
7:        Freeze $\beta^{(t+1)}$, Unfreeze $\mathbf{W}^{(t)}$
8:        $\hat{\mathbf{y}} = \psi(\phi(\mathbf{X}) - \mathbf{M}\beta^{(t+1)})$ ▷ Forward pass
9:        $\mathbf{W}^{(t+1)} = \mathbf{W}^{(t)} - \eta_2 \nabla_{\mathbf{W}^{(t)}} \mathcal{L}(\hat{\mathbf{y}}, \mathbf{y})$
10:   **end for**
11: **end procedure**

---

3 Experiments

We apply the method to a synthetic and an MRI dataset with both continuous and discrete metadata. For each experiment, we investigate the effect of metadata on a variety of architectures including a baseline CNN, the baseline network with MDN as described in Sect. 2.1, and the baseline network with PMDN as described in Sect. 2.2. The code is available at https://github.com/vento99/PMDN.

3.1 Synthetic Dataset Experiments

Data. The synthetic dataset [17] consisted of 2000 32×32 images subdivided into two groups of 1,000 images. The first group consisted of images where quadrants two and four are Gaussians with a variance sampled from the uniform distribution $\mathcal{U}(1,4)$. The second group consisted of images where quadrants two and four are Gaussians with a variance from $\mathcal{U}(3,6)$. We introduce metadata into the third quadrant of the images. In the first group, quadrant three also consists of a Gaussian with a variance from $\mathcal{U}(1,4)$ while in the second group, quadrant three consists of a Gaussian with a variance from $\mathcal{U}(3,6)$. Theoretically, complete removal of the metadata effect will lead to a maximum model accuracy of 83.33%.

Implementation. The baseline is a simple CNN of two standard blocks. The first block consists of two convolution layers with 16 and 32 filters and a ReLU activation after each convolution layer. The second block incorporates a fully connected layer of size 84 with ReLU activation followed by another fully connected layer. We use binary cross entropy loss as $\mathcal{L}$. For all other methods, we add a normalization layer (one of BatchNorm [11], MDN, or PMDN) after the convolution and first fully connected layers (before ReLU activations).

Note that we also insert a LayerNorm layer [4] before each PMDN operation in order to stabilize the input features and to enable smoother gradients, faster training, and better generalization accuracy. Similar to the setup in [17], the metadata variable is colinear with the group labels. Thus, the labels were

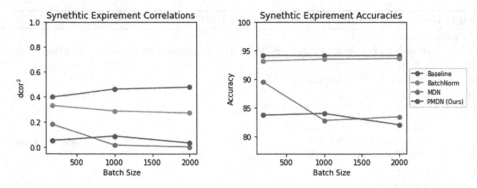

Fig. 2. Effect of batch size on different normalization strategies. Left: dcor^2 values for different batch sizes. Right: accuracy for different batch sizes. Theoretically, the maximum accuracy should be 83.3% and higher values reflect prediction error resulting from the influence of metadata.

included as an additional column in the metadata matrix $\mathbf{M}$ during training to remove the metadata effect while preserving group differences. During inference, we remove the label column from $\mathbf{M}$ and the last component from β as implemented in [17].

Evaluation. To examine whether the metadata is removed from the learned features, we calculate the squared distance correlation (dcor^2) between the output of the first FC layer and the metadata of each group separately and report the average of the two dcor^2 values. Unlike linear correlation, dcor^2 examines relationships between high-dimensional variables. Lower dcor^2 values reflect independence from metadata confounding and thus, better normalization of the feature distribution due to PMDN.

Figure 2 summarizes the results. Results for the baseline, BatchNorm, and standard MDN layers are adopted from [17]. We see that the baseline and Batch-Norm have an accuracy much greater than 83.3% (the theoretical optimum), which means that the metadata effect has not been removed from the features. Instead, the model may have learned spurious associations between metadata and labels as the dcor^2 values are much higher than those for MDN and PMDN.

For MDN, we see an inconsistency of results across different batch sizes. When the batch size is large, the batch-level closed-form solution of β approximates the true β, so that MDN successfully normalizes the metadata effect. However, for a batch size of 200, MDN performance is significantly reduced. On the other hand, PMDN shows consistent results across all batch sizes supporting our hypothesis that the penalty-based approach is impartial to batch size.

Table 1. 3D MRI dataset statistics

Site	CTRL	MCI	HIV	HAND	F/M	Age Mean ± Std
UCSF	156	148	37	145	97/389	67.00 ± 6.47
ADNI	229	397	–	–	253/373	75.16 ± 6.61
SRI	75	–	75	–	44/106	50.72 ± 11.33

3.2 Multi-label Multi-site MRI Dataset Experiments

Data. The dataset consists of 1,262 T1-Weighted MRIs from three brain studies, where each MRI was bias field corrected, skull stripped, affinely registered to a template of $64 \times 64 \times 64$ resolution. The three studies (see Table 1 for summary) were performed by (1) the Memory and Aging Center, University of California - San Francisco (UCSF) [26], (2) the Neuroscience Program, SRI International [1], and (3) the public Alzheimer's Disease Neuroimaging Initiative (ADNI1) [20]. The participants of the three studies were divided into four cohorts: healthy older adults (no neurological/psychiatric diagnosis) (CTRL; $N = 460$), adults infected with human immunodeficiency virus (HIV) without cognitive impairment (HIV; $N = 112$), HIV-infected individuals with cognitive impairment that were diagnosed with HIV-Associated Neurocognitive Disorder (HAND; $N = 145$) and HIV-negative adults diagnosed with mild cognitive impairment (MCI) but no HIV (MCI; $N = 545$). MCI is a heterogeneous condition that reflects impairment in memory and other cognitive abilities [7]. By definition, individuals with HAND meet the criteria for both MCI and HIV. Thus, this problem is formulated as a two-label classification problem: predicting whether or not individuals are infected with HIV and predicting whether or not individuals are diagnosed with cognitive impairment. For this dataset, the metadata includes the acquisition site (one-hot encoded), participant age (z-score) and participant self-identified sex (male/female).

Implementation and Baseline Models. The baseline consists of a 3D-ResNet [10] followed by a series of fully connected (FC) blocks. The 3D-ResNet consists of four standard residual blocks. Each block incorporates a 3D Conv with ReLU activation and a skip connection. The number of filters for the standard convolutions in each block are 3, 6, 9, and 6, respectively. All use kernel size 3 and padding size 1. The output of the 3D-ResNet (flattened size 2048) is passed through a FC-ReLU-FC-ReLU-FC architecture. The FC outputs are of size 128, 16, and 2, respectively. The loss ($\mathcal{L}$) we use is the focal loss [14] to combat the class imbalance. For MDN, the layers are added after each FC-ReLU and after the final FC layer. As before with the Synthetic Dataset, for PMDN, we add a LayerNorm before the first two PMDN layers and include the labels as metadata. We also examine a BatchNorm architecture where we insert BatchNorms (BNs) after each FC-ReLU. Finally, since most of the previous work focused on domain-adversarial methods for learning confounder-invariant features, we examine an adversarial training method similar to [27]. After the 3D-ResNet, we add an

Table 2. Correlations with metadata variables and accuracies of all comparison methods for the 3D Multi-label MRI dataset across different batch sizes.

|Batch|	Metric	Baseline	BN	Adversarial	MDN	PMDN (Ours)
20	Age Corr ↓	0.431	0.382	0.408	0.235	**0.213**
	Sex Corr ↓	0.209	0.237	0.154	0.172	**0.141**
	Site Corr ↓	0.388	0.312	**0.086**	0.132	0.155
	Accuracy ↑	48.8%	44.0%	26.5%	41.2%	**51.3%**
80	Age Corr ↓	0.461	0.374	0.385	0.225	**0.208**
	Sex Corr ↓	0.259	0.220	0.214	0.189	**0.187**
	Site Corr ↓	0.402	0.285	0.127	**0.126**	0.172
	Accuracy ↑	49.7%	42.1%	25.8%	41.2%	**50.7%**
160	Age Corr ↓	0.488	0.384	0.543	0.241	**0.199**
	Sex Corr ↓	0.199	0.268	**0.094**	0.174	0.188
	Site Corr ↓	0.431	0.293	0.166	0.160	**0.149**
	Accuracy ↑	45.0%	45.1%	26.8%	45.6%	**50.7%**
240	Age Corr ↓	0.488	0.369	0.382	0.185	**0.180**
	Sex Corr ↓	0.226	0.225	0.172	0.173	**0.166**
	Site Corr ↓	0.456	0.275	**0.110**	0.137	0.158
	Accuracy ↑	43.5%	42.2%	27.8%	46.2%	**51.9%**

additional head of a FC-ReLU-FC, which attempts to predict the confounding variables. The correlation loss [27] from this head is adversarially subtracted from the classification loss when updating the weights of the 3D-ResNet.

Evaluation. We perform 5-fold cross validation and report the results in Table 2. We note that $N = 240$ is the largest batch size feasible with our resource constraints. Based on the type of metadata variables (continuous, binary, or categorical), we choose different metric to investigate the metadata effect in the features. For age, we take the magnitude of the Pearson's correlation between the ages and each of the two output logits and report the average. For sex, we report on the average magnitude of the point-biserial correlation between the sexes and each of the two output logits. For site, we compute the average $dcor^2$ correlation between the site (one-hot encoded) and each of the two output logits. Finally, we calculate the accuracy for each of the four groups separately and report the average.

For each batch size, PMDN achieves the highest accuracy. This highlights that PMDN mitigates the confounding effect and produces a less biased distribution. This is also evident in the low $dcor^2$ values for PMDN. As expected, small batch size significantly compromises MDN model performance. Additionally, we see that the adversarial training method [27] removed the confounding effects for sex and site but the confounding effect of age remained as noted by the correlations. This observation underscores the inherent limitation of the adver-

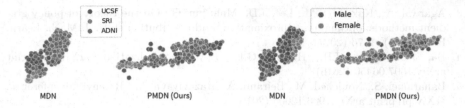

Fig. 3. tSNE visualization of features for MDN and PMDN (Left: site, Right: sex).

sarial method in controlling multiple confounds because each metadata variable requires a new adversarial component in the network. Training multiple adversarial components sacrificed model accuracy.

Figure 3 visualizes the feature space (via tSNE [18]) after removing metadata effects by MDN and PMDN for the small batch size of 80. As can be seen, the embedding space does not show a clear pattern with respect to sex differences (i.e., it is independent from the sex variable). For the case of site variable, PMDN illustrates less clustering effect compared to MDN's embedding. However, note that the three sites (UCSF, ADNI, SRI) are considerably different with respect to their class label distributions (see Table 1), which explains moderate clustering in the embedding space. In conclusion, this analysis qualitatively illustrates that of the two methods, PMDN is better at removing the confounding factor.

4 Conclusion

Herein, we introduce PMDN, our novel penalty method for removing bias in model training due to confounding factors. PMDN can be plugged into any neural network architecture and is independent from batch size. By removing the effects from confounding relationships between training and target outputs, PMDN minimizes the bias in the learned features. We show improvement of PMDN, a layer with trainable parameters, when compared to MDN, a layer with a closed-form solution, on a synthetic and a neuroimaging dataset. The improvement in accuracy and confounder independence from PMDN represent an important step towards neuroscience, imaging, or clinical applications of machine learning prediction models.

Acknowledgements. This study was partially supported by NIH Grants (AA017347, MH113406, and MH098759) and Stanford Institute for Human-Centered AI (HAI) Google Cloud Platform (GCP) Credit.

References

1. Adeli, E., et al.: Chained regularization for identifying brain patterns specific to HIV infection. Neuroimage **183**, 425–437 (2018)
2. Adeli, E., et al.: Deep learning identifies morphological determinants of sex differences in the pre-adolescent brain. Neuroimage, **223**, 117293 (2020)

3. Agarwal, A., Kakade, S.M., Lee, J.D., Mahajan, G.: On the theory of policy gradient methods: optimality, approximation, and distribution shift. J. Mach. Learn. Res. **22**(98), 1–76 (2021)

4. Ba, J.L., Kiros, J.R., Hinton, G.E.: Layer normalization. arXiv preprint arXiv:1607.06450 (2016)

5. Baharlouei, S., Nouiehed, M., Beirami, A., Razaviyayn, M.: R\'enyi fair inference. arXiv preprint arXiv:1906.12005 (2019)

6. Chen, J., et al.: Transunet: transformers make strong encoders for medical image segmentation. arXiv preprint arXiv:2102.04306 (2021)

7. Delano-Wood, L., et al.: Heterogeneity in mild cognitive impairment: differences in neuropsychological profile and associated white matter lesion pathology. J. Int. Neuropsychol. Soc. **15**(6), 906–914 (2009)

8. Deshmukh, S., Khaparde, A.: Faster region-convolutional neural network oriented feature learning with optimal trained recurrent neural network for bone age assessment for pediatrics. Biomed. Signal Process. Control, **71**, 103016 (2022)

9. Dosovitskiy, A., et al.: An image is worth 16×16 words: transformers for image recognition at scale. In: International Conference on Learning Representations (2021). https://openreview.net/forum?id=YicbFdNTTy

10. Hara, K., Kataoka, H., Satoh, Y.: Learning spatio-temporal features with 3d residual networks for action recognition. In: Proceedings of the IEEE International Conference on Computer Vision Workshops, pp. 3154–3160 (2017)

11. Ioffe, S., Szegedy, C.: Batch normalization: accelerating deep network training by reducing internal covariate shift. In: International Conference on Machine Learning, pp. 448–456. PMLR (2015)

12. Kingma, D.P., Ba, J.: Adam: a method for stochastic optimization. arXiv preprint arXiv:1412.6980 (2014)

13. Lahiri, A., Alipour, K., Adeli, E., Salimi, B.: Combining counterfactuals with shapley values to explain image models. arXiv preprint arXiv:2206.07087 (2022)

14. Lin, T.Y., Goyal, P., Girshick, R., He, K., Dollár, P.: Focal loss for dense object detection. In: Proceedings of the IEEE International Conference on Computer Vision, pp. 2980–2988 (2017)

15. Liu, T.Y., Kannan, A., Drake, A., Bertin, M., Wan, N.: Bridging the generalization gap: Training robust models on confounded biological data. arXiv preprint arXiv:1812.04778 (2018)

16. Liu, X., Li, B., Bron, E.E., Niessen, W.J., Wolvius, E.B., Roshchupkin, G.V.: Projection-wise disentangling for fair and interpretable representation learning: application to 3D facial shape analysis. In: de Bruijne, M., et al. (eds.) MICCAI 2021. LNCS, vol. 12905, pp. 814–823. Springer, Cham (2021). https://doi.org/10.1007/978-3-030-87240-3_78

17. Lu, M., et al.: Metadata normalization. In: Proceedings of the IEEE/CVF Conference on Computer Vision and Pattern Recognition, pp. 10917–10927 (2021)

18. Van der Maaten, L., Hinton, G.: Visualizing data using t-sne. J. Mach. Learn. Res. **9**(11), 2579–2605 (2008)

19. Neto, E.C.: Causality-aware counterfactual confounding adjustment for feature representations learned by deep models. arXiv preprint arXiv:2004.09466 (2020)

20. Petersen, R.C., et al.: Alzheimer's disease neuroimaging initiative (ADNI): clinical characterization. Neurology **74**(3), 201–209 (2010)

21. Robbins, H., Monro, S.: A stochastic approximation method. Ann. Math. Stat. **22**(3), 400–407 (1951)

22. Tartaglione, E., Barbano, C.A., Grangetto, M.: End: entangling and disentangling deep representations for bias correction. In: Proceedings of the IEEE/CVF Conference on Computer Vision and Pattern Recognition, pp. 13508–13517 (2021)
23. Vaswani, A., et al.: Attention is all you need. Advances in neural information processing systems 30 (2017)
24. Yao, Z., Cao, Y., Lin, Y., Liu, Z., Zhang, Z., Hu, H.: Leveraging batch normalization for vision transformers. In: Proceedings of the IEEE/CVF International Conference on Computer Vision, pp. 413–422 (2021)
25. Yong, H., Huang, J., Meng, D., Hua, X., Zhang, L.: Momentum batch normalization for deep learning with small batch size. In: Vedaldi, A., Bischof, H., Brox, T., Frahm, J.-M. (eds.) ECCV 2020. LNCS, vol. 12357, pp. 224–240. Springer, Cham (2020). https://doi.org/10.1007/978-3-030-58610-2_14
26. Kwon, D., et al.: Extracting patterns of morphometry distinguishing HIV associated neurodegeneration from mild cognitive impairment via group cardinality constrained classification. Hum. Brain Mapp. **37**(12), 4523–4538 (2016)
27. Zhao, Q., Adeli, E., Pohl, K.M.: Training confounder-free deep learning models for medical applications. Nat. Commun. **11**(1), 1–9 (2020)
28. Zhong, G., Wang, L.N., Ling, X., Dong, J.: An overview on data representation learning: from traditional feature learning to recent deep learning. J. Finan. Data Sci. **2**(4), 265–278 (2016)

mulEEG: A Multi-view Representation Learning on EEG Signals

Vamsi Kumar[1]([✉]), Likith Reddy[1], Shivam Kumar Sharma[1], Kamalaker Dadi[1], Chiranjeevi Yarra[1], Raju S. Bapi[1], and Srijithesh Rajendran[2]

[1] International Institute of Information Technology, Hyderabad, India
vamsi81523@gmail.com, likith.reddy@ihub-data.iiit.ac.in
[2] National Institute of Mental Health and Neuro Sciences, Bangalore, India
http://ihub-data.iiit.ac.in

Abstract. Modeling effective representations using multiple views that positively influence each other is challenging, and the existing methods perform poorly on Electroencephalogram (EEG) signals for sleep-staging tasks. In this paper, we propose a novel multi-view self-supervised method (mulEEG) for unsupervised EEG representation learning. Our method attempts to effectively utilize the complementary information available in multiple views to learn better representations. We introduce diverse loss that further encourages complementary information across multiple views. Our method with no access to labels, beats the supervised training while outperforming multi-view baseline methods on transfer learning experiments carried out on sleep-staging tasks. We posit that our method was able to learn better representations by using complementary multi-views (Code Available at: https://github.com/likith012/mulEEG).

Keywords: Multi-view learning · Self-supervised · Sleep staging

1 Introduction

Sleep is an important part of daily routine. Getting the right quantity and quality of sleep at the right time is essential for one's well-being. As identification of sleep stages plays a crucial role in determining the quality of sleep, it can also help identify sleep-related or other mental disorders like Obstructive Sleep Apnea, depression, schizophrenia, and dementia [4,11,13].

With the advent of the large volume of devices that can monitor Electroencephalogram(EEG) signals, the amount of data piling up is enormous. Many studies have proposed exploiting such EEG data for automated sleep stage classification using supervised methods [9,22,23,29]. However, such methods rely on

---

V. Kumar and L. Reddy—Equal Contribution.

Supplementary Information The online version contains supplementary material available at https://doi.org/10.1007/978-3-031-16437-8_38.

© The Author(s), under exclusive license to Springer Nature Switzerland AG 2022
L. Wang et al. (Eds.): MICCAI 2022, LNCS 13433, pp. 398–407, 2022.
https://doi.org/10.1007/978-3-031-16437-8_38

massive amounts of annotated data. Annotating EEG data is costly and time-consuming for a physician. The inter-rater agreement on annotations for different sleep stages is low [26], making annotations unreliable. A model trained on EEG data annotated by one expert will be biased towards their annotations and may not generalize well. Recently, self-supervised learning (SSL) has been used to learn effective representations from unlabeled data. Various self-supervised methods have been proposed on natural image datasets that rely on contrastive loss [7,16,20], clustering [5,6], or distillation [14] utilizing massive amounts of unlabeled data. Limited work has been done on self-supervised learning for EEG signals, [2] proposed SSL tasks that apply relative positioning and contrastive predictive coding on time-series EEG signals. [10] used contrastive loss while addressing the temporal dependencies in time-series more effectively. [19] extended the simCLR [7] framework to time-series EEG signals while introducing new augmentations. [25] reduced the impact of negative sampling usually caused due to contrastive loss, on spectrograms from EEG signals.

Multi-view SSL jointly trains all the views influencing each other in a self-supervised way. On natural image datasets, [24] captured the shared information across multiple views by maximizing mutual information between them. [15] focused on learning representations for videos in a self-supervised way by using complementary information from two different views (RGB and optical flow). [27] learned visual representations by using multi-modal data with a combination of inter- and intra-modal similarity preservation objectives. The same multi-view methods cannot be applied directly to EEG signals as augmentation strategies are different from images. Moreover, extracting multiple views for EEG signals differs from that on natural images. To the best of our knowledge, on physiological signals no work has been done on multi-view SSL, which aims to learn effective representations by training multiple views jointly.

Our objective is to learn effective representations from multiple views by training them jointly in a self-supervised way. We aim to utilize the complementary information in multiple views to influence each other positively during training. The main contributions of this work are as follows: 1) We design an EEG augmentation strategy for multi-view SSL. 2) We illustrate that existing multi-view self-supervised methods perform poorly and inconsistently on EEG data. 3) We propose a novel multi-view SSL method that effectively utilizes the complementary information available in multiple views to learn better representations. 4) We introduce an additional diverse loss function that further encourages the complementary information across multiple views.

2   Methods

Data Augmentation: The contrastive learning methods are heavily influenced by the stochastic data augmentations used [14], and it is important to get the right match of augmentations. We use a different family of augmentations as it produces strong variations between the two augmented views and is known to work better when a shared encoder is used [7]. Similar augmentation approaches

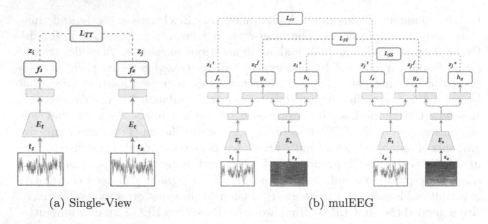

(a) Single-View   (b) mulEEG

Fig. 1. Architecture overview of Single-View and our proposed method mulEEG

are used in [10, 19] for EEG signals. We use jittering, where random uniform noise is added to the EEG signal depending on its peak-to-peak values, along with masking, where signals are masked randomly as one family of augmentations T_1. Flipping, where the EEG signal is horizontally flipped randomly, and scaling, where EEG signal is scaled with Gaussian noise are used sequentially as the second family of augmentations T_2.

Problem Formulation: The input is a continuous recording of an EEG signal from a subject and is segmented into non-overlapping 30 s EEG signal called an epoch. Each epoch is categorized into five sleep stages: Wake, REM, N1, N2, N3 according to American Academy of Sleep Medicine [3] guidelines. And each subject's EEG recording epochs belongs to one of the pretext/train/test groups where the pretext group contains a relatively large number of unlabeled subjects compared to the less number of labeled subjects in train and test groups. Throughout our study, time-series and spectrogram are used as multiple views of the same EEG signal. For each EEG epoch, we obtain two time-series augmentations denoted as t_1-T_1 and t_2-T_2. The time-series augmentations are converted into their respective spectrograms s_1, s_2. The augmentations are passed into their respective encoders, which are time-series encoder, E_t, and spectrogram encoder, E_s, to extract their high dimensional latent representations. A separate projection head is used for the time-series and spectrogram encoder, which takes input as these high dimensional latent representations and maps them to the space where contrastive loss is applied on $\mathbf{z_i}$, $\mathbf{z_j}$. Applying the contrastive loss directly on the projection head outputs gives better results [7]. We use a variant of contrastive loss called NT-Xent [7, 20] which maximizes the similarity between two augmented views while minimizing its similarity with other samples. Here, N is the batch size, τ is the temperature parameter, and a cosine similarity is used in the contrastive loss function given in (2). The function $\mathbb{1}_{[k\neq i]}\epsilon\{0,1\}$

evaluates to 1 if $k \neq i$, otherwise gives 0.

$$\ell(i,j) = -\log \frac{exp(cosine(\mathbf{z_i}, \mathbf{z_j})/\tau)))}{\sum_{k=1}^{2N} \mathbb{1}_{[k \neq i]} exp(cosine(\mathbf{z_i}, \mathbf{z_k})/\tau))}, \qquad (1)$$

$$L(\mathbf{z_i}, \mathbf{z_j}) = \frac{1}{2N} \sum_{k=1}^{N} \ell(2k-1, 2k) + \ell(2k, 2k-1)), \qquad (2)$$

Single-View and Multi-View: In single-view learning Fig. 1(a), we learn the representations for time-series, E_t, and spectrogram encoder, E_s, by training them independently using unlabeled pretext group. Inspired from [7], the encoder is shared and is trained on a pair of augmentations. For pre-training time series encoder, E_t, the time-series augmentations t_1, t_2 are passed to E_t and then its output to the time-series projection heads f_1, f_2 to obtain $\mathbf{z_i}$ and $\mathbf{z_j}$. The contrastive loss L_{TT} in (2) is applied on $\mathbf{z_i}$, $\mathbf{z_j}$ and similarly, the spectrogram encoder, E_s, is pre-trained.

In a multi-view setup, we learn the representations of time-series and spectrogram encoder by pre-training them jointly, thereby influencing each other. We present two baseline multi-view methods: 1) **CMC:** Inspired from Contrastive Multiview Coding [24], we jointly train both the encoders by sending t_1 to E_t and s_2 to E_s, and the outputs from encoders are sent to their projection heads f_1, f_2 to obtain $\mathbf{z_i}$ and $\mathbf{z_j}$. Compared to [24] we apply augmentations in our method and use a projection head. It brings both the encoders latent features $\mathbf{z_i}$, $\mathbf{z_j}$ into the same feature space and compares them by aiming for information maximization between the views. 2) **Simple Fusion:** We experiment with a setup where we jointly train both the encoders by considering the concatenation of features obtained from E_t and E_s. The augmentations t_1, s_1 are passed through E_t, E_s, and the concatenated outputs are passed through a projection head f_1 to obtain $\mathbf{z_i}$. Similarly, for augmented views t_2, s_2, we extract $\mathbf{z_j}$ and contrastive loss is applied on concatenated feature space.

mulEEG: In a supervised setup, when both views (time-series & spectrogram) are trained jointly, predicting on concatenated outputs does not always guarantee a better performance when compared to a setup where a single-view is trained individually for sleep-stage classification [21]. It is observed that such a setup behaves differently under different datasets and modalities. This behavior is somewhat surprising and is not well understood. Also, it is observed on multiple datasets with multiple modality combinations that, along with encoders, time-series and spectrogram views are favorable for identifying different sleep stages. But this favourability on sleep stages is not constant and changes from dataset to dataset. From [21], we cannot say that time-series or spectrogram views are better than others, but instead, one should aim to use both views to complement each other when trained jointly, trying to optimize the task at hand.

In our method shown in Fig. 1(b), we jointly train the encoders E_t, E_s such that the complementary information of both views is utilized optimally and positively influences each other. The time, spectrogram, and concatenation of time and spectrogram features are used in our method along with the shared encoders E_t, E_s. The augmented views of time-series and spectrogram t_1, s_1 are passed through their respective encoders. Apart from obtaining the time and spectrogram features, we also concatenate time and spectrogram features to be further given as input to three different projection heads f_1, g_1, h_1. Similarly for other augmented views t_2, s_2 output from the encoders is sent to f_2, g_2, h_2. It uses three contrastive losses, each loss working on the time-series feature space L_{TT}, spectrogram feature space L_{SS} and concatenated features space L_{FF}. Such a setup introduces flexibility to optimize between the time, spectrogram, and concatenated features during the self-supervised training.

We introduce an additional loss called Diverse Loss L_D given in (4) that further encourages the complementary information across time-series and spectrogram views. Due to contrastive loss on concatenated features, the time-series and spectrogram representations can tend to maximize the mutual information between them. Such a process can ignore the complementary information inherent to the views. Different from the contrastive loss used above, here the contrastive loss is applied on features obtained from only time-series and spectrogram projection heads $z_k = [\mathbf{z_i^t}, \mathbf{z_j^t}, \mathbf{z_i^s}, \mathbf{z_j^s}]$ $\forall i = j$ on a single sample k instead on entire batch N. The contrastive loss here tries to pull time-series features closer while pushing away spectrogram features from time-series features for a single sample. Similarly, spectrogram features are pushed closer while pushing away time-series features for a sample. This allows the representations learned by both time-series and spectrogram views to have diverse information from each other for a single sample. The total loss given in (5), is a combination of contrastive loss on time-series, spectrogram, and concatenated features that in turn combined with diverse loss. Our method can be extended to use along with recent SSL strategies, using moving average encoder [14,16], negative sampling strategies [25], etc.

$$\ell_d(z_k, a, b) = -\log \frac{exp(cosine(z_k[a], z_k[b])/\tau_d)}{\sum_{i=1}^{4} \mathbb{1}_{[i \neq a]} exp(cosine(z_k[a], z_k[i])/\tau_d))}, \tag{3}$$

$$L_D = \frac{1}{4N} \sum_{k=1}^{N} \ell_d(z_k, 1, 2) + \ell_d(z_k, 2, 1) + \ell_d(z_k, 3, 4) + \ell_d(z_k, 4, 3), \tag{4}$$

$$L_{tot} = \lambda_1 (L_{TT} + L_{FF} + L_{SS}) + \lambda_2 L_D \tag{5}$$

3   Experiments

Datasets: To evaluate our proposed method, we consider two popular publicly available sleep-staging datasets: 1) **SleepEDF:** The sleep-EDF [12] database contains a collection of 78 subjects to understand the age effects on sleep in healthy Caucasians. Each subject contains two full night recordings, with a few subjects having only one, with a total of 153 full night recordings. We randomly shuffle subjects and select 58 subjects as a pretext and remaining 20 subjects for cross-validation (5-fold). We use a single-channel EEG (Fpz-Cz) sampled 100 Hz. 2) **SHHS:** The Sleep Heart Health Study (SHHS) [28] is a multi-center cohort study comprising 6,441 subjects with a single full night recording for each subject. We have selected 326 subjects from the total subjects based on the criteria that the selected subjects are close to having a regular sleep cycle similar to [9]. A single-channel EEG (C4-A1) at a sampling rate 125 Hz is used, we convert the signals 100 Hz to keep the experiments consistent with the sleepEDF dataset. We then randomly shuffle the subjects and use a data split of 264 (pretext) and remaining 62 subjects for cross-validation (5-fold).

Implementation Details: In all our experiments, we use ResNet-50 [17] with 1D-convolutions as a time series encoder with 0.6 million parameters. For the first 1D-convolution layer a kernel size of 71 is used and for the rest, a kernel size of 25 is used which outputs a 256-dimensional feature vector. For the spectrogram encoder, [25] is used which takes in a spectrogram as input and outputs a 256-dimensional feature vector. For converting the augmented time-series EEG signal into a spectrogram, Short Time Fourier Transform is used with the number of FFT points set to 256 and hop length to 64. A non-linear two-layer MLP projection head is used as it performs better compared to a linear projection head [7], and we include Batch Norm as it improves the performance drastically [8]. The projection head architecture is Linear > BatchNorm > ReLU > Linear outputting a 128-dimensional vector to be used in the contrastive loss space. During the self-supervised pre-training, Adam optimizer with an initial learning rate of 3e−4 with $\beta_1 = 0.9$, $\beta_2 = 0.99$, and a weight decay of 3e−5 is used. It is trained for a total of 140 epochs with a batch size of 256 and a learning scheduler is used which reduces the learning rate by 1/5th with a patience of 5. For training the encoders in a supervised way, similar optimization parameters are used, but it is trained for a total of 300 epochs, batch size of 256 with learning scheduler reducing the learning rate by 1/5th with patience of 10. The temperature τ used for our model losses L_{SS}, L_{TT}, and L_{FF} is 1, 1, 1, respectively. For the diverse loss L_D the temperature τ_d used is 10 with $\lambda_1 = 1$ and $\lambda_2 = 1$ in the final loss.

Evaluation: The encoders are pre-trained using the unlabeled pretext group and evaluated on standard linear benchmarking evaluation protocol [7,8,14,16]. In linear evaluation, a linear classifier is attached on top of the frozen pre-trained encoder and only the linear classifier is trained on the train group, and the metrics are evaluated on the test group. We evaluate the experiments on datasets in two ways: 1) **Within Dataset**: One is to obtain the pretext/train/test groups within the dataset itself as done in [10,19,25] and perform a 5-fold evaluation on train and test groups. 2) **Transfer Learning**: Another is [7,16], where we pre-train on larger dataset SHHS with 264 subjects as a pretext group and evaluate on another dataset, SleepEDF with 58 subjects (pretext group in *within dataset*) as train and 20 subjects (cross-validation group in *within dataset*) as a test group.

We use the metrics Cohen's kappa (κ), accuracy, and macro-averaged F1 (MF1) score to evaluate the performance. Macro-averaged F1 score considers the class-wise performance and performs better when each class performs better compared to other metrics. We also compare our method with supervised baselines: 1) **Supervised**: The encoder is trained in a supervised way on the pretext group and evaluated on train/test groups. 2) **Randomly Initialized**: The encoder is randomly initialized and isn't trained on the pretext group. It is then further evaluated on train/test groups. These two baselines are commonly treated as lower and upper bounds to measure the quality of the learned representations using self-supervised learning [1,18].

4   Results

Within Dataset: After jointly pre-training both the encoders using the multi-view setup, the representations learned by the time-series and spectrogram encoder can be evaluated individually or on concatenated features. In our experiments, we only evaluate the time-series encoder similar to [24,27]. The performance metrics for linear evaluation protocol are shown in Table 1, comparing our method with others. All the self-supervised methods outperform the Randomly Initialized model by a drastic margin, indicating that the encoders learn useful representations from unlabeled data using self-supervised methods. Simple Fusion tends to perform worse than the single-view method on both the datasets even though the encoders are pre-trained jointly, implying the spectrogram encoder wasn't able to influence the time-series encoder positively. But the performance of CMC gives inconsistent results and performs much worse on sleepEDF but performs better on SHHS dataset compared to the single-view. Our method without the diverse loss performs consistently better on both datasets when compared to the single-view and CMC methods. But our model with diverse loss seems to perform much better consistently compared to all the self-supervised methods for both the datasets by a good margin. Compared to

Table 1. Linear evaluation performance on our proposed method and baseline models for within dataset and transfer learning

Method	Within dataset						Transfer learning		
	SleepEDF			SHHS			SHHS > SleepEDF		
	Acc	κ	MF1	Acc	κ	MF1	Acc	κ	MF1
Single-view	77.58	0.6773	66.74	79.56	0.7145	64.71	76.73	0.6669	66.42
Simple fusion	77.01	0.6683	65.71	79.39	0.7122	64.56	76.75	0.6658	65.78
CMC	74.21	0.6300	62.81	80.29	0.7215	65.93	75.84	0.6520	64.4
Ours	77.84	0.6806	67.04	80.13	0.7223	65.57	78.18	0.6869	67.88
Ours + diverse loss	**78.06**	**0.6850**	**67.82**	**81.21**	**0.7366**	**66.58**	**78.54**	**0.6914**	**68.10**
Randomly initialized	40.52	0.1189	17.04	44.75	0.0894	19.39	38.68	0.1032	16.54
Supervised	**79.08**	**0.7014**	**69.78**	**82.62**	**0.7569**	**71.41**	77.88	0.6838	67.84

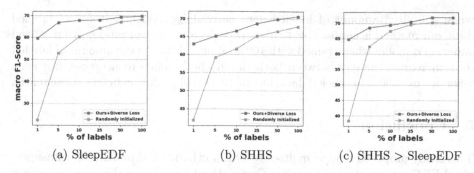

(a) SleepEDF

Fig. 2. Semi-supervised performance comparison between our method and randomly initialized on MF1 for within dataset and transfer learning

the single-view method, we observe an improvement of 3.1% for κ and 2.9% for MF1 on SHHS dataset. This shows in our method when both encoders are pre-trained jointly, spectrogram encoder was able to influence the time-series encoder positively. The supervised model seems to outperform all methods on linear evaluation.

Transfer Learning: The results for transfer learning are shown in Table 1, on linear evaluation. Similarly, we can observe all the self-supervised methods outperforming the Randomly Initialized model by a large margin. The single-view performs better than multi-view methods: Simple Fusion and CMC. Our model without the diverse loss performs better than the supervised baseline. When the diverse loss is included in our method, even better performance can be observed. When compared to the supervised baseline which has access to all the labels in pretext group, we can see both our methods with and without diverse loss, which were pre-trained with no labels, performing better than the supervised baseline. Compared to the supervised baseline, we see an increase of 1.1% on κ and 0.85% on accuracy, on conducting statistical analysis, using t-test with unequal variance, we observed that these improvements are statistically significant with p-values of 0.029 (for kappa) and 0.024 (for Acc). We can see this improvement, as supervised learning is biased to annotations generated by a physician on the SHHS dataset and couldn't generalize well to the SleepEDF dataset. This shows that the representations learned by the time-series encoder using our method are better than those learned using supervised learning on a sleep-staging task.

Semi-supervised: To evaluate the effectiveness of our method under semi-supervised settings, we finetune with different amounts of training data and compare the results of time-series encoder pre-trained using our method with the supervised model initialized randomly. We randomly select 1%, 5%, 10%, 25%, 50% and 100% of samples from train group and show the results in Fig. 2. When only 1% of labeled data is given, the supervised method performs poorly while our method performs better by a significant margin for both within dataset and transfer learning experiments. This shows the effectiveness of our method when a limited amount of labeled data is available. With only 5% of labeled data, our method is almost able to match the performance of randomly initialized supervised model when trained with 100% labeled data. As the amount of labeled data increases, the gap between both the methods tends to decrease, but even when we use 100% of labeled data, our method still performs better consistently.

5   Conclusion

This study proposed a novel multi-view SSL method called mulEEG for unsupervised EEG representation learning. Our method bootstraps the complementary information available in multiple views to learn better representations. While our method is general, specific application on sleep-stage classification is demonstrated. We show that on linear evaluation, our method was able to beat the supervised training on transfer learning and shows high efficiency on few-labeled scenarios.

References

1. Arandjelovic, R., Zisserman, A.: Look, listen and learn. In: Proceedings of the IEEE International Conference on Computer Vision, pp. 609–617 (2017)
2. Banville, H., Chehab, O., Hyvärinen, A., Engemann, D.A., Gramfort, A.: Uncovering the structure of clinical EEG signals with self-supervised learning. J. Neural Eng. 18(4), 046020 (2021)
3. Berry, R.B., et al.: Aasm scoring manual updates for 2017 (version 2.4) (2017)
4. Bianchi, M.T., Cash, S.S., Mietus, J., Peng, C.K., Thomas, R.: Obstructive sleep apnea alters sleep stage transition dynamics. PLoS One, 5(6), e11356 (2010)
5. Caron, M., Bojanowski, P., Joulin, A., Douze, M.: Deep clustering for unsupervised learning of visual features. In: Proceedings of the European conference on Computer Vision (ECCV), pp. 132–149 (2018)
6. Caron, M., Misra, I., Mairal, J., Goyal, P., Bojanowski, P., Joulin, A.: Unsupervised learning of visual features by contrasting cluster assignments. Adv. Neural. Inf. Process. Syst. 33, 9912–9924 (2020)
7. Chen, T., Kornblith, S., Norouzi, M., Hinton, G.: A simple framework for contrastive learning of visual representations. In: International Conference on Machine Learning, pp. 1597–1607. PMLR (2020)
8. Chen, X., He, K.: Exploring simple siamese representation learning. In: Proceedings of the IEEE/CVF Conference on Computer Vision and Pattern Recognition, pp. 15750–15758 (2021)
9. Eldele, E., et al.: An attention-based deep learning approach for sleep stage classification with single-channel EEG. IEEE Trans. Neural Syst. Rehabil. Eng. 29, 809–818 (2021)
10. Eldele, E., et al.: Time-series representation learning via temporal and contextual contrasting. In: Proceedings of the Thirtieth International Joint Conference on Artificial Intelligence, IJCAI-21, pp. 2352–2359 (2021)
11. Freeman, D., Sheaves, B., Waite, F., Harvey, A.G., Harrison, P.J.: Sleep disturbance and psychiatric disorders. Lancet Psychiatry 7(7), 628–637 (2020)

12. Goldberger, A., et al.: Physiobank, physiotoolkit, and physionet: components of a new research resource for complex physiologic signals. Circulation **101**(23), e215–e220 (2000)
13. Gottesmann, C., Gottesman, I.: The neurobiological characteristics of rapid eye movement (REM) sleep are candidate endophenotypes of depression, schizophrenia, mental retardation and dementia. Prog. Neurobiol. **81**(4), 237–250 (2007)
14. Grill, J.B., et al.: Bootstrap your own latent-a new approach to self-supervised learning. Adv. Neural. Inf. Process. Syst. **33**, 21271–21284 (2020)
15. Han, T., Xie, W., Zisserman, A.: Self-supervised co-training for video representation learning. Adv. Neural. Inf. Process. Syst. **33**, 5679–5690 (2020)
16. He, K., Fan, H., Wu, Y., Xie, S., Girshick, R.: Momentum contrast for unsupervised visual representation learning. In: Proceedings of the IEEE/CVF Conference on Computer Vision and Pattern Recognition, pp. 9729–9738 (2020)
17. He, K., Zhang, X., Ren, S., Sun, J.: Deep residual learning for image recognition. In: Proceedings of the IEEE Conference on Computer Vision and Pattern Recognition, pp. 770–778 (2016)
18. Korbar, B., Tran, D., Torresani, L.: Cooperative learning of audio and video models from self-supervised synchronization. Advances in Neural Information Processing Systems 31 (2018)
19. Mohsenvand, M.N., Izadi, M.R., Maes, P.: Contrastive representation learning for electroencephalogram classification. In: Machine Learning for Health, pp. 238–253. PMLR (2020)
20. Oord, A.v.d., Li, Y., Vinyals, O.: Representation learning with contrastive predictive coding. arXiv preprint arXiv:1807.03748 (2018)
21. Phan, H., Chén, O.Y., Tran, M.C., Koch, P., Mertins, A., De Vos, M.: Xsleepnet: multi-view sequential model for automatic sleep staging. In: IEEE Transactions on Pattern Analysis and Machine Intelligence (2021)
22. Sors, A., Bonnet, S., Mirek, S., Vercueil, L., Payen, J.F.: A convolutional neural network for sleep stage scoring from raw single-channel EEG. Biomed. Signal Process. Control **42**, 107–114 (2018)
23. Supratak, A., Dong, H., Wu, C., Guo, Y.: Deepsleepnet: a model for automatic sleep stage scoring based on raw single-channel EEG. IEEE Trans. Neural Syst. Rehabil. Eng. **25**(11), 1998–2008 (2017)
24. Tian, Y., Krishnan, D., Isola, P.: Contrastive multiview coding. In: Vedaldi, A., Bischof, H., Brox, T., Frahm, J.-M. (eds.) ECCV 2020. LNCS, vol. 12356, pp. 776–794. Springer, Cham (2020). https://doi.org/10.1007/978-3-030-58621-8_45
25. Yang, C., Xiao, D., Westover, M.B., Sun, J.: Self-supervised eeg representation learning for automatic sleep staging. arXiv preprint arXiv:2110.15278 (2021)
26. Younes, M., et al.: Reliability of the american academy of sleep medicine rules for assessing sleep depth in clinical practice. J. Clin. Sleep Med. **14**(2), 205–213 (2018)
27. Yuan, X., et al.: Multimodal contrastive training for visual representation learning. In: Proceedings of the IEEE/CVF Conference on Computer Vision and Pattern Recognition, pp. 6995–7004 (2021)
28. Zhang, G.Q., et al.: The national sleep research resource: towards a sleep data commons. J. Am. Med. Inform. Assoc. **25**(10), 1351–1358 (2018)
29. Zhu, G., Li, Y., Wen, P.: Analysis and classification of sleep stages based on difference visibility graphs from a single-channel EEG signal. IEEE J. Biomed. Health Inform. **18**(6), 1813–1821 (2014)

Automatic Detection of Steatosis in Ultrasound Images with Comparative Visual Labeling

Güinther Saibro[1,2,4], Michele Diana[2,4,5], Benoît Sauer[3], Jacques Marescaux[1,2], Alexandre Hostettler[1,2], and Toby Collins[1,2(✉)]

[1] Research Institute Against Digestive Cancer (IRCAD) Africa, Kigali, Rwanda
guinther.saibro@ircad.africa, toby.collins@ircad.fr
[2] Research Institute Against Digestive Cancer (IRCAD) France, Strasbourg, France
[3] Medical Imaging Group (MIM) Group, Clinique Sainte Anne, Strasbourg, France
[4] ICube Photonics Instrumentation for Health, University of Strasbourg, Strasbourg, France
[5] Department of Surgery, University Hospital of Strasbourg, Strasbourg, France

Abstract. A common difficulty in computer-assisted diagnosis is acquiring accurate and representative labeled data, required to train, test and monitor models. Concerning liver steatosis detection in ultrasound (US) images, labeling images with human annotators can be error-prone because of subjectivity and decision boundary biases. To overcome these limits, we propose comparative visual labeling (CVL), where an annotator labels the relative degree of a pathology in image pairs, that is combined with a RankNet to give per-image diagnostic scores. In a multi-annotator evaluation on a public steatosis dataset, CVL+RankNet significantly improves label quality compared to conventional single-image visual labeling (SVL) (0.97 versus 0.87 F1-score respectively, 95% CI significance). This is the first application of CVL for diagnostic medical image labeling, and it may stimulate more research for other diagnostic labeling tasks. We also show that Deep Learning (DL) models trained with CVL+RankNet or histopathology labels attain similar performance.

1 Introduction and Related Works

Non-Alcoholic Fatty Liver disease (NAFLD), also called liver steatosis, affects approximately 25% of population worldwide [1], with a higher prevalence in developing countries [26]. NAFLD patients are more likely to develop chronic diseases such as type 2 diabetes, obesity and metabolic syndrome [2]. If not identified and treated early, steatosis can lead to liver fibrosis, cirrhosis and liver cancer [7,19]. Liver biopsy is currently the gold standard for steatosis grading and quantification [3,24]. However, it is costly, invasive, requires a histopathologist, and has complication risks. Magnetic Resonance Imaging (MRI) can be

---

Supplementary Information The online version contains supplementary material available at https://doi.org/10.1007/978-3-031-16437-8_39.

© The Author(s), under exclusive license to Springer Nature Switzerland AG 2022
L. Wang et al. (Eds.): MICCAI 2022, LNCS 13433, pp. 408–418, 2022.
https://doi.org/10.1007/978-3-031-16437-8_39

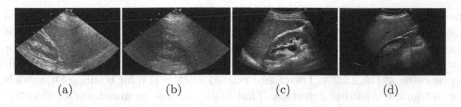

(a)   (b)   (c)   (d)

Fig. 1. Sample images from dataset 1 (a: healthy, b: pathological), and dataset 2 (c: healthy, d: pathological). The image brightness has been increased by 150% for better clarity using GIMP's exposure filter with GIMP 2.10.28 [25].

used to quantify liver fat infiltration [9], however it is not widely available for screening [6,13]. B-mode ultrasound (US) is the preferred imaging modality for screening and early steatosis detection [16], because of its convenience, low cost and non-invasiveness [5]. However, US image interpretation requires significant training and is operator dependent. Detection specificity in the literature ranges from 77.0% to 93.1% [13]. Operators can also have varying decision thresholds, and a 20% underestimation of steatosis has been reported [10]. To overcome the limits of US-based diagnosis, computer assisted diagnosis (CAD) systems have been proposed. State-of-the-art approaches use supervised deep learning with labels from histopathology or visual image labeling [5,6,14,18,28]. Histopathology provides gold standard labels, however, it has major practical and ethical difficulties associated with biopsy. In contrast, visual labeling is non-invasive, faster and cheaper. However, it requires skilled annotators, and label errors may occur because of human subjectivity, which could lead to sub-optimal CAD models, and unreliable model performance evaluation and monitoring (critical aspects when deploying CAD models as medical devices).

The main contribution of this work is comparative visual labeling with RankNet (CVL+RankNet), to significantly improve label quality in a diagnostic task. In this study, labels are binary and denote which of two images has a higher perceived degree of steatosis. Our hypothesis is that because CVL involves annotating the relative severity of the disease in image pairs, rather than absolute disease presence or absence, it can be more objective and reduce label variability and detection sensitivity biases. The CVL labels are converted into sorted real-valued scores (one per patient) using a RankNet [4]. These score can then be used to train and evaluate CAD classifiers, including regression models, which is an advantage over SVL. An experiment involving three annotators is presented using a US liver steatosis public dataset with histopathology labels as ground truth [5]. The CVL+RankNet labels of each annotator, and their fused labels, are much more accurate compared to conventional single-image visual labeling (SVL). The results indicate that CVL+RankNet may be used to create substantially higher quality labeled datasets. We also present experiments with classification and regression DL models trained with CVL+RankNet labels, which achieve similar accuracy as models trained with histopathology labels.

To the best of our knowledge, this is the first work using CVL to annotate a medical imaging diagnosis dataset and show superior label quality compared to conventional labeling. This work tackles one of the central open problems in medical imaging today (high quality labelled data acquisition). CVL has been used previously in other applications with success, especially when the task is continuous and discretized with subjective boundaries. CVL has been used for surgical kill assessment [8,27], recommendation systems [12,17] and image quality assessment [15,23,29]. The closest related work is [14] for training DL models for automatic steatosis detection. That work models an annotator's subjectivity with an annotator-aware encoder. In contrast, our work is about reducing labeling errors with a different approach to annotation, to create higher quality datasets and achieve more reliable model performance evaluation.

2   Methods

2.1   Datasets and Labeling

The only public available dataset is the dataset from Byra *et al.* [5], used in this work and [18,28], and referred to as 'Dataset 1' (Figs. 1a and 1b). This consists of 55 patients, 17 healthy and 38 pathological (mild, moderate or severe steatosis). Each patient has a pathology score (% fatty hepatocytes) where $\geq 5\%$ is the standard pathological threshold. Despite its small size, CNN-based CAD models with biopsy labels have been successfully trained on this dataset with transfer learning and strong data augmentation (ROC-AUC of 0.97 [5]). A second anonymous steatosis dataset of 54 US images (one per patient) was made available retrospectively from our partner MIM Group ('Dataset 2'), shown in Figs. 1c and 1d. The images are taken in the liver-kidney sagittal plane using a Canon Aplio a450 system (1280×960 resolution). Labeling was performed independently by 3 experienced radiographers using the CVAT annotation tool [20]. They performed SVL (healthy vs. pathological) according to standard guidelines [16]. Dataset 2 has 36 healthy and 18 pathological patients according to the annotators' majority vote. CVL was performed using a 4-way label ("first higher", "second higher", "similar normal" and "similar pathological"). The first two labels are used when one image was considered to have higher degree of pathology. The other two labels were used when both images were considered as healthy or pathological. Image pairs were generated for each dataset by uniform random allocation, resulting in an average of 17.85 comparisons per image. From this pool of image pairs, random subsets were drawn with fewer pairwise comparisons to investigate performance with less labeled data. Algorithm pseudo-code for generating image pairs is provided in the supplementary material.

2.2   Converting Comparative Visual Labels to Pathological Scores

CVL data is converted to real-valued per-image (one per patient) pathological scores using Learning To Rank (LTR), which finds a score function that orders items from comparison data. A variant of RankNet [4,11] was used because of its robustness to noise [11]. Each image is uniquely identified using one-hot encoding, denoted as $\mathbf{v}_i \in \{0,1\}^N$, where N is the total number of images. The

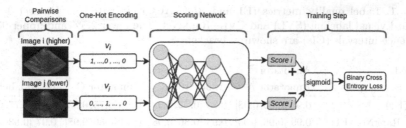

Fig. 2. Scheme of our implementation of RankNet [4], a LTR neural network trained with pairwise comparison data.

RankNet takes as input $\mathbf{v}_i$ and it outputs a ranking score $s(\mathbf{v}_i) : \{0,1\}^N \to \mathbb{R}^+$. During training, the labeled one-hot encoded image pairs $(\mathbf{v}_i, \mathbf{v}_j)$ are forward-passed through the scoring network, and a predicted pairwise label is generated as $S\left(s(\mathbf{v}_i) - s(\mathbf{v}_j)\right)$ where S is the sigmoid function (Fig. 2). The network is trained by minimizing the binary cross-entropy loss (BCE) using all pairs with a comparison differences (labeled "first higher" or "second higher"). A 4-layer RankNet with 2 hidden layers is used of dimensions 55 and 32 with a dropout rate of 0.9. The RankNet performance was not found to be strongly sensitive to the architecture (discussed in Sect. 3.1). Training typically requires less than a minute on a standard workstation PC. The CVL+RankNet scores are continuous and can be converted to binary labels using a threshold. Each annotator specified a threshold associated to their pairwise annotations as follows. Firstly, the images were presented to the annotator in order of CVL+RankNet score. The annotator then selected a contiguous range of images that they believed were on the boundary between healthy and pathological cases. The CVL+RankNet's score of the midpoint image was used as the label threshold.

2.3   Classification and Regression

Two types of models have been trained: regression models (using the real-valued CVL+RankNet scores as targets) and a binary classification models (using binary labels as targets). They are implemented as CNNs using a state-of-the-art backbone (Inception-ResNet-v2 [22]) pretrained on ImageNet. The classification model is trained with BCE loss and inverse class frequency weighting to handle the class imbalance. The regression model has three fully connected final layers of 512, 16 and 1 neurons, that is trained with the smooth L1 loss with $\beta = 1$. The models are implemented in Pytorch 1.7.1, and on a standard workstation computer with a GeForce RTX 3090 GPU, training requires approximately 2 h and 20 min for the regression and classification models respectively. All training parameters are provided in the supplementary material.

3   Experimental Results

3.1   Label Quality

The F1 and ROC area-under-curve (ROC-AUC) scores are used to assess label quality (Table 1), compared to the ground truth binary histopathology labels.

Table 1. Label quality metrics (F1 and ROC-AUC) evaluated on Dataset 1 with standard visual labeling (SVL) and CVL+RankNet. The lower 2.5% and upper 97.5% confidence intervals (CIs) are shown in brackets.

Method	Annotator			
	Annotator A	Annotator B	Annotator C	Fused labels
SVL (F1)	0.92 [0.85, 0.98]	0.83 [0.72, 0.92]	0.85 [0.75, 0.93]	0.87 [0.77, 0.94]
CVL+RankNet (F1)	0.99 [0.96, 1.00]	0.93 [0.86, 0.98]	0.93 [0.86, 0.98]	0.97 [0.93, 1.00]
CVL+RankNet (AUC)	0.99 [0.90, 1.00]	0.97 [0.88, 0.99]	0.95 [0.88, 0.99]	0.99 [0.89, 1.00]

ROC-AUC assessment is only possible with CVL+RankNet using its continuous scores. The fused labels represent the annotators' majority vote labels. Considering single-image visual labeling (SVL), F1-scores varied considerably between annotator A and the other two annotators, and the fused labels had a lower F1 score than Annotator A. F1-scores using CVL+RankNet labels were substantially higher for all annotators and the fused labels. Statistical significance was assessed using the 95% CI of paired differences between SVL and CVL+RankNet labels (bootstrap resampling with 5000 samples). Significance was found for Annotators A, B and the fused labels. McNemar's test was also performed between SVL and CVL+RankNet labels ($\alpha = 0.05$), with $p = 0.059$, $p = 0.034$ (significant), $p = 0.096$ and $p = 0.020$ (significant) for Annotators A, B, C and fused labels respectively. High ROC-AUC scores were attained for all annotators and fused labels, without significant differences between them. Furthermore, the Fleiss' Kappa (the degree of annotator agreement over that which would be expected by chance) improved from 0.75 ('substantial agreement') to 0.84 ('almost perfect agreement').

The distribution of label errors using SVL is illustrated in Fig. 3a, showing the percentage of mis-labeled images, separated into four image groups: healthy ($\leq 5\%$ fatty hepatocytes), mild (grade 1, 5–33%), moderate (grade 2, 33–66%) and severe (grade 3, $> 66\%$) steatosis. For every annotator, the majority of errors occur in the mild group, indicating the difficulty in labelling mild cases, which are also the most clinically relevant cases for early disease detection. The results show a tendency of annotators B and C to label mild cases as healthy with single-image annotation, reflecting previous findings showing steatosis is often underestimated by approximately 20% [10, 13]. This effect is also what led to the fused labels having a lower F1 score than Annotator A in Table 1 (SVL).

The distribution of comparative label errors is shown in Fig. 3c. These have been grouped into five bands that represent image pairs with different steatosis degrees from the histopathology scores. Highest error rates occur for pairs with small difference in steatosis level, which is expected, but the error rate is relatively low ($< 3.5\%$ for all annotators). Figure 3b shows the relationship between steatosis severity and CVL+RankNet scores using the fused labels. There are 4 horizontal bands representing normal and 3 pathological grades. The horizontal line shows the annotators' CVL+RankNet score threshold (39.78). Strong cor-

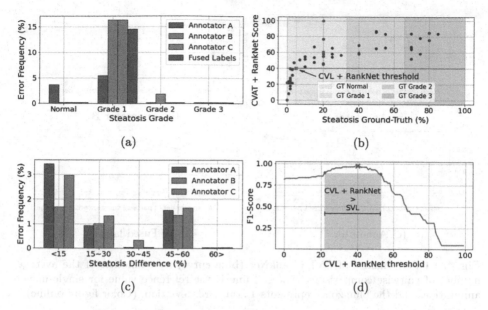

Fig. 3. (a) Shows SVL errors grouped by steatosis grade. The thin black bars indicate no errors. (b) shows the relationship between CVL+RankNet scores and steatosis score from histopathology (results with fused labels). (c) shows the distribution of pairwise label errors using CVL grouped into five pairwise-difference bands. (d) shows the range of CVL+RankNet thresholds that would attain a higher F1-score compared to SVL (fused label results are shown and the cross shows the selected threshold).

relation was found with Spearman's rank correlation ($\rho = 0.87$, $p = 7.6e^{-18}$). The CVL+RankNet scores are clearly able to distinguish well healthy cases from pathological (grades 1, 2 and 3) with almost perfect separation at the annotators' threshold (1 false positive, 1 false negative). The CVL+RankNet scores were not able to clearly separate grades, however, in practice, this not clinically relevant for early detection. The performance of CVL+RankNet for binary classification depends on the selected threshold. Figure 3d shows, using the fused labels, the range of CVL+RankNet thresholds that yield a higher F1 score compared to SVL with fused labels. The marked cross shows the threshold selected by the annotators, lying very close to the optimal threshold. In addition, this threshold may be adjusted to achieve a desired label sensitivity/specificity without relabeling images.

An evaluation of RankNet parameter sensitivity was conducted using fused labels with a grid sampling of 3 architecture parameters (number of hidden layers: $1 \rightarrow 5$, number of neurons per hidden layer: $32 \rightarrow 512$, dropout: $0 \rightarrow 0.9$). Dropout had the strongest influence, and when kept fixed to 0.7, the F1-score across architectures was very stable (mean 0.976 and 0.0089 standard deviation).

A drawback of CVL+RankNet compared to SVL is the increased number of labels. The influence of the number of pairwise comparisons per image on label quality was therefore also studied. Pairwise comparisons from each annotator

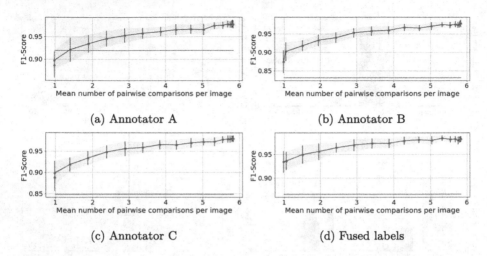

(a) Annotator A   (b) Annotator B

(c) Annotator C   (d) Fused labels

Fig. 4. F1 Performance of CVL+RankNet (blue curve) as a function of the average number of pairwise comparisons. The red line is the reference value for single-image annotations and the blue zone represents 1 standard deviation. (Color figure online)

were randomly added, until an average number of pairwise labels per image was reached. Additions were performed to ensure the images were connected by a single connected component (to ensure the ranking problem is theoretically solvable). This process was repeated 20 times, to estimate of the distribution of F1-scores as a function of number of pairwise comparisons per image. The results are shown in Fig. 4. Because it was unfeasible to make the annotators select thresholds for these RankNet models, the optimal RankNet threshold was selected (the one that maximized the F1 score). Consequently, the F1-scores in Fig. 4 represent performance using CVL+RankNet, if the annotator had selected the optimal threshold in terms of F1-score. The horizontal lines in each plot show the F1-scores of each annotator and the fused labels using SVL. We notice three trends. Firstly, with only 2 pairwise comparisons per image, CVL+RankNet exceeds SVL performance in all cases. Secondly, CVL+RankNet performance saturates at approximately 5.5 comparisons per image. Thirdly, the CVL+Ranknet F1-scores tend to a similar value at approximately 0.975 for all annotators and the fused lables. This implies that SVL reduces the performance gap between annotators, and they can all accurately rank the images in terms of pathological degree using SVL. In practice, this may significantly reduce the need for using multiple annotators to control label quality.

3.2   CNN Model Performance

The CNN models were trained and tested on Dataset 1 using leave-one-out cross-validation (LOOCV) and performance was evaluated with the ROC-AUC metric. Results are presented in Table 2, where each row represents a model configuration and each column represents the source of training labels. There are 4 configu-

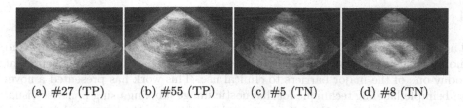

(a) #27 (TP)   (b) #55 (TP)   (c) #5 (TN)   (d) #8 (TN)

Fig. 5. Representative explanation maps using Grad-CAM [21]. showing two true positives (TPs) and two true negatives (TNs) with anonmyous patient IDs.

Table 2. ROC-AUC scores of classification and regression model configurations.

Config.	Training labels				
	Annot. A	Annot. B	Annot. C	Fused labels	GT
Classification: SVL	0.93	0.93	0.95	0.93	0.93
Classification: CVL+RankNet	0.92	0.95	0.92	0.91	0.93
Regression: GT	-	-	-	-	0.92
Regression: CVL+RankNet	0.92	0.94	0.89	0.91	0.92

rations: **Classification: SVL** (classification trained with SVL labels), **Classification: CVL+RankNet** (classification trained with CVL+RankNet labels), **Regression: GT** (regression trained with histopathology scores) and **Regression: CVL+RankNet** (regression trained with CVL+RankNet scores). There are 5 training label sources (3 individual annotators, fused labels and histopathological scores (GT)). One can see that all configurations achieved similar ROC-AUC scores. Although the CVL+RankNet labels are more accurate than SVL labels (Table 1), when used to train the classifier, the improved labels did not translate to a significant better ROC-AUC score. This indicates that the CNNs have some inherent robustness to training label errors in this task. However, the limited size of dataset 1 (17 images with mild steatosis) makes it difficult to draw firm conclusions about label errors on model performance, requiring further research. The regression and classification models also performed similarly.

Additionally, two classification CNNs were trained on Dataset 2 using SVL and CVL+RankNet labels, and tested on Dataset 1. We found that histogram equalization helped to significantly reduce the domain gap [18]. ROC-AUCs were 0.89 (CVL+RankNet) and 0.86 (SVL), and the difference was not statistically significant ($p = 0.34$). To verify that a CAD model learns appropriate task features to make correct predictions, visual explanation maps generated by Grad-CAM [21] have been produced for **Classification CVL+RankNet** (Fig. 5). Pixels with higher influence on the classification decision are illustrated in red (positive influence) and blue (negative influence). Higher influence is generally given to liver paranchyma and the liver/kidney interface (as with human radiologists), giving evidence that the CNN had correctly learned the task.

4 Conclusions

The challenges of obtaining high-quality labeled data for training, testing and monitoring deep learning models for medical image-based diagnosis represent today one of the major barriers to clinical use. This work has presented a novel and simple labeling technique for diagnostic image labeling using pairwise visual comparisons and RankNet, demonstrated to significantly improve label quality for early detection of steatosis (an important global health problem) in US images. Although we have not yet seen a significant improvement when used to train CNNs, the technique already has a high value for accurate model testing and monitoring, with many practical, economical, and ethical advantages compared to labeling with histology. In future work, we aim to gather a larger datasets and test on different pathologies, and to explore techniques such as active learning for efficient pairwise label selection.

Acknowledgements. This work has received funding from France's Région Grand Est. We also greatly thank the annotators for their invaluable participation and work on this study.

References

1. Araújo, A.R., et al.: Global epidemiology of non-alcoholic fatty liver disease/non-alcoholic steatohepatitis: what we need in the future. Liver Int. **38**, 47–51 (2018)
2. Bang, K.B., Cho, Y.K.: Comorbidities and metabolic derangement of NAFLD. J. Lifestyle Med. **5**(1), 7–13 (2015)
3. Bedossa, P.: Pathology of non-alcoholic fatty liver disease. Liver Int. **37**(Suppl. 1), 85–89 (2017)
4. Burges, C., et al.: Learning to rank using gradient descent. In: Proceedings of the 22nd International Conference on Machine Learning, ICML 2005, Bonn, Germany, pp. 89–96. ACM, New York (2005). https://doi.org/10.1145/1102351.1102363. ISBN 1595931805
5. Byra, M., et al.: Transfer learning with deep convolutional neural network for liver steatosis assessment in ultrasound images. Int. J. Comput. Assist. Radiol. Surg. **41**(1), 175–184 (2018)
6. Byra, M., et al.: Liver fat assessment in multiview sonography using transfer learning with convolutional neural networks. J. Ultrasound Med. **41**(1), 175–184 (2021)
7. Chalasani, N., et al.: The diagnosis and management of non-alcoholic fatty liver disease: practice guideline by the American association for the study of liver diseases, American college of gastroenterology, and the American gastroenterological association. Hepatology **55**(6), 2005–2023 (2012)
8. Doughty, H., Damen, D., Mayol-Cuevas, W.: Who's better? Who's best? pairwise deep ranking for skill determination. In: 2018 IEEE/CVF Conference on Computer Vision and Pattern Recognition (CVPR), pp. 6057–6066. IEEE Computer Society, Los Alamitos (2018). https://doi.org/10.1109/CVPR.2018.00634. https://doi.ieeecomputersociety.org/10.1109/CVPR.2018.00634
9. Dulai, P.S., Sirlin, C.B., Loomba, R.: MRI and MRE for non-invasive quantitative assessment of hepatic steatosis and fibrosis in NAFLD and NASH: clinical trials to clinical practice. J. Hepatol. **65**(5), 1006–1016 (2016)

10. Khov, N., Sharma, A., Riley, T.R.: Bedside ultrasound in the diagnosis of nonal-coholic fatty liver disease. World J. Gastroenterol. WJG **20**(22), 6821–6825 (2014)
11. Köppel, M., Segner, A., Wagener, M., Pensel, L., Karwath, A., Kramer, S.: Pair-wise learning to rank by neural networks revisited: reconstruction, theoretical anal-ysis and practical performance. In: Brefeld, U., Fromont, E., Hotho, A., Knobbe, A., Maathuis, M., Robardet, C. (eds.) ECML PKDD 2019. LNCS (LNAI), vol. 11908, pp. 237–252. Springer, Cham (2020). https://doi.org/10.1007/978-3-030-46133-1_15
12. Kou, N.M., Li, Y., Wang, H., Leong Hou, U., Gong, Z.: Crowdsourced top-k queries by confidence-aware pairwise judgments. In: Proceedings of the 2017 ACM Inter-national Conference on Management of Data, SIGMOD 2017, Chicago, Illinois, USA, pp. 1415–1430. ACM, New York (2017). https://doi.org/10.1145/3035918.3035953. ISBN 9781450341974
13. Lee, S.S., Park, S.H.: Radiologic evaluation of nonalcoholic fatty liver disease. World J. Gastroenterol. WJG **20**(23), 7392–7402 (2014)
14. Li, B., et al.: Learning from subjective ratings using auto-decoded deep latent embeddings. In: de Bruijne, M., et al. (eds.) MICCAI 2021. LNCS, vol. 12905, pp. 270–280. Springer, Cham (2021). https://doi.org/10.1007/978-3-030-87240-3_26. ISBN 978-3-030-87239-7
15. Lyu, J., et al.: Ultrasound volume projection image quality selection by ranking from convolutional ranknet. Comput. Med. Imaging Graph. **89**, 7392–7402 (2021)
16. Petzold, G., et al.: Diagnostic accuracy of b-mode ultrasound and hepatorenal index for graduation of hepatic steatosis in patients with chronic liver disease. PLoS One **15**(5), e0231044 (2020)
17. Prathama, F., et al.: Personalized recommendation by matrix cofactorization with multiple implicit feedback on pairwise comparison. Comput. Ind. Eng. **152**, 295–318 (2021)
18. Rhyou, S.-Y., Yoo, J.-C.: Cascaded deep learning neural network for automated liver steatosis diagnosis using ultrasound images. Sensors **21**(16), 5304 (2021)
19. Rinella, M.E., Sanyal, A.J.: Management of NAFLD: a stage-based approach. Nat. Rev. Gastroenterol. Hepatol. **13**(4), 196–205 (2016)
20. Sekachev, B., et al.: opencv/cvat: v1.1.0. https://doi.org/10.5281/zenodo.4009388
21. Selvaraju, R.R., et al.: Grad-cam: visual explanations from deep networks via gradient-based localization. In: Proceedings of the IEEE International Conference on Computer Vision, pp. 618–626 (2017)
22. Szegedy, C., Ioffe, S., Vanhoucke, V., Alemi, A.A.: Inception-v4, inception-resnet and the impact of residual connections on learning. In: Proceedings of the Thirty-First AAAI Conference on Artificial Intelligence, AAAI 2017, San Francisco, Cal-ifornia, USA, pp. 4278–4284. AAAI Press (2017)
23. Talebi, H., et al.: Rank-smoothed pairwise learning in perceptual quality assess-ment. In: 2020 IEEE International Conference on Image Processing (ICIP), pp. 3413–3417 (2020)
24. Tapper, E.B., Lok, A.S.-F.: Use of liver imaging and biopsy in clinical practice. N. Engl. J. Med. **377**(8), 756–768 (2017)
25. The GIMP Development Team. GIMP. Version 2.10.12, 12 June 2019. https://www.gimp.org
26. Younossi, Z.M., et al.: Global epidemiology of nonalcoholic fatty liver disease-meta-analytic assessment of prevalence, incidence, and outcomes. Hepatology **64**(4), 1388–1389 (2016)

27. Yu, X., et al.: Group-aware contrastive regression for action quality assessment. In: Proceedings of the IEEE/CVF International Conference on Computer Vision, pp. 7899–7908 (2021)

28. Zamanian, H., et al.: Implementation of combinational deep learning algorithm for non-alcoholic fatty liver classification in ultrasound images. J. Biomed. Phys. Eng. **11**(1), 73–84 (2021)

29. Zhang, Z., et al.: An improved pairwise comparison scaling method for subjective image quality assessment. In: 2017 IEEE International Symposium on Broadband Multimedia Systems and Broadcasting (BMSB), pp. 1–6 (2017)

Automation of Clinical Measurements on Radiographs of Children's Hips

Peter Thompson[1(✉)], Medical Annotation Collaborative[2], Daniel C. Perry[2,3], Timothy F. Cootes[1], and Claudia Lindner[1]

[1] The University of Manchester, Manchester, UK
peter.thompson-2@manchester.ac.uk
[2] University of Liverpool, Liverpool, UK
[3] Alder Hey Children's Hospital, Liverpool, UK

Abstract. Developmental dysplasia of the hip (DDH) and cerebral palsy (CP) related hip migration are two of the most common orthopaedic diseases in children, each affecting around 1–2 in 1000 children. For both of these conditions, early detection is a key factor in long term outcomes for patients. However, early signs of the disease are often missed and manual monitoring of routinely collected radiographs is time-consuming and susceptible to inconsistent measurement. We propose an automatic system for calculating acetabular index (AcI) and Reimer's migration percentage (RMP) from paediatric hip radiographs. The system applies Random Forest regression-voting to fully automatically locate the landmark points necessary for the calculation of the clinical metrics. We show that the fully automatically obtained AcI and RMP measurements are in agreement with manual measurements obtained by clinical experts, and have replicated these findings in a clinical dataset. Such a system allows for the reliable and consistent monitoring of DDH and CP patients, aiming to improve patient outcomes through hip surveillance programmes.

Keywords: Clinical decision support system · Automated radiographic measurement · Acetabular index · Reimer's migration percentage

1 Introduction

We present a system for the automatic derivation of clinically applicable metrics on paediatric hip radiographs, specifically acetabular index (AcI) and Reimer's migration percentage (RMP). These metrics are commonly used in the diagnosis of developmental dysplasia of the hip (DDH) and monitoring cerebral palsy (CP) related hip migration. AcI attempts to characterise the dysplasia of the joint by estimating the angle of the acetabular roof, and RMP estimates the proportion of the femoral head that is not covered by the acetabulum.

Supplementary Information The online version contains supplementary material available at https://doi.org/10.1007/978-3-031-16437-8_40.

© The Author(s), under exclusive license to Springer Nature Switzerland AG 2022
L. Wang et al. (Eds.): MICCAI 2022, LNCS 13433, pp. 419–428, 2022.
https://doi.org/10.1007/978-3-031-16437-8_40

AcI is a long-standing metric in the determination of hip dysplasia and is considered the most useful for children under the age of 8 years old [3,15]. The value of AcI for a healthy hip is expected to decrease significantly in the first few years of life, typically falling below 20° by 2 years of age, with 30° sometimes used as a rough upper bound for a healthy hip [20]. The reliability of the metric has been called into question, with small variations in the orientation of the pelvis [2] and the interpretation of the relevant anatomical features [9] having been identified as significant sources of error. However, AcI remains the preferred radiographic metric for assessing hip dysplasia in children.

RMP is considered a reliable and repeatable metric for assessing the severity of CP related hip migration [18]. Values greater than 33% are usually used to distinguish cases of subluxation, with 100% to distinguish dislocation [19,20]. RMP is monitored over time to assess the progression of hip migration [7].

There exist many techniques for detecting landmarks in medical images. Several contributions have used Random Forest based techniques [8], sometimes in combination with Markov random fields (MRF) [5,21].

More recently, convolutional neural networks (CNN) have been used to directly regress on the landmark coordinates. These systems are generally improved by the inclusion of a model capable of representing shape constraints, such as SSMs [1], coupled shape models [23], or a graphical component in the CNN itself [11]. Another common CNN approach is to use fully-connected networks to predict heatmaps [16], displacement-maps [25] or both [4].

We use Random Forest regression-voting in the Constrained Local Model (RFRV-CLM) framework [12,14] to automatically locate landmark points which are used to calculate AcI and RMP. RFRV-CLM has been used successfully and achieved state-of-the-art performance in several domains of medical imaging, including bone segmentation [13], automatic cephalometry [22], and wrist fracture detection [6]. Studies of the feet and paediatric hips demonstrated that the technology outperforms deep learning networks on smaller datasets [4,10].

Pham et al. [17] used CNNs to directly regress landmark coordinates in hip radiographs and used those landmarks to calculate RMP. This is the most closely related work to ours, insofar as automatically located points are used to calculate RMP. Xu et al. [24] used CNNs to estimate several clinical indicators, including AcI. These are, to our knowledge, the only prior works that perform these tasks.

Contributions: We present a fully automated system to calculate AcI and RMP from hip radiographs. The system obtains measurements in agreement with clinical experts and has been validated on an independent clinical dataset of 200 images.

2   Methods

2.1   Images

The initial dataset consisted of 450 pelvic radiographs of children aged 2–18 years-old enrolled in the Outcomes Research in Children's Hip Disease (ORCHID) study

(IRAS 227197). This was a challenging dataset containing occlusions and cases of severe disease. One third of cases had DDH, one third had no pathological condition, and one third had Perthes disease. Of the 450 images, 50 were randomly selected to be manually measured by 9 clinicians (8 trainee and 1 senior). Of these 50 cases, 30 had DDH, 10 no condition, and 10 Perthes. Of the remaining 400 images, one was excluded for image quality reasons (parts of the pelvis cropped). One hip per image was used for each image in DataI.

An independent replication dataset (DataR) was also produced, consisting of 200 DDH cases of varying severity. The children were aged 2–18 years and one image per child was collected from Alder Hey Children's Hospital. These images were each measured by 5 clinicians of a similar level of expertise to the trainees who annotated DataI. For DataR, both hips in each image were used, resulting in a dataset of 400 hips.

2.2   Manual Measurements

Both AcI and RMP require a horizontal pelvis line, the Hilgenreiner line, to be found in the image. The Hilgenreiner line passes through the triradiate cartilage on both sides of the pelvis (see Fig. 1). AcI is defined as the angle between the Hilgenreiner line and the acetabular roof. Perkin's line is perpendicular to the Hilgenreiner line and is drawn from the end of the acetabular roof. RMP is defined as the proportion of the projection of the femoral head onto the Hilgenreiner line that is on the lateral side of the Perkin's line.

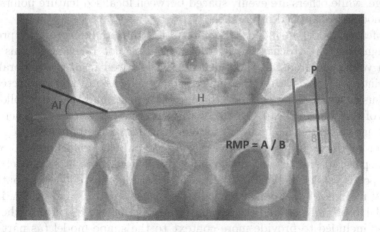

Fig. 1. Diagram showing measurement of AcI and RMP. Line H is the Hilgenreiner line, line P is Perkin's line. AcI is the angle shown on the left. RMP is distance A divided by distance B. Best viewed in colour.

We report the intraclass correlation coefficient type 2 (ICC2; considering both the images and the observers as random effects) and the proportion of variance explained by the differences between observers (PoV) to assess the manual interobserver agreement for the AcI and RMP measurements.

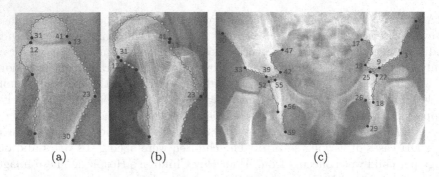

Fig. 2. The femur model for (a) a 2-year-old and (b) an 18-year-old subject, and (c) the pelvis model. Localised feature points are shown in red and labelled with their index.

2.3 Automatic Search Models

Our technique uses the RFRV-CLM framework [12,14] which enables landmark points to be automatically located in images, taking into account both local image features and global shape.

We develop two RFRV-CLM based fully automatic search models, one outlining the proximal femur and the other outlining parts of the pelvis. The models are defined as a set of points, where some points represent localised features in the image, while others are evenly spaced between localised feature points. Both search models are applied in parallel.

The femur model included 42 points along the boundary of the proximal femur, ignoring the trochanters as their shape is not relevant in this study. Since in young children the femoral head is separated from the femoral neck, fixed points were placed on either side of the epiphysis gap and connected by a continuous curve. In older children, the gap disappears and the curve follows the contour of the femoral head. This way, the same model can be used across the range of ages despite the change in topology of the visible bone. Figure 2 shows the femur model as applied to (a) a 2-year-old patient, for whom the femoral head is separated and (b) an 18-year-old patient, for whom it is connected.

The pelvis model included 60 points in total and was defined to outline at least the regions of the pelvis relevant to measuring AcI and RMP, i.e. the acetabular roof and triradiate cartilage. Other distinctive features of the pelvis were also included to provide more context to the shape model (as part of the RFRV-CLM framework). The impact of including these extraneous features is demonstrated in Sect. 3.2. The pelvis model is shown in Fig. 2c.

For each of the 449 images in DataI, we manually placed point positions as shown in Fig. 2 to create the ground truth. Based on the latter, we ran three-fold cross-validation experiments (using randomised folds) to assess the performance of the search models.

Two separate models were used to represent the femur and pelvis because the position of an extremely migrated proximal femur is highly variable, and thus the statistical modes of shape variation of a single model may not be sufficient to capture the range of position in these outlying cases. A comparison of the performance of two separate models with a single model is given in Sect. 3.2.

The search models were assessed using the point-to-point and point-to-curve errors between the manual ground truth annotations and the automatically located point positions from the three-fold cross-validation experiments.

2.4 Automatic Measurements

The automatic point positions used to calculate AcI and RMP for DataI were obtained from the three-fold cross-validation experiments. To automatically calculate AcI and RMP for the replication dataset DataR, we retrained the proximal femur and pelvis search models using all 449 images and manual annotations from DataI.[1] We then ran these models over all images in DataR, and calculated AcI and RMP based on the automatically located point positions.

As in Fig. 2c, points 9 and 39 were used to define the Hilgenreiner line, points 5 and 8 to define the acetabular roof, and point 3 to define the Perkin's line. The bounds of the femoral head were taken to be the points of maximum separation in points 31–41 in Figs. 2a and 2b when projected onto the Hilgenreiner line. AcI and RMP were then calculated as described in Sect. 2.2.

3 Results

3.1 Manual Measurements

The agreement of the manual measurements is given in Table 1. The ICC2 scores for AcI suggest good agreement on DataI and moderate agreement on DataR. For RMP, the ICC2 scores suggest good agreement for both DataI and DataR. Similarly, the PoV explained by the difference between observers suggests good agreement between observers in all experiments. Overall, the manual inter-observer agreement was slightly better on DataI compared to DataR.

3.2 Landmark Detection

The results of the three-fold cross-validation experiments to assess the landmark detection performance of the search models are shown in Table 2.

The performance of the landmark detection when two separate models were used for the femur and pelvis, compared to using a single model incorporating all femur and pelvis points, is shown in Fig. 3a. It can be seen that the two separate models significantly outperform the combined model.

Figure 3b compares the results of the complete pelvis model including all points, as shown in Fig. 2c, with a reduced pelvis model including only the parts of the pelvis relevant to measuring AcI and RMP (points 3–12 and 33–42 as in Fig. 2c). For a fair comparison, only the points in the reduced model were

---

[1] The search models are available for research purposes via www.bone-finder.com.

Table 1. Inter-observer agreement metrics for the manual measurements. (AcI = acetabular index; RMP = Reimer's migration percentage; ICC2 = intraclass correlation coefficient type 2; PoV = proportion of variance explained by the differences between observers).

Dataset	AcI		RMP	
	ICC2	PoV (%)	ICC2	PoV (%)
DataI	0.77	3.6	0.88	4.1
DataR	0.66	9.4	0.83	5.7

Table 2. Performance of the automatic landmark detection, including all femur and pelvis points and the subset of points critical to the measurements (based on applying both search models in parallel).

Measure	Mean	Med.	95th %ile	99th %ile
Point-to-curve all (mm)	1.0	0.8	1.8	3.0
Point-to-point all (mm)	2.7	2.3	4.9	7.2
Point-to-point AcI (mm)	2.5	2.0	5.8	7.9
Point-to-point RMP (mm)	2.9	1.9	7.2	11.4

included when analysing the results for the complete model in Fig. 3b. The results show that the complete pelvis model has a slight but consistent advantage over the reduced pelvis model.

3.3   Automatic Measurements

Table 3 gives the results of comparing the fully automatically obtained measurements to those made manually by the clinical experts. These show good agreement between the automatic results and the manual measurements for both AcI and RMP across the original and replication datasets. The linear mixed-effects model results demonstrate that the automatic measurements give an unbiased estimate of the manual measurements (given the p-values are not significant), and are thus in agreement with the clinicians.

Figure 4 shows scatter plots of the automatic measurements for DataI and DataR alongside all of the manual measurements, demonstrating the agreement between the manual and automatic measurements for both AcI and RMP.

Pham et al. [17] report RMP mean absolute errors of 4.5 ± 4.3 and $4.9 \pm 3.9\%$ points between their automatic RMP measurement system and their two raters. The equivalent values for our system are comparable, being 4.5 ± 6.5 on DataI and 3.3 ± 3.7 on DataR, indicating similar performance. However, these values are not directly comparable as the mean absolute error metric also depends on the distribution of RMP values for each of the datasets. Similarly, Xu et al.[24] do not report results in a way that would allow for a direct comparison, though

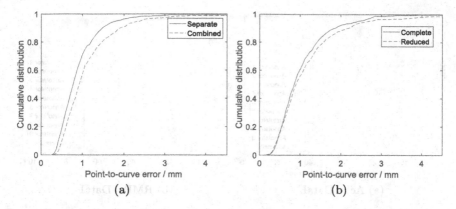

Fig. 3. Cumulative distribution functions (CDFs) showing the point-to-curve results comparing (a) two separate to a single combined model, and (b) the complete pelvis model to a reduced version including only a minimal set of points.

Table 3. Comparison of automatic and mean manual measurements (averaged over all clinical experts) showing percentage-in-range (PiR), type 2 intraclass correlation coefficient (ICC2) with its 95% confidence interval, and bias estimates with associated p-values based on a linear mixed-effects model.

Dataset	AcI				RMP			
	PiR	ICC2	Bias	p	PiR	ICC2	Bias	p
DataI	84	.86 (.78–.91)	−0.70	0.46	86	.92 (.87–.95)	−0.66	0.71
DataR	77	.86 (.84–.88)	0.03	0.97	82	.90 (.87–.92)	−1.57	0.26

they do report a mean landmark detection accuracy for AcI critical landmarks in the range of 4.69–5.37 mm, which is significantly higher than ours (see Table 2).

4   Discussion and Conclusions

We have presented an automatic system to calculate AcI and RMP, two radiographic measurements that are considered among the most important in the diagnosis and monitoring of hip diseases in children. As far as can be determined, the system performs comparably or favourably to the only previously reported systems to automatically calculate AcI or RMP. For both AcI and RMP, we have replicated our findings in an independent dataset and demonstrated that our automatically obtained measurements are in agreement with clinical experts. The system could also be used to derive other common radiographic measurements, such as acetabular depth ratio, Wilberg's angle, and neck shaft angle.

The system was developed and evaluated using a challenging dataset containing (i) occlusions; (ii) cases of severe DDH and Perthes disease; and (iii) a range of ages with a large variation in anatomy. The initial dataset also contained a

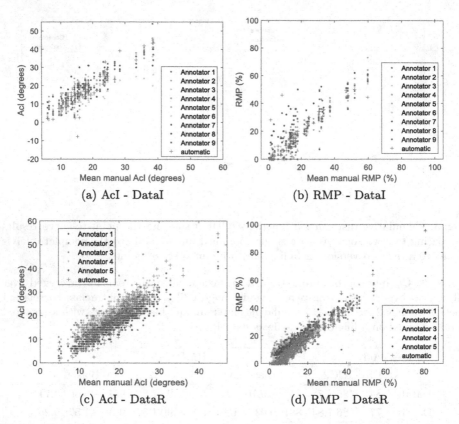

Fig. 4. Plots showing the automatic measurements in the context of all manual measurements for each dataset, sorted by the mean manual measurement.

number of post-operative cases in which the appearance of the acetabular roof was ambiguous. These cases account for the largest differences in AcI between the automatic and manual measurements, including the outlier in Fig. 4a. A limitation of this work is that, although RMP is primarily used to monitor the progression of CP, the dataset did not include any CP patients. Being able to automatically measure RMP has important implications for national surveillance systems of hip subluxation in patients with CP. The application of our system to images of children with CP is an important direction for future work. A more general problem is the fairly large spread in the manual measurements. This makes the evaluation of an automatic system challenging but also highlights the need for a more consistent method of making these radiographic measurements.

Acknowledgements. Peter Thompson was supported by the Wellcome Trust University of Manchester TPA [209741/Z/17/Z] and an Artificial Intelligence in Health and Care Award (AI_AWARD02268). Claudia Lindner was funded by the Medical Research Council, UK (MR/S00405X/1) as well as a Sir Henry Dale Fellowship jointly funded by the Wellcome Trust and the Royal Society (223267/Z/ 21/Z). Daniel C. Perry is funded by the National Institute for Health Research (NIHR) through a NIHR Research Professorship. Manual measurements were provided by the Medical Annotation Collaborative (Grace Airey, Mohammed Ali, Bishoy Bessada, Benjamin Gompels, Tom Hughes, Mustafa Javaid, Zarmina Kakakhel, Mohammed Khattak, James Redfern, David Samy, Myles Simmons, Lucy Stead). This report is independent research funded by the National Institute for Health Research (Artificial Intelligence, An Automated System for Measuring Hip Dysplasia in Children with Cerebral Palsy, AI_AWARD02268). The views expressed in this publication are those of the author(s) and not necessarily those of the National Institute for Health Research or the Department of Health and Social Care.

References

1. Arik, S.Ö., Ibragimov, B., Xing, L.: Fully automated quantitative cephalometry using convolutional neural networks. J. Med. Imaging 4(1), 014501 (2017)
2. Van der Bom, M., Groote, M., Vincken, K., Beek, F., Bartels, L.: Pelvic rotation and tilt can cause misinterpretation of the acetabular index measured on radiographs. Clin. Orthop. Relat. Res.® 469(6), 1743–1749 (2011)
3. Broughton, N., Brougham, D., Cole, W., Menelaus, M.: Reliability of radiological measurements in the assessment of the child's hip. J. Bone Joint Surg. Br. Vol. 71(1), 6–8 (1989)
4. Davison, A.K., Lindner, C., Perry, D.C., Luo, W., Cootes, T.F.: Landmark localisation in radiographs using weighted heatmap displacement voting. In: Vrtovec, T., Yao, J., Zheng, G., Pozo, J.M. (eds.) MSKI 2018. LNCS, vol. 11404, pp. 73–85. Springer, Cham (2019). https://doi.org/10.1007/978-3-030-11166-3_7
5. Donner, R., Menze, B.H., Bischof, H., Langs, G.: Global localization of 3d anatomical structures by pre-filtered hough forests and discrete optimization. Med. Image Anal. 17(8), 1304–1314 (2013)
6. Ebsim, R., Naqvi, J., Cootes, T.: Fully automatic detection of distal radius fractures from posteroanterior and lateral radiographs. In: Cardoso, M.J., et al. (eds.) CARE/CLIP -2017. LNCS, vol. 10550, pp. 91–98. Springer, Cham (2017). https://doi.org/10.1007/978-3-319-67543-5_8
7. Hägglund, G., Lauge-Pedersen, H., Persson, M.: Radiographic threshold values for hip screening in cerebral palsy. J. Child. Orthop. 1(1), 43–47 (2007). https://doi.org/10.1007/s11832-007-0012-x
8. Han, D., Gao, Y., Wu, G., Yap, P.T., Shen, D.: Robust anatomical landmark detection with application to MR brain image registration. Comput. Med. Imaging Graph. 46, 277–290 (2015)
9. Kay, R.M., Watts, H.G., Dorey, F.J.: Variability in the assessment of acetabular index. J. Pediatr. Orthop. 17(2), 170–173 (1997)
10. Lauder, J., et al.: A fully automatic system to assess foot collapse on lateral weight-bearing foot radiographs: a pilot study. Comput. Methods Program. Biomed. 213(106507), 106507 (2022). https://doi.org/10.1016/j.cmpb.2021.106507
11. Li, W., et al.: Structured landmark detection via topology-adapting deep graph learning. In: Vedaldi, A., Bischof, H., Brox, T., Frahm, J.-M. (eds.) ECCV 2020. LNCS, vol. 12354, pp. 266–283. Springer, Cham (2020). https://doi.org/10.1007/978-3-030-58545-7_16
12. Lindner, C., Bromiley, P.A., Ionita, M.C., Cootes, T.F.: Robust and accurate shape model matching using random forest regression-voting. IEEE Trans. Pattern Anal. Mach. Intell. 37(9), 1862–1874 (2014)

13. Lindner, C., Thiagarajah, S., Wilkinson, J.M., Wallis, G.A., Cootes, T.F.: Accurate fully automatic femur segmentation in pelvic radiographs using regression voting. In: Ayache, N., Delingette, H., Golland, P., Mori, K. (eds.) MICCAI 2012. LNCS, vol. 7512, pp. 353–360. Springer, Heidelberg (2012). https://doi.org/10.1007/978-3-642-33454-2_44

14. Lindner, C., et al.: Fully automatic segmentation of the proximal femur using random forest regression voting. IEEE Trans. Med. Imaging **32**(8), 1462–1472 (2013)

15. Parrott, J., et al.: Hip displacement in spastic cerebral palsy: repeatability of radiologic measurement. J. Pediatr. Orthop. **22**(5), 660–667 (2002)

16. Payer, C., Štern, D., Bischof, H., Urschler, M.: Regressing heatmaps for multiple landmark localization using CNNs. In: Ourselin, S., Joskowicz, L., Sabuncu, M.R., Unal, G., Wells, W. (eds.) MICCAI 2016. LNCS, vol. 9901, pp. 230–238. Springer, Cham (2016). https://doi.org/10.1007/978-3-319-46723-8_27

17. Pham, T.-T., Le, M.-B., Le, L.H., Andersen, J., Lou, E.: Assessment of hip displacement in children with cerebral palsy using machine learning approach. Med. Biol. Eng. Comput. **59**(9), 1877–1887 (2021). https://doi.org/10.1007/s11517-021-02416-9

18. Pons, C., Rémy-Néris, O., Médée, B., Brochard, S.: Validity and reliability of radiological methods to assess proximal hip geometry in children with cerebral palsy: a systematic review. Dev. Med. Child Neurol. **55**(12), 1089–1102 (2013)

19. Reimers, J.: The stability of the hip in children: a radiological study of the results of muscle surgery in cerebral palsy. Acta Orthop. Scand. **51**(sup184), 1–100 (1980)

20. Scrutton, D., Baird, G.: Surveillance measures of the hips of children with bilateral cerebral palsy. Arch. Dis. Child. **76**(4), 381–384 (1997)

21. Urschler, M., Ebner, T., Štern, D.: Integrating geometric configuration and appearance information into a unified framework for anatomical landmark localization. Med. Image Anal. **43**, 23–36 (2018)

22. Wang, C.W., et al.: A benchmark for comparison of dental radiography analysis algorithms. Med. Image Anal. **31**, 63–76 (2016). https://doi.org/10.1016/j.media.2016.02.004

23. Wirtz, A., Mirashi, S.G., Wesarg, S.: Automatic teeth segmentation in panoramic X-ray images using a coupled shape model in combination with a neural network. In: Frangi, A.F., Schnabel, J.A., Davatzikos, C., Alberola-López, C., Fichtinger, G. (eds.) MICCAI 2018. LNCS, vol. 11073, pp. 712–719. Springer, Cham (2018). https://doi.org/10.1007/978-3-030-00937-3_81

24. Xu, W., et al.: A deep-learning aided diagnostic system in assessing developmental dysplasia of the hip on pediatric pelvic radiographs. Front. Pediatr. **9** (2021)

25. Zhang, J., Liu, M., Shen, D.: Detecting anatomical landmarks from limited medical imaging data using two-stage task-oriented deep neural networks. IEEE Trans. Image Process. **26**(10), 4753–4764 (2017)

Supervised Contrastive Learning to Classify Paranasal Anomalies in the Maxillary Sinus

Debayan Bhattacharya[1,2]($\boxtimes$), Benjamin Tobias Becker[2], Finn Behrendt[1],
Marcel Bengs[1], Dirk Beyersdorff[3], Dennis Eggert[2], Elina Petersen[4],
Florian Jansen[2], Marvin Petersen[5], Bastian Cheng[5], Christian Betz[2],
Alexander Schlaefer[1], and Anna Sophie Hoffmann[2]

[1] Insititute of Medical Technologies and Intelligent Systems,
Hamburg University of Technology, Hamburg, Germany
debayan.bhattacharya@tuhh.de, d.bhattacharya@uke.de
[2] Department of Otorhinolaryngology, Head and Neck Surgery and Oncology,
University Medical Center Hamburg-Eppendorf, Hamburg, Germany
[3] Clinic and Polyclinic for Diagnostic and Interventional Radiology and Nuclear
Medicine, University Medical Center Hamburg-Eppendorf, Hamburg, Germany
[4] Population Health Research Department, University Heart and Vascular Center,
University Medical Center Hamburg-Eppendorf, Hamburg, Germany
[5] Clinic and Polyclinic for Neurology,
University Medical Center Hamburg-Eppendorf, Hamburg, Germany

Abstract. Using deep learning techniques, anomalies in the paranasal sinus system can be detected automatically in MRI images and can be further analyzed and classified based on their volume, shape and other parameters like local contrast. However due to limited training data, traditional supervised learning methods often fail to generalize. Existing deep learning methods in paranasal anomaly classification have been used to diagnose at most one anomaly. In our work, we consider three anomalies. Specifically, we employ a 3D CNN to separate maxillary sinus volumes without anomaly from maxillary sinus volumes with anomaly. To learn robust representations from a small labelled dataset, we propose a novel learning paradigm that combines contrastive loss and cross-entropy loss. Particularly, we use a supervised contrastive loss that encourages embeddings of maxillary sinus volumes with and without anomaly to form two distinct clusters while the cross-entropy loss encourages the 3D CNN to maintain its discriminative ability. We report that optimising with both losses is advantageous over optimising with only one loss. We also find that our training strategy leads to label efficiency. With our method, a 3D CNN classifier achieves an AUROC of 0.85 ± 0.03

A. Schlaefer and A. S. Hoffmann—These authors contributed equally to this work.

Supplementary Information The online version contains supplementary material available at https://doi.org/10.1007/978-3-031-16437-8_41.

© The Author(s), under exclusive license to Springer Nature Switzerland AG 2022
L. Wang et al. (Eds.): MICCAI 2022, LNCS 13433, pp. 429–438, 2022.
https://doi.org/10.1007/978-3-031-16437-8_41

while a 3D CNN classifier optimised with cross-entropy loss achieves an AUROC of 0.66 ± 0.1. Our source code is available at https://github. com/dawnofthedebayan/SupConCE_MICCAI_22.

Keywords: Self-supervised learning · Paranasal pathology · Nasal pathology · Magnetic resonance images

1   Introduction

Paranasal sinus anamolies are common incidental findings reported in patients who undergo diagnostic imaging of the head [21] for neuroradiological assessment. Understanding the different opacifications of the paranasal sinuses is very useful, because the frequency of these findings represent clinical challenges [5], and little is known about the incidence and significance of these morphological changes in the general population. There have been numerous studies on analysing the occurrence and progression of these incidental findings [2, 16–19], but mostly in patients with sinunasal symptoms.

In our study, elderly people (45–74 years) received an MRI for neuroradiological assessment [7] in the city of Hamburg. The purpose of our study is to find out what percentage of patients, who do not have sinunasal symptoms, show findings in the paranasal sinus system in MRI images and if it is possible to detect anomalies in the paranasal sinus system using deep learning techniques and further analyze and classify based on their volume, shape and other parameters like local contrast.

A three year retrospective study showed malignant tumors were misdiagnosed as nasal polyps with a misdiagnosis rate of 5.63% while inverted papilloma were misdiagnosed as nasal polyps with a rate of 8.45% [12]. As a first step, it would be beneficial to separate MRI with any paranasal anomaly from normal MRI using Computer Aided Diagnostics (CAD). This would allow the physicians to closely inspect the MRIs containing paranasal anomaly with finer detail. This can reduce chances of misdiagnosis while decreasing the workload of physicians from having to see normal MRIs.

There have been works of paranasal sinus anomaly diagnosis using supervised learning [8, 10, 11]. However, all these works have been used to classify at most one anomaly. Our work considers three anomalies namely: (i) mucosal thickening (ii) polyps and (iii) cysts. The anomalies are differently located within the maxillary sinus and therefore we use a 3D volume of the maxillary sinus as input to our 3D CNN. Additionally, existing works train on large datasets to achieve good performance. However, labelling is a time consuming task and in some scenarios it requires the supervision of Ears, Nose and Throat (ENT) specialised radiologists who may not be immediately available. In our case, we have a small dataset of 199 MRI volumes. To mitigate the limitations of our small labelled dataset, we use contrastive learning [14] to learn robust representations that does not overfit on the training set. In contrastive learning, an encoder learns to map positive pairs close together in the embedding space while pushing away negative pairs. In SimCLR [1], an image is transformed twice through random transformations. The transformed "views" of the image constitute a positive pair and every other image in the mini-batch is used to construct negative pairs with the reference image. There has been significant research in finding the best transformations as the chosen transformations dictate the quality of learnt representation [1]. The underlying assumption is that transformations augment the image while preserving the semantic information. However, in our case two or more images in the mini-batch can be semantically similar as they belong to the same class. Particularly, a maxillary sinus of one patient can be semantically similar to another patient's maxillary sinus if both patients do not exhibit any paranasal anomalies. Furthermore, they can be different as well due to the anatomical variations of the maxillary sinus [17,19]. Since our dataset contains anomalies, it is also important that the classifier does not overfit to a particular anomaly. Therefore, it is important for a classifier to learn anatomically invariant representation of the maxillary sinus for volumes that contain no anomaly and learn anomaly invariant representation of the maxillary sinus for volumes exhibiting one of the three anomalies. This can reduce the chances of overfitting on the training set. This motivates us to employ a supervised contrastive learning approach [9] that brings embeddings of maxillary sinus volumes without anomaly closer together while pushing away embeddings of maxillary sinus with anomaly in the embedding space and vice versa. Compared to the original method [9], we propose a training strategy that simultaneously trains our 3D CNN using the supervised contrastive loss and regular cross-entropy loss using two different projection networks. The reasoning behind this is that minimizing the contrastive loss encourages the 3D CNN to learn representations that are robust to anatomical and anomaly variations while the cross-entropy enforces the 3D CNN to preserve discriminative ability.

In summary, our contributions are three-fold. First, we demonstrate the feasibility of a deep learning approach to classify between normal and anomalous maxillary sinuses. Second, we demonstrate through extensive experiments that combining supervised contrastive loss and cross-entropy loss is the better approach to improve the discriminative ability of the 3D CNN classifier. Third, we empirically show that our method is the most label efficient.

2   Method

Our method is shown in Fig. 1. In global contrastive learning methods such as SimCLR [1], we learn a parametric function $F_\theta : X \to \mathbb{R}^D$ where $\mathbb{R}^D$ is a unit hypersphere. F_θ is trained to map semantically similar samples closer together and semantically dissimilar samples further apart through the InfoNCE loss [20]. The underlying assumption here is that images in the mini-batch are semantically dissimilar. This is untrue in our case as we have maxillary sinus volumes belonging to one of the two classes and there are semantic similarities and dissimilarities in the intra and inter class samples. Therefore, it is not possible to form meaningful clusters from the global contrastive loss described in SimCLR

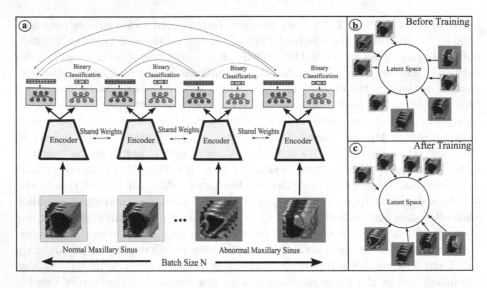

Fig. 1. (a) Our proposed method. The curved green and red lines represent similar and dissimilar representations respectively. (b)–(c) Illustration of the latent space embedding of normal and anomalous maxillary sinuses before and after the encoder is trained respectively. (Color figure online)

[1]. Therefore, to incorporate the class priors, we sample volumes from the two classes explicitly. Given an input mini-batch $B = \{x_1, x_2, ..., x_N\}$ where x_i represent the input 3D volume, a random transformation set T is used to form a pair of augmented volumes. Instead of randomly sampling N samples, we randomly sample $N/2$ maxillary sinus volumes without anomaly and $N/2$ maxillary sinus volumes with anomaly from our dataset. Each of these $N/2$ subsets undergo random transformation twice using T. Let us denote the set containing all the augmented volumes of the C classes as $\mathbb{M} = \bigcup_{c=1}^{C} M_c$. In our case, $C = 2$. Here, M_c is the subset of augmented volumes belonging to a single class and $|M_c| = N$ is its cardinality. Let $m_i, i \in \mathbb{I} = \{1, 2, ..., 2N\}$ represent the augmented volumes in set $\mathbb{M}$ and $m_{k(i)}$ is its corresponding volume augmented from the same volume in B. Furthermore, let I_c represent indices of all the augmented volumes belonging to class c such that $\mathbb{I} = \bigcup_{c=1}^{C} I_c$. Using the above stated assumptions, the InfoNCE loss that takes into consideration the class priors when making positive and negative pairs can be written as:

$$L_{simclr} = -\sum_{c=1}^{C} \frac{1}{|M_c|} \sum_{i \in I_c} log \frac{e^{sim(Z_i, Z_{k(i)})/\tau}}{e^{sim(Z_i, Z_{k(i)})/\tau} + \sum_{j \in \mathbb{I} \setminus I_c} e^{sim(Z_i, Z_j)/\tau}} \quad (1)$$

where $\tau \in \mathbb{R}^+$ is the scalar temperature parameter, $Z_i = F_\theta^{con}(m_i)$ is the normalised feature vector such that $F_\theta^{con}(.) = Proj_1(Enc(.))$, $k(i)$ is the index of the corresponding volume in $\mathbb{M}$ augmented from the same volume in B and $sim(.)$ is the cosine similarity function. Although Eq. 1 only constructs negative pairs such that the two volumes are from different classes, it still constructs positive pairs by augmenting the same volume using the random augmentation set T. As a result, we are reliant on the transformations to learn meaningful representations. However, the transformations do not guarantee anatomical and anomaly invariance as the encoder is not incentivised to produce similar representations for volumes belonging to the same class in the mini-batch. Therefore, we use a supervised contrastive loss [9] that constructs arbitrary number of positive pairs where each volume in the pair is from the same class but unique in the mini-batch. Formally, the supervised contrastive loss can be described as shown below:

$$L_{sc} = -\sum_{c=1}^{C} \frac{1}{|M_c|} \sum_{i \in I_c} log \frac{\sum_{j \in I_c \backslash \{i\}} e^{sim(Z_i, Z_j)/\tau}}{\sum_{j \in I_c \backslash \{i\}} e^{sim(Z_i, Z_j)/\tau} + \sum_{j \in \mathbb{I} \backslash I_c} e^{sim(Z_i, Z_j)/\tau}} \quad (2)$$

The main differences of Eq. 2 compared to Eq. 1 is that numerous positive pairs are constructed in the numerator by matching every volume with every other volume belonging to the same class in the mini-batch. In this case, $|M_c|$ = $N/2$ as we do not use T and $\mathbb{I} = \{1, 2, ..., N\}$. This incentivises the encoder to give similar representations for volumes belonging to the same class. This leads to learning anatomical and anomaly invariant representations. Apart from using L_{sc} we also use regular cross-entropy loss to preserve the discrimintative ability of our 3D CNN. The cross-entropy loss is formalised as follows:

$$L_{ce} = -\frac{1}{N} \sum_{i \in \mathbb{I}} y_i log(F_\theta^{class}(m_i)) \quad (3)$$

Here, $F_\theta^{class}(.)$ can be decomposed into $Proj_2(Enc(.))$ and y_i is the class label of m_i such that $y_i \in \{0, 1\}$. Therefore, our combined loss function is

$$L_{ours} = L_{sc} + \lambda L_{ce} \quad (4)$$

In our case, we set $\lambda = 1$. In summary, we train models using only L_{ce} and set this as the baseline. We then train models using L_{simclr} to show that transformation invariance does not help in our downstream classification task. Next, we train our models using L_{sc} to show the benefit of clustering based on class priors and the importance of anatomical and anomaly invariant representation learning. Finally, we train our models using L_{ours} to show that contrastive loss and cross-entropy loss improve the discriminative ability of the models and overfit the least on the training set.

2.1 Dataset

As part of the population study [7], MRI of the head and neck area of participants based in the city of Hamburg, Germany were recorded. The MRIs were recorded at University Medical Center Hamburg-Eppendorf. The age group of the

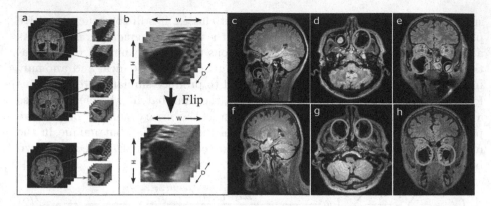

Fig. 2. [LEFT] Pre-processing steps involve (a) Extraction of 3D sub-volumes of left and right maxillary sinus from FLAIR MRI samples. (b) Flipping the coronal plane slices of right maxillary sinus sub-volume to give it the appearance of left maxillary sinus sub-volume. [RIGHT] FLAIR-MRI slices from the sagittal, axial and coronal views illustrating the difference between anomaly and normal class. The red circles denote anomalies and green circles denote normal maxillary sinus. (c) Cyst observed in right maxillary sinus. (d) Polyp observed in the left maxillary sinus (e) Cyst observed in left maxillary sinus. (f)–(h) FLAIR-MRI slices with no pathology. (Color figure online)

participants were between 45 and 74 years. Each participant had T1-weighted and fluid attenuated inversion recovery (FLAIR) sequences stored in the NIfTI[1] format. FLAIR-MRIs were chosen as the imaging modality as the incidental findings are more visible due to the higher contrast relative to T1 weighted MRI. The labelled dataset consists of 199 FLAIR-MRI volumes of which 106 patients exhibit normal maxillary sinuses and 93 patients have maxillary sinuses with anomaly in at least one maxillary sinus. The diagnosis of the observed pathology in 199 FLAIR-MRIs was confirmed by two ENT surgeons and one ENT specialised radiologist. The incidental findings are categorised and defined as follows: (i) mucosal thickening (ii) polyps (iii) cysts. The statistics of the pathology observed is reported in the supplementary material. In this work, all the anomalies are grouped into a single class called "anomaly" and all the normal maxillary sinuses are grouped into a class called "normal". Altogether, there are 269 maxillary sinus volumes without anomaly and 130 abnormal maxillary sinus volumes with anomaly. Each MRI has a resolution of $173 \times 319 \times 319$ voxels along the saggital, coronal and axial directions respectively. The voxel size is 0.53 mm $\times$ 0.75 mm $\times$ 0.75 mm.

Preprocessing: We performed rigid registration by randomly selecting one FLAIR-MRI sample as a fixed volume followed by resampling to a dimension of $128 \times 128 \times 128$. Of the resampled volumes, we extracted two sub-volumes, one for each maxillary sinus from a single patient. We made sure that the extracted sub-volumes subsumed the maxillary sinus.

---

[1] https://nifti.nimh.nih.gov/.

Since the maxillary sinus are symmetric, we horizontally flipped the coronal planes of right maxillary sinus volumes to make it look like left maxillary sinus volumes. The decision of which maxillary sinus to flip was arbitrary. Ultimately, these sub-volumes were reshaped to a standard size of $32 \times 32 \times 32$ voxels for the 3D CNN. Finally, all the maxillary sinus volumes were normalised to the range of -1 to 1. Our preprocessing step is shown in Fig. 2.

Training, Validation and Test Split: We perform a nested stratified K-fold with 5 inner and 5 outer folds. The inner fold was used to choose the best hyperparameters. In summary, each experiment has 80 volumes in test set, 64 volumes in cross validation set and 255 samples in training set. The folds are constructed by preserving the percentage of samples for the two classes.

3 Experiments and Discussion

3.1 Implementation Details

All of our experiments are implemented in PyTorch [15] and PyTorch Lightning [4]. We use a batch size of 128 for all our experiments based on hyperparameter tuning (See supplementary material). We use Adam Optimization with a learning rate of 1e-4. Similar to Chen et al. [1], we fix $\tau = 0.1$. Our encoder $Enc(.)$ is the 3DResNet18 [6]. Our projection layer $Proj_1(.)$ is a fully connected layer with input dimension 512 and output dimension 128. A ReLU activation is placed in between the layers. Projection layer $Proj_2(.)$ is a linear fully connected layer with input and output dimensions of 512 and 2 respectively. All our models are trained for 200 epochs.

3.2 Evaluation of Learnt Representations

Table 1. Evaluation of our representations

Method	L_{ce}	L_{simclr}	L_{sc}	Accuracy	F1 (weighted)	AUROC	AUPRC
Scratch	✓			0.68 ± 0.03	0.57 ± 0.04	0.66 ± 0.10	0.53 ± 0.12
Scratch (with aug)	✓			0.69 ± 0.03	0.58 ± 0.05	0.68 ± 0.10	0.56 ± 0.10
SimCLR		✓		0.65 ± 0.03	0.58 ± 0.04	0.54 ± 0.09	0.41 ± 0.09
SupCon [9]			✓	0.72 ± 0.05	0.70 ± 0.06	0.73 ± 0.08	0.61 ± 0.08
Ours	✓		✓	$\mathbf{0.80 \pm 0.02}$	$\mathbf{0.78 \pm 0.03}$	$\mathbf{0.85 \pm 0.03}$	$\mathbf{0.78 \pm 0.03}$

The metrics we have used to evaluate our representations are accuracy, F1 weighted, Area Under Receiver Operator Characteristics (AUROC) and Area Under Precision Recall Curve (AUPRC). We chose F1 weighted and AUPRC as they give a fair assessment of the performance of the models in the presence of class imbalance. Our random transformation set T consists of random affine, flip and gaussian noise as these are semantic preserving transforms. For training models with L_{simclr}, L_{sc} and L_{ours} we followed the training strategy followed by Khosla et al. [9]. The inference is performed using $Proj_2(.)$ for all the experiments. We test for statistically significant difference in our performance metrics using a permutation test with $nP = 10000$ samples and a significance level of $\alpha = 0.05$ [3]. The difference in the AUROC, AUPRC, F1 weighted and accuracy of our method is significant ($p < 0.05$) compared to the other methods. From the results in Table 1 we observe that the models trained with only L_{ce} overfit and do not generalize well. Models trained with L_{sc} achieve a significant boost in all metrics compared to models trained using L_{ce} and L_{simclr}. We conjecture this to be the case because the supervised contrastive loss clusters the maxillary sinus volumes representations based on its class leading to invariant representations and less overfitting. The absence of L_{ce} causes the clusters to be more spread out (See Fig. 3 SupCon). Our loss L_{ours} shows the best performance due to the increased discriminative ability of the classifier which is reflected by the formation of compact clusters (See Fig. 3 Ours). SimCLR performs the worst as they fail to form meaningful clusters (See Fig. 3 SimCLR).

3.3   Label Efficient Representation Learning

We evaluated the performance of models trained with the different loss functions by supplying 60% and 80% of training samples. We excluded SimCLR approach because it performed very poorly on 60% and 80% of training set. We performed

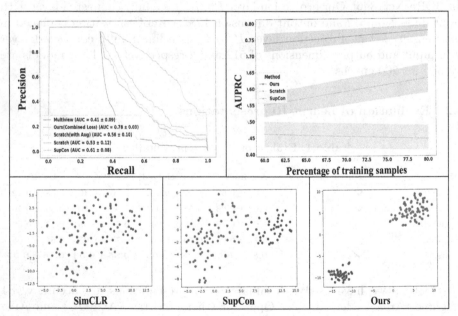

Fig. 3. (UPPER LEFT) Precision Recall Curve. (UPPER RIGHT) Illustrating the label efficiency of our method by plotting the AUPRC against the percentage of training dataset (BOTTOM LEFT, MIDDLE, RIGHT) t-SNE [13] with perplexity = 30, learning rate = 200, iterations=1000 used to visualise the representations learnt by various contrastive losses. The red dots denote normal class and purple dots denote anomaly class.

a five-fold cross validation experiment with the same training strategy. The test set is the same for all the experiments. The AUPRC of the models are displayed in the upper right graphic of Fig. 3. We observe that models trained using L_{sc} outperform the baseline by a significant margin. Even with limited data, both the losses (L_{sc} and L_{ours}) show improved performance with the injection of more training data. We also observe that L_{ours} almost achieves AUPRC of 100% training set and it is also the most label-efficient. These results reveal that our approach can reduce labelling effort of physicians to an extent thereby allowing them to invest more time in the diagnosis and evaluation of difficult clinical cases.

4   Conclusion

Previous works on population studies [2,16–19] have relied on manual analysis by physicians. Our work is a first step towards bringing automation in such studies. We show the benefit of contrastive loss in the classification of anomalies in the maxillary sinus from limited labelled dataset. Furthermore, we report the performance improvements relative to regular cross-entropy. Specifically, clustering based on class priors is helpful to learn representations that overfit less on the training set. We also show that a combinination of cross-entropy loss and supervised contrative loss improve the performance of the model. Finally, compared to other works that use deep learning for paranasal inflammation study [8,10,11], ours is the first work that tries to achieve label efficiency and thus attempts to reduce the workload of physicians. A limitation of our work is that the classification accuracy is still not satisfactory. As future work, we plan to label more MRIs and even perform classification of the type of anomaly observed in the maxillary sinus.

References

1. Chen, T., Kornblith, S., Norouzi, M., Hinton, G.: A simple framework for contrastive learning of visual representations. https://arxiv.org/pdf/2002.05709
2. Cooke, L.D., Hadley, D.M.: MRI of the paranasal sinuses: incidental abnormalities and their relationship to symptoms. J. Laryngol. Otol. **105**(4), 278–281 (1991). https://doi.org/10.1017/s0022215100115609
3. Efron, B., Tibshirani, R.: An Introduction to the Bootstrap, Monographs on Statistics and Applied Probability, vol. 57, [nachdr.] edn. Chapman & Hall, Boca Raton (1998)
4. Falcon, F.N., et al.: Pytorch lightning, vol. 3. GitHub (2019). https://github.com/PyTorchLightning/pytorch-lightning
5. Hansen, A.G., et al.: Incidental findings in MRI of the paranasal sinuses in adults: a population-based study (hunt MRI). BMC Ear Nose Throat Disord. **14**(1), 13 (2014). https://doi.org/10.1186/1472-6815-14-13
6. Hara, K., Kataoka, H., Satoh, Y.: Learning spatio-temporal features with 3D residual networks for action recognition. http://arxiv.org/pdf/1708.07632v1

7. Jagodzinski, A., et al.: Rationale and design of the Hamburg city health study. Eur. J. Epidemiol. **35**(2), 169–181 (2019). https://doi.org/10.1007/s10654-019-00577-4

8. Jeon, Y., et al.: Deep learning for diagnosis of paranasal sinusitis using multiview radiographs. Diagnost. (Basel Switz.) **11**(2) (2021). https://doi.org/10.3390/diagnostics11020250

9. Khosla, P., et al.: Supervised contrastive learning. https://arxiv.org/pdf/2004.11362

10. Kim, Y., et al.: Deep learning in diagnosis of maxillary sinusitis using conventional radiography. Invest. Radiol. **54**(1), 7–15 (2019). https://doi.org/10.1097/RLI.0000000000000503

11. Liu, G.S., et al.: Deep learning classification of inverted papilloma malignant transformation using 3d convolutional neural networks and magnetic resonance imaging. Int. Forum Allergy Rhinol. (2022). https://doi.org/10.1002/alr.22958

12. Ma, Z., Yang, X.: Research on misdiagnosis of space occupying lesions in unilateral nasal sinus. Lin chuang er bi yan hou tou jing wai ke za zhi = J. Clin. Otorhinolaryngol. Head Neck Surg. **26**(2), 59–61 (2012). https://doi.org/10.13201/j.issn.1001-1781.2012.02.005

13. van der Maaten, L., Hinton, G.: Visualizing data using t-SNE. J. Mach. Learn. Res. **9**(86), 2579–2605 (2008). http://jmlr.org/papers/v9/vandermaaten08a.html

14. van den Oord, A., Li, Y., Vinyals, O.: Representation learning with contrastive predictive coding. CoRR abs/1807.03748 (2018). http://arxiv.org/abs/1807.03748

15. Paszke, A., et al.: PyTorch: an imperative style, high-performance deep learning library. https://arxiv.org/pdf/1912.01703

16. Rak, K.M., Newell, J.D., Yakes, W.F., Damiano, M.A., Luethke, J.M.: Paranasal sinuses on MR images of the brain: significance of mucosal thickening. AJR Am. J. Roentgenol. **156**(2), 381–384 (1991). https://doi.org/10.2214/ajr.156.2.1898819

17. Rege, I.C.C., Sousa, T.O., Leles, C.R., Mendonça, E.F.: Occurrence of maxillary sinus abnormalities detected by cone beam CT in asymptomatic patients. BMC Oral Health **12**, 30 (2012). https://doi.org/10.1186/1472-6831-12-30

18. Stenner, M., Rudack, C.: Diseases of the nose and paranasal sinuses in child. GMS Curr. Top. Otorhinolaryngol. Head Neck Surg. **13**, Doc10 (2014). https://doi.org/10.3205/cto000113

19. Tarp, B., Fiirgaard, B., Christensen, T., Jensen, J.J., Black, F.T.: The prevalence and significance of incidental paranasal sinus abnormalities on MRI. Rhinology **38**(1), 33–38 (2000)

20. den van Oord, A., Li, Y., Vinyals, O.: Representation learning with contrastive predictive coding. https://arxiv.org/pdf/1807.03748

21. Wilson, R., Kuan Kok, H., Fortescue-Webb, D., Doody, O., Buckley, O., Torreggiani, W.C.: Prevalence and seasonal variation of incidental MRI paranasal inflammatory changes in an asymptomatic irish population. Ir. Med. J. **110**(9), 641 (2017)

Show, Attend and Detect: Towards Fine-Grained Assessment of Abdominal Aortic Calcification on Vertebral Fracture Assessment Scans

Syed Zulqarnain Gilani[1,2,3(✉)], Naeha Sharif[1,2,3], David Suter[1,2,3], John T. Schousboe[4], Siobhan Reid[5], William D. Leslie[6], and Joshua R. Lewis[1]

[1] Nutrition and Health Innovation Research Institute, Edith Cowan University, Joondalup, Australia
s.gilani@ecu.edu.au
[2] Centre for AI&ML, School of Science, Edith Cowan University, Joondalup, Australia
[3] Computer Science and Software Engineering, The University of Western Australia, Perth, Australia
[4] Park Nicollet Clinic and HealthPartners Institute, HealthPartners, Minneapolis, USA
[5] Department of Electrical and Computer Engineering, University of Manitoba, Winnipeg, Canada
[6] Departments of Medicine and Radiology, University of Manitoba, Winnipeg, Canada

Abstract. More than 55,000 people world-wide die from Cardiovascular Disease (CVD) each day. Calcification of the abdominal aorta is an established marker of asymptomatic CVD. It can be observed on scans taken for vertebral fracture assessment from Dual Energy X-ray Absorptiometry machines. Assessment of Abdominal Aortic Calcification (AAC) and timely intervention may help to reinforce public health messages around CVD risk factors and improve disease management, reducing the global health burden related to CVDs. Our research addresses this problem by proposing a novel and reliable framework for automated "fine-grained" assessment of AAC. Inspired by the vision-to-language models, our method performs sequential scoring of calcified lesions along the length of the abdominal aorta on DXA scans; mimicking the human scoring process.

Keywords: Abdominal Aortic Calcification · Sequential prediction · Dual-energy xray

S. Z. Gilani and N. Sharif—Joint First Authors.

Supplementary Information The online version contains supplementary material available at https://doi.org/10.1007/978-3-031-16437-8_42.

© The Author(s), under exclusive license to Springer Nature Switzerland AG 2022
L. Wang et al. (Eds.): MICCAI 2022, LNCS 13433, pp. 439–450, 2022.
https://doi.org/10.1007/978-3-031-16437-8_42

1 Introduction

Cardiovascular Disease (CVD) is the leading cause of death globally, and a significant contributor to disability worldwide [16]. Vascular calcification, a stable marker of asymptomatic CVD, occurs when calcium builds up within the walls of the arteries undergoing the atherosclerotic process, and often begins decades before clinical events such as heart attacks or strokes [13]. The abdominal aorta is one of the first vascular beds where calcification is seen, and is a marker for generalised atherosclerosis at other vascular beds [11,21]. The presence and extent of Abdominal Aortic Calcification (AAC) is associated with increased risk of future cardiovascular hospitalizations and death [8]. Given that AAC often occurs well before clinical events, this paper provides a window of opportunity to identify people at risk and intervene in a timely manner before they suffer cardiovascular events such as heart attacks or strokes [14].

The extent and severity of AAC can be assessed using lateral-lumbar radiographs, lateral spine Vertebral Fracture Assessment (VFA) Dual-energy X-ray Absorptiometry (DXA) and Quantitative Computed Tomography (QCT). Out of these, VFA DXA scans have the least amount of radiation but are of lower resolution and contain more noise. These scans can be used to semi-quantify AAC using the widely adopted Kauppila 24-point scoring method (See Sect. 2.1 for more details), which measures the calcification along the length of abdominal aorta from L1-L4. However, acquiring manual assessments for DXA images is not only time-consuming and expensive but also subjective [17].

The published methods [3,4,15] predict an overall AAC-24 score for each scan. A brief summary of these methods is given in Sect. 2.2. For the sake of reliability and explainablility, it is pertinent that the overall AAC-24 assessment mimics the human scoring process. While a single AAC-24 score provides clinically useful information on levels of cardiovascular risk, more granular (location of calcification) scoring is required to provide results to general practitioners and patients. Additionally, more granular scoring may provide better understanding of why and how AAC develops and progresses. Finally, AAC in different parts of the abdominal aorta may be more or less important for clinical outcomes such as heart attacks, stroke or death.

We address the shortcomings of the existing methods by proposing an effective framework to generate the fine-grained scores in a sequential manner without the need for ground truth annotations for the lumbar regions. We draw our inspiration from the vision-to-language domain, in particular image captioning [2], where the task is to transform visual information into a sequence of words. We intend to transform the images into a sequence of fine-grained AAC-24 scores. Our attention based encoder-decoder network is a step towards mimicking the human-like AAC-24 scoring method (see Sect. 2.1 for more details). Though such models are quite popular in the language/vision-to-language domain, this is the first time they have been used to address the particular problem of AAC-24 scoring. However, this domain adaptation comes with its own challenges. Language has a syntax and a structure which is comparatively easier to learn by the sequential models, provided they are trained on a large corpus of data. Moreover,

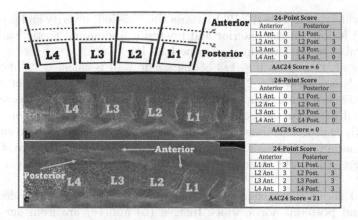

Fig. 1. AAC-24 scoring to quantify the severity of AAC. The scores of all eight segments along with the AAC-24 scores are given in the tables alongside each image.

language is flexible as there can be a number of plausible solutions, and successive words in a sentence are predictable. In contrast, AAC-24 scores are rigid (only three possibilities per segment per vertebrae), random, and the margin of error is small. We also lack the advantage of having large annotated datasets.

Our model focuses on the most salient aortic regions while generating a sequence of scores. Moreover, our algorithm can classify patients into the three risk categories of low, medium and high, with an accuracy, sensitivity, and specificity of 82%, 74% and 80% respectively on the test set. The AAC-24 scores generated by our algorithm are highly correlated (>80%) with human assessments.

In this context our contributions are as follows:

- For the first time, we frame the problem of generating fine-grained AAC assessments as *translating DXA scans to sequential scores*.
- We propose an effective framework to generate fine-grained AAC-24 scores. Our scoring process is more understandable because it is easier to compare with the way humans score, and can point out regions adjacent to particular vertebrae that are highly calcified.
- Our attention-based model has independent decoders for assessing the anterior and posterior aortic walls, which leads to better performance when compared to a single decoder for assessing both aortic walls.
- We show that despite the limited size of our dataset and severe class imbalance, our model achieves a high level of correlation with human assessment.

2 Related Work

2.1 Kauppila Scale and AAC Classes

We have used the 24-point semi-quantitative scale [6] (commonly known as the AAC-24 scale), to quantify the extent of calcification in abdominal aorta. This is the most widely used scale [17,19] to assess the location, severity and progression of calcified lesions on the anterior and posterior abdominal aortic walls in the region parallel to the lower lumbar spine L1 - L4.

A score of '1' is given if $\leq 1/3$ of the aortic wall is calcified, '2' if $> 1/3$ to $\leq 2/3$ is calcified, or '3' if $> 2/3$ is calcified. The anterior and posterior walls and segments are then summed up, for a possible score of up to 24. The scores for the anterior and posterior wall segments are summed (for a possible score of up to 24). The whole process is time consuming ($\sim$10 min), needs specialised equipment (radiology monitor) and is subjective (depends on training/experience of reader). Furthermore, the aorta is not visible in the scans if there is no calcification (Fig. 1(b)).

Figure 1 shows three examples of the AAC-24 scoring, Fig. 1(a) depicts the anterior and posterior aortic walls. Images (b) and (c) are from our dataset, where (b) reflects the difficulty in localizing the aorta when there is no calcification. Figure 1(c) shows a severe case of AAC: calcific deposits can be observed in each segment. We use the established severity categories: low (AAC-24 score 0 or 1), moderate (score 2–5) or high-risk (score $\geq$6) [9,10,18].

2.2   Automatic AAC Classification

We now summarize three relevant pieces of work available in the literature, that perform automatic AAC classification based on the overall AAC-24 score per image. Elmasri et al. [4] trained an Active Appearance Model on 20 DXA VFA images to localize the vertebra L1-L4 and the part of aorta adjacent to these vertebra. Next, they fitted this model to 53 test scans to localize the aorta and extract visual features to perform three-class classification using SVM and KNN techniques. They used non-standard class boundaries to classify the test images with an average accuracy of 92.9%. Chaplin et al. [3] followed a similar process as [4] on 195 DXA VFA scans to extract the Region of Interest (ROI), except that, they used a statistical shape model. The ROI in each scan was then warped to straighten the spine which generally leads to loss of information in a calcified aorta. Two separate U-net architectures were then used to segment the calcification in the ROI as a whole, as well as segment-wise (anterior and posterior). They report R^2 coefficient of only 0.68 between ground truth scores and segment-wise scores and $R^2 = 0.58$ with predicted AAC-24 scores for the whole image. Finally, Reid et al. [15] used a battery of CNNs to classify 1100 DXA VFA scans into three classes. They do not perform cross-validation, rather they reported their results from a single train/validate/test run and selected the network that gave them the best results. Their R^2 coefficient between the ground truth and predicted AAC-24 scores (for the whole scan) was 0.86 with an accuracy of 88.1%. It is important to note that none of the methods discussed above produce fine-grained AAC-24 scores.

3   Proposed Framework

Figure 2(a) shows our proposed framework. It starts with an image pre-processing module (for details see Sect. 4.2) which crops and resizes the images. Once the image is pre-possessed, it is passed on to the visual encoder to extract visual feature maps. We choose a pre-trained Resnet152v2 as the encoder. However, this is not a rigid choice and in future this could be replaced by other/better models. We extract feature maps from the last convolutional layer without using the classification layer of the pre-trained CNN. The feature maps are fed as input to two individual decoders, each of which independently maximizes the log likelihood over the parameter space:

$$\theta^* = \arg\max_{\theta} \sum_{(V,y)} \log p(y|V; \theta) \tag{1}$$

where θ represents the model parameters, $V = [v^1, v^2, ..., v^n]$ represents the visual feature maps extracted from the pre-processed image, and $y = \{y^1, y^2..., y^t\}$ is the sequence of segmented scores. The log-likelihood of the joint probability distribution $\log p(y|V; \theta)$ can be decomposed as:

$$\log p(y|V; \theta) = \sum_{(t=1)}^{T} \log p(y^t|y^1, ..., y^{t-1}, V; \theta) \tag{2}$$

We use a Long-Short Term Memory (LSTM) module to generate y, therefore the conditional probability $\log p(y|V)$ (dropping θ for convenience) can be modeled as $\log p(y^t|y^1, ..., y^{t-1}, V) = g(h^t, c^t)$ where g is a nonlinear function, c^t is the context vector and h^t is the hidden state of the LSTM at time t. We can model h^t as $h^t = LSTM(s^t, h^{t-1}, m^{t-1})$ where s^t is the input vector, and h^{t-1} and m^{t-1} are hidden state and memory cell vectors at time $t - 1$, respectively.

To compute the context vector c^t we use an attention module, such that the context is dependent on specific regions in the image (via image feature maps)

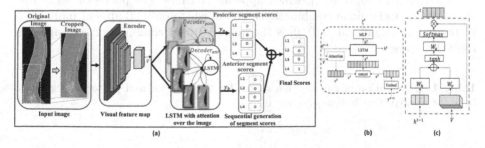

Fig. 2. Our proposed (a) framework for automatic fine-grained AAC scoring, (b) detailed schematic of our attention-based decoders, and (c) attention module. Note that $Decoder_{ant}$ and $Decoder_{post}$ have the same architecture.

as well as the decoder outputs. Therefore, c^t can be defined as $c^t = q(V, h^{t-1})$ where q is the attention function, and h^t is the hidden state of the LSTM at time t. The distribution of attention over the feature maps V (corresponding to various regions of the image) is computed using a feed-forward network and can be formalized as $z^t = W^a \tanh(W^v V + W^h h^{t-1})$ and $\beta^t = softmax(z^t)$ where W^a, W^v and W^h are the learnable parameters, and β is the attention weight over the feature maps V. Finally, c^t can be computed as:

$$c^t = \sum_{i=1}^{n} \beta^{ti} v^{ti} \tag{3}$$

We train our model using weighted cross-entropy loss, where we set the weights for each class based on the data distribution. We do not fine-tune our encoder due the limited size of our dataset. Our two decoders, $Decoder_{ant}$ and $Decoder_{post}$ (see Fig. 2(b) and (c)), have similar architecture and are trained independently to maximize the objective function given in Eq. 2.

4   Experiments

4.1   Dataset

Our dataset is comprised of randomly selected 1,916 bone-density machine-derived lateral-spine scans, obtained using iDXA GE machines [15] with a resolution of at least 1600×300 pixels. The disease severity distribution of the 1,916 scan is: *low risk* 829, *moderate risk* 445 and *high risk* 642. Although, these scans come with expert annotated AAC-24 scores [6], the location of calcified pixels is not annotated on the scans. The distribution of AAC-24 scores in the dataset (see Table 1) is very challenging as it has severe class imbalance. Specifically, this distribution of zero scores is highly skewed for L1 and L2 perhaps because vascular calcification usually starts around L4 and L3 and then progresses upwards [12]. In terms of sequences for anterior segments, our data has 176 unique (out of $4^4 = 256$ possible) combinations but only 29 of them appear more than 10 times. The most frequent sequence is $[0, 0, 0, 0]$, which appears 904 times followed by $[0, 0, 0, 1]$, which appears 77 times. For posterior segments, our data has 190 unique combinations, out of which only 30 appear more than 10 times. Once again, $[0, 0, 0, 0]$ is the most frequent combination and appears 786 (41%) times.

Table 1. Distribution of calcification scores for the anterior and posterior segments. Note the skewed distribution of score '0'.

Segment	Score			
	0	1	2	3
Anterior	5390	1255	518	501
Posterior	5098	1251	685	630

4.2 Pre-processing

To obtain the ROI i.e., the area around the lower lumbar vertebrae, we follow the pre-processing guidelines of Reid et al. [15] and crop 50% from the top, 40% from the left and 10% from the right side of each scan. Then the cropped images are resized to 900×300 pixels using the nearest neighbor interpolation, and re-scaled to values between 0 and 1. We augment the dataset by applying various affine transformations to the images, such as translation [+20, −20], scaling [+20, −20], shear [0.01°, 0.05°] and rotation [+10°, −10°]. We used the TorchVision library for data augmentation and the PyTorch Machine Learning Library for model training and evaluation.

4.3 Model and Training Parameters

We train our model M_{fgs} ("fgs" stands for fine-grained scoring model) with stochastic gradient descent using the Adam optimizer [7]. The initial learning rate and batch size were set to $1e^{-4}$ and 10 respectively. We use ResNet152v2 [5] pretrained on ImageNet as an encoding model (to extract the visual feature maps), but do not fine-tune it on our data. For an input image size of 900×300, the size of the extracted feature map is $29 \times 10 \times 2048$. We flatten the feature maps to 290×2048 and feed them individually to the two decoding networks, which we term as $Decoder_{ant}$, and $Decoder_{post}$.

The two decoders, are trained independently with sequences of anterior and posterior segment ground truth scores, respectively. Furthermore, after training is complete, the output scores of both decoders (for a given test image) are summed to get a single score corresponding to each lumbar vertebrae. Finally, the scores of L1–L4 are summed to obtain the AAC-24 scores. Both decoding pipelines are comprised of an LSTM, with a hidden size of 512, and based on an attention module, where the output sequence length is 4. We perform 10-fold stratified cross validation (where the data is split based on the distribution of AAC-24 scores, such that this distribution is maintained across all splits). In each fold 1,724 examples are used to train the network and 192 for validation. We perform early stopping based on the average Pearson correlation between the predicted and ground truth segment scores. We also use dropout (first after the hidden layer of LSTM (alpha = 0.5), then another (alpha = 0.4) before the last FC layer) as a regularization strategy [20]. Our trained network and scripts are publicly available [1].

4.4 Evaluation

To the best of our knowledge, this is the first paper that predicts fine-grained AAC-24 scores for each vertebrae; instead of a single score for L1–L4 lumbar regions. Therefore, to compare our results with the state-of-the-art we use the sum of all individual granular scores. Since Reid et al. [15] have analysed the

Table 2. Performance comparison of our model with the baseline [15] (NPV is Negative Predictive Value and PPV is Positive Predictive Value) in one-vs-rest setting using the cumulative AAC-24 predicted scores.

	Low ($n = 829$)		Moderate ($n = 445$)		High ($n = 642$)		Mean	
	M_{base}	M_{fgs}	M_{base}	M_{fgs}	M_{base}	M_{fgs}	M_{base}	M_{fgs}
Accuracy	71.14	**82.52**	62.06	**75.52**	79.12	**87.89**	70.77	**81.98**
Sensitivity	55.49	**86.37**	**59.33**	37.53	**54.83**	80.22	56.55	**68.04**
Specificity	**83.07**	79.58	62.88	**87.02**	91.37	**91.76**	79.11	**86.12**
NPV	70.99	**88.45**	**83.63**	82.16	80.06	**90.20**	78.23	**86.93**
PPV	71.43	**76.33**	32.59	**46.65**	76.19	**83.06**	60.07	**68.68**

same dataset as ours, we implement their pipeline (albeit with minor modifications) to compare with our results. Following [15], we train a baseline CNN with Resnet152v2 as its encoder. The decoder consists of a global pooling layer, followed by a dense layer with Relu activation, and another dense layer with a linear activation, the same as in [15]. The generated AAC-24 scores are classified into three risk levels, based on the thresholds discussed in Sect. 2. Note that, for fairness and transparency, we do not report results directly from [15] as we could not obtain their train/validation split and they did not perform ten-fold cross validation. The baseline model M_{base} follows the same stratified cross validation strategy as our proposed M_{fgs} model. As stated (see Sect. 4.3 and Fig. 2), our model has two decoders, for anterior and posterior segments of the lumbar regions L1–L4. It would be natural to ask whether predicting AAC scores in two segments is better than predicting them horizontally across each lumbar region e.g. L1 or L2. To ascertain this, we train a variant of our model (call it M_{fgs}^*) with a single decoder to predict a sequence of scores for each lumbar vertebra, L1–L4, where the score for L1, would be the sum of $L1_{ant}$ and $L1_{post}$. We report our results in the following section.

4.5   Results and Discussion

Table 2 reports one-vs-rest performance of our model M_{fgs} compared to the baseline M_{base} [15] after 10-fold cross validation. Our average classification accuracy $81.98 \pm 2.5\%$ is significantly better than the base line accuracy of $70.77 \pm 3.2\%$. Similarly, our average 3-class classification accuracy is $72.8 \pm 2.9\%$ while that of the baseline is $55.8 \pm 3.2\%$. Note that our AAC-24 scores are obtained by summing up the individual fine-grained scores. Our model predicts AAC-24 scores more accurately compared to the baseline model [15].

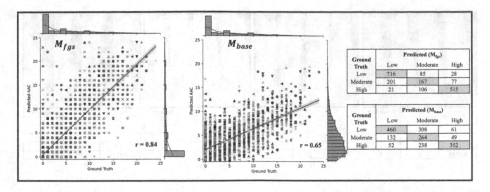

Fig. 3. Scatter plots and confusion matrix of fine-grained ground truth scores vs predicted scores for our proposed model M_{fgs} and ground truth vs the baseline M_{base} [15] overall AAC-24 score per scan.

Table 3. Correlation and error metrics between the ground truth and predicted scores for each lumbar segment. Note that $p \ll 0.001$ for both correlation metrics.

	L1		L2		L3		L4	
	M^*_{fgs}	M_{fgs}	M^*_{fgs}	M_{fgs}	M^*_{fgs}	M_{fgs}	M^*_{fgs}	M_{fgs}
Pearson correlation ↑	0.40	**0.49**	0.56	**0.64**	0.56	**0.70**	0.64	**0.69**
Kendall correlation ↑	0.34	**0.44**	0.49	**0.56**	0.47	**0.60**	0.54	**0.58**
Mean absolute error ↓	0.96	**0.60**	0.73	**0.67**	1.20	**0.88**	1.13	**1.10**

To assess the efficacy of our model at a more granular level, we compare the predicted scores with fine-grained ground truth scores. The scatter plots and confusion matrix are shown in Fig. 3. Since the baseline model [15] does not have the capability to perform fine-grained scoring, we compare its output of a single AAC-24 score (for all lumbar regions) with the corresponding ground truth scores. Note that our model is very good at classifying low and high risk patients. The figure provides evidence that fine-grained scoring results in significantly ($p \ll 0.01$) better prediction and higher correlation with human-scores.

Table 3 shows the comparison between predicting AAC scores horizontally across each vertebrae vs predicting the scores vertically for each segment (anterior and posterior), i.e. comparison between M^*_{fgs} and M_{fgs}. It makes sense that the two decoders in our model M_{fgs} 'attend' to the two vertical segments and perform better than a model that looks at each vertebrae horizontally. Thus the correlation between human annotated scores of those predicted by M_{fgs} is significantly better ($p<0.01$) than the correlation produced by our variant M^*_{fgs}.

Figure 4 shows some examples where our model succeeds (a–b) or fails (c–d). The four sub-figures in each section are from four different time stamps of our sequential attention model. The model "sees" a particular vertebrae at a given time stamp, "attends" to it and "detects" the amount of calcification. It then moves on to the next vertebrae in the sequence. Figure 4(a–b) show

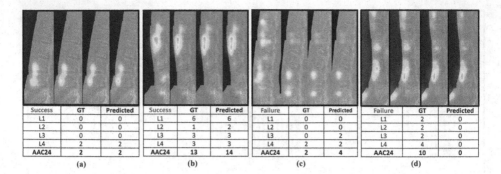

Success	GT	Predicted
L1	0	0
L2	0	0
L3	0	0
L4	2	2
AAC24	2	2

(a)

Success	GT	Predicted
L1	6	6
L2	1	2
L3	3	3
L4	3	3
AAC24	13	14

(b)

Failure	GT	Predicted
L1	0	0
L2	0	0
L3	0	2
L4	2	2
AAC24	2	4

(c)

Failure	GT	Predicted
L1	2	0
L2	2	0
L3	2	0
L4	4	0
AAC24	10	0

(d)

Fig. 4. Our qualitative results show the attention maps generated by our decoding pipeline by combining the weights of $Decoder_{ant}$ and $Decoder_{post}$ for simplicity.

how the model attends to each vertebrae and correctly scores the calcification. Figure 4(c) shows failure cases where the model over-estimates the score of L3 while (d) portrays a case where it totally fails to identify the heavy calcification. However, Fig. 4(d) is very interesting as the aorta in the DXA scan produced by the GE iDXA machine is masked for radiation dose reduction. The human experts have not scored L2 and L3 anterior sections of this scan because they are not visible. Our model is unable to "see" the aorta and hence outputs a zero score. (This is good because a higher predicted score would have meant that the model is not paying attention to the aorta in the score generation process).

5   Conclusion

This is the first work to adapt sequential attention-based models from vision-language domain to address the challenge of fine-grained AAC-24 scoring. This preliminary study on a dataset of 1,916 low resolution DXA scans, not only overcomes the bottlenecks of domain adaptation, but also provides evidence that sequential "fine-grained" scoring yields higher agreement (correlation) with expert human annotated scores. Furthermore, it highlights the necessity of developing larger LFA DXA scan datasets with granular ground truth scores to validate this technique in large population based studies.

Acknowledgement. The work was funded by a National Health and Medical Research Council (NH&MRC) of Australia Ideas grants (APP1183570). The salary of JRL is supported by a National Heart Foundation of Australia Future Leader Fellowship (ID: 102817). The study was approved by the Health Research Ethics Board for the University of Manitoba (HREB H2004:017L, HS20121). The Manitoba Health Information Privacy Committee approved access to the Manitoba data (HIPC 2016/2017-29). The results and conclusions are those of the authors and no official endorsement by Manitoba Health and Seniors Care, or other data providers is intended or should be inferred.

References

1. https://github.com/NaehaSharif/Show-Attend-and-Detect
2. Bernardi, R., et al.: Automatic description generation from images: a survey of models, datasets, and evaluation measures. J. Artif. Intell. Res. **55**, 409–442 (2016)
3. Chaplin, L., Cootes, T.: Automated scoring of aortic calcification in vertebral fracture assessment images. In: Medical Imaging 2019: Computer-Aided Diagnosis, vol. 10950, pp. 811–819. SPIE (2019)
4. Elmasri, K., Hicks, Y., Yang, X., Sun, X., Pettit, R., Evans, W.: Automatic detection and quantification of abdominal aortic calcification in dual energy X-ray absorptiometry. Proc. Comput. Sci. **96**, 1011–1021 (2016)
5. He, K., Zhang, X., Ren, S., Sun, J.: Identity mappings in deep residual networks. In: Leibe, B., Matas, J., Sebe, N., Welling, M. (eds.) ECCV 2016. LNCS, vol. 9908, pp. 630–645. Springer, Cham (2016). https://doi.org/10.1007/978-3-319-46493-0_38
6. Kauppila, L.I., Polak, J.F., Cupples, L.A., Hannan, M.T., Kiel, D.P., Wilson, P.W.: New indices to classify location, severity and progression of calcific lesions in the abdominal aorta: a 25-year follow-up study. Atherosclerosis **132**(2), 245–250 (1997)
7. Kingma, D.P., Ba, J.: Adam: a method for stochastic optimization. arXiv preprint arXiv:1412.6980 (2014)
8. Leow, K., et al.: Prognostic value of abdominal aortic calcification: a systematic review and meta-analysis of observational studies. J. Am. Heart Assoc. **10**(2), e017205 (2021)
9. Lewis, J.R., et al.: Association between abdominal aortic calcification, bone mineral density, and fracture in older women. J. Bone Miner. Res. **34**(11), 2052–2060 (2019)
10. Lewis, J.R., et al.: Long-term atherosclerotic vascular disease risk and prognosis in elderly women with abdominal aortic calcification on lateral spine images captured during bone density testing: a prospective study. J. Bone Miner. Res. **33**(6), 1001–1010 (2018)
11. Lewis, J.R., et al.: Abdominal aortic calcification identified on lateral spine images from bone densitometers are a marker of generalized atherosclerosis in elderly women. Arterioscler. Thromb. Vasc. Biol. **36**(1), 166–173 (2016)
12. Lillemark, L., Ganz, M., Barascuk, N., Dam, E.B., Nielsen, M.: Growth patterns of abdominal atherosclerotic calcified deposits from lumbar lateral x-rays. Int. J. Cardiovasc. Imaging **26**(7), 751–761 (2010)
13. Pickhardt, P.J., et al.: Automated CT biomarkers for opportunistic prediction of future cardiovascular events and mortality in an asymptomatic screening population: a retrospective cohort study. Lancet Digit. Health **2**(4), e192–e200 (2020)
14. Radavelli-Bagatini, S., et al.: Modification of diet, exercise and lifestyle (model) study: a randomised controlled trial protocol. BMJ Open **10**(11), e036366 (2020)
15. Reid, S., Schousboe, J.T., Kimelman, D., Monchka, B.A., Jozani, M.J., Leslie, W.D.: Machine learning for automated abdominal aortic calcification scoring of DXA vertebral fracture assessment images: a pilot study. Bone **148**, 115943 (2021)
16. Roth, G.A., et al.: Global burden of cardiovascular diseases and risk factors, 1990–2019: update from the GBD 2019 study. J. Am. Coll. Cardiol. **76**(25), 2982–3021 (2020)
17. Schousboe, J.T., Lewis, J.R., Kiel, D.P.: Abdominal aortic calcification on dual-energy X-ray absorptiometry: methods of assessment and clinical significance. Bone **104**, 91–100 (2017)

18. Schousboe, J.T., Taylor, B.C., Kiel, D.P., Ensrud, K.E., Wilson, K.E., McCloskey, E.V.: Abdominal aortic calcification detected on lateral spine images from a bone densitometer predicts incident myocardial infarction or stroke in older women. J. Bone Miner. Res. **23**(3), 409–416 (2008)
19. Schousboe, J.T., Wilson, K.E., Kiel, D.P.: Detection of abdominal aortic calcification with lateral spine imaging using DXA. J. Clin. Densitom. **9**(3), 302–308 (2006)
20. Srivastava, N., Hinton, G., Krizhevsky, A., Sutskever, I., Salakhutdinov, R.: Dropout: a simple way to prevent neural networks from overfitting. J. Mach. Learn. Res. **15**(1), 1929–1958 (2014)
21. Strong, J.P., et al.: Prevalence and extent of atherosclerosis in adolescents and young adults: implications for prevention from the pathobiological determinants of atherosclerosis in youth study. Jama **281**(8), 727–735 (1999)

Overlooked Trustworthiness of Saliency Maps

Jiajin Zhang[1], Hanqing Chao[1], Giridhar Dasegowda[2], Ge Wang[1],
Mannudeep K. Kalra[2(✉)], and Pingkun Yan[1(✉)]

[1] Department of Biomedical Engineering and Center for Biotechnology
and Interdisciplinary Studies, Rensselaer Polytechnic Institute, Troy, NY, USA
yanp2@rpi.edu
[2] Department of Radiology, Massachusetts General Hospital,
Harvard Medical School, Boston, MA, USA
mkalra@mgh.harvard.edu

Abstract. Various saliency visualization methods have been proposed
to explain artificial intelligence (AI) models towards building the trustworthiness of AI-driven medical image computing applications. However,
an important question has yet to be answered - *are the saliency maps
themselves trustworthy?* The trustworthiness of saliency methods has
largely been overlooked. This paper first proposes the criteria and methods to evaluate the trustworthiness of saliency maps. Then, a series of
systematic studies are performed on a large-scale dataset with a commonly adopted deep neural network. The results show that: (i) *Saliency
maps may not be relevant to the model outputs*; (ii) *Saliency maps lack
resistance and can be tampered without changing the model output*. By
demonstrating these risks of the current saliency methods, we suggest
the community using saliency maps with caution when explaining AI
models. Our source code is available at https://github.com/DIAL-RPI/
Trustworthiness-of-Medical-XAI.

Keywords: Explainability · Saliency maps · Trustworthiness ·
Medical image analysis · Deep learning

1 Introduction

Explainability is a pillar in supporting the trustworthy applications of artificial
intelligence and machine learning (AI/ML) in medical applications [1,4,9,19,
26,29]. Understanding how and why medical AI models make particular predictions is critical for building the trustworthiness of any AI-driven applications.
Towards this goal, a large body of explainable AI (XAI) methods [14,17,18,20–
24] have been proposed to reveal how AI model works by visualizing the relevance attribution from each image feature to the overall model prediction result.

J. Zhang and H. Chao—are co-first authors.

Supplementary Information The online version contains supplementary material
available at https://doi.org/10.1007/978-3-031-16437-8_43.

© The Author(s), under exclusive license to Springer Nature Switzerland AG 2022
L. Wang et al. (Eds.): MICCAI 2022, LNCS 13433, pp. 451–461, 2022.
https://doi.org/10.1007/978-3-031-16437-8_43

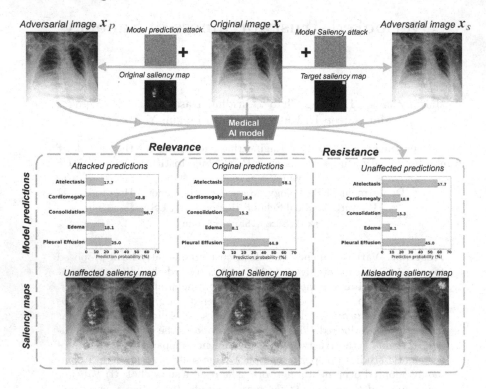

Fig. 1. Overview of our proposed methods. The trustworthiness of saliency maps can be examined from two aspects: relevance and resistance.

Saliency maps, also commonly referred to as heat maps, are the most commonly used method for explaining medical imaging AI models [14, 18, 21–24]. While the research community appreciates the development of saliency methods for XAI, an important question has yet to be answered - *can the saliency maps be trusted?*

The medical imaging research community is increasingly aware that the output of an AI model could abruptly change with subtle perturbations to the input [3, 27], especially human imperceptible adversarial attacks [6, 8, 10, 25, 28, 30]. Although efforts have been made in robustifying deep neural networks in medical image domains [6, 8, 10, 28], the reliability of saliency maps is generally overlooked [5, 11]. Little research [2, 15] has been done to understand the impact of adversarial inputs on the explainability of medical AI, which leaves potential risks to all the AI applications in the field. These previous works used intuitive approaches, such as randomizing models and labels [2] or adding constant shift to images [15], to show the irrelevance of saliency maps. Therefore, it is imperative to pay attention to systematically quantifying the trustworthiness of explainability in medical AI.

To determine the extent of these concerns, we argue that a trustworthy saliency map should have two fundamental properties: **1) Relevance.** A saliency map for an input image should change accordingly when the model's prediction for that image substantially alters due to modifications to the input; **2) Resistance.** A saliency map should stay consistent when the model's prediction of an input image remains unchanged after the image being perturbed by noise. We then develop a set of novel tools based on the adversarial robustness techniques to evaluate these two properties of the saliency map generation methods. Our experimental results underline worrisome phenomena that both properties of saliency maps may break when imperceptible perturbations are added to the input images, *i.e.*, **1) Lack of relevance.** *Saliency maps may remain the same when the model predictions change drastically.* **2) Lack of resistance.** *The generated saliency maps can be tampered into arbitrary target shape while keeping the model predictions unchanged.* These situations are illustrated in Fig. 1.

The main contributions of this paper are two-fold. **1)** Our work reveals the generally overlooked trustworthiness of saliency maps by systematically examining the bilateral relationship (*relevance* & *resistance*) between model prediction and saliency map. The two proposed properties are the necessary conditions that a trustworthy saliency method should satisfy. **2)** We propose two adversarial attack-based methods to efficiently decouple saliency maps from model predictions. Our findings suggest that the popular saliency maps could be unreliable and thus should be used with caution.

2   Method

Let's consider a deep neural network $f(x) : \mathbb{R}^N \to \mathbb{R}^K$, which takes a medical image $x \in \mathbb{R}^N$ as input, and outputs a prediction probabilities of K classes. The network is trained by minimizing a loss function $L(f(x), y) \in \mathbb{R}$, where $y \in \mathbb{R}^K$ represents the label. In this paper, without loss of generality, we focus on classification tasks with the commonly used loss functions, such as the cross-entropy (CE) loss and the binary cross-entropy (BCE). A saliency map $s_k \in \mathbb{R}^N$ for explaining a classification as the k_{th} class can be obtained as

$$s_k = S(f(x), y_k, x), \tag{1}$$

where the function $S(\cdot)$ represents a saliency generation method. The value of each element s_k^i specifies the contribution of the input feature x_i to the k_{th} class.

2.1   Relevance of Saliency Maps

We evaluate the relevance of a saliency generation method $S(\cdot)$ by quantifying the change of saliency maps when the output predictions are modified, as shown in the left of Fig. 1. Inspired by the adversarial attack methods [6,8,10,25,28], we hypothesize that we can generate an adversarial image x_p, which can fool the model into making wrong predictions without influencing the associated saliency map. We thus propose to generate such images x_p by maximizing the classification loss $L(f(x'), y)$ while minimizing the differences between the saliency maps of the generated image x' and the original image x:

$$x_p = \operatorname*{argmin}_{x'} \{ -L(f(x'), y) + \gamma_1 \cdot \|s_k(x') - s_k(x)\|^2 \}, \qquad (2)$$

$$\text{s.t. } \|x_p - x\|_\infty \le \epsilon, \qquad (3)$$

where γ_1 is a positive factor for balancing the magnitude between the terms. Eq. 3 bounds the l_∞ norm of the added perturbation to be smaller than ϵ, so that the adversarial image x_p tends to keep the appearance of the original image x. To quantitatively evaluate the change of saliency maps for the original and the adversarial image, three image similarity metrics are adopted, including the Structural Similarity Index Measure (SSIM), the Pearson Correlation Coefficient (PCC) and the Mean Squared Error (MSE).

2.2 Resistance of Saliency Maps

As shown in the right part of Fig. 1, we evaluate the resistance of saliency maps by examining the existence of an adversarial image x_s with a saliency map towards a target pattern irrelevant to the model predictions while the model predictions remain unaffected. Without loss of generality to illustrate the scenario, as shown in Fig. 1, the target pattern was designed as a highlighted square region on the top right corner of the target saliency map s_{target}. Specifically, we obtain an adversarial image x_s by pushing its saliency map $s_k(x')$ towards the target saliency map s_{target} while keeping the model predictions of the generated image x' and the original image x close by regularizing their KL-Divergence as

$$x_s = \operatorname*{argmin}_{x'} \{ \|s_k(x') - s_{target}\|^2 + \gamma_2 \cdot KL(f(x'), f(x)) \}, \qquad (4)$$

where γ_2 is a weighting factor. We constrain the bound $\|x_s - x\|_\infty \le \epsilon$ for each optimization step. The same similarity metrics (SSIM, PCC and MSE) are used to evaluate the differences between images pairs and between their corresponding saliency maps.

3   Experiments

In this paper, we examine the trustworthiness of seven popular saliency methods, including Vanilla Grad (VG) [22], Vanilla Grad×Image (VG×Image) [21], GradCAM [18], Guided-GradCAM [18], IntegratedGrad (IG) [24], SmoothGrad (SG) [23] and XRAI [14]. Please see supplementary material for more details of these saliency methods. These representative methods have been widely adopted for model explanation in many medical AI works [1,4,9,19,26].

The rest of this section is organized as follows. Section 3.1 presents the datasets and network details. The results of systematically evaluating the relevance and resistance of those seven saliency map methods are provided in Sect. 3.2 and 3.3, respectively. Since adversarial training [16] is often used to improve the model robustness against adversarial attacks, it is interesting to assess whether such a robust model can help provide resistance for the saliency maps. In Sect. 3.4, we explore the existence of universal saliency map attack by evaluating the transferability of the adversarial images across different saliency methods.

3.1 Experimental Details

Datasets. In our work, we demonstrate the discovered issues of saliency maps on a multi-class classification task, CheXpert [13], which is a chest radiograph dataset. All the images are resized into 320×320 pixels. Following the original work in [13], we focused on five major observations: Atelectasis, Cardiomegaly, Consolidation, Edema and Pleural Effusion. Since the test set of the CheXpert dataset is not yet publicly available, the original training set including $224,316$ images were randomly shuffled and split into training and validation sets with a ratio of $6:1$. The original validation set containing 234 images was used as our test set.

AI model Training. We trained two Densenet-121 [12] networks (one regular model and one adversarial robust model) as our baseline models. The model is trained with BCE loss and evaluated by the area under the receiver operating characteristic curve (AUC). Both models are trained for 10 epochs with batch size $= 64$. We used an Adam optimizer with learning rate $= 10^{-4}$. The best performed models on the validation set are selected for test. Following the previous adversarial training method [16], the robust model is trained with adversarial images generated by the PGD-5 attack (step number $= 5$, step size $= 0.01$ with bound $\epsilon = 0.03$). The model robustness was evaluated using AUC on adversarial images generated with the same setting as in the training process. The regular DenseNet achieved a mean AUC of 0.895, on par with the performance reported in the previous papers [7,13]. The adversarial robust DenseNet achieved a mean AUC of 0.865 on the original images and 0.820 on the adversarial images.

Evaluation of Relevance and Resistance. Adversarial images for evaluating the relevance of the saliency maps are optimized following Eq. 2 & 3, where we set $\gamma_1 = 10^8$ and $\epsilon = 0.03$. Each adversarial image x_p is is first initialized with the original image and optimized with an Adam optimizer for 5 steps with learning rate $= 0.01$. Adversarial images x_s for evaluating resistance is updated by an Adam optimizer with learning rate $= 2 \times 10^{-4}$. Each adversarial image sample is first initialized with the original image and optimized iteratively for 10^3 steps using Eq. 4. In our experiments, we set $\gamma_2 = 10^{-8}$ and $\epsilon = 0.03$.

3.2 Saliency Maps May Not Correlate with Model Outputs

Figure 2 shows an example case for evaluating the relevance of different saliency map generation methods. The first row exhibits the saliency maps generated with the original image as input, where the highlighted regions correspond to the observation of Atelectasis. The second row presents the generated adversarial

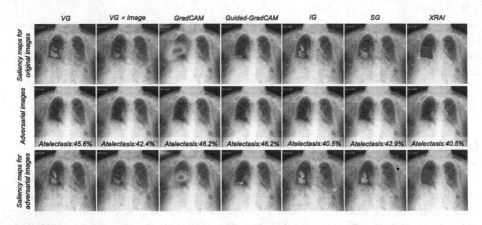

Fig. 2. An example of saliency maps remaining unchanged with altered model predictions. The original prediction probability of Atelectasis is 82.4%.

Table 1. Model performance (AUC) and AUC change on the adversarial images.

	VG	VG × Image	GradCAM	Guided-GradCAM	IG	SG	XRAI
AUC	0.406	0.402	0.470	0.470	0.442	0.427	0.442
(AUC change)	(−0.489)	(−0.493)	(−0.425)	(−0.425)	(−0.453)	(−0.468)	(−0.453)

images for different saliency methods following Eq. 2. Although the adversarial images look identical to the original image, the model prediction probabilities of Atelectasis (as shown at the bottom of each adversarial image) dropped from 82.4% (original image) to below 47.0%. The third row shows the saliency maps generated on the adversarial images. Comparing with the first row, the highlighted regions for the adversarial images did not present significant shifts. The results indicate that the saliency maps failed to reflect the changes of the model predictions.

Quantitative evaluation over the entire test set was performed to verify our findings. We first evaluated the model performance (AUC) on the adversarial images generated with Eq. 2. As shown Table 1, the AUC degraded drastically from 0.895 to below 0.470 on the adversarial images. Then, the similarity between the original and adversarial images and their corresponding saliency maps is computed as shown in Fig. 3. Counterintuitively, compared with the saliency maps for the original images, red squares in Fig. 3 demonstrate that the counterpart saliency maps for the adversarial images stay almost the same with mean SSIM > 0.85, PCC > 0.90 and MSE < 0.25. At the same time, the blue dots show that the adversarial images are almost identical to the original images with mean SSIM and PCC > 0.99, MSE < 0.01. Thus, it is evident that *the saliency maps may not correlate with the model outputs.*

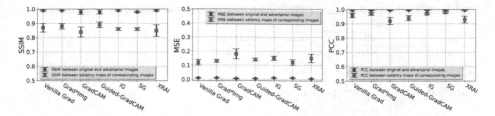

Fig. 3. Evaluation results of the relevance of popular saliency map methods in SSIM, MSE and PCC. The dots and squares in the plots denote the mean values, and the bars indicate the standard deviations.

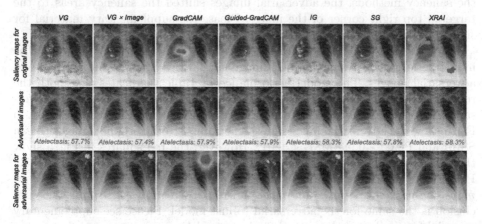

Fig. 4. An example of tampering with saliency maps without changing model predictions. The original prediction probabilities of Atelectasis is 57.7%.

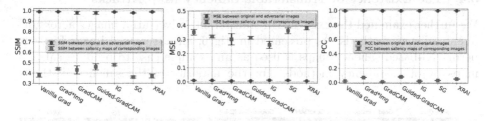

Fig. 5. Evaluation of metrics on the resistance of saliency map methods. The notations are the same as Fig. 3.

3.3   Saliency Maps Can be Tampered Without Changing Prediction

Following Eq. 4 in Sect. 2.2, we generated adversarial images for evaluating the resistance of the saliency maps. An example case is presented in Fig. 4. The first row exhibits the saliency maps generated on the original images. The highlighted regions indicate the observations of Atelectasis. As shown in the second row, not only the adversarial images look identical to the original images, their predic-

Table 2. Model performance (AUC) and AUC change on the adversarial images.

		VG	VG × Image	GradCAM	Guided-GradCAM	IG	SG	XRAI
Normal	AUC	0.888	0.882	0.870	0.870	0.882	0.887	0.882
Model	(AUC change)	(−0.007)	(−0.013)	(−0.025)	(−0.025)	(−0.013)	(−0.008)	(−0.013)
Robust	AUC	0.860	0.861	0.852	0.852	0.858	0.857	0.858
Model	(AUC change)	(−0.005)	(−0.004)	(−0.013)	(−0.013)	(−0.007)	(−0.008)	(−0.007)

tions are also consistent with the original prediction (57.7%). However, in the third row, the adversarial images successfully misled the saliency maps. For all the saliency methods, the adversarial images shifted the saliency areas to the targeted top right corner of the image. Please see supplementary material for similar visualization with the robust model.

Same as Sect. 3.2, to verify the findings in Fig. 4, we further evaluated the model performance change and the image similarities between the original and adversarial images. The model performance in Table 2 indicates that both the regular and robust model performed consistently on the original and the adversarial images, with the AUC change less than 0.025. In addition, the blue dots in Fig. 5 show the adversarial images look very similar to the original images (mean SSIM and PCC > 0.99, MSE < 0.01). However, the red squares in Fig. 5 show that the manipulated saliency maps are vastly different (mean SSIM < 0.50, PCC < 0.20) from the original saliency maps. The results indicate that *saliency maps can be tampered even the model predictions remain consistent*. The conclusion is the same for the adversarial robust model. Please see supplementary material for similar results with the robust model.

3.4   The Existence of Transferable Saliency Map Attacks

We further examined the existence of transferable adversarial images across different saliency methods. As presented in Fig. 6, an adversarial image for tampering IG-based saliency maps is used as an example case to produce saliency maps by other saliency methods. For the majority of the saliency methods,

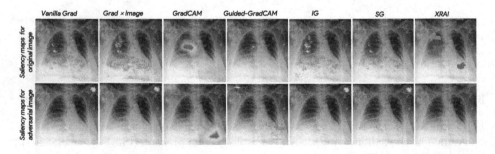

Fig. 6. The existence of transferable saliency attack across different methods. The adversarial image is generated by attacking IG but applied to other methods.

the attention area shifted towards the targeted top-right corner of the image. The saliency maps of adversarial images for GradCAM and Guided-GradCAM, although do not successfully focus on the targeted region, are still drastically different from the original ones. Quantitatively, we observed over 14% of these adversarial images can successfully alter the saliency maps of other methods towards the target map with SSIM > 0.6. These observations indicate that *even if multiple saliency maps produce consistent explanation, they may still be misleading.*

4 Conclusion

Many saliency methods have been proposed for explaining medical AI models. However, we point out that the trustworthiness of saliency methods in medical AI has been generally overlooked. Following this direction, we empirically demonstrate two potential risks when using the popular saliency methods. Our findings suggest that AI researchers and authoritative agencies in the medical domain should use saliency maps with caution when explaining AI models predictions. The propose properties of the *relevance* and *resistance* of saliency maps should be taken into consideration in the future works for validating the trustworthiness of medical AI explainability.

Acknowledgements. This research was partially supported by the National Science Foundation (NSF) under the CAREER award OAC 2046708.

References

1. Abitbol, J.L., Karsai, M.: Interpretable socioeconomic status inference from aerial imagery through urban patterns. Nat. Mach. Intell. **2**(11), 684–692 (2020)
2. Adebayo, J., Gilmer, J., Muelly, M., Goodfellow, I., Hardt, M., Kim, B.: Sanity checks for saliency maps. Advances in Neural Inf. Process. Syst. **31**, 9525–9536 (2018)
3. Antun, V., Renna, F., Poon, C., Adcock, B., Hansen, A.C.: On instabilities of deep learning in image reconstruction and the potential costs of AI. Proc. Natl. Acad. Sci. **117**(48), 30088–30095 (2020)
4. Arnaout, R., Curran, L., Zhao, Y., Levine, J.C., Chinn, E., Moon-Grady, A.J.: An ensemble of neural networks provides expert-level prenatal detection of complex congenital heart disease. Nat. Med. **27**(5), 882–891 (2021)
5. Arun, N., et al.: Assessing the trustworthiness of saliency maps for localizing abnormalities in medical imaging. Radiol.: Artif. Intell. **3**(6), e200267 (2021)
6. Bortsova, G., et al.: Adversarial attack vulnerability of medical image analysis systems: unexplored factors. Med. Image Anal. **73**, 102141 (2021)
7. Chen, B., Li, J., Lu, G., Yu, H., Zhang, D.: Label co-occurrence learning with graph convolutional networks for multi-label chest x-ray image classification. IEEE J. Biomed. Health Inform. **24**(8), 2292–2302 (2020)
8. Daza, L., Pérez, J.C., Arbeláez, P.: Towards robust general medical image segmentation. In: de Bruijne, M., et al. (eds.) MICCAI 2021. LNCS, vol. 12903, pp. 3–13. Springer, Cham (2021). https://doi.org/10.1007/978-3-030-87199-4_1

9. DeGrave, A.J., Janizek, J.D., Lee, S.I.: AI for radiographic COVID-19 detection selects shortcuts over signal. Nat. Mach. Intell. **3**(7), 610–619 (2021)
10. Finlayson, S.G., Bowers, J.D., Ito, J., Zittrain, J.L., Beam, A.L., Kohane, I.S.: Adversarial attacks on medical machine learning. Science **363**(6433), 1287–1289 (2019)
11. Ghorbani, A., Abid, A., Zou, J.: Interpretation of neural networks is fragile. In: Proceedings of the AAAI Conference on Artificial Intelligence, vol. 33, pp. 3681–3688 (2019)
12. Iandola, F., Moskewicz, M., Karayev, S., Girshick, R., Darrell, T., Keutzer, K.: DenseNet: implementing efficient convnet descriptor pyramids. arXiv preprint arXiv:1404.1869 (2014)
13. Irvin, J., Rajpurkar, P., Ko, M., et al.: CheXpert: a large chest radiograph dataset with uncertainty labels and expert comparison. In: Proceedings of the AAAI Conference on Artificial Intelligence, vol. 33, pp. 590–597 (2019)
14. Kapishnikov, A., Bolukbasi, T., Viégas, F., Terry, M.: XRAI: better attributions through regions. In: Proceedings of the IEEE/CVF ICCV, pp. 4948–4957 (2019)
15. Kindermans, P.-J., et al.: The (un)reliability of saliency methods. In: Samek, W., Montavon, G., Vedaldi, A., Hansen, L.K., Müller, K.-R. (eds.) Explainable AI: Interpreting, Explaining and Visualizing Deep Learning. LNCS (LNAI), vol. 11700, pp. 267–280. Springer, Cham (2019). https://doi.org/10.1007/978-3-030-28954-6_14
16. Madry, A., Makelov, A., Schmidt, L., Tsipras, D., Vladu, A.: Towards deep learning models resistant to adversarial attacks. In: ICLR (2018)
17. Ribeiro, M.T., Singh, S., Guestrin, C.: "Why should I trust you?" explaining the predictions of any classifier. In: Proceedings of ACM SIGKDD International Conference on Knowledge Discovery and Data Mining, pp. 1135–1144 (2016)
18. Selvaraju, R.R., Cogswell, M., Das, A., Vedantam, R., Parikh, D., Batra, D.: Grad-CAM: visual explanations from deep networks via gradient-based localization. In: Proceedings of the IEEE ICCV, pp. 618–626 (2017)
19. Shen, Y., Shamout, F.E., Oliver, J.R., et al.: Artificial intelligence system reduces false-positive findings in the interpretation of breast ultrasound exams. Nat. Commun. **12**(1), 1–13 (2021)
20. Shrikumar, A., Greenside, P., Kundaje, A.: Learning important features through propagating activation differences. In: International Conference on Machine Learning, pp. 3145–3153. PMLR (2017)
21. Shrikumar, A., Greenside, P., Shcherbina, A., Kundaje, A.: Not just a black box: learning important features through propagating activation differences. arXiv preprint arXiv:1605.01713 (2016)
22. Simonyan, K., Vedaldi, A., Zisserman, A.: Deep inside convolutional networks: visualising image classification models and saliency maps. arXiv preprint arXiv:1312.6034 (2013)
23. Smilkov, D., Thorat, N., Kim, B., Viégas, F., Wattenberg, M.: SmoothGrad: removing noise by adding noise. arXiv preprint arXiv:1706.03825 (2017)
24. Sundararajan, M., Taly, A., Yan, Q.: Axiomatic attribution for deep networks. In: Proceedings of ICML, pp. 3319–3328. PMLR (2017)
25. Szegedy, C., et al.: Intriguing properties of neural networks. In: ICLR (2014)
26. Tang, Z., et al.: Interpretable classification of Alzheimer's disease pathologies with a convolutional neural network pipeline. Nat. Commun. **10**(1), 1–14 (2019)
27. Wu, W., Hu, D., Cong, W., et al.: Stabilizing deep tomographic reconstruction. arXiv preprint arXiv:2008.01846 (2020)

28. Xu, M., Zhang, T., Li, Z., Liu, M., Zhang, D.: Towards evaluating the robustness of deep diagnostic models by adversarial attack. Med. Image Anal. **69**, 101977 (2021)
29. Zhang, J., Chao, H., Xu, X., Niu, C., Wang, G., Yan, P.: Task-oriented low-dose CT image denoising. In: de Bruijne, M., et al. (eds.) MICCAI 2021. LNCS, vol. 12906, pp. 441–450. Springer, Cham (2021). https://doi.org/10.1007/978-3-030-87231-1_43
30. Zhang, J., Chao, H., Yan, P.: Robustified domain adaptation. arXiv preprint arXiv:2011.09563 (2020)

Flexible Sampling for Long-Tailed Skin Lesion Classification

Lie Ju[1,2,3] , Yicheng Wu[3,4], Lin Wang[1,3,5], Zhen Yu[1,2,3], Xin Zhao[1],
Xin Wang[1], Paul Bonnington[2,3], and Zongyuan Ge[1,2,3(✉)]

[1] Monash-Airdoc Research, Monash University, Melbourne, Australia
julie@airdoc.com, zongyuan.ge@monash.edu
[2] eResearch Centre, Monash University, Melbourne, Australia
[3] Monash Medical AI Group, Monash University, Melbourne, Australia
[4] Faculty of Information Technology, Monash University, Melbourne, Australia
[5] Harbin Engineering University, Harbin, China
https://www.monash.edu/mmai-group

Abstract. Most of the medical tasks naturally exhibit a long-tailed distribution due to the complex patient-level conditions and the existence of rare diseases. Existing long-tailed learning methods usually treat each class equally to re-balance the long-tailed distribution. However, considering that some challenging classes may present diverse intra-class distributions, re-balancing all classes equally may lead to a significant performance drop. To address this, in this paper, we propose a curriculum learning-based framework called Flexible Sampling for the long-tailed skin lesion classification task. Specifically, we initially sample a subset of training data as anchor points based on the individual class prototypes. Then, these anchor points are used to pre-train an inference model to evaluate the per-class learning difficulty. Finally, we use a curriculum sampling module to dynamically query new samples from the rest training samples with the learning difficulty-aware sampling probability. We evaluated our model against several state-of-the-art methods on the ISIC dataset. The results with two long-tailed settings have demonstrated the superiority of our proposed training strategy, which achieves a new benchmark for long-tailed skin lesion classification.

Keywords: Long-tailed classification · Skin lesion · Flexible sampling

1 Introduction

In the real world, since the conditions of patients are complex and there is a wide range of disease categories [4], most of the medical datasets tend to exhibit a long-tailed distribution characteristics [10]. For instance, in dermatology, skin lesions can be divided into different subtypes according to their particular clinical

---

Supplementary Information The online version contains supplementary material available at https://doi.org/10.1007/978-3-031-16437-8_44.

© The Author(s), under exclusive license to Springer Nature Switzerland AG 2022
L. Wang et al. (Eds.): MICCAI 2022, LNCS 13433, pp. 462–471, 2022.
https://doi.org/10.1007/978-3-031-16437-8_44

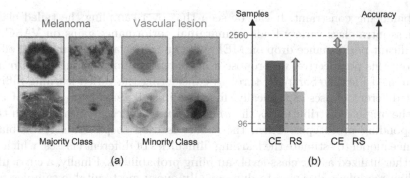

Fig. 1. An illustration of **learning difficulty**. RS denotes re-sampling. The sub-figure (a) shows two kinds of lesions melanoma and vascular lesion from the majority and minority classes respectively. The sub-figure (b) shows that naively over-sampling those minority classes can only achieve marginal performance gains on VASC but seriously hurts the recognition accuracy on MEL.

characteristics [5]. On the popular ISIC dataset [9], the ratio of majority classes and minority classes exceeds one hundred, which presents a highly-imbalanced class distribution. Deep classification models trained on such a long-tailed distribution might not recognize those minority classes well, since there are only a few samples available [18]. Improving the recognition accuracy of those rare diseases is highly desirable for the practical applications.

Most of existing methods for long-tailed learning can be roughly divided into three main categories: information augmentation, module improvement [23] and class-balancing. The information augmentation-based [18,22] methods focused on building a shared memory between head and tail but failed to tackle extremely imbalanced conditions since there are still more head samples for a mini-batch sampling. The module improvement-based methods [10,20] required multiple experts for the final decision and introduces higher computational costs. The class-balancing methods [12,24] attempted to re-balance the class distribution, which becomes the mainstream model for various long-tailed classification tasks. However, most existing methods ignored or underestimated the differences of the class-level learning difficulties. In other words, we believe that it is crucial to dynamically handle different classes according to their learning difficulties.

In Fig. 1, we illustrate our motivation with a pilot experiment on ISIC. *Melanoma (MEL)* and *vascular lesion (VASC)* are two representative skin lesions from the majority and minority classes, respectively. The former class has 2560 images for training while the latter only contains 96 training samples. It can be seen that most samples from VASC resemble one another and present a pink color. Intuitively, this kind of lesions is easier to be classified by deep models since the intra-class diversity is relatively small. By contrast, although MEL has more samples for training, various shapes and textures make the learning difficulty higher. Figure 1(b) shows the performance of two models trained with different loss functions (cross-entropy (CE) loss vs. re-sampling (RS)) for the class-balancing constraint. It can be seen that over-sampling the tailed classes with less intra-class variety leads to marginal performance gains on VASC but a significant performance drop on MEL, also known as a severe marginal effect.

From this perspective, we propose a novel curriculum learning-based framework named Flexible Sampling through considering the varied learning difficulties of different classes. Specifically, first, we initially sample a subset of data from the original distribution as the *anchor points*, which are nearest with their corresponding prototypes. Then, the subset is used to pre-train a less-biased inference model to estimate the learning difficulties of different classes, which can be further utilized as the class-level sampling probabilities. Finally, a curriculum sampling module is designed to dynamically query most suitable samples from the rest data with the uncertainty estimation. In this way, the model is trained via the easy examples first and the hard samples later, preventing the unsuitable guidance at the early training stage [17,21].

Overall, our contributions can be summarized as follows: (1) We advocate that the training based on the *learning difficulty* can filter the sub-optimal guidance at the early stage, which is crucial for the long-tailed skin lesion classification task. (2) Via considering the varied class distribution, the proposed flexible sampling framework achieves dynamic adjustments of sampling probabilities according to the class-level learning status, which leads to better performance. (3) Extensive experimental results demonstrate that our proposed model achieves new state-of-the-art long-tailed classification performance and outperforms other public methods on the ISIC dataset.

2   Methodology

The overall pipeline of our proposed flexible sampling framework is shown in Fig. 2. First, we use self-supervised learning techniques to train a model to obtain balanced representation from original long-tailed distribution. Then, we sample a subset from the training data using the prototypes (mean class features) as anchor points. Next, this subset is used to train an inference model for the estimation of learning difficulties. Finally, a novel curriculum sampling module is proposed to dynamically query new instances from the unsampled pool according to the current learning status. An uncertainty-based selection strategy is used to help find the most suitable instances to query.

2.1   SSL Pre-training for Balanced Representations

Training from the semantic labels as supervision signal on long-tailed distribution can bring significant bias since majority classes can occupy the dominant portion of the feature space [12]. A straightforward solution is to leverage self-supervised learning (SSL) technique to obtain category distribution-agnostic features. Different from the supervised learning methods, SSL techniques do not require annotations but learn representations via maximizing the instance-wise discriminativeness [11,13]. The contrastive learning loss [19] adopted in this study can be formulated as:

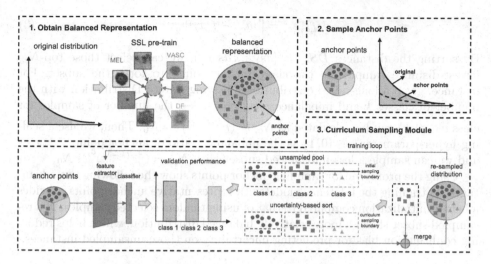

Fig. 2. The overview of our proposed framework.

$$L_{CL} = \frac{1}{N} \sum_{i=1}^{N} -log \frac{exp(v_i \cdot \frac{v_i^+}{T})}{exp(v_i \cdot \frac{v_i^+}{T}) + \Sigma_{v_i^- \in V^-} exp(v_i \cdot \frac{v_i^-}{T})}, \tag{1}$$

where v_i^+ is a positive sample for x_i after the data augmentation, v^- is a negative sample drawn from any other samples, T is the temperature factor, exp is the exponential function. SSL can learn stably well feature spaces which is robust to the underlying distribution of a dataset, especially in a strongly-biased long-tailed distribution [11].

2.2 Sample Anchor Points Using the Prototype

To sample anchor points for training the inference model, we first comupte the mean features for each class as the prototype. Concretely, after pre-training a CNN model M_{pre} with feature representation layer $f(\cdot)$ and classifier $h(\cdot)$, we can obtain features $[f(x_1), ..., f(x_i), ..., f(x_n)]$ for n instances from original distribution $[x_1, ..., x_i, ..., x_n]$ by forward feeding into $f(\cdot)$. Thus, for sorted class index $[c_0, ..., c_j, ..., c_{k-1}]$ where $N_{c_0} > ... > N_{c_{k-1}}$, the prototype m_{c_j} of class c_j can be calculated by averaging all the features from the same class:

$$m_{c_j} = \frac{\sum_{i=1, x_i \in X_{c_j}}^{N_{c_j}} f(x_i)}{N_{c_j}}, \tag{2}$$

where N_{cj} denotes the number of instances of class j. Then, we find those representative samples which have the lowest distances from prototypes by calculating their Euclidean Distance:

$$ds_{i,x_i \in c_j} = \left\| m_{c_j} - f(x_i) \right\|_2. \tag{3}$$

By sorting the distances $DS_{c_k} = [ds_1, ..., ds_{N_{c_k}}]$, we can select those top-$\hat{N}_{c_k}$ lowest-distance samples as the class anchor points to form the subset. For instance, given a long-tailed distribution following Pareto Distribution with the number of classes k and imbalance ratio $r = \frac{N_0}{N_{k-1}}$, the number of samples for class $c \in [0, k)$ can be calculated as $N_{c_j} = (r^{-(k-1)})^{c_j} * N_0$. Then, we use a scaling hyper-parameter $s \in (0, 1)$ as the imbalance ratio of anchor points $\hat{r} = r \cdot s$ and we can sample a less-imbalanced subset with $\hat{N}_{c_j} = (\hat{r}^{-(k-1)})^{c_j} * N_0$.

Using the prototype for sampling anchor points show the following two advantages: 1) training the inference model with representative anchor points provides a better initialization compared to that of using random selected samples; 2) a re-sampled subset shows a more relatively balanced distribution, which helps reducing confirmation bias for predicting uncertainty on those unsampled instances.

2.3   Curriculum Sampling Module

Class-Wise Sampling Probability. After obtaining anchor points $\hat{X}$, we use them to train the inference model for querying new samples from the unsampled instances in the training pool $U_X = C_X \hat{X}$. As we claimed that there exist various learning difficulties and some tail classes with less intra-class variety can converge well in an early stage. Over-sampling such classes may bring obvious marginal effects and do harm to other classes. Hence, a more flexible sampling strategy is required with estimating the learning difficulty of each class. Intuitively, this can be achieved by evaluating the competence of the current model. Thus, we calculate the validation accuracy of different classes after every epoch as the sampling probability:

$$p_{c_j} = 1 - a_e(c_j), \tag{4}$$

where e denotes the e_{th} epoch and $a_e(c_j)$ is the corresponding validation accuracy on j_{th} class. Then, the p_{c_j} can be normalized as $\hat{p}_{c_j} = \delta \cdot p_{c_j}$, where $\delta = K / \sum_{j=0}^{K-1} p_{c_j}$. It should be noticed that when there are limited samples for validation, the $a_e(c_j)$ can also be replaced by training accuracy with only one-time forward pass. The results evaluated on validation set can provide more unbiased reference with less risk of over-fitting since it exhibits different distribution from training set.

Instance-Wise Sample Selection. With learning difficulty-aware sampling probability at the categorical level, we need to determine which instances should be sampled for more efficient querying in an instance difficulty-aware manner. Here, we proposed to leverage uncertainty for the selection of unlabeled samples. Those samples with higher uncertainty have larger entropy since CNN model always give less-confident predictions on hard samples [15]. Training from high-uncertainty samples can well rich the representation space with discriminative features. Here, we compute the mutual information between predictions and model posterior for uncertainty measurement [8]:

$$\mathbb{I}(y_i; \omega | x_i, X) = \mathbb{H}(y_i | x_i, X) - \mathbb{E}_{p(\omega | X)}[\mathbb{H}(y_i | x_i, X)]. \tag{5}$$

where $\mathbb{H}$ is the entropy of given instance point and $\mathbb{E}$ denotes the expectation of the entropy of the model prediction over the posterior of the model parameter. A high uncertainty value implies that the posterior draws are disagreeing among themselves [14]. Then, with sorted instances $x_1, ..., x_{N_{c_j}}$ for class j, the number of newly sampling instances could be $\hat{p_{c_j}} \cdot N_{c_j}$. Finally, we merge the anchor points and newly-sampled instances into the training loop for a continual training.

3 Experiments

3.1 Datasets

We evaluate the competitive baselines and our methods on two dermatology datasets from ISIC [9] with different number of classes. *ISIC-2019-LT* is a long-tailed version constructed from ISIC-2019 Challenge[1], which aims to classify **8** kinds of diagnostic categories. We follow [18] and sample a subset from a Pareto distribution. With k classes and imbalance ratio $r = \frac{N_0}{N_{k-1}}$, the number of samples for class $c \in [0, k)$ can be calculated as $N_c = (r^{-(k-1)})^c * N_0$. For this setting, we set $r = \{100, 200, 500\}$. We select 50 and 100 images from the remained samples as validation set and test set. The second dataset is augmented from ISIC-2019 to increase the length of tail with extra 6 classes (**14** in total) added from the ISIC Archive[2] and named as *ISIC-Archive-LT*[3]. This dataset is more challenging with longer length of tail and larger imbalance ratio $r \approx 1000$. The $ISIC - Archive - LT$ dataset is divided into train: val: test = 7:1:2.

3.2 Implementation Details

All images were resized into 224×224 pixels. We used ResNet-18 and ResNet-50 [7] as our backbones for *ISIC-2019-LT* and *ISIC-Archive-LT* respectively. For the SSL pre-training, we used Adam optimizer with a learning rate of 2×10^{-4} with a batch size of 16. For the classification task, we used Adam optimizer with a learning rate of 3×10^{-4} and a batch size of 256. Some augmentations techniques were applied such as random crop, flip and colour jitter. An early stopping monitor with the patience of 20 was set when there was no further increase on the validation set with a total of 100 epochs.

For our proposed methods, we warmed up the model with plain cross-entropy loss for 30 epochs. The scaling value s for sampling anchor points was set as 0.1. We queried 10% of the samples in the unsampled pool when there was no

---

[1] https://challenge2019.isic-archive.com/.

[2] https://www.isic-archive.com/.

[3] Please see our supplementary document for more details.

Table 1. The comparative results on three different imbalance ratios {100, 200, 500}.

ISIC-2019-LT			
Methods	Top-1 @ Ratio 100	Top-1 @ Ratio 200	Top-1 @ Ratio 500
CE	55.09 (±4.10)=0.00	53.40 (±0.82)=0.00	44.40 (±1.80)=0.00
RS	61.05 (±1.26)+4.96	55.10 (±1.29)+1.70	46.82 (±1.33)+2.82
RW	61.07 (±1.73)+5.98	58.03 (±1.24)+4.63	50.55 (±1.19)+6.15
MixUp [22]	57.25 (±0.71)+2.16	52.05 (±1.71)−1.35	42.78 (±1.28)−1.62
OLTR [18]	60.95 (±2.09)+5.86	56.77 (±1.99)+3.37	48.75 (±1.52)+4.35
Focal Loss [16]	60.62 (±1.04)+5.53	53.88 (±1.19)+0.48	44.38 (±1.44)−0.02
CB Loss [2]	59.65 (±1.52)+4.56	57.75 (±1.39)+4.35	49.45 (±1.11)+5.05
LDAM [1]	58.62 (±1.46)+3.53	54.92 (±1.24)+1.52	44.90 (±2.60)+0.50
CN [6]	56.41 (±1.62)+1.32	53.55 (±0.96)+0.15	44.92 (±0.31)+0.52
ELF [3]	61.94 (±1.53)+6.85	54.60 (±1.78)+1.20	43.19 (±0.31)−1.21
Ours	**63.85 (±2.10)**+8.76	**59.38 (±1.90)**+5.98	**50.99 (±2.02)**+6.59

Table 2. The comparative results on *ISIC-Archive-LT*.

ISIC-Archive-LT				
Methods	Head	Medium	Tail	All
CE	71.23 (±1.33)	48.73 (±4.24)	36.94 (±7.17)	52.30 (±4.25)
RS	70.26 (±1.21)	55.98 (±1.63)	37.14 (±5.44)	54.46 (±2.76)
RW	59.70 (±2.84)	61.53 (±3.69)	63.64 (±3.11)	61.62 (±3.21)
CBLoss	65.27 (±5.73)	56.97 (±6.64)	66.74 (±2.94)	62.89 (±5.10)
ELF	68.81 (±3.69)	50.17 (±4.12)	60.77 (±4.99)	59.92 (±4.27)
Ours	68.26 (±2.01)	57.72 (±2.98)	67.01 (±4.12)	**64.33 (±3.04)**

further improvement with the patience of 10. Then, the optimizer will be re-initialized with a learning rate of 3×10^{-4}. For a fair comparison study, we kept all basic hyper-parameters the same on all the comparative methods. We reported the average of 5-trial running in all presented results. All experiments are implemented by PyTorch 1.8.1 platform on a single NVIDIA RTX 3090.

3.3 Comparison Study

The comparative methods selected in this study can be briefly introduced as follow: (1) plain class-balancing re-sampling (RS) and re-weighting (RW) strategies; (2) information augmentation: MixUp [22] and OLTR [18]; (3) Modified re-weighting loss functions: Focal Loss [16], Class-Balancing (CB) Loss [2] and label-distribution-aware margin (LDAM) Loss [1]; (4) Curriculum learning-based methods: Curriculum Net [6] and EarLy-exiting Framework (ELF) [3].

Performance on *ISIC-2019-LT*. The overall results of the comparative methods and our proposed framework on *ISIC-2019-LT* are summarized in Table 1

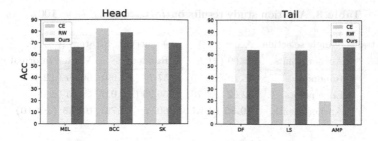

Fig. 3. The performance on several classes from head/tail classes on three methods.

using the mean and standard deviation (Std.) in terms of top-1 accuracy. Here are some findings: (1) Re-sampling strategy obtained little improvements under an extremely imbalanced condition, e.g., r = 500. Repeatedly sampling those relatively easy-to-learn classes brings obvious marginal effects. (2) CB Loss showed superior performance with 4.56%, 4.35% and 5.05% improvements on the baseline trained from cross-entropy, since it tended to sample more effective samples. (3) The proposed flexible sampling effectively outperformed other methods with 63.85%, 59.38% and 50.99% top-1 accuracy over three different imbalance ratios.

Performance on *ISIC-Archive-LT*. For *ISIC-Archive-LT* dataset with imbalanced test set, we reported the accuracy on shot-based division, e.g., head, medium, tail and their average results as most existing works [18,23]. The overall results are shown in Table 2. RS did not obtain satisfactory results. It is due to that the same training samples from tail classes can be occurred repeatedly in a mini-batch during the iteration under an extremely imbalanced condition, resulting in learning from an incomplete distribution. Compared with CBLoss which sacrificed head classes performance, our methods evenly improved the overall performance (+12.03%) with the medium (+8.99%) and tail classes (+30.07%) but only a little performance loss on head classes (−2.97%). In Fig. 3 we present the performance of several selected classes evaluated on three competitors. It is noticed that flex sampling can also improve those head classes with more learning difficulty, e.g., MEL, while RW severely hurts the recognition performance especially for those high-accuracy classes, e.g., BCC.

3.4   Ablation Study

To investigate what makes our proposed methods performant, we conduct ablation studies and report the results in Table 3. We first evaluate the selection of pre-training methods. It can be found that using SSL pre-training with anchor points can well boost the accuracy of the inference model, which shows less confirmation bias and benefits the evaluation of learning difficulties. *Random* denotes the random selection of samples and *Edge Points* denotes those samples away from the prototypes. Finally, we validated the effect of different querying strategies. Compared with random querying, uncertainty-based mutual information can help introduce samples more efficiently in an instance-aware manner.

Table 3. Ablation study results on *ISIC-2019-LT* ($r = 100$).

Pre-train	Sample selection	Acc.	+Querying Strategy	
			Random	Mutual information
CE	Anchor points	58.09 (±0.95)		
SSL	Random	56.19 (±2.20)	57.77 (±3.09)	60.99 (±2.41)
	Edge points	52.62 (±1.76)	56.95 (±1.04)	58.63 (±1.07)
	Anchor points	**59.32 (±1.58)**	60.26 (±1.73)	**63.85 (±2.10)**

4 Conclusion

In this work, we present a flexible sampling framework with taking into consideration the learning difficulties of different classes. We advocate training from a relatively balanced and representative subset and dynamically adjusting the sampling probability according to the learning status. The comprehensive experiments demonstrate the effectiveness of our proposed methods. In future work, more medical datasets and a multi-label setting can be further explored.

References

1. Cao, K., Wei, C., Gaidon, A., Arechiga, N., Ma, T.: Learning imbalanced datasets with label-distribution-aware margin loss. Adv. Neural Inf. Process. Syst. **32** (2019)
2. Cui, Y., Jia, M., Lin, T.Y., Song, Y., Belongie, S.: Class-balanced loss based on effective number of samples. In: Proceedings of the IEEE/CVF Conference on Computer Vision and Pattern Recognition, pp. 9268–9277 (2019)
3. Duggal, R., Freitas, S., Dhamnani, S., Chau, D.H., Sun, J.: ELF: an early-exiting framework for long-tailed classification. arXiv preprint arXiv:2006.11979 (2020)
4. England, N., Improvement, N.: Diagnostic imaging dataset statistical release. Department of Health 421, London (2016)
5. Esteva, A., et al.: Dermatologist-level classification of skin cancer with deep neural networks. Nature **542**(7639), 115–118 (2017)
6. Guo, S., et al.: CurriculumNet: weakly supervised learning from large-scale web images. In: Proceedings of the European Conference on Computer Vision (ECCV), pp. 135–150 (2018)
7. He, K., Zhang, X., Ren, S., Sun, J.: Deep residual learning for image recognition. In: Proceedings of the IEEE Conference on Computer Vision and Pattern Recognition, pp. 770–778 (2016)
8. Houlsby, N., Huszár, F., Ghahramani, Z., Lengyel, M.: Bayesian active learning for classification and preference learning. arXiv preprint arXiv:1112.5745 (2011)
9. ISIC: ISIC archive (2021). https://www.isic-archive.com/
10. Ju, L., et al.: Relational subsets knowledge distillation for long-tailed retinal diseases recognition. In: de Bruijne, M., et al. (eds.) MICCAI 2021. LNCS, vol. 12908, pp. 3–12. Springer, Cham (2021). https://doi.org/10.1007/978-3-030-87237-3_1
11. Kang, B., Li, Y., Xie, S., Yuan, Z., Feng, J.: Exploring balanced feature spaces for representation learning. In: International Conference on Learning Representations (2020)
12. Kang, B., et al.: Decoupling representation and classifier for long-tailed recognition. arXiv preprint arXiv:1910.09217 (2019)

13. Khosla, P., et al.: Supervised contrastive learning. Adv. Neural. Inf. Process. Syst. **33**, 18661–18673 (2020)
14. Kirsch, A., Van Amersfoort, J., Gal, Y.: BatchBALD: efficient and diverse batch acquisition for deep Bayesian active learning. Adv. Neural Inf. Process. Syst. **32** (2019)
15. Leipnik, R.: Entropy and the uncertainty principle. Inf. Control **2**(1), 64–79 (1959)
16. Lin, T.Y., Goyal, P., Girshick, R., He, K., Dollár, P.: Focal loss for dense object detection. In: Proceedings of the IEEE International Conference on Computer Vision, pp. 2980–2988 (2017)
17. Liu, F., Ge, S., Wu, X.: Competence-based multimodal curriculum learning for medical report generation. In: Proceedings of the 59th Annual Meeting of the Association for Computational Linguistics and the 11th International Joint Conference on Natural Language Processing (Volume 1: Long Papers), pp. 3001–3012 (2021)
18. Liu, Z., Miao, Z., Zhan, X., Wang, J., Gong, B., Yu, S.X.: Large-scale long-tailed recognition in an open world. In: Proceedings of the IEEE/CVF Conference on Computer Vision and Pattern Recognition, pp. 2537–2546 (2019)
19. Van den Oord, A., Li, Y., Vinyals, O.: Representation learning with contrastive predictive coding. arXiv e-prints, p. arXiv-1807 (2018)
20. Wang, X., Lian, L., Miao, Z., Liu, Z., Yu, S.X.: Long-tailed recognition by routing diverse distribution-aware experts. arXiv preprint arXiv:2010.01809 (2020)
21. Wang, Y., Gan, W., Yang, J., Wu, W., Yan, J.: Dynamic curriculum learning for imbalanced data classification. In: Proceedings of the IEEE/CVF International Conference on Computer Vision, pp. 5017–5026 (2019)
22. Zhang, H., Cisse, M., Dauphin, Y.N., Lopez-Paz, D.: Mixup: beyond empirical risk minimization. arXiv preprint arXiv:1710.09412 (2017)
23. Zhang, Y., Kang, B., Hooi, B., Yan, S., Feng, J.: Deep long-tailed learning: a survey. arXiv preprint arXiv:2110.04596 (2021)
24. Zhou, B., Cui, Q., Wei, X.S., Chen, Z.M.: BBN: bilateral-branch network with cumulative learning for long-tailed visual recognition. In: Proceedings of the IEEE/CVF Conference on Computer Vision and Pattern Recognition, pp. 9719–9728 (2020)

A Novel Deep Learning System for Breast Lesion Risk Stratification in Ultrasound Images

Ting Liu, Xing An, Yanbo Liu, Yuxi Liu, Bin Lin, Runzhou Jiang, Wenlong Xu, Longfei Cong, and Lei Zhu[✉]

Shenzhen Mindray BioMedical Electronics, Co., Ltd., Shenzhen, China
zhulei@mindray.com

Abstract. This paper presents a novel deep learning system to classify breast lesions in ultrasound images into benign and malignant and into Breast Imaging Reporting and Data System (BI-RADS) six categories simultaneously. A multitask soft label generating architecture is proposed to improve the classification performance, in which task-correlated labels are obtained from a dual-task teacher network and utilized to guide the training of a student model. In student model, a consistency supervision mechanism is embedded to constrain that a prediction of BI-RADS is consistent with the predicted pathology result. Moreover, a cross-class loss function that penalizes different degrees of misclassified items with different weights is introduced to make the prediction of BI-RADS closer to the annotation. Experiments on our private and two public datasets show that the proposed system outperforms current state-of-the-art methods, demonstrating the great potential of our method in clinical diagnosis.

Keywords: Classification · Breast ultrasound image · Multitask soft label · Consistency supervision · Cross-class loss

1 Introduction

Breast cancer has become the most commonly diagnosed cancer in women with a critical mortality [1]. In clinic, ultrasound has been widely used for breast screening and lesion diagnosis. Experienced radiologists can recognize the malignant risk of lesions from their appearance in ultrasound, and issue reports referring to the Breast Imaging Reporting and Data System (BI-RADS) [2] guideline to advise treatments. The BI-RADS guideline includes several essential terms for describing breast lesions and criteria for classifying the risk. Specifically, BI-RADS divides the malignancy likelihood of a lesion into 6 categories, i.e. BI-RADS 2, 3, 4a, 4b, 4c and 5. A higher BI-RADS grade means a larger malignant risk [2] as shown in Fig. 1. Patients with any suspicious mass that is assessed as category 4a, 4b, 4c or 5 will normally be suggested to undergo a preliminary biopsy or surgical excision [2].

Over the last decade, various computer-aided diagnosis (CAD) systems have been developed to assess the malignant risk of breast lesions in ultrasound, which can effectively relieve the workload of physicians and improve their diagnostic performance [3].

© The Author(s), under exclusive license to Springer Nature Switzerland AG 2022
L. Wang et al. (Eds.): MICCAI 2022, LNCS 13433, pp. 472–481, 2022.
https://doi.org/10.1007/978-3-031-16437-8_45

Han et al. [4] adopted a pretrained convolutional neural network (CNN) model to categorize lesions in ultrasound as benign and malignant. Qian et al. [5] developed a combined ultrasonic B-mode and color Doppler system to classify the malignant risk of a lesion into four BI-RADS categories, i.e. BI-RADS 2, 3, 4 (including 4a, 4b and 4c) and 5, and the sum of the probabilities of BI-RADS 2 and 3 was considered as the benign probability while that of BI-RADS 4 and 5 was regard as malignant likelihood. Liu et al. [6] and Xing et al. [7] developed deep learning models with two branches, one for classifying lesions into BI-RADS six categories and the other for pathologic binary classification (benign and malignant).

Although many researches have achieved good performances, two main issues remain. First, several studies demonstrated that combining BI-RADS stratifications and pathology classification could improve the classification performance, but they achieved the combination by simply adding two branches following the CNN, ignoring the relevance between them [7, 8]. Second, no clear distinction between adjacent BI-RADS categories exists in BI-RADS guideline, and the general rule is that a higher BI-RADS grade represents a larger malignant likelihood [2]. Therefore, minor differences between BI-RADS assessments of a lesion given by different physicians or CADs exist and are acceptable, while huge distinctions are unreasonable. However, to the best of our knowledge, few studies focused on minimizing the gap between their predicted BI-RADS categories and the annotations.

To solve the problems above, an automatic breast lesion risk stratification system that classifies lesions into benign and malignant and into BI-RADS categories simultaneously is presented. we propose a multitask soft label generating architecture, in which task-correlated soft labels are generated from a dual-task teacher network and utilized to train a student model. Moreover, a consistency supervision mechanism is proposed to constrain that the prediction of BI-RADS category is consistent with the predicted pathology result, and a cross-class loss function that penalizes the different degrees of misclassification items with different weights is introduced to make the prediction of BI-RADS categories as closer as the annotations.

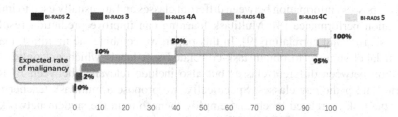

Fig. 1. Expected malignant rate of BI-RADS categories [2].

2 Methodology

The overall architecture is illustrated in Fig. 2. The teacher model is a dual-task network, one task is to classify lesions into BI-RADS six categories, the other is for benign and malignant classification. Task-correlated soft labels are obtained from the teacher network and utilized to train the student model. In student model, consistency supervision mechanism (CSM) constrains that a lesion predicted as BI-RADS 2 or 3 (BI-RADS 4c or 5) is categorized as benign (malignant), thus making the predictions of two branches consistent. The cross-class loss function (CCLF) penalizes different degrees of mis-classified items of BI-RADS categories with different weights to make the prediction closer to the annotation. Details are described as follows.

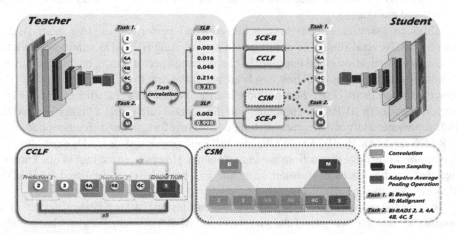

Fig. 2. Illustration of our method. *SLB:* Soft label of BI-RADS. *SLP:* Soft label of pathology. *SCE-B:* Soft label cross entropy of BI-RADS. *SCE-P:* Soft label cross entropy of pathology. *CCLF:* Cross-class loss function. *CSM:* Consistency supervision mechanism.

2.1 Multitask Label Generating Architecture

Soft labels possess information between different classes and are usually used to improve classification performance [8]. Multitask learning can improve prediction results by message sharing and correlating [9]. In this paper, we combine the multitask learning with soft label strategy to obtain task-correlated soft labels which not only contain the relation between different classes but also include relevance between BI-RADS categories and pathology classes. Specifically, we propose a multitask teacher model to learn the task-correlated labels and employ them to train the student network. The teacher model consists of a CNN module, an adaptive average pooling operation, and two classification branches at the end. RepVGG-A2 [10] is adopted as the backbone network to extract features. Hard labels of BI-RADS categories and pathologic information are utilized to train the teacher model using two normal cross entropy loss functions.

The soft labels of BI-RADS (SLB_i) and pathology (SLP_j) are expressed as:

$$\begin{cases} SLB_i = \frac{1}{N_i} \sum tb'(x_{ij}), \ tbc'(x_{ij}) = i \ \& \ tpc'(x_{ij}) = j \\ SLP_j = \frac{1}{N_j} \sum tp'(x_{ij}), \ tpc'(x_{ij}) = j \ \& \ tbc'(x_{ij}) = i \end{cases} \tag{1}$$

where x_{ij} denotes an input image belonging to the ith BI-RADS and the jth pathology. $i \in \{0, 1, 2, 3, 4, 5\}$ and j is 0 or 1. We ran the trained teacher model on the training set, and obtained predicted results. $tb'(x)$ and $tbc'(x)$ represent the output probability vector and predicted category of BI-RADS, and $tp'(x)$ and $tpc'(x)$ are that of pathology respectively. To compute soft label of the ith BI-RADS (SLB_i), a predicted probability vector of BI-RADS is summed up if the predicted BI-RADS result is i and the pathology result equals to the annotation ($tbc'(x_{ij}) = i$ & $tpc'(x_{ij}) = j$). N_i is the number of all qualified cases. The calculation of SLP_j is similar to SLB_i.

The task-correlated labels are then used to train the student model that has the same structure as the teacher network. The loss function is defined as:

$$L_{sl} = -\sum b \cdot logb'(x) - \sum p \cdot logp'(x) \qquad (2)$$

where x denotes the input image. b and p represent the labels from SLB and SLP respectively. $b'(x)$ and $p'(x)$ are the prediction probabilities of BI-RADS and pathology from the student model.

2.2 Consistency Supervision Mechanism

According to [2], a lesion annotated as BI-RADS 2 or 3 is more likely to be benign while a lesion with BI-RADS 4c or 5 is more likely to be malignant. We propose a consistency supervision mechanism (CSM) to constrain the above relevance to make the predictions of two branches consistent. That is, the CSM restricts that a lesion with prediction of BI-RADS 2 or 3 is predicted as benign while a lesion predicted as BI-RADS 4c or 5 is classified as malignant. The consistency loss function is defined as:

$$L_c = -\sum \begin{cases} p_B \cdot log(1 - \sum_{4c}^{5} b'(x)), & p_B \geq 0.5 \\ p_M \cdot log(1 - \sum_{2}^{3} b'(x)), & else \end{cases} \qquad (3)$$

where $p_B + p_M = 1$. p_B and p_M represent the benign and malignant value in the soft label from SLP. $p_B \geq 0.5$ means the input lesion is benign while $p_B < 0.5$ represents that the lesion is malignant. $b'(x)$ is the predicted probability of BI-RADS. $\sum_{4c}^{5} b'(x)$ represents the sum of predicted probabilities of BI-RADS 4c and 5. $\sum_{2}^{3} b'(x)$ is the sum of probabilities of BI-RADS 2 and 3. By optimizing the L_c, the predicted probability of BI-RADS 2 or 3 (BI-RADS 4c or 5) is positively associated with that of benign (malignant). Blurred operation was made for BI-RADS 4a and 4b, since the malignant probability of them is not significant (varies from 2% to 50%) [2].

2.3 Cross-Class Loss Function

As mentioned above, minor differences between BI-RADS assessments given by different physicians or CADs are acceptable, while huge distinctions are unreasonable. Hence, we propose a cross-class loss function that penalizes the different degrees of misclassification items with different weights to make the prediction of BI-RADS closer to the annotation. The cross-class loss function is defined as:

$$L_{cc} = e^{\|n-m\|} \cdot \left(b'_m(x) - b_m\right)^2 \qquad (4)$$

where n and m represent the nth category annotated by radiologists and m^{th} category predicted by the model respectively. The prediction is wrong when n is not equal to m. $b'_m(x)$ and b_m are the mth value in the predicted probability vector and the soft label from SLB respectively. $e^{\|n-m\|}$ is the penalty coefficient, and a larger misclassified degree results in a greater punishment coefficient.

The total loss in the student network is summarized as:

$$\mathcal{L} = a \cdot L_{sl} + \beta \cdot L_c + \gamma \cdot L_{cc} \qquad (5)$$

where α, β and γ are balance parameters, and are set to 1, 0.5 and 0.5 respectively.

3 Experiments

Dataset. Pathologically confirmed breast lesions in women were collected from 32 hospitals and one lesion per patient was included. BI-RADS assessment categories were annotated by several high-experienced radiologists according to the BI-RADS lexicon. A total of 5012 patients with 3220 benign and 1792 malignant lesions were collected, including 14042 images (7793 benign and 6249 malignant images). The dataset was divided at patient-level. The patients were divided into training set (4010 lesions with 11258 images), validation set (501 lesions with 1382 images) and testing set (501 lesions with 1402 images) according to 8:1:1. The average patient age was 43.8 ± 13.2 years for the training set, 44.0 ± 12.5 years for the validation set and 42.8 ± 13.1 for the testing set. The details of each set are described in Table 1 (B: benign, M: malignant) and Fig. 3.

Table 1. Benign and malignant distribution at patient-level.

	Training		Validation		Testing	
	B	M	B	M	B	M
Patients	2564	1446	329	172	327	174
Images	6207	5051	798	584	788	614
Average age (years)	43.8 ± 13.2		44.0 ± 12.5		42.8 ± 13.1	
Average size (cm)	1.86 ± 1.00		1.82 ± 1.03		1.80 ± 0.94	

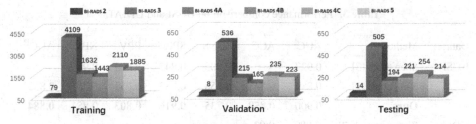

Fig. 3. BI-RADS distribution at image-level.

Implementation Details. The region of a lesion with its 60-pixel border was cropped to reduce the irrelevant background. The cropped images were converted to square by adding 0 to shorter edges, and then resized to 224×224. Both teacher and student models used the same dataset distributions and parameter settings. The pretrained RepVGG-A2 was adopted and the optimizer was set as SGD with a learning rate of 0.01 in the training stage. Random horizontal flipping, scaling, and brightness and contract transformation were utilized as data augmentation. We trained the model for 100 epochs with a batch size of 128. The model that performed with the highest AUC on validation set was chosen as the final model for testing.

Results Analysis. We measured the AUC (AUC^P), Accuracy (ACC), Sensitivity (SENS), Specificity (SPEC), positive predictive value (PPV), and negative predictive value (NPV) of pathology classification as well as AUC (AUC^B) and Kappa of BI-RADS categorization to evaluate the classification performance of the proposed method. The novel VGG19 [11], ResNet18 [12], EfficientNet [13], DenseNet [14] and RepVGG-A2 were re-implemented with two classification branches and evaluated on the same dataset for comparison. Results are shown in Table 2. Our method outperforms all these algorithms in all above metrics.

Table 2. Performance comparison on our test set (%).

Method	AUC^P	ACC	SENS	SPEC	PPV	NPV	AUC^B	Kappa
VGG19	0.897	0.809	0.750	0.867	0.847	0.780	0.880	0.544
ResNet18	0.893	0.803	0.736	0.873	0.858	0.760	0.861	0.577
EfficientNet	0.876	0.799	0.738	0.859	0.839	0.768	0.842	0.548
DenseNet	0.905	0.833	0.819	0.843	0.795	0.863	0.863	0.579
RepVGG-A2	0.908	0.835	0.814	0.850	0.809	0.855	0.868	0.622
Ours	**0.958**	**0.897**	**0.865**	**0.921**	**0.895**	**0.897**	**0.931**	**0.677**

We also test our method on BUSI [15] and UDIAT [16] to compare with others. The results are shown in Table 3 (Italics: test in a subset of the dataset). Our method outperforms [3, 4, 7], and [17] in most metrics on both datasets.

Table 3. Performance comparison on BUSI and UDIAT.

Dataset	Method	AUC^P	ACC	SENS	SPEC	PPV	NPV	AUC^B
BUSI	Shen et al. [3]	0.927	-	0.905	0.842	0.672	0.949	-
	Xing et al. [7]	0.889	0.843	0.758	0.883	0.751	-	0.832
	Ours	0.900	**0.859**	0.735	**0.916**	**0.803**	0.881	**0.884**
UDIAT	*Zhang et al.* [17]	*0.889*	*0.92*	-	-	-	-	-
	Byra et al. [4]	*0.893*	*0.840*	*0.851*	*0.834*	-	-	-
	Xing et al. [7]	0.870	0.859	0.685	0.945	0.860	-	0.872
	Ours	**0.905**	0.877	0.685	**0.972**	**0.925**	**0.862**	**0.916**

Ablation Study. A series of ablation experiments were conducted to validate the effectiveness of the proposed methods. The results are shown in Table 4 and Fig. 4. Comparing with teacher model (RepVGG-A2 in Table 2), the task-correlated soft labels make the prediction results improved about 2% over all metrics on both classification tasks (S). Based on S, the AUC^P and AUC^B increase by 1.2% (p = 0.0160) and 3.7% (p < 0.0001) respectively with the help of CSM. After using the CCLF, not only AUC^P and AUC^B but also accuracy (increased by around 4%) and sensitivity (grown by over 2.5%) are risen comparing to S + CSM. The final student network (S + CSM + CCLF) outperforms the teacher model in a huge margin over most metrics demonstrating the advantage of our method.

Table 4. Results of ablation studies.

Method	AUC^P	ACC	SENS	SPEC	PPV	NPV	AUC^B	Kappa
Teacher	0.908	0.835	0.814	0.850	0.809	0.855	0.868	0.622
Student (S)	0.920	0.852	0.832	0.867	0.830	0.869	0.882	0.635
P value	0.0318	-	-	-	-	-	0.1186	-
S + CSM	0.932	0.859	0.839	0.876	0.840	0.875	0.919	0.642
P value	0.0160	-	-	-	-	-	< 0.0001	-
S + CSM + CCLF	**0.958**	**0.897**	**0.865**	**0.921**	**0.895**	**0.897**	**0.931**	**0.677**
P value	< 0.0001	-	-	-	-	-	0.0991	-

Figure 5 compares the class activation maps [18] (CAMs) between teacher model and final student model. The CAMs of two tasks in the final model (the last two columns) are more consistent than in the teacher model (the second and third column), displaying that the proposed method successfully utilized the relevance between the two tasks. The first two rows show that the teacher model misclassifies a benign lesion with annotation of BI-RADS 3 and a malignant lesion labeled as BI-RADS 4c in term of pathologic

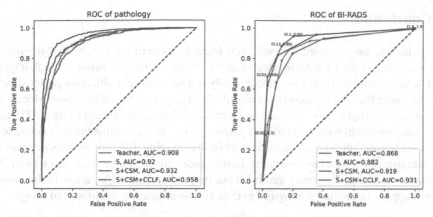

Fig. 4. ROC comparison among ablation studies.

classes but the final student model does not, which demonstrates that our method is able to correct misclassification by constraining the correlation between the pathologic classes and BI-RADS categories. The third and fourth rows present that the BI-RADS category predicted by student model is closer to the annotation than that by teacher model indicating the effectiveness of cross-class function.

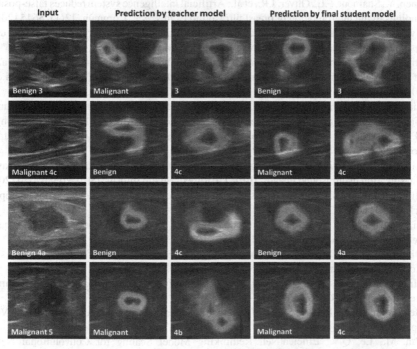

Fig. 5. Comparison of class activation maps.

4   Conclusion

In conclusion, we proposed a novel deep learning system for benign and malignant classification and for BI-RADS categorization simultaneously to assist clinical diagnosis. The task-correlated soft labels successfully improved the classification performance, demonstrating the effectiveness of the multitask label generating architecture. Moreover, the consistency supervision mechanism guaranteed that the prediction of BI-RADS category was consistent with the predicted pathology result, meanwhile the cross-class loss function improved the classification accuracies by utilizing different weights to penalize different degrees of misclassified items. Furthermore, the experiment results on two public datasets indicated the great potential of our method in clinical diagnosis. In the future, we will apply our method for thyroid nodule risk stratification in ultrasound images.

References

1. Sung, H., et al.: Global cancer statistics 2020: GLOBOCAN estimates of incidence and mortality worldwide for 36 cancers in 185 countries. CA Cancer J Clin **71**(3), 209–249 (2021)
2. Medelson, E.B., Böhm-Véle,z M., Berg, W.A., et al.: ACR BI-RADS® ultrasound. In: ACR BI-RADS® Atlas, Breast Imaging Reporting and Data System. American College of Radiology, Reston (2013)
3. Shen, Y., Shamout, F.E., Oliver, J.R., et al.: Artificial intelligence system reduces false-positive findings in the interpretation of breast ultrasound exams. Nat. Commun. **12**(1), 5645 (2021)
4. Byra, M., et al.: Breast mass classification in sonography with transfer learning using a deep convolutional neural network and color conversion. Med. Phys. **46**(2), 746–755 (2019)
5. Qian, X., et al.: A combined ultrasonic B-mode and color Doppler system for the classification of breast masses using neural network. Eur. Radiol. **30**(5), 3023–3033 (2020). https://doi.org/10.1007/s00330-019-06610-0
6. Liu, J., et al.: Integrate domain knowledge in training CNN for ultrasonography breast cancer diagnosis. In: Frangi, A.F., Schnabel, J.A., Davatzikos, C., Alberola-López, C., Fichtinger, G. (eds.) MICCAI 2018. LNCS, vol. 11071, pp. 868–875. Springer, Cham (2018). https://doi.org/10.1007/978-3-030-00934-2_96
7. Xing, J., et al.: Using BI-RADS stratifications as auxiliary information for breast masses classification in ultrasound images. IEEE J. Biomed. Health Inform. **25**(6), 2058–2070 (2021)
8. Hinton, G., Vinyals, O., Dean, J.: Distilling the knowledge in a neural network. arXiv preprint arXiv:1503.02531 2.7 (2015)
9. Li, Y., Kazameini, A., Mehta, Y., et al.: Multitask learning for emotion and personality detection. arXiv preprint arXiv:2101.02346 (2021)
10. Ding, X., Zhang, X., Ma, N., Han, J., Ding, G., Sun, J.: RepVGG: making VGG-style ConvNets great again. In: CVPR, pp. 13728–13737 (2021)
11. Simonyan, K., Zisserman, A.: Very Deep Convolutional Networks for Large-Scale Image Recognition. CoRR, abs/1409.1556 (2015)
12. He, K., Zhang, X., Ren, S., Sun, J.: Deep residual learning for image recognition. In: CVPR, pp. 770–778 (2016)
13. Tan, M., Le, Q.V.: EfficientNet: Rethinking Model Scaling for Convolutional Neural Networks. arXiv:1905.11946 (2019)
14. Huang, G., Liu, Z., Van Der Maaten, L., Weinberger, K.Q.: Densely connected convolutional networks. In: CVPR, pp. 2261–2269 (2017)

15. Al-Dhabyani, W., Gomaa, M., Khaled, H., Fahmy, A.: Dataset of breast ultrasound images. Data in brief **28**, 104863 (2020)
16. Yap, M.H., Pons, G., et al.: Automated breast ultrasound lesions detection using convolutional neural networks. IEEE J. Biomed. Health Inform. **22**(4), 1218–1226 (2017)
17. Zhang, E., Seiler, S., Chen, M., Lu, W., Gu, X.: Boundary-aware semi-supervised deep learning for breast ultrasound computer-aided diagnosis. In: Annual International Conference of the IEEE Engineering in Medicine and Biology Society (EMBC), vol. 2019, pp. 947–950 (2019)
18. Selvaraju, R.R., Cogswell, M., et al.: Grad-CAM: visual explanations from deep networks via gradient-based localization. In: IEEE International Conference on Computer Vision, pp. 618–626 (2017)

Deep Treatment Response Assessment and Prediction of Colorectal Cancer Liver Metastases

Mohammad Mohaiminul Islam[1,2], Bogdan Badic[1,3], Thomas Aparicio[4],
David Tougeron[5], Jean-Pierre Tasu[1,5], Dimitris Visvikis[1],
and Pierre-Henri Conze[1,6(✉)]

[1] Inserm, LaTIM UMR 1101, Brest, France
[2] Maastricht University, Maastricht, The Netherlands
[3] University Hospital of Brest, Brest, France
[4] Saint-Louis Hospital, AP-HP, Paris, France
[5] University Hospital of Poitiers, Poitiers, France
[6] IMT Atlantique, Brest, France
pierre-henri.conze@imt-atlantique.fr

Abstract. Evaluating treatment response is essential in patients who develop colorectal liver metastases to decide the necessity for second-line treatment or the admissibility for surgery. Currently, RECIST1.1 is the most widely used criteria in this context. However, it involves time-consuming, precise manual delineation and size measurement of main liver metastases from Computed Tomography (CT) images. Moreover, an early prediction of the treatment response given a specific chemotherapy regimen and the initial CT scan would be of tremendous use to clinicians. To overcome these challenges, this paper proposes a deep learning-based treatment response assessment pipeline and its extension for prediction purposes. Based on a newly designed 3D Siamese classification network, our method assigns a response group to patients given CT scans from two consecutive follow-ups during the treatment period. Further, we extended the network to predict the treatment response given only the image acquired at first time point. The pipelines are trained on the PRODIGE20 dataset collected from a phase-II multi-center clinical trial in colorectal cancer with liver metastases and exploit an in-house dataset to integrate metastases delineations derived from a U-Net inspired network as additional information. Our approach achieves overall accuracies of 94.94% and 86.86% for treatment response assessment and early prediction respectively, suggesting that both treatment response assessment and prediction issues can be effectively solved with deep learning.

Keywords: Treatment response · Siamese network · Colorectal cancer · Liver metastases · Longitudinal analysis

---

Supplementary Information The online version contains supplementary material available at https://doi.org/10.1007/978-3-031-16437-8_46.

© The Author(s), under exclusive license to Springer Nature Switzerland AG 2022
L. Wang et al. (Eds.): MICCAI 2022, LNCS 13433, pp. 482–491, 2022.
https://doi.org/10.1007/978-3-031-16437-8_46

1   Introduction

The management of patients with colorectal cancer, the second leading cause of cancer death [1], is a major public health issue. Approximately half of the patients with colorectal cancer develop a distant recurrence. The liver, through the development of liver metastases, is the most common spread site, accounting for 15–25% of patients at diagnosis and a further 18–25% of patients within 5 years [2]. With an estimated 5-year survival rate from 37% to 58%, hepatic resection consists of the complete removal of lesions, leaving at least 30% of the parenchyma. Surgery is currently the only curative treatment. When it is not an option due to the tumor burden, the treatment regimen consists of palliative oncological treatments. In this context, Computed Tomography (CT) image analysis is a critical step in assessing the response to treatments. Most evaluation methods are based on measures related to lesion size. In particular, the Response Evaluation Criteria In Solid Tumours (RECIST) 1.1 [3] criterion consists of a unidimensional assessment of lesions which are classified into 3 categories: target, non-target or new lesions. The existence of progression can be affirmed by a progression of the diameter of target lesions of 20% or more, in case of indisputable increase of non-target lesions or when at least one new lesion has appeared. However, morphologic criteria are not suited for treatments that give tumor necrosis (anti-angiogenic agents) or that provide an immune response (immunotherapy) since density information is not taken into account. Moreover, these criteria require localizing, identifying and segmenting hepatic metastases before measurement. These tasks are still largely performed manually by clinicians, hence time-consuming and prone to high intra and inter-expert variability [4]. This finding also limits the option of extracting quantitative indices from CT scans to determine the treatment response group in a *radiomics* fashion [5].

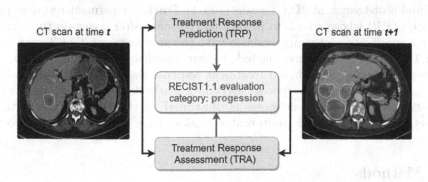

Fig. 1. Overview of the proposed systems along side follow-up CT examinations from the same patient. Liver and liver metastases are resp. with pink and yellow boundaries. (Color figure online)

Deep learning (DL) has been widely used in medical image analysis for a variety of diagnosis purposes [6] but fairly few in the domain of treatment response evaluation. The majority of studies [7–9] concentrate on disease identification and evaluation via the analysis of images obtained at a single time point. For example, Maaref et al. [10] proposed a two-step approach for predicting treatment response of liver metastases from CT scans. First, they segmented and identified untreated liver lesions and then applied an Inception network to predict the treatment response. This method is fundamentally constrained in terms of evolution estimation since it does not account for therapy-induced changes. DL has recently been applied to the analysis of longitudinal clinical factors with the aim of assessing disease risk or development [11–13]. Given the complexity of 3D volumetric data, there remains a great demand for a DL approach that effectively captures the dynamic information from baseline and follow-up images.

More recent studies are taking the longitudinal fact into consideration. Pre- and post-treatment Magnetic Resonance (MR) images were employed by Zhu et al. to predict tumor response to chemotherapy in patients with colorectal liver metastases through a 3D multi-stream deep convolutional network [14]. Both works from Maaref et al. [10] and Zhu et al. [14] treated the response prediction as a binary classification task by creating two synthetic groups (response versus non-response), out of the 4 groups (complete response, partial response, progressive disease, stable disease) originally defined by RECIST 1.1 [3]. This is prohibitive since in this scenario, clinicians would not have the fine-grained division into multiple response groups, which is essential for patient-specific management. One of the comprehensive studies on this topic came from Jin et al. who proposed a multi-task DL approach for both tumor segmentation and response prediction [12]. They designed two Siamese sub-networks linked at multiple layers to enable an in-depth comparison between pre- and post-treatment images.

However, none of the studies address treatment response assessment (TRA) which aims to evaluate the response (considering all response groups from RECIST1.1) of a certain time frame given a specific chemotherapy regimen with pre- and post-treatment 3D CT scans (Fig. 1). Further, treatment response prediction (TRP) whose goal is to predict the response after a certain time frame given a specific chemotherapy regimen with the pre-treatment 3D CT scan only (Fig. 1) has never been investigated, to our knowledge. In this direction, we propose to develop a DL-based treatment response assessment pipeline and its extension for prediction purposes based on a newly designed 3D Siamese classification network. Its effectiveness is illustrated on the PRODIGE20 dataset collected from a phase-II multi-center clinical trial in metastatic colorectal cancer.

2 Methods

We propose a DL based end-to-end pipeline addressing treatment response assessment and prediction. The pipeline utilizes pair of volumetric CT scans of consecutive examinations taken at two different time points, i.e. pre-treatment t and post-treatment $t + 1$, to assess the response of the applied chemotherapy treatment (Sect. 2.2). Further, we extend the pipeline for predicting the treatment response group at time $t + 1$ given only the CT scan acquired at time t

(Sect. 2.3). A liver metastasis segmentation task has also been tackled (Sect. 2.1) as it helps to improve both treatment assessment and prediction results.

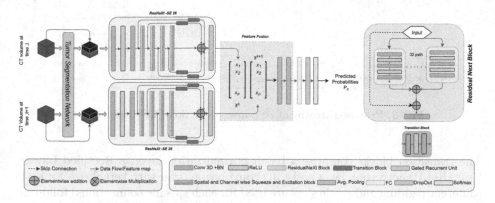

Fig. 2. Proposed treatment response assessment (TRA) pipeline.

2.1 Liver Metastasis Segmentation

The pipeline starts with segmenting liver metastases from volumetric CT scans. For this task, we employ a 2D U-Net [15] whose encoding part is supplanted by a pre-trained VGG-19 trained on ImageNet [16] since considerable gains over randomly weighted encoders have been revealed for many applications [17,18]. We stack the segmentation mask with the original CT scan in the channel direction. The idea is to use the metastasis segmentation as a guiding signal for treatment response assessment and prediction, as an early attention mechanism.

2.2 Treatment Response Assessment

As core of our treatment response assessment (TRA) pipeline, we propose a Siamese network [19] built upon a modified 3D ResNeXt-26 [20] architecture. The overall design of the pipeline is shown in Fig. 2. Each residual block is fitted with a 3D concurrent spatial and channel squeeze-and-excitation (SE) layer [21] from standard SE modules. We also introduced long skip-connections that transfer features arising from different network depths (first convolutional layer and three subsequent ResNeXt blocks) to the very end of the pipeline by aggregating them with the final output of the ResNeXt blocks. We perform an element-wise addition operation among all the feature maps from skip-connections and feature maps to the last ResNeXt block. This aggregated feature map is then fed into an average pooling layer. However, one of the inherent problem with skip-connections and feature map transfer is that feature map dimensions are different at various network depths. We therefore introduced a learned down-sampling step referred as transition block and inspired by the DenseNet [22] architecture.

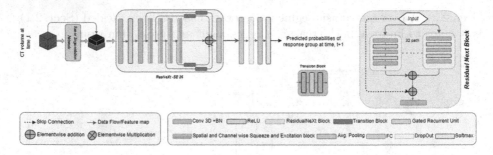

Fig. 3. Proposed treatment response prediction (TRP) pipeline.

The transition blocks comprise a batch normalization layer, a Rectified Linear Unit (ReLU) activation layer and a 3D convolution layer with kernel size $1 \times 1 \times 1$ (different padding and stride sizes are used for each specific connections) followed by a max-pooling layer. The transition block used in DenseNet [22] employs an average pooling layer at the end instead of the max-pooling layer. Nevertheless, we found max-pooling to yield better performance in terms of accuracy and loss.

We consider now the features extracted by each stream of the Siamese network from CT scans acquired at both time stamps. Feature fusion is formulated as a sequence classification problem with Gated Recurrent Units (GRU) [23]. Features of both volumes are used as input sequences of two successive GRU. Finally, two Fully-Connected (FC) layers followed by a Softmax yields the final prediction of belonging to each response group (complete response, partial response, progression, stable disease). The progression class integrates both the increase in (non-)target lesion size and the appearance of new lesions. The main architectural differences with [12] is the use of ResNeXt blocks with SE, short skip-connections additionally to long ones and successive GRU for feature fusion (Fig. 2).

2.3 Treatment Response Prediction

To predict the early treatment response (TRP), we adopted the TRA pipeline (Sect. 2.2) by modifying it. In the case of early prediction, we predict the response at time $t+1$ by only feeding the model with the CT scan acquired at time t. Now, we only require a single 3D ResNeXt branch of the network instead of a Siamese style architecture. The feature fusion mechanism (GRU) was stripped off and we added two FC layer and Softmax activation to reach final class predictions.

3 Experiments and Results

3.1 Imaging Datasets

Experiments were performed on the PRODIGE20 dataset [24] collected for a phase-II multi-center clinical trial in colorectal cancer with liver metastasis that evaluated chemotherapy alone or combined with Bevacizumab in elderly patient (more than 75 years) during follow-up using specific endpoints. A total of 400 consecutive CT scan pairs from 102 patients were considered for our experiments. It is worth noting here that, due to the scarcity of data, multiple CT scans (acquired at different times after contrast injection) from the same examination were occasionally included. Images pairs were divided into 4 response groups: Complete Response (CR), Partial Response (PR), Progression (P) and Stable (S) in accordance with RECIST [3]. Hence, our problem is a 3D image classification problem with 4 classes. We split the dataset into training (60%), validation (15%) and test (25%) sets, corresponding to 241, 60 and 99 CT scan pairs. The splitting process was done such that CT scans from the same patient belong to the same subset and considering almost similar class distributions among subsets. Each class represents the following dataset portions: 63.67% for S, 20.61% for PR, 9.59% for P and 6.12% for CR. To train the segmentation network, we employed an in-house dataset derived from a multi-center study of metastatic colorectal cancer. 171 patients participated in the study, and several CT examination series were available for each patient (245 CT in total). Ground truth liver and liver metastasis contours were established by an experienced clinician. This in-house dataset was split into training, validation and testing sets of 70%, 15% and 15%.

3.2   Implementation Details

Regarding segmentation tasks, we used a hybrid loss combining Dice and binary cross-entropy and relied on random search [25] for hyper-parameter optimization. We fine-tuned the model with the entire segmentation dataset including validation/test sets. Next, we blindly ran inference on the PRODIGE20 dataset for getting the liver segmentation to extract the 3D region of interest and obtaining metastasis delineations to complement the inputs of the proposed pipelines. We finally trained the TRA and TRP models on PRODIGE20. Here, we selected only the region that contains the liver based on the obtained liver masks. This area was resized to $176 \times 176 \times 128$ to deal with GPU memory limitations.

The CT scans utilized in this study came from a variety of scanners. Hence, the images in the dataset had varying spatial and voxel resolutions. The original resolutions varied from $0.67 \times 0.67 \times 1$ to $2.1 \times 2.1 \times 2\,mm^3$. As a result, all images were resampled to $1.0 \times 1.0 \times 1.0\,mm^3$ using cubic Spline interpolation to ensure that the whole dataset had the same voxel spacing and slice thickness. During training, we also applied data augmentation by introducing random rotations (between -20 to $+20°$ where randomness and selection of axes follow a normal distribution), flips (along x, y and z-axis), contrast adjustment (defined as $c = \mu + \alpha \times (v - \mu)$ where μ is the mean voxel value and α a scaling factor between 0.5 and 1.5) as well as noise addition.

Since our dataset has a large class imbalance problem, it is desirable to make the influence of minor classes (especially PR, P and CR) more important during training. One simple way of achieving this is by weighting the loss function. Hence, we employed the weighted cross-entropy loss to reduce the negative effects of class imbalance. First, we define a weight vector $W \in \mathcal{R}^k$ with elements $w_k > 0$

Table 1. Performance of treatment response assessment and prediction pipelines.

	acc	sens	spec	F1	AUC
TRA	94.94	94.89	97.22	94.72	95.56
TRP	86.86	83.22	94.17	83.02	89.25
Model A	92.42	93.17	97.22	92.30	94.17
Model B	89.39	86.89	96.46	82.23	90.39
Model C	87.87	89.17	95.51	84.66	92.34

defined over the range of class labels $i \in \{1, 2.., K\}$ where $K = 4$. Based on the initial definition of cross-entropy, its weighted counterpart is defined as follows:

$$\mathcal{L}_{WCE} = -\sum_{i=1}^{K} w_i \, y_i \, \log p_i \tag{1}$$

where y_i is the ground truth label and p_i the predicted confidence score. In practice, we calculated the w_i as $1/f_i$, where f_i is the normalized frequency of the ith class. Then, we re-normalized the w_i weights to make the sum equals to 1. The Adam optimizer was employed to optimize both TRA and TRP pipelines.

3.3   Performance

TRA and TRP pipelines were evaluated in a one-versus-all fashion using metrics including overall accuracy (acc), sensitivity (sens), specificity (spec), F1 score (F1) and area under the curve (AUC). Scores are reported for test data in Table 1. Receiver operating characteristic (ROC) analysis was performed separately to choose the optimal threshold value that maximizes the Youden index [26]. Our pipeline for TRA achieved 94.94% overall accuracy with high F1-score indicating a good balance between precision and recall as well as a well-functioning classification model despite of the class imbalance issue. This can also be realized from the confusion matrix for the final TRA model (Fig. 4b) which shows that the model can classify samples from dominant and minor classes almost alike. On the other hand, our TRP pipeline achieved 86.86% overall accuracy. Even if both pipelines have an architectural similarity, the score is much less in the case of TRP due to the increased task complexity of predicting treatment response using the pre-treatment CT only. GradCAM [27] was used to see what the model is focusing on to make the decision (see Appendix). We noticed that the model learned to focus on the region of the largest metastasis cluster inside the liver. The size of the primary metastasis cluster as well as darker regions (necrotic tissues) seem to play a significant role in producing higher levels of activation.

3.4 Ablation Study

This section describes the ablation study that compares the TRA pipeline to 3 striped versions of it to identify the importance of its main components. In the first ablated version (Model A), we removed the GRU-based feature fusion and replaced it with simple feature concatenation followed by FC layer. Then, we trained and tested the model in the exact same way. We kept all the settings and hyper-parameters the same to conduct a fair comparison. Next, we experimented a second version (Model B) by removing the long skip-connections and transition blocks. The third version (Model C) works on the source images only, without segmentation masks. Receiver operating characteristic (ROC) curve provided in Fig. 4a shows the performance of each model on a wide range of threshold values. Table 1 summarizes the performance of each ablated version. Our study shows that striping off the GRU-based feature fusion mechanism reduces the overall accuracy by 2.52%. Further, omitting the skip connection and transition block reduces 3.03% more. Finally, supplying only the input images without the segmentation masks further reduces the accuracy by 1.52%.

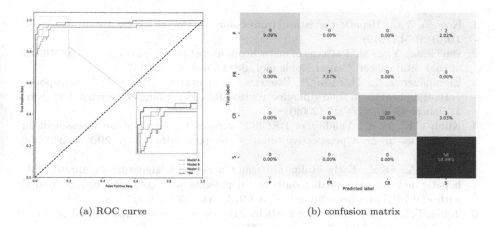

(a) ROC curve   (b) confusion matrix

Fig. 4. Receiver operating characteristic (ROC) curve and confusion matrix on test set for the treatment response assessment (TRA) model. Additional results in Appendix.

4 Conclusion

In this paper, we proposed a deep learning based treatment response assessment and prediction pipeline for liver metastases that explores the assessment and prediction consistency arising from longitudinal 3D CT scans acquired for patients with metastatic colorectal cancer. The primary contribution is incorporating the longitudinal CT scan information within the deep framework to assess and early predict chemotherapy response, which has been under-investigated in the field of medical imaging. Our contributions significantly alleviate the burden of manual liver metastases segmentation required by RECIST1.1, which is beneficial to the development of computer-aided tools. A future possible extension is to build a complete chemotherapy regimen recommendation system able to predict the best treatment for each patient and which would include other organ assessment such as lungs, lymph nodes, peritoneal cavity and bones.

Acknowledgments. This work was partially funded by Ligue contre le Cancer. The PRODIGE20 dataset was provided with the support from FFCD (Fédération Franco-phone de Cancérologie Digestive). The authors would like to thank all the PRODIGE20 investigators as well as A. Dohan from AP-HP for fruitful discussions.

Compliance with Ethical Standards. CT data acquisition was performed in line with the principles of the Declaration of Helsinki. Ethical approval was provided by the Ethics Committee CPP EST I DIJON n°100109 in Jan, 26 2010 and registered in clinicaltrials.gov with number NCT01900717. Authors declare that they do not have any conflicts of interest.

References

1. Kow, A.W.C.: Hepatic metastasis from colorectal cancer. J. Gastrointest. Oncol. **10**(6), 1274 (2019)
2. van Gestel, Y., et al.: Patterns of metachronous metastases after curative treatment of colorectal cancer. Cancer Epidemiol. **38**(4), 448–454 (2014)
3. Eisenhauer, E.A., Therasse, P., Bogaerts, J., Schwartz, L.H., et al.: New response evaluation criteria in solid tumours: revised RECIST guideline (version 1.1). Eur. J. Cancer **45**(2), 228–247 (2009)
4. Kuhl, C.K., et al.: Validity of RECIST version 1.1 for response assessment in metastatic cancer: a prospective, multireader study. Radiology **290**(2), 349–356 (2019)
5. Dohan, A., et al.: Early evaluation using a radiomic signature of unresectable hepatic metastases to predict outcome in patients with colorectal cancer treated with FOLFIRI and bevacizumab. Gut **69**(3), 531–539 (2020)
6. Topol, E.J.: High-performance medicine: the convergence of human and artificial intelligence. Nat. Med. **25**(1), 44–56 (2019)
7. Esteva, A., et al.: Dermatologist-level classification of skin cancer with deep neural networks. Nature **542**(7639), 115–118 (2017)
8. Gulshan, V., et al.: Development and validation of a deep learning algorithm for detection of diabetic retinopathy in retinal fundus photographs. JAMA **316**(22), 2402–2410 (2016)
9. Lee, H., et al.: An explainable deep-learning algorithm for the detection of acute intracranial haemorrhage from small datasets. Nat. Biomed. Eng. **3**(3), 173–182 (2019)
10. Maaref, A., et al.: Predicting the response to FOLFOX-based chemotherapy regimen from untreated liver metastases on baseline CT: a deep neural network approach. J. Digit. Imaging **33**(4), 937–945 (2020)
11. Graffy, P.M., Liu, J., Pickhardt, P.J., Burns, J.E., Yao, J., Summers, R.M.: Deep learning-based muscle segmentation and quantification at abdominal CT: application to a longitudinal adult screening cohort for sarcopenia assessment. Br. J. Radiol. **92**(1100), 20190327 (2019)

12. Jin, C., et al.: Predicting treatment response from longitudinal images using multi-task deep learning. Nat. Commun. **12**(1), 1–11 (2021)
13. Cao, Y., et al.: Longitudinal assessment of COVID-19 using a deep learning-based quantitative CT pipeline: illustration of two cases. Radiol.: Cardiothorac. Imaging **2**(2) (2020)
14. Zhu, H.-B., et al.: Deep learning-assisted magnetic resonance imaging prediction of tumor response to chemotherapy in patients with colorectal liver metastases. Int. J. Cancer **148**, 1717–1730 (2021)
15. Ronneberger, O., Fischer, P., Brox, T.: U-net: convolutional networks for biomedical image segmentation. In: Navab, N., Hornegger, J., Wells, W.M., Frangi, A.F. (eds.) MICCAI 2015. LNCS, vol. 9351, pp. 234–241. Springer, Cham (2015). https://doi.org/10.1007/978-3-319-24574-4_28
16. Deng, J., Dong, W., Socher, R., Li, L.-J., Li, K., Fei-Fei, L.: ImageNet: a large-scale hierarchical image database. In: IEEE Conference on Computer Vision and Pattern Recognition, pp. 248–255 (2009)
17. Conze, P.-H., Brochard, S., Burdin, V., Sheehan, F.T., Pons, C.: Healthy versus pathological learning transferability in shoulder muscle MRI segmentation using deep convolutional encoder-decoders. Comput. Med. Imaging Graph. **83**, 101733 (2020)
18. Conze, P.-H., et al.: Abdominal multi-organ segmentation with cascaded convolutional and adversarial deep networks. Artif. Intell. Med. **117**, 102109 (2021)
19. Yan, Y., et al.: Longitudinal detection of diabetic retinopathy early severity grade changes using deep learning. In: Fu, H., Garvin, M.K., MacGillivray, T., Xu, Y., Zheng, Y. (eds.) OMIA 2021. LNCS, vol. 12970, pp. 11–20. Springer, Cham (2021). https://doi.org/10.1007/978-3-030-87000-3_2
20. Xie, S., Girshick, R., Dollár, P., Tu, Z., He, K.: Aggregated residual transformations for deep neural networks. In: IEEE Conference on Computer Vision and Pattern Recognition, pp. 1492–1500 (2017)
21. Roy, A.G., Navab, N., Wachinger, C.: Concurrent spatial and channel 'squeeze & excitation' in fully convolutional networks. In: Frangi, A.F., Schnabel, J.A., Davatzikos, C., Alberola-López, C., Fichtinger, G. (eds.) MICCAI 2018. LNCS, vol. 11070, pp. 421–429. Springer, Cham (2018). https://doi.org/10.1007/978-3-030-00928-1_48
22. Huang, G., Liu, Z., Van Der Maaten, L., Weinberger, K.Q.: Densely connected convolutional networks. In: IEEE Conference on Computer Vision and Pattern Recognition, pp. 4700–4708 (2017)
23. Cho, K., et al.: Learning phrase representations using RNN encoder-decoder for statistical machine translation. arXiv preprint arXiv:1406.1078 (2014)
24. Aparicio, T., et al.: Bevacizumab+chemotherapy versus chemotherapy alone in elderly patients with untreated metastatic colorectal cancer: a randomized phase II trial-PRODIGE 20 study results. Ann. Oncol. **29**(1), 133–138 (2018)
25. Bergstra, J., Bengio, Y.: Random search for hyper-parameter optimization. J. Mach. Learn. Res. **13**(2), 281–305 (2012)
26. Rücker, G., Schumacher, M.: Summary ROC curve based on a weighted Youden index for selecting an optimal cutpoint in meta-analysis of diagnostic accuracy. Stat. Med. **29**(30), 3069–3078 (2010)
27. Selvaraju, R.R., Cogswell, M., Das, A., Vedantam, R., Parikh, D., Batra, D.: Grad-CAM: visual explanations from deep networks via gradient-based localization. In: IEEE International Conference on Computer Vision, pp. 618–626 (2017)

CACTUSS: Common Anatomical CT-US Space for US Examinations

Yordanka Velikova[1]($\boxtimes$), Walter Simson[1], Mehrdad Salehi[1],
Mohammad Farid Azampour[1,3], Philipp Paprottka[2], and Nassir Navab[1,4]

[1] Computer Aided Medical Procedures, Technical University of Munich,
Munich, Germany
dani.velikova@tum.de
[2] Interventional Radiology, Klinikum rechts der Isar, Munich, Germany
[3] Department of Electrical Engineering, Sharif University of Technology,
Tehran, Iran
[4] Computer Aided Medical Procedures, John Hopkins University, Baltimore, USA

Abstract. Abdominal aortic aneurysm (AAA) is a vascular disease in
which a section of the aorta enlarges, weakening its walls and potentially rupturing the vessel. Abdominal ultrasound has been utilized for
diagnostics, but due to its limited image quality and operator dependency, CT scans are usually required for monitoring and treatment planning. Recently, abdominal CT datasets have been successfully utilized to
train deep neural networks for automatic aorta segmentation. Knowledge
gathered from this solved task could therefore be leveraged to improve
US segmentation for AAA diagnosis and monitoring. To this end, we
propose CACTUSS: a common anatomical CT-US space, which acts as
a virtual bridge between CT and US modalities to enable automatic
AAA screening sonography. CACTUSS makes use of publicly available
labelled data to learn to segment based on an intermediary representation that inherits properties from both US and CT. We train a segmentation network in this new representation and employ an additional
image-to-image translation network which enables our model to perform
on real B-mode images. Quantitative comparisons against fully supervised methods demonstrate the capabilities of CACTUSS in terms of
Dice Score and diagnostic metrics, showing that our method also meets
the clinical requirements for AAA scanning and diagnosis.

Keywords: Ultrasound · Computer aided intervention · Abdominal
Aortic Aneurysm · Domain Adaptation

1 Introduction

Abdominal aortic aneurysm (AAA) is a life-threatening disease of the main blood
vessel in the human body, the aorta, where an aneurysm, or expansion, occurs
thereby weakening the aorta walls. AAA can lead to a high risk of a rupturing
of the aorta with an overall incidence rate of 1.9% to 18.5%, in males age 60+
years of age and an average subsequent mortality rate 60% [19].

© The Author(s), under exclusive license to Springer Nature Switzerland AG 2022
L. Wang et al. (Eds.): MICCAI 2022, LNCS 13433, pp. 492–501, 2022.
https://doi.org/10.1007/978-3-031-16437-8_47

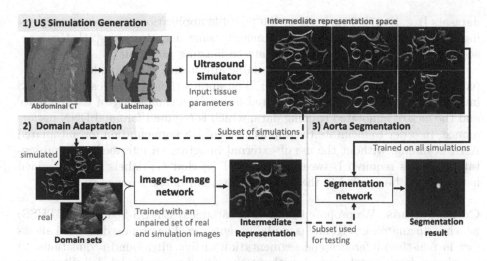

Fig. 1. Overview of the proposed framework. In phase one, an established ultrasound simulator is re-purposed and parameterized to define an intermediate representation, between the ultrasound and CT space. In phase two, an unsupervised network is trained separately in an isolated fashion to translate clinical ultrasound images to the intermediate representation defined in phase one. In phase three, a segmentation network is trained on the segmentation task using only samples generated with the ultrasound simulator. At inference time, real ultrasound images are passed to the image-to-image network, translated to the intermediate representation and segmented. This is the first time that the segmentation network has seen the intermediate representation from real ultrasound images.

Abdominal ultrasound has been recommended as an initial examination modality for asymptomatic patients with a high risk of AAA. There is evidence of a significant reduction of premature death from AAA in men aged 65 and above who undergo ultrasound screening [19]. Per definition, the aorta is considered aneurysmatic when the absolute anterior to posterior diameter is larger than 3 cm, independently of the relative body size of the patient. However, because the interpretation of the US image is heavily based on the sonographer's experience, the resulting diagnosis is largely operator-dependent, as reported in [15].

To overcome the challenge of reproducible ultrasound screening, robotic ultrasound (RUS) imaging has been proposed to offer reproducible ultrasound scans independent of operator skill [7,8,11]. Specifically for screening of AAA, this has required an external camera and MRI atlas to locate and track the trajectory of the aorta, which reduces the usability and subsequent acceptance of the methods [9,20]. Furthermore, ultrasound image quality has been criticized for not offering the resolution needed to make an accurate measurement [4].

Computed tomography (CT) scans are used in clinical practice to assess, manage, and monitor AAA after an initial discovery during screening [3]. In recent years segmentation models based on deep neural networks have demonstrated great performance for automatized CT aorta segmentation, and numerous studies have been trained on large, expert annotated, and publicly available datasets [1,2,10,17]. This leads to the possible application of automatic screening and monitoring of AAA in CT imaging using deep learning [21]. However, acquiring a CT scan exposes the patient to ionizing radiation.

Ultrasound imaging can serve as a viable alternative to CT and help to reduce patient exposure to ionizing radiation. The application of deep learning for US image segmentation has been hampered due to the complexity of the modality and the lack of annotated training data, which is required for good DNN performance. In order to facilitate the applications of US segmentation for automated AAA scanning, without the use of external imaging, an intermediate representation (IR) is required between US and CT so that CT labels and pretrained networks can be applied to the task of ultrasound image segmentation.

Contributions. We propose Common Anatomical CT-US Space (CACTUSS) which is an anatomical IR and is modality agnostic. The proposed method allows for: 1) real-time inference and segmentation of live ultrasound acquisitions, 2) training a deep neural network without the use of manually labeled ultrasound images, 3) reproducible and interpretable screening and monitoring of AAA.

We evaluate the results from the proposed approach by comparing it to a fully supervised segmentation network and investigate the use of the proposed method for measuring the anterior-posterior aortic diameter compared to the current clinical workflow. In total, the proposed method meets the clinical requirements associated with AAA screening and diagnosis. The source code for our method is publicly available[1].

2   Method and Experimental Setup

CACTUSS (c.f. Fig. 1) consists of three main phases: (1) Joint anatomical IR generator, (2) Domain Adaptation network, (3) Aorta Segmentation network.

Ultrasound Simulation Parametrization: Ultrasound Simulation has been an active research topic in the last two decades [6,16,18]. Since ultrasound data is limited in quantity and difficult to acquire, simulated ultrasound images are used to define an intermediate ultrasound representation. To help define a common anatomical space, we take advantage of a hybrid US simulator introduced by [16]. In CACTUSS, the hybrid ray-tracing convolutional ultrasound simulator is used to define an anatomical IR with anisotropic properties, preserving the direction-dependent nature of US imaging while also having modality-specific artifacts of ultrasound imaging and well-defined contrast and resolution of CT. This anatomical IR should reside on the joint domain boundary of US and CT distributions. This has the benefit that from a single CT scan, a large number of samples of the IR can be created. Simulation parameters are listed in Table 1 and Table 2 and describe the characteristics of the US machine, which allow for direct spatial mapping from the CT domain to the ultrasound domain. Input to the simulator is a three-dimensional label map where each voxel is assigned

---

[1] https://github.com/danivelikova/cactuss.

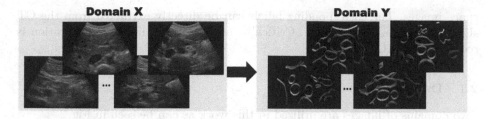

Fig. 2. Images from both image domains

six acoustic parameters that describe the tissue characteristics - speed of sound c, acoustic impedance Z, and attenuation coefficient α, which are used to compute the acoustic intensity at each point along the travel path of the ray. In this way, we create a virtual modality that provides important characteristics from ultrasound, such as tissue interfaces, while learning from annotated CT.

Table 1. Ultrasound scan parameters for ISS

Parameter	Value
Probe width	59 mm
Probe angle	40°
Image depth	100 mm
Focus depth	50 mm
Scan lines	196
Axial resolution	1024

Table 2. Ultrasound simulation parameters for ISS

Parameter	Value
Elevational rays	10
RF noise	0
Scale exponent 1	1.0
Scale exponent 2	0.2
TGC alpha	0.65
TGC scale	0.2

Domain Adaptation. Since there is a domain shift between the IR and real ultrasound B-modes, we learn a mapping between them while preserving the patient-specific anatomical characteristics of each image. In order to translate real ultrasound images into the IR we employ a recent Contrastive Learning for Unpaired image-to-image translation network (CUT) [13]. The CUT network assumes a maximum correlation between the content information of a patch of the target image with the spatially corresponding patch of the source image vs. any other patches in the source image. The network generator function $G : \mathcal{X} \mapsto \mathcal{Y}$ translates input domain images $\mathcal{X}$ to look like an output domain images $\mathcal{Y}$, with unpaired samples from source $X = x \in \mathcal{X}$ and target $Y = y \in \mathcal{Y}$ respectively. The generator G is composed of an encoder G_{enc} and a decoder G_{dec}, which are applied consecutively $\widehat{y} = G(z) = G_{dec}(G_{enc}(x))$. G_{enc} is restricted to extracting content characteristics, while G_{dec} learns to create the desired appearance using a patch contrastive loss [13]. The generated sample Y is stylized, while preserving the structure of the input x. Thus, samples can have the appearance of the IR while maintaining the anatomical content of the US image.

Aorta Segmentation: In the last phase, a segmentation network is trained on the samples from phase 1 to perform aorta segmentation on intermediate space images. The corresponding labels can be directly extracted from the CT slices, saving manual labelling. Critically, no ultrasound image segmentation is required for CACTUSS.

2.1 Data

Two domains of images are utilized in this work as can be seen in Fig. 2.

Intermediate Space: Eight partially labeled CT volumes of men and women were downloaded from a publicly available dataset Synapse[2]. These labels were augmented with labels of bones, fat, skin and lungs to complete the label map. The CTs were used to generate 5000 simulated intermediate space samples with a size of 256×256 pixels. From those simulated data, a subset of 500 images was used for domain Y for the CUT network training. This dataset will be referred to as intermediate space set (ISS).

In-vivo Images: Ten US abdominal sweeps were acquired of the aortas of ten volunteers (m:6/f:4), age $= 26 \pm 3$ with a convex probe (CPCA19234r55) on a cQuest Cicada US scanner (Cephasonics, Santa Clara, CA, US). Per sweep, 50 frames were randomly sampled, each with size 256×256 pixels, for a total of 500 samples and used for domain X of the CUT network. For testing the segmentation network, which was trained only on IRs, a subset of 100 frames, with 10 random frames per volunteer was labelled by a medical expert and used as a test set. For the purpose of comparing against a supervised approach, additional images were annotated to train a patient-wise split 8-fold-cross validation network with 50 images per fold from 8 subjects. Additionally, 23 images from a volunteer not from the existing datasets were acquired from ACUSON Juniper (Siemens Healthineers, Erlangen, Germany) with a 5C1 convex probe and annotated for further evaluation.

2.2 Training

For phase 2 we train the CUT network for 70 epochs with a learning rate of 10^{-5} and default hyperparameters. For phase 3 we train a U-Net [14] for 50 epochs with a learning rate of 10^{-3}, batch size of 64, Adam optimizer and DSC loss. Both models were implemented in PyTorch 1.8.1 and trained on a Nvidia GeForce RTX 3090 using Polyaxon[3]. Phase 3 training includes augmentations with rotation, translation, scaling and noise and is randomly split in 80–20% ratio for training and validation, respectively. For testing, the test set, consisting of 100 images from the in-vivo images, is inferred through the CUT network and translated into the common anatomical representation before being inferred with the phase 3 network.

---

[2] https://www.synapse.org//#!Synapse:syn3193805/wiki/89480.
[3] https://polyaxon.com/.

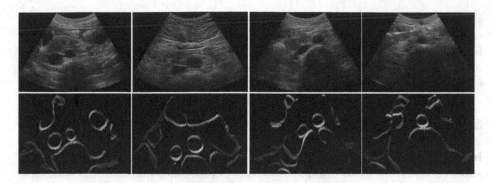

Fig. 3. Examples of B-mode images after inference through the CUT network to the IR. Top row: input B-mode. Bottom row: result after inference.

2.3   Evaluation Metrics

We use the following metrics to quantitatively evaluate our method: For CUT we use the Fréchet inception distance (FID) [5] for performance evaluation and early stopping regularization. FID quantifies the difference in feature distribution between two sets of images e.g. real and IR, using feature vectors from the Inception network. As proposed in [5] we use the second layer for FID calculation and consider the epochs with the top 3 FID scores and qualitatively select based on the desired appearance. For the segmentation model, we report the average Dice Score (DSC) and mean absolute error (MAE) of the diameter of the resulting segmentation as proposed in [12].

2.4   Experiments

We test the proposed framework quantitatively wrt. the following experiments:

Imaging Metrics: We evaluate the accuracy of the proposed method by comparing it to a supervised network. For this, we train an 8-fold cross-validation U-Net, where each fold contains 50 in-vivo images from one subject. We test on 3 hold-out subjects and report the average DSC.

Clinical Applicability: We measure the anterior-posterior diameter of the aorta in mm, according to current clinical practice [4], and report the MAE and standard deviation compared to ground truth labels for both CACTUSS and the supervised segmentation. Clinically, an error of less than 8 mm is considered acceptable for a medical diagnosis of AAA [4].

Robustness: We evaluate against images of a patient scanned with a second US machine as described in Sect. 2.1. Thus we show how robust is the method to domain shift and again evaluate against the supervised network.

Different Intermediate Representation: We replace the proposed common anatomical IR with two alternatives to test the sensitivity of the proposed method's IR choice and specification. The first alternative tested processes CT slices with a Canny edge detector, bilaterial filter, and subsequent convex mask with shape from a convex US probe. The second is a realistic ultrasound simulation from the same label map as the ISS. These alternative IRs were evaluated on a DCS score on 100 in-vivo frames passed through the trained model, with expert annotation ground truth.

3   Results and Discussion

In Tables 3 and 4, we present the DSC and MAE values of CACTUSS and a supervised U-Net when evaluated on the Cephasonics and Siemens scanners, respectively. This evaluation is performed on two real-world scanners, while the CACTUSS phase 3 segmentation network has only been trained on synthetic samples from the ISS. Remarkably, on the Cephasonics scanner, CACTUSS achieves a higher DSC in aortic segmentation and lower mean absolute error of aorta diameter measurements, a key metric in AAA diagnosis. On the Siemens scanner, CACTUSS has a slightly higher MAE, but still exhibits a lower standard deviation. Furthermore, CACTUSS diameter measurement results are still within the clinically accepted range. For this particular experiment, four out of 23 images were wrongly predicted and only from the supervised method.

Table 3. Evaluation of CACTUSS on Cicada samples.

	Supervised	CACTUSS
DSC	85.7 ± 0.02	**90.4 ± 0.003**
MAE	4.3 ± 1.9	**2.9 ± 1.9**

Table 4. Evaluation of CACTUSS on Juniper samples.

	Supervised	CACTUSS
DSC	81.3	**88.0**
MAE	**4.9 ± 7.0**	7.6 ± 1.5

Alternative Intermediate Representations: Result from evaluating alternative IRs on the are reported in Table 5. The proposed IR in CACTUSS outperforms both edge detection and realistic simulated ultrasound images.

Table 5. Comparison of DSC of segmentation given alternative IRs.

	Edges	US Simulation	CACTUSS
DSC	66.1	72.7	**89.3**

3.1   Discussion

CACTUSS was able to not only successfully segment real B-mode images while being trained only on IR data but was able to surpass the supervised U-Net as depicted in Table 3. Furthermore, CACTUSS was able to achieve an aortic diameter measure accuracy of 2.9 ± 1.9 compared to 4.3 ± 1.9 for the supervised

U-Net on the Cephasonics machine. For both ultrasound devices, the diameter accuracy is within the accuracy required for clinical diagnosis AAA [4]. The results showed that it performed well independently from the machine used; however, the performance may vary due to different preprocessing steps and filters in each US machine.

Our testing of alternative intermediate representations showed the unique advantage that CACTUSS offers. By embedding the anatomical layout in a space between US and CT, i.e. with the contrast of CT and attenuation and reflectively of ultrasound, the greatest segmentation performance was displayed. By testing ultrasound simulation representations and an edge detection representation with high contrast, we show that the choice of representation is not arbitrary. In the case of the US simulations, the reduction of performance is likely due to the increased complexity and the lower SNR of the image due to the addition of ultrasound-specific features such as shadows, reflections and speckle. Alternative representations are also possible, but our testing shows that including fundamental physical assumptions of both modalities enhances model performance.

One challenge of using CUT for domain adaptation is the possibility of hallucinations. Those networks are prone to hallucinate incorrect characteristics with the increasing number of training loops. However, we mitigate this issue by integrating FID to select the most performant CUT model (Fig. 3). This approach can remain challenging for complex outputs; however, the CACTUSS IR is simplified in structure and is cleared from features such as speckle noise or reflections, thus improving trainability.

The reproducibility of diagnostic measurements between sonographers, which is heavily dependent on their expertise, can lead to large inter as well as intraobserver variability. In particular, the differences in measurements between sonographers lie between 0.30–0.42 cm, and the mean repeatability among technicians is 0.20 cm [4]. Neural network-based methods provide a standardized computer-aided diagnostic approach that improves the reproducibility of results since models are deterministic. In this way, CACTUSS shows reproducible deterministic results, which are within the clinically accepted ranges, and shows stability in evaluation results.

Additionally, CACTUSS shows reproducible results on images from different US machines, which is a positive indication that the algorithm can be machine agnostic. Moreover, CACTUSS can also be referred to as modality agnostic since intermediate representation images can also be generated from other medical modalities such as MRI. Initial experimental results on AAA sample images show that CACTUSS is able to successfully generate an IR for AAA B-mode images independently of anatomical size and shape[4]. The desired segmentation performance can be achieved by re-training the segmentation network on any in-distribution data. This shows that CACTUSS is applicable to AAA cases and has the potential to generalize to other applications and anatomies. This demonstrates the adaptivity of the proposed method.

---

[4] https://github.com/danivelikova/cactuss.

4   Conclusion

In this work, we presented CACTUSS, a common anatomical CT-US space, which is generated out of physics-based simulation systems to address better the task of aorta segmentation for AAA screening and monitoring. We successfully show that US segmentation networks can be trained with existing labeled data from other modalities and effectively solve clinical problems. By better utilizing medical data, we show that the problem of aorta segmentation for AAA screening can be performed within the rigorous standards of current medical practice. Furthermore, we show the robustness of this work by evaluating CACTUSS on data from multiple scanners. Future work includes the integration of CACTUSS in robotic ultrasound platforms for automatic AAA screening and clinical feasibility studies of the method.

Acknowledgements. This work was partially supported by the department of Interventional Radiology in Klinikum rechts der Isar, Munich. We would also like to thank Teresa Schäfer for helping us with the data annotations and Dr. Magdalini Paschali for revising the manuscript.

References

1. Brutti, F., et al.: Deep learning to automatically segment and analyze abdominal aortic aneurysm from computed tomography angiography. Cardiovas. Eng. Technol. 1–13 (2021). https://doi.org/10.1007/s13239-021-00594-z
2. Cao, L., et al.: Fully automatic segmentation of type b aortic dissection from CTA images enabled by deep learning. Eur. J. Radiol. **121**, 108713 (2019)
3. Chaikof, E.L., et al.: The society for vascular surgery practice guidelines on the care of patients with an abdominal aortic aneurysm. J. Vasc. Surg. **67**(1), 2–77.e2 (2018). https://www.sciencedirect.com/science/article/pii/S0741521417323698
4. Hartshorne, T., McCollum, C., Earnshaw, J., Morris, J., Nasim, A.: Ultrasound measurement of aortic diameter in a national screening programme. Eur. J. Vasc. Endovasc. Surg. **42**(2), 195–199 (2011)
5. Heusel, M., Ramsauer, H., Unterthiner, T., Nessler, B., Hochreiter, S.: Gans trained by a two time-scale update rule converge to a local nash equilibrium (2017)
6. Jensen, J.A., Nikolov, I.: Fast simulation of ultrasound images. In: 2000 IEEE Ultrasonics Symposium. Proceedings. An International Symposium (Cat. No. 00CH37121), vol. 2, pp. 1721–1724. IEEE (2000)
7. Jiang, Z., et al.: Autonomous robotic screening of tubular structures based only on real-time ultrasound imaging feedback. IEEE Trans. Ind. Electron. **69**(7), 7064–7075 (2021)
8. Kojcev, R., et al.: On the reproducibility of expert-operated and robotic ultrasound acquisitions. Int. J. Comput. Assist. Radiol. Surg. **12**(6), 1003–1011 (2017). https://doi.org/10.1007/s11548-017-1561-1
9. Langsch, F., Virga, S., Esteban, J., Göbl, R., Navab, N.: Robotic ultrasound for catheter navigation in endovascular procedures. In: 2019 IEEE/RSJ International Conference on Intelligent Robots and Systems (IROS), pp. 5404–5410 (2019). https://doi.org/10.1109/IROS40897.2019.8967652, ISSN: 2153-0866

10. López-Linares, K., et al.: Fully automatic detection and segmentation of abdominal aortic thrombus in post-operative CTA images using deep convolutional neural networks. Med. Image Anal. **46**, 202–214 (2018)
11. Merouche, S., et al.: A robotic ultrasound scanner for automatic vessel tracking and three-dimensional reconstruction of b-mode images. IEEE Trans. Ultrason. Ferroelectr. Freq. Control **63**(1), 35–46 (2015)
12. Milletari, F., Navab, N., Ahmadi, S.A.: V-net: fully convolutional neural networks for volumetric medical image segmentation. In: 2016 Fourth International Conference on 3D Vision (3DV), pp. 565–571. IEEE (2016)
13. Park, T., Efros, A.A., Zhang, R., Zhu, J.Y.: Contrastive learning for unpaired image-to-image translation. https://arxiv.org/abs/2007.15651
14. Ronneberger, O., Fischer, P., Brox, T.: U-Net: convolutional networks for biomedical image segmentation. In: Navab, N., Hornegger, J., Wells, W.M., Frangi, A.F. (eds.) MICCAI 2015. LNCS, vol. 9351, pp. 234–241. Springer, Cham (2015). https://doi.org/10.1007/978-3-319-24574-4_28
15. Rumack, C.M., Levine, D.: Diagnostic Ultrasound E-Book. Elsevier Health Sciences, Amsterdam (2017)
16. Salehi, M., Ahmadi, S.-A., Prevost, R., Navab, N., Wein, W.: Patient-specific 3D ultrasound simulation based on convolutional ray-tracing and appearance optimization. In: Navab, N., Hornegger, J., Wells, W.M., Frangi, A.F. (eds.) MICCAI 2015. LNCS, vol. 9350, pp. 510–518. Springer, Cham (2015). https://doi.org/10.1007/978-3-319-24571-3_61
17. Shen, D., Wu, G., Suk, H.I.: Deep learning in medical image analysis. Annu. Rev. Biomed. Eng. **19**(1), 221–248 (2017)
18. Treeby, B.E., Tumen, M., Cox, B.T.: Time domain simulation of harmonic ultrasound images and beam patterns in 3d using the k-space pseudospectral method. In: Fichtinger, G., Martel, A., Peters, T. (eds.) MICCAI 2011. LNCS, vol. 6891, pp. 363–370. Springer, Heidelberg (2011). https://doi.org/10.1007/978-3-642-23623-5_46
19. Ullery, B.W., Hallett, R.L., Fleischmann, D.: Epidemiology and contemporary management of abdominal aortic aneurysms. Abdom. Radiol. **43**(5), 1032–1043 (2018). https://doi.org/10.1007/s00261-017-1450-7
20. Virga, S., et al.: Automatic force-compliant robotic ultrasound screening of abdominal aortic aneurysms. In: 2016 IEEE/RSJ International Conference on Intelligent Robots and Systems (IROS), pp. 508–513. IEEE (2016)
21. Yamashita, R., Nishio, M., Do, R.K.G., Togashi, K.: Convolutional neural networks: an overview and application in radiology. Insights Imaging **9**(4), 611–629 (2018). https://doi.org/10.1007/s13244-018-0639-9

INSightR-Net: Interpretable Neural Network for Regression Using Similarity-Based Comparisons to Prototypical Examples

Linde S. Hesse[1,2(✉)] and Ana I. L. Namburete[2,3]

[1] Institute of Biomedical Engineering, Department of Engineering Science, University of Oxford, Oxford, UK
linde.hesse@seh.ox.ac.uk
[2] Oxford Machine Learning in Neuroimaging (OMNI) Laboratory, Department of Computer Science, University of Oxford, Oxford, UK
[3] Wellcome Centre for Integrative Neuroscience (WIN), Nuffield Department of Clinical Neurosciences, University of Oxford, Oxford, UK

Abstract. Convolutional neural networks (CNNs) have shown exceptional performance for a range of medical imaging tasks. However, conventional CNNs are not able to explain their reasoning process, therefore limiting their adoption in clinical practice. In this work, we propose an inherently interpretable CNN for regression using similarity-based comparisons (*INSightR-Net*) and demonstrate our methods on the task of diabetic retinopathy grading. A prototype layer incorporated into the architecture enables visualization of the areas in the image that are most similar to learned *prototypes*. The final prediction is then intuitively modeled as a mean of prototype labels, weighted by the similarities. We achieved competitive prediction performance with our *INSightR-Net* compared to a ResNet baseline, showing that it is not necessary to compromise performance for interpretability. Furthermore, we quantified the quality of our explanations using sparsity and diversity, two concepts considered important for a good explanation, and demonstrated the effect of several parameters on the latent space embeddings.

Keywords: Interpretability · Regression · Diabetic retinopathy

1 Introduction

Deep learning methods are able to achieve exceptional performance on several regression tasks in medical imaging. However, a key limitation of most of these methods is that they cannot explain their reasoning process, hampering their

Supplementary Information The online version contains supplementary material available at https://doi.org/10.1007/978-3-031-16437-8_48.

© The Author(s), under exclusive license to Springer Nature Switzerland AG 2022
L. Wang et al. (Eds.): MICCAI 2022, LNCS 13433, pp. 502–511, 2022.
https://doi.org/10.1007/978-3-031-16437-8_48

adoption in clinical practice [10]. As clinicians are responsible for providing optimal patient care, it is crucial that they trust the model and are able to verify the model's decision. For this reason, it is important to develop models that are interpretable and can provide an explanation for their prediction.

In interpretability research, *post-hoc* methods that attempt to explain a trained black box model have been the most popular approach due to the often assumed trade-off between performance and interpretability [10]. However, post-hoc explanations have shown to not always be an accurate representation of the model's decision-making process and can therefore be problematic [13]. For example, saliency maps can resemble edge maps rather than being dependent on the trained model [1]. On the other hand, explanations obtained from an inherently interpretable model are by design an accurate representation, but developing interpretable convolutional neural network (CNN)-based architectures is a challenging task [3,9,13].

A popular interpretable CNN for the classification of natural images is *ProtoPNet* [3], which achieved comparable accuracy to several non-interpretable baselines. This method is based on learning representative examples from the training set, referred to as *prototypes*, and classifies new images by computing their similarity to each of the learned prototypes. This method was later extended to breast lesion classification by incorporating part annotations [2], and has been applied to a few other classification tasks in medical imaging [11,14].

However, many medical imaging tasks are intrinsically regression problems, such as grading the severity of disease progression. While these often can be modeled as classification tasks, this ignores the linear inter-dependence of the classes, i.e. predicting a grade 4 or 2 for a true grade of 1 is penalized equally. Additionally, a human observer is likely to only fully understand an explanation if it consists of a limited number of concepts [4]. For this reason, sparsity is considered to be important for an effective explanation [13]. Providing an explanation for each class separately, as done in [13], is thus undesirable, especially for regression tasks with a large range of possible values.

In this work, we propose an Interpretable Neural Network using Similarity-based comparisons for Regression (*INSightR-Net*). Our network incorporates a prototype layer [3], providing insight into which image parts the network considers to be similar to a set of learned prototypes. The final predictions are modeled as a weighted mean of prototype labels, therefore providing an intuitive explanation for a regression task. We also propose to use a new similarity function in our prototype layer and show that this results in a sparser explanation, while maintaining the same prediction performance.

In addition to sparsity, a good explanation should also be specific to a certain sample, i.e. not all samples should have the same explanation. For this reason, we quantitatively assess both *explanation sparsity* and *diversity* to assess the quality of our explanations. Furthermore, in contrast to previous work [2,3,11,14], we provide an analysis of the latent space representations in the prototype layer and study the effect of each of the loss components in an ablation study. We demonstrate the efficacy of our proposed *INSightR-Net* on a large publicly available dataset of diabetic retinopathy grading [5].

2   Methods

Architecture. The network architecture consists of a ResNet-based CNN back-bone (having a sigmoid as last activation), denoted by f, followed by a proto-typical layer g_p and one fully connected layer h_{fc} (Fig. 1). Let $X \in \mathbb{R}^{w \times h \times 3}$ be a three-channel input image of size $w \times h$ with ground-truth label y. The feature extractor f then extracts a latent representation of X, given by $Z \in \mathbb{R}^{w_z \times h_z \times c_z} : f(X)$, in which c_z represents the number of output channels of f, and w_z and h_z are the spatial dimensions of the latent representation. By definition, both w_z and $h_z > 1$, meaning that Z can be split into latent *patches* each of size $(1, 1, c_z)$. Such a latent patch will be denoted by $\tilde{Z}$, and can be interpreted as the latent representation of an *image part*. In the case of retinal images, $\tilde{Z}$ could for example encode optic disc information, or the presence of micro-aneurysms.

Prototype Layer. In g_p, m prototypes are learned, denoted by $P = \{P_j \in \mathbb{R}^{1 \times 1 \times c_z} \; \forall j \in [1, m]\}$. Both $\tilde{Z}$ and P_j can be considered as points in the same latent space and, P_j can thus also be interpreted as the latent representation of an image part. The prototype layer g_p aims to learn representative P_j's so that for a new image X, a prediction can be made based on the L_2 distances of its latent patches, $\tilde{Z}$, to each of the prototypes in P. While the prototypes can initially be located anywhere in the latent space, a *projection* stage in training moves each P_j to the closest $\tilde{Z}$ from the training set (see *Training Algorithm*).

As shown in Fig. 1, for image X with latent representation Z, g_p computes the squared L_2 distances between all $\tilde{Z}$ and P_j, followed by a min-pool opera-tion to obtain the minimum distance of Z to each of the prototypes. The dis-tance computation is implemented using a generalized convolution with the L_2 norm instead of the conventional inner product [12]. Effectively, this means that the learned convolutional filters of this layer represent P, and can be jointly optimized with the convolutional parameters of f. The min-distances are subse-quently converted to similarities, $\mathbf{s}$, using a similarity function ($\phi(\cdot)$). Mathemati-cally, this layer can thus be described by: $\mathbf{s} = g_p(Z) = \phi(\mathbf{d})$, with $\mathbf{d}$ the minimum squared L_2 distances between each P_j and Z, defined by: $\mathbf{d} = \min_{\tilde{Z} \in Z} \|\tilde{Z} - P\|_2^2$.

To induce sparsity in the model predictions, we changed the logarithmic similarity function used in [3] to the following function:

$$\phi(\mathbf{d}) = \frac{1}{(\mathbf{d}/d_{max}) + \epsilon} \tag{1}$$

where d_{max} is the maximum distance possible in latent space (exists because the last activation of f is a sigmoid), and ϵ a small number to prevent division by zero. As the slope of this function increases with decreasing distance, it will amplify the differences in distances. This ultimately results in very high similarity values for some prototypes that will dominate the prediction and as such produce a sparse solution rather than one in which all prototypes contribute equally to the final prediction.

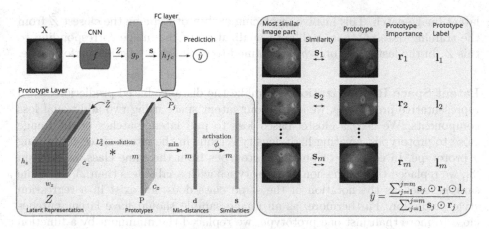

Fig. 1. Overview of INSightR-Net (left) and an example prediction (right). Variables next to the boxes indicate the dimensions, unannotated sides have a dimension of 1.

In order to make a prediction using the computed similarities, each P_j is assigned a continuous label, l_j, and should thus encode an image part that is representative for that label. In contrast to [3], where a predefined number of prototypes per class is used, the m prototypes in *INSightR-Net* are continuous and span the label range of the dataset.

Fully Connected Layer. The final prediction is made by passing the similarity scores, $\mathbf{s}$, through the fully connected layer, h_{fc}, resulting in a single predicted value for each image, denoted by $\hat{y}$. As $\mathbf{s}$ consists of m similarities, h_{fc} learns m weights, denoted by $\boldsymbol{\theta}_h$. In order to enable modeling of the final prediction as a weighted mean of prototype labels, we generate the prediction as follows:

$$\hat{y} = \frac{\sum_{j=1}^{j=m} \mathbf{s}_j \odot \boldsymbol{\theta}_{h,j}^2}{\sum_{j=1}^{j=m} \mathbf{s}_j \odot \boldsymbol{\theta}_{h,j}^2 / l_j} \tag{2}$$

where l_j is the label of prototype P_j, and $\odot$ denotes element-wise multiplication. Given that we can break $\boldsymbol{\theta}_h^2$ down into the prototype label and a weighting vector $\mathbf{r}$ as $\boldsymbol{\theta}_{h,j}^2 = \mathbf{r}_j \odot l_j$, Eq. 2 can be rewritten into a weighted mean, with the weights consisting of the prototype similarity $\mathbf{s}$ and a weighting vector $\mathbf{r}$ as: $\mathbf{w} = \mathbf{s} \odot \mathbf{r}$. Intuitively, $\mathbf{r}$ can now be considered as a *prototype importance score*. Squared weights, $\boldsymbol{\theta}_h^2$, were applied in the last layer to enforce $\mathbf{w}$ to be strictly positive, which is desired in a weighted mean computation.

Training Algorithm. We train our network with the same three-stage protocol as introduced in [3]. In the first stage, the weights of h_{fc} are frozen, and both the weights of f and g_p are trained for a number of epochs. Next, the learned proto-types, which are the convolutional weights of g_p, are *projected* onto the closest latent image patch. This involves replacing each prototype by the closest $\tilde{Z}$ from the training dataset, thus enabling visualization of the image corresponding to this $\tilde{Z}$. In the last stage, only h_{fc} is trained to optimize prediction performance.

Latent Space Regularization. To make the distance-based predictions learn representative prototypes, we shaped our latent space using two additional loss components. We used a cluster-based loss to pull latent patches of the image close to prototypes, creating high-density regions in the latent space containing a prototype. In contrast to pulling prototypes from the same class, as done in [3], we replaced this expression by prototypes with a label less than Δ_l from the sample label, as the notation of the same class does not exist in a regression style prediction. Furthermore, as an image sample should have latent patches close to more than just one prototype, we replaced the minimum by a function that takes the average of the k minimum values, denoted by min_k, resulting in:

$$\mathcal{L}_{Clst} = \frac{1}{n} \sum_{i=1}^{n} min_k(\mathbf{d}_{i,j}) \quad \forall j : ||\mathbf{l}_j - y_i||_1 < \Delta_l \qquad (3)$$

where n is the number of samples in the batch and $\mathbf{d}_{i,j}$ the minimum distance between sample i and prototype j. To facilitate prototype projection, it is desired that each prototype has at least one latent image patch nearby in latent space. As the cluster loss only attracts the k closest prototypes within a certain label range, some prototypes can become outliers lacking image patches in their vicinity. For this reason, an additional prototype sample distance loss [8] was included:

$$\mathcal{L}_{PSD} = -\frac{1}{m} \sum_{j=1}^{m} log(1 - \min_{i \in [1,n]} (\frac{\mathbf{d}_{i,j}}{d_{max}})) \qquad (4)$$

The total training loss is then defined by: $\mathcal{L} = \alpha_{MSE}\mathcal{L}_{MSE} + \alpha_{Clst}\mathcal{L}_{Clst} + \alpha_{PSD}\mathcal{L}_{PSD}$, with $\mathcal{L}_{MSE}$ the mean squared error loss between y and $\hat{y}$, and $\alpha_{(\bullet)}$ the weighting parameters of the loss components.

3   Experimental Setup

Dataset. For this study we used the publicly available EyePACS dataset [5], which was previously used for a Kaggle challenge on diabetic retinopathy (DR) detection. The dataset consists of a large number of RGB retina images in the macula-centered field, that were labeled on the presence of DR on a scale from 0 (healthy) to 4 (most severe). The dataset consists of a training (used for training and validation) and test subset (reserved for final evaluation) and is highly unbalanced. As we considered unbalanced data to be out of scope for this study, we sampled a subset of the data ensuring label balance. Specifically, a maximum of 2443 images per grade was used from the train set, and label balance was achieved by oversampling the minority classes. In the test set we used 1206 images per grade, resulting in a total test set of 6030 images. Even though the labels in this dataset are categorical, the underlying disease progression can be considered as a regression problem. To prevent problems with negative labels, we shifted the labels to a range from 1–5 instead of 0–4, however, all results are reported without this shift.

The retina images were preprocessed using the technique proposed by the top entry of the 2015 Kaggle challenge [6]. This consisted of re-scaling the retina images to have a radius of 300 pixels, subtracting the local average color and clipping with a circular mask with a radius of 270 pixels to remove boundary effects. Finally, the pre-processed images were center-cropped to 512×512.

Implementation. We used a ResNet-18 as the CNN backbone [7] that was pretrained as a regression task on our dataset, with the addition of one more convolutional block (Conv+ReLu+Conv+Sigmoid) to reduce the latent image size to 9×9 ($w_z = w_h = 9$). The number of prototypes, m, was set to 50 and the depth of Z, c_z, to 128. Based on cross-validation, in the loss function α_{MSE} and α_{Clst} were set to 1 and α_{PSD} to 10. The k in $\mathcal{L}_{Clst}$ was set to 3 and Δ_l to 0.5.

We trained our network with two subsequent cycles of the three-step protocol, each cycle consisting of 20 epochs of the joint training stage followed by 10 epochs of last layer training. In the first 5 epochs of the first joint stage, only g_p and the additional convolutional block were trained. The prototype labels were fixed at the beginning of training with equal intervals between 0.1 and 5.9 and did not change during prototype projection. The prototype label range was set slightly larger than the ground-truth labels to make sure that the network could predict values at the boundaries of the range. θ_h^2 was initialized to 1, effectively setting the importance of each prototype ($\mathbf{r}_j$) to 1. We performed 5-fold cross-validation using the train data and selected the best model of each fold with the validation loss. We report the average results of these models on the held-out test set.

We used the Adam optimizer with a learning rate of $1e-5$ for f, and $1e-3$ for both g_p and h_{fc}. The batch size was set to 30 and we applied random rotation (between 0 and 360°) and scaling (between 0.9 and 1.1) as augmentation. The same pretrained ResNet-18 backbone was used as baseline (without the additional block), adding an average pool and fully connected layer. This baseline was finetuned for 30 epochs with a learning rate of $1e-4$, decaying by a factor 2 every 5 epochs. All experiments were run on a GeForce GTX 1080 GPU using Python 3.7 and Pytorch 1.7. One cross-validation fold took 2.5 h to complete. All code is available at https://github.com/lindehesse/INSightR-Net.

4   Results and Discussion

To evaluate the prediction performance we used the mean absolute error (MAE) and accuracy. The prediction performance of *INSightR-Net* and the ResNet-18 baseline are shown in Table 1. It is evident that the performance of *INSightR-Net* is almost equal to that of the baseline, demonstrating that it is not necessary to sacrifice prediction performance to gain interpretability.

Table 1. Quantitative results of model predictions. Each arrow indicates the optimum of the metric. Five-class accuracy was computed from rounding regression scores to the closest integer. Shown standard deviations are across the five folds.

	MAE ↓	Accuracy ↑	s_{spars} ↓	Diversity ↑
ResNet-18 baseline	**0.59 ± 0.004**	**0.52 ± 0.003**	–	–
(1a) with Log-Similarity [3]	0.61 ± 0.003	0.51 ± 0.003	19.2 ± 1.15	35.0 ± 4.24
(1b) w/o $\mathcal{L}_{PSD}$	0.60 ± 0.004	0.51 ± 0.005	15.2 ± 0.82	31.2 ± 1.47
(1c) w/o $\mathcal{L}_{Clst}$	0.60 ± 0.001	**0.52 ± 0.003**	14.8 ± 1.37	**40.8 ± 6.18**
(1d) w/o $\mathcal{L}_{Clst}$ and $\mathcal{L}_{PSD}$	0.60 ± 0.003	0.51 ± 0.012	18.7 ± 0.93	37.6 ± 4.45
(1e) w/o min-k in $\mathcal{L}_{Clst}$	0.60 ± 0.004	**0.52 ± 0.008**	**9.9 ± 1.10**	32.0 ± 1.67
(1) INSightR (ours)	**0.59 ± 0.004**	**0.52 ± 0.005**	14.6 ± 0.73	35.8 ± 2.64

In Fig. 2 the top-3 most contributing prototypes (with the highest $\mathbf{w}_j$) are shown for a representative example from the test set. The activations overlaid on top of the image sample indicate the similarity of each latent patch to the respective prototype. In a similar way, the activation maps overlaid on the prototypes represent the activation of the image that the prototype was projected on. Both of these activation maps can thus be interpreted as the regions that the model looked at while computing the similarity score. A more detailed description of the visualization of activation maps can be found in [3].

By including a prototypical layer in our architecture, our model's decision process is thus inherently interpretable and provides not only local attention on the input image, but also similar examples (prototypes) from the training set. This contrasts with many other interpretable methods that are typically post-hoc, and provide only local attention [15]. Furthermore, our method is specifically tailored for regression and by necessity provides a single explanation for the prediction as opposed to a separate explanation for each class [3].

As it is challenging to assess the quality of the prediction from visual examples, we introduced two new metrics to quantify the quality of a prediction: explanation sparsity and model diversity. We express sparsity as the number of prototypes required to explain 80% of the prediction (the s_{spars} for which $\sum_{j=1}^{j=s_{spars}} \mathbf{w}_j = 0.8 \cdot \sum_{j=1}^{j=m} \mathbf{w}_j$). In this definition, a lower s_{spars} results in a sparser solution. While sparsity is a desired quality, a low sparsity could also be a result of the same few prototypes being used for each explanation. For this reason, we also quantify model diversity, which we define by the number of prototypes that are in the top-5 contributing prototypes for at least 1% of the test set images. Ideally, the explanation of each sample is different (high *diversity*) and can mostly be captured by a minimal number of prototypes (low s_{spars}).

It can be seen in Table 1 that replacing the log-activation as the similarity function improves sparsity ($p < 0.001$ for all folds using a Wilcoxon signed-rank test) while preserving model diversity. This confirms our hypothesis that the new similarity function increases the sparsity of the explanation. Furthermore, this similarity function also increases prediction performance by a small amount.

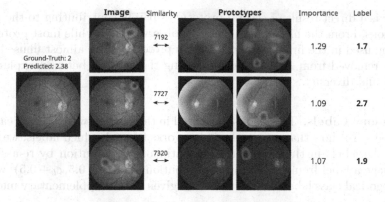

Fig. 2. Example prediction showing the top-3 prototypes and their similarity scores. These top-3 prototypes constitute 33% of the weights for the final prediction. The activation maps were upsampled from 9×9, and the shown prototype labels were shifted to match the true label range (resulting in a prototype label range from -0.9 to 4.9).

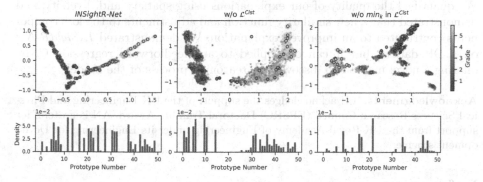

Fig. 3. Top row: 2D PCA of latent space embeddings ($\tilde{Z}$) of the test set samples (stars) and learned prototypes (circles). For each sample only the 5 embeddings closest to a prototype are shown. Bottom row: probability density histograms of the occurrence of a certain prototype in the top-5 most contributing prototypes of a test set sample.

Ablation Studies. The effect of removing loss components is shown in Table 1. It can be seen that removing $\mathcal{L}_{PSD}$ results in worse s_{spars} and model diversity, clearly demonstrating the need for this component. However, removing $\mathcal{L}_{clust}$ shows a less clear effect, improving model diversity while maintaining a similar s_{spars}. To demonstrate the effect of this component in more detail, we visualized their latent representations in Fig. 3. It is evident that without $\mathcal{L}_{clust}$, the samples are spread out in latent space with prototypes scattered in these representations. On the other hand, with $\mathcal{L}_{clust}$, the samples are clustered around prototypes, thus encouraging each prototype to display a representative concept.

Removing the min_k averaging (replacing it with a minimum) in $\mathcal{L}_{Clust}$ reduces s_{spars} as well as model diversity. It can be seen in Fig. 3 that the improved sparsity is a direct result of several prototypes barely contributing to the model predictions. From the histograms it can be observed that while most prototypes are being used in the first two panels, many prototypes are almost unused when min_k is removed from $\mathcal{L}_{Clust}$, demonstrating the balance between explanation sparsity and diversity.

Continuous Labels. The results presented in this section were on the (categorical) labels. To show that our method also works for real-valued labels, we transformed the labels in the train set to a continuous distribution by re-assigning each image a label from the uniform distribution $\sim U(c_l - 0.5, c_l + 0.5)$, with c_l the categorical class label. These results are given in the supplementary material.

5   Conclusion

In this work we showed that *INSightR-Net* is able to provide an intuitive explanation for a regression task while maintaining baseline prediction performance. We quantified the quality of our explanations using sparsity and diversity, and demonstrated that a new similarity function and adjustments to the loss components contributed to an improved explanation. We demonstrated *INSightR-Net* on a DR dataset but it can be applied to any feedforward regression CNN, offering a novel method to discover the reasoning process of the model.

Acknowledgments. LH acknowledges the support of the UK Engineering and Physical Sciences Research Council (EPSRC) Doctoral Training Award. AN is grateful for support from the UK Royal Academy of Engineering under its Engineering for Development scheme.

References

1. Adebayo, J., Gilmer, J., Muelly, M., Goodfellow, I., Hardt, M., Kim, B.: Sanity checks for saliency maps. Adv. Neural Inf. Process. Syst. **31** (2018)
2. Barnett, A.J., et al.: A case-based interpretable deep learning model for classification of mass lesions in digital mammography. Nat. Mach. Intell. **3**(12), 1061–1070 (2021)
3. Chen, C., Li, O., Tao, D., Barnett, A., Rudin, C., Su, J.K.: This looks like that: deep learning for interpretable image recognition. Adv. Neural Inf. Process. Syst. **32** (2019)
4. Cowan, N.: The magical mystery four: how is working memory capacity limited, and why? Curr. Dir. Psychol. Sci. **19**(1), 51–57 (2010)
5. EyePACS: Diabetic retinopathy detection (2015). https://www.kaggle.com/c/diabetic-retinopathy-detection/data
6. Graham, B.: Kaggle diabetic retinopathy detection competition report, pp. 24–26. University of Warwick (2015)
7. He, K., Zhang, X., Ren, S., Sun, J.: Deep residual learning for image recognition. In: Proceedings of the IEEE Conference on Computer Vision and Pattern Recognition, pp. 770–778 (2016)

8. Kraft, S., et al.: SPARROW: semantically coherent prototypes for image classification. In: The 32nd British Machine Vision Conference (BMVC) 2021 (2021)

9. Li, O., Liu, H., Chen, C., Rudin, C.: Deep learning for case-based reasoning through prototypes: a neural network that explains its predictions. In: Proceedings of the AAAI Conference on Artificial Intelligence, vol. 32 (2018)

10. Markus, A.F., Kors, J.A., Rijnbeek, P.R.: The role of explainability in creating trustworthy artificial intelligence for health care: a comprehensive survey of the terminology, design choices, and evaluation strategies. J. Biomed. Inform. **113**, 103655 (2021)

11. Mohammadjafari, S., Cevik, M., Thanabalasingam, M., Basar, A., Initiative, A., et al.: Using protopnet for interpretable Alzheimer's disease classification. In: Proceedings of the Canadian Conference on Artificial Intelligence (2021)

12. Nalaie, K., Ghiasi-Shirazi, K., Akbarzadeh-T, M.R.: Efficient implementation of a generalized convolutional neural networks based on weighted Euclidean distance. In: 2017 7th International Conference on Computer and Knowledge Engineering (ICCKE), pp. 211–216. IEEE (2017)

13. Rudin, C.: Stop explaining black box machine learning models for high stakes decisions and use interpretable models instead. Nat. Mach. Intell. **1**(5), 206–215 (2019)

14. Singh, G., Yow, K.C.: These do not look like those: an interpretable deep learning model for image recognition. IEEE Access **9**, 41482–41493 (2021)

15. van der Velden, B.H., Kuijf, H.J., Gilhuijs, K.G., Viergever, M.A.: Explainable artificial intelligence (XAI) in deep learning-based medical image analysis. Med. Image Anal. 102470 (2022)

Disentangle Then Calibrate: Selective Treasure Sharing for Generalized Rare Disease Diagnosis

Yuanyuan Chen[1], Xiaoqing Guo[2], Yong Xia[1(✉)], and Yixuan Yuan[2(✉)]

[1] National Engineering Laboratory for Integrated Aero-Space-Ground-Ocean Big Data Application Technology, School of Computer Science and Engineering, Northwestern Polytechnical University, Xi'an 710072, China
yxia@nwpu.edu.cn
[2] Department of Electrical Engineering, City University of Hong Kong, Hong Kong SAR, China
yxyuan.ee@cityu.edu.hk

Abstract. Annotated images for rare disease diagnosis are extremely hard to collect. Therefore, identifying rare diseases based on scarce amount of data is of far-reaching significance. Existing methods target only at rare diseases diagnosis, while neglect to preserve the performance of common disease diagnosis. To address this issue, we first disentangle the features of common diseases into a disease-shared part and a disease-specific part, and then employ the disease-shared features alone to enrich rare-disease features, without interfering the discriminability of common diseases. In this paper, we propose a new setting, *i.e.*, generalized rare disease diagnosis to simultaneously diagnose common and rare diseases. A novel selective treasure sharing (STS) framework is devised under this setting, which consists of a gradient-induced disentanglement (GID) module and a distribution-targeted calibration (DTC) module. The GID module disentangles the common-disease features into disease-shared channels and disease-specific channels based on the gradient agreement across different diseases. Then, the DTC module employs only disease-shared channels to enrich rare-disease features via distribution calibration. Hence, abundant rare-disease features are generated to alleviate model overfitting and ensure a more accurate decision boundary. Extensive experiments conducted on two medical image classification datasets demonstrate the superior performance of the proposed STS framework.

1 Introduction

Deep learning has achieved remarkable success in medical image analysis for computer-aided diagnosis (CAD) of various diseases [4,6,14]. However, the automated diagnosis of rare diseases using medical imaging remains challenging,

---

Y. Chen—This work was done when Yuanyuan Chen was a visiting student at the Department of Electrical Engineering, City University of Hong Kong.

© The Author(s), under exclusive license to Springer Nature Switzerland AG 2022
L. Wang et al. (Eds.): MICCAI 2022, LNCS 13433, pp. 512–522, 2022.
https://doi.org/10.1007/978-3-031-16437-8_49

since deep learning usually requires large amounts of labeled training data [3,24], which is extremely difficult to be collected for rare diseases [9,10]. Therefore, studying automated rare disease diagnosis using an extraordinarily scarce amount of medical images is of far-reaching significance and clinical values for assisting medical professionals in making better decisions.

Rare disease diagnosis can be treated as a few-shot learning (FSL) task [9,15,25], where rare diseases are regarded as few-shot classes and relevant common diseases with abundant samples are considered as many-shot classes. Currently, FSL-based rare disease diagnosis is mainly based on self-supervised learning (SSL), meta-learning, or metric-learning. In SSL-based methods [10,18], a feature extractor is first trained on common-disease data using contrastive learning, and then fine-tuned on rare-disease data for rare disease diagnosis. Meta-learning based techniques [9,15] simulate the few-shot setting by iteratively sampling small datasets from common-disease data to train a model, which is able to adapt quickly to few-shot classes. Metric-learning based methods [2,25] learn appropriate comparison metrics, based on which each query sample can be classified by comparing it with the support samples of each class. Overall, these methods focus on training a model that can perform well on rare-disease data with common-disease knowledge. Besides, there exists another kind of FSL methods, which solve the few-shot problem through data augmentation and have been widely used for natural image classification. Specifically, they augment the features of few-shot classes by explicitly transferring either the intra-class variance [5,7,17] or feature distribution [1,21] from many-shot to few-shot classes.

Despite achieving good performance, these FSL-based methods still face two challenges in their efforts to be a clinically demanding CAD tool. On one hand, these methods [2,9,10,15,18,25] target only at rare diseases diagnosis, while neglecting to preserve the performance of common disease diagnosis. In clinical practice, each case could belong to either a rare disease or a related common disease. Thus, a CAD tool should be able to diagnose both common and rare diseases, instead of just differentiating rare diseases. On the other hand, data augmentation-based FSL methods [5,7,17] directly transfer the knowledge embedded in all features from many-shot classes to few-shot classes. Although these methods perform well in classifying few-shot classes, they may achieve less-optimal performance on simultaneous common and rare diseases diagnosis, since they neglect the distinctions of features from common and rare diseases.

To address both issues, we advocate to first disentangle the features of common-disease data into a disease-shared part and a disease-specific part, and then employ the disease-shared features alone to enrich rare-disease features for better diagnosis. The disease-shared features represent the knowledge contained in all diseases and thus can be transferred among diseases. Transferring these features from common diseases to rare diseases can expand the feature distribution and provide rich information for rare diseases without interfering their discriminability. In contrast, the disease-specific features include unique knowledge of each disease, and should not be transferred among different diseases. Such feature disentanglement and selective feature enrichment provide a potential for simultaneous diagnosis of common and rare diseases.

In this paper, we propose a new setting, *i.e.*, generalized rare disease diagnosis, aiming to simultaneously diagnose common and rare diseases. We devise a novel selective treasure sharing (STS) framework under this new setting, which consists of a gradient-induced disentanglement (GID) module and a distribution-targeted calibration (DTC) module. Specifically, the GID module disentangles the common-disease features into disease-shared channels and disease-specific channels based on the gradient agreement across different diseases. Then, the DTC module utilizes disease-shared channels to enrich the features of rare diseases by distribution calibration. The distribution of rare-disease features is calibrated at the lesion region and normal region, respectively. Thus, abundant features can be generated to alleviate model overfitting and derive a better decision boundary. Finally, two well-performing diagnosis models constructed for common and rare disease, respectively, are concatenated to form a generalized diagnosis framework, which is then fine-tuned using the samples of all diseases. We evaluated the proposed STS framework on two medical image datasets, and it achieves superior performance over state-of-the-art approaches.

2   Method

Given a dataset with image-label pairs $\mathcal{D} = \{\mathbf{x}_i, \mathbf{y}_i\}$, where $\mathbf{x}_i$ is an input image and $\mathbf{y}_i$ is the corresponding one-hot label, the class set is divided into common diseases $\mathcal{D}_c$ with abundant training samples and rare diseases $\mathcal{D}_r$ with only few-shot training samples. Note that there is no overlap between the label spaces of $\mathcal{D}_c$ and $\mathcal{D}_r$. The goal of the proposed selective treasure sharing (STS) framework is to train a generalized model which can simultaneously diagnose all diseases. The STS framework is composed of a common branch, a rare branch, a gradient-induced disentanglement (GID) module, and a distribution-targeted calibration (DTC) module (see Fig. 1). The training process of the STS framework consists of three steps. 1) We utilize all training data from $\mathcal{D}_c$ to train the common branch with the standard cross entropy loss. During this process, GID module disentangles the convolutional kernels of the common-branch into disease-shared channels and disease-specific channels. 2) We sample rare-disease data from $\mathcal{D}_r$ to train the rare branch. At each iteration, common-disease data and rare-disease data are sent to the common branch and rare branch, respectively. Then, DTC module utilizes the disease-shared channels of common-disease features to calibrate the distribution of rare-disease features for feature enrichment. The generated rare-disease features are fed forward for rare-disease diagnosis. 3) We concatenate both common and rare branches at the last fully connected layer to form a generalized diagnosis framework, and then fine-tune it using the samples of all diseases to perform simultaneous common and rare diseases diagnosis. During the inference process, testing data of all diseases are directly fed into the generalized diagnosis framework without the operations in two modules.

2.1   Gradient-Induced Disentanglement (GID)

Existing FSL-based methods [5,7,17] utilize all features of many-shot classes to enrich features of few-shot classes. However, they are infeasible for simultaneous

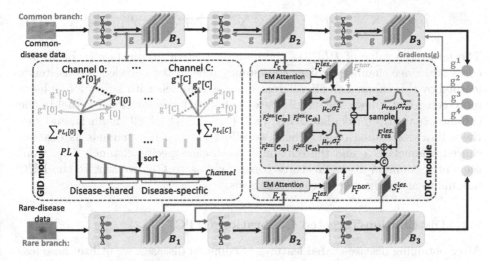

Fig. 1. Architecture of the proposed *Selective Treasure Sharing (STS)* framework, including (1) a gradient-induced disentanglement (GID) module for disease-shared channel disentanglement and (2) a distribution-targeted calibration (DTC) module for rare-disease feature enriching. GID module and DTC module are applied to all network blocks (B_1, B_2 and B_3) of each branch.

common and rare disease diagnosis, since they neglect the distinctions of common and rare diseases. To handle this issue, we devise a GID module to disentangle the common-disease features into disease-shared and disease-specific channels, and only utilize the disease-shared channels to enrich rare-disease features. As network gradients can reflect the consistency of knowledge learned from different aspects [12,19], GID module innovatively performs feature disentanglement by measuring channel gradient agreement across different common diseases.

Given a set of training data composed of N common-diseases D_c, we sample a mini-batch training samples from each disease and utilize them to perform a forward propagation for the common branch. Then, we obtain the cross-entropy loss $\mathcal{L}_i(\theta)$ ($i \in \{1, \cdots, N\}$) of each disease and the averaged loss $\mathcal{L}(\theta) = \frac{1}{N}\sum_{i=1}^{N}\mathcal{L}_i(\theta)$ of all diseases. In order to figure out the disease-shared channels and disease-specific channels, we record the optimization direction indicated by different diseases for each channel. Specifically, we perform loss back propagation using the loss $\mathcal{L}_i(\theta)$ of each disease and the averaged $\mathcal{L}(\theta)$, respectively. After that, we obtain a gradient vector $g^i = \nabla\theta\mathcal{L}_i(\theta)$ of each disease and a consensus gradient vector $g^\star = \nabla\theta\mathcal{L}(\theta)$ of all diseases. Let $g^i[j]$ and $g^\star[j]$ denote the j_{th} channel of $g^{(i)}$ and $g^\star$. If channel gradients of two diseases are similar in both magnitude and direction, this channel learns similar knowledge on two corresponding diseases.

To calculate the channel gradient consistency for all commo diseases, we project the gradient $g^i[j]$ of each disease to the consensus gradient $g^\star[j]$, and calculate the sum of projection lengths by Eq. 1.

$$PL[j] = \sum_{i=1}^{N} PL_{g^i[j] \to g^*[j]} = \sum_{i=1}^{N} |g^i[j]| cos < g^i[j], g^*[j] > \qquad (1)$$

The larger the value of $PL[j]$ is, the more consistent gradients $g^i[j]$ of all diseases are. Since channel gradient consistency reflects the knowledge consistency learned by this channel on all different disease, channels with larger $PL[j]$ learn more consistent knowledge on all diseases. For each convolutional kernel in the common branch, we calculate $PL[j]$ ($j \in \{1, \cdots, C\}$) for all channels, and sort them in a descending order. Then, we select the top-K_1 channels as disease-shared channels $\mathbf{C}_{sh}$ and the remaining as disease-specific channels $\mathbf{C}_{sp}$. Consequently, common-disease features corresponding to $\mathbf{C}_{sh}$ are sent to the DTC module to assist the rare diseases diagnosis.

2.2   Distribution-Targeted Calibration (DTC)

After obtaining disease-shared features of common diseases, we utilize these features alone to enrich the rare-disease features by channel-wise distribution calibration. Due to the various feature distribution in lesion and normal regions, directly performing channel-wise calibration to each whole feature map may lead to biased information. Therefore, we design a DTC module, which calibrates the lesion-region distribution and normal-region distribution for each feature map, respectively, thus deriving abundant features to alleviate the model overfitting.

Region Detection. Given feature maps F_c and F_r from a common disease and a rare disease, we first utilize the off-the-shelf EM attention module [8] to obtain attention maps of each feature map. EM attention module learns a compact set of bases through expectation-maximization algorithms, and multiple attention maps are generated to highlight different feature components. For each feature map, we select two attention maps which best highlight the lesion region and normal region, and obtain a lesion mask and a normal mask based on these two attention maps. Then, based on these masks, we crop the lesion (normal) region $F_c^{les.}$ ($F_c^{nor.}$) from common-disease feature map F_c and the lesion (normal) region $F_r^{les.}$ ($F_r^{nor.}$) from rare-disease feature map F_r, respectively.

Distribution Calibration. Considering features learned from abundant training samples of common diseases represent richer disease-shared information, we utilize the distribution statistics of common-disease features to calibrate the distribution of rare-disease features at those disease-shared channels. Taking the lesion region as an example, we denote the disease-shared feature channels of $F_c^{les.}$ and $F_r^{les.}$ as $F_c^{les.}[\mathbf{C}_{sh}]$ and $F_r^{les.}[\mathbf{C}_{sh}]$, respectively. Assuming each feature channel in $F_c^{les.}[\mathbf{C}_{sh}]$ follows a Gaussian distribution, intensity mean μ_c and intensity variance σ_c can be calculated for each channel. Similarly, μ_r and σ_r can be obtained for each channel in $F_r^{les.}[\mathbf{C}_{sh}]$. Then, we calculate the residual distribution between each pair of channels in $F_c^{les.}[\mathbf{C}_{sh}]$ and $F_r^{les.}[\mathbf{C}_{sh}]$ as follows

$$\mu_{res} = \alpha(\mu_c - \mu_r), \quad \sigma_{res}^2 = \beta(\sigma_c^2 - \sigma_r^2), \qquad (2)$$

where α and β determines the calibration degree.

Feature Enrichment. After obtaining the residual distribution for each channel in $F_r^{les.}[\mathbf{C}_{sh}]$, an adequate number of residual features can be sampled from it to diversify the original channel. To make the feature sampling operation differentiable, we incorporate the re-parameterization technique into the sampling process (Eq. 3). Specifically, for each channel j ($j \in \mathbf{C}_{sh}$), we first sample the residual features from a normal Gaussian distribution, and then modify the features' distribution based on the statistics μ_{res} and σ_{res}^2 we learned before.

$$F_{res}^{les.}[j] = \mu_{res} + \sigma_{res}^2 * \epsilon_t, \quad \epsilon_t \in \mathcal{N}(0,1) \tag{3}$$

The sampled residual features $F_{res}^{les.}[j]$ contain rich disease-shared knowledge that has not yet been learned from few-shot samples of rare diseases. Hence, for each channel j in $\mathbf{C}_{sh}$, we add the sampled residual features $F_{res}^{les.}[j]$ to the original features $F_r^{les.}[j]$ to form a new complemented channel (Eq. 4).

$$S_r^{les.}[j] = \begin{cases} F_r^{les.}[j] + F_{res}^{les.}[j]), & if \ c \in \mathbf{C}_{sh}, \\ F_r^{les.}[j], & if \ c \in \mathbf{C}_{sp}. \end{cases} \tag{4}$$

Similar to the lesion region, we perform distribution calibration and feature enrichment to the normal region and obtain abundant normal-region features $S_r^{nor.}$ for each feature map. Calibrated features $S_r^{les.}$ and $S_r^{nor.}$ are then fed forward to next stage of the rare branch for the final rare-disease diagnosis.

3   Experiments

Dataset: We evaluate the performance of the proposed STS framework on two medical image classification datasets. (1) The ISIC 2018 Skin Lesion Dataset has a total of 10,015 skin lesion images from seven skin diseases: melanocytic nevus (6,705), benign keratosis (1,099), melanoma (1,113), basal cell carcinoma (514), actinic keratosis (327), dermatofibroma (115), and vascular lesion (142). We follow the few-shot classification setting of [9] and utilize the four classes with largest amount of cases as common diseases and the left three classes as rare diseases. (2) The Kvasir dataset [16] consists of eight classes showing anatomical landmarks, phatological findings or endoscopic procedures. Each class contains 500 images. We selected the three phatological classes *i.e.*, Esophagus, Polyps, Ulcerative Colitis, as rare diseases and the rest as common diseases.

Implementation: We utilized WideResNet [22] as backbones of both branches following [11,21]. The Adam optimizer is utilized with learning rate of 0.001. The size of input images is 112×112. The batch-size of the common and rare branch is set to 16 and 8, respectively. When optimizing the common and rare branch, the number of iterations is set to 4000. The hyper-parameters K_1, α and β were empirically set to 150, 0.5 and 0.5, respectively. For the common branch, we randomly select 80% data for training and the remaining 20% are used for testing. For the rare branch, each time we randomly selected K ($K = 1$ or 5) samples from each rare disease as few-shot training data and the remaining

Table 1. Performance ((mean±std)/%) of seven methods on ISIC 2018 dataset and Kvasir dataset under settings of 3-way-1-shot and 3-way-5-shot.

Task	Met.	Method						
		FCICL [13]	CRCKD [20]	D-EMD [23]	DC [21]	DAML [9]	URL-PSS [18]	**STS (ours)**
$ISIC$	AUC	$54.38_{\pm 3.22}$	$54.97_{\pm 3.03}$	$58.25_{\pm 3.06}$	$60.05_{\pm 3.21}$	$57.66_{\pm 2.99}$	$61.04_{\pm 2.67}$	$61.82_{\pm 2.37}$
1_{shot}	ACC	$53.16_{\pm 3.17}$	$53.62_{\pm 2.98}$	$57.31_{\pm 2.89}$	$58.94_{\pm 2.97}$	$56.23_{\pm 3.02}$	$59.39_{\pm 2.87}$	$60.05_{\pm 3.07}$
$Rare$	$F1$	$53.60_{\pm 2.96}$	$54.01_{\pm 3.07}$	$57.88_{\pm 3.01}$	$59.21_{\pm 2.96}$	$56.53_{\pm 2.98}$	$59.75_{\pm 3.02}$	$60.36_{\pm 2.99}$
$ISIC$	AUC	$80.92_{\pm 0.88}$	$82.35_{\pm 0.93}$	$79.62_{\pm 1.01}$	$77.24_{\pm 1.38}$	$80.17_{\pm 1.22}$	$81.73_{\pm 1.17}$	$83.47_{\pm 1.23}$
1_{shot}	ACC	$79.61_{\pm 0.73}$	$81.88_{\pm 0.80}$	$79.06_{\pm 1.23}$	$76.37_{\pm 0.97}$	$78.92_{\pm 1.04}$	$80.64_{\pm 1.31}$	$82.25_{\pm 1.31}$
$Com.$	$F1$	$80.07_{\pm 0.88}$	$82.04_{\pm 1.16}$	$79.12_{\pm 1.33}$	$77.01_{\pm 1.01}$	$77.83_{\pm 0.94}$	$80.75_{\pm 1.22}$	$82.09_{\pm 1.02}$
$ISIC$	AUC	$70.63_{\pm 3.02}$	$71.87_{\pm 2.71}$	$75.68_{\pm 1.93}$	$76.59_{\pm 2.29}$	$73.37_{\pm 2.87}$	$76.94_{\pm 2.36}$	$78.96_{\pm 2.17}$
5_{shot}	ACC	$69.46_{\pm 2.76}$	$70.04_{\pm 2.69}$	$74.28_{\pm 2.06}$	$75.54_{\pm 2.31}$	$72.13_{\pm 2.77}$	$75.25_{\pm 2.32}$	$77.87_{\pm 2.08}$
$Rare$	$F1$	$69.22_{\pm 2.76}$	$69.88_{\pm 2.81}$	$74.06_{\pm 2.31}$	$74.94_{\pm 2.17}$	$72.07_{\pm 2.28}$	$74.94_{\pm 2.42}$	$77.64_{\pm 1.95}$
$ISIC$	AUC	$80.36_{\pm 1.02}$	$81.41_{\pm 0.88}$	$78.49_{\pm 1.21}$	$76.91_{\pm 1.58}$	$79.62_{\pm 1.42}$	$80.79_{\pm 1.32}$	$82.31_{\pm 1.38}$
5_{shot}	ACC	$79.74_{\pm 0.98}$	$80.87_{\pm 0.79}$	$78.06_{\pm 1.08}$	$76.54_{\pm 1.21}$	$79.34_{\pm 1.39}$	$79.93_{\pm 1.17}$	$81.87_{\pm 1.25}$
$Com.$	$F1$	$78.62_{\pm 1.28}$	$80.14_{\pm 1.06}$	$77.31_{\pm 0.99}$	$76.33_{\pm 0.89}$	$77.57_{\pm 1.29}$	$78.75_{\pm 1.33}$	$80.62_{\pm 1.49}$
$Kvas.$	AUC	$63.37_{\pm 3.05}$	$64.08_{\pm 2.98}$	$66.22_{\pm 2.82}$	$66.83_{\pm 2.96}$	$65.34_{\pm 2.82}$	$68.38_{\pm 2.91}$	$70.25_{\pm 2.93}$
5_{shot}	ACC	$62.43_{\pm 2.99}$	$63.82_{\pm 3.01}$	$65.47_{\pm 3.09}$	$66.26_{\pm 2.99}$	$64.62_{\pm 2.98}$	$67.49_{\pm 2.86}$	$69.33_{\pm 2.88}$
$Rare$	$F1$	$63.30_{\pm 3.02}$	$64.39_{\pm 2.91}$	$65.96_{\pm 3.03}$	$67.01_{\pm 3.12}$	$64.75_{\pm 3.23}$	$68.41_{\pm 2.99}$	$70.07_{\pm 2.94}$
$Kvas.$	AUC	$83.81_{\pm 0.87}$	$84.08_{\pm 0.88}$	$81.59_{\pm 1.21}$	$77.46_{\pm 1.38}$	$79.82_{\pm 1.42}$	$81.87_{\pm 1.32}$	$85.29_{\pm 1.21}$
5_{shot}	ACC	$83.34_{\pm 1.22}$	$83.54_{\pm 0.98}$	$80.94_{\pm 1.17}$	$77.25_{\pm 1.25}$	$79.63_{\pm 1.28}$	$81.25_{\pm 1.19}$	$83.97_{\pm 0.93}$
$Com.$	$F1$	$83.62_{\pm 1.08}$	$84.04_{\pm 1.16}$	$81.12_{\pm 1.06}$	$78.31_{\pm 0.77}$	$79.42_{\pm 0.98}$	$82.96_{\pm 1.23}$	$84.72_{\pm 1.02}$

samples as test data. We run the whole framework 20 times in total and reported the averaged performance over 20 runs. Evaluation metrics include the area under receiver operating characteristic (AUC), accuracy (ACC) and F1-score.

Comparing to Existing Methods: We compared the proposed STS framework against six state-of-the-art methods, including two conventional medical image classification methods [13,20], two FSL classification methods [21,23] and two rare disease diagnosis methods [9,18]. For a fair comparison, we implemented these methods using the hyper-parameter settings reported in original papers under the same data split as the proposed STS. Besides, for methods [9,18,21,23] which are designed for only rare disease diagnosis, we add a common baseline model for each method and then fine-tune both the common model and the rare model together using the same way as the proposed STS, in order to achieve simultaneous common and rare diseases diagnosis. To make better comparison among these methods, we reported the diagnosis performance of common and rare diseases under each setting, respectively. *Results on ISIC 2018:* The performance of each method on the ISIC 2018 Skin Lesion dataset is recorded in Table 1. To validate the effectiveness of the proposed STS with different numbers of rare-disease training samples, we trained the rare branch with 5 training images per rare disease (*i.e.*, the 3-way-5-shot setting) or only 1 training image per rare disease (*i.e.*, the 3-way-1-shot setting), respectively. Under the 3-way-5-shot setting, the proposed STS ranks the first on all metrics for both common and rare disease diagnosis. It is observed that although methods [13,20] perform well on common disease diagnosis, their rare-disease diagnosis performance is quite poor, since these methods are designed for many-shot classes and can

Table 2. Results ((mean ± std)/%) of ablation study on ISIC dataset.

D	M	GID	DTC	AUC (%)	ACC (%)	F1 (%)
Rare	I			70.94 ± 2.82	69.81 ± 2.73	69.49 ± 2.54
	II		√	75.34 ± 2.51	74.97 ± 2.60	74.27 ± 2.36
	STS	√	√	**78.96 ± 2.17**	**77.87 ± 2.08**	**77.64 ± 1.95**
Common	I			80.65 ± 1.23	80.04 ± 0.99	79.21 ± 1.17
	II		√	77.06 ± 1.42	76.56 ± 1.33	75.02 ± 1.07
	STS	√	√	**82.31 ± 1.38**	**81.87 ± 1.25**	**80.62 ± 1.49**

not tackle with the few-shot problem for rare disease diagnosis. Comparing with FSL-based methods [9,18,21,23], the proposed STS achieves better performance on both common and rare disease diagnosis, indicating that the distribution calibration to those disease-shared channels is beneficial to the simultaneous diagnosis of common and rare diseases. Under the 3-way-1-shot setting, the proposed STS still achieves the highest performance for both common and rare disease diagnosis. It reveals the proposed STS is also effective in rare diagnosis based on extremely scarce amount of data, *i.e.*, only 1 image per class. **Results on Kvasir:** To verify the generalization ability of the proposed STS framework, we chose the common and rare disease diagnosis under 3-way-5-shot setting as a case study and tested the performance of each method on the Kvasir dataset in Table 1. The proposed STS still obtains the highest results on all four metrics for both common and rare disease diagnosis, further indicating the generalization ability of the proposed STS over different datasets.

Ablation Study: To evaluate the effectiveness of GID module and DTC module, we chose the classification of common diseases and rare diseases (under the 3-way-5-shot setting) on ISIC dataset as a case study, and tested the performance of the following model variants. Model I is the baseline which trains a classification model using images from common diseases and then directly fine-tunes it using images from rare diseases. Model II merely incorporates DTC module to the baseline, *i.e.*, performing distribution calibration for all channels without channel disentanglement by GID module. The diagnosis performance of these variants is shown in Table 2. It reveals that Model I has the lowest performance on rare diseases diagnosis, indicating fine-tuning the model with few-shot training samples of rare diseases leads to model overfitting and thus deriving biased decision boundary. The proposed STS achieves an AUC improvement of 3.62% over Model II on rare diseases diagnosis, indicating that utilizing only disease-shared channels of common diseases to assist rare diseases diagnosis is better than using all channels. For common diseases diagnosis, Model II performs worse than Model I, suggesting transferring all features from common diseases to rare diseases interferes the discriminability of common diseases. Besides, the proposed STS outperforms Model I with an AUC improvement of 1.66%, demonstrating transferring only disease-shared channels from common diseases to rare diseases is beneficial to simultaneous common and rare disease diagnosis.

4   Conclusion

We propose a generalized rare disease diagnosis framework STS for both common and rare disease diagnosis, which consists of a GID module and a DTC module. The GID module disentangles the common-disease knowledge into disease-shared channels and disease-specific channels, and only disease-shared channels will be transferred to assist the rare disease diagnosis in DTC module based on distribution calibration. Thus, abundant rare-disease features can be generated to alleviate model overfitting without interfering the discriminability of common-disease diagnosis. Extensive experiments conducted on two medical image classification datasets demonstrate the superior performance of the proposed STS framework.

Acknowledgement. This work was supported in part by the National Natural Science Foundation of China under Grants 62171377, in part by the Key Research and Development Program of Shaanxi Province under Grant 2022GY-084, in part by Hong Kong Research Grants Council (RGC) Early Career Scheme grant 21207420 (CityU 9048179), and in part by Hong Kong RGC Collaborative Research Fund grant C4063-18G (CityU 8739029).

References

1. Afrasiyabi, A., Lalonde, J.-F., Gagné, C.: Associative alignment for few-shot image classification. In: Vedaldi, A., Bischof, H., Brox, T., Frahm, J.-M. (eds.) ECCV 2020. LNCS, vol. 12350, pp. 18–35. Springer, Cham (2020). https://doi.org/10.1007/978-3-030-58558-7_2
2. Ali, S., Bhattarai, B., Kim, T.-K., Rittscher, J.: Additive angular margin for few shot learning to classify clinical endoscopy images. In: Liu, M., Yan, P., Lian, C., Cao, X. (eds.) MLMI 2020. LNCS, vol. 12436, pp. 494–503. Springer, Cham (2020). https://doi.org/10.1007/978-3-030-59861-7_50
3. Cai, J., et al.: Deep lesion tracker: monitoring lesions in 4D longitudinal imaging studies. In: Proceedings of the IEEE/CVF Conference on Computer Vision and Pattern Recognition (CVPR), pp. 15159–15169 (2021)
4. Chen, Y., Xia, Y.: Iterative sparse and deep learning for accurate diagnosis of Alzheimer's disease. Pattern Recogn. **116**, 107944 (2021)
5. Gao, H., Shou, Z., Zareian, A., Zhang, H., Chang, S.F.: Low-shot learning via covariance-preserving adversarial augmentation networks. In: In Advances in Neural Information Processing Systems (NeurIPS), pp. 981–991 (2018)
6. Guo, X., Yang, C., Liu, Y., Yuan, Y.: Learn to threshold: thresholdnet with confidence-guided manifold mixup for polyp segmentation. IEEE Trans. Med. Imaging **40**(4), 1134–1146 (2021)
7. Li, K., Zhang, Y., Li, K., Fu, Y.: Adversarial feature hallucination networks for few-shot learning. In: Proceedings of the IEEE/CVF Conference on Computer Vision and Pattern Recognition (CVPR), pp. 13470–13479 (2020)

8. Li, X., Zhong, Z., Wu, J., Yang, Y., Lin, Z., Liu, H.: Expectation-maximization attention networks for semantic segmentation. In: Proceedings of the IEEE/CVF International Conference on Computer Vision (ICCV), pp. 9167–9176 (2019)

9. Li, X., Yu, L., Jin, Y., Fu, C.-W., Xing, L., Heng, P.-A.: Difficulty-aware meta-learning for rare disease diagnosis. In: Martel, A.L., et al. (eds.) MICCAI 2020. LNCS, vol. 12261, pp. 357–366. Springer, Cham (2020). https://doi.org/10.1007/978-3-030-59710-8_35

10. Mai, S., Li, Q., Zhao, Q., Gao, M.: Few-shot transfer learning for hereditary retinal diseases recognition. In: de Bruijne, M., et al. (eds.) MICCAI 2021. LNCS, vol. 12908, pp. 97–107. Springer, Cham (2021). https://doi.org/10.1007/978-3-030-87237-3_10

11. Mangla, P., Kumari, N., Sinha, A., Singh, M., Krishnamurthy, B., Balasubramanian, V.N.: Charting the right manifold: manifold mixup for few-shot learning. In: Proceedings of the IEEE/CVF Winter Conference on Applications of Computer Vision, pp. 2218–2227 (2020)

12. Mansilla, L., Echeveste, R., Milone, D.H., Ferrante, E.: Domain generalization via gradient surgery. In: Proceedings of the IEEE/CVF International Conference on Computer Vision, pp. 6630–6638 (2021)

13. Marrakchi, Y., Makansi, O., Brox, T.: Fighting class imbalance with contrastive learning. In: de Bruijne, M., et al. (eds.) MICCAI 2021. LNCS, vol. 12903, pp. 466–476. Springer, Cham (2021). https://doi.org/10.1007/978-3-030-87199-4_44

14. Pan, Y., Liu, M., Xia, Y., Shen, D.: Disease-image-specific learning for diagnosis-oriented neuroimage synthesis with incomplete multi-modality data. IEEE Trans. Pattern Anal. Mach. Intell. (2021). https://doi.org/10.1109/TPAMI.2021.3091214

15. Paul, A., Tang, Y., Shen, T.C., Summers, R.M.: Discriminative ensemble learning for few-shot chest x-ray diagnosis. Med. Image Anal. **68**, 101911 (2021)

16. Pogorelov, K., et al: A multi-class image dataset for computer aided gastrointestinal disease detection. In: Proceedings of the 8th ACM on Multimedia Systems Conference, pp. 164–169 (2017)

17. Schwartz, E., et al.: δ-encoder: an effective sample synthesis method for few-shot object recognition. In: International Conference on Neural Information Processing Systems (NIPS), pp. 2850–2860 (2018)

18. Sun, J., Wei, D., Ma, K., Wang, L., Zheng, Y.: Unsupervised representation learning meets pseudo-label supervised self-distillation: a new approach to rare disease classification. In: de Bruijne, M., et al. (eds.) MICCAI 2021. LNCS, vol. 12905, pp. 519–529. Springer, Cham (2021). https://doi.org/10.1007/978-3-030-87240-3_50

19. Wang, Y., Zhang, R., Zhang, S., Li, M., Xia, Y., Zhang, X., Liu, S.: Domain-specific suppression for adaptive object detection. In: Proceedings of the IEEE/CVF Conference on Computer Vision and Pattern Recognition, pp. 9603–9612 (2021)

20. Xing, X., Hou, Y., Li, H., Yuan, Y., Li, H., Meng, M.Q.-H.: Categorical relation-preserving contrastive knowledge distillation for medical image classification. In: de Bruijne, M., et al. (eds.) MICCAI 2021. LNCS, vol. 12905, pp. 163–173. Springer, Cham (2021). https://doi.org/10.1007/978-3-030-87240-3_16

21. Yang, S., Liu, L., Xu, M.: Free lunch for few-shot learning: distribution calibration. In: International Conference on Learning Representations (ICLR) (2021)

22. Zagoruyko, S., Komodakis, N.: Wide residual networks. arXiv preprint arXiv:1605.07146 (2016)

23. Zhang, C., Cai, Y., Lin, G., Shen, C.: Deepemd: few-shot image classification with differentiable earth mover's distance and structured classifiers. In: Proceedings of the IEEE/CVF conference on computer vision and pattern recognition (CVPR), pp. 12203–12213 (2020)

24. Zhang, J., Xie, Y., Xia, Y., Shen, C.: Dodnet: learning to segment multi-organ and tumors from multiple partially labeled datasets. In: Proceedings of the IEEE/CVF Conference on Computer Vision and Pattern Recognition (CVPR), pp. 1195–1204 (2021)
25. Zhu, W., Li, W., Liao, H., Luo, J.: Temperature network for few-shot learning with distribution-aware large-margin metric. Pattern Recogn. **112**, 107797 (2021)

Learning Robust Representation for Joint Grading of Ophthalmic Diseases via Adaptive Curriculum and Feature Disentanglement

Haoxuan Che[1,3(✉)], Haibo Jin[1], and Hao Chen[1,2,3]

[1] Department of Computer Science and Engineering,
The Hong Kong University of Science and Technology, Kowloon, Hong Kong
{hche,hjinag,jhc}@cse.ust.hk
[2] Department of Chemical and Biological Engineering,
The Hong Kong University of Science and Technology, Kowloon, Hong Kong
[3] Center for Aging Science, The Hong Kong University of Science and Technology,
Kowloon, Hong Kong

Abstract. Diabetic retinopathy (DR) and diabetic macular edema (DME) are leading causes of permanent blindness worldwide. Designing an automatic grading system with good generalization ability for DR and DME is vital in clinical practice. However, prior works either grade DR or DME independently, without considering internal correlations between them, or grade them jointly by shared feature representation, yet ignoring potential generalization issues caused by difficult samples and data bias. Aiming to address these problems, we propose a framework for joint grading with the dynamic difficulty-aware weighted loss (DAW) and the dual-stream disentangled learning architecture (DETACH). Inspired by curriculum learning, DAW learns from simple samples to difficult samples dynamically via measuring difficulty adaptively. DETACH separates features of grading tasks to avoid potential emphasis on the bias. With the addition of DAW and DETACH, the model learns robust disentangled feature representations to explore internal correlations between DR and DME and achieve better grading performance. Experiments on three benchmarks show the effectiveness and robustness of our framework under both the intra-dataset and cross-dataset tests.

1 Introduction

Diabetic retinopathy (DR) and diabetic macular edema (DME) are complications caused by diabetes, which are the most common leading cause of the visual loss and blindness worldwide [1]. It is vital to classify stages of DR and DME in clinical practice because treatments are more effective for those diseases at early stages, and patients could receive tailored treatments based on severity. Physicians grade DR into multiple stages according to the severity of retinopathy lesions like hemorrhages, hard and soft exudates, etc., while they classify DME into three stages via occurrences of hard exudates and the shortest distances of hard exudates to the macula center [2,3], as shown in Fig. 1.

© The Author(s), under exclusive license to Springer Nature Switzerland AG 2022
L. Wang et al. (Eds.): MICCAI 2022, LNCS 13433, pp. 523–533, 2022.
https://doi.org/10.1007/978-3-031-16437-8_50

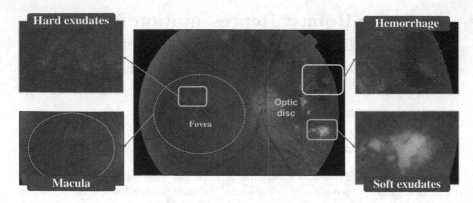

Fig. 1. Symptoms of DR and DME are different in fundus images [1–4]. Physicians grade DR via soft and hard exudates, hemorrhage, etc., yet determine DME via occurrences of hard exudates and the shortest distances of hard exudates to the macula center.

Recently, large progress has been made by deep learning based methods on the grading of DR [5–8] and DME [9,10]. For example, He et al. [6] proposed a novel category attention block to explore discriminative region-wise features for each DR grade, and Ren et al. [9] presented a semi-supervised method with vector quantization for DME grading. However, DR and its associated DME were treated as separate diseases in these methods. Later, several works proposed to conduct the grading jointly [11–13]. Among them, Gulshan et al. [11], and Krasuse et al. [12] only leveraged the relationship implicitly by treating them as a multi-task problem. Subsequently, Li et al. [13] explored the internal correlation between DR and DME explicitly by designing a cross-disease attention network (CANet).

Despite the fruitful effectiveness of existing works on DR and DME joint grading, two vital issues remain unsolved. First, there is no proper solution for difficult samples, even though they have raised challenges for DR and DME grading tasks. For example, some previous works re-modify the 4-class DR grading task as a binary classification due to challenging samples [13–15]. In clinical settings, however, fine-grained grading is necessary due to the need for tailored treatments to avoid visual loss of patients. Second, the entangling feature representations are prone to emphasize the potential bias caused by the class imbalance and stereotyped disease context [16], where severe DR always accompanies DME in existing public datasets [2,3], however, DME can occur at any stage of DR [4]. The bias can harm the generalization ability of the model, leading to subpar performance on cross-domain data. Models with robust generalization ability are especially significant for clinical application [17,18]. However, existing works do not explore how to avoid the potential bias and mitigate the degeneration of generalization performance.

Therefore, we propose a difficulty-aware and feature disentanglement-based framework for addressing difficult samples and potential bias to jointly grade

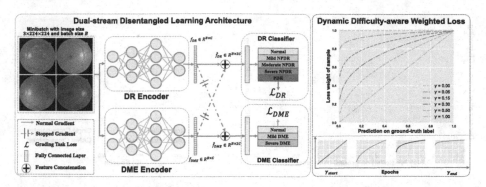

Fig. 2. The overview of our proposed framework. The framework uses two encoders to extract disentangled features f_{DR} and f_{DME} for DR and DME grading tasks, respectively. The concatenated features, $\hat{f}_{DR}$ and $\hat{f}_{DME}$, flow into classifiers to learn correlations between tasks, while the model stops gradients of DR features back-propagating to the DME encoder, and vice versa. The dynamic difficulty-aware weighted loss (DAW) weights samples adaptively via predictions on true labels with γ adjusting weights dynamically during training.

DR and DME. Firstly, inspired by curriculum learning (CL) [19], we propose the dynamic difficulty-aware weighted loss (DAW) to learn robust feature representations. DAW weights samples via the evaluation of their difficulties adaptively by the consistency between model predictions and ground-truth labels. Moreover, it focuses on learning simple samples at early training, and then gradually emphasizes difficult samples, like learning curriculum from easy to challenging. Such a dynamic and adaptive curriculum helps models build robust feature representations and speed up training [19]. Meanwhile, given the success of feature disentanglement in generalization [20,21], we design a dual-stream disentangled learning architecture (DETACH). DETACH prevents potential emphasized bias by disentangling feature representations to improve the robustness and generalization ability of models. It enables encoders to receive supervision signals only from their tasks to avoid entangling feature representations while retaining the ability to explicitly learn internal correlations of diseases for the joint grading task. Experiments show that our framework improves performance under both intra-dataset and cross-dataset tests against state-of-the-art (SOTA) approaches.

2   Methodology

An overview of our framework is shown in Fig. 2. Our framework disentangles feature representations of DR and DME via cutting off gradient back-propagation streams, and learns from simple to difficult samples via controlling the difficulty-aware parameter. In this section, we first show clues of difficult samples and generalization performance degeneration through a preliminary analysis on the Messidor dataset [2], and then introduce the proposed methods to address these issues.

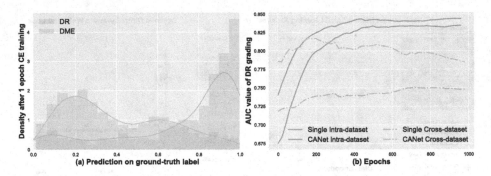

Fig. 3. (a) is a probability distribution histogram of model predictions on ground-truth labels after one epoch cross-entropy loss (CE) training, which shows divergent learning difficulties among samples and between tasks. (b) shows increasing performance in the intra-dataset test yet different trends in the cross-dataset test, implying potential emphasized bias and generalization performance degeneration.

Preliminary Analysis. Figure 3(a) implies divergent difficulties among samples and between grading tasks, as also reported in [22], which claims a phenomenon of low agreement on grading for experts even adhering to a strict protocol. Such grading difficulty may be introduced by the ambiguity of samples near decision boundaries [23]. Unlike standard samples, difficult samples with ambiguity require special learning strategies [24]. Besides, as shown in Fig. 3(b), an entangling representation-based model, CANet, improves its DR grading performance in the intra-dataset test yet performs continual-decreasingly in the cross-dataset test. On the contrary, a vanilla ResNet50 trained only for DR grading (denoted as Single) with cross-entropy loss (CE) improves performance in both tests. A possible explanation [16,20] is that the model with entangled features may emphasize potential class imbalance and stereotyped correlations with DR and DME during intra-domain training, and thus suffers from performance decrease on cross-domain data. Our preliminary analysis implies the existence of difficult samples and generalization ability degeneration in DR and DME joint grading. Thus, reasonable learning strategies and disentangled feature representations are necessary for DR and DME grading.

Weighting Samples via Difficulty Adaptively. As illustrated previously, CL can help handle difficult samples [19]. However, one critical puzzle for CL is how to measure the difficulties of samples reasonably. Instead of requiring extra models or the knowledge of experts to measure difficulty, we alternatively use consistency between model predictions and labels to evaluate the difficulties of samples for the model adaptively. It is inspired by the implicit weighting scheme of CE [25]. For notation convenience, we denote y as the one-hot encoding label of the sample, p as the softmax output of the model, and p_t as the probability of the ture class where the sample belongs. We define the loss as $CE(p, y) = -log(p_t)$

and show its gradient as:

$$\frac{\partial CE(p, y)}{\partial \theta} = -\frac{1}{p_t} \nabla_\theta p_t \qquad (1)$$

where θ denotes parameters of the model. As Eq. 1 shows, CE implicitly weights more on the samples whose model predictions are less congruent with labels, i.e., those difficult samples with smaller p_t and hence larger $1/p_t$, than samples whose model predictions are more consistent with labels, i.e., simple samples [25].

Inspired by the implicit effects of p_t on gradients of samples shown in Eq. 1, we design a dynamic difficulty-adaptive weighted loss (DAW) to explicitly mitigate the enhancement on difficult samples. Specifically, we propose to add a difficulty-adaptive weight α^γ with a tunable difficulty-aware parameter $\gamma \in [0,1]$ to CE, and we define DAW as:

$$\mathcal{L}_{DAW}(p, y, \gamma) = -\alpha^\gamma log(p_t) \qquad (2)$$

where α is the numeric value of p_t, and thus α^γ also acts on gradients directly. DAW weights samples by difficulty adaptively measured via p_t, which is continually updated during the training process. Such adaptive consideration of the keep-evolving model delivers a proper curriculum, where difficulties of samples are continual-changing for the model [26]. Moreover, DAW can emphasize or reduce the effects of p_t via controlling γ. As shown in the right of Fig. 2, larger γ, i.e., more sensitive difficulty-awareness, means that the model applies smaller weights on difficult samples, and vice versa.

Learning from Simple to Difficult Dynamically. DAW learns from simple to difficult samples by setting γ initially as γ_{start} and gradually decreasing it to γ_{end}, and the range $[\gamma_{end}, \gamma_{start}]$ denotes the dynamic difficulty-aware range. A larger γ_{start} means that the model learns less on difficult samples at the beginning, and a smaller γ_{end} means that the model treats samples more equally in the end. Such learning dynamics help the model to establish common features first, then try to learn distinctive features via its current knowledge during the training, and finally build robust feature representations [19]. After reaching γ_{end}, γ stops decreasing and the model continues to learn with $\gamma = \gamma_{end}$. In this paper, we adopt a linear decreasing strategy for γ, i.e., γ decreases equally every epoch, to verify the effectiveness of DAW in the experiment section, but more sophisticated algorithms can be used to boost the performance further.

Compared to existing CL methods, DAW handles samples adaptively and dynamically. To be specific, the existing works need extra networks or expert knowledge to evaluate the difficulty of samples as [26,27] or sample difficulties are usually predefined and fixed during training as [19]. In contrast, for one thing, DAW uses continual-updating predictions to measure sample difficulties for the model at that moment adaptively; for another, decreasing γ focuses more on learning difficult samples than before dynamically. As a result, DAW could set a reasonable and effective curriculum for the model.

Disentangling Feature Representations of DR and DME. To improve the generalization ability of our model, we propose a feature disentanglement module, named dual-stream disentangled learning architecture (DETACH), shown in Fig. 2. It prevents the potential decrease of generalization performance by disentangling feature representations of DR and DME grading, while exploring internal correlations between DR and DME grading. Specifically, DETACH assigns two encoders with linear classifiers to DR and DME grading tasks. Encoders extract latent task features f_{DR} and f_{DME} individually. Classifiers use the concatenation of those features, denoted as $\hat{f}_{DR}$ and $\hat{f}_{DME}$, for downstream DR and DME grading tasks, respectively. The core of DETACH is the feature detachment, which cuts off gradient computing of features, and thus, leads to independent back-propagation streams and disentangled features of DR and DME. For example, $\hat{f}_{DR}$, the concatenation of f_{DR} and detached f_{DME}, will flow into the DR classifier for grading. Thus, the DME encoder does not receive any supervision signals from the DR grading task, and vice versa. DETACH has the following strengths for DR and DME joint grading: a) encoders learn DR and DME features independently, considering different symptoms and grading criteria of DR and DME as shown in Fig. 1, and b) classifiers explicitly explore internal correlations between DR and DME, and c) disentangled feature representations will prevent potential bias being emphasized by entangling feature representations [16].

3   Experiments

Datasets and Implementation Settings. We evaluate the proposed framework in both intra-dataset and cross-dataset tests on three fundus benchmarks: DeepDRiD [28], Messidor [2], and IDRiD [3], where the first one focuses on DR grading, and the last two have both DR and DME grading tasks. For the intra-dataset test, we split Messidor into five folds for cross-validation, and we use the train and test sets provided by the organizers for DeepDRiD and IDRiD. For the cross-dataset generalization test, we use Messidor as the train set, IDRiD as the test set, and group DR grade 3 and 4 in IDRiD as a new grade 3 to align DR severity levels with Messidor. We report the accuracy (ACC), area under the ROC curve (AUC), and macro F1-score (F1) of DR and DME grading tasks for the intra-dataset test, and additionally, report recall (REC) and precision (PRE) for the cross-dataset test. We adopt two ImageNet pre-trained ResNet50 as encoders, two fully connected layers with input size 4096 as linear classifiers for DR and DME, respectively, Adam as the optimizer, and set the initial learning rate as $1e^{-4}$, the batch size as 16, the dynamic difficulty-aware range as $[0.15, 1]$ with γ decreasing in the first 400 epochs, and the training ends at the 500th epoch.

Comparisons with State-of-the-Art Approaches. To the best of our knowledge, there is only one previous work [13] explicitly exploring the DR and DME joint grading. It re-modified two competitive multi-task learning methods,

Table 1. Comparisons with SOTA approaches in the intra-dataset test.

Method	Messidor						IDRID					
	DME			DR			DME			DR		
	AUC	F1	ACC	AUC	F1	ACC	AUC	F1	ACC	AUC	F1	ACC
CANet [13]	90.5	66.6	89.3	84.8	61.6	**71.4**	87.9	66.1	78.6	78.9	42.3	57.3
Multi-task net [30]	88.7	66.0	88.8	84.7	**61.7**	69.4	86.1	60.3	74.8	78.0	43.9	59.2
MTMR-net [29]	89.2	64.1	89.0	84.5	60.5	70.6	84.2	61.1	79.6	79.7	45.3	**60.2**
Ours	**92.6**	**70.9**	**90.3**	**86.6**	61.6	70.6	**89.5**	**72.3**	**82.5**	**84.8**	**49.4**	59.2

Table 2. Ablation study in the intra-dataset test.

Method	Messidor						IDRID					
	DME			DR			DME			DR		
	AUC	F1	ACC	AUC	F1	ACC	AUC	F1	ACC	AUC	F1	ACC
Joint Training	90.0	64.6	88.4	84.6	59.0	69.7	84.0	61.1	75.7	82.3	41.9	57.3
DETACH w/ CE	91.7	69.7	90.1	85.5	60.2	**70.7**	87.7	66.1	**84.3**	80.2	37.9	48.5
Ours	**92.6**	**70.9**	**90.3**	**86.6**	**61.6**	70.6	**89.5**	**72.3**	82.5	**84.8**	**49.4**	**59.2**

MTMR-Net [29] and Multi-task net [30] for the joint grading task. We compare our proposed framework with the above approaches. Note that the previous joint grading work re-modified DR grading into two categories [13], i.e., the referable and non-referable. However, we consider the original fine-grained grading problem for DR and DME. Table 1 shows the results of the intra-dataset experiment on the Messidor and IDRiD dataset. Our method consistently outperforms other SOTA approaches under the AUC metric. The results indicate that our framework learns more robust feature representations.

Ablation Study. To better analyze the effects of components of our proposed method, we conduct an ablation study. Table 2 shows the result of the ablation study on Messidor and IDRiD dataset in the intra-dataset test. The joint training is an ordinary multi-task learning model containing one ResNet50 with ImageNet pre-training and two linear classifiers for DR and DME grading. The performance improves with the addition of DETACH and DAW, showing the positive effect of our proposed components. It also illustrates that the model could learn better correlations via disentangled feature representations.

The Cross-Dataset Generalization Test. A recent study suggests evaluating models in the cross-dataset test to measure generalization capacity [17]. To further investigate the generalization and robustness of our method, we conduct experiments with the above approaches in the cross-dataset test. Table 3 shows the results of those approaches and ablation study under the cross-dataset test setting. The results show that our method is more robust and generalized than existing approaches. Besides, the considerable gap between the joint training and

Table 3. Comparisons and ablation study in the cross-dataset test.

Methods	DME					DR				
	AUC	F1	ACC	REC	PRE	AUC	F1	ACC	REC	PRE
CANet [13]	83.5	58.2	78.5	59.8	60.7	78.6	30.2	45.0	37.5	36.5
Multi-task net [30]	85.5	56.9	78.5	59.0	57.7	79.2	**32.2**	44.6	38.1	**39.0**
MTMR-net [29]	81.8	61.5	76.3	61.3	67.3	79.5	28.7	46.0	37.3	31.9
Joint Training	83.4	59.1	78.5	60.2	61.8	75.2	27.7	43.3	35.5	30.3
DETACH w/CE	87.3	64.4	**84.3**	65.1	68.1	79.5	31.9	**49.2**	**39.2**	34.9
Ours	**87.7**	**70.0**	80.9	**69.0**	**73.4**	**79.7**	30.8	42.9	37.0	36.5

Table 4. Comparisons on different loss functions.

Loss	DME, IDRiD			DR, IDRiD			DR, DeepDRiD		
	AUC	F1	ACC	AUC	F1	ACC	AUC	F1	ACC
CE	84.2	64.7	79.6	76.5	47.8	57.3	83.1	51.9	61.0
FL [31]	86.2	65.4	80.6	75.3	40.5	**58.3**	83.2	**53.0**	**63.0**
GCE [25]	88.6	71.3	80.6	77.4	40.3	52.4	82.9	52.2	60.0
$\mathcal{L}_{DAW}$	**90.2**	**74.2**	**83.5**	**82.9**	**50.2**	55.3	**84.6**	52.0	60.1

DETACH implies our disentangled learning strategy has better generalization ability than entangling feature representations.

Loss Study. Finally, to verify the effectiveness of DAW, we conduct experiments comparing it with three popular loss functions, including Cross-Entropy Loss (CE), Focal Loss (FL) [31], and Generalized Cross-Entropy Loss (GCE) [25]. To avoid the influence of entangling feature representation in joint grading, we evaluate the above loss functions by individual DR or DME grading tasks on the IDRiD and DeepDRiD. We adopt a ResNet50 as the backbone and a fully connected layer with input size 2048 as the classifier, set the training epochs as 1000, and the batch size as 8. We adopt default parameter settings for the above loss functions for fairness, i.e., $\gamma = 2$ for FL, $q = 0.7$ for GCE, and dynamic difficulty-aware range $[0, 1]$ for DAW. The result in Table 4 illustrates the superiority of DAW on the grading task, where difficult samples require a reasonable learning strategy. This result also implies that DAW helps the model learn robust feature representations for grading tasks.

4   Conclusion

In this paper, we focus on the joint grading task of DR and DME, which suffers from difficult samples and potential generalization issues. To address them, we propose the dynamic difficulty-aware weighted loss (DAW) and dual-stream disentangled learning architecture (DETACH). DAW measures the difficulty of samples adaptively and learns from simple to difficult samples dynamically. DETACH builds disentangled feature representations to explicitly learn internal correlations between tasks, which avoids emphasizing potential bias and mitigates the generalization ability degeneration. We validate our methods on three benchmarks under both intra-dataset and cross-dataset tests. Potential future works include exploring the adaptation of the proposed framework in other medical applications for joint diagnosis and grading.

Acknowledgments. This work was supported by funding from Center for Aging Science, Hong Kong University of Science and Technology, and Shenzhen Science and Technology Innovation Committee (Project No. SGDX20210823103201011).

References

1. Cho, N., et al.: IDF diabetes atlas: global estimates of diabetes prevalence for 2017 and projections for 2045. Diabetes Res. Clin. Pract. **138**, 271–281 (2018)
2. Decencière, E., et al.: Feedback on a publicly distributed image database: the messidor database. Image Anal. Stereol. **33**(3), 231–234 (2014)
3. Porwal, P., et al.: Indian diabetic retinopathy image dataset (IDRID): a database for diabetic retinopathy screening research. Data **3**(3), 25 (2018)
4. Das, A., McGuire, P.G., Rangasamy, S.: Diabetic macular edema: pathophysiology and novel therapeutic targets. Ophthalmology **122**(7), 1375–1394 (2015)
5. Zhou, K., et al.: Multi-cell multi-task convolutional neural networks for diabetic retinopathy grading. In: 2018 40th Annual International Conference of the IEEE Engineering in Medicine and Biology Society (EMBC), pp. 2724–2727. IEEE (2018)
6. He, A., Li, T., Li, N., Wang, K., Fu, H.: CABNet: category attention block for imbalanced diabetic retinopathy grading. IEEE Trans. Med. Imaging **40**(1), 143–153 (2020)
7. Liu, S., Gong, L., Ma, K., Zheng, Y.: GREEN: a graph REsidual rE-ranking network for grading diabetic retinopathy. In: Martel, A.L., et al. (eds.) MICCAI 2020. LNCS, vol. 12265, pp. 585–594. Springer, Cham (2020). https://doi.org/10.1007/978-3-030-59722-1_56
8. Tian, L., Ma, L., Wen, Z., Xie, S., Xu, Y.: Learning discriminative representations for fine-grained diabetic retinopathy grading. In: 2021 International Joint Conference on Neural Networks (IJCNN), pp. 1–8. IEEE (2021)
9. Ren, F., Cao, P., Zhao, D., Wan, C.: Diabetic macular edema grading in retinal images using vector quantization and semi-supervised learning. Technol. Health Care **26**(S1), 389–397 (2018)
10. Syed, A.M., Akram, M.U., Akram, T., Muzammal, M., Khalid, S., Khan, M.A.: Fundus images-based detection and grading of macular edema using robust macula localization. IEEE Access **6**, 58784–58793 (2018)
11. Gulshan, V., et al.: Development and validation of a deep learning algorithm for detection of diabetic retinopathy in retinal fundus photographs. JAMA **316**(22), 2402–2410 (2016)
12. Krause, J., et al.: Grader variability and the importance of reference standards for evaluating machine learning models for diabetic retinopathy. Ophthalmology **125**(8), 1264–1272 (2018)

13. Li, X., Hu, X., Yu, L., Zhu, L., Fu, C.W., Heng, P.A.: CANet: cross-disease attention network for joint diabetic retinopathy and diabetic macular edema grading. IEEE Trans. Med. Imaging **39**(5), 1483–1493 (2019)

14. Wang, Z., Yin, Y., Shi, J., Fang, W., Li, H., Wang, X.: Zoom-in-net: deep mining lesions for diabetic retinopathy detection. In: Descoteaux, M., Maier-Hein, L., Franz, A., Jannin, P., Collins, D.L., Duchesne, S. (eds.) MICCAI 2017. LNCS, vol. 10435, pp. 267–275. Springer, Cham (2017). https://doi.org/10.1007/978-3-319-66179-7_31

15. Vo, H.H., Verma, A.: New deep neural nets for fine-grained diabetic retinopathy recognition on hybrid color space. In: 2016 IEEE International Symposium on Multimedia (ISM), pp. 209–215. IEEE (2016)

16. Chu, S., Kim, D., Han, B.: Learning debiased and disentangled representations for semantic segmentation. In: Advances in Neural Information Processing Systems, vol. 34 (2021)

17. Geirhos, R., et al.: Shortcut learning in deep neural networks. Nat. Mach. Intell. **2**(11), 665–673 (2020)

18. Chen, C., Dou, Q., Jin, Y., Chen, H., Qin, J., Heng, P.-A.: Robust multimodal brain tumor segmentation via feature disentanglement and gated fusion. In: Shen, D., et al. (eds.) MICCAI 2019. LNCS, vol. 11766, pp. 447–456. Springer, Cham (2019). https://doi.org/10.1007/978-3-030-32248-9_50

19. Bengio, Y., Louradour, J., Collobert, R., Weston, J.: Curriculum learning. In: Proceedings of the 26th Annual International Conference on Machine Learning, pp. 41–48 (2009)

20. Träuble, F., et al.: On disentangled representations learned from correlated data. In: International Conference on Machine Learning, pp. 10401–10412. PMLR (2021)

21. Montero, M.L., Ludwig, C.J., Costa, R.P., Malhotra, G., Bowers, J.: The role of disentanglement in generalisation. In: International Conference on Learning Representations (2020)

22. Sánchez, C.I., Niemeijer, M., Dumitrescu, A.V., Suttorp-Schulten, M.S., Abramoff, M.D., van Ginneken, B.: Evaluation of a computer-aided diagnosis system for diabetic retinopathy screening on public data. Investig. Ophthalmol. Vis. Sci. **52**(7), 4866–4871 (2011)

23. Toneva, M., Sordoni, A., des Combes, R.T., Trischler, A., Bengio, Y., Gordon, G.J.: An empirical study of example forgetting during deep neural network learning. In: International Conference on Learning Representations (2018)

24. Yang, X., Dong, M., Guo, Y., Xue, J.-H.: Metric learning for categorical and ambiguous features: an adversarial method. In: Hutter, F., Kersting, K., Lijffijt, J., Valera, I. (eds.) ECML PKDD 2020. LNCS (LNAI), vol. 12458, pp. 223–238. Springer, Cham (2021). https://doi.org/10.1007/978-3-030-67661-2_14

25. Zhang, Z., Sabuncu, M.: Generalized cross entropy loss for training deep neural networks with noisy labels. In: Advances in Neural Information Processing Systems, vol. 31 (2018)

26. Jiang, L., Zhou, Z., Leung, T., Li, L.J., Fei-Fei, L.: Mentornet: learning data-driven curriculum for very deep neural networks on corrupted labels. In: International Conference on Machine Learning, pp. 2304–2313. PMLR (2018)

27. Kumar, M., Packer, B., Koller, D.: Self-paced learning for latent variable models. In: Advances in Neural Information Processing Systems, vol. 23 (2010)

28. DeepDRiD: The DeepDR diabetic retinopathy image dataset (DeepDRiD) website. https://isbi.deepdr.org. Accessed 20 Feb 2022

29. Liu, L., Dou, Q., Chen, H., Qin, J., Heng, P.A.: Multi-task deep model with margin ranking loss for lung nodule analysis. IEEE Trans. Med. Imaging **39**(3), 718–728 (2019)

30. Chen, Q., Peng, Y., Keenan, T., Dharssi, S., Agro, E., et al.: A multi-task deep learning model for the classification of age-related macular degeneration. AMIA Transl. Sci. Proc. **2019**, 505 (2019)

31. Lin, T.Y., Goyal, P., Girshick, R., He, K., Dollár, P.: Focal loss for dense object detection. In: Proceedings of the IEEE International Conference on Computer Vision, pp. 2980–2988 (2017)

Spatiotemporal Attention for Early Prediction of Hepatocellular Carcinoma Based on Longitudinal Ultrasound Images

Yiwen Zhang[1,2], Chengguang Hu[3], Liming Zhong[1,2], Yangda Song[4],
Jiarun Sun[4], Meng Li[4], Lin Dai[4], Yuanping Zhou[4(✉)], and Wei Yang[1,2(✉)]

[1] School of Biomedical Engineering, Southern Medical University, Guangzhou, China
weiyanggm@gmail.com
[2] Guangdong Provincial Key Laboratory of Medical Image Processing,
Southern Medical University, Guangzhou, China
[3] Department of Gastroenterology, Nanfang Hospital, Southern Medical University,
Guangzhou, China
[4] Department of Infectious Diseases and Hepatology Unit, Nanfang Hospital,
Southern Medical University, Guangzhou, China
yuanpingzhou@163.com

Abstract. Early screening is an important way to reduce the mortality
of hepatocellular carcinoma (HCC) and improve its prognosis. As a non-
invasive, economic, and safe procedure, B-mode ultrasound is currently
the most common imaging modality for diagnosing and monitoring HCC.
However, because of the difficulty of extracting effective image features
and modeling longitudinal data, few studies have focused on early prediction of HCC based on longitudinal ultrasound images. In this paper,
to address the above challenges, we propose a spatiotemporal attention network (STA-HCC) that adopts a convolutional-neural-network–
transformer framework. The convolutional neural network includes a
feature-extraction backbone and a proposed regions-of-interest attention block, which learns to localize regions of interest automatically and
extract effective features for HCC prediction. The transformer can capture long-range dependencies and nonlinear dynamics from ultrasound
images through a multihead self-attention mechanism. Also, an age-based
position embedding is proposed in the transformer to embed a more-
appropriate positional relationship among the longitudinal ultrasound
images. Experiments conducted on our dataset of 6170 samples collected
from 619 cirrhotic subjects show that STA-HCC achieves impressive performance, with an area under the receiver-operating-characteristic curve
of 77.5%, an accuracy of 70.5%, a sensitivity of 69.9%, and a specificity
of 70.5%. The results show that our method achieves state-of-the-art
performance compared with other popular sequence models.

Keywords: Hepatocellular carcinoma · Early prediction · Deep
learning · Attention · Longitudinal data

Y. Zhang and C. Hu are the co-first authors.

© The Author(s), under exclusive license to Springer Nature Switzerland AG 2022
L. Wang et al. (Eds.): MICCAI 2022, LNCS 13433, pp. 534–543, 2022.
https://doi.org/10.1007/978-3-031-16437-8_51

1 Introduction

Hepatocellular carcinoma (HCC) is the most common primary malignancy of the liver and one of the most common cause of cancer deaths in the world [8,24]. The poor prognosis and high mortality of HCC are related to the lack of effective monitoring and the low accuracy of early diagnosis. Most patients' HCC is caused by the worsening of liver cirrhosis, while patients with liver cirrhosis are often asymptomatic [1], and patients with sensible hepatic symptoms are considered to have advanced-stage HCC [11]. Currently, various treatments have been shown to be effective for treating early HCC, so it is very important to identify potential high-risk patients, which could help physicians to improve the efficiency and implementation of screening surveillance strategies.

Previous studies have developed some machine-learning models for early HCC prediction. Used widely is the risk-prediction model based on Cox proportional-hazards regression and logistic regression [12,14,22]. Wu et al. [23] characterized the performances of 14 HCC risk-prediction models through meta-analysis followed by external validation; although the results showed that all the models gave generally acceptable discrimination with pooled area under receiver-operating-characteristic curve (AUC) ranging from 70% to 83% for three-year prediction, the calibration performance was poorly reported in most external-validation studies. Ioannou et al. [7] developed a recurrent neural network (RNN) model based on longitudinal data for HCC prediction; compared with the conventional logistic-regression models, their model achieved better performance with AUC of 75.9%. However, most existing HCC prediction models use structured demographic information and laboratory variables instead of images, whereas imaging examination is an indispensable part of HCC screening in practice [10]. As a noninvasive, economic, and safe procedure, conventional B-mode ultrasound (US) is currently the most common imaging modality for diagnosing and monitoring HCC [25], as shown in Fig. 1. However, few studies have focused on the early prediction of HCC based on longitudinal US images.

One of the challenges for early prediction of HCC based on longitudinal US images is that it is difficult to define regions of interest (ROIs). Not all patients with liver cirrhosis have nodules, so it can only be analyzed based on the whole US image. Although some visualization technologies [16,26] have shown that convolutional neural networks (CNNs) based on supervised training

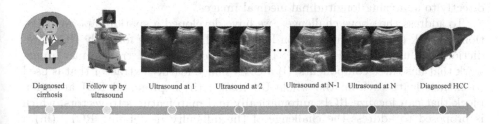

Fig. 1. Timeline of ultrasound examination before diagnosis in a HCC patient.

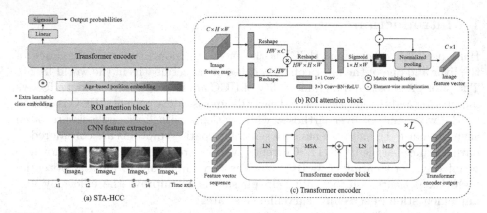

Fig. 2. (a) Overview of proposed spatiotemporal attention network (STA-HCC) for early prediction of hepatocellular carcinoma (HCC) based on longitudinal B-mode ultrasound (US) images. Visual illustration of (b) region of interest (ROI) attention block and (c) transformer encoder.

can localize focal areas automatically and extract task-specific features from images, the performance of vanilla CNNs is limited. However, a concept that has recently become popular in computer vision is that of attention mechanisms [9,21], which are effective at capturing useful areas in images by constructing long-range dependencies. Therefore, attention mechanisms offer a solution for localizing ROIs automatically in liver US images.

Another challenge is how to deal with sequential images and irregular temporal components. Because of the lack of temporal component embedding, conventional sequence models such as Markov models and conditional random fields deal with sequential data but have poor ability to learn long-range dependencies. Recently, RNNs have achieved great success in disease prediction [17], genomic analysis [20], and patient trajectory modeling [13]. The serial propagation structure of RNNs makes it easy for them to capture sequence order, but they lack the awareness of sequence interval. Transformers [19] constitute another framework for processing sequential data, this being the current state-of-the-art framework in natural language processing. A transformer has a more-efficient parallel feedforward structure with a multihead attention mechanism, allowing it to consider the full sequence simultaneously. However, these methods cannot be applied directly to analyzing longitudinal medical images.

To address the above challenges, we have developed a spatiotemporal attention network (STA-HCC) based on longitudinal US images for early HCC prediction, as illustrated in Fig. 2(a). STA-HCC adopts a CNN–transformer framework that has three components: (i) a CNN image feature extractor that is used to extract features from the whole US image; (ii) a proposed ROI attention block that can localize ROIs automatically and map features to vectors, which is proposed to address the challenge of the difficulty in defining ROIs; (iii) a transformer sequence model that can establish the relationship among longitudinal US images by capturing long-range dependencies and nonlinear dynamics. Moreover, we propose an age-based position embedding (PE) in the transformer to establish a more-appropriate positional relationship among longitudinal US images, which solves the problem of embedding irregular temporal components.

2   Methodology

2.1   Problem Formulation

Our aim is to predict whether a patient with cirrhosis will develop HCC within N years based on past US images. In practice, patients may receive US examinations of various times, and a given patient may have different results in different periods. Therefore, the collected data must be resampled into samples (input–label pairs) of various lengths. We denote the data of each patient $p \in \{1, 2, \ldots, P\}$ as $D^p = \{\mathbf{v}^{(p,1)}, \mathbf{v}^{(p,2)}, \mathbf{v}^{(p,3)}, \ldots, \mathbf{v}^{(p,s_p)}, g^p_{HCC}\}$, where s_p is the number of US examinations in patient p's data, $\mathbf{v}^{(p,j)} = \{x^{(p,j)}, g^{(p,j)}\}$ is the record containing a US image and the patient age in examination j, and g^p_{HCC} is the patient age at diagnosis of HCC (which is set to "None" if HCC is not diagnosed during the follow-up). The subset $V^{(p,j)} = \{\mathbf{v}^{(p,1)}, \mathbf{v}^{(p,2)}, \ldots, \mathbf{v}^{(p,j)}\} \subseteq D^p$ denotes all the records of patient p collected until time j. The term $y^{(p,j)}$ is the label corresponding to $V^{(p,j)}$ and is defined as

$$y^{(p,j)} = \begin{cases} 0, & g^p_{HCC} = \text{None} \; ; \\ 0, & g^p_{HCC} - g^{(p,j)} > N; \\ 1, & g^p_{HCC} - g^{(p,j)} \leq N. \end{cases} \tag{1}$$

Finally, the original dataset is resampled into sample set $X = \{V^{(p,j)}\}$ and label set $Y = \{y^{(p,j)}\}$, where $p \in [1, P]$ and $j \in [1, s_p]$. The aim of this study is to seek a function f that learns the mapping between X and Y ($f : X \rightarrow Y$).

2.2   CNN Image Feature Extractor

Feature extraction is an important part of image analysis. Without ROIs, the feature extractor must analyze the whole image and then localize the potential regions associated with HCC and extract effective features, which challenges the capability of the extractor. Because of the powerful feature-extraction ability of CNNs, various CNN-based backbone networks have been used widely in medical image analysis [4,6,18], and we also use a CNN-based backbone network to extract features from a single US image. The high-dimensional feature map generated by the backbone network can be obtained by

$$h^{(p,j)} = G(x^{(p,j)}), \tag{2}$$

where G denotes the pre-trained CNN-based backbone network and $h^{(p,j)} \in \mathbb{R}^{C \times H \times W}$ is an image feature map.

2.3 ROI Attention Block

Although the CNN image feature extractor itself has a certain ability to localize focal areas, its ones are usually inaccurate and poorly interpreted. Instead, we propose a novel spatial attention block called ROI attention to localize ROIs more accurately, as shown in Fig. 2(b). Inspired by nonlocal attention [21], we adopt dot-product similarity with embedded version to construct long-range dependencies and compute attention maps. The difference between our method and the nonlocal one is that we further reduce the dimension of the attention map followed by a gating mechanism with a sigmoid activation. We define the attention map of $h^{(p,j)}$ as $m^{(p,j)} \in \mathbb{R}^{1 \times H \times W}$, which can be formulated by

$$m^{(p,j)} = \sigma(\beta(\theta(h^{(p,j)})^T \otimes \phi(h^{(p,j)}))), \tag{3}$$

where $\theta(\cdot)$ and $\phi(\cdot)$ are linear embeddings that are 1×1 convolutions in space. We use a Conv(3×3)-BN-ReLU-Conv(1×1) combination as the dimension-reduction function $\beta(\cdot)$. The symbol $\otimes$ defines matrix multiplication, and $\sigma(\cdot)$ is a sigmoid activation function. The learned attention map is element-wise multiplied with the feature map $h^{(p,j)}$ to reweight the feature pixels: $\tilde{h}^{(p,j)} = h^{(p,j)} \odot \lambda(m^{(p,j)})$, where $\odot$ denotes element-wise multiplication and $\lambda(\cdot)$ is the broadcasting function. Then the reweighted feature map $\tilde{h}^{(p,j)}$ must be transformed into a feature vector to adapt to the input form of the transformer encoder. Because the feature map is reweighted by the attention map of $(0,1)$, the conventional average pooling reduces the absolute value of the feature and makes it greatly influenced by the area of the focal areas in the attention map. Therefore, we propose a normalized pooling that is defined as

$$q_c^{(p,j)} = \text{Pool}_{norm}(\tilde{h}_c^{(p,j)}, m^{(p,j)}) = \frac{\sum_{h=1}^H \sum_{w=1}^W \tilde{h}_{c,h,w}^{(p,j)}}{\sum_{h=1}^H \sum_{w=1}^W m_{h,w}^{(p,j)}}, \tag{4}$$

where $q_c^{(p,j)}$ is the c-th value of the US image feature vector $q^{(p,j)} \in \mathbb{R}^C$.

2.4 Transformer for HCC Prediction

The aforementioned CNN feature extractor and ROI attention block only extract features from a single US image, while the transformer can establish the relationship among longitudinal US images by means of its multihead attention mechanism. Because the parallel structure cannot capture the sequence order and interval, we must explicitly embed the position information into the input of the transformer. Conventional position embedding adopts a sinusoidal function with the position order of tokens as input. However, for longitudinal medical data, patient age is obviously a more-appropriate factor to measure the positional relationship among tokens. Therefore, we propose age-based position embedding with a sinusoidal function:

$$
\begin{aligned}
PE_{(g^{(p,j)},2i)} &= \sin(g^{(p,j)}/10000^{2i/C}), \\
PE_{(g^{(p,j)},2i+1)} &= \cos(g^{(p,j)}/10000^{2i/C}),
\end{aligned}
\tag{5}
$$

where i is the dimension and C is the length of feature vector $q^{(p,j)}$. For given longitudinal data $V^{(p,j)}$, the image feature vector and the position embedding in each record can be obtained by Eqs. (4) and (5), respectively. The image feature sequence $Q^{(p,j)} \in \mathbb{R}^{j \times C}$ and the position embedding sequence $E^{(p,j)} \in \mathbb{R}^{j \times C}$ are defined as

$$Q^{(p,j)} = [q^{(p,1)} \ q^{(p,2)} \ \cdots \ q^{(p,j)}],$$
$$E^{(p,j)} = [e^{(p,1)} \ e^{(p,2)} \ \cdots \ e^{(p,j)}]. \tag{6}$$

The position embedding sequence is added to the image feature sequence to embed position information: $Z^{(p,j)} = Q^{(p,j)} + E^{(p,j)}$. We then prepend a learnable [class] token embedding vector $z_{cls} \in \mathbb{R}^C$ to the projected feature sequence for classification: $\tilde{Z}^{(p,j)} = [z_{cls} \ Z^{(p,j)}]$.

The embedding $\tilde{Z}^{(p,j)}$ is input into the transformer encoder. The transformer encoder includes L transformer encoder blocks, and each encoder block used in our model consists of a multihead self-attention (MSA), a multilayer perceptron (MLP), two layer normalizations (LNs), and two residual connections, as shown in Fig. 2(c). The output of the transformer encoder can be formulated by

$$\hat{Z}^{(p,j)} = F(\tilde{Z}^{(p,j)}), \tag{7}$$

where $\hat{Z}^{(p,j)} = [\hat{z}_{cls}^{(p,j)} \ \hat{z}^{(p,1)} \ \cdots \ \hat{z}^{(p,j)}] \in \mathbb{R}^{(j+1) \times C}$ and F denotes the transformer encoder. Finally, $\hat{z}_{cls}$ is input to a linear classifier followed by a sigmoid function to obtain the prediction probability:

$$\hat{y}^{(p,j)} = \sigma(\psi(\hat{z}_{cls}^{(p,j)}) + b), \tag{8}$$

where $\psi(\cdot)$ is a linear function and b is the trainable bias in the classifier.

3 Experiments

3.1 Experimental Settings

Dataset and Evaluation. The dataset was collected from a local hospital comprised longitudinal US examination records of 619 cirrhotic subjects from January 2011 to December 2020, of which 89 were diagnosed with HCC. All subjects with undiagnosed HCC were followed up for more than three years, and the longitudinal US examination record intervals of each subject were more than two months and less than seven months. The number of US images per subject, s_p, ranged from two to 22 with an average of ten, and 6170 samples were included in our experiment in total. The prediction period N was set to three, so the ratio of positive to negative samples was 492:5678. Four metrics were used to measure the performances of the models: (i) area under the receiver-operating-characteristic curve (AUC), (ii) accuracy (ACC), (iii) sensitivity (SEN), and (iv) specificity (SPE). We divided the subjects into five folds for cross-validation, while all quantitative metrics were calculated based on resampled samples.

Implementation Details. We chose SE-ResNet50 [6] as the CNN image feature extractor, so the channel number C of the feature map was 2048. The transformer encoder consisted of six transformer encoder blocks, and the number of MSA heads and the MLP dimension were four and 512, respectively. The CNN image feature extractor with ROI attention block and the transformer encoder were trained step by step. First, the CNN image feature extractor with ROI attention block connected to a linear classifier was trained to predict HCC from a single US image. Then the complete STA-HCC was trained to predict HCC based on longitudinal US images, and only the transformer encoder and last linear classifier were optimized. They both used class-balanced focal loss [3] as the loss function. We used a stochastic-gradient-descent optimizer with a poly learning-rate decay schedule to train both networks. The initial learning rate and batch size were set to 0.001 and 48, respectively. All images were resized to 384×512 and uniformly linearly normalized to $[0, 1]$ for training.

3.2   Evaluations

Ablation Study. We conducted ablation studies to analyze the contribution of each part in our STA-HCC, and the quantitative results are shown in Fig. 3. The main components of our method are the image feature extractor, the ROI attention block, the transformer, and the age-based PE, of which the image feature extractor is indispensable. Therefore, the ablation studies included (i) STA-HCC without the ROI attention block (w/o ROI attention), (ii) STA-HCC without the transformer (w/o transformer), (iii) STA-HCC without the age-based PE (w/o age-based PE), and (iv) our STA-HCC. The results show that STA-HCC w/o transformer shows significant decreases of 7.3% in AUC, 6.2% in ACC, 4.9% in SEN, and 6.2% in SPE; this reveals that the association among longitudinal data constructed by the transformer plays an important role in HCC prediction. Meanwhile, STA-HCC w/o ROI attention block leads to decreases of 3.1% in AUC, 1.6% in ACC, 1.2% in SEN, and 1.5% in SPE, which shows that our ROI attention block is helpful for extracting more-effective image features. Removing the age-based PE reduces all metrics by ~2%, which demonstrates the importance of age information in longitudinal medical data analysis.

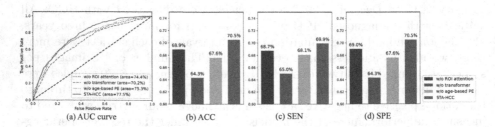

Fig. 3. Results of ablation study on proposed STA-HCC. All results are the average of five-fold cross-validation.

Table 1. Quantitative comparison of different sequence models. The feature vector sequences extracted by SE-ResNet50 with ROI attention block were used as the input for all models.

Sequence model	AUC [%]	ACC [%]	SEN [%]	SPE [%]
LSTM [5]	74.2	66.0	68.3	65.9
BiLSTM [15]	74.4	66.8	68.1	66.7
GRU [2]	74.3	67.3	67.1	67.4
Transformer [19]	76.1	68.7	68.9	68.8
Transformer with age-based PE (ours)	**77.5**	**70.5**	**69.9**	**70.5**

Compared with Other Sequence Models. A sequence model is an important component of longitudinal data analysis. To show that our method is effective, we compare it with other popular sequence models. With the same feature vector sequences extracted by SE-ResNet50 with the ROI attention block as input, we compare our method with long short-term memory (LSTM) [5], bidirectional LSTM (BiLSTM) [15], a gated recurrent unit (GRU) [2], and a transformer with conventional PE [19]. Table 1 shows that the conventional transformer outperforms the popular RNNs, being better by more than 1% in each metric, which reflects the superiority of its parallel structure and attention mechanism. Our transformer with age-based PE leads to further increases of 1.4% in AUC, 1.7% in ACC, 1.0% in SEN, and 1.7% in SPE, which shows that our method is more powerful in longitudinal data modeling.

Visualization of ROI Attention. Our ROI attention block is aimed at localizing focal areas automatically to address the challenge that it is difficult to define ROIs manually in liver US images. To show its effectiveness, the attention maps in our ROI attention block are visualized in Fig. 4. The visualization

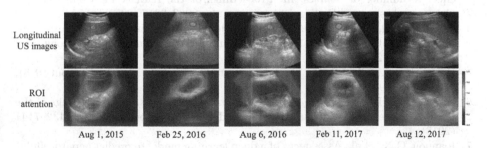

Fig. 4. Visualization of attention in proposed ROI attention block. The longitudinal US images are from a subject with liver cirrhosis, for whom a liver cirrhosis nodule was found on January 31, 2015; US re-examination was then performed regularly, and HCC was confirmed finally on August 28, 2018. The red circle/ellipse marks the lesion area. (Color figure online)

results show that ROI attention can track and focus on the same lesion areas and ignore other irrelevant areas in the longitudinal US image, which provides an effective and highly relevant feature sequence for the subsequent transformer. However, because there is no extra constraint on the ROI attention map, its focal areas are not sufficiently local. In the future, we will improve the precision positioning ability of the proposed approach.

4   Conclusion

We have presented the STA-HCC CNN–transformer framework for early HCC prediction based on longitudinal US images. The proposed ROI attention block in the STA-HCC can produce reliable attention maps that focus automatically on ROIs. The longitudinal feature sequences extracted based on ROI attention are input into the transformer to construct long-range dependencies. Moreover, an age-based PE is used to establish an appropriate positional relationship among longitudinal US images. Consequently, the proposed STA-HCC achieves state-of-the-art performance for early HCC prediction based on longitudinal US images.

Acknowledgements. This research is supported by the National Natural Science Foundation of China (No. 82172020 and 81772923) and Guangdong Provincial Key Laboratory of Medical Image Processing (No. 2020B1212060039).

References

1. Bruix, J., Reig, M., Sherman, M.: Evidence-based diagnosis, staging, and treatment of patients with hepatocellular carcinoma. Gastroenterology **150**(4), 835–853 (2016)
2. Cho, K., et al.: Learning phrase representations using RNN encoder-decoder for statistical machine translation. arXiv preprint arXiv:1406.1078 (2014)
3. Cui, Y., Jia, M., Lin, T.Y., Song, Y., Belongie, S.: Class-balanced loss based on effective number of samples. In: Proceedings of the IEEE/CVF Conference on Computer Vision and Pattern Recognition, pp. 9268–9277 (2019)
4. He, K., Zhang, X., Ren, S., Sun, J.: Deep residual learning for image recognition. In: Proceedings of the IEEE Conference on Computer Vision and Pattern Recognition, pp. 770–778 (2016)
5. Hochreiter, S., Schmidhuber, J.: Long short-term memory. Neural Comput. **9**(8), 1735–1780 (1997)
6. Hu, J., Shen, L., Sun, G.: Squeeze-and-excitation networks. In: Proceedings of the IEEE Conference on Computer Vision and Pattern Recognition, pp. 7132–7141 (2018)
7. Ioannou, G.N., et al.: Assessment of a deep learning model to predict hepatocellular carcinoma in patients with hepatitis c cirrhosis. JAMA Netw. Open **3**(9), e2015626–e2015626 (2020)
8. Kanda, T., Goto, T., Hirotsu, Y., Moriyama, M., Omata, M.: Molecular mechanisms driving progression of liver cirrhosis towards hepatocellular carcinoma in chronic hepatitis b and c infections: a review. Int. J. Mol. Sci. **20**(6), 1358 (2019)

9. Li, X., Zhong, Z., Wu, J., Yang, Y., Lin, Z., Liu, H.: Expectation-maximization attention networks for semantic segmentation. In: Proceedings of the IEEE/CVF International Conference on Computer Vision, pp. 9167–9176 (2019)

10. Marrero, J.A., et al.: Diagnosis, staging, and management of hepatocellular carcinoma: 2018 practice guidance by the American association for the study of liver diseases. Hepatology **68**(2), 723–750 (2018)

11. Obi, S., et al.: Combination therapy of intraarterial 5-fluorouracil and systemic interferon-alpha for advanced hepatocellular carcinoma with portal venous invasion. Cancer Interdisc. Int. J. Am. Cancer Soc. **106**(9), 1990–1997 (2006)

12. Papatheodoridis, G., et al.: PAGE-B predicts the risk of developing hepatocellular carcinoma in Caucasians with chronic hepatitis B on 5-year antiviral therapy. J. Hepatol. **64**(4), 800–806 (2016)

13. Pham, T., Tran, T., Phung, D., Venkatesh, S.: Predicting healthcare trajectories from medical records: a deep learning approach. J. Biomed. Inform. **69**, 218–229 (2017)

14. Poh, Z., et al.: Real-world risk score for hepatocellular carcinoma (RWS-HCC): a clinically practical risk predictor for HCC in chronic hepatitis B. Gut **65**(5), 887–888 (2016)

15. Schuster, M., Paliwal, K.K.: Bidirectional recurrent neural networks. IEEE Trans. Signal Process. **45**(11), 2673–2681 (1997)

16. Selvaraju, R.R., Cogswell, M., Das, A., Vedantam, R., Parikh, D., Batra, D.: Gradcam: visual explanations from deep networks via gradient-based localization. In: Proceedings of the IEEE International Conference on Computer Vision, pp. 618–626 (2017)

17. Sharma, D., Xu, W.: phyLoSTM: a novel deep learning model on disease prediction from longitudinal microbiome data. Bioinformatics **37**(21), 3707–3714 (2021)

18. Tan, M., Le, Q.: Efficientnet: rethinking model scaling for convolutional neural networks. In: International Conference on Machine Learning, pp. 6105–6114. PMLR (2019)

19. Vaswani, A., et al.: Attention is all you need. In: Advances in Neural Information Processing Systems, pp. 5998–6008 (2017)

20. Wang, H., Li, C., Zhang, J., Wang, J., Ma, Y., Lian, Y.: A new LSTM-based gene expression prediction model: L-GEPM. J. Bioinform. Comput. Biol. **17**(04), 1950022 (2019)

21. Wang, X., Girshick, R., Gupta, A., He, K.: Non-local neural networks. In: Proceedings of the IEEE Conference on Computer Vision and Pattern Recognition, pp. 7794–7803 (2018)

22. Wong, G.L.H., et al.: Liver stiffness-based optimization of hepatocellular carcinoma risk score in patients with chronic hepatitis B. J. Hepatol. **60**(2), 339–345 (2014)

23. Wu, S., et al.: Hepatocellular carcinoma prediction models in chronic hepatitis B: a systematic review of 14 models and external validation. Clin. Gastroenterol. Hepatol. **19**(12), 2499–2513 (2021)

24. Yang, J.D., Hainaut, P., Gores, G.J., Amadou, A., Plymoth, A., Roberts, L.R.: A global view of hepatocellular carcinoma: trends, risk, prevention and management. Nat. Rev. Gastroenterol. Hepatol. **16**(10), 589–604 (2019)

25. Yao, Z., et al.: Preoperative diagnosis and prediction of hepatocellular carcinoma: radiomics analysis based on multi-modal ultrasound images. BMC Cancer **18**(1), 1–11 (2018)

26. Zhou, B., Khosla, A., Lapedriza, A., Oliva, A., Torralba, A.: Learning deep features for discriminative localization. In: Proceedings of the IEEE Conference on Computer Vision and Pattern Recognition, pp. 2921–2929 (2016)

NVUM: Non-volatile Unbiased Memory for Robust Medical Image Classification

Fengbei Liu[1(✉)], Yuanhong Chen[1], Yu Tian[1], Yuyuan Liu[1], Chong Wang[1], Vasileios Belagiannis[2], and Gustavo Carneiro[1]

[1] Australian Institute for Machine Learning, University of Adelaide, Adelaide, Australia
fengbei.liu@adelaide.edu.au
[2] Otto von Guericke University Magdeburg, Magdeburg, Germany

Abstract. Real-world large-scale medical image analysis (MIA) datasets have three challenges: 1) they contain noisy-labelled samples that affect training convergence and generalisation, 2) they usually have an imbalanced distribution of samples per class, and 3) they normally comprise a multi-label problem, where samples can have multiple diagnoses. Current approaches are commonly trained to solve a subset of those problems, but we are unaware of methods that address the three problems simultaneously. In this paper, we propose a new training module called Non-Volatile Unbiased Memory (NVUM), which non-volatility stores running average of model logits for a new regularization loss on noisy multi-label problem. We further unbias the classification prediction in NVUM update for imbalanced learning problem. We run extensive experiments to evaluate NVUM on new benchmarks proposed by this paper, where training is performed on noisy multi-label imbalanced chest X-ray (CXR) training sets, formed by Chest-Xray14 and CheXpert, and the testing is performed on the clean multi-label CXR datasets OpenI and PadChest. Our method outperforms previous state-of-the-art CXR classifiers and previous methods that can deal with noisy labels on all evaluations. Our code is available at https://github.com/FBLADL/NVUM.

Keywords: Chest X-ray classification · Multi-label classification · Imbalanced classification

1 Introduction and Background

The outstanding results shown by deep learning models in medical image analysis (MIA) [14,15] depend on the availability of large-scale manually-labelled training

This work was supported by the Australian Research Council through grants DP180103232 and FT190100525.

Supplementary Information The online version contains supplementary material available at https://doi.org/10.1007/978-3-031-16437-8_52.

© The Author(s), under exclusive license to Springer Nature Switzerland AG 2022
L. Wang et al. (Eds.): MICCAI 2022, LNCS 13433, pp. 544–553, 2022.
https://doi.org/10.1007/978-3-031-16437-8_52

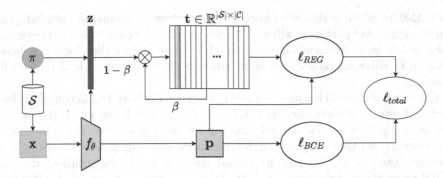

Fig. 1. NVUM training algorithm: 1) sample input image $\mathbf{x}$ from training set $\mathcal{S}$ and calculate label distribution prior π; 2) train model f_θ and get sample logits $\mathbf{z}$ and prediction $\mathbf{p}$; 3) update memory $\mathbf{t}$ with (3); and 4) minimise the loss that comprises $\ell_{BCE}(.)$ in (1) and $\ell_{REG}(.)$ in (2).

sets, which is expensive to obtain. As a affordable alternative, these manually-labelled training sets can be replaced by datasets that are automatically labelled by natural language processing (NLP) tools that extract labels from the radiologists' reports [9,29]. However, the use of these alternative labelling processes often produces unreliably labelled datasets because NLP-extracted disease labels, without verification by doctors, may contain incorrect labels, which are called *noisy labels* [21,22]. Furthermore, differently from computer vision problems that tend to be multi-class with a balanced distribution of samples per class, MIA problems are usually multi-label (e.g., a disease sample can contain multiple diagnosis), with severe class imbalances because of the variable prevalence of diseases. Hence, robust MIA methods need to be flexible enough to work with *noisy multi-label* and *imbalanced* problems.

State-of-the-art (SOTA) noisy-label learning approaches are usually based on noise-cleaning methods [5,13,16]. Han et al. [5] propose to use two DNNs and use their disagreements to reject noisy samples from the training process. Li et al. [13] rely on semi-supervised learning that treats samples classified as noisy as unlabelled samples. Other approaches estimate the label transition matrix [4,30] to correct model prediction. Even though these methods show state-of-the-art (SOTA) results in noisy-label problems, they have issues with imbalanced and multi-label problems. First, noise-cleaning methods usually rely on detecting noisy samples by selecting large training loss samples, which are either removed or re-labelled. However, in imbalanced learning problems, such training loss for clean-label training samples, belonging to minority classes, can be larger than the loss for noisy-label training samples belonging to majority classes, so these SOTA noisy-label learning approaches may inadvertently remove or re-label samples belonging to minority classes. Furthermore, in multi-label problems, the same sample can have a mix of clean and noisy labels, so it is hard to adapt SOTA noisy-label learning approaches to remove or re-label particular labels of each sample. Another issue in multi-label problems faced by transition matrix methods is that they are designed to work for multi-class problems, so their adaptation to multi-label problems will need to account for the correlation between the multiple labels. Hence, current noisy-label learning approaches have not been designed to solve all issues present in noisy multi-label imbalanced real-world datasets.

Current imbalanced learning approaches are usually based on decoupling classifier and representation learning [11,28]. For instance, Kang et al. [11] notice that learning with an imbalanced training set does not affect the representation learning, so they only adjust for imbalanced learning when training the classifier. Tang et al. [28] identify causal effect in stochastic gradient descent (SGD) momentum update on imbalanced datasets and propose a de-confounded training scheme. Another type of imbalanced learning is based on loss weighting [2,27] that up-weights the minority classes [2] or down-weights the majority classes [27]. Furthermore, Menon et al. [20] discover that decoupling approach that based on correlation between classifier weight norm and data distribution is only applicable for SGD optimizer, which is problematic for MIA methods that tend to rely on other optimizers, such as Adam, that show better training convergence. Even though the papers above are effective for imbalanced learning problems, they do not consider the combination of imbalanced and noisy multi-label learning.

To address the noisy multi-label imbalanced learning problems present in real-world MIA datasets, we introduce the **Non-volatile Unbiased Memory (NVUM)** training module, which is described in Fig. 1. Our contributions are:

- NVUM that stores a non-volatile running average of model logits to explore the multi-label noise robustness of the early learning stages. This memory module is used by a new regularisation loss to penalise differences between current and early-learning model logits;
- The NVUM update takes into account the class prior distribution to unbias the classification predictions estimated from the imbalanced training set;
- Two new noisy multi-label imbalanced evaluation benchmarks, where training is performed on chest X-ray (CXR) training sets from Chest Xray14 [29] and CheXpert [9], and testing is done on the clean multi-label CXR datasets OpenI [3] and PadChest [1].

2   Method

We assume the availability of a noisy-labelled training set $\mathcal{S} = \{(\mathbf{x}^i, \tilde{\mathbf{y}}^i)\}_{i=1}^{|\mathcal{S}|}$, where $\mathbf{x}^i \in \mathcal{X} \subset \mathbb{R}^{H \times W \times R}$ is the input image of size $H \times W$ with R colour channels, and $\tilde{\mathbf{y}}^i \in \{0,1\}^{|\mathcal{C}|}$ is the noisy label with the set of classes denoted by $\mathcal{C} = \{1, ..., |\mathcal{C}|\}$ (note that $\tilde{\mathbf{y}}_i$ represents a binary vector in multi-label problems, with each label representing one disease).

2.1   Non-volatile Unbiased Memory (NVUM) Training

To describe the NVUM training, we first need to define the model, parameterised by θ and represetned as a deep neural network, with $\mathbf{p} = \sigma(f_\theta(\mathbf{x}))$, where $\mathbf{p} \in$

$[0,1]^{|C|}$, $\sigma(.)$ denotes the sigmoid activation function and $\mathbf{z} = f_\theta(\mathbf{x})$, with $\mathbf{z} \in \mathcal{Z} \in \mathbb{R}^{|C|}$ representing a logit. The training of the model $f_\theta(\mathbf{x})$ is achieved by minimising the following loss function:

$$\ell_{total}(\mathcal{S}, \mathbf{t}, \theta) = \frac{1}{|\mathcal{S}|} \sum_{(\mathbf{x}^i, \tilde{\mathbf{y}}^i) \in \mathcal{S}} \ell_{BCE}(\tilde{\mathbf{y}}^i, \mathbf{p}^i) + \ell_{REG}(\mathbf{t}^i, \mathbf{p}^i), \tag{1}$$

where ℓ_{BCE} denotes the binary cross-entropy loss for handling multi-label classification and ℓ_{REG} is a regularization term defined by:

$$\ell_{REG}(\mathbf{t}^i, \mathbf{p}^i) = \log(1 - \sigma((\mathbf{t}^i)^\top \mathbf{p}^i)). \tag{2}$$

here $\mathbf{t} \in \mathbb{R}^{|\mathcal{S}| \times |C|}$ is our proposed memory module designed to store an unbiased multi-label running average of the predicted logits for all training samples and $\mathbf{t}$ uses the class prior distribution π for updating, denoted by $\pi(c) = \frac{1}{|\mathcal{S}|} \sum_{i=1}^{|\mathcal{S}|} \tilde{\mathbf{y}}(c)$ for $c \in \{1, ..., |C|\}$. The memory module $\mathbf{t}$ is initialised with zeros, as in $\mathbf{t}_0 = \mathbf{0}^{|\mathcal{S}| \times |C|}$, and is updated in every epoch $k > 0$ with:

$$\mathbf{t}_k^i = \beta \mathbf{t}_{k-1}^i + (1 - \beta)(\mathbf{z}_k^i - \log \pi), \tag{3}$$

where $\beta \in [0,1]$ is a hyper-parameter controlling the volatility of the memory storage, with β set to larger value representing a non-volatile memory and $\beta \approx 0$ denoting a volatile memory that is used in [6] for contrastive learning. To explore the early learning phenomenon, we set $\beta = 0.9$ so the regularization can enforce the consistency between the current model logits and the logits produced at the beginning of the training, when the model is robust to noisy label. Furthermore, to make the training robust to imbalanced problems, we subtract the log prior of the class distributions, which has the effect of increasing the logits with larger values for the classes with smaller prior. This counterbalances the issue faced by imbalanced learning problems, where the logits for the majority classes can overwhelm those from the minority classes, to the point that logit inconsistencies found by the regularization from noisy labels of the majority classes may become indistinguishable from the clean labels from minority classes.

The effect of Eq. (2) can be interpreted by inspecting the loss gradient, which is proved in the supplementary material. The gradient of (1) is:

$$\nabla_\theta \ell_{total}(\mathcal{S}, \theta) = \frac{1}{|\mathcal{S}|} \sum_{i \in \mathcal{S}} \mathbf{J}_{\mathbf{x}^i}(\theta)(\mathbf{p}^i - \tilde{\mathbf{y}}^i + \mathbf{g}^i), \tag{4}$$

$$\text{where} \quad \mathbf{g}_c^i = -\sigma((\mathbf{t}^i)^\top (\mathbf{p}^i))\mathbf{p}_c^i(1 - \mathbf{p}_c^i)\mathbf{t}_c$$

where $\mathbf{J}_{\mathbf{x}^i}(\theta)$ is the Jacobian matrix w.r.t. θ for the ith sample. Assume $\mathbf{y}_c$ is the hidden true label of the sample $\mathbf{x}^i$, then the entry $\mathbf{t}_c^i > 0$ if $\mathbf{y}_c = 1$, and $\mathbf{t}_c^i < 0$ if $\mathbf{y}_c = 0$ at the early stages of training. During training, we consider four conditions explained below, where we assume that $\sigma((\mathbf{t}^i)^\top(\mathbf{p}^i))\mathbf{p}_c^i(1 - \mathbf{p}_c^i) > 0$. When the training sample has clean label:

- if $\tilde{\mathbf{y}}_c = \mathbf{y}_c = 1$, the gradient of the BCE term, $\mathbf{p}_c^i - \tilde{\mathbf{y}}_c^i \approx 0$ given that the model is likely to fit clean samples. With $\mathbf{t}_c^i > 0$, the sign of $\mathbf{g}_c^i$ is negative, and the model keeps training for these positive labels even after the early-training stages.
- if $\tilde{\mathbf{y}}_c = \mathbf{y}_c = 0$, the gradient of the BCE term, $\mathbf{p}_c^i - \tilde{\mathbf{y}}_c^i \approx 0$ given that the model is likely to fit clean samples. Given that $\mathbf{t}_c^i < 0$, we have $\mathbf{g}_c^i > 0$, and the model keeps training for these negative labels even after the early-training stages.

Therefore, adding $\mathbf{g}_c^i$ in total loss ensures that clean samples gradient magnitudes remains relatively high, encouraging a continuing optimisation using the clean label samples. For a noisy-label sample, we have:

- if $\tilde{\mathbf{y}}_c = 0$ and $\mathbf{y}_c = 1$, the gradient of the BCE loss is $\mathbf{p}_c^i - \tilde{\mathbf{y}}_c^i \approx 1$ because the model will not fit noisy label during early training stages. With $\mathbf{t}_c^i > 0$, we have $\mathbf{g}_c^i < 0$, which reduces the gradient magnitude from the BCE loss.
- if $\tilde{\mathbf{y}}_c = 1$ and $\mathbf{y}_c = 0$, $\mathbf{p}_c^i - \tilde{\mathbf{y}}_c^i \approx -1$. Given that $\mathbf{t}_c^i < 0$, we have $\mathbf{g}_c^i > 0$, which also reduces the gradient magnitude from the BCE loss.

Therefore, for noisy-label samples, $\mathbf{g}_c^i$ will counter balance the gradient from the BCE loss and diminish the effect of noisy-labelled samples in the training.

3   Experiment

Datasets. For the experiments below, we use the NIH Chest X-ray14 [29] and CheXpert [9] as noisy multi-label imbalanced datasets for training and Indiana OpenI [3] and PadChest [1] datasets for clean multi-label testing sets.

For the noisy sets, **NIH Chest X-ray14 (NIH)** contains 112,120 CXR images from 30,805 patients. There are 14 labels (each label is a disease), where each patient can have multiple diseases, forming a multi-label classification problem. For a fair comparison with previous papers [7,25], we adopt the official train/test data split [29]. **CheXpert (CXP)** contains 220k images with 14 different diseases, and similarly to NIH, each patient can have multiple diseases. For pre-processing, we remove all lateral view images and treat uncertain and empty labels as negative labels. Given that the clean test set from CXP is not available and the clean validation set is too small for a fair comparison, we further split the training images into 90% training set and 10% *noisy* validation set with no patient overlapping. For the clean sets, **Indiana OpenI (OPI)** contains 7,470 frontal/lateral images with manual annotations. In our experiments, we only use 3,643 frontal view of images for evaluation. **PadChest (PDC)** is a large-scale dataset containing 158,626 images with 37.5% of images manually labelled. In our experiment, we only use the manually labelled samples as the clean test set. To keep the number of classes consistent between different datasets, we trim the training and testing sets based on the shared classes between these datasets[1].

---

[1] We include a detailed description and class names in the supplementary material.

Table 1. Class-level and mean testing AUC on OPI [3] and PDC [1] for the experiment based on training on NIH [29]. Best results for OPI/PDC are in **bold**/<u>underlined</u>.

Models	ChestXNet [25]		Hermoza et al. [7]		Ma et al. [17]		DivideMix [13]		Ours	
Datasets	OPI	PDC	OPI	PDC	OPI	PDC	OPI	PDC	OPI	PDC
Atelectasis	86.97	84.99	86.85	83.59	84.83	79.88	70.98	73.48	**88.16**	<u>85.66</u>
Cardiomegaly	89.89	92.50	89.49	91.25	90.87	91.72	74.74	81.63	**92.57**	<u>92.94</u>
Effusion	94.38	96.38	95.05	96.27	94.37	96.29	84.49	<u>97.75</u>	**95.64**	96.56
Infiltration	76.72	70.18	**77.48**	64.61	71.88	73.78	84.03	<u>81.61</u>	72.48	72.51
Mass	53.65	75.21	95.72	<u>86.93</u>	87.47	85.81	71.31	77.41	**97.06**	85.93
Nodule	86.34	75.39	82.68	<u>75.99</u>	69.71	68.14	57.45	63.89	**88.79**	75.56
Pneumonia	**91.44**	76.20	88.15	75.73	84.79	76.49	64.65	72.32	90.90	<u>82.22</u>
Pneumothorax	80.48	79.63	75.34	74.55	82.21	<u>79.73</u>	71.56	75.46	**85.78**	79.50
Edema	83.73	<u>98.07</u>	85.31	97.78	82.75	96.41	80.71	91.81	**86.56**	95.70
Emphysema	82.37	79.10	83.26	<u>79.81</u>	79.38	75.11	54.81	59.91	**83.70**	79.38
Fibrosis	90.53	96.13	86.26	96.46	83.17	93.20	76.98	84.71	**91.67**	<u>98.40</u>
Pleural Thickening	81.58	72.29	77.99	69.95	77.59	67.87	63.98	58.25	**84.82**	<u>74.80</u>
Hernia	89.82	86.72	93.90	89.29	87.37	86.87	66.34	72.11	**94.28**	<u>93.02</u>
Mean AUC	83.69	83.29	86.01	83.25	82.80	82.41	70.92	76.18	**88.65**	<u>85.55</u>

Implementation Details. We use the ImageNet [26] pre-trained DenseNet 121 [8] as the backbone model for $f_\theta(.)$ on NIH and CXP. We use Adam [12] optimizer with batch size 16 for NIH and 64 for CXP. For NIH, we train for 30 epochs with a learning rate of 0.05 and decay with 0.1 at 70% and 90% of the total of training epochs. Images are resized from 1024×1024 to 512×512 pixels. For data augmentation, we employ random resized crop and random horizontal flipping. For CXP, we train for 40 epochs with a learning rate of $1e^{-4}$ and follow the learning rate decay policy as on NIH. Images are resized to 224×224. For data augmentation, we employ random resized crop, $10°$ random rotation and random horizontal flipping. For both datasets, we use $\beta = 0.9$ and normalized by ImageNet mean and standard deviation.

All classification results are reported using area under the ROC curve (AUC). To report performance on clean test sets OPI and PDC, we adopt a common noisy label setup [5,13] that selects the best performance checkpoint on noisy validation, which is the noisy test set of NIH and the noisy validation set of CXP. All experiments are implemented with Pytorch [23] and conducted on an NVIDIA RTX 2080ti GPU. The training takes 15 h on NIH and 14 h on CXP.

3.1 Experiments and Results

Baselines. We compared NVUM with several methods, including the CheXNet baseline [25], Ma et al.'s approach [17] based on a cross-attention network, the current SOTA for NIH on the official test set is the model by Hermoza et al. [7] that is a weakly supervised disease classifier that combines region proposal and saliency detection. We also show results from DivideMix [13], which uses a noisy-label learning algorithm based on small loss sample selection and semi-supervised

Table 2. Class-level and mean testing AUC on OPI [3] and PDC [1] for the experiment based on training on CXP [9]. Best results for OPI/PDC are in **bold**/<u>underlined</u>.

Methods	CheXNet [25]		Hermoza et al. [7]		Ma et al. [17]		DivideMix [13]		Ours	
Datasets	OPI	PDC	OPI	PDC	OPI	PDC	OPI	PDC	OPI	PDC
Cardiomegaly	84.00	80.00	87.01	87.20	82.83	85.89	71.14	66.51	**88.86**	<u>88.48</u>
Edema	88.16	98.80	87.92	98.72	86.46	97.47	75.36	95.51	**88.63**	<u>99.60</u>
Pneumonia	**65.82**	58.96	65.56	53.42	61.88	54.83	57.65	40.53	64.90	<u>67.89</u>
Atelectasis	77.70	72.23	78.40	<u>75.33</u>	80.13	72.87	73.65	64.12	**80.81**	75.03
Pneumothorax	77.35	<u>84.75</u>	62.09	78.65	51.08	71.57	68.75	54.05	**82.18**	83.32
Effusion	85.81	91.84	87.00	<u>93.44</u>	**88.43**	92.92	78.60	79.89	83.54	89.74
Fracture	57.64	60.26	57.47	53.77	59.92	60.44	**60.35**	59.43	57.02	<u>62.67</u>
Mean AUC	76.64	78.12	75.06	77.29	72.96	76.57	69.36	65.72	**77.99**	<u>80.96</u>

Table 3. Pneumothorax and Mass/Nodule AUC using the manually labelled clean test from [18]. Baseline results obtained from [31]. Best results are in **bold**.

	BCE	F-correction [24]	MentorNet [10]	Decoupling [19]	Co-teaching [5]	ELR [16]	Xue et al. [31]	Ours
Pneu	87.0	80.8	86.6	80.1	87.3	87.1	**89.1**	88.9
M/N	84.3	84.8	83.7	84.3	82.0	83.2	84.6	**85.5**

learning. DivideMix has the SOTA results in many noisy-label learning benchmarks. All methods are implemented using the same DenseNet121 [8] backbone.

Quantitative Comparison. Table 1 shows the class-level AUC result for training on NIH and testing on OPI and PDC. Our approach achieves the SOTA results on both clean test sets, consistently outperforming the baselines [7,17,25], achieving 2% mean AUC improvement on both test sets. Compared with the current SOTA noisy-label learning DivideMix [13], our method outperforms it by 18% on OPI and 9% on PDC. This shows that for noisy multi-label imbalanced MIA datasets, noisy multi-class balanced approaches based on small-loss selection is insufficient because they do not take into account the multi-label and imbalanced characteristics of the datasets. Table 2 shows class-level AUC results for training on CXP and testing on OPI and PDC. Similarly to the NIH results on Table 1, our approach achieves the best AUC results on both test sets with at least 3% improvement on OPI and 3% on PDC. In addition, DivideMix [13] shows similar results compared with NIH. Hence, SOTA performance on both noisy training sets suggests that our method is robust to different noisy multi-label imbalanced training sets.

Additional Benchmark. Using the recently proposed noisy label benchmark by Xue et al. [31], we further test our approach against the SOTA in the field. The benchmark uses a subset of the official NIH test set [18], with 1,962 CXR images manually re-labelled by at least three radiologists per image. For the results, we follow [31] and consider the AUC results only for Pneumothorax (Pneu) and average of Mass and Nodule (M/N). We use the same hyperparameters as

$\ell_{BCE}\ \pi$	$\ell_{REG}\ \pi$	OPI	PDC
✓		82.91±0.78	82.27±1.02
✓	✓	85.80±0.04	84.35±0.35
✓		✓	85.24±0.70	84.39±0.21
✓	✓	✓	85.36±0.11	83.04±0.79
✓	✓	✓	✓	86.68±0.16	85.02±0.18
NVUM		**88.17±0.48**	**85.49±0.06**

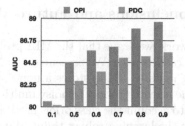

Fig. 2. (Left) Mean AUC results of training on NIH using the class prior distribution π applied to different components of NVUM. (Right) Mean AUC results on OPI (blue) and PDC (green) of training on NIH with different β values for the NVUM memory update in (3). (Color figure online)

above. The results in Table 3 shows that our method outperforms most noisy label methods and achieves comparable performance to [31] on Pneumothorax (88.9 vs 89.1) and better performance on Mass/Nodule (85.5 vs 84.6). However, it is important to mention that differently from [31] that uses two models, we use only one model, so our method requires significantly less training time and computation resources. Furthermore, the clean test set from [18] is much smaller than OPI and PDC with only two classes available, so we consider results in Table 1 and 2 more reliable than Table 3.

3.2 Ablation Study

Different Components of NVUM with π. We first study in Fig. 2 (left) how results are affected by the prior added on different components of NVUM. We run each experiment three times and show mean and standard deviation of AUC results. By adding the class prior π to ℓ_{BCE} [20], we replace the BCE term in (1) with $\ell_{BCE}(\tilde{\mathbf{y}}^i, \sigma(f_\theta(\mathbf{x}^i + \log \pi)))$. We can also add the class prior π to ℓ_{REG} by replacing the regularization term in (1) with $\ell_{REG}(\mathbf{t}^i, \sigma(f_\theta(\mathbf{x} + \log \pi)))$. We observe a 2% improvement for OPI and PDC for both modifications compared to ℓ_{BCE} baseline, demonstrating that it is important to handle imbalanced learning in MIA problems. Furthermore, we combine two modifications together and achieve additional 1% improvement. However, instead of directly working on the loss functions, as suggested in [20], we work on the memory module given that it also enforces the early learning phenomenon, addressing the combined noisy multi-label imbalanced learning problem.

Different β. We also study different values for β in (3). First, we test a volatile memory update with $\beta = 0.1$, which shows a significantly worse performance because the model is overfitting the noisy multi-label of the training set. This indicates traditional volatile memory [6] cannot handle noisy label learning. Second, the non-volatile memory update with $\beta \in \{0.5, ..., 0.9\}$ shows a performance that improves consistently with larger β. Hence, we use $\beta = 0.9$ as our default setup.

4   Conclusions and Future Work

In this work, we argue that the MIA problem is a problem of *noisy multi-label* and *imbalanced learning*. We presented the Non-volatile Unbiased Memory (NVUM) training module, which stores a non-volatile running average of model logits to make the learning robust to noisy multi-label datasets. Furthermore, The NVUM takes into account the class prior distribution when updating the memory module to make the learning robust to imbalanced learning. We conducted experiments on proposed new benchmark and recent benchmark [31] and achieved SOTA results. Ablation study shows the importance of carefully accounting for imbalanced and noisy multi-label learning. For the future work, we will explore an precise estimation of class prior π during the training for accurate unbiasing.

References

1. Bustos, A., Pertusa, A., Salinas, J.M., de la Iglesia-Vayá, M.: PadChest: a large chest x-ray image dataset with multi-label annotated reports. Med. Image Anal. **66**, 101797 (2020)
2. Cao, K., Wei, C., Gaidon, A., Arechiga, N., Ma, T.: Learning imbalanced datasets with label-distribution-aware margin loss. In: Advances in Neural Information Processing Systems, pp. 1567–1578 (2019)
3. Demner-Fushman, D., et al.: Preparing a collection of radiology examinations for distribution and retrieval. J. Am. Med. Inform. Assoc. **23**(2), 304–310 (2016)
4. Goldberger, J., Ben-Reuven, E.: Training deep neural-networks using a noise adaptation layer (2016)
5. Han, B., et al.: Co-teaching: robust training of deep neural networks with extremely noisy labels. In: Advances in Neural Information Processing Systems, vol. 31 (2018)
6. He, K., Fan, H., Wu, Y., Xie, S., Girshick, R.: Momentum contrast for unsupervised visual representation learning. arXiv preprint arXiv:1911.05722 (2019)
7. Hermoza, R., et al.: Region proposals for saliency map refinement for weakly-supervised disease localisation and classification. arXiv preprint arXiv:2005.10550 (2020)
8. Huang, G., et al.: Densely connected convolutional networks. In: CVPR, pp. 4700–4708 (2017)
9. Irvin, J., et al.: CheXpert: a large chest radiograph dataset with uncertainty labels and expert comparison. In: Proceedings of the AAAI Conference on Artificial Intelligence, vol. 33, pp. 590–597 (2019)
10. Jiang, L., Zhou, Z., Leung, T., Li, L.J., Fei-Fei, L.: MentorNet: learning data-driven curriculum for very deep neural networks on corrupted labels. In: ICML (2018)
11. Kang, B., et al.: Decoupling representation and classifier for long-tailed recognition. arXiv preprint arXiv:1910.09217 (2019)
12. Kingma, D.P., Ba, J.: Adam: a method for stochastic optimization. arXiv preprint arXiv:1412.6980 (2014)
13. Li, J., Socher, R., Hoi, S.C.: DivideMix: learning with noisy labels as semi-supervised learning. arXiv preprint arXiv:2002.07394 (2020)
14. Litjens, G., et al.: A survey on deep learning in medical image analysis. Med. Image Anal. **42**, 60–88 (2017)

15. Liu, F., Tian, Y., Chen, Y., Liu, Y., Belagiannis, V., Carneiro, G.: ACPL: anti-curriculum pseudo-labelling forsemi-supervised medical image classification. arXiv preprint arXiv:2111.12918 (2021)
16. Liu, S., Niles-Weed, J., Razavian, N., Fernandez-Granda, C.: Early-learning regularization prevents memorization of noisy labels. arXiv preprint arXiv:2007.00151 (2020)
17. Ma, C., Wang, H., Hoi, S.C.H.: Multi-label thoracic disease image classification with cross-attention networks. In: Shen, D., et al. (eds.) MICCAI 2019. LNCS, vol. 11769, pp. 730–738. Springer, Cham (2019). https://doi.org/10.1007/978-3-030-32226-7_81
18. Majkowska, A., et al.: Chest radiograph interpretation with deep learning models: assessment with radiologist-adjudicated reference standards and population-adjusted evaluation. Radiology **294**(2), 421–431 (2020)
19. Malach, E., Shalev-Shwartz, S.: Decoupling "when to update" from "how to update". In: Advances in Neural Information Processing Systems, vol. 30 (2017)
20. Menon, A.K., Jayasumana, S., Rawat, A.S., Jain, H., Veit, A., Kumar, S.: Long-tail learning via logit adjustment. arXiv preprint arXiv:2007.07314 (2020)
21. Oakden-Rayner, L.: Exploring the chestxray14 dataset: problems. Wordpress: Luke Oakden Rayner (2017)
22. Oakden-Rayner, L.: Exploring large-scale public medical image datasets. Acad. Radiol. **27**(1), 106–112 (2020)
23. Paszke, A., et al.: Pytorch: an imperative style, high-performance deep learning library. In: Advances in Neural Information Processing Systems, vol. 32 (2019)
24. Patrini, G., Rozza, A., Krishna Menon, A., Nock, R., Qu, L.: Making deep neural networks robust to label noise: a loss correction approach. In: Proceedings of the IEEE Conference on Computer Vision and Pattern Recognition, pp. 1944–1952 (2017)
25. Rajpurkar, P., et al.: CheXNet: radiologist-level pneumonia detection on chest X-rays with deep learning. arXiv preprint arXiv:1711.05225 (2017)
26. Russakovsky, O., et al.: Imagenet large scale visual recognition challenge. IJCV **115**(3), 211–252 (2015)
27. Tan, J., et al.: Equalization loss for long-tailed object recognition. In: Proceedings of the IEEE/CVF Conference on Computer Vision and Pattern Recognition, pp. 11662–11671 (2020)
28. Tang, K., Huang, J., Zhang, H.: Long-tailed classification by keeping the good and removing the bad momentum causal effect. In: NeurIPS (2020)
29. Wang, X., Peng, Y., Lu, L., Lu, Z., Bagheri, M., Summers, R.M.: ChestX-ray8: hospital-scale chest X-ray database and benchmarks on weakly-supervised classification and localization of common thorax diseases. In: CVPR, pp. 2097–2106 (2017)
30. Xia, X., et al.: Are anchor points really indispensable in label-noise learning? arXiv preprint arXiv:1906.00189 (2019)
31. Xue, C., Yu, L., Chen, P., Dou, Q., Heng, P.A.: Robust medical image classification from noisy labeled data with global and local representation guided co-training. IEEE Trans. Med. Imaging **41**(6), 1371–1382 (2022)

Local Graph Fusion of Multi-view MR Images for Knee Osteoarthritis Diagnosis

Zixu Zhuang[1,3], Sheng Wang[1,3], Liping Si[4], Kai Xuan[1], Zhong Xue[3], Dinggang Shen[2,3], Lichi Zhang[1], Weiwu Yao[4], and Qian Wang[2(✉)]

[1] School of Biomedical Engineering, Shanghai Jiao Tong University, Shanghai, China
[2] School of Biomedical Engineering, ShanghaiTech University, Shanghai, China
wangqian2@shanghaitech.edu.cn
[3] Shanghai United Imaging Intelligence Co., Ltd., Shanghai, China
[4] Department of Imaging, Tongren Hospital, Shanghai Jiao Tong University School of Medicine, Shanghai, China

Abstract. Magnetic resonance imaging (MRI) has become necessary in clinical diagnosis for knee osteoarthritis (OA), while deep neural networks can contribute to the computer-assisted diagnosis. Recent works prove that instead of only using a single-view MR image (e.g., sagittal), integrating multi-view MR images can boost the performance of the deep network. However, existing multi-view networks typically encode each MRI view to a feature vector, fuse the feature vectors of all views, and then derive the final output using a set of shallow computations. Such a *global fusion* scheme happens at a coarse granularity, which may not effectively localize the often tiny abnormality related to the onset of OA. Therefore, this paper proposes a *Local Graph Fusion Network* (LGF-Net), which implements graph-based representation of knee MR images and multi-view fusion for OA diagnosis. We first model the multi-view MR images to a unified knee graph. Then, the patches of the same location yet from different views are encoded to one-dimensional features and are exchanged mutually during fusing. The *local fusion* of the features further propagates following edges by Graph Transformer Network in the LGF-Net, which finally yields the grade of OA. The experimental results show that the proposed framework outperforms state-of-the-art methods, demonstrating the effectiveness of local graph fusion in OA diagnosis.

Keywords: Knee OA diagnosis · Graph representation · Multi-view MRI

1 Introduction

Knee osteoarthritis (OA) is one of the most common musculoskeletal diseases and a major cause of disability in older people, which brings enormous medical costs and social burdens [14,20]. Magnetic resonance imaging (MRI) has become essential for assessing knee OA. It is advantageous to early OA diagnosis and

© The Author(s), under exclusive license to Springer Nature Switzerland AG 2022
L. Wang et al. (Eds.): MICCAI 2022, LNCS 13433, pp. 554–563, 2022.
https://doi.org/10.1007/978-3-031-16437-8_53

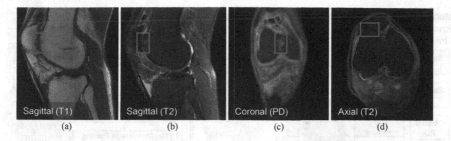

Fig. 1. In clinical routines, the knee MR images are often scanned in 2D of multiple views (sagittal, coronal, and axial) and modalities (T1, T2, and PD). The red boxes indicate the same OA-related abnormality in individual images. (Color figure online)

intervention due to the excellent contrast of MRI on soft tissues and the rendering of the whole joint [6]. In clinical routines, the knee MR images are often scanned in 2D of multiple views (sagittal, coronal, and axial) and modalities (T1, T2, and PD) [9,18], with examples provided in Fig. 1. While the sagittal view is pivotal, radiologists need to read other views in diagnosing, particularly to establish the association for a specific lesion area across different views.

Recently, many researchers have applied deep learning methods in MRI-based computer-aided diagnosis (CAD) of knee disease and achieved promising progress. However, most of them address the scenario where only a single view of MRI is available [11,12,15,24]. The idea to utilize multi-view MR images for knee disease diagnosis has thus drawn a lot of attention [4,5,22]. For instance, Bien et al. [5] proposed MRNet that encoded all the 2D MRI slices and merged them by max-pooling in each view. They then fused the disease probabilities from three views with logistic regression.

Several other multi-view approaches have similar pipelines yet vary on the fusion strategy. Azcona et al. [3] replaced the backbone in MRNet with ResNet18 [10] and fused the multi-view features by max-pooling. Belton et al. [4] encoded all the views and their slices and max-pooling them all together. Wilhelm et al. [17] chose the recurrent neural network (RNN) for slice features fusion. In this paper, we call these methods that fuse the encoded features of individual MRI slices or volumes *global fusion* methods.

Despite getting better performance by utilizing multi-view MR images instead of a single view, there are some disadvantages in the current global fusion methods.

- **Fine-Granularity Spatial Association for Lesions**. Radiologists usually read MR images in a *local fusion* manner. That is, they often check all three views for a certain lesion. However, the feature encoding each slice/volume lacks spatial localization in the existing methods. Therefore, in global fusion, it is difficult to establish a cross-view association for the feature representation of a specific lesion, which may reduce the diagnostic performance.
- **Distortion from Large Background**. The lesion areas contributing to knee OA diagnosis are distributed near the bone-cartilage interface. Concerning the

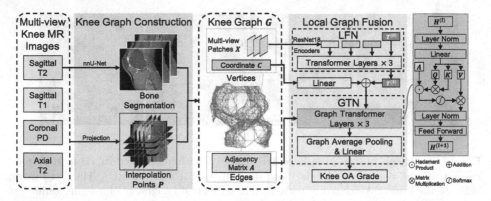

Fig. 2. Our proposed framework is composed of two major parts: (1) Knee Graph Construction, which embeds the multi-view MR images to a unified graph representation; (2) Local Graph Fusion, which first fuses the multi-view knee MR patches by Local Fusion Network (LFN), and then derives the knee OA grade by Graph Transformer Network (GTN).

curved thin-layer shapes of the cartilages, the MR images may have too much information from the background that is irrelevant to OA diagnosis.

We propose a local graph fusion network (LGF-Net) to address these challenges. First, we model the multi-view MR images to a unified knee graph. Then, the features of the graph vertices are encoded from the patches of the same location in the knee joint yet different views/modalities by our Local Fusion Network. Next, we propagate the vertex features following the edges of the knee graph and finally yield the grade of OA using a Graph Transformer Network. Compared with the global fusion methods, our framework fuses multi-view knee MR images locally, establishing an association for the lesion-specific features and suppressing irrelevant background.

2   Method

As illustrated in Fig. 2, the proposed framework consists of two parts: (1) Knee Graph Construction, which uses the MRI localization and knee bone segmentation to build the knee graph representation for all multi-view images of a subject; (2) Local Graph Fusion, which encodes the multi-view patches from vertices of the knee graph via Local Fusion Network (LFN) first, and then derives the OA grade for the subject by ensembling through the Graph Transformer Network (GTN).

2.1   Knee Graph Construction

The knee graph is defined as $G = \{V, A\}$, which is composed of the vertices V and the edges A, to represent the knee of each subject. Since the OA-related imaging biomarkers are distributed near the bone-cartilage interface, the vertices V need to cover these regions, and the edges A provide the pathways for message passing between vertices.

Vertex Collection. To represent a multi-view knee MRI by graph G, it is essential to collect a set of vertices V for the graph. The vertices need to cover the OA-related regions in all multi-view MR images. An MRI file stores world coordinate origin, orientation, and voxel sizes, which allows individual planes to be projected into the world coordinate system. Therefore, given sagittal, coronal, and axial slices, their intersection yields a set of intersection points P, as shown in the gray part in Fig. 2. From P the vertices V will be extracted. Note that, for each intersection point in P, the 2D patches can be naturally center-cropped from the MRI planes of respective views. Meanwhile, the sagittal T1 and T2 images in this work share the same coordinate system, including their origin, orientation, and spacing.

Next, we segment the femur, tibia, and patella in each sagittal-view T2 MRI by an in-house nnU-Net tool [13]. The OA-related knee joint region is cropped by localizing with the bone segmentation [21]: the upper bound of the cropped region is 5 mm above the patella; the inferior bound is 80 mm below the upper bound; the anterior bound is 3 mm in front of the patella; and the posterior bound is 120 mm behind the anterior bound. The bone surface is then reconstructed as S in the cropped region.

The vertices V is collected from those points located in P that are near the bone surface. By defining $V = \{X, C\}$, we have the coordinates $C = \{p | p \in P, \text{Distance}(p, S) < t\}$. Here the threshold t is the half the mean of the inter-slice spacing in input multi-view MR images to ensure that the vertices are uniformly distributed near the bone-cartilage interface.

Each vertex is signified by multiple 2D patches corresponding to individual views with the same center, denoted by $X \in \mathbb{R}^{N \times m \times p \times p}$. N is the number of vertices, related to the inter-slice spacing and the number of slices in different views of each subject. And $p = 64$ is the size of the patches. In addition, we have $m = 4$ since there are three views and two modalities are available in sagittal view (cf. Fig. 1).

Edge Connection. We further establish edge connection, allowing adjacent vertices to exchange their features along the bone-cartilage interface. The edges are represented by the adjacency matrix A: $A[i, j] = 1$, if $j \in \mathcal{N}_i$. Because the distribution of the vertices is relatively uniform, the edges can be established by simply connecting each vertex with its nearest $|\mathcal{N}_i|$ neighbors include itself ($|\mathcal{N}_i| = 9$ in our implementation).

The process embeds the multi-view knee MR images into a unified knee graph representation. It helps the subsequent computation focus on local lesion-specific visual cues and avoids being distorted by irrelevant background information.

2.2 Local Fusion Network (LFN)

LFN is developed for encoding the multi-view local patches and fusing them into a feature vector in each vertex. As shown in Fig. 2, LFN consists of CNN encoders for embedding patches in each view and a multi-layer Transformer for fusing multi-view patch embeddings.

Particularly we have three CNN encoders for three views separately. The sagittal-view encoder supports two-channel input since this view has dual input modalities. The other two views handle single-channel inputs. Due to the small patch size, a shallow encoder such as ResNet18 [10] is sufficient according to our experiments.

Referring to Vision Transformer [8], an auxiliary token $T^{(0)}$ that has the same shape with the patch embedding is added to the 3-layer Transformer. It is randomly initialized and then learned in the training stage. This token will calculate the attention with the patch embeddings of the three views and fuse them in each vertex, eventually yielding a one-dimensional vector as the fused local feature $T^{(3)}$ representing each vertex.

It is demonstrated in the literature that the model performance can be further improved by pre-training its encoder [16]. This paper uses patch-level OA grading as the pre-training task for LFN. We will also verify the performance gain of the pre-training.

2.3 Graph Transformer Network (GTN)

We propose the GTN to aggregate multi-view patch features along edges in the knee graph. The GTN includes three Graph Transformer Layers (GTLs), a graph average pooling, and a linear layer. More details are provided in the right-most of Fig. 2.

The GTL is designed to update vertex features based on the local attention mechanism. For the l-th GTL, the input $H^{(l)}$ generates the three embeddings Q, K, and V by a linear layer. The adjacency matrix A constrains that each vertex only computes the attention $\hat{A}$ with the adjacent vertices including itself:

$$\hat{A} = \text{Softmax}(A \odot \frac{QK^T}{\sqrt{d}}), \tag{1}$$

where $\odot$ is the Hadamard product, and d is the embedding length. We take the local attention $\hat{A}$ as the edge weights to update the vertex feature:

$$H^{(l+1)} = \text{FeedForward}(\hat{A}V). \tag{2}$$

Note that the initial $H^{(0)}$ sums up $T^{(3)}$ and the projected coordinates C.

The graph average pooling is applied to acquire the graph embedding H_g based on all vertex features $H^{(3)}$ from the last GTL: $H_g = \sum_i^N H^{(4)}[i]/N$. Finally, the knee OA grade can be computed by a linear projection of H_g.

3   Experimental Results

3.1   Data and Experimental Settings

The knee MRI dataset used in our experiments has been collected from Anonymous Institute. There are 1251 subjects, while each subject is scanned by a T1-weighted sequence (T1) and a T2-weighted sequence with fat suppression (T2) in the sagittal view, a proton-density-weighted sequence (PD) in the coronal view, and a T2 in axial view. The examples of the images can be found in Fig. 1.

Referring to the Whole-Organ MRI Score (WORMS) [1], and Astuto *et al.* [2], we classify knee OA into three grades based on the severity: Grade 0, no knee OA, with intact cartilage (corresponding to Score 0 and 1 in WORMS); Grade 1, mild knee OA, with partial cartilage defects (Score 2, 3, and 4 in WORMS); and Grade 2, severe knee OA, with complete cartilage defects (Score 2.5, 5, and 6 in WORMS). In addition to the subject grades, two radiologists also label the lesion areas to facilitate our pre-training of the LFN. In summary, the subject numbers of Grade 0, 1, and 2 are 518, 391, and 342, respectively. We evaluate the grading performance with 5-fold cross-validation by the following metrics: accuracy (ACC), sensitivity (SEN), specificity (SPE), precision (PRE), and F1 score (F1S).

3.2   Comparison with State-of-the-Art Methods

We compare the proposed LGF-Net with state-of-the-art methods to demonstrate its graph-based multi-view fusion superiority. We particularly compare with the best performing multi-view models, namely MRNet [5], MPFuseNet [4], and Azcona *et al.* [3]. These works demonstrate that the multi-view fusion strategy works powerfully in various volumetric medical images tasks. In addition, we use ELNet [22] and 3D ResNet18 [7] as the baselines for the single-view model.

Our method owns the advantages in that the graph representation naturally supports heterogeneous knee MR images as the input. However, many existing methods, including the above, require homogeneous inputs. To this end, we pad all the MR images to the same number of slices, resample them to a fixed spacing, and crop them to the same size, if needed by the comparing methods.

The comparisons are shown in Table 1. Note that while our method can benefit from fine-granularity supervision induced to pre-train the LFN by task of knee OA classification in patch level, the comparing methods cannot utilize this information. Therefore, we temporarily disable the pre-training and denote our method as "LGF (No PT)" in the table for a fair comparison. Still, with all three MRI views as the input, our method achieves 75.5% in ACC, significantly higher than early multi-view methods. In addition, the single-view methods perform consistently worse than multi-view methods except for the MRNet with an old backbone. The results validate the merit of using multi-view MR images for CAD of knee OA. Further, our method best utilizes the multi-view MR images even though the pre-training is disabled.

Table 1. Comparing LGF-Net with other methods for knee OA grading.

	Views			Performance(%)				
	Sagittal	Coronal	Axial	ACC	SEN	SPE	PRE	F1S
MRNet [5]	✓	✓	✓	71.8	71.0	85.8	71.5	71.0
MPFuseNet [4]	✓	✓	✓	73.4	73.0	86.8	73.6	72.7
Azcona et al. [3]	✓	✓	✓	73.4	73.1	86.8	73.8	73.1
ELNet [22]		✓		71.2	71.7	85.7	73.4	71.1
3D ResNet18 [7]	✓			66.1	66.5	83.2	67.8	66.4
3D ResNet18 [7]		✓		64.6	64.2	82.0	65.8	64.3
3D ResNet18 [7]			✓	63.8	63.9	82.0	63.5	63.1
LGF-Net (No PT)	✓	✓	✓	75.5	75.0	87.7	75.4	74.6
LGF-Net (Proposed)	✓	✓	✓	**81.5**	**81.1**	**90.7**	**81.5**	**81.0**

Further, our method is featured by pre-training the LFN with patch-level knee OA classification in the same dataset and the same data division, which brings significant improvement as demonstrated by the last two rows in Table 1. Specifically, the radiologists have labeled the lesion areas related to OA when grading the image dataset. We then crop the patches and train the LFN to map the patches to the corresponding grades. Considering the imbalance of patch numbers across grades, we restrict the number of Grade-0 patches to 10830, in addition to 8679 Grade-1 and 10830 Grade-2 patches for the pre-training.

Notice that our method performs much better when the pre-training is enabled (81.5% in ACC with pre-training, vs. 75.5% without pre-training). This indicates that the accuracy of knee OA diagnosis can be significantly improved by mining fine-granularity lesion visual cues and adding the pre-training task for stronger supervision.

3.3   Contribution of Multiple Views

This section shows the contributions of individual views toward the final grading in our method. After the knee graph construction, we train LGF-Net in selected views and their combinations. The comparing results are presented in Table 2.

Table 2. Contributions of individual views and their combinations in LGF-Net.

Views			Performance(%)				
Sagittal	Coronal	Axial	ACC	SEN	SPE	PRE	F1S
✓	✓	✓	**81.5**	**81.1**	**90.7**	**81.5**	**81.0**
✓	✓		79.9	79.2	89.9	79.3	79.1
✓		✓	79.5	78.6	89.6	78.7	78.4
	✓	✓	79.7	78.6	89.7	79.9	79.0
✓			79.5	77.9	89.4	80.9	78.5
	✓		78.4	77.2	89.0	78.4	77.4
		✓	75.6	73.9	87.3	75.5	73.8

One can notice that adding more views improves the knee OA diagnosis performance constantly. For example, when only utilizing the sagittal view (of both T1 and T2), the SEN is 77.9%. Then, adding the coronal view improves the SEN to 79.2%. The highest SEN can be achieved at 81.1% when the axial view is further incorporated. The results also underscore the role of the sagittal view. That is, as a single view, the sagittal T1 and T2 combined together yields better performance than the other two views alone. The finding may support that the dual-modal input brings in higher performance.

3.4 Ablation Study on the Design of LGF-Net

Past works have shown that multi-view fusion can be done using simple concatenation [22] and pooling [5]. Therefore, we replace the Transformer Layers (TLs) with concatenation, mean-pooling, and max-pooling. We also replace the Graph Transformer Layers (GTLs) with widely adopted GCN [19], GAT [23], and identity layer. The results for the ablation study are presented in Table 3.

Table 3. Controlled ablation study of LFN and GTN in the proposed LFN-Net.

Alternative		Performance (%)				
TLs	GTLs	ACC	SEN	SPE	PRE	F1S
Concatenation	GTLs	78.5	77.7	89.3	77.3	77.1
Mean-Pooling	GTLs	79.3	78.7	89.6	78.8	78.5
Max-Pooling	GTLs	78.3	77.6	89.0	77.4	77.3
TLs	GCN	77.8	77.3	88.7	77.9	77.1
TLs	GAT	80.0	78.7	89.7	79.8	78.9
TLs	Identity	76.5	76.2	88.2	76.4	75.8
Ours (TLs+GTLs)		**81.5**	**81.1**	**90.7**	**81.5**	**81.0**

Compared to TLs, the other patch fusion methods show lower performance (ACC: 81.5% vs. 78.3%–79.3%). It demonstrates that using the attention mechanism to fuse multi-view patches in TLs helps the network better focus on the lesion regions.

Replacing GTLs with other networks also leads to degradation in performance (ACC: 81.5% vs. 76.5%–80.0%). It proves that GTLs can update features better along the bone-cartilage interface, and thus GTLs give the model a stronger capability of identifying OA.

4   Conclusion

In this paper, we have proposed a local graph fusion framework for automatic grading of knee OA based on multi-view MR images, consisting of the unified knee graph construction and the local graph fusion. It allows information from multiple views to fuse in a local manner, which is enabled by the unified knee graph representation. While we achieve superior performance compared to many state-of-the-art methods, we want to emphasize that our method well fits the heterogeneous imaging data in the clinical scenario. Therefore, our method has a high potential for CAD of knee OA.

Acknowledgement. This work was supported by the National Key Research and Development Program of China (2018YFC0116400), National Natural Science Foundation of China (NSFC) grants (62001292) and Interdisciplinary Program of Shanghai Jiao Tong University (YG2019QNA17).

References

1. Alizai, H., et al.: Cartilage lesion score: comparison of a quantitative assessment score with established semiquantitative MR scoring systems. Radiology **271**(2), 479–487 (2014)
2. Astuto, B., et al.: Automatic deep learning-assisted detection and grading of abnormalities in knee MRI studies. Radiol. Artif. Intell. **3**(3), e200165 (2021)
3. Azcona, D., McGuinness, K., Smeaton, A.F.: A comparative study of existing and new deep learning methods for detecting knee injuries using the MRNet dataset. In: 2020 International Conference on Intelligent Data Science Technologies and Applications (IDSTA), pp. 149–155. IEEE (2020)
4. Belton, N., et al.: Optimising knee injury detection with spatial attention and validating localisation ability. In: Papież, B.W., Yaqub, M., Jiao, J., Namburete, A.I.L., Noble, J.A. (eds.) MIUA 2021. LNCS, vol. 12722, pp. 71–86. Springer, Cham (2021). https://doi.org/10.1007/978-3-030-80432-9_6
5. Bien, N., et al.: Deep-learning-assisted diagnosis for knee magnetic resonance imaging: development and retrospective validation of mrnet. PLoS Med. **15**(11), e1002699 (2018)
6. Calivà, F., Namiri, N.K., Dubreuil, M., Pedoia, V., Ozhinsky, E., Majumdar, S.: Studying osteoarthritis with artificial intelligence applied to magnetic resonance imaging. Nat. Rev. Rheumatol. **18**, 1–10 (2021)
7. Chen, S., Ma, K., Zheng, Y.: Med3D: transfer learning for 3D medical image analysis. arXiv preprint arXiv:1904.00625 (2019)
8. Dosovitskiy, A., et al.: An image is worth 16x16 words: transformers for image recognition at scale. In: International Conference on Learning Representations (2020)
9. Garwood, E.R., Recht, M.P., White, L.M.: Advanced imaging techniques in the knee: benefits and limitations of new rapid acquisition strategies for routine knee MRI. Am. J. Roentgenol. **209**(3), 552–560 (2017)
10. He, K., Zhang, X., Ren, S., Sun, J.: Deep residual learning for image recognition. In: Proceedings of the IEEE Conference on Computer Vision and Pattern Recognition, pp. 770–778 (2016)
11. Huo, J., et al.: Automatic grading assessments for knee MRI cartilage defects via self-ensembling semi-supervised learning with dual-consistency. Med. Image Anal. **80**, 102508 (2022)

12. Huo, J., et al.: A self-ensembling framework for semi-supervised knee cartilage defects assessment with dual-consistency. In: Rekik, I., Adeli, E., Park, S.H., Valdés Hernández, M.C. (eds.) PRIME 2020. LNCS, vol. 12329, pp. 200–209. Springer, Cham (2020). https://doi.org/10.1007/978-3-030-59354-4_19
13. Isensee, F., Jaeger, P.F., Kohl, S.A., Petersen, J., Maier-Hein, K.H.: nnU-Net: a self-configuring method for deep learning-based biomedical image segmentation. Nat. Methods 18(2), 203–211 (2021)
14. Jamshidi, A., Pelletier, J.P., Martel-Pelletier, J.: Machine-learning-based patient-specific prediction models for knee osteoarthritis. Nat. Rev. Rheumatol. 15(1), 49–60 (2019)
15. Liu, F., et al.: Fully automated diagnosis of anterior cruciate ligament tears on knee MR images by using deep learning. Radiol. Artif. Intell. 1(3), 180091 (2019)
16. Liu, F., et al.: Deep learning approach for evaluating knee MR images: achieving high diagnostic performance for cartilage lesion detection. Radiology 289(1), 160–169 (2018)
17. Nikolas, W., Jan, L., Carina, M., von Eisenhart-Rothe, R., Rainer, B.: Maintaining the spatial relation to improve deep-learning-assisted diagnosis for magnetic resonace imaging of the knee. Zeitschrift für Orthopädie und Unfallchirurgie 158(S 01), DKOU20-670 (2020)
18. Peterfy, C., Gold, G., Eckstein, F., Cicuttini, F., Dardzinski, B., Stevens, R.: MRI protocols for whole-organ assessment of the knee in osteoarthritis. Osteoarthr. Cartil. 14, 95–111 (2006)
19. Scarselli, F., Gori, M., Tsoi, A.C., Hagenbuchner, M., Monfardini, G.: The graph neural network model. IEEE Trans. Neural Networks 20(1), 61–80 (2008)
20. Si, L., et al.: Knee cartilage thickness differs alongside ages: a 3-T magnetic resonance research upon 2,481 subjects via deep learning. Front. Med. 7, 1157 (2021)
21. Suzuki, T., Hosseini, A., Li, J.S., Gill, T.J., IV., Li, G.: In vivo patellar tracking and patellofemoral cartilage contacts during dynamic stair ascending. J. Biomech. 45(14), 2432–2437 (2012)
22. Tsai, C.H., Kiryati, N., Konen, E., Eshed, I., Mayer, A.: Knee injury detection using MRI with efficiently-layered network (ELNet). In: Medical Imaging with Deep Learning, pp. 784–794. PMLR (2020)
23. Veličković, P., Cucurull, G., Casanova, A., Romero, A., Liò, P., Bengio, Y.: Graph attention networks. In: International Conference on Learning Representations (ICLR), pp. 1–12 (2017)
24. Zhuang, Z., et al.: Knee cartilage defect assessment by graph representation and surface convolution. arXiv preprint arXiv:2201.04318 (2022)

DeepCRC: Colorectum and Colorectal Cancer Segmentation in CT Scans via Deep Colorectal Coordinate Transform

Lisha Yao[1,2], Yingda Xia[3(✉)], Haochen Zhang[1,2], Jiawen Yao[3], Dakai Jin[3],
Bingjiang Qiu[1], Yuan Zhang[1], Suyun Li[1,2], Yanting Liang[1,2],
Xian-Sheng Hua[3], Le Lu[3], Xin Chen[2,4], Zaiyi Liu[1,2], and Ling Zhang[3]

[1] Guangdong Provincial People's Hospital, Guangzhou, China
[2] South China University of Technology, Guangzhou, China
[3] Alibaba Group, Hangzhou, China
yingda.xia@alibaba-inc.com
[4] Guangzhou First People's Hospital, Guangzhou, China

Abstract. We propose DeepCRC, a topology-aware deep learning-based approach for automated colorectum and colorectal cancer (CRC) segmentation in routine abdominal CT scans. Compared with MRI and CT Colonography, regular CT has a broader application but is more challenging. Standard segmentation algorithms often induce discontinued colon prediction, leading to inaccurate or completely failed CRC segmentation. To tackle this issue, we establish a new 1D colorectal coordinate system that encodes the position information along the colorectal elongated topology. In addition to the regular segmentation task, we propose an auxiliary regression task that directly predicts the colorectal coordinate for each voxel. This task integrates the global topological information into the network embedding and thus improves the continuity of the colorectum and the accuracy of the tumor segmentation. To enhance the model's architectural ability of modeling global context, we add self-attention layers to the model backbone, and found it complementary to the proposed algorithm. We validate our approach on a cross-validation of 107 cases and outperform nnUNet by an absolute margin of 1.3% in colorectum segmentation and 8.3% in CRC segmentation. Notably, we achieve comparable tumor segmentation performance with the human inter-observer (DSC: 0.646 vs. 0.639), indicating that our method has similar reproducibility as a human observer.

Keywords: Colorectal cancer · Colorectal coordinate · Segmentation

1 Introduction

Colorectal cancer (CRC) is the third most common cancer and the second leading cause of cancer-related death worldwide [16]. Computed tomography (CT),

---

L. Yao and Y. Xia—Equal contribution.

© The Author(s), under exclusive license to Springer Nature Switzerland AG 2022
L. Wang et al. (Eds.): MICCAI 2022, LNCS 13433, pp. 564–573, 2022.
https://doi.org/10.1007/978-3-031-16437-8_54

Table 1. A conceptual comparison between our work and existing studies on colorectal cancer segmentation.

Studies	Modality	Scan range	2D/3D	Preparation	Cost	Challenge
[7,24]	MRI	rectum	2D	No	High	Low
[5,8,15]	MRI	rectum	3D	No	High	Low
[10,13]	CTC	colon+rectum	2D	Yes	Medium	Medium
Decathlon [2] Task08 [6,17]	CT	colon+rectum	3D	Yes	Low	High
Ours	CT	colon+rectum	3D	No	Low	High

which provides location and morphology information of CRC, is routinely performed for cancer detection, diagnosis, staging, surgical planning, and treatment response monitoring [3], as well as structural evaluation of the entire colorectum. Automatic segmentation of colorectum, especially CRC, is the key to achieving automation of these clinical tasks, which can greatly increase the efficiency, reproducibility, and potentially accuracy of the entire clinical workflow.

In this paper, we aim to provide the first exploration of automated segmentation of the colorectum (colon and rectum) and CRC in routine abdominal CT scans (without bowel preparation). The comparison between our work and existing studies in CRC segmentation is shown in Table 1. Compared with MRI [5,7,8,15,24], CT is much less costly and is generally superior to MRI for the hollow viscera (colon). In clinical practice, MRI mainly allows staging rectal cancer only. For CT Colonography (CTC), it is still not widely implemented because bowel preparation and distention is time-consuming and sometimes causes adverse events, such as examination-related pain, and vasovagal syncope or presyncope [21]. The CRC segmentation task in Medical Segmentation Decathlon [2] uses CT but still after bowel preparation with barium coating for contrast enhancement. As a result, the successful segmentation of the colorectum and tumor in routine abdominal CT without bowel preparation will have a broader application and impact.

However, this task is also more challenging due to the following reasons. (i) The colorectum takes up a large space in the abdomen, and it is always hard to be continually traced caused by the mixing of other organs such as the small intestine. (ii) CRCs are usually small and hard to distinguish from the contents of colorectum in regular CT scans (without bowel preparation), compared with a) CT Colonography, which has clean and adequate colon distention by laxative purgation and colonic insufflation, and b) pelvic MRI, which only contains the rectum structure. Furthermore, the discontinued segmentation of colorectum might eventually cause the misdetection of CRC.

To tackle these two challenges, we propose *Deep Colorectal Coordinate Transform* to improve the accuracy of colorectum and CRC segmentation at the same time. Topologically, the colorectum has a single-path and continual structure that extends between the caecum and the rectum. This special pattern motivates us to propose an auxiliary voxel-wise regression problem to improve segmentation continuity. As shown in an illustrative example in Fig. 1, we set up a new one-dimensional coordinate system based on the centerline of the colorectum. Technically, we transform the 3D voxel space into 1D colorectal coordinate space by projecting each foreground colorectum voxel into the 1D coordinate system, thus obtaining a normalized scalar for each voxel. In addition to the voxel-wise classification in conventional segmentation task formulation, we directly regress the relative position, *i.e.* the coordinate value in this coordinate system, of each foreground colorectum voxel. This auxiliary task forces the network to learn the global structural information of the colorectum and the positional information of CRCs, and thus helps the network to achieve better segmentation performance. In terms of network architecture design, we add self-attention layers with positional embedding to enhance the ability to model global context. This design not only provides the architectural basis for the proposed coordinate regression task, but improves the ability of tumor and non-tumor distinction as well.

Our approach is related to recent advance in 3D medical image segmentation with deep networks [6,11,14] and approaches beyond the voxel space, such as boundary projection [12], distance transform [20], and mesh models [22]. For example, Ni et al. proposed a 3D segmentation framework with an elastic boundary projection to obtain the 2D surface of subjects [12]. Yao et al. learned 3D geometry of organs using a 3D mesh representation for better segmentation performance [22]. These methods take full advantage of topology information and achieve promising segmentation accuracy. Moreover, recent works in attention models [4,18,19] also motivate us to enhance the global context of 3D CNNs with self-attention layers.

We validate our proposed method on an in-house dataset, including 107 CT scans with manual colorectum and CRC annotations. Our approach outperforms a strong baseline nnUNet [6] by an absolute DSC margin of 1.3% in colorectum segmentation and 8.3% in CRC segmentation. Moreover, the CRC segmentation performance of our approach (DSC = 0.646) is comparable with the inter-observer variability (DSC = 0.635), illustrating a strong potential for clinical application.

2   Method

Problem Statement. We aim at colorectum and tumor (CRC) segmentation in contrast-enhanced CT scan. For each patient, we have an image $\mathbf{X}$ and its corresponding label $\mathbf{Y}$ in the venous phase. We denote the whole dataset as $S = \{(\mathbf{X_i}, \mathbf{Y_i})|i = 1, 2, ..M\}$, where $\mathbf{X_i} \in \mathbb{R}^{H_i \times W_i \times D_i}$ is a 3D volume representing the CT scan of the i-th patient. $\mathbf{Y_i} \in \mathcal{L}^{H_i \times W_i \times D_i}$ is a voxel-wise annotated label with the same three dimensional size (H_i, W_i, D_i) as $\mathbf{X_i}$. $\mathcal{L} = \{0, 1, 2\}$ represents our segmentation targets, *i.e.*, background, colorectum, and tumor (CRC).

2.1   Colorectal Coordinate Transform

In this section, we discuss how to set up the colorectal coordinate system and how to transform each voxel in $\mathbf{Y}$ into the coordinate system. The output of this

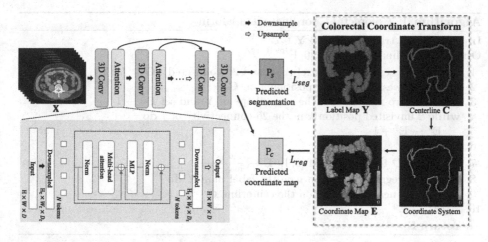

Fig. 1. The pipeline of the proposed method. Our network has an encoder-decoder architecture with self-attention layers during downsampling. In the training phase, we transform the label map $\mathbf{Y}$ into the groundtruth coordinate map $\mathbf{E}$ via "Colorectal Coordinate Transform". Our network takes the input of 3D CT images and outputs the segmentation $\mathbf{P}_s$ (supervised by $\mathcal{L}_{seg}$) and the voxel-wise coordinate map $\mathbf{P}_c$ (supervised by $\mathcal{L}_{reg}$). In the testing phase, we only use $\mathbf{P}_s$ as the segmentation prediction.

process is the groundtruth coordinate map $\mathbf{E}$ which is only used in the training phase. The overall algorithm is described in Algorithm 1 and visually illustrated in Fig. 1 "Colorectal Coordinate Transform".

First, we extract the centerline $\mathbf{C}$ of the colorectum based on the ground truth label map $\mathbf{Y}$. This process is denoted as $\mathbf{C} = f_{cl}(\mathbf{Y})$, where f_{cl} is a centerline extraction algorithm and $\mathbf{C}$ is the extracted 3D centerline image of the same size as $\mathbf{X}$ and $\mathbf{Y}$. We use a robust centerline extraction algorithm [9] to avoid the false centerlines easily produced by the irregular colorectum boundaries. $\mathbf{C}$ is also guaranteed to be one-voxel thick with 26-connectivity. The automated algorithm might fail to extract the correct centerline of some regions where the bowel has severe adhesion. We correct these centerlines semi-automatically by erasing to split these adhesion in $\mathbf{Y}$ and rerunning the algorithm. An example of the extracted $\mathbf{C}$ is shown in Fig. 1 (Centerline $\mathbf{C}$).

Second, we build a colorectal coordinate map $\mathbf{E}$ which is initialized as a all-zero matrix with the same shape as $\mathbf{Y}$. Then we find the lowest foreground position j on the centerline $\mathbf{C}$. We use this position as the starting point to trace the centerline and mark it incrementally on $\mathbf{E}$, which is further normalized to the range of $[0, 1]$ (Fig. 1 Coordinate system).

Finally, we propagate the coordinates along the centerline to the foreground voxels in $\mathbf{Y}$. For each foreground position p, we find the nearest point q on the centerline, and update $\mathbf{E}^p$ with the same coordinate of q on the centerline. After this step, we transform the groundtruth label $\mathbf{Y} \in$ into a coordinate map $\mathbf{E} \in [0, 1]^{H \times W \times D}$ (Fig. 1 Coordinate map $\mathbf{E}$).

Algorithm 1. Colorectal Coordinate Transform

Input: Ground truth label map $\mathbf{Y} \in \mathcal{L}^{H \times W \times D}$
Output: Coordinate map $\mathbf{E} \in [0,1]^{H \times W \times D}$

1: Extract the centerline (1-voxel thick and 26-connected): $\mathbf{C} = f_{cl}(\mathbf{Y})$
2: Find the lowest foreground position j on $\mathbf{C}$
3: Initialize zero map $\mathbf{E}$ with the same shape of $\mathbf{Y}$ and set $\mathbf{E}^j \leftarrow 1$
4: **while** $\exists$ unvisited position k in the 26-connectivity of j **do**
5:     $\mathbf{E}^k \leftarrow \mathbf{E}^j + 1$ ▷ Trace the centerline and mark it incrementally
6:     $j \leftarrow k$
7: Normalize D to the range of $[0,1]$: $\mathbf{E} \leftarrow \frac{\mathbf{E}}{\max_i(\mathbf{E}^i)}$.   ▷ Coordinate system
8: **for** each foreground position p on the label map $\mathbf{Y}$ **do**
9:     Find its nearest point q on the centerline C
10:     $\mathbf{E}^p \leftarrow \mathbf{E}^q$ ▷ Project the coordinates to each foreground voxel
11: **return** $\mathbf{E}$

2.2 Network Training

Figure 1 shows the training diagram of the proposed method. Our network has two outputs, i.e., the regular segmentation prediction $\mathbf{P}_s$ and the coordinate map prediction $\mathbf{P}_c$. Following nnUNet [6], the segmentation loss (denoted as $\mathcal{L}_{seg}$) is defined as a summation of cross-entropy loss and Dice loss [11]. Additionally, we define a regression loss $\mathcal{L}_{reg}$ that minimizes the difference between the predicted coordinate map $\mathbf{P}_c$ and the generated coordinate map $\mathbf{E}$.

$$\mathcal{L}_{reg} = \sum_j ||\mathbf{P}_c^j - \mathbf{E}^j||^2 \tag{1}$$

where j is the j-th voxel of $\mathbf{E}$. With a controlling loss scaler α, the final loss function $\mathcal{L}$ is defined as:

$$\mathcal{L} = \mathcal{L}_{seg} + \alpha \mathcal{L}_{reg} \tag{2}$$

In the testing phase, the trained network takes only an image $\mathbf{X}$ as the input and simply keeps the segmentation prediction $\mathbf{P}_s$ as the final output. The process of colorectal coordinate transform is not needed.

2.3 Network Architecture

In terms of architectural improvement, we integrate the global self-attention layer to enhance the model's ability to model global context. Our proposed auxiliary coordinate regression loss (Eq. 1) requires the network to understand of the colorectum's topology globally. However, vanilla UNet-based segmentation networks heavily rely on local textual change and have a limited receptive field. Thus, an architectural improvement in the global context is desired.

As shown in Fig. 1, our network has an encoder-decoder architecture with skip connections. We add self-attention layers after each downsampling block in the encoder. In each self-attention layer, we first downsample the feature map to a fixed spatial size (H_t, W_t, D_t) and reshape the feature map to obtain N tokens $(N = H_t \times W_t \times D_t)$. We then add a learnable positional embedding to the tokens before forwarding the tokens to multi-head attention layers. Finally, we reshape the tokens back to (H_t, W_t, D_t) and upsample it to the original size of the feature map.

3   Experiments

3.1   Dataset and Annotation

In this study, we retrospectively collected 3D volumetric venous phase CT from 107 patients (including 75 males and 32 females, aged from 30 to 89 years old) with colorectal cancers. The patients were injected with the contrast agents at the rate of 2.5–3.5 ml/s and scanned by a GE or Philips CT scanner operated at 120 kVp and 130–250 mAs. The median voxel spacing of the dataset is $0.78 \times 0.78 \times 5$ mm^3. The colorectum and CRC were manually segmented by an experienced radiologist (10-yr) using ITK-SNAP software [23]. More specifically, the colorectum was carefully traced slice-by-slice and annotated from scratch manually, which took about one hour for each volume on average. The CRC was annotated by referring to the corresponding clinical and pathological reports, as well as other CT phases if necessary, and further checked by a senior radiologist specialized in CRC imaging for 23 years. Additionally, to test the inter-observer variability of CRC segmentation, all 107 cases were detected and delineated by another medical student (2-yr in CRC imaging) with only venous phase CT provided.

3.2   Implementation Details

We conduct five-fold cross-validation on 107 cases. In the training phase, we resample all images to the spacing of (2 mm, 2 mm, 5 mm) and randomly crop patches of $(160, 160, 80)$ voxels for the network input. Our approach is built on the nnUNet [6] framework with 5 downsampling blocks and 5 upsampling blocks with skip connections. The self-attention layers are appended after each downsampling block where the spatial sizes are reduced to $(10, 10, 10)$, and thus, the number of tokens is 1000 for the multi-head attention. We use RAdam optimizer with an initial learning rate of 0.001 and polynomial learning decay. The batch size is 2 and the total iteration is 75000. Standard augmentations including random flip, random rotation, and 3D elastic transformation are utilized. We set the regression loss scaler $\alpha = 50$ for the best model performance. We also set $\mathbf{E} \leftarrow \mathbf{E}+1$ and thus $\mathbf{E} \in [1, 2]$ to increase the coordinate value difference between the coordinate starting point (i.e., 1) and the background voxels (i.e., 0). In the testing phase, we resample the test sample to the training spacing and resample the model prediction to the original spacing. We use the sliding-window testing scheme with a stride of $(80, 80, 40)$ voxels. The average testing time for each volume is approximately 10 s on an Nvidia Tesla V100 16G GPU.

Table 2. Results on 5-fold cross-validation of the dataset. We compare the results of nnUNet, nnUNet with self-attention blocks (denoted as "nnUNet+self-attn."), nnUNet with regression loss ("nnUNet+$\mathcal{L}_{reg}$"), and our DeepCRC (combined). Results are reported as mean±std. The last two rows demonstrate the leading performances on the Decathlon [2] challenge (CRC segmentation on CT with bowel preparation).

	Methods	DSC↑	MSD(mm)↓	HD$_{95}$(mm)↓	TDR↑
Colorectum	nnUNet	0.849 ± 0.072	1.19 ± 1.13	6.08 ± 5.56	–
	nnUNet+self-attn	0.851 ± 0.066	1.29 ± 1.14	6.40 ± 5.92	–
	nnUNet+$\mathcal{L}_{reg}$	0.864 ± 0.060	1.08 ± 1.02	5.03 ± 4.91	–
	DeepCRC	0.862 ± 0.058	1.07 ± 0.96	5.07 ± 4.74	–
Tumor	nnUNet	0.563 ± 0.306	14.6 ± 38.4	29.7 ± 46.5	0.841
	nnUNet+self-attn.	0.616 ± 0.287	20.4 ± 49.3	33.0 ± 55.2	0.869
	nnUNet+$\mathcal{L}_{reg}$	0.617 ± 0.289	17.6 ± 42.5	30.4 ± 49.9	0.869
	DeepCRC	0.646 ± 0.275	12.5 ± 36.5	23.3 ± 43.4	0.879
	Inter-observer	0.639 ± 0.280	15.6 ± 40.5	26.1 ± 48.0	0.879
Decathlon-Task08	nnUNet [1]	0.583	–	–	–
Colon cancer [2]	Swin UNETR [1]	0.595	–	–	–

Note: The median MSD and HD$_{95}$ of DeepCRC are 0.62 mm and 4.12 mm; the inter-observer variability are 1.38 mm and 5.75 mm. The large mean values of MSD and HD$_{95}$ are mainly caused by a small proportion of segmentations with larger errors.

3.3   Results

Quantitative Results. Dice-Sørensen coefficient (DSC), mean surface distance (MSD), Hausdorff distance (HD$_{95}$), and tumor detection rate (TDR) are used to evaluate the colorectum and tumor segmentation, and results over the 5-fold cross-validation of the whole dataset are reported. TDR measures the ratio of the cases that tumor is correctly detected in the dataset, where we use DSC > 0.1 as the criterion for successful detection. Table 2 compares our approach (DeepCRC) with nnUNet [6] and two ablation configurations of our contributions, denoted as "nnUNet+self-attention" and "nnUNet+$\mathcal{L}_{reg}$" respectively. In terms of the colorectum segmentation, the performances with the regression loss (the last two approaches) are higher than the segmentation-loss-only counterpart (the first two approaches), mainly because our auxiliary task to predict the colorectal coordinate provides another topology and location information for colorectum segmentation. As for the tumor segmentation, both self-attention and regression loss improve the segmentation performance (about 5% absolute improvement each in the DSC score). Moreover, these two contributions are complementary to each other, resulting in an even better performance (another 3% absolute improvement) when combined together, reaching a DSC score of 0.646.

In addition, for the state-of-the-art colon cancer segmentation results on a different but comparable dataset (decathlon challenge Task08 [2]), nnUNet is still among the top performers, with a DSC score of 0.583 [1] in CT imaging with bowel preparation (126 training data), which is slightly higher than nnUNet's DSC score of 0.563 on our CT imaging without bowel preparation (107 training

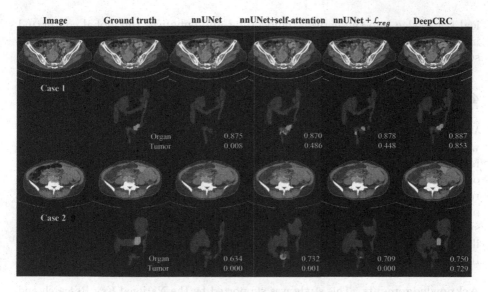

Fig. 2. The visual illustration of colorectum (red) and colorectal tumor (green) segmentation. The DSC scores of the organ and the tumor are marked at the bottom-right of each result. We compare our method (DeepCRC) with the nnUNet baseline and two other ablation experiments in Table 2. In the first case, our approach improves the continuity of the colorectum segmentation and the quality of the tumor segmentation. In the second case, our approach successfully detects the tumor while the other approaches fail, because we correctly predict the colon topology around the tumor. (Color figure online)

data). Our approach (DSC = 0.646) evidently outperforms the strong baseline nnUNet by over 8% in tumor DSC and has reached within the human interobserver variability (0.639).

Qualitative Analysis. We analyze the reason for our improvements by visual illustration of two cases segmented by different method configurations, i.e., nnUNet, +self-attention, +$\mathcal{L}_{reg}$, and DeepCRC (Fig. 2). In the first case, the nnUNet baseline predicts discontinuous colorectum while our DeepCRC improves the integrity of the colorectum prediction. The tumor predictions of the two ablation configurations are only partially correct (about 0.4 in DSC score) due to the discontinuity. Our approach predicts a relatively intact colorectum and an accurate tumor extent. The second case is more challenging due to the irregular colorectum shape. The baseline approaches all fail to detect the tumor, but our approach succeeds because our proposed method can reliably predict a continued topology around the tumor region and distinguishe the tumor from the organ and the background.

From the illustrated qualitative results, we hypothesize that the tumor will affect the appearance of the colorectum and make the affected region harder to be distinguished from non-colorectum regions. Our approach integrates the topological knowledge into the network embedding and enhances the global contextual information. This is beneficial to the improvement of continuity in colorectum segmentation and especially the ability to detect the colorectal tumor.

4   Conclusion

We propose DeepCRC, a colorectum and colorectal tumor segmentation framework in regular contrast-enhanced CT. We introduce an additional auxiliary regression task to directly predict the relative position of each voxel in the colorectal topology and self-attention layers to model global context. Experimental results show that when trained on only small-sized ($n < 100$) data, our approach outperforms nnUNet with improved integrity in colorectum segmentation and substantially better accuracy in tumor segmentation, achieving an accuracy similar to the inter-observer variability. Our approach could serve as the core for the fully-automated diagnosis, treatment, and follow-up of CRC using CT imaging.

Acknowledgements. This study was supported by the National Key Research and Development Program of China [grant number 2021YFF1201003], the National Science Fund for Distinguished Young Scholars [grant number 81925023], the National Natural Scientific Foundation of China [grant number 82072090].

References

1. Medical Segmentation Decathlon, Challenge Leaderboard. https://decathlon-10. grand-challenge.org/evaluation/challenge/leaderboard/
2. Antonelli, M., et al.: The medical segmentation decathlon. arXiv preprint arXiv:2106.05735 (2021)
3. Argilés, G., et al.: Localised colon cancer: ESMO clinical practice guidelines for diagnosis, treatment and follow-up. Ann. Oncol. **31**(10), 1291–1305 (2020)
4. Dosovitskiy, A., et al.: An image is worth 16x16 words: transformers for image recognition at scale. ICLR (2021)
5. Huang, Y.J., et al.: 3-D Roi-aware U-net for accurate and efficient colorectal tumor segmentation. IEEE Trans. Cybern. **51**(11), 5397–5408 (2020)
6. Isensee, F., Jaeger, P.F., Kohl, S.A., Petersen, J., Maier-Hein, K.H.: nnU-Net: a self-configuring method for deep learning-based biomedical image segmentation. Nat. Methods **18**(2), 203–211 (2021)
7. Jian, J., et al.: Fully convolutional networks (FCNs)-based segmentation method for colorectal tumors on T2-weighted magnetic resonance images. Australas. Phys. Eng. Sci. Med. **41**(2), 393–401 (2018). https://doi.org/10.1007/s13246-018-0636-9
8. Jiang, Y., et al.: ALA-Net: adaptive lesion-aware attention network for 3D colorectal tumor segmentation. IEEE Trans. Med. Imaging **40**(12), 3627–3640 (2021)
9. Jin, D., Iyer, K.S., Chen, C., Hoffman, E.A., Saha, P.K.: A robust and efficient curve skeletonization algorithm for tree-like objects using minimum cost paths. Pattern Recogn. Lett. **76**, 32–40 (2016)
10. Liu, X., et al.: Accurate colorectal tumor segmentation for CT scans based on the label assignment generative adversarial network. Med. Phys. **46**(8), 3532–3542 (2019)

11. Milletari, F., Navab, N., Ahmadi, S.A.: V-net: fully convolutional neural networks for volumetric medical image segmentation. In: 2016 Fourth International Conference on 3D Vision (3DV), pp. 565–571. IEEE (2016)

12. Ni, T., Xie, L., Zheng, H., Fishman, E.K., Yuille, A.L.: Elastic boundary projection for 3D medical image segmentation. In: Proceedings of the IEEE/CVF Conference on Computer Vision and Pattern Recognition, pp. 2109–2118 (2019)

13. Pei, Y., Mu, L., Fu, Y., He, K., Li, H., Guo, S., Liu, X., Li, M., Zhang, H., Li, X.: Colorectal tumor segmentation of CT scans based on a convolutional neural network with an attention mechanism. IEEE Access **8**, 64131–64138 (2020)

14. Ronneberger, O., Fischer, P., Brox, T.: U-Net: convolutional networks for biomedical image segmentation. In: Navab, N., Hornegger, J., Wells, W.M., Frangi, A.F. (eds.) MICCAI 2015. LNCS, vol. 9351, pp. 234–241. Springer, Cham (2015). https://doi.org/10.1007/978-3-319-24574-4_28

15. Soomro, M.H., et al.: Automated segmentation of colorectal tumor in 3D MRI using 3D multiscale densely connected convolutional neural network. J. Healthc. Eng. 2019 (2019)

16. Sung, H., et al.: Global cancer statistics 2020: GLOBOCAN estimates of incidence and mortality worldwide for 36 cancers in 185 countries. CA Cancer J. Clin. **71**(3), 209–249 (2021)

17. Tang, Y., et al.: Self-supervised pre-training of swin transformers for 3D medical image analysis. arXiv preprint arXiv:2111.14791 (2021)

18. Vaswani, A., et al.: Attention is all you need. arXiv preprint arXiv:1706.03762 (2017)

19. Wang, X., Girshick, R., Gupta, A., He, K.: Non-local neural networks. In: Proceedings of the IEEE Conference on Computer Vision and Pattern Recognition, pp. 7794–7803 (2018)

20. Wang, Y., et al.: Deep distance transform for tubular structure segmentation in CT scans. In: Proceedings of the IEEE/CVF Conference on Computer Vision and Pattern Recognition, pp. 3833–3842 (2020)

21. Wolf, A.M., et al.: Colorectal cancer screening for average-risk adults: 2018 guideline update from the American cancer society. CA Cancer J. Clin. **68**(4), 250–281 (2018)

22. Yao, J., Cai, J., Yang, D., Xu, D., Huang, J.: Integrating 3D geometry of organ for improving medical image segmentation. In: Shen, D., et al. (eds.) MICCAI 2019. LNCS, vol. 11768, pp. 318–326. Springer, Cham (2019). https://doi.org/10.1007/978-3-030-32254-0_36

23. Yushkevich, P.A., et al.: User-guided 3D active contour segmentation of anatomical structures: significantly improved efficiency and reliability. Neuroimage **31**(3), 1116–1128 (2006)

24. Zheng, S., et al.: MDCC-Net: multiscale double-channel convolution U-Net framework for colorectal tumor segmentation. Comput. Biol. Med. **130**, 104183 (2021)

Graph Convolutional Network with Probabilistic Spatial Regression: Application to Craniofacial Landmark Detection from 3D Photogrammetry

Connor Elkhill[1,2,3]([✉]), Scott LeBeau[3], Brooke French[3], and Antonio R. Porras[1,2,3,4,5]

[1] Department of Biostatistics and Informatics, Colorado School of Public Health, University of Colorado Anschutz Medical Campus, Aurora, CO 80045, USA
connor.2.elkhill@cuanschutz.edu
[2] Computational Bioscience Program, School of Medicine, University of Colorado Anschutz Medical Campus, Aurora, CO 80045, USA
[3] Department of Pediatric Plastic and Reconstructive Surgery, Children's Hospital Colorado, Aurora, CO 80045, USA
[4] Department of Pediatric Neurosurgery, Children's Hospital Colorado, Aurora, CO 80045, USA
[5] Department of Pediatrics, School of Medicine, University of Colorado Anschutz Medical Campus, Aurora, CO 80045, USA

Abstract. Quantitative evaluation of pediatric craniofacial anomalies relies on the accurate identification of anatomical landmarks and structures. While segmentation and landmark detection methods in standard clinical images are available in the literature, image-based methods are not directly applicable to 3D photogrammetry because of its unstructured nature consisting in variable numbers of vertices and polygons. In this work, we propose a graph-based convolutional neural network based on Chebyshev polynomials that exploits vertex coordinates, polygonal connectivity, and surface normal vectors to extract multi-resolution spatial features from the 3D photographs. We then aggregate them using a novel weighting scheme that accounts for local spatial resolution variability in the data. We also propose a new trainable regression scheme based on the probabilistic distances between each original vertex and the anatomical landmarks to calculate coordinates from the aggregated spatial features. This approach allows calculating accurate landmark coordinates without assuming correspondences with specific vertices in the original mesh. Our method achieved state-of-the-art landmark detection errors.

Keywords: Graph convolutional neural network · Anatomical landmark detection · 3D photogrammetry

1 Introduction

Quantitative assessment of craniofacial anomalies plays an important role both in the diagnosis and in the pre- and post-treatment evaluation of several pediatric developmental pathologies [1–3]. During the last decade, three-dimensional (3D) photogrammetry has emerged as a popular medical imaging modality for pediatric craniofacial evaluation [4, 5] because of its low cost, non-invasive nature and its fast acquisition. Most

© The Author(s), under exclusive license to Springer Nature Switzerland AG 2022
L. Wang et al. (Eds.): MICCAI 2022, LNCS 13433, pp. 574–583, 2022.
https://doi.org/10.1007/978-3-031-16437-8_55

data-driven craniofacial analysis methods rely on the segmentation of specific regions or the accurate identification of anatomical landmarks [6, 7]. While segmentation and landmark detection methods in traditional medical images have been widely explored in the literature [8–10] they are not applicable to 3D photogrammetry because it provides a graph-like surface representation with vertices and edges instead of a traditional voxel-based image representation.

Several approaches have been presented for segmentation and landmark identification in point-based data representations. Spectral domain filters in graph-based representations suggested by Kipf and Welling [11] or Defferrard et al. [12] have been successfully used in a variety of segmentation tasks in the biomedical domain [13–15] but they require edges connecting nodes. Since not all data types have existing edges, point cloud-based approaches were developed to consider each node in the dataset independently, as seen in PointNet [16] and PointCNN [17]. However, these methods do not incorporate any connections between nodes and are therefore unable to consider the local geometry of the dataset. This was improved with the subsequent PointNet++ [18], which estimated spatial features by partitioning points within a given radius to construct a hierarchical neural network for node segmentation tasks. Although this network structure provides robust state-of-the-art performance on various segmentation tasks, it is only applicable to point clouds and cannot exploit explicit data structure information in the form of point connections or polygonal connectivity. For further information on state-of-the-art point cloud based learning methods, please refer to [19].

In this work, we propose a novel graph-based landmark detection architecture that exploits vertex coordinates and their connectivity available from clinical 3D photographs to avoid hierarchical neighborhood partitioning as in PointNet++. We use Chebyshev polynomials [12] to quantify multi-resolution spatial features at every vertex in our mesh only considering its connected neighborhood. Then, we use a weighting scheme dependent on the local data density at every surface location to aggregate the spatial features at each vertex. Finally, we propose a novel probabilistic regression framework that uses these aggregated spatial features to calculate landmark locations without assuming correspondences between landmarks and vertices. Our architecture allows leveraging the inherent structure of the mesh while enabling landmark coordinates to be identified at any location in the original mesh.

2 Materials and Methods

In this section we first describe the dataset used in our work. Then, we provide details about our 3D photogrammetry pre-processing and graph construction. Finally, we present the novel deep learning architecture that we propose for anatomical landmark identification.

2.1 Data

After IRB approval by the local ethics review committee at the University of Colorado Anschutz Medical Campus (#20-1563), we collected retrospectively 982 3D photographs acquired at Children's Hospital Colorado between 2014 and 2021 using the 3DMD-head System (3dMD, Atlanta, GA). These patients had a diagnosis of craniosynostosis (70.4%), cleft lip or palate (2.4%), deformational plagiocephaly (19.8%), or other craniofacial anomalies (7.4%). The average age was 3.53 ± 2.35 years. All photographs were manually annotated by an expert radiographer with the location of the 13 craniofacial landmarks represented in Fig. 1.

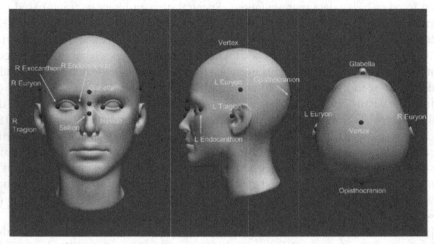

Fig. 1. Landmark locations. L and R indicate left and right, respectively. Red landmarks indicate Bookstein type I or II, whereas blue landmarks indicate Bookstein type III [20]. (Color figure online)

2.2 Graph Construction

To reduce the number of vertices and polygons in our dataset, we decimated each original 3D photograph using quadratic error estimation [21]. We dynamically adjusted the target reduction ratio to approximate the final number of vertices to no less than 7,000. Then, we constructed a graph from the vertices and polygons so that each graph node corresponds to one vertex in the mesh and each graph edge connects nodes that share at least one polygon. Our average graph size was $7,998 \pm 4,096$ nodes and six edges per node. For each graph node, we stored its coordinates and the unit surface normal vector at its position.

2.3 Graph Convolutional Network

We represented each graph $G(V, E)$ with N nodes using a feature matrix $B \in \mathbb{R}^{N \times 6}$ containing the coordinates and normal vectors at each vertex, and a binary adjacency matrix $A \in \mathbb{R}^{N \times N}$ representing node connections or edges. Then, we designed a graph convolutional neural network (GCN) that operates on both matrices and that is represented in Fig. 2.

We used spectral graph convolutions in the Fourier domain estimated using Chebyshev polynomials to extract spatial features at each graph node [12]. Given a graph $G(V, E)$ we first compute the normalized graph Laplacian $\tilde{L} = D^{-\frac{1}{2}} A D^{-\frac{1}{2}}$, where A is the adjacency matrix of the graph and D is the degree matrix computed by taking the row-wise sum of the adjacency matrix. The output spatial features from a given convolutional layer i with input features b_i is then computed as:

$$h_i = \sum_{k=0}^{K-1} \theta_k T_k\left(\tilde{L}\right) b_i, \tag{1}$$

where $T_k\left(\tilde{L}\right)$ is the kth order Chebyshev polynomial evaluated using the normalized Laplacian $\tilde{L}$ and θ_k are the trainable weights for the polynomial of order k.

The order of the Chebyshev polynomials indicates the maximum degree of separation between neighboring nodes used to approximate convolutional features. Since the distance between neighboring nodes (the local spatial resolution of the graph) may vary at different spatial location, the same polynomial degree may extract features at different spatial resolutions depending on the local data density. Hence, we propose a graph-based convolutional architecture that extracts spatial features at each node using Chebyshev polynomials of different orders. As presented in Fig. 2, we propose two parallel convolutional networks that quantify features at two different resolution levels: a high spatial resolution using low-order polynomials ($k = 3$) and a low spatial resolution using high-order polynomials ($k = 7$). The output of each layer is then activated using a rectified linear unit (ReLU) and followed by graph normalization as detailed in [22].

After multi-resolution feature quantification, we propose a new weighting scheme that acts as a feature selector and integrates the spatial features accounting for the local node density observed in the data. Given the average pairwise Euclidean distance $\overline{d}_n$ between a node n and its neighboring nodes, the weight of the high-resolution features extracted using low-order Chebyshev polynomials was computed as

$$w_n = \frac{\overline{d}_n - \min(\overline{d}_G)}{\max\left(\overline{d}_G\right) - \min(\overline{d}_G)}, \tag{2}$$

where $\overline{d}_G$ contains the average distance of each graph node to its neighbors. Given the low- and high-order spatial features calculated after a graph-convolution h_i^{low} and h_i^{high}, respectively, we aggregate them at each node as

$$h_i = h_i^{low} w_n + h_i^{high}(1 - w_n). \tag{3}$$

Note that this aggregation scheme links the confidence or importance of node features calculated at different spatial resolutions with the local resolution in the original dataset. The aggregated features h_i at each node are then fed to two fully connected layers, the first one followed by ReLU activation and graph normalization. The final layer calculates L features from each node, where L represents the number of output landmarks. We used softmax activation to convert these output features h_c^{FC} to an estimation of the probability

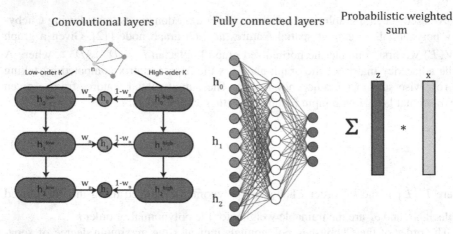

Fig. 2. Proposed network architecture. The spatial features are calculated at different resolutions (using different Chebyshev filter orders k) by the convolutional layers and are aggregated using the data density weights w_n at each node n. The aggregated features h_0, h_1, h_2 at each node are then used by fully connected layers (in white) that perform linear operations on them. All network parameters are shared across all graph nodes. A graph-wise softmax function produces a probabilistic distance matrix z that is used to regress landmark coordinates based on the original graph node coordinates x.

of each node representing each anatomical landmark. Hence, our network calculates at each node

$$z_c = \frac{\exp\left(h_c^{FC}\right)}{\sum_{\forall j}\exp\left(h_j^{FC}\right)}. \tag{4}$$

Finally, instead of using these probabilities to identify the most likely node to represent each landmark (as in most segmentation tasks [16, 17]), we propose to interpret these values as probabilistic distances between nodes and landmarks. This different perspective compared to other state-of-the-art works [14, 18, 23] enables the regression of the Euclidean coordinates of each landmark $\hat{y}_c$ as

$$\hat{y}_c = \sum_{n=1}^{N} x_n z_c, \tag{5}$$

where x_n are the Euclidean coordinates of each node of the graph. Note that this new approach enables the calculation of landmark coordinates that don't necessarily correspond with the coordinates of any graph node. We optimized our network parameters using the following loss function, where L is the number of landmarks, $\hat{y}_c$ and y_c are the estimated and ground-truth landmark coordinates, respectively:

$$\mathcal{L} = \frac{1}{L}\sum_{c}^{L} \left(\hat{y}_c - y_c\right)^2. \tag{6}$$

3 Experiments and Results

3.1 Training Details

We randomly divided our dataset into training (80%, N = 785), validation (10%, N = 98) and test (10%, N = 98) photographs. We performed data augmentation by randomly rotating each image between -30 and $30°$ on each axis a total of five times for each image, resulting in a dataset with 4,710, 588, and 588 training, validation and test images, respectively. Our network was trained with 128 hidden features (h_i in Fig. 2.) using the Adam optimizer with the AMSGrad algorithm [24]. We used a minibatch size of 5, learning rate 0.001, and $\beta_1 = 0.9$ and $\beta_2 = 0.99$. The network was trained for no more than 400 epochs and training was stopped upon convergence of the validation loss. Our network was implemented on a NVIDIA Quadro RTX 4000 GPU system using Python 3.9 and Pytorch Geometric [25]. Model parameters were estimated to utilize 1.558 MB of memory while training and the final model file size is 1.54 MB on the disk. Model evaluation execution time using our testing dataset was 15.07 min for 588 images (1.53 s per image). The trained model can be found at the following link: https://github.com/cuMIP/3dphoto.

3.2 Performance Evaluation

To evaluate the performance of our method, we divided our 13 landmarks into two groups following the Bookstein classification [26]. The first group of landmarks contained the Bookstein types I and II landmarks whose placement is defined by clear anatomical criteria: nasion, sellion, left and right tragion, left and right endocanthion, and left and right exocanthion. The second group included the Bookstein type III landmarks whose placement is arbitrary and highly dependent on the observer: glabella, left and right euryon, opsithocranion, and vertex. Figure 1 shows the classification of each craniofacial landmark. Our network achieved an average landmark detection error in the test dataset of 3.39 ± 3.76 mm for the first landmark group, and 11.30 ± 9.88 mm for the second landmark group (p < 0.001 estimated using a Mann-Whitney U-test).

Table 1. Summary of landmark detection errors (in mm) for the proposed method and all other networks trained in this work. P-values calculated from Mann-Whitney U-test between the proposed method and evaluated alternatives.

Method	Group 1 landmarks error	Group 2 landmarks error
Proposed method	3.39 ± 3.76	11.30 ± 9.88
Single resolution (K = 7)	3.66 ± 4.12 (p < 0.001)	11.46 ± 10.17 (p = 0.358)
Single resolution (K = 3)	4.80 ± 4.96 (p < 0.001)	14.70 ± 10.79 (p < 0.001)
Unweighted aggregation	3.40 ± 3.32 (p = 0.246)	11.61 ± 10.20 (p = 0.151)
PointNet++	6.06 ± 10.89 (p < 0.001)	12.47 ± 15.08 (p = 0.223)
PointNet++ Weighted Regression	4.26 ± 3.03 (p < 0.001)	9.03 ± 8.49 (p < 0.001)

Table 1 shows the comparison of the performance of our network with four other alternative architectures: one that calculates single resolution spatial features (Single resolution in Table 1), one that does not use weighted aggregation when combining spatial features after the convolutional layers (Unweighted aggregation in Table 1), the most popular and best-performing state-of-the-art method for node classification - PointNet++ [18], and a PointNet++ architecture incorporating a probabilistic distance weighted regression scheme for landmark placement (PointNet++ Weighted Regression in Table 1).

4   Discussion

We present a graph convolutional neural network to automatically identify anatomical landmarks from 3D photogrammetry. Our network uses Chebyshev convolutional filters to derive spatial features at different resolutions, and we propose a novel weighting scheme that aggregates those features accounting for the local node density of the data. Our experiments show that our multi-resolution weighting scheme improves landmark placement accuracy as compared to the use of single-resolution features (Single resolution models in Table 1) or simple feature aggregation (Unweighted aggregation model in Table 1). Improvement using weighted as opposed to unweighted aggregation was more substantial in Bookstein type III landmarks located in smoother areas of the head surface represented with lower point density in the 3D photographs, suggesting the importance of accounting for local spatial resolution variability.

We also propose a new probabilistic framework to regress landmark coordinates as a combination of graph node coordinates. This regression does not assume any correspondences between the landmarks and the vertices in the original mesh, which provides an increased landmark placement accuracy as compared to the state-of-the-art method PointNet++ that employs a more traditional node segmentation approach (PointNet++ in Table 1). We also show that our probabilistic framework to regress landmark coordinates improved the landmark placement accuracy when we incorporated it in PointNet++ (PointNet++ Weighted Regression in Table 1).

Our results indicate that our proposed network outperforms the results of PointNet++ in identifying Bookstein type I and II landmarks, which are located at clearly defined anatomical locations. The average errors reported for these landmarks is also within the suggested clinically acceptable accuracy range (<4 mm) [27]. Unlike PointNet++, our network explicitly leverages the data structure in the original 3D photographs represented by polygon connections between vertices instead of performing arbitrary spatial samplings to estimate the local data structure. However, PointNet++ with weighted regression provided more accurate results locating Bookstein type III landmarks whose placement is highly arbitrary and dependent on the observer. Given such variability, interpretation of these results is subjective and may be only explained by random variability in the ground truth. We also speculate that this may be related to global contextual features learned by PointNet++ from the training dataset that may help with landmark placement in areas without distinctive anatomical information.

One limitation of the presented method is its direct dependence from the spatial data resolution derived from the use of Chebyshev filters. While our new multi-resolution spatial feature calculation and aggregation schemes were designed to mitigate this dependency, our method may lose accuracy in datasets with highly variable spatial resolution. Although aggregation of features calculated at additional spatial resolutions could potentially reduce the impact of these challenging scenarios, we will evaluate in our future work how a high variability in spatial resolution may affect the accuracy of our method. An additional limitation of this work is the use of a single expert to define the ground truth landmark locations.

Methods for automatic landmark detection are essential for efficiency and repeatability. While different approaches for craniofacial landmark detection on traditional 2D photographs (or 2D projections of 3D data) are available in the literature [10, 28], there has been a paucity of works for automatic landmark placement in 3D photogrammetry. Compared to traditional images, this modality presents additional challenges related to the variable number of data points between patients and the difficulties establishing anatomical correspondences in large datasets to train accurate models. Our graph-based method was trained and tested on a large dataset of 3D photographs with different numbers of vertices and polygons, and it does not require the establishment of any anatomical correspondences between subjects.

During the last decade, there have been many efforts towards the clinical use of 3D photogrammetry because of its low cost, non-invasive and radiation-free nature [3, 5]. Those efforts have been hindered by the lack of automatic and reproducible analysis methods. This is of particular importance in pediatric populations such as the one used in this study, as they are especially vulnerable to the potentially harmful effects of radiation and/or sedation required in other imaging modalities such as computed tomography and magnetic resonance imaging. The landmark detection model presented in this work can be leveraged to establish anatomical correspondences between subjects in large datasets that are required for statistical analysis and modeling. It also enables the automatic quantification of clinically relevant craniofacial features that can be used for patient screening and diagnosis [3], surgical planning [29], or post-surgical evaluation [6].

5   Conclusion

We presented a novel graph convolutional neural network for the automatic identification of craniofacial landmarks in pediatric clinical 3D photographs. Our method leverages the data structure to extract multi-resolution features, which are combined using a new aggregation scheme that adapts to the local data resolution. In addition, we propose a regression scheme that enables landmark placements at any location in the input surface without assuming landmark correspondences with specific vertices. Our method showed state-of-the-art performance and could be leveraged for a fully automated quantitative clinical assessment of pediatric 3D photographs.

Acknowledgements. CE is supported by the National Library of Medicine (NLM) under project number T15LM009451. ARP was supported by the National Institute of Dental & Craniofacial Research of the National Institutes of Health under Award Number R00DE027993. The content is solely the responsibility of the authors and does not necessarily represent the official views of the National Institutes of Health.

References

1. Trainor, P.A., Richtsmeier, J.T.: Facing up to the challenges of advancing craniofacial research. Am. J. Med. Genet. A **167**(7), 1451–1454 (2015). https://doi.org/10.1002/ajmg.a.37065
2. Wang, J.Y., Dorafshar, A.H., Liu, A., Groves, M.L., Ahn, E.S.: The metopic index: an anthropometric index for the quantitative assessment of trigonocephaly from metopic synostosis. J. Neurosurg. Pediatr. **18**(3), 275–280 (2016). https://doi.org/10.3171/2016.2.PEDS15524
3. Mathijssen, I.M.J.: Updated guideline on treatment and management of craniosynostosis. J. Craniofac. Surg. **32**(1), 371–450 (2021). https://doi.org/10.1097/SCS.0000000000007035
4. Schweitzer, T., Böhm, H., Meyer-Marcotty, P., Collmann, H., Ernestus, R.-I., Krauß, J.: Avoiding CT scans in children with single-suture craniosynostosis. Childs Nerv. Syst. **28**(7), 1077–1082 (2012). https://doi.org/10.1007/s00381-012-1721-0
5. Tanikawa, C., Akcam, M.O., Takada, K.: Quantifying faces three-dimensionally in orthodontic practice. J. Cranio-Maxillofac. Surg. **47**(6), 867–875 (2019). https://doi.org/10.1016/j.jcms.2019.02.012
6. Porras, A.R., et al.: Quantification of head shape from three-dimensional photography for presurgical and postsurgical evaluation of craniosynostosis. Plast. Reconstr. Surg. **144**(6), 1051e–1060e (2019). https://doi.org/10.1097/PRS.0000000000006260
7. Cho, M.-J., et al.: Quantifying normal craniofacial form and baseline craniofacial asymmetry in the pediatric population. Plast. Reconstr. Surg. **141**(3), 380e–387e (2018). https://doi.org/10.1097/PRS.0000000000004114
8. Ma, Q., et al.: Automatic 3D landmarking model using patch-based deep neural networks for CT image of oral and maxillofacial surgery. Int. J. Med. Robot. **16**(3), e2093 (2020). https://doi.org/10.1002/rcs.2093
9. Torosdagli, N., Liberton, D.K., Verma, P., Sincan, M., Lee, J.S., Bagci, U.: Deep geodesic learning for segmentation and anatomical landmarking. IEEE Trans. Med. Imaging **38**(4), 919–931 (2019). https://doi.org/10.1109/TMI.2018.2875814
10. Song, Y., Qiao, X., Iwamoto, Y., Chen, Y.: Automatic cephalometric landmark detection on x-ray images using a deep-learning method. Appl. Sci. **10**(7), 2547 (2020). https://doi.org/10.3390/app10072547
11. Kipf, T.N., Welling, M.: Semi-supervised classification with graph convolutional networks. ArXiv160902907 Cs Stat (2017). Accessed 21 Feb 2022. http://arxiv.org/abs/1609.02907
12. Defferrard, M., Bresson, X., Vandergheynst, P.: Convolutional neural networks on graphs with fast localized spectral filtering. ArXiv160609375 Cs Stat (2017). Accessed 08 Dec 2021. http://arxiv.org/abs/1606.09375
13. Soberanis-Mukul, R.D., Navab, N., Albarqouni, S.: An uncertainty-driven GCN refinement strategy for organ segmentation. Machine Learning for Biomedical Imaging MELBA (2020). arXiv:2012.03352
14. Wolterink, J.M., Leiner, T., Išgum, I.: Graph convolutional networks for coronary artery segmentation in cardiac CT angiography. In: Zhang, D., Zhou, L., Jie, B., Liu, M. (eds.) GLMI 2019. LNCS, vol. 11849, pp. 62–69. Springer, Cham (2019). https://doi.org/10.1007/978-3-030-35817-4_8
15. Parisot, S., et al.: Spectral graph convolutions for population-based disease prediction. In: Descoteaux, M., Maier-Hein, L., Franz, A., Jannin, P., Collins, D.L., Duchesne, S. (eds.) MICCAI 2017. LNCS, vol. 10435, pp. 177–185. Springer, Cham (2017). https://doi.org/10.1007/978-3-319-66179-7_21
16. Qi, C.R., Su, H., Mo, K., Guibas, L.J.: PointNet: deep learning on point sets for 3D classification and segmentation. ArXiv161200593 Cs (2017). Accessed 12 Oct 2021. http://arxiv.org/abs/1612.00593

17. Li, Y., Bu, R., Sun, M., Wu, W., Di, X., Chen, B.: PointCNN: convolution on $\mathcal{X}$-transformed points. ArXiv180107791 Cs (2018). Accessed 21 Feb 2022. http://arxiv.org/abs/1801.07791

18. Qi, C.R., Yi, L., Su, H., Guibas, L.J.: PointNet++: deep hierarchical feature learning on point sets in a metric space. ArXiv170602413 Cs (2017). Accessed 25 Oct 2021. http://arxiv.org/abs/1706.02413

19. Guo, Y., Wang, H., Hu, Q., Liu, H., Liu, L., Bennamoun, M.: Deep learning for 3D point clouds: a survey. IEEE Trans. Pattern Anal. Mach. Intell. **43**(12), 4338–4364 (2021). https://doi.org/10.1109/TPAMI.2020.3005434

20. Caple, J., Stephan, C.N.: A standardized nomenclature for craniofacial and facial anthropometry. Int. J. Legal Med. **130**(3), 863–879 (2015). https://doi.org/10.1007/s00414-015-1292-1

21. Garland, M., Heckbert, P.S.: Surface simplification using quadric error metrics. In: Proceedings of the 24th Annual Conference on Computer Graphics and Interactive Techniques - SIGGRAPH 1997, pp. 209–216 (1997). https://doi.org/10.1145/258734.258849

22. Cai, T., Luo, S., Xu, K., He, D., Liu, T.-Y., Wang, L.: GraphNorm: a principled approach to accelerating graph neural network training. ArXiv200903294 Cs Math Stat (2021). Accessed 17 Feb 2022. http://arxiv.org/abs/2009.03294

23. Li, W., et al.: Structured landmark detection via topology-adapting deep graph learning. In: Vedaldi, A., Bischof, H., Brox, T., Frahm, J.-M. (eds.) ECCV 2020. LNCS, vol. 12354, pp. 266–283. Springer, Cham (2020). https://doi.org/10.1007/978-3-030-58545-7_16

24. Reddi, S.J., Kale, S., Kumar, S.: On the convergence of Adam and beyond. ArXiv190409237 Cs Math Stat (2019). Accessed 24 Feb 2022. http://arxiv.org/abs/1904.09237

25. Fey, M., Lenssen, J.E.: Fast graph representation learning with PyTorch geometric. ArXiv190302428 Cs Stat (2019). Accessed 15 Dec 2021. http://arxiv.org/abs/1903.02428

26. Bookstein, F.L.: "Landmarks," in Morphometric Tools for Landmark Data: Geometry and Biology, pp. 55–87. Cambridge University Press, Cambridge (1992)

27. Yue, W., Yin, D., Li, C., Wang, G., Xu, T.: Automated 2-D cephalometric analysis on X-ray images by a model-based approach. IEEE Trans. Biomed. Eng. **53**(8), 1615–1623 (2006). https://doi.org/10.1109/TBME.2006.876638

28. Torres, H.R., et al.: Anthropometric landmark detection in 3D head surfaces using a deep learning approach. IEEE J. Biomed. Health Inform. **25**(7), 2643–2654 (2021). https://doi.org/10.1109/JBHI.2020.3035888

29. García-Mato, D., et al.: Effectiveness of automatic planning of fronto-orbital advancement for the surgical correction of metopic craniosynostosis. Plast. Reconstr. Surg. - Glob. Open **9**(11), e3937 (2021). https://doi.org/10.1097/GOX.0000000000003937

Dual-Distribution Discrepancy for Anomaly Detection in Chest X-Rays

Yu Cai[1,2(✉)], Hao Chen[3], Xin Yang[2], Yu Zhou[2], and Kwang-Ting Cheng[1,3]

[1] Department of Electronic and Computer Engineering,
The Hong Kong University of Science and Technology, Hong Kong, China
yu.cai@connect.ust.hk
[2] School of Electronic Information and Communications,
Huazhong University of Science and Technology, Wuhan, China
[3] Department of Computer Science and Engineering,
The Hong Kong University of Science and Technology, Hong Kong, China

Abstract. Chest X-ray (CXR) is the most typical radiological exam for diagnosis of various diseases. Due to the expensive and time-consuming annotations, detecting anomalies in CXRs in an unsupervised fashion is very promising. However, almost all of the existing methods consider anomaly detection as a one-class classification (OCC) problem. They model the distribution of only known normal images during training and identify the samples not conforming to normal profile as anomalies in the testing phase. A large number of unlabeled images containing anomalies are thus ignored in the training phase, although they are easy to obtain in clinical practice. In this paper, we propose a novel strategy, Dual-distribution Discrepancy for Anomaly Detection (DDAD), utilizing both known normal images and unlabeled images. The proposed method consists of two modules. During training, one module takes both known normal and unlabeled images as inputs, capturing anomalous features from unlabeled images in some way, while the other one models the distribution of only known normal images. Subsequently, inter-discrepancy between the two modules, and intra-discrepancy inside the module that is trained on only normal images are designed as anomaly scores to indicate anomalies. Experiments on three CXR datasets demonstrate that the proposed DDAD achieves consistent, significant gains and outperforms state-of-the-art methods. Code is available at https://github.com/caiyu6666/DDAD.

Keywords: Anomaly detection · Chest X-ray · Deep learning

1 Introduction

Thanks to the cost-effectiveness and low radiation dose, combined with a reasonable sensitivity to a wide variety of pathologies, chest X-ray (CXR) is the most commonly performed radiological exam [4]. To alleviate radiologists' reading burden and improve diagnosis efficiency, automatic CXR analysis using deep

© The Author(s), under exclusive license to Springer Nature Switzerland AG 2022
L. Wang et al. (Eds.): MICCAI 2022, LNCS 13433, pp. 584–593, 2022.
https://doi.org/10.1007/978-3-031-16437-8_56

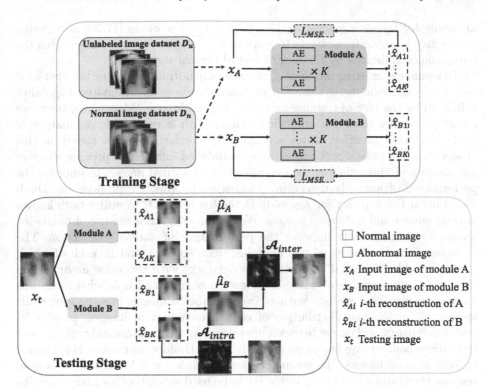

Fig. 1. Illustration of the proposed DDAD. To simplify the visualization, we utilize the AE as the backbone in both modules A and B here.

learning is becoming popular [11,12]. However, annotations of CXRs are difficult to obtain because of the expertise requirements and time-consuming reading, which motivates the development of anomaly detection that requires few or even no annotations.

Previously, most existing methods consider anomaly detection as a one-class classification (OCC) problem [15], where only normal images are utilized for training and samples not conforming to normal profiles are identified as anomalies in the testing phase. Under this setting, extensive methods based on reconstruction [2] or self-supervised learning [7] have been proposed for anomaly detection. Reconstruction-based methods [1,3,6,13,16] have been proven effective. They train the reconstruction networks like variants of autoencoder (AE) to minimize the reconstruction error on normal images, while unseen abnormal images are assumed not able to be reconstructed, and in turn yield larger reconstruction errors. To avoid the reconstruction of anomaly and reduce miss detection, some methods [5,14] utilize Variational AE (VAE) [8] to approximate the normative distribution and perform image restoration iteratively to ensure the output is anomaly-free, yielding higher difference with the abnormal input. However, the iterative restoration process is computationally complex and time-consuming. Recently, some self-supervised methods [10,18,19,22] try to synthesize defects manually for training models to detect irregularities. Some [17,20] also design contrastive learning frameworks to learn more effective representations. But the performance is limited by the lack of real abnormal samples.

In summary, existing methods either use only normal images or use synthetic abnormal images on top of them during training, whose discriminative capability is limited by the lack of training on real abnormal images. Meanwhile, there are plenty of readily available unlabeled images with a reasonable anomaly rate (AR) in clinical practice, which are ignored by existing works. Based on this observation, we raise a problem: whether unlabeled images can provide effective information of abnormalities as a complement of normal images to improve the performance of anomaly detection? Motivated by this, we propose the Dual-distribution Discrepancy for Anomaly Detection (DDAD) to utilize both known normal images and unlabeled images. We demonstrate that the use of unlabeled images can significantly improve the performance of anomaly detection. The proposed DDAD consists of two modules, denoted as A and B, each of which is an ensemble of several reconstruction networks with the same architecture. During training, module A takes both known normal and unlabeled images as inputs, capturing anomalous features from unlabeled images in some way, while module B models the distribution of only known normal images. Intuitively, high discrepancy will derive between modules A and B in abnormal regions, thus inter-discrepancy is applied as an anomaly score. Besides, as module B is trained on only normal images, the reconstructions' variance will be high in abnormal regions, thus intra-discrepancy inside B can be used as another anomaly score. To the best of our knowledge, it is the first time that unlabeled images are utilized for anomaly detection. Experiments show that our method achieves significant improvement and obtains state-of-the-art performance on three CXR datasets.

2   Method

In this section, we introduce the proposed DDAD. Previously, most of the existing works formulate the anomaly detection as an OCC problem. That is, given a normal image dataset $D_n = \{x_{ni}, i = 1, ..., N\}$ with N normal images, and a test dataset $D_t = \{(x_{ti}, y_i), i = 1, ..., T\}$ with T annotated normal or abnormal images, where $y_i \in \{0, 1\}$ is the image label (0 for normal image and 1 for abnormal image), the goal is to train a model based on the normal image set D_n which can identify anomalies in the test dataset D_t during inference. Different from previous works, our proposed DDAD makes full use of unlabeled images in clinical practice. Specifically, except for the normal image dataset D_n, we also utilize a readily available unlabeled image dataset $D_u = \{x_{ui}, i = 1, ..., M\}$ with M unlabeled images including both normal and abnormal images, to improve the performance of anomaly detection.

2.1   Dual-Distribution Modeling

In the proposed DDAD as shown in Fig. 1, we use two modules, denoted as A and B, trained on different datasets to model the dual-distribution. Each module is an ensemble of K reconstruction networks with the same architecture but different random initialization of parameters and random shuffling of training samples, trained by the Mean Squared Error (MSE) Loss to minimize reconstruction errors on the training set. Specifically, module A is trained on both normal image dataset D_n and unlabeled image dataset D_u as:

$$\mathcal{L}_A = \frac{1}{N+M} \sum_{\boldsymbol{x}_A \in D_n \cup D_u} \sum_{i=1}^{K} \|\boldsymbol{x}_A - \hat{\boldsymbol{x}}_{Ai}\|^2, \tag{1}$$

where N and M are sizes of datasets D_n and D_u respectively, $\boldsymbol{x}_A$ is the input image of module A, and $\hat{\boldsymbol{x}}_{Ai}$ is the reconstruction of $\boldsymbol{x}_A$ from i-th network in module A. Similarly, the loss function of module B trained on only normal image dataset D_n can be written as:

$$\mathcal{L}_B = \frac{1}{N} \sum_{\boldsymbol{x}_B \in D_n} \sum_{i=1}^{K} \|\boldsymbol{x}_B - \hat{\boldsymbol{x}}_{Bi}\|^2. \tag{2}$$

Through this way, module A captures effective information of abnormalities from the unlabeled dataset as a complement of normal images. Therefore, high discrepancy between A and B will derive at abnormal regions as module B never sees anomalies during training. Besides, based on the theory of Deep Ensemble [9], networks in module B will also show high uncertainty (i.e., intra-discrepancy) on unseen anomlies. Our anomaly scores (described in Sec. 2.2) are designed based on the above analysis to indicate abnormal regions subsequently.

2.2   Dual-Distribution Discrepancy-Based Anomaly Scores

Given a testing image $\boldsymbol{x}_t$, the pixel-wise reconstruction error $\mathcal{A}_{rec}^p = (x_t^p - \hat{x}_t^p)^2$ has been widely used as anomaly score previously. Base on the proposed DDAD and above analysis, we propose to use intra- and inter-discrepancy to indicate anomalies as following:

$$\mathcal{A}_{intra}^p = \sqrt{\frac{1}{K} \sum_{i=1}^{K} (\hat{\mu}_B^p - \hat{x}_{Bi}^p)^2}; \quad \mathcal{A}_{inter}^p = |\hat{\mu}_A^p - \hat{\mu}_B^p|. \tag{3}$$

Here p is the index of pixels, $\hat{\mu}_A$ and $\hat{\mu}_B$ are average maps of reconstructions from modules A and B, respectively. As shown in Fig. 1, our discrepancy maps can indicate potential abnormal regions based on the pixel-wise anomaly scores. The image-level anomaly score is obtained by averaging the pixel-level scores in each image.

Compared with $\mathcal{A}_{rec}$, our anomaly scores for the first time consider the discrepancy between different distributions, leading to stronger discriminative capability. Intuitively, higher AR in unlabeled dataset will lead to greater difference between two distributions on abnormal regions, deriving more competitive $\mathcal{A}_{inter}$. Fortunately, experiments in Fig. 2 demonstrate that even if AR is 0 we can still achieve a consistent improvement compared with the reconstruction baseline, while a low AR can lead to significant boost. Besides, our discrepancies are all computed among reconstructions, rather than between the input and reconstruction as $\mathcal{A}_{rec}$ does. This can reduce the false positive detection caused by reconstruction ambiguity of AE around high frequency regions [2,13].

2.3   Uncertainty-Refined Dual-Distribution Discrepancy

Due to the reconstruction ambiguity of AE, high reconstruction errors often appear at high frequency regions, e.g., around normal region boundaries, leading to false positive detection. To address this problem, AE-U [13] proposed to refine the $\mathcal{A}_{rec}$ using estimated pixel-wise uncertainty. It generates the reconstruction $\hat{x}_i$ and corresponding uncertainty $\sigma^2(x_i)$ for each input x_i, trained by:

$$\mathcal{L} = \frac{1}{NP} \sum_{i=1}^{N} \sum_{p=1}^{P} \{ \frac{(x_i^p - \hat{x}_i^p)^2}{\sigma_p^2(x_i)} + \log\sigma_p^2(x_i) \} \tag{4}$$

Training on normal images, the numerator of the first term is an *MSE* loss to minimize the reconstruction error, while the $\sigma_p^2(x_i)$ at the denominator will be learned automatically to be large at pixels with high reconstruction errors to minimize the first term. Besides, the second term drives the predicted uncertainty to be small at other regions. The two loss terms together ensures that the predicted uncertainty will be larger at only normal regions with high reconstruction errors, thus it can be used to refine the anomaly score at pixel-level.

   In this work, we design a strategy similar to AE-U while adapting to DDAD well. We use AE-U as the backbone of DDAD, and utilize the uncertainty predicted by our module B, which is trained on only normal image dataset, to refine our intra- and inter-discrepancy at p-th pixel as:

$$\mathcal{A}_{intra}^p = \frac{\sqrt{\frac{1}{K} \sum_{i=1}^{K} (\hat{\mu}_B^p - \hat{x}_{Bi}^p)^2}}{\sigma_p}; \quad \mathcal{A}_{inter}^p = \frac{|\hat{\mu}_A^p - \hat{\mu}_B^p|}{\sigma_p}. \tag{5}$$

Here σ_p is the average uncertainty predicted by AE-Us in module B.

3   Experiments

3.1   Datasets

We conduct extensive experiments on three CXR datasets: 1) RSNA Pneumonia Detection Challenge dataset[1], 2) VinBigData Chest X-ray Abnormalities Detection dataset[2], 3) Chest X-ray Anomaly Detection (CXAD) dataset. The performance is assessed with area under the ROC curve (AUC). **RSNA dataset** contains 8,851 normal and 6,012 lung opacity images. In experiments, we use

---

[1] https://www.kaggle.com/c/rsna-pneumonia-detection-challenge.
[2] https://www.kaggle.com/c/vinbigdata-chest-xray-abnormalities-detection.

3,851 normal images as the normal dataset D_n, 4,000 images with different ARs as the unlabeled dataset D_u, and 1,000 normal and 1,000 lung opacity images as the test dataset D_t. **VinBigData dataset** contains 10,606 normal and 4,394 abnormal images that include 14 categories of anomalies in total. In experiments, we use 4,000 normal images as D_n, 4,000 images as D_u, and 1000 normal and 1000 abnormal images as D_t. **CXAD dataset**, collected by us for this study, contains 3,299 normal and 1,701 abnormal images that include 18 categories of anomalies in total. In experiments, we use 2,000 normal images as D_n, 2,000 images as D_u, and 499 normal and 501 abnormal images as D_t.

3.2 Implementation Details

The AE in our experiments contains an encoder and a decoder. The encoder contains 4 convolutional layers with kernel size 4 and stride 2, whose channel sizes are 16–32–64–64. The decoder contains 4 deconvolutional layers with the same kernel size and stride as the encoder, and the channel sizes are 64–32–16–1. The encoder and decoder are connected by 3 fully connected layers. All layers except the ouput layer are followed by batch normalization and ReLU. For fair comparison, MemAE [6] and AE-U [13] in our experiments are modified based on this AE. All the input images are resized to 64×64. K is set to 3. Each model is trained for 250 epochs using the Adam optimizer with a learning rate of 5e-4. All experiments were run on a single NVIDIA TITAN Xp GPU.

3.3 Ablation Study

DDAD with Different ARs. In clinical practice, the AR of unlabeled dataset D_u is unknown. In order to simulate the real scenario, we evaluate the proposed DDAD based on AE on RSNA dataset with AR of D_u varying from 0 to 100%. We use the reconstruction-based method with AE as the baseline for comparison. As shown in Fig. 2, the proposed DDAD method, especially when using $\mathcal{A}_{inter}$, achieves consistent and significant improvement compared with baseline, while DDAD using $\mathcal{A}_{inter}$ performs better with the increasing AR of D_u. Note that $\mathcal{A}_{intra}$ is computed inside module B, thus irrelevant to AR.

More specifically, several observations from Fig. 2 demonstrate the superiority of our method. First, even in the extreme situation (i.e., AR is 0), the DDAD method can still achieve better performance than baseline. That's to say, we can apply the DDAD strategy in any situations and get improvement consistently regardless of AR. Intuitively, when AR is 0, dataset $D_n \cup D_u$ only contains normal images, thus module A degenerates to the same as B. However, in this situation module A is trained on a larger normal dataset than baseline, which leads to more robust models and supports the consistent improvement. Second, even if AR is low (e.g., 20%), the DDAD can achieve a significant improvement. That means the proposed DDAD can improve the performance considerably in clinical practice as there are always some abnormal cases.

Reference to ARs of several public large-scale CXR datasets (e.g., 71% in RSNA and 46% in ChestX-ray8 [21]), we generally assume an AR of 60% for

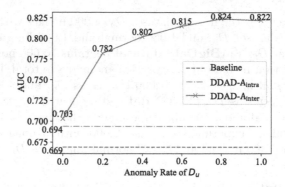

Fig. 2. Performance of DDAD (AE) on RSNA dataset with varying AR of D_u compared with the reconstruction baseline using AE.

D_u in following experiments. We visualize the histograms in this setting using AE as the backbone for qualitative analysis in Fig. 3. The overlap of normal and abnormal histograms in DDAD is significantly less than the reconstruction method, suggesting stronger discriminative capability for identifying anomalies.

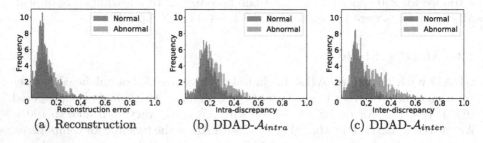

(a) Reconstruction   (b) DDAD-$\mathcal{A}_{intra}$ (c) DDAD-$\mathcal{A}_{inter}$

Fig. 3. Histograms of anomaly score for normal (blue) and abnormal (red) images in the test set of RSNA. The backbone is AE. Scores are normalized to [0,1]. (Color figure online)

DDAD with Different Backbones. Our proposed DDAD method can use any AEs' variants as the backbones. In order to further explore the advantages, DDAD built on different backbones are compared with corresponding reconstruction baselines (Rec.) in Table 1. AR of D_u is set to 60% for all DDAD methods. The results show that DDAD based on AE, MemAE [6] and AE-U [13] can all outperform corresponding baselines on three CXR datasets by a large margin. In terms of AUC, DDAD-$\mathcal{A}_{inter}$ improves baselines of AE, MemAE and AE-U by 14.6%, 10.8% and 4.3% on RSNA dataset, by 15.1%, 13.2% and 12.1% on VinBigData dataset, by 6.5%, 3.9% and 5.0% on CXAD dataset, respectively. DDAD-$\mathcal{A}_{intra}$ improves baselines of AE, MemAE and AE-U by 2.5%, 4.9% and

0.6% on RSNA dataset, by 4.2%, 3.7% and 0.5% on VinBigData dataset, by 4.2%, 3.4% and 2.8% on CXAD dataset, respectively.

We also test the ensemble of K reconstruction models using $\mathcal{A}_{rec}$, demonstrating that simple ensemble has no significant improvement.

Table 1. Performance of different methods built on three backbones. Bold numbers indicate the best results for each baseline.

Method	Anomaly score	AUC								
		RSNA			VinBigData			CXAD		
		AE	MemAE	AE-U	AE	MemAE	AE-U	AE	MemAE	AE-U
Rec	$\mathcal{A}_{rec}$	0.669	0.680	0.867	0.559	0.558	0.738	0.556	0.560	0.664
Rec. (ensemble)		0.669	0.670	0.866	0.555	0.553	0.731	0.550	0.552	0.659
DDAD	$\mathcal{A}_{intra}$	0.694	0.729	0.873	0.601	0.595	0.743	0.598	0.594	0.692
	$\mathcal{A}_{inter}$	**0.815**	**0.788**	**0.910**	**0.710**	**0.690**	**0.859**	**0.621**	**0.599**	**0.714**

3.4   Comparison with State-of-the-Art Methods

We compare our method with four state-of-the-art (SOTA) methods, including AE, MemAE [6], f-AnoGAN [16] and AE-U [13]. Results in Table 2 show that while all DDAD methods outperform the reconstruction methods using the same backbones, the DDAD built on AE-U outperforms all other methods and achieves state-of-the-art performance.

Table 2. Comparison with SOTA methods. Bold face with underline indicates the best, and bold face for the second best.

Method	Anomaly score	AUC		
		RSNA	VinBigData	CXAD
AE	$\mathcal{A}_{rec}$	0.669	0.559	0.556
MemAE		0.680	0.558	0.560
f-AnoGAN[†]		0.798	**0.763**	0.619
AE-U		0.867	0.738	0.664
Ours (AE)	$\mathcal{A}_{intra}$	0.694	0.601	0.589
	$\mathcal{A}_{inter}$	0.815	0.710	0.621
Ours (MemAE)	$\mathcal{A}_{intra}$	0.729	0.595	0.594
	$\mathcal{A}_{inter}$	0.788	0.690	0.599
Ours (AE-U)	$\mathcal{A}_{intra}$	**0.873**	0.743	**0.692**
	$\mathcal{A}_{inter}$	<u>**0.910**</u>	<u>**0.859**</u>	<u>**0.714**</u>

[†] Consistent with [16], we combine pixel-level and feature-level reconstruction errors as $\mathcal{A}_{rec}$ for f-AnoGAN.

4   Conclusion

In this paper, we propose the Dual-distribution Discrepancy for Anomaly Detection (DDAD), which fully utilizes both known normal and unlabeled images. Two new anomaly scores, intra- and inter-discrepancy, are designed based on DDAD for identifying abnormalities. Experiments on three CXR datasets demonstrate that the proposed DDAD can achieve consistent and significant gains using any reconstruction networks as backbones. The performance reveals an increasing trend with the increasing of AR in the unlabeled dataset, while it also outperforms the reconstruction method when AR is 0. The state-of-the-art performance is achieved by DDAD method built on AE-U. In conclusion, DDAD is the first method that utilizes readily available unlabeled images to improve performance of anomaly detection. We hope this work will inspire researchers to explore anomaly detection in a more effective way.

Acknowledgement. This work was supported in part by the National Key Research and Development Program of China (grant No. 2018AAA0100400), the National Natural Science Foundation of China (grant No. 62176098, 61872417 and 62061160490), the Natural Science Foundation of Hubei Province of China (grant No. 2019CFA022), and the UGC Grant (grant No. BGF.005.2021).

References

1. Akcay, S., Atapour-Abarghouei, A., Breckon, T.P.: GANomaly: semi-supervised anomaly detection via adversarial training. In: Jawahar, C.V., Li, H., Mori, G., Schindler, K. (eds.) ACCV 2018. LNCS, vol. 11363, pp. 622–637. Springer, Cham (2019). https://doi.org/10.1007/978-3-030-20893-6_39
2. Baur, C., Denner, S., Wiestler, B., Navab, N., Albarqouni, S.: Autoencoders for unsupervised anomaly segmentation in brain MR images: a comparative study. Med. Image Anal. **69**, 101952 (2021)
3. Baur, C., Wiestler, B., Albarqouni, S., Navab, N.: Scale-space autoencoders for unsupervised anomaly segmentation in brain MRI. In: Martel, A.L., et al. (eds.) MICCAI 2020. LNCS, vol. 12264, pp. 552–561. Springer, Cham (2020). https://doi.org/10.1007/978-3-030-59719-1_54
4. Erdi Çallı, Ecem Sogancioglu, Bram van Ginneken, Kicky G van Leeuwen, and Keelin Murphy. Deep learning for chest x-ray analysis: a survey. Med. Image Anal. **72**,102125 (2021)
5. Chen, X., Pawlowski, N., Glocker, B., Konukoglu, E.: Normative ascent with local gaussians for unsupervised lesion detection. Med. Image Anal. **74**, 102208 (2021)
6. Gong, D., et al.: Memorizing normality to detect anomaly: memory-augmented deep autoencoder for unsupervised anomaly detection. In: Proceedings of the IEEE/CVF International Conference on Computer Vision, pp. 1705–1714 (2019)
7. Jing, L., Tian, Y.: Self-supervised visual feature learning with deep neural networks: a survey. IEEE Trans. Pattern Anal. Mach. Intell. **43**(11), 4037–4058 (2020)
8. Kingma, D.P., Welling, M.: Auto-encoding variational bayes. arXiv preprint arXiv:1312.6114 (2013)
9. Lakshminarayanan, B., Pritzel, A., Blundell, C.: Simple and scalable predictive uncertainty estimation using deep ensembles. Adv. Neural Inf. Process. Syst. 30 (2017)

10. Li, C.L., Sohn, K., Yoon, J., Pfister, T.: Cutpaste: self-supervised learning for anomaly detection and localization. In: Proceedings of the IEEE/CVF Conference on Computer Vision and Pattern Recognition, pp. 9664–9674 (2021)
11. Luo, L., Chen, H., Zhou, Y., Lin, H., Heng, P.-A.: OXnet: deep omni-supervised thoracic disease detection from chest X-rays. In: de Bruijne, M., et al. (eds.) MICCAI 2021. LNCS, vol. 12902, pp. 537–548. Springer, Cham (2021). https://doi.org/10.1007/978-3-030-87196-3_50
12. Luo, L., et al.: Deep mining external imperfect data for chest x-ray disease screening. IEEE Trans. Med. Imaging **39**(11), 3583–3594 (2020)
13. Mao, Y., Xue, F.-F., Wang, R., Zhang, J., Zheng, W.-S., Liu, H.: Abnormality detection in chest x-ray images using uncertainty prediction autoencoders. In: Martel, A.L., et al. (eds.) MICCAI 2020. LNCS, vol. 12266, pp. 529–538. Springer, Cham (2020). https://doi.org/10.1007/978-3-030-59725-2_51
14. Marimont, S. N., Tarroni, G.: Anomaly detection through latent space restoration using vector quantized variational autoencoders. In: 2021 IEEE 18th International Symposium on Biomedical Imaging (ISBI), pp. 1764–1767. IEEE (2021)
15. Ruff, L., et al.: Deep one-class classification. In: International Conference on Machine Learning, pp. 4393–4402. PMLR (2018)
16. Schlegl, T., Seeböck, P., Waldstein, S.M., Langs, G., Schmidt-Erfurth, U.: f-anogan: fast unsupervised anomaly detection with generative adversarial networks. Med. Image Anal. **54**, 30–44 (2019)
17. Sohn, K., Li, C.L., Yoon, J., Jin, M., Pfister, T.: Learning and evaluating representations for deep one-class classification. arXiv preprint arXiv:2011.02578 (2020)
18. Tan, J., Hou, B., Batten, J., Qiu, H., Kainz, B.: Detecting outliers with foreign patch interpolation. arXiv preprint arXiv:2011.04197 (2020)
19. Tan, J., Hou, B., Day, T., Simpson, J., Rueckert, D., Kainz, B.: Detecting outliers with poisson image interpolation. In: de Bruijne, M., et al. (eds.) MICCAI 2021. LNCS, vol. 12905, pp. 581–591. Springer, Cham (2021). https://doi.org/10.1007/978-3-030-87240-3_56
20. Tian, Y., et al.: Constrained contrastive distribution learning for unsupervised anomaly detection and localisation in medical images. In: de Bruijne, M., et al. (eds.) MICCAI 2021. LNCS, vol. 12905, pp. 128–140. Springer, Cham (2021). https://doi.org/10.1007/978-3-030-87240-3_13
21. Wang, X., Peng, Y., Lu, L., Lu, Z., Bagheri, M., Summers, R.M.. Chestx-ray8: hospital-scale chest x-ray database and benchmarks on weakly-supervised classification and localization of common thorax diseases. In: Proceedings of the IEEE Conference on Computer Vision and Pattern Recognition, pp. 2097–2106 (2017)
22. Zavrtanik, V., Kristan, M., Skočaj, D.: Draem-a discriminatively trained reconstruction embedding for surface anomaly detection. In: Proceedings of the IEEE/CVF International Conference on Computer Vision, pp. 8330–8339 (2021)

Skin Lesion Recognition with Class-Hierarchy Regularized Hyperbolic Embeddings

Zhen Yu[1,6,7], Toan Nguyen[5], Yaniv Gal[3], Lie Ju[6,7], Shekhar S. Chandra[4], Lei Zhang[1], Paul Bonnington[6], Victoria Mar[2], Zhiyong Wang[5], and Zongyuan Ge[6,7,8(✉)]

[1] Central Clinical School, Faculty of Medicine, Nursing and Health Sciences, Monash University, Melbourne, Australia
[2] Victorian Melanoma Service, Alfred Health, Melbourne, Australia
[3] Khu, Auckland, New Zealand
[4] School of Information Technology and Electrical Engineering, The University of Queensland, Brisbane, QLD, Australia
[5] School of Computer Science, The University of Sydney, Sydney, Australia
[6] eResearch Centre, Monash University, Melbourne, Australia
[7] Monash Airdoc Research, Monash University, Melbourne, Australia
[8] Faculty of Engineering, Monash University, Melbourne, Australia
zongyuan.ge@monash.edu
https://mmai.group

Abstract. In practice, many medical datasets have an underlying taxonomy defined over the disease label space. However, existing classification algorithms for medical diagnoses often assume semantically independent labels. In this study, we aim to leverage class hierarchy with deep learning algorithms for more accurate and reliable skin lesion recognition. We propose a hyperbolic network to jointly learn image embeddings and class prototypes. The hyperbola provably provides a space for modeling hierarchical relations better than Euclidean geometry. Meanwhile, we restrict the distribution of hyperbolic prototypes with a distance matrix which is encoded from the class hierarchy. Accordingly, the learned prototypes preserve the semantic class relations in the embedding space and we can predict label of an image by assigning its feature to the nearest hyperbolic class prototype. We use an in-house skin lesion dataset which consists of ∼230k dermoscopic images on 65 skin diseases to verify our method. Extensive experiments provide evidence that our model can achieve higher accuracy with less severe classification errors compared to that of models without considering class relations.

Keywords: Skin lesion recognition · Class hierarchy · Deep learning · Hyperbolic geometry

1 Introduction

Recent advances in deep learning have greatly improved the accuracy of classification algorithms for medical image diagnosis. Typically, these algorithms

© The Author(s), under exclusive license to Springer Nature Switzerland AG 2022
L. Wang et al. (Eds.): MICCAI 2022, LNCS 13433, pp. 594–603, 2022.
https://doi.org/10.1007/978-3-031-16437-8_57

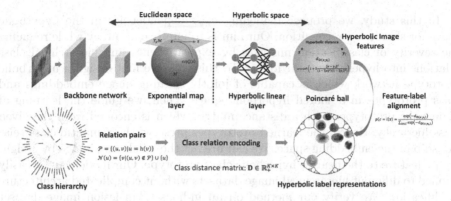

Fig. 1. The proposed hyperbolic model with class hierarchy for skin lesion recognition.

assume mutually exclusive and semantically independent labels [3,4,15]. The classification performance is evaluated by treating all classes other than the true class as equally wrong. However, many medical datasets have an underlying class hierarchy defined over the label space. Accordingly, different disease categories can be organized from general to specific in the semantic concepts. Diseases from a same super-class often share similar clinical characteristics. Incorporating the constraint of class relations in a diagnostic algorithm has at least two benefits. First, a class hierarchy defines a prior knowledge on the structure of disease labels. Learning model with such knowledge would facilitate the model training and boost the performance compared with that of using semantic-agnostic labels. Second, a class tree indicates the semantic similarity between each class pair and a model can be optimized with the semantic metric to reduce the severity of prediction errors [2,6,8]. Take the example of a diagnostic model in dermatology: the common non-cancerous melanocytic lesion has at least two sub-classes: lentigo and benign nevus. Undoubtedly, mistaking a lentigo for a begin nevus is more tolerable than of mistaking a malignant melanoma for a begin nevus. By taking mistake severity into consideration, we can somewhat preclude models from making a egregious diagnostic error which is crucial in deploying the model in real world scenarios.

However, relatively few works use hierarchical class clues in the context of medical image analysis [1,14]. Although these methods report promising results, they require the network architecture to be adapted for a specific hierarchy and they neglect semantic measurements in evaluating the performance of the algorithms. Besides, these models are built in Euclidean space while study [5] shows that Euclidean space suffer from heavy volume intersection and points arranged with Euclidean distances would no longer be capable of persevering the structure of the original tree. By contrast, approaches with hyperbolic geometry for modelling symbolic data have demonstrated to be more effective in representing hierarchical relations [5,12]. The hyperbolic space can reflect complex structural patterns inherent in taxonomic data with a low dimensional embedding.

In this study, we propose modelling class dependencies in the hyperbolic space for skin lesion recognition. Our aim is to improve accuracy while reducing the severity of classification mistakes by explicitly encoding hierarchical class relations into hyperbolic embeddings. To this end, We first design a hyperbolic prototype network which is capable of jointly learning image embeddings and class prototypes in a shared hyperbolic space. Then, we guide the learning of hyperbolic prototypes with a distance matrix which is encoded from the given class hierarchy. Hence, the learned prototypes preserve the semantic class relationship in the embedding space. We can predict the label of an image by assigning its feature to the nearest hyperbolic class prototype. Our model can be easily applied to different hierarchical image datasets without complicated architecture modification. We verify our method on an in-house skin lesion image dataset which consists of approximately 230k dermoscopic images organized in three-level taxonomy of 65 skin diseases. Extensive experiments prove that our model can achieve higher classification accuracy with less severe classification errors than models without considering class relations. Moreover, we also conducted an ablation study by comparing hyperbolic space trained hierarchical models to those trained in Euclidean space.

2   Method

2.1   Hyperbolic Geometry

The hyperbolic space $\mathbb{H}^n$ is a homogeneous, simply connected Riemannian manifold with constant negative curvature. There exist five insightful models of $\mathbb{H}^n$ and they are conformal to the Euclidean space. Following [12], we use the Poincaré ball model because it can be easily optimized with gradient-based methods. The Poincaré ball model $(\mathbb{D}^n, g^{\mathbb{D}})$ is defined by the manifold $\mathbb{D}^n = \{x \in \mathbb{R}^n : c\,\|x\| < 1, c \geq 0\}$ endowed with the Riemannian metric $g_x^{\mathbb{D}} = \lambda_x^{2c} g^E$, where c denotes the curvature, $\lambda_x^c = \frac{2}{1-c\|x\|^2}$ is the conformal factor and $g^E = \mathbf{I}^n$ is the Euclidean metirc tensor. The hyperbolic space has very different geometric properties than that in the Euclidean space. We introduce basic hyperbolic operations involved in this study as following:

Poincaré Distance. The distance between two points $\mathbf{x}_1, \mathbf{x}_2 \in \mathbb{D}_c^n$ is calculated as:

$$d_c(\mathbf{x}_1, \mathbf{x}_2) = \frac{2}{\sqrt{c}}\operatorname{arctanh}\left(\sqrt{c}\,\|-\mathbf{x}_1 \oplus_c \mathbf{x}_2\|\right) \qquad (1)$$

$$\mathbf{x}_1 \oplus_c \mathbf{x}_2 = \frac{\left(1 + 2c\langle \mathbf{x}_1, \mathbf{x}_2\rangle + c\,\|\mathbf{x}_2\|^2\right)\mathbf{x}_1 + \left(1 - c\,\|\mathbf{x}_1\|^2\right)\mathbf{x}_2}{1 + 2c\langle \mathbf{x}_1, \mathbf{x}_2\rangle + c^2\,\|\mathbf{x}_1\|^2\,\|\mathbf{x}_2\|^2} \qquad (2)$$

Exponential Map. The *exponential map* defines a projection from tangent space $T_{\mathbf{x}}\mathbb{D}_c^n$ of a Riemannian manifold $\mathbb{D}_c^n$ to itself which enables us to map a vector in Euclidean space $\mathbb{R}^n \cong T_{\mathbf{x}}\mathbb{D}_c^n$ to the hyperbolic manifold. The mathematical definition of *exponential map* is given by:

$$\exp_{\mathbf{x}}^c(\mathbf{v}) = \mathbf{x} \oplus_c \left(\tanh\left(\sqrt{c}\frac{\lambda_{\mathbf{x}}^c \|\mathbf{v}\|}{2} \right) \frac{\mathbf{v}}{\sqrt{c}\|\mathbf{v}\|} \right) \tag{3}$$

where $\mathbf{x}$ denotes the reference point and default $\mathbf{0}$ are used if not specified.

Hyperbolic Linear Projection. The linear projection in hyperbolic space is based on the Möbius matrix-vector multiplication. Let $\psi : \mathbb{R}^n \to \mathbb{R}^m$ be a linear map defined in Euclidean space. Then, for $\forall \mathbf{x} \in \mathbb{D}_c^n$, if $\psi(\mathbf{x}) \neq \mathbf{0}$, the calculation of the linear map is defined as:

$$\psi^{\otimes_c}(\mathbf{x}) = \frac{1}{\sqrt{c}}\tanh\left(\frac{\|\psi(\mathbf{x})\|}{\|\mathbf{x}\|}\tanh^{-1}\left(\sqrt{c}\|\mathbf{x}\| \right) \right) \frac{\psi(\mathbf{x})}{\|\psi(\mathbf{x})\|} \tag{4}$$

If we consider a bias of $\mathbf{b} \in \mathbb{D}_c^n$ in the linear map, then the $\mathbf{x}$ in the above equation should be replaced with the Möbius sum: $\mathbf{x} \leftarrow \mathbf{x} \oplus_c \mathbf{b}$.

2.2   Hyperbolic Prototype Network for Image Classification

As shown in Fig. 1, the proposed hyperbolic prototype network (HPN) consists of a backbone network, a *exponential map* layer, a hyperbolic linear layer and a classification layer. First, the backbone network extracts image representations from the Euclidean space and then the *exponential map* layer projects it into the Poincaré ball. After that, we use a hyperbolic linear layer to transform the projected hyperbolic image embeddings so that their dimensions are fitted with that of class prototypes in the shared hyperbolic space. Finally, the classification layer performs matching between hyperbolic image embeddings with respect to the corresponding class prototypes.

Formally, consider a dataset $\mathcal{N}$ consists of m samples $\{(\mathbf{x}_i, \mathbf{y}_i)\}^m$ from K classes. Let $\{\mathbf{a}_1, ..., \mathbf{a}_K\}$ be the hyperbolic prototypes for the K classes. Then, the hyperbolic network computes probability distributions over all the class prototypes for each input image as:

$$p(\hat{\mathbf{y}}_i = k|\mathbf{x}_i) = \frac{\exp(-d_c(\mathbf{z}_i, \mathbf{a}_k))}{\sum_{j=1}^K \exp(-d_c(\mathbf{z}_i, \mathbf{a}_j))}, \forall k \in K \tag{5}$$

$$\mathbf{z}_i = \psi^{\otimes_c}(\exp_{\mathbf{0}}^c(f(\mathbf{x}_i)) \oplus_c \mathbf{b}) \tag{6}$$

where $f(\cdot)$ denotes the function of backbone network. The network can be directly optimized with the cross-entropy loss on the hyperbolic distance-based logits:

$$\mathcal{L}_{DCE} = \frac{1}{m}\sum_{i \in m}\left(\frac{1}{T}d_c(\mathbf{z}_i, \mathbf{a}_{\mathbf{y}_i}) + \log\left(\sum_{j=1}^K \exp\left(-\frac{1}{T}d_c(\mathbf{z}_i, \mathbf{a}_j) \right) \right) \right) \tag{7}$$

where T is a temperature factor for scaling the distance logits which is fixed as 0.1.

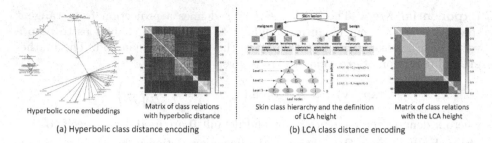

(a) Hyperbolic class distance encoding   (b) LCA class distance encoding

Fig. 2. Illustration of class relation encoding methods. Both of the approaches produce a matrix which indicates the dissimilarity of class pairs.

2.3   Incorporating Constraint of Class Relations

Hierarchical Class Relations Encoding: Let's assume that the label of dataset $\mathcal{N}$ can be organized into a class tree $\mathcal{H} = (V, E)$ with h hierarchical levels, where V and E denote nodes and edges, respectively. Each node corresponds to a class label while an edge $(u, v) \in E$ indicates entailment relation[1] between the node pair. We exploit two methods for encoding class relations from the hierarchical tree $\mathcal{H}$: 1) hyperbolic class distance (HCD) encoding, and 2) low common ancestor (LCA) class distance (LCD) encoding. Both of the methods produce a matrix of class distance $\mathbf{D} \in \mathbb{R}_+^{K \times K}$ which indicates the dissimilarity between each pair of leaf-classes (see Fig. 2). For the HCD encoding, we first learn hyperbolic representations for all class labels following [5] and then compute pairwise hyperbolic distances for leaf classes. Similar to [2], in the LCD encoding, we directly define the distance between two classes as the height of their LCA in the hierarchy.

Class Distance Guided Hyperbolic Prototype Learning: To introduce such class relations into class prototypes, we propose to guide the prototype learning by constraining the distance between prototypes to be consistent with the class distance in the $\mathbf{D}$. As described by Sala et al. [13], the distortion of a mapping between the finite metric space of hierarchical class distance $\mathbf{D}[i, j]$ and the continuous metric space of prototype distance $d_c(\mathbf{a}_i, \mathbf{a}_j)$ can be defined as:

$$\text{disto}(\mathbf{d}, \mathbf{D}) = \frac{1}{K(K-1)} \sum_{i,j \in K^2, i \neq j} \frac{|d_c(\mathbf{a}_i, \mathbf{a}_j) - \mathbf{D}[i, j]|}{\mathbf{D}[i, j]} \tag{8}$$

A low-distortion mapping means that the learned prototypes preserve well the relations between classes defined by the $\mathbf{D}$. However, achieving low-distortion mapping requires the prototypes to be arranged in the embedding space with the specific distance constraint, and this may conflict with the cross-entropy loss (7) which encourages the distance between an embedding to negative class prototypes to be as large as possible. Hence, we introduce a scale factor in the

---

[1] Namely, v is a sub-concept of u.

formulation of the distortion for removing the discrepancy between the scale of the prototype distance and the hierarchical class distance:

$$s = \sum_{i,j \in K^2} \frac{d_c\left(\mathbf{a}_i, \mathbf{a}_j\right)}{\mathbf{D}\left[i, j\right]} \Big/ \sum_{i,j \in K^2} \frac{d_c\left(\mathbf{a}_i, \mathbf{a}_j\right)^2}{\mathbf{D}\left[i, j\right]^2} \tag{9}$$

The s is dynamically changed depending on the value of $\mathbf{d}$ and $\mathbf{D}$. Then, we obtain the following smooth surrogate of the disto^s for optimization:

$$\mathcal{L}_{disto} = \frac{1}{K\left(K-1\right)} \min_{s \in \mathbb{R}_+} \sum_{i,j \in K^2, i \neq j} \left(\frac{s \cdot d_c\left(\mathbf{a}_i, \mathbf{a}_j\right) - \mathbf{D}\left[i, j\right]}{\mathbf{D}\left[i, j\right]} \right)^2 \tag{10}$$

Finally, we optimize the proposed hyperbolic network by combining $\mathcal{L}_{DCE}$ and $\mathcal{L}_{disto}$. The $\mathcal{L}_{DCE}$ enables us to jointly learn hyperbolic image embeddings and class prototypes, while the $\mathcal{L}_{disto}$ forces the class prototypes following the semantic distribution defined by the given class hierarchy:

$$\mathcal{L} = \mathcal{L}_{DCE} + \mathcal{L}_{disto} \tag{11}$$

3   Experiment and Results

3.1   Dataset and Implementation

We evaluate the proposed method on an in-house dataset. We denote the dataset as $molemap^+$ as it is bigger and more diverse compared to the first version used in [7]. The $molemap^+$ includes 235,268 tele-dermatology verified dermoscopic images organized in three-level tree-structured taxonomy of 65 skin conditions (shown in Fig. 3). We split the dataset into training, validation and testing set with a ratio of 7:1:2. The standard data augmentation techniques such as random resized cropping, colour transformation, and flipping are equally used in all experiments. Each dermoscopic image is resized to a fixed input size of 320×320. We use ReseNet-34 [9] as the backbone for all models and train them using ADAM optimizer with a batch size of 100 and a training epoch of 45. The initial learning rates is set to 1×10^{-5} and 3×10^{-4} for the backbone layers and new added layers, respectively. We adjust learning rate with a step decay schedule. The decay factor is set to 0.1 and associated with a decay epoch of 15.

3.2   Evaluation Metrics

We consider accuracy and two semantic measures for assessing performance of models: **Mistake severity (MS):** Inspired by [2], we measure the severity of a mis-classification with the height of the LCA between the class of incorrect prediction and the true class. **Hierarchical distance (HD) of top-k^2:** This metric computes the mean hierarchical class distance between the true class and the top-k predicted classes. The measurement is meaningful for assessing the reliance of a model in assisting clinicians making diagnosis.

---

2 We set k as 5 in this study.

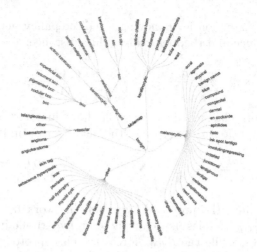

Fig. 3. Structure of the skin disease taxonomy.

3.3 Quantitative Results

Ablation Study: Here, we give ablation results of our model to illustrate how different settings affect the final performance. Figure 4(a) shows the effect of curvature which determines the distortion of the hyperbolic ball. It can be seen that a small c produces better performance and the accuracy drops 1.6% when increasing c from 0.01 to 1. As demonstrated by [10], this is because large curvatures could bring numerical instability in hyperbolic operations. Therefore, we set c as 0.01 in the following experiments. Then, we report accuracy by varying the hyperbolic dimension in Fig. 4(b). It can be noted that the model with dimension of 320 achieves highest accuracy of 60.94%. However, the performance gap is not significant compared to other models with lower dimension settings. Even reducing the dimension from 320 to 16, the accuracy only decreases ~0.5%. This result verifies the efficiency of hyperbolic embeddings in representing imaging data. In Table 1, we compare the performance of the HPN trained with and without class hierarchy. It can be seen that both the HCD and LCD encoding boost the performance compared to the baseline hyperbolic network. Among them, the HPN trained with the class relation matrix derived from LCA encoding gives best accuracy and semantic metrics. Since our dataset is highly imbalanced, in Fig. 4(c), we further give the detailed performance improvement on head classes, middle class and tail classes separately. When using the class hierarchy, the accuracy increases 0.9% for tail classes which is higher than that of 0.2% for both middle and head classes.

Comparative Study: We then compare our model with that of models trained with and without class hierarchy in Euclidean space and hyperbolic space, respectively. The details of those model are described in the Appendix. From Table 1, we can observe that all class-hierarchy trained models apart from the

Table 1. Comparison of the proposed model with other methods.

Models	Class hierarchy	Accuracy	Semantic metrics	
			Mistake severity	Mean HD@5
Baseline CNN	✗	58.47	1.95	1.65
Soft-label [2]	✓	58.88	1.95	1.46
Multi-branch CNN [16]	✓	60.28	1.92	1.61
Fixed-hyperbolic embeddings [11]	✓	56.87	1.85	1.40
Euclidean prototype net [6]	✓ (LCD)	60.76	1.90	1.55
Hyperbolic prototype net	✗	60.68	1.91	1.63
	✓ (HCD)	60.94	1.83	1.43
	✓ (LCD)	61.04	1.82	1.41

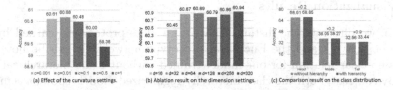

Fig. 4. Ablation results on the curvature setting and the dimension setting.

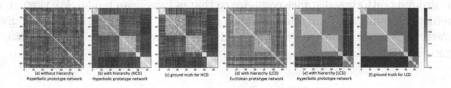

Fig. 5. Pairwise class prototype distance learned with and without class hierarchy.

method with fixed-hyperbolic embeddings show higher accuracy compared to the baseline CNN. While the semantic metric of all models learned with class hierarchy are better than that of the baseline CNN. Certainly, this result highlights the value of incorporating class relations for skin lesion recognition. For the fixed-hyperbolic embeddings that achieves best semantic measurements with the lowest accuracy, we attribute this discrepancy to the softmax-based cross-entropy optimization. Because it is hard to minimize the loss on the probability distribution of distance (Eq. (5)) between an image representation and fixed prototypes with a semantic distance constraint. Noticeably, when using class hierarchy, our model outperforms the Euclidean prototype net which has best performance among all Euclidean models.

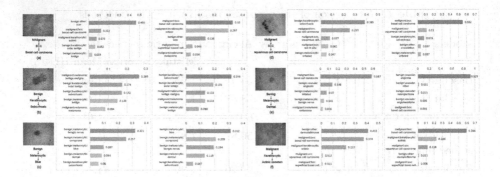

Fig. 6. Examples of predictions results from our model trained with and without the class hierarchy. (a)–(c) are incorrectly predicted samples by both model. (d)–(f) are samples correctly predicted by hierarchy-aware HPN but mis-classified by hierarchy-agnostic HPN (best viewed in zoom in mode).

3.4   Visualization Results

In Fig. 5, we illustrate the distance matrix for class prototypes learned by our hyperbolic network and the Euclidean prototype network. It can be note that there is no clear relation for our model trained without using hierarchical class clues. By contrast, both the class-hierarchy regularized models shows a semantic connection in the prototypical class distance matrix. From Fig. 5(d) and Fig. 5(e), we can see the pairwise prototypes distance of the hyperbolic network is closer to the ground truth class distance compare to that of Euclidean model. We then give prediction results for individual samples in Fig. 6. It can be seen the hierarchy-aware HPN gives more reasonable predictions compared with the hierarchy-agnostic HPN.

4   Conclusion

In this study, we present a hyperbolic network with class hierarchy for skin lesion recognition. Our model is capable of jointly learning hyperbolic image embeddings and class prototypes while preserve class relations from the hierarchy. We evaluate the proposed method on a large-scale in-house skin lesion dataset by reporting both accuracy and semantic measurements derived from the class tree. Experiments demonstrate that our model can capture well class relations and the hyperbolic network outperforms other Euclidean models.

References

1. Barata, C., Marques, J.S., Celebi, M.E.: Deep attention model for the hierarchical diagnosis of skin lesions. In: 2019 IEEE/CVF Conference on Computer Vision and Pattern Recognition Workshops, pp. 2757–2765 (2019)

2. Bertinetto, L., Mueller, R., Tertikas, K., Samangooei, S., Lord, N.A.: Making better mistakes: leveraging class hierarchies with deep networks. In: Proceedings of the IEEE/CVF Conference on Computer Vision and Pattern Recognition, pp. 12506–12515 (2020)

3. Brinker, T.J., et al.: Deep learning outperformed 136 of 157 dermatologists in a head-to-head dermoscopic melanoma image classification task. Eur. J. Cancer **113**, 47–54 (2019)

4. Esteva, A., et al.: Dermatologist-level classification of skin cancer with deep neural networks. Nature **542**(7639), 115–118 (2017)

5. Ganea, O., Bécigneul, G., Hofmann, T.: Hyperbolic entailment cones for learning hierarchical embeddings. In: International Conference on Machine Learning, pp. 1646–1655 (2018)

6. Garnot, V.S.F., Landrieu, L.: Leveraging class hierarchies with metric-guided prototype learning (2021)

7. Ge, Z., Demyanov, S., Chakravorty, R., Bowling, A., Garnavi, R.: Skin disease recognition using deep saliency features and multimodal learning of dermoscopy and clinical images. In: Descoteaux, M., Maier-Hein, L., Franz, A., Jannin, P., Collins, D.L., Duchesne, S. (eds.) MICCAI 2017. LNCS, vol. 10435, pp. 250–258. Springer, Cham (2017). https://doi.org/10.1007/978-3-319-66179-7_29

8. Goyal, P., Choudhary, D., Ghosh, S.: Hierarchical class-based curriculum loss. In: Proceedings of the Thirtieth International Joint Conference on Artificial Intelligence, pp. 2448–2454 (2021)

9. He, K., Zhang, X., Ren, S., Sun, J.: Deep residual learning for image recognition. In: Proceedings of the IEEE Conference on Computer Vision and Pattern Recognition, pp. 770–778 (2016)

10. Khrulkov, V., Mirvakhabova, L., Ustinova, E., Oseledets, I., Lempitsky, V.: Hyperbolic image embeddings. In: Proceedings of the IEEE/CVF Conference on Computer Vision and Pattern Recognition, pp. 6418–6428 (2020)

11. Long, T., Mettes, P., Shen, H.T., Snoek, C.G.: Searching for actions on the hyperbole. In: Proceedings of the IEEE/CVF Conference on Computer Vision and Pattern Recognition, pp. 1141–1150 (2020)

12. Nickel, M., Kiela, D.: Poincaré embeddings for learning hierarchical representations. In: Advances in Neural Information Processing Systems, vol. 30 (2017)

13. Sala, F., De Sa, C., Gu, A., Ré, C.: Representation tradeoffs for hyperbolic embeddings. In: International Conference on Machine Learning, pp. 4460–4469. PMLR (2018)

14. Yang, J., et al.: Hierarchical classification of pulmonary lesions: a large-scale radiopathomics study. In: International Conference on Medical Image Computing and Computer-Assisted Intervention, pp. 497–507 (2020)

15. Yu, L., Chen, H., Dou, Q., Qin, J., Heng, P.A.: Automated melanoma recognition in dermoscopy images via very deep residual networks. IEEE Trans. Med. Imaging **36**(4), 994–1004 (2016)

16. Zhu, X., Bain, M.: B-CNN: branch convolutional neural network for hierarchical classification. arXiv preprint arXiv:1709.09890 (2017)

Reinforcement Learning Driven Intra-modal and Inter-modal Representation Learning for 3D Medical Image Classification

Zhonghang Zhu[1], Liansheng Wang[1(✉)], Baptiste Magnier[2], Lei Zhu[3,4], Defu Zhang[1], and Lequan Yu[5]

[1] Department of Computer Science at School of Informatics, Xiamen University, Xiamen, China
zzhonghang@stu.xmu.edu.cn, {lswang,dfzhang}@xmu.edu.cn
[2] Euromov Digital Health in Motion, University Montpellier, IMT Mines Ales, Ales, France
baptiste.magnier@mines-ales.fr
[3] The Hong Kong University of Science and Technology (Guangzhou), Guangzhou, China
leizhu@ust.hk
[4] The Hong Kong University of Science and Technology, Hong Kong, SAR, China
[5] Department of Statistics and Actuarial Science, The University of Hong Kong, Hong Kong, SAR, China
lqyu@hku.hk

Abstract. Multi-modality 3D medical images play an important role in the clinical practice. Due to the effectiveness of exploring the complementary information among different modalities, multi-modality learning has attracted increased attention recently, which can be realized by Deep Learning (DL) models. However, it remains a challenging task for two reasons. First, the prediction confidence of multi-modality learning network cannot be guaranteed when the model is trained with weakly-supervised volume-level labels. Second, it is difficult to effectively exploit the complementary information across modalities and also preserve the modality-specific properties when fusion. In this paper, we present a novel Reinforcement Learning (RL) driven approach to comprehensively address these challenges, where two Recurrent Neural Networks (RNN) based agents are utilized to choose reliable and informative features within modality (intra-learning) and explore complementary representations across modalities (inter-learning) with the guidance of dynamic weights. These agents are trained via Proximal Policy Optimization (PPO) with the confidence increment of the prediction as the reward. We take the 3D image classification as an example and conduct experiments on a multi-modality brain tumor MRI data. Our approach outperforms other methods when employing the proposed RL-based multi-modality representation learning.

Keywords: Multi-modality learning · 3D medical images · Reinforcement learning · Classification

© The Author(s), under exclusive license to Springer Nature Switzerland AG 2022
L. Wang et al. (Eds.): MICCAI 2022, LNCS 13433, pp. 604–613, 2022.
https://doi.org/10.1007/978-3-031-16437-8_58

1   Introduction

Multi-modality images, *e.g.*, different MRI, are widely used in medical applications [2,18]. Integrating the strengths of multiple modalities by exploring their rich information and discovering the underlying correlations among them is an effective manner to improve the diagnosis and prognosis tasks. In other aspects, many medical images involve 3D format, therefore, multi-modality 3D image classification is important in medical image computing field. However, it is challenging to develop such multi-modality learning algorithms for several reasons. On the one hand, for learning within a single modality (*i.e.*, intra-modal learning), since obtaining slice-level labels of a 3D volume via manual labeling is tedious and time-consuming [6]; it is difficult to identify features containing modality-specific information without precise instructions. On the other hand, for learning among different modalities (*i.e.*, inter-modal learning), since the underlying correlations among them are unclear, exploiting complement yet discriminative representations from each modality is also non-trivial.

Recently, an increasing number of studies have been investigated for multi-modality learning. As an example, to obtain complementary features of different modalities, Canonical Correlation Analysis (CCA) [3] projects the features of each modality to a new robust space. Multiple Kernel Learning (MKL) [7] utilizes a set of predefined kernels from multi-view data to integrate these modalities using the optimized weights. In addition, there are several works that applied DL networks for multi-modal learning [9,11,13,15,20]. These methods can be roughly categorised into two branches: 2D-based methods [10,19] and 3D-based methods [5,8,14]. For the first line of methods, 3D volumes are firstly projected into 2D images and then are integrated for the final prediction. However, these methods are insufficient in capturing the complicated spatial characteristics of 3D volumes. In contrast, 3D-based methods can work well in capturing spatial relations between different volumes for learning more complementary multi-modality representations. However, there still exist several limitations for 3D-based methods. First, the particular use of the 3D fusion models can only be supervised by the volume-level labels with limited information, which leads to high uncertainty prediction [1]. While the risk-sensitive tasks, like medical diagnosis, require high prediction confidence for the purposes of avoiding critical mistakes. Second, to explore complementary representation in multi-modality learning, the weighted fusion method with fixed weight is widely applied in these 3D fusion researches. However, it is unreasonable to merely assign specific weights to different modalities, besides, the weights should be dynamically allocated by data-driven rather than artificial.

In this paper, a novel Reinforcement Learning (RL) driven approach for effective multi-modality 3D medical image analysis is presented, where two RL-based agents are learned for dynamical intra-modal and inter-modal features enhancement to learn latent modality representation and underlying correlations among different modalities. Specifically, to enable such intra-modal and inter-modal feature enhancement, we explore two key techniques based on the characteristics of multi-modality medical images. (1) We propose an iterative hybrid-enhancement

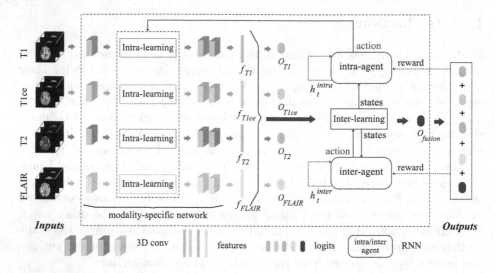

Fig. 1. Illustrations of the proposed method: given a group of 3D volumes in different modalities *i.e.*, T1, T1ce, T2, FLAIR, the model iteratively processes a sequence of multi-modality images with intra-learning and inter-learning modules. The proposal agents (intra-agent and inter-agent) are supervised by the reward function designed as confidence increment of the outputs. Hidden states h_t^{intra} and h_t^{inter} of both agents help them exploit information from previous inputs and produce promising actions for intra-modality learning and inter-modality learning, respectively.

network to integrate intra-features and inter-features, where the enhanced intra-features in each iteration are regarded as the state of two designed agents which generate strategies for intra-enhance and inter-enhance in the next training iteration. (2) We take the prediction confidence increment as the supervision of the agents, which encourages the agents to customize the enhanced strategies that can promote the prediction confidence. Finally, the whole framework is trained in an end-to-end manner for hybrid multi-modality learning.

2   Method

The detailed architecture of our proposed framework is shown in Fig. 1. Given an input volume, in order to capture intra-modality representations, a 3D convolution layer in the modality-specific network is first employed to generate shallow modality features which are enhanced by the RL-based intra-learning module to focus on the salient part. Then, modality-specific features are passed to another RL-based inter-modal learning module to conduct multi-modality learning for prediction promotion. Specifically, the output of the agent is taken as the state of the intra-agent and inter-agent to determine actions (enhancement weights) for intra-/inter- modality learning in the next iteration. The increments of the softmax prediction are used as reward function to train the agents for facilitating the agents to propose actions that enable the network to produce correct predictions in high confidence. The detailed design of RL-based intra-/inter- modality learning process is introduced in the following subsections.

2.1 RL-Based Intra-modality Learning

In the proposed method, the RL based intra-modality learning is an alternating decision making process. Specifically, at the beginning of a training iteration, 3D volumes sized in $B \times 1 \times d \times w \times h$ are sent to individual 3D convolution layers to get the modality-specific primary representations sized in $B \times c \times d \times w \times h$, respectively. The B, c denotes the batch size and channel number, while d, w, h indicates the depth, width and height of the input 3D volumes, respectively. Intra-modality learning is then implemented by multiplying the intra-modality enhancement weights $W_{s,t} \in \mathbb{R}^{B \times c}$ with modality-specific primary representations, where t denotes the number of iterations.

We extract the feature from the shallow layer of the modality-specific network to conduct intra-modality learning with the assumption that these features keep the modality-specific spatial relationship. Moreover, different modalities share the same intra-modality enhancement weights determined by the intra-agent whose initial state is designed as the sum of different high-level modality-specific representations, i.e., $f_{T1} \in \mathbb{R}^{B \times 8c}$, $f_{T1ce} \in \mathbb{R}^{B \times 8c}$, $f_{T2} \in \mathbb{R}^{B \times 8c}$, $f_{FLAIR} \in \mathbb{R}^{B \times 8c}$, aiming that the proposed intra-enhancement action can facilitate sharing of intra-modality spatial relationships among different modalities. In our experiments, intra-modality enhancement weights $W_{s,t}$ in the first iteration are initialized as 1, and updated by the intra-agent in the following iterations.

2.2 RL-Based Inter-modality Learning

In inter-modality learning, different high-level modality-specific representations f_{T1}, f_{T1ce}, f_{T2}, f_{FLAIR} are fused to generate a $B \times 32c$ feature forwarded to two independent RNN termed intra-agent and inter-agent. Inter-modality enhancement weights $W_{f,tk} \in \mathbb{R}^{B \times 1}$, $k \in (0, 1, 2, 3)$ are involved in the inter-modality learning which can be formulated as:

$$f_{fusion} = Concat(f_{T1} \times W_{f,t0}, f_{T1ce} \times W_{f,t1}, f_{T2} \times W_{f,t2}, f_{FLAIR} \times W_{f,t3}), \quad (1)$$

where the $Concat$ represents the concatenation operation, and $W_{f,t0}$, $W_{f,t1}$, $W_{f,t2}$, $W_{f,t3}$ are different weights for different modalities. Same as intra-modality enhancement weights $W_{s,t}$, the inter-modality enhancement weights $W_{f,tk}$ are also set as 1 when $t = 0$, and subsequently updated by the inter-agent.

With the fusion feature f_{fusion}, a Fully Connected (FC) layer is adopted to generate a primary fusion representation. The primary fusion representation includes comprehensive features of all modalities, which is suitable to be set as the initial state of the inter-agent. While if we split the fusion feature f_{fusion} into individual modality-specific features and add them together, the voting information of each modality-specific feature for the current predicted state can be synthesized to the greatest extent, without harming characteristics specific of each modality, which can be regarded as states of the intra-agent. with the initial states, the intra-agent and inter-agent will be triggered to propose the new $W_{s,t}$ and $W_{f,tk}$, $k \in (0, 1, 2, 3)$ for next iteration procedure.

Note that several iterations are conducted in a training epoch. For each iteration, the hidden state h_t^{intra} and h_t^{inter} which aggregates the information of past states are maintained within intra-agent and inter-agent, respectively. In addition, modality-specific logit outputs O_{T1}, O_{T1ce}, O_{T2} and O_{FLAIR}, *i.e.*, the output of each modality-specific network followed a FC layer and a soft-max layer, with modality fusion logit O_{fusion} are saved in a memory bank until the end of the epoch. These sequential logit outputs will be used for calculating reward of the agents and the training loss.

2.3   Reward Function and Training Procedure

Operating under the intuition that the normalized prediction probability (*i.e.*, normalized from 0 to 1) reflects model confidence of the prediction. In this study, we define the reward as increments of the soft-max prediction probability on the ground truth labels for the optimization of two agents. Specifically, the reward function to train the agent is designed as $r_t = p_t - p_{t-1}$, where p_t is the soft-max prediction probability with the ground truth label at the tth iteration process, which is derived from the tth logit outputs. Then both the inter-agent and intra-agent are trained simultaneously by maximizing discounted reward $r_{all} = \sum_{t=1}^{3} \gamma^{t-1} \cdot r_t$ where the γ is set to 0.1.

The network of our RL-based multi-modality learning model includes two components: modality-specific network consisted of four cascaded 3D encoders which are used for representation extraction of each single-modality, and a RL module includes two agents for intra-/inter- modality enhancement weights proposal. Each 3D CNN extractor contains six 3D blocks followed the avgpool layer and FC layer. Besides, the inter-agent and intra-agent share a similar structure including a RNN followed a FC layer, but the FC layer has different nodes number which is set as c for intra-agent and 4 for inter-agent, respectively.

To optimize the whole framework, we collect modality-specific logits and fusion logits of each iteration to calculate the training loss. Subsequently, the total loss of our approach can be computed as:

$$L = \sum_{t=0}^{3} L_{O_{T1,t}} + L_{O_{T1ce,t}} + L_{O_{T2,t}} + L_{O_{FLAIR,t}} + \lambda \cdot L_{O_{fusion,t}}, \quad (2)$$

where $L_{O_{.,t}}$ denotes loss calculated by cross entropy loss at the tth iteration and $L_{O_{T1,.}}$, $L_{O_{T1ce,.}}$, $L_{O_{T2,.}}$, $L_{O_{FLAIR,.}}$ and $L_{O_{fusion,.}}$ represent the loss calculated between outputs of each iteration and the labels. The λ is set to 4 as a trade-off parameters to balance the influence of modality-specific loss and fusion loss. In the test phase, we only keep the sum of the outputs of the last iteration as the final outputs which can be denoted as $Outputs = O_{T1,3} + O_{T1ce,3} + O_{T2,3} + O_{FLAIR,3} + O_{fusion,3}$.

Table 1. Quantitative results (mean ± standard deviation) of different methods on BraTS18.

Method	Acc	Precision	Recall	F1 score
2D CNN	0.613 ± 0.178	0.617 ± 0.259	0.581 ± 0.190	0.555 ± 0.224
TPCNN [17]	0.636 ± 0.000	–	–	–
M^{2}Net [19]	0.664 ± 0.061	0.574 ± 0.141	0.613 ± 0.075	0.589 ± 0.102
3D CBAM [16]	0.650 ± 0.087	0.603 ± 0.082	0.516 ± 0.077	0.518 ± 0.084
3D MFB [18]	0.575 ± 0.018	0.586 ± 0.022	0.540 ± 0.017	0.521 ± 0.037
Ours	**0.692 ± 0.189**	**0.675 ± 0.251**	**0.648 ± 0.205**	**0.622 ± 0.242**

3 Experiment

3.1 Experiment Setup

Datasets. Experiments are carried out on the BraTS 2018: a multi-modality MRI dataset in which each patient includes T1, T1ce, T2 and FLAIR volumes. In this study, we also focus on the overall survival prediction task defined in [19], finally, we have 165 subjects with survival information for this dataset. Our prediction task is constructed following [19], in which patients are divided into three classes: (1) low survival risk, (2) middle survival risk, (3) high survival risk.

Implementation Details. For experiments of BraTS 2018, we first locate the tumor region according to the tumor mask and extract an image volume that is centered on the tumor region. Then we resize the extracted image volume of each subject to a predefined size (*i.e.*, $d = 64$, $w = 64$, $h = 64$). Three individual Adam optimizers are taken to train the feature extraction backbone and two agents, respectively. The learning rate of the feature extraction backbone is set as 1e−4 while the learning rate of agents is set as 1e−2 and the weight decays are set to 1e−5. In every training epoch, the agent will iterate three times (t = 3). The batch size is set to 10 and we adopt a 10-fold cross-validation and report the average performance of 10 folds. For each fold, we further divide the data (the other 9 folds) into training set (80%) and validation set (20%) and take the best model on validation part for evaluation.

Evaluation Metric. We evaluate our method with Accuracy (Acc), Precision, Recall and F1 score. The precision and recall are calculated with one-class-versus-all-other-classes and then calculate F1 score $\left(F1 = \frac{2*Precision*Recall}{Precision+Recall}\right)$.

3.2 Experimental Results

Comparisons with the State of the Art. The proposed method is also compared with other fusion methods, results are reported in Table 1. For comparisons, we choose the following methods: 1) <u>2D CNN fusion:</u> we project the input 3D volume onto 2D images along the vertical axis by averaging the sum

Table 2. Quantitative results (mean ± standard deviation) of ablation studies on BraTS18.

Method	Acc	Precision	Recall	F1 score
Baseline	0.631 ± 0.081	0.546 ± 0.087	0.546 ± 0.043	0.574 ± 0.088
Ours w/o Inter	0.681 ± 0.194	0.666 ± 0.259	**0.660 ± 0.210**	**0.627±0.234**
Ours (1 modal)	0.600 ± 0.071	0.462 ± 0.095	0.429 ± 0.063	0.364 ± 0.033
Ours (2 modal)	0.614 ± 0.065	0.558 ± 0.026	0.496 ± 0.015	0.431 ± 0.035
Ours (3 modal)	0.650 ± 0.079	0.620 ± 0.062	0.575 ± 0.019	0.528 ± 0.039
Ours	**0.692 ± 0.189**	**0.675 ± 0.251**	0.648 ± 0.205	0.622 ± 0.242

of all slices for each modality. Then, we use the ResNet34 [4] (replace the input channel from three to one) as the modality-specific feature extractor, and then fuse the outputs using the concatenation operation. 2) TPCNN [17]: this method uses a CNN model to extract features from multi-modal data and then employs XGBoost to build the regression model. 3) M^{2}Net [19]: a multi-modal shared network to fuse modality-specific features using a bilinear pooling model, exploiting their correlations to provide complementary information. 4) 3D CBAM [16]: using the CBAM [16] module to adaptively generate enhancement weights for intra-/inter- modalities. 5) 3D MFB [18]: a hybrid-fusion network with Mixed Fusion Block (MFB) to adaptively weight different fusion strategies. We retain the encoder of the modality-specific network with the MFB block in [18] to match our classification task. T1 and FLAIR modalities are taken as inputs.

Note that we have the same setting as the TPCNN and M^{2}Net, and take the results reported in [19]. As shown in Table 1, the 3D CBAM [16] which fuses features under channel attention mechanism in the training process, achieves worse performance than the proposed method. In addition, the 3D CBAM shares the same baseline network with ours, indicating that the accuracy boost is due to the RL-agent module not the increasing of backbone size. Moreover, we find that the agent tends to pay more attention to the FLAIR modality from the learned weights for inter-modality fusion. It proves that the proposed RL-based hierarchic multi-modality learning method can provide more effective feature enhancement weights by iteratively learning from previous actions aim to achieve higher prediction confidence, further promoting the classification accuracy.

Ablation Study. We first conduct ablation experiments to validate the design of our proposed different components. We compare the following different settings. (1) Baseline: the 3D network described in Sect. 2.3 without agent modules. We train the model with $Loss = L_{O_{T1}} + L_{O_{T1ce}} + L_{O_{T2}} + L_{O_{FLAIR}} + 4 \times L_{O_{fusion}}$, and the prediction is produced as $Outputs = O_{T1} + O_{T1ce} + O_{T2} + O_{FLAIR} + O_{fusion}$. (2) Ours w/o Inter: the proposed method with RL-based intra-learning, while the modality-specific features are concatenated without inter-learning. (3) Ours (1 modal): the proposed method using one modality, $i.e.$T1, without inter-learning. (4) Ours (2 modal): the proposed method using two modalities, $i.e.$T1

Table 3. Evaluations (mean ± standard deviation) of proposed methods on LNDb.

Method	Acc	Sen	Spe	F1 score
Baseline	0.718 ± 0.007	0.507 ± 0.013	0.480 ± 0.006	0.473 ± 0.006
3D CBAM [16]	0.732 ± 0.027	0.496 ± 0.019	0.498 ± 0.021	0.490 ± 0.020
Baseline w/ia	**0.744 ± 0.012**	**0.520 ± 0.019**	**0.502 ± 0.010**	**0.496 ± 0.010**

and FLAIR. (5) Ours (3 modal): the proposed method using three modalities, *i.e.*T1, T2 and FLAIR. (6) Ours: the proposed method with RL-based intra-/inter- modality learning using four modalities.

Table 2 shows the ablation results. It is observed that the proposed method improves the performance of classification by adopting the proposed RL-based intra-/inter- modality learning. Especially, RL-based intra-modality learning contributes an accuracy improvement of 5% as the input 3D volumes are enhanced by the RL-based proposed weights, while the RL-based inter-modality enhancement gets an improvement of 1.1% in accuracy on this dataset. It is worth noting that our method has better performance than other attention-based fusion methods (compared to 3D CBAM [16] and 3D MFB [18]), which demonstrates that our design of RL-based modality enhancement is better. In addition, from the results, it can be seen that the prediction performance of our method improves when using more modalities, which also verifies the effectiveness of multi-modality learning. We adopt the T1 and FLAIR for Ours (2 modal) and 3D MFB [18] to demonstrate that our method can get better performance using the same modalities.

Validation of the RL-Based Enhancement. To demonstrate the effectiveness of our method, we further explore the performance of RL-based intra-modality learning on LNDb [12] which contains a total of 229 lung nodule CT images from the training set and distinguish lung nodules into three texture classes (solid, sub-solid, and GGO). For experiments of LNDb, we first extract $96 \times 96 \times 96$ cubes from the whole CT scans according to the given center location of a lung nodule. Only one agent is used for intra-modality learning in the LNDb classification model. The batch size is set to 4 and we adopt a 5-fold cross-validation strategy for performance evaluation.

As shown in Table 3, RL-based enhancement also promotes the classification accuracy of LNDb by about 2.6% compared to the baseline, which demonstrates the efficiency of our method on another imaging data. Moreover, we also compare the performance of CBAM [16] on the intra-modality learning on the LNDb, proving that the proposed method helps the model to learn better intra-modal representations with higher metrics.

4   Conclusion

This paper innovatively introduces the RL strategy into the intra-modality learning and inter-modality learning and proposes a novel hierarchic feature enhancement framework for multi-modality learning. With the purpose of exploiting complementary inter-modality features while preserving intra-modality features, the multi-modality learning problem is modeled as a dynamic hierarchic feature enhancement issue. In addition, the proposed RL-based multi-modality learning method is general and has great potential to boost the performance of various medical image tasks. Our future work will focus on different more effective training strategies and extend our framework to other multi-modality medical image analysis problems.

Acknowledgement. This work was supported by the National Key Research and Development Program of China (2019YFE0113900).

References

1. Browning, J., et al.: Uncertainty aware deep reinforcement learning for anatomical landmark detection in medical images. In: de Bruijne, M., et al. (eds.) MICCAI 2021. LNCS, vol. 12903, pp. 636–644. Springer, Cham (2021). https://doi.org/10.1007/978-3-030-87199-4_60
2. Fan, J., Cao, X., Yap, P.T., Shen, D.: Birnet: brain image registration using dual-supervised fully convolutional networks. Med. Image Anal. **54**, 193–206 (2019)
3. Hardoon, D.R., Szedmak, S., Shawe-Taylor, J.: Canonical correlation analysis: an overview with application to learning methods. Neural Comput. **16**(12), 2639–2664 (2004)
4. He, K., Zhang, X., Ren, S., Sun, J.: Deep residual learning for image recognition. In: CVPR, pp. 770–778 (2016)
5. Khvostikov, A., Aderghal, K., Benois-Pineau, J., Krylov, A., Catheline, G.: 3D cnn-based classification using smri and md-dti images for alzheimer disease studies. arXiv preprint arXiv:1801.05968 (2018)
6. Lei, W., et al.: One-shot weakly-supervised segmentation in medical images. arXiv preprint arXiv:2111.10773 (2021)
7. Lin, Y.Y., Liu, T.L., Fuh, C.S.: Multiple kernel learning for dimensionality reduction. IEEE Trans. Pattern Anal. Mach. Intell. **33**(6), 1147–1160 (2010)
8. Morani, K., Unay, D.: Deep learning based automated covid-19 classification from computed tomography images. arXiv preprint arXiv:2111.11191 (2021)
9. Ngiam, J., Khosla, A., Kim, M., Nam, J., Lee, H., Ng, A.Y.: Multimodal deep learning. In: ICML (2011)
10. Nie, D., et al.: Multi-channel 3d deep feature learning for survival time prediction of brain tumor patients using multi-modal neuroimages. Sci. Rep. **9**(1), 1–14 (2019)
11. Nie, D., Zhang, H., Adeli, E., Liu, L., Shen, D.: 3D deep learning for multi-modal imaging-guided survival time prediction of brain tumor patients. In: Ourselin, S., Joskowicz, L., Sabuncu, M.R., Unal, G., Wells, W. (eds.) MICCAI 2016. LNCS, vol. 9901, pp. 212–220. Springer, Cham (2016). https://doi.org/10.1007/978-3-319-46723-8_25
12. Pedrosa, J., et al.: Lndb: a lung nodule database on computed tomography. arXiv preprint arXiv:1911.08434 (2019)
13. Peng, Y., Huang, X., Qi, J.: Cross-media shared representation by hierarchical learning with multiple deep networks. In: IJCAI, pp. 3846–3853 (2016)

14. Saha, A., et al.: Weakly supervised 3D classification of chest CT using aggregated multi-resolution deep segmentation features. In: Medical Imaging 2020: Computer-Aided Diagnosis, vol. 11314, p. 1131408. International Society for Optics and Photonics (2020)
15. Wang, A., Lu, J., Cai, J., Cham, T.J., Wang, G.: Large-margin multi-modal deep learning for RGB-D object recognition. IEEE Trans. Multimedia **17**(11), 1887–1898 (2015)
16. Woo, S., Park, J., Lee, J.Y., Kweon, I.S.: Cbam: convolutional block attention module. In: Proceedings of the European Conference on Computer Vision (ECCV), pp. 3–19 (2018)
17. Zhou, F., Li, T., Li, H., Zhu, H.: TPCNN: two-phase patch-based convolutional neural network for automatic brain tumor segmentation and survival prediction. In: Crimi, A., Bakas, S., Kuijf, H., Menze, B., Reyes, M. (eds.) BrainLes 2017. LNCS, vol. 10670, pp. 274–286. Springer, Cham (2018). https://doi.org/10.1007/978-3-319-75238-9_24
18. Zhou, T., Fu, H., Chen, G., Shen, J., Shao, L.: Hi-net: hybrid-fusion network for multi-modal mr image synthesis. IEEE Trans. Med. Imaging **39**(9), 2772–2781 (2020)
19. Zhou, T., et al.: M^2Net: multi-modal multi-channel network for overall survival time prediction of brain tumor patients. In: Martel, A.L., et al. (eds.) MICCAI 2020. LNCS, vol. 12262, pp. 221–231. Springer, Cham (2020). https://doi.org/10.1007/978-3-030-59713-9_22
20. Zhou, T., Thung, K.H., Zhu, X., Shen, D.: Effective feature learning and fusion of multimodality data using stage-wise deep neural network for dementia diagnosis. Hum. Brain Mapp. **40**(3), 1001–1016 (2019)

A New Dataset and a Baseline Model for Breast Lesion Detection in Ultrasound Videos

Zhi Lin[1], Junhao Lin[1], Lei Zhu[2,3(✉)], Huazhu Fu[4], Jing Qin[5],
and Liansheng Wang[1]

[1] Department of Computer Science, School of Informatics, Xiamen University,
Xiamen, China
[2] ROAS Thrust, System Hub, The Hong Kong University of Science and Technology
(Guangzhou), Guangzhou, China
[3] Department of Electronic and Computer Engineering, The Hong Kong University
of Science and Technology, Hong Kong SAR, China
`leizhu@ust.hk`
[4] Institute of High Performance Computing, Agency for Science, Technology
and Research, Fusionopolis, Singapore
[5] School of Nursing, The Hong Kong Polytechnic University, Kowloon, Hong Kong

Abstract. Breast lesion detection in ultrasound is critical for breast
cancer diagnosis. Existing methods mainly rely on individual 2D ultrasound images or combine unlabeled video and labeled 2D images to
train models for breast lesion detection. In this paper, we first collect and annotate an ultrasound video dataset (188 videos) for breast
lesion detection. Moreover, we propose a clip-level and video-level feature aggregated network (CVA-Net) for addressing breast lesion detection in ultrasound videos by aggregating video-level lesion classification
features and clip-level temporal features. The clip-level temporal features
encode local temporal information of ordered video frames and global
temporal information of shuffled video frames. In our CVA-Net, an inter-
video fusion module is devised to fuse local features from original video
frames and global features from shuffled video frames, and an intra-video
fusion module is devised to learn the temporal information among adjacent video frames. Moreover, we learn video-level features to classify the
breast lesions of the original video as benign or malignant lesions to further enhance the final breast lesion detection performance in ultrasound
videos. Experimental results on our annotated dataset demonstrate that
our CVA-Net clearly outperforms state-of-the-art methods. The corresponding code and dataset are publicly available at https://github.com/
jhl-Det/CVA-Net.

Keywords: Breast lesion detection in ultrasound videos · Inter-video
fusion module · Intra-video fusion module · Clip/video aggregation

---

Z. Lin and J. Lin—Equal contribution.

© The Author(s), under exclusive license to Springer Nature Switzerland AG 2022
L. Wang et al. (Eds.): MICCAI 2022, LNCS 13433, pp. 614–623, 2022.
https://doi.org/10.1007/978-3-031-16437-8_59

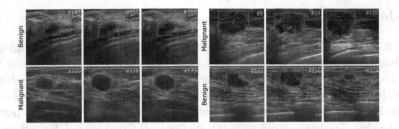

Fig. 1. Examples of our annotated ultrasound video dataset for breast lesion detection.

1   Introduction

Breast cancer is a leading cause of death for women worldwide [16]. Currently, ultrasound imaging is the most commonly used and effective technique for breast cancer detection due to its versatility, safety, and high sensitivity [13]. Detecting breast lesions in ultrasound is often taken as an important step of computer-aided diagnosis systems to assist radiologists in the ultrasound-based breast cancer diagnosis [1,20,23]. However, accurate breast lesion detection in ultrasound videos is challenging due to blurry breast lesion boundaries, inhomogeneous distributions, changeable breast lesion sizes and positions in dynamic videos.

Existing methods [9,10,18,19,21] mainly performed the breast lesion segmentation or detection in 2D ultrasound images, or fused unlabeled videos with labeled 2D images for ultrasound video breast lesion detection. With the dominated results of convolutional neural networks on medical imaging, it is highly desirable to extend deep-learning-based breast lesion detection from the image level to video level, since the latter can leverage temporal consistency to address many in-frame ambiguities. The major obstacle for this extension is the lack of an ultrasound video dataset with appropriate annotations for breast lesion segmentation, both of which are essential for training deep models for breast lesion segmentation in ultrasound videos.

To do so, we in this work first provide a video dataset for breast lesion detection in ultrasound; see Fig. 1 for several examples of annotated videos. Then, we present a novel network for boosting breast lesion detection in ultrasound videos by aggregating video-level classification features and clip-level temporal features, which contains a local temporal feature from the input video frames and a global temporal information from shuffled video frames. The contributions of this work could be summarised as: 1) We develop a novel network to learn clip-level temporal features and video-level lesion classification features for boosting breast lesion detection in ultrasound videos. 2) We collect and annotate a video dataset (118 videos) for breast lesion detection in ultrasound videos. 3) An inter-video fusion module is devised to attentively aggregate local features from the original video frames and global features from the shuffled video frames, while an intra-video fusion module is developed to fuse temporal features encoded among adjacent video frames. 4) Experimental results on our annotated dataset demonstrate that our network sets a new state-of-the-art performance on breast lesion detection in ultrasound videos.

2   Method

Figure 2 shows the schematic illustration of the developed clip-level and video-level feature aggregation network (CVA-Net). The motivation behind our CVA-Net is to integrate video-level lesion classification features and clip-level features from adjacent video frames, and such clip-level features include local temporal information from the original video and global temporal information from the shuffled video. To do so, given an input ultrasound video with T frames, we first shuffle the ordered frames index sequence $\{1, \cdots, T\}$ of the input video to obtain a new index sequence, which is then used to generate a shuffled video. For a current video frame (I_k), our CVA-Net takes three neighboring images (denoted as I_k, I_{k-1}, and I_{k+1}) of the input video, and then passes I_k, I_{k-1}, and I_{k+1} into a feature extraction backbone (i.e., ResNet50 [5]) with three convolutional layers to obtain three features, which are denoted as L_k, L_{k-1}, and L_{k+1}). Apparently, L_k, L_{k-1}, and L_{k+1} contain three CNN feature maps with different spatial resolutions. Meanwhile, we take three corresponding images (denoted as S_k, S_{k-1}, and S_{k+1}) of the shuffled video, apply data augmentation techniques (e.g., horizontal flip, random crop, random pepper) on them, and pass augmented images into the feature extraction backbone to produce another three features G_k, G_{k-1}, and G_{k+1}. Afterwards, we devise an inter-video fusion module to integrate local temporal features (L_k, L_{k-1}, and L_{k+1}) from input ultrasound video, and global features (G_k, G_{k-1}, and G_{k+1}) from the shuffled video to obtain three features, which are denoted as P_k, P_{k-1}, and P_{k+1}. Then, we devise an intra-video fusion to fuse these three features P_k, P_{k-1}, and P_{k+1} to produce a new feature map Q_k, which is then passed into a classifier to predict whether the breast lesions are benign or malignant, and a bounding box predictor to produce the final breast lesion detection of the current video frame I_k.

2.1   Inter-video Fusion Module

Existing breast lesion detection in ultrasound video methods [1] extracted the temporal information encoded in ordered video frames sampled from the input video [2]. Unlike this, we devise an inter-video fusion module to utilize three frames from a shuffled video to remove the temporal order and enhance the global semantic information for detecting breast lesions. As shown in Fig. 2, our inter-video fusion module has three inter-video fusion blocks to fuse three pairs of local features and global features, which are L_k and G_k, L_{k-1} and G_{k-1}, as well as L_{k+1} and G_{k+1}, respectively. By doing so, we can fuse the local information from the input video and the global information from the shuffled video, thereby improving the video breast lesion detection.

Figure 3(a) depicts the schematic illustration of the inter-video fusion block, which integrates the local feature $L_k=(L_k^1, L_k^2, L_k^3)$ and the global feature $G_k=(G_k^1, G_k^2, G_k^3)$. The inter-video fusion block utilizes three attention blocks to integrate three features of L_k and G_k, and the i-th attention is for L_k^i and G_k^i (i=1, 2, 3); see Fig. 3(b). Specifically, we first apply two different linear layers on L_k^i, reshape both of them into two $c \times wh$ matrices, and also reshape L_k^i into a

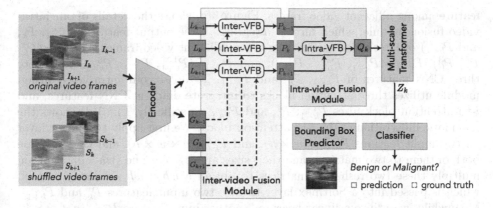

Fig. 2. Schematic illustration of our clip/video-level feature aggregation network (CVA-Net) for breast lesion detection network in ultrasound videos. "Inter-VFB" and "Intra-VFB" denotes a inter-video fusion block and a intra-video fusion block.

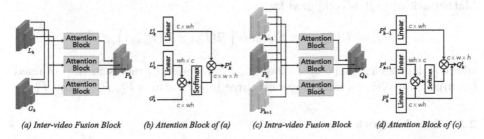

(a) Inter-video Fusion Block   (b) Attention Block of (a)   (c) Intra-video Fusion Block   (d) Attention Block of (c)

Fig. 3. Schematic illustration of the inter-video fusion module and intra-video fusion module of our network (see Fig. 2).

$wh \times c$ matrix. Then we multiply two reshaped matrices to obtain a similarity matrix ($wh \times wh$), followed by a softmax layer. Then, we multiple the result with another reshaped matrix of L_k^i, and reshape the multiplication output to obtain the output feature H_k^i ($c \times w \times h$). Mathematically, P_k^i is computed by:

$$P_k^i = \hat{L}_k^i \times Softmax\left(\mathcal{R}(G_k^i) \times \mathcal{R}(\tilde{L}_k^i)\right), \quad (1)$$

where $\mathcal{R}$ denotes a reshape operation. $\tilde{L}_k^i$ and $\hat{L}_k^i$ represents the obtained features via utilizing two different linear layers on the input feature L_k^i.

2.2 Intra-video Fusion Module

Although our inter-video fusion module is capable of integrating several paired of local features from the original video and the global features of the shuffled video, it neglects the temporal information among the adjacent video frames. In this regard, we develop an intra-video fusion module to capture the temporal feature among adjacent video frames. Figure 3(c) shows the details of our intra-video fusion module, which further integrates three output features (P_{k-1}, P_k, and P_{k+1}) of inter-video fusion from three adjacent video frames. Let $P_k = (P_k^1, P_k^2, P_k^3)$, $P_{k-1} = (P_{k-1}^1, P_{k-1}^2, P_{k-1}^3)$, and $P_{k+1} = (P_{k+1}^1, P_{k+1}^2, P_{k+1}^3)$ to denote three CNN features of P_{k-1}, P_k, and P_{k+1}. To do so, our intra-video fusion module utilizes three attention blocks to integrate different CNN features, and i-th attention block fuses P_{k-1}^i, P_k^i, and P_{k+1}^i. Specifically, Fig. 3(d) shows the schematic illustration of the i-th attention block. We first apply two linear layer on two input features P_k^i ($c \times w \times h$), and P_{k+1}^i ($c \times w \times h$), and then reshape both of them to two matrices, and their sizes are ($c \times wh$) and ($wh \times c$). We then multiply these two reshaped matrices to obtain a $wh \times wh$ similarity matrix, which is passed into a Softmax layer to fuse two input features P_k^i and P_{k+1}^i. Meanwhile, we utilize a linear layer on another input feature P_{k-1}^i ($c \times w \times h$) and reshape it into $c \times wh$. To further fuse P_{k-1}^i, we multiply its reshaped matrix with the similarity matrix from P_k^i and P_{k+1}^i to obtain a new feature ($c \times wh$), which is then reshaped into a 3D feature map ($c \times w \times h$); see Q_k^i of Fig. 3(d). Mathematically, Q_k^i is computed by:

$$Q_k^i = \mathcal{R}(\hat{P}_{k-1}^i) \times Softmax\left(\mathcal{R}(\hat{P}_k^i) \times \mathcal{R}(\hat{P}_{k+1}^i)\right) , \qquad (2)$$

where $\mathcal{R}$ denotes a reshape operation. $\hat{P}_k^i$, $\hat{P}_{k+1}^i$, and $\hat{P}_{k-1}^i$ represent the obtained features via utilizing a linear layer on three input features P_k^i, P_{k-1}^i, and P_{k+1}^i.

2.3   Our Network

Apparent from learning clip-level features (i.e., $P_k = (P_k^1, P_k^2, P_k^3)$) via fusing local and global temporal features via our inter-video fusion modules and our intra-video fusion module, we further learn a video-level information, which represents the classification result (benign or maligant) on each breast lesion of the input ultrasound video. To do so, we first pass three features (i.e., P_k^1, P_k^2, and P_k^3) of P_k into a multi-scale transformer (i.e., deformable-DETR [25]) to learn a long-range dependency feature map Z. We then apply a linear layer on Z to classify the breast lesions as benign lesions or malignant lesions. Furthermore, we predict bounding boxes of breast lesion as the final breast lesion detection result of the current video frame I_k (see Fig. 2). We utilize a negative log-likelihood to compute the lesion classification loss, and a linear combination of the ℓ_1 loss and the generalized IoU loss [12] for computing the breast lesion detection loss.

Implementation Details.   We utilize ResNet-50 [5] pre-trained on ImageNet [3] to initialize the backbone of our network, and deformable DETR [25] pre-trained on COCO 2017 [8] to initialize the multi-scale transformer module of our network, while other network parameters are initialized by a normal distribution. All training videos are randomly horizontally flipped, resized, and cropped for data augmentation. Our network is implemented on Pytorch and trained using an Adam with 50 epochs, an initial learning rate of 2×10^{-4}, and a weight decay of 1×10^{-4}. The whole architecture is trained on three GeForce RTX 2080 Ti GPUs, and each GPU has a batch-size of 1.

Table 1. Quantitative comparisons of our network and compared methods on our annotated video dataset.

Method	Type	Backbone	AP	AP_{50}	AP_{75}
GFL [6]	Image	ResNet-50	23.4	46.3	22.2
Cascade RPN [14]	Image	ResNet-50	24.8	42.4	27.3
Faster R-CNN [11]	Image	ResNet-50	25.2	49.2	22.3
VFNet [22]	Image	ResNet-50	28.0	47.1	31.0
RetinaNet [7]	Image	ResNet-50	29.5	50.4	32.4
DFF [26]	Video	ResNet-50	25.8	48.5	25.1
FGFA [24]	Video	ResNet-50	26.1	49.7	27.0
SELSA [17]	Video	ResNet-50	26.4	45.6	29.6
Temporal ROI Align [4]	Video	ResNet-50	29.0	49.9	33.1
MEGA [2]	Video	ResNet-50	32.3	57.2	35.7
CVA-Net (ours)	Video	ResNet-50	**36.1**	**65.1**	**38.5**

3 Experiments and Results

Dataset. Note that there is no public ultrasound video breast lesion detection benchmark dataset with annotations. To evaluate the effectiveness of the developed network, we collect a breast lesion ultrasound video dataset with 188 videos, which has 113 malignant videos and 75 benign videos. These 118 videos has 25, 272 images in total, and the number of ultrasound images at each video varied from 28 to 413. Each video has a complete scan of the tumor, from its appearance to the largest section, then to its disappearance. All videos are acquired by LOGIQ-E9 and PHILIPS TIS L9-3. Two pathologists with eight years of experience in breast pathology were invited to manually annotate the breast lesion rectangles inside of each video frame and give the corresponding classification label for the breast lesions of the video. We further randomly and evenly select 38 videos from the dataset as a testing set (about 20% of the dataset) and the remaining videos as the training set.

Evaluation Metrics. We utilize three widely-used metrics for quantitatively comparing different ultrasound video breast lesion detection methods. They are average precision (AP), AP_{50}, and AP_{75}. Please refer to [15] for the definitions of AP, AP_{50}, and AP_{75}.

3.1 Comparisons with State-of-the-Arts

We compare our network against ten state-of-the-art methods, including five image-based methods and five video-based methods. The image-based methods include, GFL [6], Cascade RPN [14], Faster R-CNN [11], VFNet [22], and RetinaNet [7]. The five video-based methods are DFF [26], FGFA [24], SELSA [17], Temporal ROI Align [4], and MEGA [2]. For providing a fair comparison, we

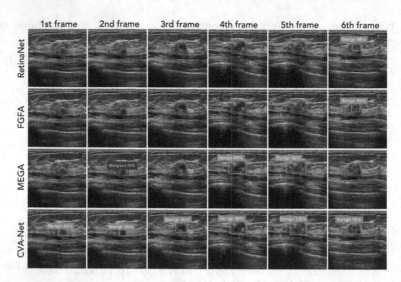

Fig. 4. Visual comparisons of video breast lesion detection results produced by our network and state-of-the-art methods (i.e., RetinaNet, FGFA, and MEGA) for multiple frames of an ultrasound video with benign breast lesions.

obtain the detection results of all compared methods by exploiting its public implementations or implementing them by ourselves. We also re-train these networks on our dataset, and fine-tune the network parameters for achieving the best detection results of ultrasound video breast lesions.

Quantitative Comparisons. Table 1 summarizes the quantitative results of our network and all ten compared breast lesion video detection methods. In general, video-based methods tend to have larger AP, AP_{50}, and AP_{75} scores than image-based ones. Specifically, among ten compared methods, MEGA has the largest AP score of 32.3, the largest AP_{50} score of 57.2, and the largest AP_{75} score of 35.7. On the contrary, our method further outperforms MEGA in terms of all three metrics (AP, AP_{50}, and AP_{75}). Compared to MEGA, our CVA-Net improves the AP score from 32.3 to 36.1, the AP_{50} score from 57.2 to 65.1, and the AP_{75} score from 35.7 to 38.5.

Visual Comparisons. Figure 4 visually compares breast lesion detection results produced by our network and four compared methods for multiple frames of an ultrasound video with benign breast lesions. As shown in the first two rows of Fig. 4, RetinaNet and FGFA fail to detect the breast lesions of the first five frames. MEGA obtains a better breast lesion detection result among three compared methods, but also fails to detect breast lesions at the 1st, 3rd, and 6th video frames. Moreover, although MEGA can detect the breast lesion of the 2nd frame, the classification result is incorrect (see the blue rectangle of Fig. 4). However, our method (see CVA-Net at last row of Fig. 4) can correctly detect the breast lesions of all video frames and has a correct breast lesion classification result for each video frame.

Table 2. Quantitative results of ablation study experiments. "Ours-w/o-cla" is to remove the lesion classification branch from our network. "Ours-w/o-aug" denotes that we do not utilize any data augmentation operation on shuffled video frames, while "Ours-VF", "Ours-RR", and "Ours-CC" denote that we utilize the vertical flip, random rotation, and center crop operation on the shuffled video frames.

	AP	AP_{50}	AP_{75}
Basic	27.8	52.8	26.4
Basic+Inter-video	33.5	55.1	38.5
Basic+Intra-video	27.8	60.9	36.3
Our method	**36.1**	**65.1**	**38.5**
Ours-w/o-cla	34.4	61.4	35.5
Ours-w/o-aug	35.6	62.1	37.8
Ours-VF	35.5	59.8	37.1
Ours-RR	35.4	62.1	38.4
Ours-CC	35.8	62.4	37.9

3.2 Ablation Study

Effectiveness of Clip-Level Features. We first construct three baselines to verify the clip-level features learned by our inter-video fusion and intra-video fusion. The first baseline network ("basic") is to remove all inter-frame fusion modules and the intra-frame fusion modules from our network. The second baseline network ("basic+inter-video") is to add inter-video fusion module into "basic", while the third one ("basic+intra-video") is to add the intra-video fusion module into "basic". As shown in Table 2, "basic+inter-video" has a larger AP, AP_{50}, and AP_{75} scores than "basic", showing that the leveraging our inter-video fusion module to fuse local information of the input video and global information of the shuffled video can enhance the breast lesion detection performance in ultrasound video. Moreover, our network achieves superior results of AP, AP_{50}, and AP_{75} over "basic+inter-video" and "basic+intra-video". It improves the AP score from 27.8 to 36.1, the AP_{50} score from 52.8 to 65.1, and the AP_{75} score from 26.5 to 38.6. It indicates that integrating the inter-video fusion module and the intra-video fusion module together in our method can further enhance the breast lesion detection accuracy in ultrasound video.

Effectiveness of Video-Level Features. We also compare our method with and without the lesion classification branch and show their results in Table 2. It shows that the video breast lesion detection accuracy of our method are reduced when we remove the lesion classification branch from our network.

Effectiveness of Data Augmentation on Shuffled Video Frames. Note that we have utilized a data augmentation technique (i.e., random pepper) on shuffled video frames before passing them into the feature extraction backbone (see Encoder of Fig. 2). Hence, we further show the video breast lesion detection results of our network without and with different data augmentation techniques, including vertical flip, random rotation, and center crop. Apparently, our method with a random pepper has the best detection performance, and thus our network empirically utilizes this data augmentation in our experiments.

4 Conclusion

This paper first collects and annotates the first ultrasound video dataset for breast lesion detection with 188 videos. Moreover, we present a clip-level and video-level feature aggregation network for boosting breast lesion detection in ultrasound videos. The main idea of our network is to combine clip-level features and video-level features for detecting breast lesions in videos, and the clip-level features are learned by devising inter-video fusion modules and intra-video fusion on the input ordered video and a shuffled video. Experimental results on our annotated dataset show that our network has obtained superior breast lesion detection performance over state-of-the-art methods. Our further work includes the collection of more video data, exploration of a more systematic approach or complicated video shuffling operations.

Acknowledgment. This work was supported by the National Natural Science Foundation of China (No. 61902275, No. 12026604), AME Programmatic Fund (A20H4b0141), and Hong Kong Research Grants Council under General Research Fund (No. 15205919).

References

1. Chen, S., et al.: Semi-supervised breast lesion detection in ultrasound video based on temporal coherence. arXiv preprint arXiv:1907.06941 (2019)
2. Chen, Y., Cao, Y., Hu, H., Wang, L.: Memory enhanced global-local aggregation for video object detection. In: Proceedings of the IEEE/CVF Conference on Computer Vision and Pattern Recognition, pp. 10337–10346 (2020)
3. Deng, J., Dong, W., Socher, R., Li, L.J., Li, F.F.: Imagenet: a large-scale hierarchical image database. In: IEEE Conference on Computer Vision and Pattern Recognition (2009)
4. Gong, T., et al.: Temporal ROI align for video object recognition. In: Proceedings of the AAAI Conference on Artificial Intelligence. vol. 35, pp. 1442–1450 (2021)
5. He, K., Zhang, X., Ren, S., Sun, J.: Deep residual learning for image recognition. In: IEEE Conference on Computer Vision and Pattern Recognition (CVPR), pp. 770–778 (2016)
6. Li, X., et al.: Generalized focal loss: learning qualified and distributed bounding boxes for dense object detection. arXiv preprint arXiv:2006.04388 (2020)
7. Lin, T.Y., Goyal, P., Girshick, R., He, K., Dollár, P.: Focal loss for dense object detection. In: Proceedings of the IEEE International Conference on Computer Vision, pp. 2980–2988 (2017)
8. Lin, T., et al.: Microsoft COCO: common objects in context. In: Fleet, D., Pajdla, T., Schiele, B., Tuytelaars, T. (eds.) ECCV 2014. LNCS, vol. 8693, pp. 740–755. Springer, Cham (2014). https://doi.org/10.1007/978-3-319-10602-1_48

9. Movahedi, M.M., Zamani, A., Parsaei, H., Tavakoli Golpaygani, A., Haghighi Poya, M.R.: Automated analysis of ultrasound videos for detection of breast lesions. Middle East J. Cancer **11**(1), 80–90 (2020)

10. Qi, X., et al.: Automated diagnosis of breast ultrasonography images using deep neural networks. Med. Image Anal. **52**, 185–198 (2019)

11. Ren, S., He, K., Girshick, R., Sun, J.: Faster R-CNN: towards real-time object detection with region proposal networks. arXiv preprint arXiv:1506.01497 (2015)

12. Rezatofighi, S.H., Tsoi, N., Gwak, J., Sadeghian, A., Reid, I., Savarese, S.: Generalized intersection over union: a metric and a loss for bounding box regression. In: IEEE/CVF Conference on Computer Vision and Pattern Recognition (CVPR), pp. 658–666 (2019)

13. Stavros, A.T., Thickman, D., Rapp, C.L., Dennis, M.A., Parker, S.H., Sisney, G.A.: Solid breast nodules: use of sonography to distinguish between benign and malignant lesions. Radiology **196**(1), 123–134 (1995)

14. Vu, T., Jang, H., Pham, T.X., Yoo, C.D.: Cascade RPN: delving into high-quality region proposal network with adaptive convolution. ArXiv abs/1909.06720 (2019)

15. Wang, X., Kong, T., Shen, C., Jiang, Y., Li, L.: SOLO: segmenting objects by locations. In: Vedaldi, A., Bischof, H., Brox, T., Frahm, J.-M. (eds.) ECCV 2020. LNCS, vol. 12363, pp. 649–665. Springer, Cham (2020). https://doi.org/10.1007/978-3-030-58523-5_38

16. Wild, C., Weiderpass, E., Stewart, B.W.: World Cancer Report: Cancer Research for Cancer Prevention. IARC Press (2020)

17. Wu, H., Chen, Y., Wang, N., Zhang, Z.: Sequence level semantics aggregation for video object detection. In: 2019 IEEE/CVF International Conference on Computer Vision (ICCV), pp. 9216–9224 (2019)

18. Xue, C., et al.: Global guidance network for breast lesion segmentation in ultrasound images. Med. Image Anal. **70**, 101989 (2021)

19. Yang, Z., Gong, X., Guo, Y., Liu, W.: A temporal sequence dual-branch network for classifying hybrid ultrasound data of breast cancer. IEEE Access **8**, 82688–82699 (2020)

20. Yap, M.H., et al.: Automated breast ultrasound lesions detection using convolutional neural networks. IEEE J. Biomed. Health Inform. **22**(4), 1218–1226 (2017)

21. Zhang, E., Seiler, S., Chen, M., Lu, W., Gu, X.: Birads features-oriented semi-supervised deep learning for breast ultrasound computer-aided diagnosis. Phys. Med. Biol. **65**(12), 125005 (2020)

22. Zhang, H., Wang, Y., Dayoub, F., Sunderhauf, N.: Varifocalnet: an IoU-aware dense object detector. In: IEEE/CVF Conference on Computer Vision and Pattern Recognition (CVPR), pp. 8510–8519 (2021)

23. Zhu, L., et al.: A second-order subregion pooling network for breast lesion segmentation in ultrasound. In: Martel, A.L., et al. (eds.) MICCAI 2020. LNCS, vol. 12266, pp. 160–170. Springer, Cham (2020). https://doi.org/10.1007/978-3-030-59725-2_16

24. Zhu, X., Wang, Y., Dai, J., Yuan, L., Wei, Y.: Flow-guided feature aggregation for video object detection. In: IEEE International Conference on Computer Vision (ICCV), pp. 408–417 (2017)

25. Zhu, X., Su, W., Lu, L., Li, B., Wang, X., Dai, J.: Deformable detr: deformable transformers for end-to-end object detection. arXiv preprint arXiv:2010.04159 (2020)

26. Zhu, X., Xiong, Y., Dai, J., Yuan, L., Wei, Y.: Deep feature flow for video recognition. In: Proceedings of the IEEE Conference on Computer Vision and Pattern Recognition, pp. 2349–2358 (2017)

RemixFormer: A Transformer Model for Precision Skin Tumor Differential Diagnosis via Multi-modal Imaging and Non-imaging Data

Jing Xu[1], Yuan Gao[1], Wei Liu[1], Kai Huang[2], Shuang Zhao[2], Le Lu[1],
Xiaosong Wang[1], Xian-Sheng Hua[1], Yu Wang[1(✉)], and Xiang Chen[2]

[1] DAMO Academy, Alibaba Group, Hangzhou, China
tonggou.wangyu@alibaba-inc.com
[2] Department of Dermatology, Xiangya Hospital Central South University,
Changsha, China

Abstract. Skin tumor is one of the most common diseases worldwide and the survival rate could be drastically increased if the cancerous lesions were identified early. Intrinsic visual ambiguities displayed by skin tumors in multi-modal imaging data impose huge amounts of challenges to diagnose them precisely, especially at the early stage. To achieve high diagnosis accuracy or precision, all possibly available clinical data (imaging and/or non-imaging) from multiple sources are used, and even the missing-modality problem needs to be tackled when some modality may become unavailable. To this end, we first devise a new disease-wise pairing of all accessible patient data if they fall into the same disease category as a remix operation of data samples. A novel cross-modality-fusion module is also proposed and integrated with our transformer-based multi-modality deep classification framework that can effectively perform multi-source data fusion (i.e., clinical images, dermoscopic images and accompanied with clinical patient-wise metadata) for skin tumors. Extensive quantitative experiments are conducted. We achieve an absolute 6.5% increase in averaged F1 and 2.8% in accuracy for the classification of five common skin tumors by comparing to the prior leading method on Derm7pt dataset of 1011 cases. More importantly, our method obtains an overall 88.5% classification accuracy using a large-scale in-house dataset of 5601 patients and in ten skin tumor classes (pigmented and non-pigmented). This experiment further validates the robustness and implies the potential clinical usability of our method, in a more realistic and pragmatic clinic setting.

Keywords: Skin tumor · Multi-modality fusion · Remix sampling

---

Supplementary Information The online version contains supplementary material available at https://doi.org/10.1007/978-3-031-16437-8_60.

© The Author(s), under exclusive license to Springer Nature Switzerland AG 2022
L. Wang et al. (Eds.): MICCAI 2022, LNCS 13433, pp. 624–633, 2022.
https://doi.org/10.1007/978-3-031-16437-8_60

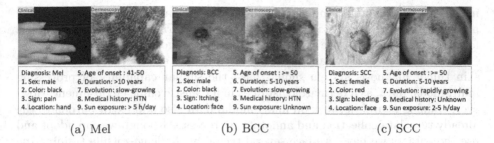

Diagnosis: Mel	5. Age of onset : 41-50
1. Sex: male	6. Duration: >10 years
2. Color: black	7. Evolution: slow-growing
3. Sign: pain	8. Medical history: HTN
4. Location: hand	9. Sun exposure: > 5 h/day

(a) Mel

Diagnosis: BCC	5. Age of onset : >= 50
1. Sex: male	6. Duration: 5-10 years
2. Color: black	7. Evolution: slow-growing
3. Sign: Itching	8. Medical history: HTN
4. Location: face	9. Sun exposure: Unknown

(b) BCC

Diagnosis: SCC	5. Age of onset : >= 50
1. Sex: female	6. Duration: 5-10 years
2. Color: red	7. Evolution: rapidly growing
3. Sign: bleeding	8. Medical history: Unknown
4. Location: face	9. Sun exposure: 2-5 h/day

(c) SCC

Fig. 1. Samples of X-SkinTumor-10 dataset

1   Introduction

Accurate early skin lesion diagnosis is crucial to prevent skin cancers and can significantly increase the 5-year survival rate of malignant tumors [10] such as Melanoma (Mel). However, it remains a challenging task even for well-trained professionals, and experienced dermatologists find clues in multi-modal data from different types of clinical sources. For example, precision diagnosis of skin tumors often involves examining lesions, analyzing details in dermatoscopic images, and referencing metadata, e.g., medical history. Figure 1 shows three skin tumor cases with different conditions, where the clinical and dermoscopic images are listed and accompanied with patients' meta information. Specifically, in Fig. 1(a), patient's non-imaging data (the age, slow progression over the years, and long duration of the sun exposure) largely support the diagnosis of malignant melanoma in addition to the observation of black plaque visually.

In recent years, computer-aided diagnosis has shown some impressive performance in supporting dermatologists' diagnoses [4–7,11–15]. Esteva et al. [4] demonstrated that the convolutional neural networks (CNN) are capable of classifying malignant and benign skin tumors with a level of competence comparable to dermatologists. Tschandl et al. [13,14] also reported that learning-based classifiers not only can outperform human experts in the diagnosis of pigmented skin lesions but can further improve the diagnostic accuracy when high quality computer-aided clinical decision-making is available to dermatologists. Such a way performs better over those by either algorithms or physicians alone. When it comes to employing multi-modal data, Ge et al. [5] proposed a deep convolutional neural network architecture to capture discriminative features from both clinical and dermoscopic images and showed that the multi-modality method significantly outperforms single-modality methods. Furthermore, Haenssle et al. [6,7] studied CNN's diagnostic performance with a large group of dermatologists, in which most dermatologists were outperformed by the CNN models when only dermoscopic images are provided. When given multi-modal information, most dermatologists only performed equivalently as the CNN-based models. Thus, one can reasonably assume that better performance can be achieved provided that the neural networks are also trained with a range of multi-modal data as dermatologists did. Most recently, Tang et al. [12] constructed a two-stage approach named FusionM4Net. It concatenates features of clinical and dermoscopic images at the first stage, and then incorporates the patient's metadata with the prediction from the first stage via SVM-based clustering. The final diagnosis is formed by the fusion of the predictions from two stages.

To achieve the performance on a par with human experts, a computer-aided skin tumor diagnosis system should have the similar capability of processing multi-modal data, as a dermatologist does in a realistic clinic environment. However, training neural network on multi-modal data also add extra burden to the already tidy data collection and annotation process. Although we can adopt and use the data as we have, we face several technical challenges while training the multi-modal neural networks. First, it is more than common in multi-modal data to have some modalities inaccessible, whenever they are not acquired or indeed missing. It is often unrealistic to ensure the completeness of every modality for each sample/patient. Furthermore, another difficulty is how to effectively merge the multiple source of information for the multi-modality models, especially together with the first challenge in the training phase. In contrast to previous methods [1,8,12], where paired data are required for the training, a novel data sampling strategy via disease-wise pairing (DWP as a remix of data samples on the disease-class level) is presented. Integrated with a scalable cross-modality-fusion module, our proposed multi-modal classification framework can better handle the incomplete data in the model training and achieve higher classification accuracy in both the publicly released dataset (covering mainly pigmented tumors) and a large-scale private skin tumor dataset with 10 categories, containing additional non-pigmented skin tumors such as basal cell carcinoma (BCC) and squamous cell carcinoma (SCC) which have higher incidence rates in clinics.

Our contributions are three-fold: (a) We propose a new transformer-based multi-modality classification framework for skin tumors, which includes the novel DWP sampling strategy for tackling the missing modality issue in training data. This disease-wise pairing augmentation process makes our framework more flexible for model training, and more importantly, provides better training generalizability for the trained model; (b) We improve the multi-modal data fusion with a more efficient cross-modality fusion module than the conventional concatenation. A leading recognition accuracy of 81.3% is achieved on Derm7pt [8] dataset; (c) We compose a large-scale multi-modality dataset with significantly more cases than the existing databases and importantly, it covers both pigmented and non-pigmented skin tumors (closer to the real data distributions of daily clinical routines). Our multi-modal framework achieves 88.5% accuracy and 98.4% AUC on this more comprehensive and realistic database.

2   Method

Our proposed framework is outlined in the Fig. 2. During training, the multi-modal samples are first grouped in a disease-wise fashion, i.e., multi-modal data can be randomly selected and paired as long as they belong to the same disease category. It can be seen as a remix of the multi-modal data on the disease level since a multi-modal tuple here will have data from different studies/patients. There are numerous possible permutations of modalities among all the data in

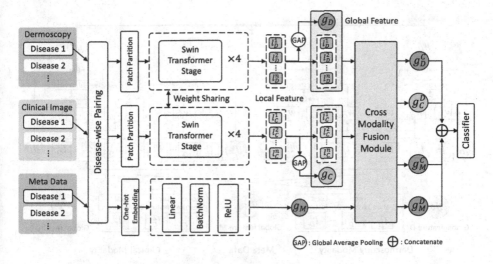

Fig. 2. The pipeline of our multi-modal cross-fusion transformer

the same disease class. It not only significantly increases the amount of training data (as a form of effective data augmentation) but also largely enhances the robustness and generalizability of the trained model.

A Swin Transformer [9] is adopted as the unified backbone to encode the clinical and dermoscopic images. Metadata is separately processed using the one-hot representation and followed by a linear embedding. The image features and metadata feature are fed into the cross-modality fusion (CMF) module to form a global representation for the final disease classification. The entire framework can be trained in an end-to-end manner using a standard cross entropy loss.

2.1 Unified Transformer Backbone

During diagnosis, experienced dermatologists usually consider various visual features including shape, size, color, texture, location and distribution, etc., of which some are localized features and some need spatial global context. Although multi-scale convolutional neural networks can be used to model these complicated local-global interactions, Transformer [3] based methods (ViTs) are more suitable choices by the virtue of their non-local modeling capabilities. Besides, Transformer is a natural choice for multi-modal feature encoding. As an improvement to ViT, Swin Transformer [9] only computes self-attention locally within non-overlapping windows and thus has less complexity. From previous work [1,8,12], separate backbones are employed for clinical and dermoscopic images. We empirically find that using the shared backbone for both clinical and dermoscopic images will not compromise the performance, and actually this simplicity design of choice makes the training and inference more efficient. Taking all these considerations into account, a unified Swin Transformer backbone is adopted into our multi-modal framework.

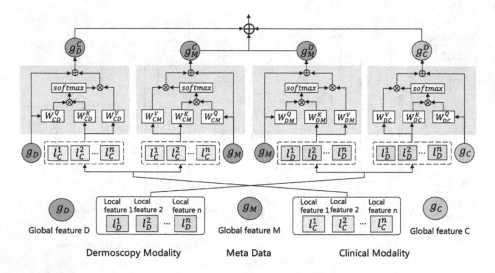

Fig. 3. The Cross-Modality-Fusion module.

2.2 Disease-Wise Pairing

Missing modality has been common when dealing with multi-modal data. Requiring the completeness of every modality in the clinical dataset adds extra burden to the already hash process of data collection and cleanse. To learn a good multi-modality representation at the class level, we integrate a disease-wise pairing scheme into our data augmentation pipeline during training to complete some of the modality missing. Unlike using instance-wise pairs for training, the DWP encourages the network to input/fuse the different modality data channels from different patients (as long as they belong under the same disease category) into the feature space. Our framework is convenient to add new training samples for any modality, which greatly reduces the cost of data collection and maximizes the data utilization. Furthermore, it serves as an additional regularization to prevent the model from learning some spurious correlation within a certain pair of modalities, and thus provides better generalizability (i.e., semantic data modality fusion at the disease level) and extra performance boosts.

Specifically, we define a sample data as $s_i = \{c_i, d_i, m_i\}$, where c_i, d_i, and m_i represent a clinical image, a dermoscopic image, and a set of metadata respectively. $S = \{s_i \mid i \in \{1, ..., I\}\}$ is used to indicate the set of sampled training data with the total amount of I. Let C, D, M be the three modalities of clinical image, dermoscopy and metadata, respectively; C^k, D^k, M^k be the grouped sets of data with the corresponding disease class k. When DWP is turned on (based on $p > T_p, p \in [0,1]$), c_i is randomly sampled from $\{c_j^k \mid \forall c_j^k \in C^k, k \in \{1, ..., K\}\}$, d_i from $\{d_j^k \mid \forall d_j^k \in D^k, k \in \{1, ..., K\}\}$, and m_i from $\{m_j^k \mid \forall m_j^k \in M^k, k \in \{1, ..., K\}\}$, where K is the number of diseases to be considered. Otherwise, $\{c_i, d_i, m_i\}$ will naturally be from the same study and patient if available (zero padding is used for missing modalities during training and testing). T_p is a cutoff threshold to control the probability of applying DWP to the input samples, where p is a random number generated by a uniform distribution on the interval $[0, 1]$. For each input sample, we apply DWP when $p > T_p$ (T_p is empirically set to 0.6).

2.3   Cross-Modality Fusion

Partially inspired by [2], the CMF module is designed to fuse the global features of each modality with the local features from another modality in a cyclic manner across all modalities. Specifically, the global features from each modality will exchange information with the local features from other modalities through a multi-head attention module. Since the global feature has already gathered the information from local features in its own modality, the multi-head attention will fuse the information from the local features of other modalities.

The detailed diagram for the fusion of three engaged modalities is shown in Fig. 3. l_C or l_D is the output feature map of the last stage in Swin Transformer. g_C or g_D are generated by applying a global average pooling (GAP) layer, and they are the class tokens for ViT backbones. g_M is the output by a linear layer after one-hot embedding with metadata. After layer normalization (LN), a cross-attention block is utilized to fuse features by taking the local features as $\mathcal{K}$ and $\mathcal{V}$ and global features as $\mathcal{Q}$:

$$g_{X'} = \begin{cases} \mathrm{GAP}(l_{X'}^1, l_{X'}^2, ..., l_{X'}^n), & X' \in \{C, D\} \\ g_M, & X' = M \end{cases} \quad (1)$$

$$f_{X'}^X = \mathrm{LN}\left(concat(g_{X'}, l_X^1, l_X^2, ..., l_X^n)\right), \quad \tilde{g}_{X'} = f_{X'}^X[0], \quad z_X = f_{X'}^X[1:] \quad (2)$$

$$\mathcal{Q} = \tilde{g}_{X'} W_{XX'}^{\mathcal{Q}}, \quad \mathcal{K} = z_X W_{XX'}^{\mathcal{K}}, \quad \mathcal{V} = z_X W_{XX'}^{\mathcal{V}} \quad (3)$$

$$M_{att} = \mathrm{softmax}(\frac{\mathcal{Q}\mathcal{K}^T}{\sqrt{F/h}}), \quad M_{cross} = M_{att}\mathcal{V} \quad (4)$$

$$g_{X'}^X = \tilde{g}_{X'} + \mathrm{linear}(M_{cross}), \quad X \in \{C, D\}, \quad X' \in \{C, D, M\} \setminus X \quad (5)$$

where X and X' are defined as two different modalities, l is the local feature and g is the global feature. $W_{XX'}^{\mathcal{Q}}, W_{XX'}^{\mathcal{K}}, W_{XX'}^{\mathcal{V}} \in \mathcal{R}^{F \times F}$ are learnable parameters, F is the dimension of features and h is the number of heads.

3   Experiment and Results

Data. We employ two datasets in this study: one public and one private dataset. The public dataset Derm7pt contains 413 training cases, 203 validation cases and 395 testing cases. Each case comprises a dermoscopic image and a clinical image, the diagnostic label is divided into 5 types: Mel, Nevus (Nev), Seborrheic Keratosis (SK), BCC, and Miscellaneous (Misc). Please refer to [8] for details, on which we will compare our method with the previous leading Fusion4MNet [12]. Our private dataset named X-SkinTumor-10 was collected from Xiangya Hospital from 2016 to 2021, and annotated by the dermatologists with at least

Table 1. Comparison of our proposed model with other methods on Derm7pt.

Method	F1						Acc
	Nev	BCC	Mel	Misc	SK	Avg.	
HcCNN [1]	—	—	0.605	—	—	—	0.699
Inception-comb [8]	0.838	0.588	0.632	0.559	**0.552**	0.634	0.742
ViT-S/16 [3]	0.855	0.414	0.723	0.706	0.429	0.625	0.777
FusionM4net [12]	0.878	0.452	0.727	0.646	0.181	0.577	0.785
RemixFormer	**0.883**	**0.595**	**0.755**	**0.743**	0.519	**0.699**	**0.813**

five years experience. The dataset contains 14,941 images and 5,601 patients, and only 2,198 patients have the three-modality paired data. The percentages of missing data are 0.4%, 33.6% and 37.7% for clinical images, dermoscopic images and metadata, respectively. The patient's metadata with 9 attributes is shown in Fig. 1. The dataset has ten types of skin tumors: SCC (8.5%), Mel (6.1%), BCC (13.4%), AK (2%), keloid (Kel, 3.4%), dermatofibroma (DF, 3.2%), sebaceous nevus (SN, 3.1%), SK (22.9%), Nev (35.7%), haemangioma (Hem, 1.6%).

Implementations. On X-SkinTumor-10, we perform 5-fold cross-validation to evaluate our method, where the entire dataset is randomly divided into 5 folds on the patient level with a 3:1:1 ratio as training, validation, and testing set. Our backbone is a regular Swin-B/384, the data augmentation includes flip, rotation, random affine transformation. We adopt SGD optimizer with cosine learning rate schedule, and the initial learning rate is 1e-4. Models are trained for 200 epochs on 4 NVIDIA Tesla V100 GPUs with batch size 64. On Derm7pt, we use exactly the same data augmentation as [12], and apply Swin-T/224 model with fewer parameters to verify the effectiveness of our method. In the CMF module, the dimensions of the local features and global features are 768 in Swin-T, 1024 in Swin-B, and the number of heads is 8. We initialize the model parameters with ImageNet pretrained weights to speed up the convergence. We utilize area under the curve, macro-averaged F1-score, sensitivity, precision, specificity, and overall accuracy as the evaluation metrics, corresponding to the abbreviations AUC, F1, Sen, Pre, Spe, and Acc respectively.

Results As listed in Table 1, we compare the proposed RemixFormer with other methods on the public dataset of Derm7pt [8]. For completeness, we use ViT-S/16 and fuse the three features (only class tokens of C and D, metadata feature) by concatenation. Our method with Swin-T backbone outperforms FusionM4Net by 12.2% and 2.8% in the average F1 and overall Acc, respectively. The relatively poor performance of SK is mainly due to the long-tail problem, which may be addressed in future work. It worth mentioning that our model has noticeably less parameters than FusionM4Net (32.3M vs. 54.4M).

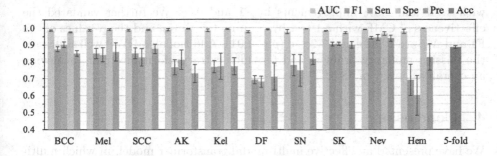

Fig. 4. 5-fold cross validation result on X-SkinTumor-10. The far right is the average accuracy of 5 folds, and error bars indicate the standard deviation.

Table 2. Ablation Study for multi-modality models.

Modality	Fusion		DWP	AUC	F1	Sen	Pre	Spe	Acc
	concat	CMF							
Clinical				0.952	0.599	0.592	0.619	0.968	0.766
Dermoscopy				0.953	0.606	0.601	0.629	0.769	0.778
C + D	✓			0.962	0.661	0.717	0.630	0.974	0.788
C + D		✓		0.966	0.681	0.729	0.652	0.975	0.805
C + D + M	✓			0.970	0.702	0.744	0.677	0.979	0.835
C + D + M		✓		0.976	0.749	0.765	0.740	0.983	0.864
C + D + M		✓	✓	**0.984**	**0.784**	**0.795**	**0.778**	**0.985**	**0.885**

Using 5-fold cross-validation, RemixFormer with Swin-B backbone achieves an overall 88.5% ± 0.8% classification accuracy on the more comprehensive X-SkinTumor-10. The average AUC, F1, Sen, Spe and Pre of the 10 conditions are 98.4% ± 0.3%, 81.2% ± 1.2%, 80.4% ± 2.2%, 98.6% ± 0.1% and 82.8% ± 1.1%, respectively, and the corresponding metrics for each condition are shown in Fig. 4. The huge performance difference between Nev and Hem is largely due to the imbalanced data distribution. For multi-instance cases, we use majority voting to select the prediction during inference.

To justify our design choices, we set aside 1034 three-modality cases from X-SkinTumor-10 as the test set to perform ablation study. Each case has one clinical and one dermoscopic image with metadata. We first conduct the modality-wise ablation experiments on this fixed test set. As listed in Table 2, model using only dermoscopic images performs better than clinical images. Combining two modalities (C+D) brings better performance, and simple features concatenation can improve F1 and Acc by 5.5% and 1% respectively. When the concatenation is replaced by CMF, the F1 and Acc are further improved by 2% and 1.7%. Increasing the modality to three (C+D+M) also brings significant performance gain. The model with CMF again outperforms the one using concatenation, with 4.7% and 2.9% improvements in F1 and Acc. We further validated the effectiveness of CMF with a t-test by running ten times, and the p-values of the F1 and Acc are 1.8e−4 and 7.2e−6 ($p < 0.01$). Additionally, DWP well addresses the missing modality problem and provides extra performance gain, achieving an overall accuracy of 88.5%.

4   Conclusion

We have presented an effective multi-modal transformer model, in which multi-modal data are properly fused by a novel cross-modality fusion module. To handle the missing modality problem, we implement a new disease-wise sampling strategy, which augments to form class-wise multi-modality image pairs (within the same class) and facilitates sufficient training. Swin transformer as an efficient image feature extractor is shared by both the dermoscopic and clinical image network streams. Through quantitative experiments, we achieve a new record on the public dataset of Derm7pt, surpassing the previous best method by 2.8% in accuracy. Our method is also validated on a large scale in-house dataset X-SkinTumor-10 where our reported quantitative performance of an overall 88.5% classification accuracy on recognizing ten classes of pigmented and non-pigmented skin tumors demonstrates excellent clinical potential.

Acknowledgement. This work was supported by National Key R&D Program of China (2020YFC2008703) and the Project of Intelligent Management Software for Multimodal Medical Big Data for New Generation Information Technology, the Ministry of Industry and Information Technology of the People's Republic of China (TC210804V).

References

1. Bi, L., Feng, D.D., Fulham, M., Kim, J.: Multi-label classification of multi-modality skin lesion via hyper-connected convolutional neural network. Pattern Recogn. **107**, 107502 (2020)
2. Chen, C.F.R., Fan, Q., Panda, R.: Crossvit: cross-attention multi-scale vision transformer for image classification. In: Proceedings of the IEEE/CVF International Conference on Computer Vision (ICCV), pp. 357–366 (2021)
3. Dosovitskiy, A., et al.: An image is worth 16×16 words: transformers for image recognition at scale. In: International Conference on Learning Representations (2021)
4. Esteva, A., et al.: Dermatologist-level classification of skin cancer with deep neural networks. Nature **542**, 115–118 (2017)
5. Ge, Z., Demyanov, S., Chakravorty, R., Bowling, A., Garnavi, R.: Skin disease recognition using deep saliency features and multimodal learning of dermoscopy and clinical images. In: Medical Image Computing and Computer Assisted Intervention (MICCAI 2017), pp. 250–258 (2017)
6. Haenssle, H.A., et al.: Man against machine: diagnostic performance of a deep learning convolutional neural network for dermoscopic melanoma recognition in comparison to 58 dermatologists. Ann. Oncol. **29**, 1836–1842 (2018)

7. Haenssle, H.A., et al.: Man against machine reloaded: performance of a market-approved convolutional neural network in classifying a broad spectrum of skin lesions in comparison with 96 dermatologists working under less artificial conditions. Ann. Oncol. **31**, 137–143 (2020)
8. Kawahara, J., Daneshvar, S., Argenziano, G., Hamarneh, G.: Seven-point checklist and skin lesion classification using multitask multimodal neural nets. IEEE J. Biomed. Health Inform. **23**(2), 538–546 (2019)
9. Liu, Z., et al.: Swin transformer: Hierarchical vision transformer using shifted windows. In: Proceedings of the IEEE/CVF International Conference on Computer Vision (ICCV), pp. 10012–10022 (2021)
10. Perera, E., Gnaneswaran, N., Jennens, R., Sinclair, R.: Malignant melanoma. Healthcare **2**(1), 1 (2013)
11. Soenksen, L.R., Kassis, T., Conover, S.T., Marti-Fuster, B., Gray, M.L.: Using deep learning for dermatologist-level detection of suspicious pigmented skin lesions from wide-field images. Sci. Transl. Med. **13**(581), eabb3652 (2021)
12. Tang, P., et al.: Fusionm4net: a multi-stage multi-modal learning algorithm for multi-label skin lesion classification. Med. Image Anal. **76**(102307), 1–13 (2022)
13. Tschandl, P., et al.: Comparison of the accuracy of human readers versus machine-learning algorithms for pigmented skin lesion classification: an open, web-based, international, diagnostic study. Lancet Oncol. **20**, 938–947 (2019)
14. Tschandl, P., et al.: Human-computer collaboration for skin cancer recognition. Nat. Med. **26**, 1229–1234 (2020)
15. Yu, Z., et al.: End-to-end ugly duckling sign detection for melanoma identification with transformers. In: de Bruijne, M., et al. (eds.) MICCAI 2021. LNCS, vol. 12907, pp. 176–184. Springer, Cham (2021). https://doi.org/10.1007/978-3-030-87234-2_17

Building Brains: Subvolume Recombination for Data Augmentation in Large Vessel Occlusion Detection

Florian Thamm[1,2(✉)], Oliver Taubmann[2], Markus Jürgens[2],
Aleksandra Thamm[1], Felix Denzinger[1,2], Leonhard Rist[1,2], Hendrik Ditt[1],
and Andreas Maier[1]

[1] Friedrich-Alexander University Erlangen-Nuremberg, Erlangen, Germany
florian.thamm@fau.de
[2] Siemens Healthcare GmbH, Forchheim, Germany

Abstract. Ischemic strokes are often caused by large vessel occlusions (LVOs), which can be visualized and diagnosed with Computed Tomography Angiography scans. As time is brain, a fast, accurate and automated diagnosis of these scans is desirable. Human readers compare the left and right hemispheres in their assessment of strokes. A large training data set is required for a standard deep learning-based model to learn this strategy from data. As labeled medical data in this field is rare, other approaches need to be developed. To both include the prior knowledge of side comparison and increase the amount of training data, we propose an augmentation method that generates artificial training samples by recombining vessel tree segmentations of the hemispheres or hemisphere subregions from different patients. The subregions cover vessels commonly affected by LVOs, namely the internal carotid artery (ICA) and middle cerebral artery (MCA). In line with the augmentation scheme, we use a 3D-DenseNet fed with task-specific input, fostering a side-by-side comparison between the hemispheres. Furthermore, we propose an extension of that architecture to process the individual hemisphere subregions. All configurations predict the presence of an LVO, its side, and the affected subregion. We show the effect of recombination as an augmentation strategy in a 5-fold cross validated ablation study. We enhanced the AUC for patient-wise classification regarding the presence of an LVO of all investigated architectures. For one variant, the proposed method improved the AUC from 0.73 without augmentation to 0.89. The best configuration detects LVOs with an AUC of 0.91, LVOs in the ICA with an AUC of 0.96, and in the MCA with 0.91 while accurately predicting the affected side.

Keywords: Computed Tomography Angiography · Stroke · Augmentation

Supplementary Information The online version contains supplementary material available at https://doi.org/10.1007/978-3-031-16437-8_61.

© The Author(s), under exclusive license to Springer Nature Switzerland AG 2022
L. Wang et al. (Eds.): MICCAI 2022, LNCS 13433, pp. 634–643, 2022.
https://doi.org/10.1007/978-3-031-16437-8_61

1   Introduction

A multitude of branches and bifurcations characterize the vascular structure in the brain. Once this system is disturbed, focal deficits may be the consequence. Ischemic stroke is the most common cerebrovascular accident, where an artery is occluded by a clot preventing blood flow. If in particular large vessels are affected, this condition is called large vessel occlusion (LVO), which primarily appears in the middle cerebral artery (MCA) and/or internal carotid artery (ICA). Computed Tomography Angiography (CTA) is commonly used for diagnosis, for which contrast agent is injected to enhance all perfused vascular structures. LVOs in the anterior circulation become visible in such images as unilateral discontinuation of a vessel. Therefore, human readers take symmetries into account and usually compare the left and right hemispheres to detect LVOs. Despite common anatomical patterns, the individual configuration and appearance of the vessel tree can differ substantially between patients, hence automated and accurate methods for LVO detection are desirable.

Amukotuwa et al. developed a pipeline consisting of 14 image processing steps to extract hand-crafted features followed by a rule-based classifier. With their commercially available product, they achieved an area under the receiver operator characteristic curve (AUC) between 0.86 and 0.94 depending on the patient cohort [1,2]. Multi-phase CTA which consists of scans acquired in the arterial, peak venous, and late venous phase has been used as well, as it offers more information about the underlying hemodynamics. Stib et al. [12] computed maximum-intensity projections of the vessel tree segmentation of each phase and predicted the existence of LVOs with a 2D-DenseNet architecture [6]. They performed an ablation study on 424 patients using the individual phases and achieved AUCs between 0.74 and 0.85. Another commercially available detection tool for LVOs utilizing a Convolutional Neural Network (CNN) has been evaluated by Luijten et al. [9], who report an AUC of 0.75 on a cohort of 646 patients.

In all studies mentioned above, large data pools have been used for development and evaluation. In contrast, Thamm et al. [14] proposed a method to counteract the problem of limited data while achieving a comparable performance as related work. They showed that the application of strong elastic deformations on vessel tree segmentation leads to significant improvements in terms of the detection of LVOs.

However, deformations only change vessel traces and do not vary the underlying tree topologies. Furthermore, vessels, especially the ICA, have asymmetric diameters in the majority of patients [3], which does not get significantly altered by deformations. Finally, the image regions relevant for strokes are spatially distant from each other. Previous work did not consider this for the development of their architectures.

Advancing the idea of data enrichment with information-preserving methods, we propose a randomized recombination of different patients as a novel augmentation method to create new synthetic data. This data is leveraged to train a 3D-DenseNet with a task-specific input feeding, which allows the network

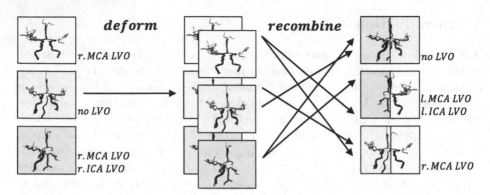

Fig. 1. Proposed augmentation method. Vessel tree segmentations are deformed and randomly recombined yielding artificial patients.

to perceive the vessel trees of both hemispheres simultaneously. This approach is extended by using two 3D-DenseNet encoding individual subregions (ICA, MCA) to further integrate prior knowledge. As the proposed augmentation neither requires additional memory nor delays training, it can be supplemented with other more computational expensive methods like the elastic deformations which in related works have shown positive impact (Fig. 1). Furthermore, we additionally predict if the left or right ICA and/or MCA vessels are occluded, which implicitly enables a coarse localization of the LVO.

2   Methods

2.1   Data

The data from [14] was used consisting of CTA scans covering the head/neck region of 151 patients, of which 44/57 suffered from left-/right-sided LVOs. 17 from original 168 cases were removed due to unsuccessful contrast agent injections, other artifacts or failed registrations with the atlas (see Sect. 2.2) that were still included in the original study of [14]. The ICA was affected in 12/24 and the MCA in 40/50 cases for the left/right side, i.e. in some cases ICA and MCA are affected. MCA LVOs are either located proximal/distal in M1 or in the transition from M1 to M2. The data was provided by one clinical site and acquired with a Somatom Definition AS+ (Siemens Healthineers, Forchheim, Germany).

2.2   Preprocessing

For preprocessing, the cerebral vasculature is segmented following the approach of Thamm et al. [13]. Then the scans are non-rigidly registered using Chefd'hotel et al.'s method [4] into a reference coordinate system of a probabilistic brain

atlas [7] with isotropic voxel spacings of 1 mm. The vessel segmentation is transformed into the reference coordinate system using the determined deformation
field. As a result, the vessel trees have a uniform orientation and sampling. The
reference coordinate system either enables an easy split of the volume into both
hemispheres or allows to extract specific subregions of the segmentation.

2.3   Hemisphere Recombination

To augment our data we leverage the fact that the human brain and therefore the
cerebral vasculature is quasisymmetric w.r.t. to the mid-plane dividing the brain
into the left and right hemisphere. We therefore split the vessel tree segmentations of P patients along the brain mid-plane. Next, we mirror all left-sided
hemisphere's vessel trees in sagittal direction such that they become congruent to their right counterparts. A "hemisphere tree" describes in this context
the segmented tree-like vasculature corresponding to one hemisphere. The set of
hemisphere tree segmentations for all patients is denoted as $\mathcal{X}^{H} = \{x_1^H, \ldots, x_N^H\}$
where $N = 2P$. Each hemisphere tree x^H of volume size $100 \times 205 \times 90$ voxels
(mm^3) (atlas space split in half) is labeled regarding the presence of an LVO
with $y^H \in \{0, 1\}$ (1 if LVO positive, 0 else). Analogously, the set of all labels is
denoted as $\mathcal{Y}^{H} = \{y_1^H, \ldots, y_N^H\}$.

New artificial patients can be generated by randomly drawing two hemisphere
trees x_i^H and x_j^H from $\mathcal{X}^{H}$ which are recombined to a new stack $s_{ij} = \{x_i^H, x_j^H\}$.
Recombinations of two LVO positive hemisphere trees lead to no meaningful
representation and are hence excluded. If r is the ratio of LVO positives cases
among the P patients, the augmentation scheme leads to

$$R = \underbrace{2r(2 - r)P^2}_{\text{LVO pos.+Mirr}} + \underbrace{(2 - r)^2 P^2}_{\text{LVO neg.+Mirr}} \tag{1}$$

possible recombinations including the mirrored representations. In this work,
approximately 81k recombinations can be achieved using P = 151 patient with
r = 67% being LVO positive.

We consider a multi-class classification problem with one-hot encoded labels
$y'^H \in \{0, 1\}^3$. The encoding comprises three exclusive classes using the following
order: No LVO, left LVO or right LVO. For two recombined hemisphere trees, i
for left and j for the right, the labels y_i^H, y_j^H relate to $y_{ij}'^H$ as

$$y_{ij}'^H = \begin{bmatrix} \neg(y_i^H \vee y_j^H) \\ y_i^H \\ y_j^H \end{bmatrix}. \tag{2}$$

By stacking randomly selected hemisphere trees creating new artificial data sets,
a neural network f can be trained for all i and j such that $f(s_{ij}^H) = \hat{y}_{ij}'^H \approx$
$y_{ij}'^H$ $\forall i, j \in \{1, \ldots N\}$. The stacking can be done sagittally or channel-wise
depending on the architecture (Sect. 2.5). The proposed augmentation scheme
is not suitable for LVOs located at the basilar artery, due to artifacts that may
occur close to the brain mid-plane.

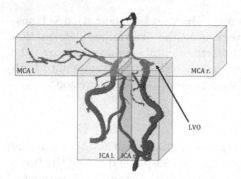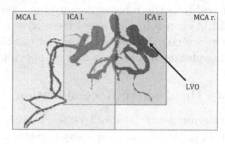

(a) ICA/MCA subvolumes in an angled perspective.

(b) ICA/MCA subvolumes viewed axially caudal.

Fig. 2. Circle of Willis and surrounding vessels including the subvolumes covering the ICAs/MCAs each left and right with on colored overlay. This example shows an LVO on the right proximal MCA visible as an interruption in the respective vessel trace. (Color figure online)

The recombination does neither have to be random and unstructured nor cover all permutations. Instead, the construction can be targeted to create specific distributions. In our case, we sample the three classes to be uniformly distributed, whereby each hemisphere appears once in one epoch.

2.4   Recombination of ICA and MCA Subvolumes

The above-mentioned recombination strategy can be extended if the each hemisphere vessel segmentation is further partitioned in two subvolumes, respectively covering the ICA and MCA regions initially defined based on the atlas. Likewise with $\mathcal{X}^\mathrm{H}$, the subvolumes of the hemisphere trees of all patients are described as the ICA set $\mathcal{X}^\mathrm{I} = \{\boldsymbol{x}_1^\mathrm{I}, \ldots, \boldsymbol{x}_N^\mathrm{I}\}$ with a size of $55 \times 121 \times 57$ voxels (mm^3) for $\boldsymbol{x}^\mathrm{I}$, and the MCA set $\mathcal{X}^\mathrm{M} = \{\boldsymbol{x}_1^\mathrm{M}, \ldots, \boldsymbol{x}_N^\mathrm{M}\}$ with x^M of size $60 \times 77 \times 76$ voxels (mm^3). Each subvolume is associated with a corresponding label $\mathcal{Y}^\mathrm{I}$ and $\mathcal{Y}^\mathrm{M}$ analogously to $\mathcal{Y}^\mathrm{H}$ described in Sect. 2.3.

For the synthesis of new artificial patients, subvolumes $\boldsymbol{x}_i^\mathrm{I}$ and $\boldsymbol{x}_j^\mathrm{I}$ are drawn from $\mathcal{X}^\mathrm{I}$ and $\boldsymbol{x}_k^\mathrm{M}$ and $\boldsymbol{x}_l^\mathrm{M}$ from $\mathcal{X}^\mathrm{M}$, forming a new stack $s_{ijkl} = \{\boldsymbol{x}_i^\mathrm{I}, \boldsymbol{x}_j^\mathrm{I}, \boldsymbol{x}_k^\mathrm{M}, \boldsymbol{x}_l^\mathrm{M}\}$ whereby asymmetric global LVOs (e.g. l. ICA and r. MCA) are excluded. A patient's one-hot encoded labels are not only represented globally by $\boldsymbol{y}^{\prime\mathrm{H}}$ anymore but may be decomposed into $\boldsymbol{y}^{\prime\mathrm{M}} \in \{0,1\}^3$ and $\boldsymbol{y}^{\prime\mathrm{I}} \in \{0,1\}^3$ identically representing the exclusive classes described above, restricted to their respective region:

$$\boldsymbol{y}_{ijkl}^{\prime\mathrm{H}} = \begin{bmatrix} \neg(y_i^\mathrm{I} \vee y_j^\mathrm{I} \vee y_k^\mathrm{M} \vee y_l^\mathrm{M}) \\ y_i^\mathrm{I} \vee y_k^\mathrm{M} \\ y_j^\mathrm{I} \vee y_l^\mathrm{M} \end{bmatrix}, \boldsymbol{y}_{ij}^{\prime\mathrm{I}} = \begin{bmatrix} \neg(y_i^\mathrm{I} \vee y_j^\mathrm{I}) \\ y_i^\mathrm{I} \\ y_j^\mathrm{I} \end{bmatrix}, \boldsymbol{y}_{kl}^{\prime\mathrm{M}} = \begin{bmatrix} \neg(y_k^\mathrm{M} \vee y_l^\mathrm{M}) \\ y_k^\mathrm{M} \\ y_l^\mathrm{M} \end{bmatrix}$$

$$(3)$$

The suggested distinction between ICA and MCA within one hemisphere increases the number of possible recombinations. Assuming in all LVO positives, the ICA and MCA are affected, we obtain

$$R = \underbrace{2r^2(2-r)^2 P^4 + 4r(2-r)^3 P^4}_{\text{LVO pos.+Mirr}} + \underbrace{(2-r)^4 P^4}_{\text{LVO neg.+Mirr}} \tag{4}$$

recombinations. Analogously to Sect. 2.5, 5730M recombinations are possible in this work.

2.5 Architectures

We make use of the 3D-DenseNet architecture [6] such that it takes advantage of the proposed data representation by design. The first variant is trained on $\mathcal{X}^{\mathrm{H}}$ and its extension is trained using both, $\mathcal{X}^{\mathrm{I}}$ and $\mathcal{X}^{\mathrm{M}}$. As baseline serves the approach by Thamm et al. [14] utilizing 3D-DenseNets trained on whole heads which were not split into two hemispheres. However, all architectures predict $\hat{\boldsymbol{y}}^{\prime \mathrm{M}}$, $\hat{\boldsymbol{y}}^{\prime \mathrm{I}}$ and $\hat{\boldsymbol{y}}^{\prime \mathrm{H}}$ (Eq. 3).

Baseline (Whole Head). We compare the suggested architectures and augmentation schemes to the methods presented in Thamm et al. [14]. Each data set is deformed 10 times with a random elastic field using the parameters described in [14] and mirrored sagittally (left/right flip). More deformed data sets did not lead to any improvements. A 3D-DenseNet [6] ($\approx$4.6 m parameters, growth rate of 32 and 32 initial feature maps) receives the vessel tree segmentations covering the entire head generated and preprocessed in the way described in Sect. 2.2. The recombination method is applicable here, if the two drawn hemispheres trees $\boldsymbol{x}_i^{\mathrm{H}}$ and $\boldsymbol{x}_j^{\mathrm{H}}$ are sagittally concatenated along the brain midplane. The indices for the labels are hence $\boldsymbol{y}_{ijij}^{\prime \mathrm{H}}$, $\boldsymbol{y}_{ij}^{\prime \mathrm{I}}$ and $\boldsymbol{y}_{ij}^{\prime \mathrm{I}}$, using the notation described in Sect. 2.4.

Hemisphere-Stack (H-Stack). The first proposed variant is based on the 3D-DenseNet architecture ($\approx$3 m parameters - larger capacity did not lead to better performances) as well and receives the samples as described in Sect. 2.3 drawn from $\mathcal{X}^{\mathrm{H}}$, but concatenated channel-wise (Fig. 3(a)). This has the advantage that the corresponding positions on both sides are spatially aligned within the same receptive fields of convolutional layers even early in the network and are carried over to deeper stages of the network due to the skip connections of the DenseNet architecture. Hence, a left/right comparison between both hemispheres is encouraged by design and does not need to be encoded over long spatial distances. The only difference between H-Stack and the baseline exists therein the feeding and composition of the data to demonstrate the impact of the suggested data representation. The label computation is identical. Furthermore, as mentioned in Sect. 2.3 we extend the recombination with the elastic deformations using RandomElasticDeformation [11] (TorchIO, max. displacement 20 voxels, 6 anchors, 10 repetitions).

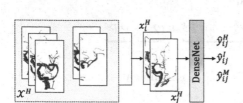

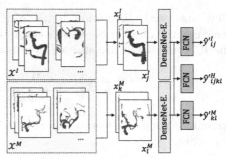

(a) H-Stack utilizing a DenseNet to predict all 9 classes

(b) IM-Stack consisting of two enconders each returning feature vectors which are processed by 3 classifiers

Fig. 3. Schematics of the H-Stack and IM-Stack variant both predicting the presence of an LVO and in the positive case its side.

ICA-MCA-Stack (IM-Stack). In the second variant, samples s_{ijkl} are drawn from $\mathcal{X}^I$ and $\mathcal{X}^M$ according to the scheme presented in Sect. 2.4. The ICA and MCA volumes are concatenated channel-wise with their side-counterpart as visualized in Fig. 3(b). Each vessel region is encoded with a DenseNet ($\approx$ 3m parameters, growth rate of 16 and 32 initial feature maps) similarly as described in the H-Stack variant. The last latent space feature vectors (globally max pooled to a length of 64) of both encoders are forwarded to three individual fully connected heads (consisting of one dense layer and one batch-norm layer each) reducing the space to the required dimension of three. The global head receives a concatenation of both encodings. LVOs can affect either the ICA or MCA or both of the vessels. By analyzing the ICA and MCA regions separately, the network is able to focus on the characteristic LVO-patterns in one vessel, independent to any occurrences of LVOs in the respective other vessel. The ICA and MCA volumes are deformed as well (max. displacement 20 voxels, 4 anchors, 10 repetitions).

2.6 Experiments

All models of Sect. 2.5 are trained by Adam [8] (learning rate 10^{-5}, batch size 6) using the BCE-Loss (PyTorch 1.6 [10] and Python 3.8). We furthermore used the 3D-Densenet implementation by [5] and, besides the changes in Sect. 2.5, kept default values for all other details concerning the DenseNets used in this work. Early stopping (p = 100 epochs) has been applied monitored on the validation AUC. To evaluate the effect of the recombination, we perform an ablation study combined with a 5-fold cross validation (3-1-1 split for training, validation and testing). Validation/Testing is done on original, not recombined, data. We applied the deformation (D) and recombination (R) on each method, including the baseline. Additionally, the data is mirrored (M) for the baseline.

Table 1. Quantitative evaluation of the 5-fold cross validation showing the ROC-AUCs of the respective class-wise prediction for the presence/absence of an LVO, and the accuracy for the affected side on LVO-pos. cases. The abbreviation "R" stands for the proposed recombination method, "D" for the deformation and "M" for mirroring.

Method	Global		ICA		MCA	
	Class AUC	Side Acc.	Class AUC	Side Acc.	Class AUC	Side Acc.
Whole Head [14]	0.79	0.84	0.82	0.86	0.79	0.84
Whole Head + D + M [14]	0.84	0.94	0.86	0.88	0.85	0.97
Whole Head + R	0.87	0.94	0.82	0.86	0.87	0.97
Whole Head + R + D	0.89	0.93	0.81	**0.94**	0.89	0.98
H Stack	0.73	0.91	0.74	0.83	0.75	0.94
H Stack + D	0.82	0.86	0.86	0.86	0.85	0.93
H Stack + R	0.87	0.93	0.95	0.86	0.89	0.94
H Stack + R + D	0.89	0.92	0.95	0.88	**0.91**	0.96
IM Stack	0.84	0.92	0.71	0.81	0.82	0.91
IM Stack + D	0.86	**0.96**	0.93	0.88	0.86	**0.99**
IM Stack + R	0.88	0.94	0.92	0.86	0.89	0.97
IM Stack + R + D	**0.91**	**0.96**	**0.96**	0.92	**0.91**	0.98

3   Results

The AUCs for the "Class" categories, depicted in Table 1, are determined by evaluating the test data according to the probability of the LVO-pos. class (sum of left and right LVO-pos. class) against the LVO-neg. class (no LVO). "Side" is measured by the accuracy of taking the argmax of the left or right class prediction on the LVO-pos. cases. Potential false negatives are taken into account. Variances and 95% confidence intervals can be found in the suppl. material. It is evident that the recombination by itself consistently leads to a better performance for the classification of LVOs than deformation. Both augmentations combined complement each other boosting every method to its individual best performance. Except for the "Whole Head" model, which seems to not properly detect ICA LVOs while being accurate in predicting the side in all three classes. H-Stack detects ICA and MCA LVOs better than the baseline "Whole Head" except on the global label, where both methods perform equally well. In contrast, IM-Stack outperforms the other methods w.r.t. the LVO classification by a significant margin, achieving an AUC of 0.91 globally, 0.96 for the ICA and 0.91 for the MCA, while for the side being superior or close to the best performer.

4   Conclusion

We present a novel and efficient augmentation technique for patient-level LVO classification based on the recombination of segmented subvolumes from multiple patients. While the data sets created in this manner may be of limited realism overall due to differences in patient anatomy, contrast bolus or image quality, their use in training consistently and significantly improved the performance of all tested models. It appears that repeatedly presenting the individual sub-volumes to the models in new contexts strongly benefits generalization, even if the artificial data sets providing that context are less representative of real samples. We evaluated the proposed recombination as well as a state-of-the-art deformation-based augmentation in a baseline architecture and two models specifically designed to exploit the inherent symmetry of the brain to detect and coarsely localize (ICA/MCA) an anterior LVO. Best results were achieved when both types of augmentation were combined with our task-specific models.

References

1. Amukotuwa, S.A., Straka, M., Dehkharghani, S., Bammer, R.: Fast automatic detection of large vessel occlusions on CT angiography. Stroke **50**(12), 3431–3438 (2019)
2. Amukotuwa, S.A., et al.: Automated detection of intracranial large vessel occlusions on computed tomography angiography: a single center experience. Stroke **50**(10), 2790–2798 (2019)
3. Caplan, L.R.: Arterial occlusions: does size matter? J. Neurol. Neurosurg. Psychiatry **78**(9), 916 (2007)
4. Chefd'Hotel, C., Hermosillo, G., Faugeras, O.: Flows of diffeomorphisms for multimodal image registration. In: Proceedings IEEE International Symposium on Biomedical Imaging, pp. 753–756 (2002). https://doi.org/10.1109/ISBI.2002.1029367
5. Hara, K., Kataoka, H., Satoh, Y.: Can spatiotemporal 3D CNNs retrace the history of 2D CNNs and imagenet? In: Proceedings of the IEEE Conference on Computer Vision and Pattern Recognition (CVPR), pp. 6546–6555 (2018)
6. Huang, G., Liu, Z., Van Der Maaten, L., Weinberger, K.Q.: Densely connected convolutional networks. In: IEEE Conference on Computer Vision and Pattern Recognition (CVPR), pp. 4700–4708 (2017). https://doi.org/10.1109/CVPR.2017.243
7. Kemmling, A., Wersching, H., Berger, K., Knecht, S., Groden, C., Nölte, I.: Decomposing the hounsfield unit. Clin. Neuroradiol. **22**(1), 79–91 (2012). https://doi.org/10.1007/s00062-011-0123-0
8. Kingma, D.P., Ba, J.: Adam: A method for stochastic optimization (2015)
9. Luijten, S.P., et al.: Diagnostic performance of an algorithm for automated large vessel occlusion detection on CT angiography. J. Neurointerventional Surg. **14**(8), 794–798 (2021). https://doi.org/10.1136/neurintsurg-2021-017842
10. Paszke, A., et al.: Pytorch: an imperative style, high-performance deep learning library. Adv. Neural Inf. Process. Syst. **32** (2019)
11. Pérez-García, F., Sparks, R., Ourselin, S.: Torchio: a python library for efficient loading, preprocessing, augmentation and patch-based sampling of medical images in deep learning. Comput. Methods Programs Biomed. **208**, 106236 (2021). https://doi.org/10.1016/j.cmpb.2021.106236
12. Stib, M.T., et al.: Detecting large vessel occlusion at multiphase CT angiography by using a deep convolutional neural network. Radiology **297**(3), 640–649 (2020)

13. Thamm, F., Jürgens, M., Ditt, H., Maier, A.: VirtualDSA++: automated segmentation, vessel labeling, occlusion detection and graph search on CT-angiography data. In: Kozlíková, B., Krone, M., Smit, N., Nieselt, K., Raidou, R.G. (eds.) Eurographics Workshop on Visual Computing for Biology and Medicine. The Eurographics Association (2020). https://doi.org/10.2312/vcbm.20201181
14. Thamm, F., Taubmann, O., Jürgens, M., Ditt, H., Maier, A.: Detection of large vessel occlusions using deep learning by deforming vessel tree segmentations. In: Bildverarbeitung für die Medizin 2022. I, pp. 44–49. Springer, Wiesbaden (2022). https://doi.org/10.1007/978-3-658-36932-3_9

Hybrid Spatio-Temporal Transformer Network for Predicting Ischemic Stroke Lesion Outcomes from 4D CT Perfusion Imaging

Kimberly Amador[1,2(✉)], Anthony Winder[2], Jens Fiehler[3], Matthias Wilms[2,4], and Nils D. Forkert[2,4]

[1] Department of Biomedical Engineering, University of Calgary, Calgary, CA, Canada
kimberlyalejandra.am@ucalgary.ca
[2] Department of Radiology and Hotchkiss Brain Institute, University of Calgary, Calgary, CA, Canada
[3] Department of Diagnostic and Interventional Neuroradiology, University Medical Center Hamburg-Eppendorf, DE Hamburg, Germany
[4] Alberta Children's Hospital Research Institute, University of Calgary, Calgary, CA, Canada

Abstract. Predicting the follow-up infarct lesion from baseline spatio-temporal (4D) Computed Tomography Perfusion (CTP) imaging is essential for the diagnosis and management of acute ischemic stroke (AIS) patients. However, due to their noisy appearance and high dimensionality, it has been technically challenging to directly use 4D CTP images for this task. Thus, CTP datasets are usually post-processed to generate parameter maps that describe the perfusion situation. Existing deep learning-based methods mainly utilize these maps to make lesion outcome predictions, which may only provide a limited understanding of the spatio-temporal details available in the raw 4D CTP. While a few efforts have been made to incorporate raw 4D CTP data, a more effective spatio-temporal integration strategy is still needed. Inspired by the success of Transformer models in medical image analysis, this paper presents a novel hybrid CNN-Transformer framework that directly maps 4D CTP datasets to stroke lesion outcome predictions. This hybrid prediction strategy enables an efficient modeling of spatio-temporal information, eliminating the need for post-processing steps and hence increasing the robustness of the method. Experiments on a multicenter CTP dataset of 45 AIS patients demonstrate the superiority of the proposed method over the state-of-the-art. Code is available on GitHub.

Keywords: Stroke · CT perfusion · Transformer · Deep learning

M. Wilms and N.D. Forkert—Shared last authorship.

Supplementary Information The online version contains supplementary material available at https://doi.org/10.1007/978-3-031-16437-8_62.

© The Author(s), under exclusive license to Springer Nature Switzerland AG 2022
L. Wang et al. (Eds.): MICCAI 2022, LNCS 13433, pp. 644–654, 2022.
https://doi.org/10.1007/978-3-031-16437-8_62

1   Introduction

Acute ischemic stroke (AIS) is caused by a decrease in blood flow to the brain due to an arterial occlusion, leading to neurological deterioration and death [23]. For the diagnosis and treatment of AIS, spatio-temporal (4D) Computed Tomography Perfusion (CTP) has become an essential tool to visualize the passage of a contrast agent through the brain [26], allowing to identify the irreversibly damaged infarct core and salvageable penumbra tissue. However, their noisy appearance and high-dimensional characteristics make it challenging to directly use baseline 4D CTP images for clinical-decision making and outcome prediction. Basic approaches, therefore, deconvolve the CTP curve for each voxel with the arterial input function (AIF), resulting in so-called perfusion parameter maps that describe the brain hemodynamics (e.g., cerebral blood flow–CBF, cerebral blood volume–CBV, mean transit time–MTT) [13]. In clinical practice, perfusion maps are thresholded to segment the infarct core and penumbra and guide treatment decisions. While computationally efficient, such a simplistic approach frequently fails to capture the complexity of the processes underlying AIS [10].

Given the success of deep learning in medical image analysis [14], several methods for predicting follow-up infarct lesions from baseline perfusion maps have been proposed. For example, in [15,27], convolutional neural networks (CNNs) are utilized to predict the final infarct core from Magnetic Resonance Imaging (MRI)-based perfusion maps that significantly outperform clinical thresholding methods. More recently, methods like [16] have been developed to leverage the spatio-temporal nature of perfusion MRI by incorporating raw 4D information into the model, obtaining better results compared to using perfusion maps alone. While most approaches rely on MRI datasets (potentially due to the better signal-to-noise ratio), only a few efforts have been made to use 4D CTP imaging for this purpose, and even fewer have tried to directly exploit its temporal properties. Notable exceptions are [1,17]. In [17], a CNN is trained on clinical parameters and raw 4D CTP measurements. However, the temporal data characteristics are rather implicitly analyzed by combining features from all time points via standard convolutions, and the approach still relies on a manually selected AIF. In contrast, [1] proposed a fully-automated and end-to-end trainable method that directly handles the temporal dimension of the 4D CTP data without the need to select an AIF. This is achieved by extracting feature maps from each 2D image of the temporal sequence using a CNN backbone. Those maps are then combined using a temporal convolutional network (TCN) [2] that explicitly respects and analyzes the temporal dimension of the perfusion data.

Both 4D CTP-based outcome prediction approaches discussed above, [17] and [1], outperform methods using standard perfusion maps. However, there is still potential for improvements: The reported mean Dice scores comparing predicted and real follow-up lesions are rather low (<0.4), and it can also be argued that only very basic temporal information processing techniques have been utilized so far. We strongly believe that an enhanced processing and analysis of the temporal dimension of the CTP data offers exciting opportunities for performance improvements. While in [1] this is handled with a hierarchical TCN, its tree-like architecture that recursively connects neighbouring points in time (and their combinations at higher levels) imposes a rather strong prior on the temporal relations the network can learn. Transformers [20], on the other hand, do not suffer from this disadvantage as their attention mechanism allows them to learn complex short- and long-term temporal relations in a data-driven way.

After their success in core machine learning and computer vision, Transformer-based networks have become increasingly popular for temporal and spatial medical image analysis, achieving state-of-the-art results on various image classification [7, 22] and segmentation tasks [3, 12]. This is due to their inherent ability to identify long-range spatial relations, which is hard to achieve using standard CNNs. However, their fully-connected architecture components usually prevent Transformer models from being directly utilized on medical image data for efficiency reasons. Existing spatial Transformer models in medical image analysis are often hybrid approaches (e.g., [5, 21, 24]) that utilize encoder-like 3D CNNs for (local) feature extraction, with the Transformer part being responsible for modeling the global (spatial) interaction between the local features. Although these methods capitalize on the strengths of both worlds, none of the previously described models allow for spatio-temporal feature learning of 4D medical images. This motivates us to propose a novel and hybrid CNN- and Transformer-based model capable of efficiently capturing spatio-temporal relationships for improved AIS lesion outcome prediction from raw 4D CTP data. In contrast to [17], the proposed network fully automatically analyzes the data without the need for post-processing steps like deconvolution or manual AIF selection. While our approach shares similarities with [1], the self-attention mechanism of the proposed Transformer part effectively overcomes the limitations of their TCN-based temporal modeling mechanism. Additional crucial differences between our approach and [1] are a shared CNN for efficient feature extraction across all time points instead of individual ones as proposed in [1], and the U-Net-like skip connections for learning-based upsampling of the final lesion masks are combined along the temporal dimension via absolute differences rather than simple concatenations. The advantages of our pipeline are highlighted by our evaluation on a challenging multicenter 4D CTP dataset consisting of 45 AIS patients.

The contributions of this paper can be summarized as follows: (1) We propose a novel hybrid CNN-Transformer model for processing spatio-temporal CTP images. (2) We are the first to explore the potential of Transformer in the context of stroke lesion outcome prediction. (3) We demonstrate the benefit of using a Transformer-based approach by evaluating it against standard CNNs and TCNs.

2   Methods

In this work, we present a novel hybrid architecture based on CNNs and Transformer to predict a binary segmentation mask of the follow-up stroke lesion directly from raw baseline 4D CTP imaging data. The proposed method is end-to-end trainable and organized into three intertwined modules: (a) A shared

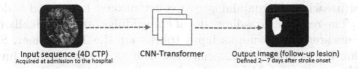

Input sequence (4D CTP)
Acquired at admission to the hospital

CNN-Transformer

Output image (follow-up lesion)
Defined 2—7 days after stroke onset

Fig. 1. Summarized workflow illustrating the stroke lesion outcome prediction problem.

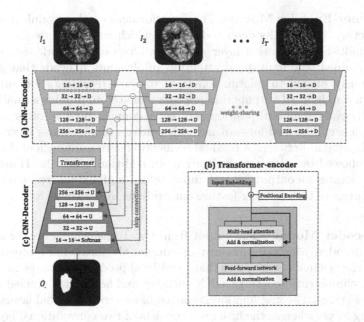

Fig. 2. The proposed model consists of: (a) shared CNN-encoder used to extract high-level spatial features from each time point, (b) Transformer-encoder that merges the temporal data, (c) and CNN-decoder that predicts the stroke lesion. Also, the CNN-encoder is directly connected to the CNN-decoder via skip connections at each level.

CNN-encoder extracts spatial features from each time point, (b) a Transformer-encoder that captures the temporal feature relations, (c) and a CNN-decoder that predicts the probability of infarction on a pixel-by-pixel basis based on the Transformer output. A workflow summarizing the lesion outcome prediction problem is shown in Fig. 1, and an overview of the model architecture in Fig. 2.

CNN-Encoder Module. To reduce the computational requirements and enable a fair comparison with the state-of-the-art, the proposed method handles each temporal sequence of axial 2D slices of the full 4D CTP sequence independently (3D+t problem reduced to multiple 2D+t problems). Thus, let $I = (I_0, I_1, \ldots, I_T)$ be an input CTP sequence I with T time points/2D images. First, T identical 2D CNN-encoders with shared weights are used in parallel to extract the spatial feature representations from each time point while continuously reducing its spatial resolution. There are five levels in each encoder (see

Fig. 2a), with each level containing two convolutional layers and a downsampling layer. The resulting low-dimensional embeddings are individually flattened and then concatenated to use as input tokens for the Transformer. Since the self-attention layers in the Transformer are order-agnostic [20], we further use an Embedding layer to add a positional encoding to the existing input token embeddings so that information about their location in the sequence is preserved.

Transformer-Encoder Module. The Transformer-encoder module follows the same structure as originally described in [20], which consists of two key components: a multi-head attention layer and a feed-forward network (see Fig. 2b). Briefly explained, the multi-head attention implements h heads that allow to simultaneously attend to information from distinct representation subspaces, resulting in h individual outputs that are concatenated to provide a final output. As a result, every temporal position is directly connected to all other positions in the sequence. The feed-forward network consists of two dense layers with a ReLU activation in between. A residual connection surrounds each of these components, followed by a layer normalization. In addition, since the Transformer-encoder generates an output for each input token, a global max-pooling layer is used to aggregate the resulting feature embeddings into a single feature embedding.

CNN-Decoder Module. The output from the Transformer-encoder is fed to the CNN-decoder, which continuously upsamples the extracted representations to restore the original resolution so that pixel-level predictions can be made. The decoder is symmetrical to a single CNN-encoder, and hence, it is divided into five levels (see Fig. 2c). The first four levels consist of two convolutional layers and an upsampling layer, whereas the fifth level consists of two convolutional layers and a final softmax layer. Moreover, in order to recover localized spatial information, the CNN-encoder is directly connected to the CNN-decoder via skip connections at each level. To do so, the absolute value of their difference is concatenated as proposed in [8], helping to further highlight the difference between time points and reduce memory usage. After applying a threshold of 0.5 to the resulting probability maps, a binary segmentation mask of the stroke lesion is generated. Full details of the proposed model can be found in the supplementary material.

3   Experiments and Results

Data. The proposed method is evaluated on a multicenter database, consisting of 45 baseline CTP scans of acute ischemic stroke patients from the ERASER [9] prospective cohort study and an in-house dataset collected from 2012 to 2016. A registered follow-up non-contrast CT scan, acquired between 2 and 7 d after stroke onset, is also available for each patient, including a ground truth lesion segmentation. These infarct lesions were segmented by experienced radiologists using a semi-automated method based on AnToNIa [11] and ITK-SNAP [28]. Both, baseline and follow-up scans, have a resolution of $512 \times 512 \times N$ (with N

slices varying from 16 to 35) and a voxel size of $0.45 \times 0.45 \times 5\,mm^3$. The temporal resolution of the baseline CTP scans ranges from 1.6 to 2.8 s, with a total number of time points between 45 and 151. Our pre-processing pipeline follows [1] to enable meaningful comparisons: We extract 32-second image sequences from the original 4D CTP scans, as this is considered to capture most of the dynamic perfusion process of the contrast agent. This time-window extraction is an automated process that measures the average 3D image intensity over time for each patient, and then aligns the center of the 32-second window with the peak of each intensity-time curve. In addition, all axial slices (=2D input images) are cropped to a size of 384×256 pixels, such that the images only include the brain hemisphere affected by the stroke lesion.

Experimental Design. We performed experiments with two additional methods to evaluate the performance gain of using raw 4D CTP data over perfusion maps to predict the stroke lesion outcome, as well as to demonstrate the advantage of using Transformer over TCNs to learn the cerebral blood flow dynamics from 4D CTP images. For this, we train a standard 2D U-Net [18] architecture that mimics [15] and uses four perfusion parameter maps (CBF, CBV, MTT, and Tmax) obtained through deconvolution of the 4D CTP datasets with an AIF using AnToNIa [11]. This method represents the state-of-the-art for perfusion maps-based prediction. Furthermore, we compare our method to the TCN-based approach from [1], which to the best of our knowledge, is the only method for fully automated lesion prediction from raw 4D CTP (see Sect. 1 for a discussion of [1]). Both, the proposed method and the TCN model, use the same pre-processed CTP sequences with $T = 32$ time points as input, while the 2D U-Net model uses perfusion maps. All three models use the follow-up lesion segmentations as ground truth. All experiments are performed via ten-fold cross-validation with stratified sampling at the patient level to ensure that follow-up lesion volumes ranging from small to large are represented in all folds. This evaluation scheme has the benefit of improving the performance estimate when limited data is available. The models are trained for 100 epochs to maximize the Dice score between the predicted and ground truth lesion segmentations using Adam with an initial learning rate of 1e−3. A batch size of 1 is used due to hardware restrictions and $h = 8$ attention heads are utilized for the Transformer. These parameters were empirically determined based on extensive experiments on a small validation subset of the data, where we found the proposed method to be insensitive to small variations with respect to its parameters. All models are implemented in Python using Keras [6] with a TensorFlow 2.0 backend and trained on a NVIDIA Tesla V100 GPU with 16GB. The baseline methods were re-implemented as described in the original papers, and our implementations are available on GitHub[1].

Evaluation Metrics. To quantitatively analyze the results, the predicted 2D binary segmentation masks for each patient are concatenated back into a 3D volume and then compared to the ground truth 3D lesion segmentations from

---

[1] https://github.com/kimberly-amador/Spatiotemporal-CNN-Transformer.

the real follow-up data using four different metrics: Dice similarity coefficient (DSC), average Hausdorff distance (HD), volume error, and absolute volume error. Moreover, a paired t-test with $p < 0.05$ is used to assess the statistical significance of differences between the proposed method and the two baselines.

Results and Discussion. The results of our evaluation are summarized in Table 1 and Fig. 3. From Fig. 3, it can be seen that the lesion outcome predictions made by the proposed method are the most similar to the ground truth regardless of the lesion size. They are also considerably smoother and less noisy compared to the results of both baseline methods. This is also confirmed by the quantitative results: In terms of lesion overlap and surface distance, the proposed method achieves the best results (DSC: 0.45, HD: 6.9 mm), which are significantly better than those obtained with the TCN model (DSC: 0.40, HD: 7.8 mm; p<0.05) and the U-Net trained on perfusion maps (DSC: 0.37, HD: 9.0 mm; p<0.001). The mean volume errors indicate that all three methods slightly overestimate the lesion sizes, although no significant differences are found between them (p>0.11). Such a divergence between DSC and volume error can be expected as the latter only evaluates size (and not localization agreement). This means that the smaller the overlap, the more likely it is that two similar-volume segments are not aligned [19]. The absolute volume error, on the other hand, significantly improves from 76 ml to 58 ml when comparing the proposed method with the TCN model (p<0.05). Furthermore, using the proposed method results in less outliers across the patient population as evidenced by the reduced volume error variance.

Table 1. Quantitative comparison of the proposed method and the two baseline methods for stroke lesion outcome prediction. All results are averaged across the ten folds and are displayed as mean (standard deviation). The best results are bolded. Paired t-tests were performed to determine statistical significance of differences to the proposed method (* and ** denoting a p-value of <0.05 and <0.001, respectively).

Methods	DSC	Volume error	Absolute Volume error	HD
TCN [1]	0.40 (0.24) **	27 (120) mL	76 (97) mL *	7.8 (7.2) mm *
U-Net [15, 18]	0.37 (0.25) **	**3 (108) mL**	73 (80) mL	9.0 (8.9) mm **
Proposed	**0.45 (0.26)**	19 (83) mL	**58 (63) mL**	**6.9 (7.3) mm**

Our results also support the findings from [1, 17] that learning from 4D CTP data is more informative and leads to better results than using perfusion maps. As shown in Fig. 3 for Patient B, this is most likely due to the lack of a real spatio-temporal context in the maps, which makes it hard for the U-Net to identify smaller lesions. The results in Table 1 and Fig. 3 also confirm our initial hypothesis that our Transformer-based setup is better suited than the TCN employed in [1] to fully utilize the complex perfusion information available in the raw 4D CTP. Our proposed approach significantly outperforms [1], generates visibly smoother predictions, and, therefore, sets a new baseline for lesion outcome

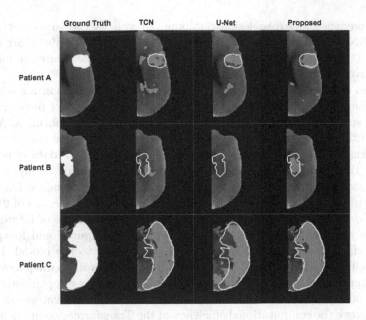

Fig. 3. Qualitative comparison of stroke lesion outcome predictions for three randomly selected patients (rows) using the proposed method (last column), TCN-based method [1] (second column), and U-Net approach [15,18] (third column). The white mask/contour corresponds to the ground truth segmentation in the follow-up image, while the blue masks represent the predictions computed by each model. (Color figure online)

prediction from raw 4D CTP data. We believe that our approach benefits from the multi-head attention mechanism of the Transformer. Unlike TCNs, where the information is only combined from neighbouring time points, our setup allows the network to compare feature activations over the entire input sequence to capture complex relationships in the data. This behaviour is likely amplified by the addition of positional encodings that help to explicitly learn content-position relationships.

4   Conclusion

In this work, we present a novel hybrid spatio-temporal Transformer network for predicting stroke lesion outcomes using raw 4D CTP imaging data. In contrast to most existing approaches, our method does not rely on any specific post-processing of the spatio-temporal image sequence such as deconvolution or AIF selection. It effectively integrates elements of CNNs and Transformer, which results in a powerful end-to-end trainable network that ensures that both local spatial context and temporal relationships of 4D CTP sequences are leveraged for an improved performance. Extensive experiments on a multicenter dataset demonstrate the benefits of using raw 4D CTP imaging over standard perfusion maps to predict the stroke lesion outcome, as well as the superiority of our hybrid CNN-Transformer method against a competing state-of-the-art method that uses TCNs. The proposed method, therefore, sets a new baseline for stroke lesion outcome prediction from raw 4D CTP sequences.

We also believe that our work provides a promising direction for advancing stroke lesion outcome prediction in general, as well as a novel perspective for analyzing spatio-temporal medical image datasets using Transformers. A disadvantage of the proposed method is the increased computational complexity and memory consumption that originates from using raw 4D CTP data in combination with Transformers. Since individual time points are used as input tokens, the dimension of the model grows quadratically with the sequence length. We partially circumvent this problem by processing temporal sequences of 2D slices of the 4D data individually and cropping the brain hemisphere of interest prior to the analysis. While this effectively improves the computational footprint of the method, it also limits the spatial context available to the model. In addition, modeling spatio-temporal information from both brain hemispheres might help to improve the lesion outcome prediction and avoid over-optimistic results regarding false positives. Future work should, therefore, investigate ways such as [25] to improve the computational efficiency of the Transformer component of our method so that full size volumes can be analyzed. Moreover, a detailed (visual) analysis of the impact the attention mechanism has on the model's performance could be carried out by employing and adapting dedicated interpretability and explainability methods for Transformer models [4].

Acknowledgements. This work is supported by the Alberta Innovates Graduate Student Scholarships, T. Chen Fong Doctoral Research Excellence Scholarship in Imaging Science, the Canada Research Chairs Program, and the River Fund at Calgary Foundation.

References

1. Amador, K., Wilms, M., Winder, A., Fiehler, J., Forkert, N.D.: Stroke lesion outcome prediction based on 4D CT perfusion data using temporal convolutional networks. In: Proceedings of the Fourth Conference on Medical Imaging with Deep Learning, pp. 22–33 (2021)
2. Bai, S., Kolter, J.Z., Koltun, V.: An empirical evaluation of generic convolutional and recurrent networks for sequence modeling. arXiv:1803.01271 (2018)
3. Cao, H., Wang, Y., Chen, J., Jiang, D., Zhang, X., et al.: Swin-unet: unet-like pure transformer for medical image segmentation. arXiv:2105.05537 (2021)
4. Chefer, H., Gur, S., Wolf, L.: Transformer interpretability beyond attention visualization. In: Proceedings of the IEEE/CVF Conference on Computer Vision and Pattern Recognition, pp. 782–791 (2021)
5. Chen, J., Lu, Y., Yu, Q., Luo, X., Adeli, E., et al.: TransUNet: transformers make strong encoders for medical image segmentation. arXiv:2102.04306 (2021)
6. Chollet, F., et al.: Keras (2015). https://keras.io
7. Dai, Y., Gao, Y., Liu, F.: TransMed: transformers advance multi-modal medical image classification. Diagnostics **11**(8), 1384 (2021)

8. Daudt, R.C., Le Saux, B., Boulch, A.: Fully convolutional siamese networks for change detection. arXiv:1810.08462 (2018)
9. Fiehler, J., Thomalla, G., Bernhardt, M., Kniep, H., Berlis, A., Dorn, F., et al.: ERASER: a thrombectomy study with predictive analytics end point. Stroke **50**(5), 1275–1278 (2019)
10. Flottmann, F., Broocks, G., Faizy, T.D., Ernst, M., Forkert, N.D., Grosser, M., et al.: CT-perfusion stroke imaging: a threshold free probabilistic approach to predict infarct volume compared to traditional ischemic thresholds. Sci. Rep. **7**(1), 1–10 (2017)
11. Forkert, N.D., Cheng, B., Kemmling, A., Thomalla, G., Fiehler, J.: ANTONIA perfusion and stroke: a software tool for the multi-purpose analysis of MR perfusion-weighted datasets and quantitative ischemic stroke assessment. Methods Inf. Med. **53**(6), 469–481 (2014)
12. Karimi, D., Vasylechko, S., Gholipour, A.: Convolution-free medical image segmentation using transformers. arXiv:2102.13645 (2021)
13. Laughlin, B., Chan, A., Tai, W.A., Moftakhar, P.: RAPID automated CT perfusion in clinical practice. Pract. Neurol. **18**(9), 38–55 (2019)
14. Lo Vercio, L., Amador, K., Bannister, J.J., Crites, S., Gutierrez, A., MacDonald, M.E., et al.: Supervised machine learning tools: a tutorial for clinicians. J. Neural Eng. **17**(6), 062001 (2020)
15. Nielsen, A., Hansen, M.B., Tietze, A., Mouridsen, K.: Prediction of tissue outcome and assessment of treatment effect in acute ischemic stroke using deep learning. Stroke **49**(6), 1394–1401 (2018)
16. Pinto, A., Pereira, S., Meier, R., Alves, V., Wiest, R., Silva, C.A., Reyes, M.: Enhancing clinical MRI perfusion maps with data-driven maps of complementary nature for lesion outcome prediction. In: International Conference on Medical Image Computing and Computer-Assisted Intervention, pp. 107–115 (2018)
17. Robben, D., Boers, A.M., Marquering, H.A., Langezaal, L.L., Roos, Y.B., van Oostenbrugge, R.J., et al.: Prediction of final infarct volume from native CT perfusion and treatment parameters using deep learning. Med. Image Anal. **59**, 101589 (2020)
18. Ronneberger, O., Fischer, P., Brox, T.: U-net: convolutional networks for biomedical image segmentation. In: International Conference on Medical Image Computing and Computer-Assisted Intervention, pp. 234–241 (2015)
19. Taha, A.A., Hanbury, A.: Metrics for evaluating 3D medical image segmentation: analysis, selection, and tool. BMC Med. Imaging **15**, 29 (2015)
20. Vaswani, A., Shazeer, N., Parmar, N., Uszkoreit, J., Jones, L., Gomez, A.N., et al.: Attention is all you need. In: Proceedings of the 31st International Conference on Neural Information Processing Systems, pp. 5998–6008 (2017)
21. Wang, W., Chen, C., Ding, M., Yu, H., Zha, S., Li, J.: TransBTS: multimodal brain tumor segmentation using transformer. In: International Conference on Medical Image Computing and Computer Assisted Intervention, pp. 109–119 (2021)
22. Wang, X., Yang, S., Zhang, J., Wang, M., Zhang, J., Huang, J., et al.: TransPath: transformer-based self-supervised learning for histopathological image classification. In: International Conference on Medical Image Computing and Computer Assisted Intervention, pp. 186–195 (2021)
23. Chan, B.P., Albers, G.W.: Acute ischemic stroke. Curr. Treat. Options. Neurol. **1**(2), 83–95 (1999). https://doi.org/10.1007/s11940-999-0009-5
24. Xie, Y., Zhang, J., Shen, C., Xia, Y.: CoTr: efficiently bridging CNN and transformer for 3D medical image segmentation. In: International Conference on Medical Image Computing and Computer Assisted Intervention, pp. 171–180 (2021)

25. Xiong, Y., Zeng, Z., Chakraborty, R., Tan, M., Fung, G., et al.: Nyströmformer: a nyström-based algorithm for approximating self-attention. In: Proceedings of the AAAI Conference on Artificial Intelligence, pp. 14138–14148 (2021)

26. Yu, Y., Han, Q., Ding, X., Chen, Q., Ye, K., Zhang, S., et al.: Defining core and penumbra in ischemic stroke: a voxel- and volume-based analysis of whole brain CT perfusion. Sci. Rep. **6**, 20932 (2016)

27. Yu, Y., Xie, Y., Thamm, T., Gong, E., Ouyang, J., Huang, C., et al.: Use of deep learning to predict final ischemic stroke lesions from initial magnetic resonance imaging. JAMA Netw. Open **3**(3), e200772 (2020)

28. Yushkevich, P.A., et al.: User-guided 3D active contour segmentation of anatomical structures: significantly improved efficiency and reliability. Neuroimage **31**(3), 1116–1128 (2006)

A Medical Semantic-Assisted Transformer for Radiographic Report Generation

Zhanyu Wang[1], Mingkang Tang[2], Lei Wang[3], Xiu Li[2], and Luping Zhou[1(✉)]

[1] University of Sydney, Sydney, NSW, Australia
{zhanyu.wang,luping.zhou}@sydney.edu.au
[2] Shenzhen International Graduate School of Tsinghua University, Shenzhen, China
tmk20@mails.tsinghua.edu.cn, li.xiu@sz.tsinghua.edu.cn
[3] University of Wollongong, Wollongong, NSW, Australia
leiw@uow.edu.au

Abstract. Automated radiographic report generation is a challenging cross-domain task that aims to automatically generate accurate and semantic-coherence reports to describe medical images. Despite the recent progress in this field, there are still many challenges at least in the following aspects. First, radiographic images are very similar to each other, and thus it is difficult to capture the fine-grained visual differences using CNN as the visual feature extractor like many existing methods. Further, semantic information has been widely applied to boost the performance of generation tasks (e.g. image captioning), but existing methods often fail to provide effective medical semantic features. Toward solving those problems, in this paper, we propose a memory-augmented sparse attention block utilizing bilinear pooling to capture the higher-order interactions between the input fine-grained image features while producing sparse attention. Moreover, we introduce a novel Medical Concepts Generation Network (MCGN) to predict fine-grained semantic concepts and incorporate them into the report generation process as guidance. Our proposed method shows promising performance on the recently released largest benchmark MIMIC-CXR. It outperforms multiple state-of-the-art methods in image captioning and medical report generation.

Keywords: Radiographic report generation · Semantic concepts · Sparse attention transformer

1 Introduction

Automated radiographic report generation aims to generate a paragraph to address the observations and findings of a given radiology image. It has significant application scenarios in reducing the workload and mitigating the diagnostic errors of radiologists who are under pressure to report increasingly complex studies in less time, especially in emergency radiology reporting.

Z. Wang and M. Tang—Equal contribution.

© The Author(s), under exclusive license to Springer Nature Switzerland AG 2022
L. Wang et al. (Eds.): MICCAI 2022, LNCS 13433, pp. 655–664, 2022.
https://doi.org/10.1007/978-3-031-16437-8_63

Due to its clinical importance, medical report generation has gained increasing attention. The encoder-decoder paradigm has prevailed in this field, motivated by its success in generic image captioning [1,5,6,10,12,17,18,25,27,28], where many approaches apply similar structures including an encoder based on convolutional neural networks (CNN), and a decoder based on recurrent neural networks (RNN) [12,25,27]. Furthermore, several research works employ carefully-designed attention mechanism to boost performance [1,10,18]. Most recently, there have been efforts starting to explore transformer-structured framework [5,6,17] in this field.

Despite these progresses, there are still some non-negligible problems in the medical report generation methods rooted in image captioning due to the non-trivial different characteristics of these two tasks. First, unlike natural images, radiographic images are very similar to each other. The medical report generation methods using the CNN-based image encoder extract image-level features due to the lack of image region annotations in such applications, which could not well cater for the local details reflecting the fine-grained image patterns of clinic importance. This problem could be somewhat mitigated by employing the recently developed Vision Transformer model [7] that explores the dependency of image regions without generating regional proposals. However, there still lacks mechanism to better treat fine-grained patterns. Second, the differences between medical reports are also fine-grained, that is, the reports are dominated by similar sentences describing the common content of the images while the disease-related words may be submerged. A possible remedy is to incorporate semantic textual concepts into the model training to guide the report generation. However, such information has not been well explored for medical report generation, especially in the recently popular transformer-structured models.

To cope with the limitations mentioned above, we propose the following solutions to advance radiographic report generation. Firstly, we carefully deal with the fine-grained differences existed in radiographic images from two aspects. On the one hand, we utilize CLIP [21] rather than ResNet as our visual feature extractor. The advantage of CLIP is two-fold: i) it is built upon Vision Transformer that extracts regional visual features and relationships, and ii) it is trained by matching image-text pairs and thus produces text-enhanced image features. On the other hand, more importantly, we introduce an memory-augmented sparse attention for high-order feature interactions and embed it into the transformer encoders and decoders. This attention makes use of bilinear pooling that proves to be effective for fine-grained visual recognition [16], memorizes historical information for long report generation, and produces sparse attention for efficient report generation. Second, to incorporate semantic textual concepts, we introduce a novel lightweight medical concepts generation network to predict fine-grained semantic concepts and incorporate them into medical report generation process, which is different from the usage of sparse medical tags in the existing methods [10]. In sum, our main contributions include: 1) We introduce a bilinear-pooling-assisted sparse attention block and embed it into a transformer network to capture the fine-grained visual difference existed between

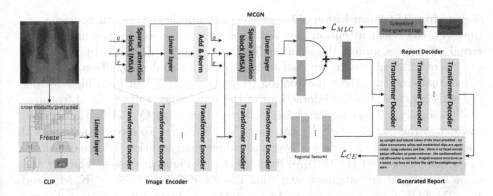

Fig. 1. An overview of the proposed framework, which comprises an Image Encoder, a Medical Concepts Generation Network (MCGN) and a Report Decoder. The transformer encoder and decoder are embedded with a Memory-augmented Sparse nonlinear Attention (MSA) to capture the higher-order interactions between the input fine-grained image features. Meanwhile, we inject a medical concepts generator to provide semantic information to facilitate report generation.

radiographic images; 2) We propose a medical concepts generation network to provide enriched semantic information to benefit radiographic report generation; 3) We extensively validate our model on the recently released largest dataset MIMIC-CXR. The results indicate that our framework outperforms multiple state-of-the-art methods in image captioning and medical report generation.

2 Methodology

The proposed framework adopts the encoder-decoder paradigm, where the transformer encoder and decoder are embedded with a Memory-augmented Sparse Attention block (MSA) to capture the higher-order interactions between the input fine-grained image features. Meanwhile, a Medical semantic Concepts Generation Network (MCGN) is incorporated into the report generation model to further improve the performance. In the following, we first introduce the MSA block in Sect. 2.1 and how the MSA block are embedded into the encoder and decoder in Sect. 2.2 and Sect. 2.4, and our proposed MCGN in Sect. 2.3 (Fig. 1).

2.1 Memory-Augmented Sparse Attention of High-Order Interaction

Given the three input matrices, queries $\mathbf{Q} \in \mathbf{R}^{N \times D_q}$, keys $\mathbf{K} \in \mathbf{R}^{N \times D_k}$ and values $\mathbf{V} \in \mathbf{R}^{N \times D_v}$, the conventional self-attention block [23] employed by transformer first computes the dot products (element-wise sum) of each query with all keys and then applies a softmax function to obtain the weights on the values. Formally, the attention is calculated by: $\text{Attention}(\mathbf{Q}, \mathbf{K}, \mathbf{V}) = \text{softmax}(\frac{\mathbf{Q}\mathbf{K}^{\mathbf{T}}}{\sqrt{d_k}})\mathbf{V}$. It exploits only 1st order interaction of the input vectors since the weight is derived from the linear fusion of the given query and keys via element-wise sum. However, the linear attention may not be sufficient to capture the fine-grained visual differences among the input radiographic images. Inspired by the recent success of bilinear pooling applied in fine-grained visual recognition, we inject bilinear-pooling into the self-attention to capture the 2nd or even higher-order (by stacking these blocks) interactions of the input fine-grained visual features.

Specifically, we assume the global feature as the query $\mathbf{Q} \in \mathbf{R}^{D_q}$, the regional features as the key $\mathbf{K} \in \mathbf{R}^{N \times D_k}$ and value $\mathbf{V} \in \mathbf{R}^{N \times D_v}$. To record the historical information, we extent the set of keys and values with additional "memory-slots" to encode and collect the features from all the previous processes. The key and value of our memory-augment attention can be defined as: $\hat{\mathbf{K}} = [\mathbf{K}, \mathbf{M}_k]$ and $\hat{\mathbf{V}} = [\mathbf{V}, \mathbf{M}_v]$, respectively, where $\mathbf{M}_k$ and $\mathbf{M}_v$ are learnable matrices with n_m rows, and $[\cdot, \cdot]$ indicates concatenation. Then, a low-rank bilinear pooling [13] is performed to obtain the joint bilinear query-key $\mathbf{B}_k$ and query-value $\mathbf{B}_v$ by

$$\mathbf{B}_k = \sigma(\mathbf{W}_k \hat{\mathbf{K}}) \odot \sigma(\mathbf{W}_q^k \hat{\mathbf{Q}}), \quad \mathbf{B}_v = \sigma(\mathbf{W}_v \hat{\mathbf{v}}) \odot \sigma(\mathbf{W}_q^v \hat{\mathbf{Q}}) \quad (1)$$

where $\mathbf{W}_k \in \mathbf{R}^{D_B \times D_k}$, $\mathbf{W}_v \in \mathbf{R}^{D_B \times D_v}$, $\mathbf{W}_q^k \in \mathbf{R}^{D_B \times D_q}$ and $\mathbf{W}_q^v \in \mathbf{R}^{D_B \times D_v}$ are learning parameters, σ denotes ReLU unit, and $\odot$ represents element-wise multiplication. Next, we use a linear layer to project $\mathbf{B}_k \in \mathbf{R}^{D_c \times D_B}$ into a intermediate representation $\mathbf{B}_k' = \sigma(\mathbf{W}_{\tilde{B}}^k \mathbf{B}_k)$, then use another linear layer to map $\mathbf{B}_k'$ from $\mathbf{D}_c$ dimension to 1 dimension to obtain the spatial-wise attention weight $\alpha_s \in \mathbb{R}^{D_c \times 1}$. Unlike [19] using softmax to normalize α_s, we utilized another ReLU unit to prune out all negative scores of low query-key relevance, automatically ensuring the sparse property of the attention weight $\beta_s = \sigma(\alpha_s)$. Meanwhile, we perform a squeeze-excitation operation [8] to $\mathbf{B}_k'$ to obtain channel-wise attention $\beta_c = \text{sigmoid}(\mathbf{W}_c)\bar{\mathbf{B}}$, where $\mathbf{W}_c \in \mathbf{R}^{D_B \times D_c}$ is learnable parameters and $\bar{\mathbf{B}} \in \mathbf{R}^{D_B \times 1}$ is an average pooling of $\mathbf{B}_k'$. The output attended features $\hat{v}$ of our memory-augmented sparse attention integrate the enhanced bilinear values with spatial and channel-wise bilinear attention $\hat{\mathbf{Q}} = \text{Attention}(\mathbf{K}, \mathbf{V}, \mathbf{Q}) = \beta_c \odot \text{LN}(\beta_s \mathbf{B}_v)$. Where $\text{LN}(\cdot)$ denotes variants of layer normalization [2].

Such structure design benefits the model in three ways. First, it can explore higher-order interactions between the input single-model (in the encoder) or multi-model (in the decoder) features, resulting in a more robust representative capacity of the output attended features. Second, the memory tokens we embed into the MSA block can record the previous generation process and collect historical information, which can be valuable for our task's long report generation. Third, we propose a softmax-free sparse attention module, pruning out all negative attention scores and producing sparse attention of transformer decoder with higher efficiency in the process of report generation. We measured the speedup per training step on 500 steps with about 16 samples per batch to compare the running efficiency of softmax-base attention and our sparse Relu-based attention. We perform three runs on a single NVIDIA TESLA V100 and report average results. Without performance degradation, our sparse attention can reduce the training time from 0.497 s per batch to 0.488 s per batch.

2.2   Image Encoder

Let's denote an input image by $\mathbf{I}$. The pre-trained CLIP model (ViT-B/16) is utilized to extract the regional features of I: $f = \text{CLIP}(\mathbf{I})$, where $f \in \mathbb{R}^{D_c \times D_f}$. We take the mean-pooled feature embedding f as the initial $\mathbf{Q}^{(0)} = \frac{1}{N_c} \sum_{i=1}^{N_c} f_i$, and f as the initial $\mathbf{K}^{(0)}$ and $\mathbf{V}^{(0)}$, and feed them into the encoding layers.

Encoding Layer. We embed our memory-augmented sparse attention into a Transformer-like layer. Formally, for the m-th transformer encoder TE_m, we take the previous output attended feature $\hat{\mathbf{Q}}^{(m-1)}$ as the input query, couple it with the current input keys $\mathbf{K}^{m-1} = \left\{ \mathbf{k}_i^{(m-1)} \right\}_{i=1}^{N}$, and values $\mathbf{V}^{(m-1)} = \{\mathbf{v}_i^{m-1}\}_{i=1}^{N}$, which could be expressed as $\mathbf{K}^{(m)}, \mathbf{V}^{(m)}, \mathbf{Q}^{(m)} = TE_m(\mathbf{K}^{(m-1)}, \mathbf{V}^{(m-1)}, \hat{\mathbf{Q}}^{(m-1)})$, with $\hat{\mathbf{Q}}^{(m)} = \text{Attention}(\mathbf{K}^{(m-1)}, \mathbf{V}^{(m-1)}, \hat{\mathbf{Q}}^{(m-1)})$ computed by MSA block (see Sect. 2.1). Then, all the keys and values are further updated conditioned on the new attended feature $\hat{\mathbf{Q}}^{(m)}$ by equation: $\mathbf{k}_i^{(m)} = \text{LN}(\sigma(\mathbf{W}_m^k[\hat{\mathbf{Q}}^{(m)}, \mathbf{k}_i^{(m-1)}]) + \mathbf{k}_i^{(m-1)})$ and $\mathbf{v}_i^{(m)} = \text{LN}(\sigma(\mathbf{W}_m^v[\hat{\mathbf{Q}}^{(m)}, \mathbf{v}_i^{(m-1)}]) + \mathbf{v}_i^{(m-1)})$, respectively, where $\mathbf{W}_m^k$ and $\mathbf{W}_m^v$ are learnable parameters. [,] indicates concatenation. This means that each key or value is concatenated with the new attended feature, followed with a residual connection and layer normalization [2]. Noted that the entire image encoder consists of $M = 6$ encoding layers in this paper. In particular, when $m = 1$, $\hat{\mathbf{Q}}^{(0)} = \mathbf{Q}^{(0)} = \frac{1}{N_c} \sum_{i=1}^{N_c} f_i$.

2.3   Medical Concepts Generation Network

In order to train the Medical Concepts Generation Network (MCGN), we utilize RadGraph [9] to extract the pseudo-medical concepts as the ground-truth for multi-label classification (MLC). Specifically, the RadGraph is a knowledge graph of clinic radiology entitles and relations based on full-text chest x-ray radiology reports. We use these entities as the medical semantic concepts and finally select 768 concepts according to their occurrence frequency.

Our MCGN consists of an MSA block for processing the intermediate features generated by the m-th Encoding layer of Image Encoder. The output of MSA is $\mathbf{V}_c = \text{Attention}(\mathbf{K}^{(m-1)}, \mathbf{V}^{(m-1)}, \hat{\mathbf{Q}}^{(m-1)})$, where $m \in [1, M]$ denotes the m-th encoding layer. Then, a linear projection maps $\mathbf{V}_c$ from D to K dimensions, where K is the number of medical concepts. The overall medical concepts classification loss can be expressed as:

$$\mathcal{L}_{MLC} = -\frac{1}{K} \cdot \sum_i y_i \cdot \log((1 + \exp(-x_i))^{-1})$$
$$+ (1 - y_i) \cdot \log\left(\frac{\exp(-x_i)}{1 + \exp(-x_i)}\right), \tag{2}$$

where x_i is the prediction for ith medical concept ($i \in \{0, 1, \cdots, K\}$), and y_i is the ground-truth label for ith medical concept where $y_i \in \{0, 1\}$, with $y_i = 1$

meaning the input image has the corresponding medical concept while $y_i = 0$ meaning the opposite.

2.4   Report Decoder

The report decoder aims to generate the output report conditioned on the attended visual embeddings $\mathbf{V}_i = \hat{\mathbf{Q}}^{(}m)_{m=0}^{M}$ from image encoder and the medical concepts embeddings $\mathbf{V}_c$ from MCGN. Therefore, we first fused those embeddings by the equation: $\mathbf{v}_f = \mathbf{W}_f[\hat{\mathbf{Q}}^{(0)}, \hat{\mathbf{Q}}^{(1)}, \cdots, \hat{\mathbf{Q}}^{(M)}] + \mathbf{V}_c$, where $\mathbf{W}_f$ is a learnable parameter. The input of transformer decoder is thus set as the concatenation of the fused feature $\mathbf{v}_f$, the regional features $\mathbf{K}^{(M)}$ and $\mathbf{V}^{(M)}$ output by image encoder, and the input word $\mathbf{w}_t$. Generally, the first decoder sub-layer takes $\mathbf{v}_f$ as the query to calculate sparse nonlinear attention with the word embeddings (taken as keys and values) of $\mathbf{w}_t$ to obtain the visual-enhanced word embeddings. Then the output of the first sub-layer will be coupled with the image encoder's output $\mathbf{K}^{(M)}$ and $\mathbf{V}^{(M)}$ as $\mathbf{Q}, \mathbf{K}, \mathbf{V}$ input to another sparse attention block and repeat this process. Note that we employ residual connections around each sub-layers similar to the encoder, followed by layer normalization. The decoder is also composed of a stack of $N = 6$ identical layers.

We train our model parameters θ by minimizing the negative log-likelihood of $\mathbf{P}(T)$ given the image features:

$$\mathcal{L}_{CE} = -\sum_{i=1}^{N} log P_\theta(\mathbf{t}_i|\mathbf{I}, \mathbf{t}_{i-1}, \cdots, \mathbf{t}_1) \tag{3}$$

where $\mathbf{P}(\mathbf{t}_i|\mathbf{I}, \mathbf{t}_{i-1}, \cdots, \mathbf{t}_1)$ represents the probability predicted by the model for the i-th word based on the information of the image $\mathbf{I}$ and the first $(i-1)$ words.

Overall Objective Function. Our overall objective integrates the two losses regarding report generation and multi-class classification, which is defined as:

$$\mathcal{L}_{all} = \lambda_{CE}\mathcal{L}_{CE} + \lambda_{MLC}\mathcal{L}_{MLC} \tag{4}$$

The hyper-parameters λ_{CE} and λ_{MLC} balance the two losses terms, and their values are given in Sect. 3.

3   Experiments

3.1   Data Collection

MIMIC-CXR [11] is the recently released largest dataset to date containing both chest radiographs and free-text reports. It consists of 377110 chest x-ray images and 227835 reports from 64588 patients of the Beth Israel Deaconess Medical Center. In our experiment, we adopt MIMIC-CXR's official split following [5,17] for a fair comparison, resulting in a total of 222758 samples for training, and 1808 and 3269 samples for validation and test, respectively. We convert all tokens to lowercase characters and remove non-word illegal characters and the words whose occurring frequency is lower than 5, counting to 5412 unique words remaining in the dataset.

Table 1. Performance comparison on MIMIC-CXR dataset.

Methods	Bleu-1	Bleu-2	Bleu-3	Bleu-4	ROUGE	METEOR	CIDEr
Show-Tell [25]	0.308	0.190	0.125	0.088	0.256	0.122	0.096
Att2in [27]	0.314	0.198	0.133	0.095	0.264	0.122	0.106
AdaAtt [18]	0.314	0.198	0.132	0.094	0.267	0.128	0.131
Transformer [23]	0.316	0.199	0.140	0.092	0.267	0.129	0.134
M2Transformer [6]	0.332	0.210	0.142	0.101	0.264	0.134	0.142
R2Gen [5]	0.353	0.218	0.145	0.103	0.277	0.142	0.141
R2GenCMN [4]	0.353	0.218	0.148	0.106	0.278	0.142	0.143
PPKED[†] [17]	0.360	0.224	0.149	0.106	0.284	0.149	0.237
Self-boost [26]	0.359	0.224	0.150	0.109	0.277	0.141	0.270
baseline	0.352	0.215	0.145	0.105	0.264	0.133	0.240
baseline+MSA	0.369	0.231	0.159	0.117	0.280	0.142	0.297
Ours (baseline+MSA+MCGN)	0.373	0.235	0.162	0.120	0.282	0.143	0.299
Ours+RL	**0.413**	**0.266**	**0.186**	**0.136**	**0.298**	**0.170**	**0.429**

3.2 Experimental Settings

Evaluation Metrics. Following the standard evaluation protocol[1], we utilise the most widely used BLEU scores [20], ROUGE-L [15], METEOR [3] and CIDER [24] as the metrics to evaluate the quality of the generated text report.

Implementation Details. We extract our pre-computed image features by a pre-trained "CLIP-B/16" [21][2] model, resulting in a regional feature matrix $f \in \mathbb{R}^{768 \times 196}$ (reshaped from $768 \times 14 \times 14$). The number of memory vectors is set to 3. The number of the layers in both the image transformer encoder and the text transformer decoder is set to 6 and the number of heads in multi-head attention is set to 8. The hyper-parameters λ_{CE}, λ_{MLC} are set as 1 and 5, respectively. We train our model using Adam optimizer [14] on eight NVIDIA TESLA V100 with a mini-batch size of 32 for each GPU. The learning rates is set to be $5e-5$, and the model is trained in a total of 60 epochs. At the inference stage, we adopt the beam search strategy and set the beam size as 3. In addition, following [22], we also apply reinforcement learning with CIDEr as the reward and further train the model for 20 epochs to boost the performance.

3.3 Results and Discussion

Comparison with SOTA. Table 1 summarizes the performance comparisons between the state-of-the-art methods and our proposed model on the MIMIC-CXR Official test split. Specifically, there are five SOTA image captioning methods in the comparison, including Show-tell [25], AdaAtt [18], Att2in [1], Transformer [23], and M2transformer [6]. Moreover, we also compare with four

---

[1] https://github.com/tylin/coco-caption.
[2] https://github.com/openai/CLIP/.

Image	Medical Concepts	Ground Truth	BL	BL+MSXA+MCGN	BL+MSXA+MCGN+RL

Fig. 2. An example report generated by the proposed model. The correctly predicted medical concepts and the key information in the report are marked red. (Color figure online)

SOTA medical report generation methods: including R2Gen [5], R2GenCMN [4], PPKED [17] and Self-boost [26]. For PPKED [17], we quote the performance from their paper (marked with † in Table 1) since this model does not release the code. For the other methods in comparison, we download the codes released publicly and re-run them on the MIMIC-CXR dataset with the same experimental setting as ours, so they are strictly comparable.

As shown in Table 1, our proposed method is the best performer over almost all evaluation metrics among the comparing methods, even without the reinforcement learning. Specifically, the two very recent medical report generation models, R2Gen [5] and Self-boost [26], perform better than other image captioning methods but still lose to ours. R2Gen adopts a memory-augmented transformer decoder, but its image encoder still relies on the CNN model, which may fail to identify fine-grained differences of radiographic images. For the Self-boost model, despite utilizing an image-text matching network to help the image encoder learn fine-grained visual differences, it builds upon CNN and LSTM and does not incorporate semantic information into the report generation process like ours. It is also noted that although performing slightly inferior than PPKED [17] on Meteor and Rouge, our models shows clear advantages over PPKED on Bleu and CIDEr metrics. PPKED is a transformer based model and encodes semantic concepts into an external knowledge graph to guide the report generation. However, this knowledge graph is sparse with only 25 nodes (concepts) and requires extra efforts to construct, while our fine-grained medical concepts can be easily picked up from the reports.

Ablation Study. Table 1 (lower part) shows the results of the ablation study to single out the contributions of each component of our model. We remove MCGN from our model and replace MSA with conventional self-attention [23] as the baseline to verify the performance improvements brought by the proposed attention and our concepts generation network. In Table 1, there are three components: MSA, MCGN, and RL, representing Memory-augmented Sparse Attention, Medical Concepts Generation Network, and Reinforcement Learning, respectively. The symbols "+" or "−" indicate the inclusion or exclusion of the following component. The benefit of using MSA block can be well reflected by the improvement from "baseline" to "baseline+MSA". As shown, the performance can be further boosted by additionally introducing medical concepts ("baseline+MSA+MCGN"). Moreover, as mentioned, even removing reinforcement learning, our model still significantly outperforms the comparing methods in Table 1. Our proposed model produces more accurate and descriptive findings compared with baseline, as the qualitative results show in Fig. 2.

4   Conclusions

We propose a medical report generation model utilizing sparse nonlinear attention in the transformer-structured encoder-decoder paradigm. Compared with prior arts, our proposed model considers higher-order interactions across the input feature vectors and thus can better cater for the fine-grained radiographic images. Our proposed Medical Concepts Generation Network introduces rich semantic concepts and encodes them as semantic information to benefit the radiographic report generation task. Extensive experimental results demonstrate that our approach surpasses the state-of-the-art with a large margin.

References

1. Anderson, P., et al.: Bottom-up and top-down attention for image captioning and visual question answering. In: CVPR (2018)
2. Ba, J.L., Kiros, J.R., Hinton, G.E.: Layer normalization. arXiv preprint arXiv:1607.06450 (2016)
3. Banerjee, S., Lavie, A.: METEOR: an automatic metric for MT evaluation with improved correlation with human judgments. In: ACL (2005)
4. Chen, Z., Shen, Y., Song, Y., Wan, X.: Cross-modal memory networks for radiology report generation. In: Proceedings of the 59th Annual Meeting of the Association for Computational Linguistics and the 11th International Joint Conference on Natural Language Processing (2021)
5. Chen, Z., Song, Y., Chang, T.H., Wan, X.: Generating radiology reports via memory-driven transformer. arXiv preprint arXiv:2010.16056 (2020)
6. Cornia, M., Stefanini, M., Baraldi, L., Cucchiara, R.: Meshed-memory transformer for image captioning. In: CVPR (2020)
7. Dosovitskiy, A., et al.: An image is worth 16x16 words: transformers for image recognition at scale. arXiv preprint arXiv:2010.11929 (2020)
8. Hu, J., Shen, L., Sun, G.: Squeeze-and-excitation networks. In: CVPR (2018)
9. Jain, S., et al.: Radgraph: extracting clinical entities and relations from radiology reports. arXiv preprint arXiv:2106.14463 (2021)
10. Jing, B., Xie, P., Xing, E.P.: On the automatic generation of medical imaging reports. In: ACL (2018)
11. Johnson, A.E.W., et al.: MIMIC-CXR: a large publicly available database of labeled chest radiographs. CoRR (2019)
12. Karpathy, A., Li, F.: Deep visual-semantic alignments for generating image descriptions. In: CVPR (2015)

13. Kim, J.H., On, K.W., Lim, W., Kim, J., Ha, J.W., Zhang, B.T.: Hadamard product for low-rank bilinear pooling. arXiv preprint arXiv:1610.04325 (2016)
14. Kingma, D.P., Ba, J.: Adam: a method for stochastic optimization. In: ICLR (2015)
15. Lin, C.Y.: ROUGE: a package for automatic evaluation of summaries. In: ACL (2004)
16. Lin, T.Y., RoyChowdhury, A., Maji, S.: Bilinear CNN models for fine-grained visual recognition. In: ICCV (2015)
17. Liu, F., Wu, X., Ge, S., Fan, W., Zou, Y.: Exploring and distilling posterior and prior knowledge for radiology report generation. In: CVPR (2021)
18. Lu, J., Xiong, C., Parikh, D., Socher, R.: Knowing when to look: adaptive attention via a visual sentinel for image captioning. In: CVPR (2017)
19. Pan, Y., Yao, T., Li, Y., Mei, T.: X-linear attention networks for image captioning. In: CVPR (2020)
20. Papineni, K., Roukos, S., Ward, T., Zhu, W.: Bleu: a method for automatic evaluation of machine translation. In: ACL (2002)
21. Radford, A., et al.: Learning transferable visual models from natural language supervision. In: ICML (2021)
22. Rennie, S.J., Marcheret, E., Mroueh, Y., Ross, J., Goel, V.: Self-critical sequence training for image captioning. In: CVPR (2017)
23. Vaswani, A., et al.: Attention is all you need. In: NIPS (2017)
24. Vedantam, R., Zitnick, C.L., Parikh, D.: CIDEr: consensus-based image description evaluation. In: CVPR (2015)
25. Vinyals, O., Toshev, A., Bengio, S., Erhan, D.: Show and tell: a neural image caption generator. In: CVPR (2015)
26. Wang, Z., Zhou, L., Wang, L., Li, X.: A self-boosting framework for automated radiographic report generation. In: CVPR (2021)
27. Xu, K., et al.: Show, attend and tell: neural image caption generation with visual attention. In: ICML (2015)
28. You, Q., Jin, H., Wang, Z., Fang, C., Luo, J.: Image captioning with semantic attention. In: CVPR (2016)

Personalized Diagnostic Tool for Thyroid Cancer Classification Using Multi-view Ultrasound

Han Huang[1,2,3], Yijie Dong[4], Xiaohong Jia[4], Jianqiao Zhou[4], Dong Ni[1,2,3], Jun Cheng[1,2,3](✉), and Ruobing Huang[1,2,3](✉)

[1] National-Regional Key Technology Engineering Laboratory for Medical Ultrasound, School of Biomedical Engineering, Health Science Center, Shenzhen University, Shenzhen, China
chengjun583@qq.com, ruobing.huang@szu.edu.cn

[2] Medical Ultrasound Image Computing (MUSIC) Lab, Shenzhen University, Shenzhen, China

[3] Marshall Laboratory of Biomedical Engineering, Shenzhen University, Shenzhen, China

[4] Department of Ultrasound Medicine, Ruijin Hospital, School of Medicine, Shanghai Jiaotong University, Shanghai, China

Abstract. Over the past decades, the incidence of thyroid cancer has been increasing globally. Accurate and early diagnosis allows timely treatment and helps to avoid over-diagnosis. Clinically, a nodule is commonly evaluated from both transverse and longitudinal views using thyroid ultrasound. However, the appearance of the thyroid gland and lesions can vary dramatically across individuals. Identifying key diagnostic information from both views requires specialized expertise. Furthermore, finding an optimal way to integrate multi-view information also relies on the experience of clinicians and adds further difficulty to accurate diagnosis. To address these, we propose a personalized diagnostic tool that can customize its decision-making process for different patients. It consists of a multi-view classification module for feature extraction and a personalized weighting allocation network that generates optimal weighting for different views. It is also equipped with a self-supervised view-aware contrastive loss to further improve the model robustness towards different patient groups. Experimental results show that the proposed framework can better utilize multi-view information and outperform the competing methods.

Keywords: Personalized diagnosis · Thyroid cancer · Ultrasound · Multi-view

1 Introduction

The world has witnessed a rapid global increase in the incidence of thyroid carcinoma during the past decades [1]. Accurate diagnosis allows access to timely

© The Author(s), under exclusive license to Springer Nature Switzerland AG 2022
L. Wang et al. (Eds.): MICCAI 2022, LNCS 13433, pp. 665–674, 2022.
https://doi.org/10.1007/978-3-031-16437-8_64

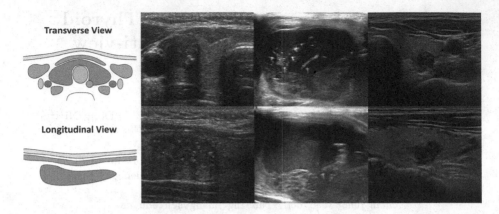

Fig. 1. Illustration of the transverse (upper row) and longitudinal (lower row) planes of the thyroid. Each row corresponds to the US images of one patient, while the anatomical diagrams are shown in the left. It can be seen that both the imaging conditions and the lesion appearance vary dramatically across individuals.

intervention and helps to prevent over-diagnosis. Ultrasound (US) is the primary technique used in thyroid cancer screening due to its wide availability, low cost and non-ionizing nature [6]. The correct interpretation of thyroid US requires spatial understanding of the anatomy (see Fig. 1) and accurate recognition of diagnostic related features. These necessitate both expertise and a high-level of experience, and may elude the less-experienced clinicians.

To alleviate this issue, many computer-aided diagnosis (CAD) tools were proposed [5,12,19]. For example, Chi et al. [5] used a GoogLeNet to extract features of US and then applied a Random Forest classifier to identify thyroid nodules. Liu et al. [12] combined deep features from CNN with conventional ones, forming a hybrid model for thyroid cancer classification. Zhao et al. [19] proposed a local and global features disentangled network to classify thyroid nodules into benign and malignant cases. These methods shed light on how to approach the problem, while they only focused on the primary transverse view of the US scans and neglected the other. However, US scanning for thyroid is often done both in transverse and longitudinal planes in clinical practice [2]. Other clinical studies also revealed that longitudinal planes help to promote diagnosis accuracy by excluding suspicious benign nodules and discovering potential malignancy markers that might not be visible in the transverse planes [14,17]. For example, the column 2 of Fig. 1 shows an example whose longitudinal view indicates existence of micro-calcification, while its transverse view fails to provide this information. Similarly, the column 4 shows another case where the lesion seems to have a regular shape when viewed from the transverse direction, while its longitudinal direction exhibits the opposite information. New diagnostic tool is needed to incorporate this rich multi-view information.

Combining information from different views has been investigated previously to analyze other medical images. Modern approaches mainly leveraged deep learning (DL) models to extract and fuse multi-view information. For example, Kyono et al. [10] presented a novel multi-view multi-task (MVMT) CNN model to fuse the four views of mammograms for breast cancer diagnosis. Pi et al. [16] developed an attention-augmented deep neural network (AADNN) to detect bone metastasis using anterior and posterior views of X-ray images. Wang et al. [18] used depth-wise separable convolution-based CNNs to organize the five views of echocardiograms. Note that, to the best of our knowledge, there also exist two multi-view studies in thyroid US [4,11]. The former [11] essentially used a three-branch CNN model to classify thyroid nodules with different resolutions, while the latter [4] referred to the deep features, statistical and texture features of US images as different "views" of thyroid lesions and combined this information through an ensemble model. Neither of them utilized both the transverse and longitudinal views of thyroid US images for lesion classification.

Despite showing promising results, these methods treated different views equally, overlooking the fact that one of the views might contain critical diagnostic information while the other should be de-emphasized. More importantly, as the lesions vary in size, shape, appearance across individuals (see Fig. 1), this disparity among the importance of different views should be customized to the particular condition of each patient. To address this, we propose a personalized diagnostic tool for automated thyroid cancer classification. Our contribution is three-fold:

- The first DL-based framework that intelligently integrates both the transverse and longitudinal views for thyroid cancer diagnosis using US.
- A personalized weighting allocation network that customizes the multi-view weighting for different patients. It can be trained end-to-end and does not require additional supervision.
- A self-supervised view-aware contrastive loss that considers intra-class variation inside patient groups and can further improve the model performance.

Experiment results showed that the trained model is able to intelligently exploit multi-view information and outperform state-of-the-art approaches in thyroid cancer diagnosis.

2 Methods

Processing information from heterogeneous sources is a non-trivial task, while combining them according to individual specifications can be more challenging and is rarely explored. To address this, we propose a framework that learns to assign customized multi-view weighting for thyroid cancer diagnosis. Figure 2 displays the overall framework. The inputs are first passed to a multi-view classification network (MVC) for feature extraction and pre-prediction. Then, the features are sent to a novel personalized weighting allocation network (PAWN) for multi-view weighting generation. The training procedure is also equipped with a view-aware contrastive loss (VACL) (lower part of Fig. 2). Details of each component is explained next.

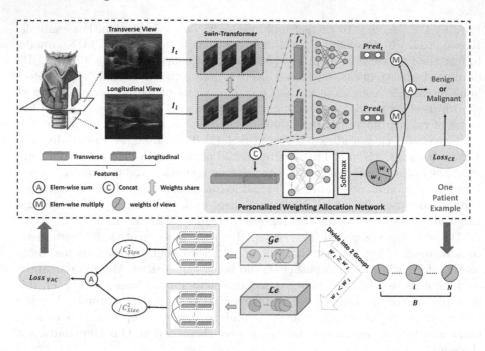

Fig. 2. Schematic of the overall framework. Multi-view thyroid US images are first processed by the swin-transformer backbone (grey block). The derived multi-level features are sent to subsequent MLPs to generate view-specific predictions, respectively. The PAWN then automatically assigns the optimal multi-view weighting for each patient (w_t, w_l). These pairs of weights are also collected to separate samples and the view-aware contrastive loss $\mathcal{L}_{VAC}$ is calculated. (Color figure online)

2.1   Multi-view Classification Model

Extracting multi-view features automatically while dealing with view-related information efficiently has always been a challenge in multi-view learning. The proposed MVC model uses a two-branch network to address this, where each branch corresponds to one anatomical view of thyroid US (orange block in Fig. 2). The swin-transformer architecture is adopted as the feature extraction backbone due to its prominent modeling capacity [13]. Note that two branches share the same parameters to reduce over-parameterization. This design can help prevent overfitting and also encourage interaction between different views. Formally, the MVC model accepts a pair of thyroid US images: I_t^p and I_l^p of the same patient p. The former is collected from the transverse direction, while the latter corresponds to the longitudinal view. The feature extraction backbone then yields corresponding high-level features: $f_t^p = ST(I_t^p), f_l^p = ST(I_l^p)$. Then, f_t^p, f_l^p are fed into unique multilayer perceptron (MLP) heads respectively to generate view-specific decisions, denoted as $pred_t^p$ and $pred_l^p$. This structure boosts the model's flexibility and allows learning view-specific representations and making view-independent decisions. The f_t^p, f_l^p are sent to the subsequent PAWN, while $pred_t^p$ and $pred_l^p$ are saved to generate the final prediction result.

Note that the MVC network is a self-contained model and can be trained independently to warm up the learning. Specifically, the output of MVC model is formulated as $\hat{pred}^p = 0.5*pred_t^p+0.5*pred_l^p$, and evaluated against the ground truth. The model can then be optimized through a standard cross-entropy loss ($\mathcal{L}_{CE}$) for a few epochs. The PWAN is then attached to the pre-trained MVC for end-to-end optimization. Empirical experiments found that synchronized optimization facilitates the learning of holistic features.

2.2 Personalized Weighting Generation

An ideal multi-view model should make predictions using all the available information but may put emphasis on a particular view based on the specific situation. To fulfill this, a lightweight network named PAWN was designed to generate the optimal weighting for different views to yield accurate diagnosis. Instead of directly processing raw images, the PAWN operates on high-level features (f_t^p, f_l^p) of different views. This helps to reduce repetitive computation and can accelerate convergence. Moreover, it also forces the PAWN to allocate weights explicitly based on diagnosis-related features and coordinates parameter updates across modules. The PAWN, therefore, only requires a simple architecture to operate, i.e., a three-layer MLP (with a hidden size of 256, 128, 32, respectively). It generates optimal weighting for each view which can be denoted as: $(w_t^p, w_l^p) = Softmax(PWAN(f_t^p, f_l^p))$, $s.t. w_t^p + w_l^p = 1$. The $Softmax()$ operation enforces the subjective constraint. The value of (w_t^p, w_l^p) controls the impact of the corresponding view in the final decision-making process. Given view-independent predictions $(pred_t^p, pred_l^p)$, the final classification result for patient p is calculated as:

$$\hat{pred}^p = w_t^p * pred_t^p + w_l^p * pred_l^p. \tag{1}$$

As the whole process is differentiable, the parameters of PWAN can then be updated using back-propagation given the ground truth classification label. This is especially beneficial as manually assigning multi-view weights is infeasible and might suffer from observer subjectivity. During the inference, the operator can inspect both the view-independent decision $pred_t^p$, $pred_l^p$ and the combined one $\hat{pred}^p$. This might also help to strengthen the interpretability of the model decision-making process.

Note that similar to the state-of-the-art subject-customized approaches (e.g. [15]), it is also possible to adopt a view selected strategy to generate the final outcome. However, this approach discards part of the inputs and inevitably loses information. We argue that all the views should be considered jointly for accurate thyroid cancer diagnosis and utilized in a weighted way. Later validation experiments also validate this conjecture (see Sect. 4).

2.3 View-Aware Contrastive Loss

Learning effective visual representations without human supervision is a long-standing problem. Recently, discriminative approaches based on contrastive learning in the latent space have shown great promise [3,7,9]. The core of contrastive learning is to penalize dis-similar features in high-level space to increase model robustness and may help to alleviate over-fitting. As lesions have large variations in size, shape, appearance, and more importantly, different diagnostic features in different anatomical planes, rashly restricting all representations may produce adverse effects and confuse the training. Therefore, we argue that applying contrastive learning to multi-view data requires further consideration. In specific, we propose a novel contrastive loss equipped with view awareness, which segregates samples based on whether they emphasize the same view in lesion classification and only constrain similarity within the same group.

Formally, given a training batch $B = \{1, ..., i, ..., N\}$, the corresponding feature set and the weight set can be defined as: $F = \{(f_t^1, f_l^1), ..., (f_t^N, f_l^N)\}$, $W = \{(w_t^1, w_l^1), ..., (w_t^N, w_l^N)\}$, respectively. B can be divided into two groups: $Ge = \{i | i \in B, w_t^i \geq w_l^i\}$, and $Le = \{i | i \in B, w_t^i < w_l^i\}$, while $Ge + Le = B$. It is clear to see that cases belonging to the same group have similar weighting distributions over different views and their features should be relatively close in high dimensional space. The VACL can be defined as follows:

$$\mathcal{L}_{VAC} = \frac{\sum_{\substack{i,j \in Ge \\ i \neq j}} Dis(i,j)}{C_{Size(Ge)}^2 + \epsilon} + \frac{\sum_{\substack{i,j \in Le \\ i \neq j}} Dis(i,j)}{C_{Size(Le)}^2 + \epsilon} \tag{2}$$

where $Dis(i,j) = \|f_t^i - f_t^j\|_1 + \|f_l^i - f_l^j\|_1$ measures the mean absolute distance between the high-level features of the patient i and j. C_{Size}^2 represents the combination formula and is used to normalize the numerator based on total number of calculations. $Size(Ge), Size(Le)$ represent the number of samples in each group. ϵ is a small positive number to avoid division by 0. $\mathcal{L}_{VAC}$ can therefore selectively encourage feature resemblance between samples with similar diagnostic patterns and avoid penalizing the rest. Finally, the overall loss function can be summarized as follows:

$$\mathcal{L}_{total} = \mathcal{L}_{CE} + \lambda * \mathcal{L}_{VAC}, \tag{3}$$

where λ is a hyper-parameter that controls the influence of the proposed contrastive constraint.

3 Materials and Experiments

We validate our approach on an in-house dataset containing 4529 sets of multi-view US images of thyroid nodules. Each set contains a pair of multi-view (a transverse and a longitudinal views) US images collected from one patient. This study was approved by the local Institutional Review Board. We randomly split the dataset at the patient level into 7:1:2 for training, validation, and test. All images are resized to 224×224. Biopsies were carried out for each patient used as the ground truth.

We compare the performance of the proposed model with that of single-view models and popular multi-view approaches (i.e., MVMT [10], AADNN [16]). The single view models use the same feature extraction backbone as the proposed model but are trained only using transverse (row 1, Table 1) or longitudinal view (row 2, Table 1). Note that some multi-modal models are closely related to this task as well. We therefore select two state-of-the-art works: AW3M [8] and AdaMML [15], as the former treats different branches differently, while the latter selects different modalities to perform classification for different patients. For ablations study, we also implement the proposed model without the PAWN and VACL (row 7, Table 1) and only without the VACL (row 8, Table 1) to verify the effectiveness of each component of the proposed model. Furthermore, we also switch the VACL with an original contrastive loss [9] (row 9, Table 1) to examine whether the view-dependent design of VACL is effective.

Different augmentation strategies are applied for all experiments, including scaling, rotation, flipping, and mixup. Weights pre-trained from ImageNet are used for initializing. We use the AdamW optimizer with a learning rate of 5e-4. The weight decay is set to 0.05. ϵ is set to 0.001. The value of λ is defined by considering the order of magnitude of the two losses and is empirically set to be 0.01. The MVC network is trained for 300 epochs together with the PWAN. Accuracy (ACC), sensitivity (SEN), precision (PRE), specificity (SPE) and F1-score are used as evaluation metrics. All experiments are implemented in PyTorch with a NVIDIA RTX A6000 GPU.

4   Results and Discussion

Table 1. Results of the comparison experiments. Each row represents one approach while each column corresponds to different evaluation metrics.

	Methods	ACC(%)	SEN(%)	SPE(%)	PRE(%)	F1-score(%)
Single-view	Transverse	79.21	85.57	66.85	83.38	84.46
	Longitudinal	79.58	85.29	68.49	84.03	84.65
Multi-view	MVMT [10]	81.00	85.38	**72.50**	**85.78**	85.58
	AADNN [16]	80.75	82.75	68.12	84.18	85.69
	AW3M [8]	80.45	90.07	61.75	82.07	85.88
	AdaMML [15]	65.90	65.14	67.40	79.52	71.61
	Ours w/o PAWN	81.87	89.41	67.21	84.13	86.69
	Ours w/o VACL	82.18	**90.72**	65.57	83.66	87.05
	Ours with OCL	80.94	89.03	65.21	83.26	86.05
	Ours	**83.29**	89.97	70.31	85.49	**87.67**

As shown in row 1 and 2 in Table 1, comparable performance was obtained for the two single-view models. This shows that both views provide informative features for cancer diagnosis. In fact, the longitudinal model scored slightly higher ACC and F1-score in our datasets, suggesting its necessity in decision making. On the other hand, most multi-view models exhibited better performance than that of single-view ones. This indicates that there exists complementary information hidden in different views of thyroid US, and their combination could lead to a more accurate diagnosis. Compared with AADNN, the MVMT obtained higher ACC and PRE. It might be caused by its larger size with greater modeling capacity. The AW3M model scored higher F1-score, displaying a balanced performance. This may stem from the fact that this model breaks the equilibrium of different branches and learns to weight different views differently. However, it overlooks that this weighting should vary according to individual specification of each patient. As a result, it performed inferiorly than the proposed method, which scored 87.67% in F1-score (last row in Table 1). Another interesting comparison method is the AdaMML model, the only approach that customizes its decision-making for different patients. Nevertheless, it exhibited inferior performance in our dataset (row 6 in Table 1). This may result from the fact that this approach was originally proposed to analyze multi-modal data for video classification and learns to discard modalities to improve accuracy. This approach could handle data redundancy and noise but inevitably loses some information as well. However, in our task, both views contain crucial information and should be jointly considered during classification. On the contrary, our personalized diagnostic tool avoids information loss by allocating different multi-view weights for different patients. It scored ACC = 83.29%, SEN = 89.97%, SPE = 70.31%, PRE = 85.49% and F1-score = 87.67%, and outperformed all competing methods, indicating that it is a suitable solution for our task.

As an ablation study, we investigated how the proposed framework would perform without the PWAN and VACL. It can be observed in Table 1 that the MVC model alone could achieve promising performance (row 7), proving itself as an effective baseline for multi-view classification. Meanwhile, the introduction of PAWN boosted the model performance (row 8) as this design enables personalized weighting for different views. Moreover, the addition of VACL continued to improve the model's accuracy (compare row 8 with 10). We conjecture that this constraint helps to harness the learning of high-level representations and increases the robustness of the model. Note that as both the design of PAWN and VACL is general, they could be easily extended to other multi-view or multi-modal problems in future applications. To fully investigate the efficacy of the VACL, we also implemented the same framework with an original contrastive loss (row 9). Results show that this variant scored inferiorly than that trained with the proposed VACL (row 10). This suggests that segregating samples based on their importance of views during contrastive learning is more beneficial than the vanilla one. Comparing row 9 and 8, the additional contrastive loss led to a decrease of 1.69% in SEN, 1.24% in ACC and 1% in F1-score. It further indicates that blindly enforcing features resemblance might be detrimental for the challenging thyroid cancer diagnosis.

5   Conclusions

In this paper, we proposed a personalized diagnostic tool for thyroid cancer diagnosis using multi-view US. Its design encourages interactions between different views via a weight-sharing multi-branch backbone, while also allows asymmetric emphasis on different views through a portable weighting allocation network. It also leverages a novel view-aware contrastive loss to further increase the model robustness. The experimental results showed that the proposed method outperformed single-view models and other state-of-the-art multi-view approaches. The design of this framework is general, and could be applied to other multi-view applications.

Acknowledgements. This work was supported by the National Natural Science Foundation of China (No. 62101342, 62171290, and 61901275); Shenzhen-Hong Kong Joint Research Program (No. SGDX20201103095613036), Shenzhen Science and technology research and Development Fund for Sustainable development project (No. KCXFZ20201221173613036); National Natural Science Foundation of China (No. 82071928); Shenzhen University Startup Fund (2019131).

References

1. Cabanillas, M.E., McFadden, D.G., Durante, C.: Thyroid cancer. The Lancet **388**(10061), 2783–2795 (2016)
2. Chaudhary, V., Bano, S.: Thyroid ultrasound. Indian J. Endocrinol. Metab. **17**(2), 219 (2013)
3. Chen, T., Kornblith, S., Norouzi, M., Hinton, G.: A simple framework for contrastive learning of visual representations. In: International Conference on Machine Learning, pp. 1597–1607. PMLR (2020)
4. Chen, Y., Li, D., Zhang, X., Jin, J., Shen, Y.: Computer aided diagnosis of thyroid nodules based on the devised small-datasets multi-view ensemble learning. Med. Image Anal. **67**, 101819 (2021)
5. Chi, J., Walia, E., Babyn, P., Wang, J., Groot, G., Eramian, M.: Thyroid nodule classification in ultrasound images by fine-tuning deep convolutional neural network. J. Digit. Imaging **30**(4), 477–486 (2017)
6. Hegedüs, L.: The thyroid nodule. N. Engl. J. Med. **351**(17), 1764–1771 (2004)
7. Hjelm, R.D., et al.: Learning deep representations by mutual information estimation and maximization. arXiv preprint arXiv:1808.06670 (2018)
8. Huang, R., et al.: Aw3m: An auto-weighting and recovery framework for breast cancer diagnosis using multi-modal ultrasound. Med. Image Anal. **72**, 102137 (2021)
9. Khosla, P., et al.: Supervised contrastive learning. Adv. Neural. Inf. Process. Syst. **33**, 18661–18673 (2020)
10. Kyono, T., Gilbert, F.J., Schaar, M.: Multi-view multi-task learning for improving autonomous mammogram diagnosis. In: Machine Learning for Healthcare Conference, pp. 571–591. PMLR (2019)
11. Liu, T., et al.: Automated detection and classification of thyroid nodules in ultrasound images using clinical-knowledge-guided convolutional neural networks. Med. Image Anal. **58**, 101555 (2019)

12. Liu, T., Xie, S., Yu, J., Niu, L., Sun, W.: Classification of thyroid nodules in ultrasound images using deep model based transfer learning and hybrid features. In: 2017 IEEE International Conference on Acoustics, Speech and Signal Processing (ICASSP), pp. 919–923. IEEE (2017)
13. Liu, Z., et al.: Swin transformer: hierarchical vision transformer using shifted windows. In: Proceedings of the IEEE/CVF International Conference on Computer Vision, pp. 10012–10022 (2021)
14. Moon, H.J., Kwak, J.Y., Kim, E.K., Kim, M.J.: A taller-than-wide shape in thyroid nodules in transverse and longitudinal ultrasonographic planes and the prediction of malignancy. Thyroid **21**(11), 1249–1253 (2011)
15. Panda, R., Chen, C.F.R., Fan, Q., Sun, X., Saenko, K., Oliva, A., Feris, R.: Adamml: Adaptive multi-modal learning for efficient video recognition. In: Proceedings of the IEEE/CVF International Conference on Computer Vision, pp. 7576–7585 (2021)
16. Pi, Y., Zhao, Z., Xiang, Y., Li, Y., Cai, H., Yi, Z.: Automated diagnosis of bone metastasis based on multi-view bone scans using attention-augmented deep neural networks. Med. Image Anal. **65**, 101784 (2020)
17. Sipos, J.A.: Advances in ultrasound for the diagnosis and management of thyroid cancer. Thyroid **19**(12), 1363–1372 (2009)
18. Wang, J., et al.: Automated interpretation of congenital heart disease from multi-view echocardiograms. Med. Image Anal. **69**, 101942 (2021)
19. Zhao, S.X., Chen, Y., Yang, K.F., Luo, Y., Ma, B.Y., Li, Y.J.: A local and global feature disentangled network: toward classification of benign-malignant thyroid nodules from ultrasound image. IEEE Trans. Med. Imaging (2022)

Morphology-Aware Interactive Keypoint Estimation

Jinhee Kim[1], Taesung Kim[1], Taewoo Kim[1], Jaegul Choo[1], Dong-Wook Kim[2],
Byungduk Ahn[3], In-Seok Song[2(✉)], and Yoon-Ji Kim[4(✉)]

[1] KAIST, Daejeon, South Korea
{seharanul17,zkm1989,special1ktu,jchoo}@kaist.ac.kr
[2] Korea University Anam Hospital, Seoul, South Korea
densis@korea.ac.kr
[3] Papa's Dental Clinic, Seoul, South Korea
[4] Asan Medical Center, Ulsan University School of Medicine, Seoul, South Korea
yn0331@gmail.com

Abstract. Diagnosis based on medical images, such as X-ray images, often involves manual annotation of anatomical keypoints. However, this process involves significant human efforts and can thus be a bottleneck in the diagnostic process. To fully automate this procedure, deep-learning-based methods have been widely proposed and have achieved high performance in detecting keypoints in medical images. However, these methods still have clinical limitations: accuracy cannot be guaranteed for all cases, and it is necessary for doctors to double-check all predictions of models. In response, we propose a novel deep neural network that, given an X-ray image, automatically detects and refines the anatomical keypoints through a user-interactive system in which doctors can fix mispredicted keypoints with fewer clicks than needed during manual revision. Using our own collected data and the publicly available AASCE dataset, we demonstrate the effectiveness of the proposed method in reducing the annotation costs via extensive quantitative and qualitative results.

Keywords: Interactive keypoint estimation · AI-assisted image analysis

1 Introduction

Keypoint-based analysis of medical images is a widely used approach in clinical treatment [1,6,7,16,24]. Anatomical points obtained from an image can be utilized to measure significant features of body parts where keypoints are located. For example, vertex points of cervical vertebrae can be used to measure the concavity and height of the vertebrae, which correlates to skeletal growth [6,16].

J. Kim and T. Kim—Both authors contributed equally.

Supplementary Information The online version contains supplementary material available at https://doi.org/10.1007/978-3-031-16437-8_65.

ⓒ The Author(s), under exclusive license to Springer Nature Switzerland AG 2022
L. Wang et al. (Eds.): MICCAI 2022, LNCS 13433, pp. 675–685, 2022.
https://doi.org/10.1007/978-3-031-16437-8_65

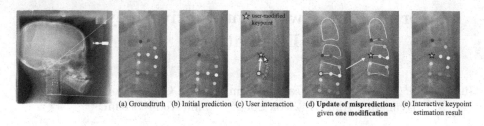

(a) Groundtruth   (b) Initial prediction   (c) User interaction   (d) **Update of mispredictions** (e) Interactive keypoint
                                                                        given **one modification**   estimation result

Fig. 1. Interactive keypoint estimation results on a cephalometric X-ray image. The goal is to estimate (a) 13 keypoints on the cervical vertebrae, each of which determines vertebrae morphology. Here, (b) the initial prediction misses one vertebra at the top, making the entire prediction wrong. A manual revision will be no better than annotating from scratch. However, in our method, if (c) a user corrects only one point, (d) the remaining points come up together. All keypoints appropriately reflect user-interaction information in (e) the final result.

Usually, doctors manually annotate such keypoints, but the process is costly, requiring significant human effort. Accordingly, studies on fully automatic keypoint estimation approaches have been widely conducted [3,10,13,15,20,24,27]. However, fully automatic methods can be inaccurate because data available for training is usually scarce in the medical domain due to privacy issues. Besides, errors of inaccurate keypoints can also cascade to subsequent feature extraction and diagnosis procedures. Thus, a thorough revision of erroneous model predictions is crucial, yet manually correcting individual errors is highly time-consuming. To assist the process, a keypoint estimation model that, after receiving user correction for a few mispredicted keypoints, automatically revises the remaining mispredictions is needed. However, to the best of our knowledge, little research has been conducted on such an approach despite its need and importance. In response, we propose a novel framework, the **interactive keypoint estimation method**, which aims to reduce the number of required user modifications compared to manual revision, thereby reducing human labor in the revision process, as shown in Fig. 1.

Recently, interactive segmentation tasks have gained much attention for their ability to reduce human effort in precise image segmentation of medical image datasets [17,21,28] as well as real-world datasets [5,11,18]. Reviving iterative training with mask guidance (RITM) [19] achieves the state-of-the-art performance in image segmentation by reactivating iterative training for multiple user revisions. Also, approaches that optimize user revision during the inference time, such as BRS [5] and f-BRS [18], have shown remarkable performance in the interactive segmentation task. Given these successes, we develop our interactive keypoint estimation model from the interactive segmentation approaches.

However, when we apply these methods to the interactive keypoint estimation task, we find that information of user modifications to keypoints does not propagate to distant points in the model; it only locally affects nearby keypoints. To address this issue, we propose the **interaction-guided gating network**, which

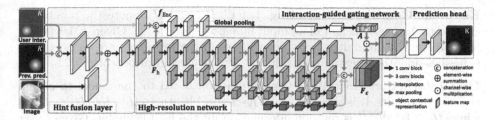

Fig. 2. Overview of proposed interactive keypoint estimation model. It receives an image, user interaction (User inter.), and its previous prediction (Prev. pred.) and outputs a heatmap of keypoint locations that reflects interactive user feedback.

can propagate user modification information across the spatial dimensions of the image. The proposed gating network better reflects user modification information in all mispredictions, including distant keypoints. We also observe that the degree of freedom between keypoints is minimal in medical images, e.g., vertex points on the cervical vertebrae have limited deformation. In Fig. 1, the cervical vertebrae consist of a series of similarly shaped polygonal vertebrae with five corner points. This indicates that the distances or the angles between keypoints may be similar across different images. If so, we can use these higher-order statistics obtained from keypoints to make the model explicitly aware of dependent relationships between keypoints. To this end, we propose **the morphology-aware loss**, which facilitates explicit learning of inter-keypoint relationships by regularizing distances and angles between predicted keypoints to be close to the groundtruth.

Consequently, our main contributions are as follows: (i) We propose a novel interactive keypoint estimation network in which users can revise inaccurate keypoints with fewer modifications than needed when manually revising all mispredictions. (ii) We introduce an interaction-guided gating network, which allows user modification information to be propagated to all inaccurate keypoints, including distant ones. (iii) We propose the morphology-aware loss, which utilizes higher-order statistics to make the model aware of inter-keypoint relationships as well as individual groundtruth locations of keypoints. (iv) We verify the effectiveness of our approach through extensive experiments on a dataset we collected on our own as well as on the public AASCE challenge dataset [24].

2   Methodology

2.1   Interactive Keypoint Estimation

Interactive keypoint estimation involves correcting mispredicted keypoints via user interaction. The task aims to reduce the number of required user modifications compared to manual revision by automatically revising all mispredictions given user feedback to a few. In our work, manual revision indicates fully-manually revised results by a user without the assistance of an interactive model.

Keypoint Coordinate Encoding. Let C, W, and H denote the number of channels, width, and height of an input image $\mathcal{I} \in \mathbb{R}^{C \times W \times H}$. Given K keypoints to estimate, the groundtruth x-y coordinates of the keypoints are encoded to a Gaussian-smoothed heatmap $\mathcal{H} \in \mathbb{R}^{K \times W \times H}$, as used in prior work [9, 22].

User-Interaction Encoding. User interaction is fed to the model to obtain the revised results. It is encoded into a K-channel heatmap $\mathcal{U} = \{\mathcal{U}_1, \mathcal{U}_2, ..., \mathcal{U}_K\}$, where $\mathcal{U}_n \in \mathbb{R}^{1 \times W \times H}$ is revision information for the n-th keypoint. Given l user modifications to a subset of keypoints, $\{c_1, c_2, ...c_l\}$, the corresponding channels $\{\mathcal{U}_{c_1}, \mathcal{U}_{c_2}, ..., \mathcal{U}_{c_l}\}$ are activated as the Gaussian-smoothed heatmap, whereas the other channels corresponding to the unmodified keypoints are filled with zeros. Formally, a user interaction heatmap for the n-th keypoint can be expressed as

$$\mathcal{U}_n(i,j) = \begin{cases} \exp\left(\frac{(i-x_n)^2 + (j-y_n)^2}{-2\sigma^2}\right), & \text{if } n \in \{c_1, c_2, ..., c_l\}. \\ 0, & \text{otherwise,} \end{cases} \tag{1}$$

where $p_n = (x_n, y_n)$ denotes user-modified coordinates of the n-th keypoint.

Keypoint Coordinate Decoding. Following Bulat et al. [2], we transform the predicted heatmaps into 2D keypoint coordinates by using differentiable local soft-argmax to reduce quantization errors caused by the heatmap.

Synthesizing User Interaction During Training. During training, we simulate user interaction by randomly sampling a subset of groundtruth keypoints as user-modified keypoints. First, we define a multinomial distribution of the number of user modifications with a range of $[0, K]$. Since our goal is to correct errors with a small number of modifications, we set the probability of obtaining a larger number of modifications to become exponentially smaller. Once the total number of modifications is determined, the keypoints to be corrected are randomly sampled from a discrete uniform distribution $\text{Unif}(1, K)$.

Network Architecture. As illustrated in Fig. 2, we employ a simple convolutional block, called hint fusion layer, to feed additional input to the model, e.g., user clicks, without any architectural changes to the backbone following RITM [19]. The hint fusion layer receives an input image, user interaction, and previous prediction of the model and returns a tensor having the same shape as the output of the first block of the backbone network. Also, we adopt an iterative training procedure so that users can repeatedly revise a keypoint prediction result. Meanwhile, the previous prediction of the model is fed to the model in the next step to make it aware of its earlier predictions [12, 19]. During the inference time, we extend this to selectively provide previous predictions just for user-modified keypoints, not all keypoints, to facilitate differentiating the mispredictions. We use the high-resolution network (HRNet)-W32 [22] with object-contextual representations (OCR) [26] as a pre-trained backbone network. The model is trained with a binary cross-entropy loss L_g between the predicted heatmap $\hat{\mathcal{H}}$ and the Gaussian-smoothed target heatmap $\mathcal{H}$. As a final prediction, we post-process the predictions so that a user-modified point stays where the user wants it.

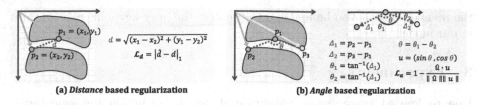

(a) *Distance* based regularization   (b) *Angle* based regularization

Fig. 3. Illustration of morphology-aware loss.

2.2   Interaction-Guided Gating Network

We propose an interaction-guided gating network, which effectively propagates user-interaction information throughout the entire spatial area of an image. Given user interaction $\mathcal{U}$ and the downsampled feature map $\mathbf{F}_h$ from the backbone network, the proposed method generates a channel-wise gating weight $\mathbf{A} \in \mathbb{R}^{k_c}$ to recalibrate the intermediate feature map $\mathbf{F}_c \in \mathbb{R}^{k_c \times W' \times H'}$ according to user feedback summarized globally from every pixel position. Inspired by Hu et al. [4], which use global average pooling to aggregate global spatial information, we employ global max-pooling to selectively retrieve significant interaction-aware features from the entire spatial area for each channel. Then, the gating weight $\mathbf{A}$ is generated by subsequent fully connected layers and a sigmoid activation function and utilized to gate the feature map F_c channel-wise. Finally, the reweighted feature map is decoded into a keypoint heatmap by our prediction head.

2.3   Morphology-Aware Loss

The morphology-aware loss aims to explicitly utilize the inter-keypoint relationships as well as the individual information of each keypoint to make the model aware of the morphological associations among keypoints. Specifically, we exploit two higher-order statistics: the distance between two keypoints and the angle among three keypoints, as shown in Fig. 3. Instead of using all possible combinations of keypoint sets, we leverage only the ones that rarely deviate across the dataset, assuming that learning significantly varying relationships can degrade the model performance. Thus, we select two subsets to apply the morphology-aware loss based on the standard deviation values of the distance and angle. Given standard deviation values as $S_d = \sqrt{\mathbb{E}[(d - \mathbb{E}[d])^2]}$ and $S_a = \sqrt{-\ln(\mathbb{E}[u_x]^2 + \mathbb{E}[u_y]^2)}$, where d and u are the distance and angle vector defined in Fig. 3, the subsets are obtained as $\mathcal{P}_d = \{(p_m, p_n) | S_d(d(p_m, p_n)) < t_d, m \neq n; m, n \in [1, K]\}$ and $\mathcal{P}_a = \{(p_m, p_n, p_l) | S_a(u(p_m, p_n, p_l)) < t_a, m \neq n \neq l; m, n, l \in [1, K]\}$, where t_d and t_a are threshold values. Here, the circular variance is computed for the angle. Finally, the morphology-aware loss is defined as $L_m = L_d + \lambda_m L_a$, where L_d is the L1 loss for distance applied to the set $\mathcal{P}_d$, and L_a is the cosine similarity loss for angle vector applied to the set $\mathcal{P}_a$. Ultimately, when a user revises one keypoint, the keypoints having regular relationships with the revised one will also be updated to preserve the inter-keypoint relationships in our model.

3   Experiments

Due to limited space, more experimental details, including implementation details and hyperparameter settings, are available on **our project webpage**.

Dataset. We collect our own dataset, the *Cephalometric X-ray* dataset, which contains 6,504 cephalometric X-ray images (677/692/5,135 for training/validation/test) of 4,280 subjects without overlapping patients in each set. We use a small training set and large test set in consideration of the medical domain, in which reliable performance validation is critical, and data available for training is scarce [8,14,23]. Each image is annotated with 13 keypoints, which include the vertex points of the cervical vertebrae. The keypoints can be used to examine the skeletal growth of a patient by calculating anatomical measurements such as the concavity of the vertebrae [6,16]. The *AASCE* dataset [24] contains spinal anterior-posterior X-ray images (352/128/128 for training/validation/test). In each image, 68 points representing the four vertices of 17 vertebrae are annotated.

Evaluation Metrics. We measure the mean radial error (MRE), which is calculated as $\text{MRE} = \frac{1}{K}\sum_{n=1}^{K}(||p_n - \hat{p}_n||_2)$. Also, we borrow the evaluation protocol of interactive segmentation tasks [11,18], number of clicks (NoC), to evaluate the performance of the proposed interactive keypoint estimation approach. In our work, NoC measures the average number of clicks required to achieve a target MRE. We set ranges of target MRE as $[0, 10]$ and $[0, 60]$ for Cephalometric X-ray and AASCE, respectively. We report selected results in Table 1, and the results for different target MRE values are provided in Fig. 6 of the supplementary material. When measuring NoC, we limit the maximum number of user modifications to a small value because we aim to revise all mispredictions with as few clicks as possible. Since Cephalometric X-ray and AASCE include 13 and 68 target keypoints, we set the limits to five and ten clicks, respectively. Finally, $\text{NoC}_\alpha @\beta$ denotes the average number of clicks to achieve the target MRE of β pixels when a prediction of an image can be maximally modified by up to α clicks. Similarly, $\text{FR}_\alpha @\beta$ counts the average rate of images that fail to achieve the target MRE of β pixels when the model is given a maximum of α clicks. Under the assumption that a doctor would correct clearly wrong keypoints first, the keypoint having the highest error for each prediction is selected to revise in all experiments.

Compared Baselines. We assess the proposed interactive keypoint estimation model by comparing it with state-of-the-art interactive segmentation approaches, which share similar concepts in the interactive system: BRS [5], f-BRS [18], and RITM [19]. Along with the ablation study, this is an essential evaluation step in our method, given the lack of research on interactive keypoint estimation. The baselines encode user feedback as distance maps [5,18] or hard masks [19]

Table 1. Comparison with baselines on Cephalometric X-ray and AASCE.

Method	Cephalometric X-ray					AASCE				
	FR_5 @3	NoC_5 @3	NoC_5 @4	NoC_5 @5	NoC_5 @6	FR_{10} @20	NoC_{10} @20	NoC_{10} @30	NoC_{10} @40	NoC_{10} @50
BRS [5]	9.72	3.18	1.65	0.78	0.36	11.72	3.02	2.41	1.91	1.59
f-BRS [18]	9.80	2.99	1.34	0.52	0.22	52.34	7.36	5.61	4.55	3.82
RITM [19]	10.69	3.12	1.48	0.60	0.25	13.28	3.56	2.73	2.14	1.56
Ours	**4.48**	**2.32**	**0.86**	**0.31**	**0.13**	**6.25**	**2.46**	**1.88**	**1.41**	**1.19**

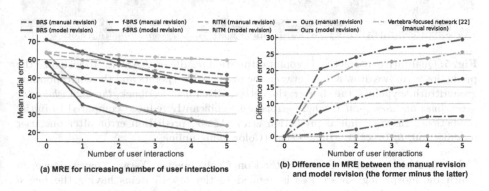

(a) MRE for increasing number of user interactions

(b) Difference in MRE between the manual revision and model revision (the former minus the latter)

Fig. 4. Comparison with manual revision on AASCE.

tailored to interactive segmentation, so we replace them with our K-channel user interaction heatmap in the experiments. Additionally, we modify the baselines to output a K-channel heatmap and to minimize the binary cross-entropy loss between the predicted and groundtruth heatmap. For BRS and f-BRS, we observe that their backpropagation refinement schemes degrade model performance in our task; we report their results without the schemes.

Quantitative Comparison with Baselines. Table 1 shows that our method consistently outperforms the baselines in terms of the number of clicks (NoC) and failure rate (FR) by a significant margin. For example, our method reduces the failure rate by about half compared to the baselines. Moreover, the mean number of clicks of our method is consistently lower than that of the baselines for both datasets. In Fig. 4, when the proposed interactive keypoint estimation approach (*model revision*) is applied, the prediction error is significantly reduced compared to the fully manually revised result (*manual revision*). The superiority of the proposed framework is also revealed by the comparison with a baseline model without the interactive framework, the Vertebra-focused network [25]. The error reduction rate of the network is notably slow when its predictions are manually revised. Also, among all models, our model achieves the largest improvement compared to manual revision and consistently outperforms RITM by a large margin, decreasing the error by five pixels on average. Altogether, our approach successfully propagates user revision information to unrevised but inaccurate keypoints, demonstrating the effectiveness of the proposed interactive framework.

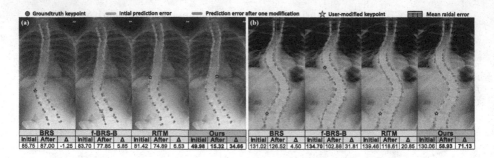

Fig. 5. Qualitative interactive keypoint estimation results on AASCE. To visualize the prediction error, we draw a line between the predicted keypoints and the corresponding groundtruth keypoints; the shorter, the better. Given user feedback, *the area where the green lines are dominant* is where errors are significantly reduced compared to initial predictions. Initial, initial prediction error; After, prediction error after one user modification; Δ, Initial minus After (Color figure online)

Table 2. Ablation study of our method on Cephalometric X-ray. Low variance, in which the morphology-aware loss is applied to the *top* 15 items having the *lowest variance* among all possible items; high variance, in which the morphology-aware loss is applied to the *top* 15 items having the *highest variance* among all; adjacent points, in which all keypoint sets comprise an *edge or internal angle of vertebrae*.

Interaction-guided gating			Morphology-aware loss		Performance			
	Pooling method	Activation function		Criterion for $\mathcal{P}_d$ and $\mathcal{P}_a$	FR$_5$ @3	NoC$_5$ @3	NoC$_5$ @4	NoC$_5$ @5
✗	-	-	✗	-	10.09	3.09	1.46	0.60
✓	max	sigmoid	✗	-	5.88	2.51	0.98	0.37
✓	**max**	**sigmoid**	✓	**low variance**	**4.48**	**2.32**	**0.86**	**0.31**
✗	-	-	✓	low variance	6.33	2.56	1.03	0.38
✓	average	sigmoid	✓	low variance	5.39	2.53	1.00	0.37
✓	max	softmax	✓	low variance	5.71	2.54	0.96	0.34
✓	max	sigmoid	✓	high variance	6.23	2.62	1.04	0.38
✓	max	sigmoid	✓	adjacent points	4.79	2.44	0.93	0.33

Qualitative Comparison with Baselines. Figure 5 illustrates the prediction results on AASCE with their mean radial errors before and after user revision. Our proposed model remarkably reduces the initial error in a much wider area than the baselines, successfully propagating user feedback to distant mispredictions.

Ablation Study. Table 2 validates each component of our method. The results demonstrate that our model outperforms the ablated versions. We ablate the distance and angle sets criterion to apply the morphology-aware loss, $\mathcal{P}_d$ and $\mathcal{P}_a$. As a result, learning highly varying inter-keypoint relationships degrades the model performance contrary to learning rarely-varying relationships. Results for different threshold values are provided in Table 4 of the supplementary material.

4   Conclusion

This paper focuses on improving the efficiency and usability of the keypoint annotation to speed up the overall process while improving diagnostic accuracy. To
this end, we introduce a novel interactive keypoint estimation network, incorporating the interaction-guided attention network and the morphology-aware
loss to revise inaccurate keypoints with a small number of user modifications.
We demonstrate the effectiveness of the proposed approach through extensive
experiments and analysis of two medical datasets. This work assumes that a
user will always provide a correct modification to the model and revise a clearly
wrong keypoint first. Thus, future work can address noisy interactive inputs by
real users or find the most effective keypoint to correct all other keypoints and
recommend users to revise it first. To the best of our knowledge, this is the first
work to propose an interactive keypoint estimation framework, and it will be
helpful to researchers working on human-in-the-loop keypoint annotation.

Acknowledgements. This work was supported by the Institute of Information &
communications Technology Planning & Evaluation (IITP) grant funded by the Korean
government(MSIT) (No. 2019-0-00075, Artificial Intelligence Graduate School Pro-
gram(KAIST)), the National Research Foundation of Korea (NRF) grant funded by
the Korean government (MSIT) (No. NRF-2019R1A2C4070420), the National Supercomputing Center with supercomputing resources including technical support (KSC-
2022-CRE-0119), and the Korea Medical Device Development Fund grant funded by
the Korea government (the Ministry of Science and ICT, the Ministry of Trade, Industry and Energy, the Ministry of Health & Welfare, the Ministry of Food and Drug
Safety) (Project Number: 1711139098, RS-2021-KD000009).

References

1. Bier, B., et al.: X-ray-transform invariant anatomical landmark detection for pelvic
   trauma surgery. In: Frangi, A.F., Schnabel, J.A., Davatzikos, C., Alberola-López,
   C., Fichtinger, G. (eds.) MICCAI 2018. LNCS, vol. 11073, pp. 55–63. Springer,
   Cham (2018). https://doi.org/10.1007/978-3-030-00937-3_7
2. Bulat, A., Sanchez, E., Tzimiropoulos, G.: Subpixel heatmap regression for facial
   landmark localization. In: The British Machine Vision Conference (BMVC) (2021)
3. Chen, R., Ma, Y., Chen, N., Lee, D., Wang, W.: Cephalometric landmark detection
   by attentive feature pyramid fusion and regression-voting. In: Shen, D., et al. (eds.)
   MICCAI 2019. LNCS, vol. 11766, pp. 873–881. Springer, Cham (2019). https://
   doi.org/10.1007/978-3-030-32248-9_97
4. Hu, J., Shen, L., Sun, G.: Squeeze-and-excitation networks. In: Proceedings of
   the IEEE Conference on Computer Vision and Pattern Recognition (CVPR), pp.
   7132–7141 (2018)
5. Jang, W.D., Kim, C.S.: Interactive image segmentation via backpropagating refinement scheme. In: Proceedings of the IEEE Conference on Computer Vision and
   Pattern Recognition (CVPR), pp. 5297–5306 (2019)
6. Kim, D.W., et al.: Prediction of hand-wrist maturation stages based on cervical
   vertebrae images using artificial intelligence. Orthod. Craniofac. Res. **24**, 68–75
   (2021)

7. Kordon, F., et al.: Multi-task localization and segmentation for X-Ray guided planning in knee surgery. In: Shen, D., et al. (eds.) MICCAI 2019. LNCS, vol. 11769, pp. 622–630. Springer, Cham (2019). https://doi.org/10.1007/978-3-030-32226-7_69

8. Lee, H.J., Kim, J.U., Lee, S., Kim, H.G., Ro, Y.M.: Structure boundary preserving segmentation for medical image with ambiguous boundary. In: Proceedings of the IEEE Conference on Computer Vision and Pattern Recognition (CVPR) (2020)

9. Li, J., Su, W., Wang, Z.: Simple pose: rethinking and improving a bottom-up approach for multi-person pose estimation. In: Proceedings the AAAI Conference on Artificial Intelligence (AAAI), pp. 11354–11361 (2020)

10. Li, W., et al.: Structured landmark detection via topology-adapting deep graph learning. arXiv preprint arXiv:2004.08190 (2020)

11. Lin, Z., Zhang, Z., Chen, L.Z., Cheng, M.M., Lu, S.P.: Interactive image segmentation with first click attention. In: Proceedings of the IEEE Conference on Computer Vision and Pattern Recognition (CVPR), pp. 13339–13348 (2020)

12. Mahadevan, S., Voigtlaender, P., Leibe, B.: Iteratively trained interactive segmentation. In: British Machine Vision Conference (BMVC) (2018)

13. Payer, C., Štern, D., Bischof, H., Urschler, M.: Integrating spatial configuration into heatmap regression based CNNs for landmark localization. Med. Image Anal. **54**, 207–219 (2019)

14. Peng, C., Lin, W.A., Liao, H., Chellappa, R., Zhou, S.K.: Saint: spatially aware interpolation network for medical slice synthesis. In: Proceedings of the IEEE Conference on Computer Vision and Pattern Recognition (CVPR) (2020)

15. Qian, J., Luo, W., Cheng, M., Tao, Y., Lin, J., Lin, H.: CephaNN: a multi-head attention network for cephalometric landmark detection. IEEE Access **8**, 112633–112641 (2020)

16. Safavi, S.M., Beikaii, H., Hassanizadeh, R., Younessian, F., Baghban, A.A.: Correlation between cervical vertebral maturation and chronological age in a group of Iranian females. Dental Res. J. **12**(5), 443 (2015)

17. Sakinis, T., et al.: Interactive segmentation of medical images through fully convolutional neural networks. arXiv preprint arXiv:1903.08205 (2019)

18. Sofiiuk, K., Petrov, I., Barinova, O., Konushin, A.: f-BRS: rethinking backpropagating refinement for interactive segmentation. In: Proceedings of the IEEE Conference on Computer Vision and Pattern Recognition (CVPR), pp. 8623–8632 (2020)

19. Sofiiuk, K., Petrov, I., Konushin, A.: Reviving iterative training with mask guidance for interactive segmentation. arXiv preprint arXiv:2102.06583 (2021)

20. Wang, C.W., et al.: A benchmark for comparison of dental radiography analysis algorithms. Med. Image Anal. **31**, 63–76 (2016)

21. Wang, G., et al.: Interactive medical image segmentation using deep learning with image-specific fine tuning. IEEE Trans. Med. Imaging **37**(7), 1562–1573 (2018)

22. Wang, J., et al.: Deep high-resolution representation learning for visual recognition. IEEE Trans. Pattern Anal. Mach. Intell. (TPAMI) **43**(10), 3349–3364 (2020)

23. Wang, S., et al.: LT-Net: label transfer by learning reversible voxel-wise correspondence for one-shot medical image segmentation. In: Proceedings of the IEEE Conference on Computer Vision and Pattern Recognition (CVPR) (2020)

24. Wu, H., Bailey, C., Rasoulinejad, P., Li, S.: Automatic landmark estimation for adolescent idiopathic scoliosis assessment using BoostNet. In: Descoteaux, M., Maier-Hein, L., Franz, A., Jannin, P., Collins, D.L., Duchesne, S. (eds.) MICCAI 2017. LNCS, vol. 10433, pp. 127–135. Springer, Cham (2017). https://doi.org/10.1007/978-3-319-66182-7_15

25. Yi, J., Wu, P., Huang, Q., Qu, H., Metaxas, D.N.: Vertebra-focused landmark detection for scoliosis assessment. In: 2020 IEEE 17th International Symposium on Biomedical Imaging (ISBI), pp. 736–740 (2020)

26. Yuan, Y., Chen, X., Wang, J.: Object-contextual representations for semantic segmentation. In: Vedaldi, A., Bischof, H., Brox, T., Frahm, J.-M. (eds.) ECCV 2020. LNCS, vol. 12351, pp. 173–190. Springer, Cham (2020). https://doi.org/10.1007/978-3-030-58539-6_11

27. Zhong, Z., Li, J., Zhang, Z., Jiao, Z., Gao, X.: An attention-guided deep regression model for landmark detection in cephalograms. In: Shen, D., et al. (eds.) MICCAI 2019. LNCS, vol. 11769, pp. 540–548. Springer, Cham (2019). https://doi.org/10.1007/978-3-030-32226-7_60

28. Zhou, T., Li, L., Bredell, G., Li, J., Konukoglu, E.: Quality-aware memory network for interactive volumetric image segmentation. In: de Bruijne, M., et al. (eds.) MICCAI 2021. LNCS, vol. 12902, pp. 560–570. Springer, Cham (2021). https://doi.org/10.1007/978-3-030-87196-3_52

GazeRadar: A Gaze and Radiomics-Guided Disease Localization Framework

Moinak Bhattacharya, Shubham Jain, and Prateek Prasanna[(⊠)]

Stony Brook University, Stony Brook, NY, USA
{moinak.bhattacharya,shubham.jain.1,prateek.prasanna}@stonybrook.edu

Abstract. We present *GazeRadar*, a novel radiomics and eye gaze-guided deep learning architecture for disease localization in chest radiographs. *GazeRadar* combines the representation of radiologists' visual search patterns with corresponding radiomic signatures into an integrated *radiomics-visual attention representation* for downstream disease localization and classification tasks. Radiologists generally tend to focus on fine-grained disease features, while radiomics features provide high-level textural information. Our framework first 'fuses' radiomics features with visual features inside a teacher block. The visual features are learned through a teacher-focal block, while the radiomics features are learned through a teacher-global block. A novel Radiomics-Visual Attention loss is proposed to transfer knowledge from this joint radiomics-visual attention representation of the teacher network to the student network. We show that *GazeRadar* outperforms baseline approaches for disease localization and classification tasks on 4 large scale chest radiograph datasets comprising multiple diseases. Code: https://github.com/bmi-imaginelab/gazeradar.

Keywords: Disease localization · Eye-gaze · Fusion · Radiomics

1 Introduction

Medical image interpretation is a complex visuo-cognitive task that requires an understanding of a disease's textural patterns and locations in the image. Previous studies have demonstrated the importance of visual search patterns, obtained from eye-gaze tracking of radiologists, in disease classification, localization [9,17,22,25,39,41] and segmentation [18]. Despite the spatially-rich information, gaze-derived attention regions do not always coincide with the actual disease regions. On the other hand, handcrafted radiomics features contain context-rich textural information that focuses on abnormalities, primarily disease-specific features, manifested both within and surrounding the dis-

---

Supplementary Information The online version contains supplementary material available at https://doi.org/10.1007/978-3-031-16437-8_66.

© The Author(s), under exclusive license to Springer Nature Switzerland AG 2022
L. Wang et al. (Eds.): MICCAI 2022, LNCS 13433, pp. 686–696, 2022.
https://doi.org/10.1007/978-3-031-16437-8_66

ease regions [30]. Several computer-aided diagnostic techniques focus on independently utilizing visual patterns and radiomics features [27,29,33]. While radiomics features have long been used for different diagnostic tasks, the concept of coupling textural features with visual attention is still unexplored. Radiologists' visual search patterns are honed through years of training, and different levels of expertise often leads to variations in these search patterns, even on the same image [36,37]. Studies have shown that disease diagnosis can be enhanced by taking advantage of gaze patterns from multiple radiologists [2,16,35]. While Bhattacharya et al. [3] show that transformer-based architectures can leverage human visual attention from a single radiologists' gaze patterns for diagnostic tasks, they do not investigate how to fuse multiple readers' visual search patterns in a deep learning setting. Simple averaging or majority voting may lead to dispersion of attention regions or losing information regarding gaze variations.

Motivation and Overview. To address the aforementioned limitations, we propose a novel approach that couples radiomics features with visual search patterns from multiple radiologists to infer the *radiomics-visual attention*, and leverages it in a deep learning framework for improved disease detection and diagnosis. The motivation for our approach stems from a) the importance of multiple radiologists' gaze patterns in medical image interpretation, b) the importance of co-learning visual attention and textural attention, and c) distillation ability of this *radiomics-visual* attention knowledge to deep learning architectures for downstream classification and localization tasks. A global-focal learning paradigm to mimic radiologists' cognitive behavior has shown promising results in disease diagnosis [3]. This learning paradigm presents the opportunity to incorporate radiomics features as complementary attributes into the global-focal framework. Radiomics features provide a representation of disease related imaging changes as a global context, while visual attention can help characterize detailed fine-grained features into a focal context. We use this as a motivation to design a modified global-focal architecture that integrates textural information and human visual cognition with self-attention-based learning of transformers. The main contributions of this paper are: (1) We present *GazeRadar*, a novel global-focal student-teacher architecture for disease localization based on radiomics information and visual search patterns. The teacher block learns a joint representation of radiomics and visual attention features. This representation is then used to train a student block for downstream classification and localization tasks. (2) We develop novel *Radiomics Attention Fusion* and *Gaze Attention Fusion* strategies to fuse radiomics features and gaze features, respectively. (3) We design a novel *Radiomics-Visual Attention Loss* for transferring the joint radiomics-visual knowledge from the teacher block to the student block. To the best of our knowledge, this is the first work that incorporates both, radiomics and radiologists' search patterns into a decision-making pipeline.

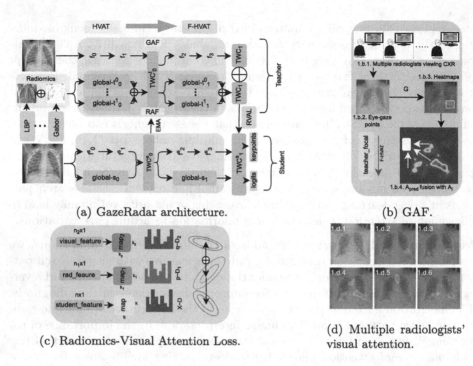

(a) GazeRadar architecture.

(b) GAF.

(c) Radiomics-Visual Attention Loss.

(d) Multiple radiologists' visual attention.

Fig. 1. GazeRadar: a radiomics-visual attention fusion architecture.

2 Methodology

Figure 1 presents an overview of the proposed *GazeRadar* architecture. There are two primary network blocks, the student and the teacher block. Each of these blocks consist of a global-focal network. The teacher-focal network learns human visual attention from the tracked eye gaze points of radiologists, and the teacher-global network learns attention from radiomics features. The teacher block comprises two sub-blocks, namely Gaze Attention Fusion (GAF) and Radiomics Attention Fusion (RAF). GAF module fuses visual attention regions from multiple radiologists to provide a consolidated visual attention region. This region is used for pre-training the teacher, referred to as Fused-Human Visual Attention Training (F-HVAT). Note that in the inference stage, we do not need eye-gaze data. The RAF module fuses different radiomics features using cross-attention. RAF attempts to learn 'radiomically relevant' regions from the fused radiomics features. The student network learns radiomics-visual attention from the teacher network. Henceforth, we use the following notations: g: global, f: focal, $\mathcal{D}$: probability distribution, $\mathcal{L}$: loss functions, $\mathcal{N}$: normal distribution, and the following abbreviations: GAF: Gaze Attention Fusion, RAF: Radiomics Attention Fusion, HVAT: Human Visual Attention Training, F-HVAT: Fused-HVAT, TWC: Two Way Cross, RVAL: Radiomics-Visual Attention Loss.

2.1   Teacher Block

The teacher network is designed to fuse the visual attention maps into a single representation and couple it with the radiomics attention map for downstream tasks. The teacher-focal block learns visual attention features from the eye gaze patterns of radiologists. The search patterns of different radiologists are non-uniform and hence the visual attention map may spread across different sections of the lungs, as shown in Fig. 1d.* in Fig. 1. Figure 1d.1 are eye-gaze points from different radiologists, shown in different colors. In Fig. 1d.2–1d.6, the visual attention maps of 5 different radiologists are shown. The teacher-global block learns the radiomics attention.

Global-Focal Network. The global and focal networks are variants of shifting window transformer architecture, inspired by [3]. In the teacher block, the global network is the RAF, and the focal network is pre-trained with GAF. The global blocks are represented as g, and the focal blocks are represented as f. There are two parallel global blocks connected with a focal block. There are k-sets of global blocks, represented as g_i^k, where $i \in \{0, 1\}$ and $k \in \{0, 1\}$, cascaded with four focal blocks represented as f_i, where $i \in \{0, 1, 2, 3\}$. Here, i is the number of shifting blocks, and k is the number of radiomics features. These blocks are connected by a Two Way Cross (TWC) module, represented as $C_i(x, y)$, which is a cross-attention block between g_0^k-f_1, and g_1^k-f_3. TWC module is a cross-attention between x and y, shown as $C_i(x, y) = MLP(LN_1(MHA(LN_0^0(x), LN_0^1(y)) + (x + y))) + z$, where $z = MHA(LN_0^0(x), LN_0^1(y)) + (x + y)$. Here, MHA is Multi-head attention, LN is Layer Norm and MLP is Multi-layered perceptron. The $C_0(\hat{g}_0, f_1)$ is TWC$_0$ layer, and $C_1(\hat{g}_1, f_3)$ is TWC$_1$ layer. Here, $\hat{g}_0 = \sum_{i=0}^{k} \lambda_i^{int} * g_0^i$, and $\hat{g}_1 = \sum_{i=0}^{k} \lambda_i^{final} * g_1^i$, where λ_k^{int} and λ_k^{final} are the intermediate and final weight parameters for k radiomics features, respectively.

Gaze Attention Fusion. The visual attention maps from n radiologists (shown in Fig. 1b.1), represented as $\mathcal{A}_i$, where $i \in \{1, 2, ..., n\}$, are first obtained as explained in Sect. 4. These maps are generally localized in different sections of the raw image. The teacher block is pre-trained with single (HVAT$_1$ and HVAT$_2$, jointly termed as HVAT) and multiple (F-HVAT) radiologists' eye gaze, also shown in Fig. 1a and Fig. 1b, and discussed in Sect. 4. Then, the raw image is fed to this pre-trained teacher to produce a predicted visual attention map, represented as $\mathcal{A}_{pred}$. The visual attention maps from multiple radiologists, $\mathcal{A}_i$, are fused with $\mathcal{A}_{pred}$ as the reference region. This can be defined as weighted linear minimization of the distances between multiple probability distributions. The predicted probability distribution is represented as $p \sim \mathcal{N}(\mu_{pred}, \sigma_{pred}^2)$, and the probability distribution of visual attention maps from radiologists are represented as $q_i \sim \mathcal{N}(\mu_i, \sigma_i^2)$. Here, $[\mu_{pred}, \sigma_{pred}]$ are the mean and standard deviation of the predicted distribution, and $[\mu_i, \sigma_i]$ are the mean and standard deviation of the distribution from visual attention regions. The n-Gaze Attention Loss (n-GAL) is the weighted distance between the p and q_i, represented as $\mathcal{L}_{n-GAL} = -\ln(d_{\mathcal{B}})$, and $d_{\mathcal{BC}} = \sum_{i=1}^{n} \alpha_i * \sqrt{p.q_i}$. Here, $d_{\mathcal{BC}}$ is a variation of Bhattacharyya coefficient [4, 5], and α_i is the parameter for weighting.

Radiomics Attention Fusion. The radiomic features are obtained from the raw images, $\mathcal{I}$, represented as $\mathcal{R}_i = \mathcal{F}_i(\mathcal{I})$, where $i \in \{0, 1, ..k\}$, as shown in Fig. 1a. Here, $\mathcal{R}_i$ are the radiomics features, $\mathcal{F}_i$ are the set of radiomic filters applied to $\mathcal{I}$. The $\mathcal{R}_i$ are fed to the global networks, shown as, $\mathcal{O}_0^i = g_0^i(\mathcal{R}_i)$. This output set is fused, as $\hat{\mathcal{O}}_0 = \sum_{i=0}^{k} \lambda_i^{int} \mathcal{O}_0^i$ and fed to $\mathcal{C}_0$ layer. The output of this $\mathcal{C}_0$ layer is represented as c_0, which is then provided to the next global set to obtain the output set, $\mathcal{O}_1^i = g_1^i(c_0)$. This output set is also fused, $\hat{\mathcal{O}}_1 = \sum_{i=0}^{k} \lambda_i^{final} \mathcal{O}_1^i$, and fed to $\mathcal{C}_1$.

2.2   Student Block

Global-Focal Network. The student block is a global-focal network where two global layers are stacked in-parallel with four focal layers. Similar to the teacher, the global and focal layers of the student are variants of shifting window transformers. The two global blocks are represented as g_i^s, $i \in \{0, 1\}$, and the focal blocks are represented as f_i^s, $i \in \{0, 1, 2, 3\}$. The g_0^s and f_1^s are fed into TWC_0^s in student block, represented as $c_0^s = \mathcal{C}_0^s(g_0^s(t_g(\mathcal{I})), f_1^s(f_0^s(t_f(\mathcal{I}))))$. Here $\{t_g, t_f\}$ are the augmentations of global and focal blocks respectively. The output c_0^s is fed to the subsequent global g_1^s, and focal f_2^s layers. The final output from TWC_1^s is represented as $c_1^s = \mathcal{C}_1^s(g_1^s(c_0^s), f_3^s(f_2^s(c_0^s)))$.

Training. The teacher block is updated with the student block using exponential moving average (EMA). The output of the student block is represented as c_1^s, and the output of the teacher block is represented as c_1. The EMA is represented as $\theta_{c_1} = \delta * \theta_{c_1}' + (1 - \delta) * \theta_{c_1^s}$. Here, δ is the smoothing coefficient, θ_{c_1}' is the parameter of the teacher block, $\theta_{c_1^s}$ is the parameter of the student block, and θ_{c_1} is the updated parameter of the teacher block. The downstream tasks are classification and localization. Consequently, classification heads and detection heads are appended to the output of the student block c_1^s. The classification head outputs predicted logits of shape $[\mathcal{B}, \mathcal{N}]$, where $\mathcal{B}$ is the batch size during training, and $\mathcal{N}$ is the number of classes in the datasets. The bounding box head outputs bounding boxes of shape $[\mathcal{B}, 4]$, where $\mathcal{B}$ is the batch size mentioned before, and 4 is the number of key-points of bounding boxes, in this case, $[x_{min}, y_{min}, x_{max}, y_{max}]$. A cross-entropy loss is applied for classification, shown as, $\mathcal{L}_{cls} = \sum_{i=0}^{\mathcal{B}} \hat{y}_i \log(y_i)$, where y is the predicted logit, and $\hat{y}$ is the ground-truth logit. For bounding box regression, a weighted addition of Generalised Intersection over Union (GIoU) and Mean Squared Error (MSE) loss is applied. This loss is represented as, $\mathcal{L}_{bbox} = \lambda_1^b * \mathcal{L}_{GIoU} + \lambda_2^b * \mathcal{L}_{MSE}$, where $(\lambda_1^b, \lambda_2^b)$ are the weights for adding the bounding box losses. Here, $\lambda_1^b + \lambda_2^b$ should be equal to 1. The final loss is represented as:

$$\mathcal{L} = \lambda_0^l * \mathcal{L}_{cls} + \lambda_1^l * \mathcal{L}_{bbox} + \lambda_2^l * \mathcal{L}_{rval} \tag{1}$$

where $(\lambda_0^l, \lambda_1^l, \lambda_2^l)$ are the weights for adding the individual loss components for final loss, and $\lambda_0^l + \lambda_1^l + \lambda_2^l = 1$. $\mathcal{L}_{rval}$ is the Radiomics-Visual Attention Loss (RVAL), described in the following subsection.

2.3   Radiomics-Visual Attention Loss

The teacher outputs a joint representation of radiomics, and visual attention features. The final TWC_1 layer C_1 takes both $\hat{g}_1$ and f_3 as input. The $\hat{g}_1$ is the fused radiomics attention representation, and f_3 is the visual attention representation. The output from this final TWC_1 layer, shown as $c_1 = C_1(\hat{g}_1, f_3)$, is the joint radiomics-visual attention feature representation. As shown in Fig. 1c, the radiomics attention features are represented as $p \sim \mathcal{D}_1$, and the visual attention features are represented as $q \sim \mathcal{D}_2$, where p and q are probability distributions. The joint representation of these distributions can be represented as $\mathcal{P} \sim \mathcal{D}_{12}$. From Fig. 1c, this output is represented as $\mathcal{X} \sim \mathcal{D}$. We propose a novel loss function $\mathcal{L}_{rval}$ that calculates the distance between these probability distributions, represented as:

$$\mathcal{L}_{rval} = -\ln \left(\int \sqrt{\mathcal{X}(\hat{c}_1^s) * \mathcal{P}(\hat{c}_1)} \right) \tag{2}$$

where $\hat{c}_1^s$ is the feature map obtained after post-processing the output from the TWC_1^s of the student block, and $\hat{c}_1$ is the feature map obtained after post-processing the output from the TWC_1 of the teacher block.

3   Datasets and Environment

We use 4 datasets for developing and validating our proposed techniques: 1) RSNA Pneumonia Detection Challenge dataset [32] consisting of radiographs with presence and absence of pneumonia, 2) SIIM-FISABIO-RSNA COVID-19 Detection dataset [19] for COVID-19 classification and localization, 3) NIH Chest X-rays [38], and 4) VinBigData Chest X-ray Abnormalities Detection dataset [28] comprising 14 common thorax diseases. The training, validation, and testing splits are provided in Table 1. For experimentation, we used Google Cloud Platform (GCP) with TPUs from TensorFlow Research Cloud (TRC). All experiments are in TensorFlow and Keras v2.8.0.

4   Results and Discussion

Implementation. The HVAT comprises 2 stages, namely $HVAT_1$ and $HVAT_2$. During $HVAT_1$, the teacher network is pre-trained on eye-gaze data from [10,15] which contains single radiologist eye-gaze points on 1083 chest x-rays from the MIMIC-CXR [10,14] dataset. For $HVAT_2$, the teacher network is fine-tuned on similar eye-gaze data from REFLACX [10,20,21] which contains single radiologist eye-gaze points on 2507 chest x-rays from the MIMIC-CXR [10,14] dataset. Finally, for F-HVAT, the teacher network is further finetuned on n-radiologists' eye gaze points (in this case $n = 5$) with GAF, as explained in Sect. 2, for 109 chest x-rays from REFLACX [10,20,21] dataset. The attention regions, shown as heatmaps (Fig. 1b.3), are human attention based diagnostically important areas. A Gaussian filter, represented as $\mathcal{G}$, with standard deviation, $\sigma = 64$, generates the attention heatmaps. The contours from these attention heatmaps

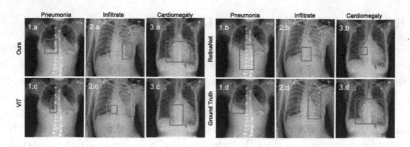

Fig. 2. Qualitative Results: Example localization results are shown on 3 disease types, namely Pneumonia (1.*), Infiltrate (2.*) and Cardiomegaly (3.*), for *GazeRadar* (*.a), RetinaNet (*.b) and ViT (*.c) architectures. Here, *.d are the ground-truth localizations.

are selected with a thresholding value of $\hat{\lambda} = 140$. Then, bounding boxes are generated from the contour with the largest area, shown in Fig. 1b.3. All the images are resized to 256×256 pixels. During $HVAT_i$, and F-HVAT, the output of the teacher network is a $[\hat{\mathcal{B}}, 2]$ tensor of probability values (representing two classes: normal and disease) and a $[\hat{\mathcal{B}}, 4]$ tensor of keypoints. Here, $\hat{\mathcal{B}}$ is the batch size during $HVAT_i$ and F-HVAT.

We use an Adam optimizer with a batch size of 64 for 50 epochs. The initial learning rate (LR), set to 1×10^{-2}, is scheduled with an exponential scheduler with decay steps $= 10^5$ and decay rate $= 0.2$. There is an early stopping criteria with patience $= 20$ to minimize the validation loss. *GazeRadar* follows similar training settings as the baselines. Also, in RAF, we have the radiomic features, $\mathcal{R}_i$ generated from $\mathcal{F}_i$. In our experiments, $i \in \{0, 1\}$ where $\mathcal{F}_0$ is Local Binary Pattern (LBP) [11] and $\mathcal{F}_1$ is an orthogonal Gabor filter [8], also shown in Fig. 1a.

Comparisons and Performance Metrics. *GazeRadar* is compared against standard localization architectures like variants of RetinaNet [24], Center-Net [40], YOLOv3 [31], YOLOv5 [13], and recent vision transformer architectures appended with a detection head. The vision transformer architectures used for comparisons are ViT [7], CCT [12], DeTR [6], and Swin Transformer [26]. To measure the performance of *GazeRadar* for both classification and localization tasks, we use Mean Squared Error (MSE) and Area-under-Curve (AUC). As shown in Table 1, *GazeRadar* outperforms state-of-the-art (SOTA) on 3 datasets (RSNA, SIIM, NIH), and achieves comparable results on VBD. We observe that *GazeRadar* outperforms the majority of vision transformer based architectures on all the datasets.

Ablation Studies. In Table 2, we show the performance of different components of GAF and RAF. F-HVAT is pre-trained with $HVAT_1$ and $HVAT_2$. GAF is a component of F-HVAT. Hence, we independently evaluate $HVAT_1$ and $HVAT_2$. From Table 2, for GAF, we see that $HVAT_2$ performs significantly better than $HVAT_1$ for classification with comparable localization performance. We therefore infer that only radiomics attention, without RVAL, is not well-suited for transferring information from teacher block to student block. We also show, in the

Table 1. Quantitative Comparisons. MSE ($\downarrow$) and AUC ($\uparrow$) are shown for RSNA (Train = 21158, Val = 3022, Test = 6045), SIIM (4433, 633, 1266), NIH (688, 98, 196) and VBD (47539, 6791, 13582) datasets. The comparisons are shown on baseline architectures and *GazeRadar*.

Datasets	RSNA [32]		SIIM [19]		NIH [38]		VBD [28]	
Baselines	MSE	AUC	MSE	AUC	MSE	AUC	MSE	AUC
RN-R50 [24]	03.38	90.18	10.83	82.89	05.67	54.13	02.17	95.98
RN-R101 [24]	02.82	94.57	10.47	82.57	05.87	63.61	01.97	**96.65**
RN-R152 [24]	03.26	91.00	10.44	83.54	07.16	62.21	**01.91**	96.51
CN-R50 [40]	24.74	60.66	27.84	69.89	19.85	51.06	35.11	91.51
CN-R101 [40]	05.86	77.83	45.94	38.98	30.84	45.73	37.43	65.33
CN-R152 [40]	06.57	80.68	28.51	68.55	21.18	50.37	43.13	79.78
YOLOv3 [31]	**02.73**	93.12	29.64	72.29	**04.45**	56.77	07.05	83.28
YOLOv5 [13]	04.15	72.63	12.93	72.29	04.50	56.62	07.05	83.28
ViT-B16+DH [7]	04.13	80.53	13.28	72.23	05.64	57.33	11.04	89.70
ViT-B32+DH [7]	05.08	75.49	13.32	72.84	05.22	55.53	14.49	92.76
ViT-L16+DH [7]	04.88	81.92	13.34	72.39	05.50	57.10	10.73	91.68
ViT-L32+DH [7]	05.12	72.42	13.30	72.43	05.12	56.54	10.30	88.90
CCT+DH [12]	03.99	79.28	12.89	73.50	04.54	56.77	03.83	91.20
DeTr [6]	04.45	68.97	12.96	72.29	04.63	56.77	07.06	83.28
SwinT0+DH [26]	06.35	93.51	07.14	74.42	04.57	64.10	14.87	92.50
SwinT1+DH [26]	07.09	93.75	06.97	75.46	**04.45**	78.70	15.41	91.44
Ours	03.56	**96.27**	**06.56**	**99.36**	08.57	**98.68**	12.12	94.26

Supplementary, different components of the *GazeRadar* architecture and observe how appending different modules affects the system performance. The ablations fundamentally explain the effects of individual global-focal components, adding a teacher network, and then further training using RVAL.

Qualitative Analysis. In Fig. 2, we show localization results of *GazeRadar* in comparison to RetinaNet and ViT for three different disease types: pneumonia, infiltrate, and cardiomegaly. 1.* represents Pneumonia, 2.* represents Infiltrate, and 3.* represents Cardiomegaly samples. *.a are the results from *GazeRadar*, *.b are the results from RetinaNet, *.c are the results from ViT, and *.d are the ground truth bounding boxes. We observe that the predicted bounding boxes overlap better with the ground truth for *GazeRadar* as compared with the baseline methods.

Table 2. GAF-RAF Ablations. MSE ($\downarrow$) and AUC ($\uparrow$) for NIH and SIIM datasets.

DS	NIH [38]		SIIM [19]		DS	NIH [38]		SIIM [19]	
GAF	MSE	AUC	MSE	AUC	**RAF**	MSE	AUC	MSE	AUC
$HVAT_1$	07.78	60.64	14.61	97.37	RAF	17.63	61.43	17.55	96.56
$HVAT_2$	08.39	89.77	18.60	98.38	S+RAF	08.17	59.86	23.64	80.18

5   Conclusion

This paper presents *GazeRadar*, a novel architecture that fuses radiomics and visual attention to learn a joint representation. This radiomics-visual attention is leveraged to train a student block for classification and localization tasks. A novel Radiomics-Visual Attention Loss (RVAL) is proposed to calculate the distance between the student block attention distribution, and the joint representation. We demonstrated the feasibility of this approach with two radiomics features; however, as described in the methodology, this may be readily extended to other features and also to 3D imaging modalities. Our results demonstrate that radiomics and radiologists' visual search patterns harbor important complementary cues regarding disease characteristics and its location; these features can be leveraged in a deep learning framework using a student-teacher architecture. Future work will involve incorporation of RVAL in lung nodule classification [1] and computer-assisted intervention tasks [23,34].

Acknowledgments. Reported research was partly supported by NIH 1R21CA2 58493-01A1, NIH 75N92020D00021 (subcontract), and the OVPR and IEDM seed grants at Stony Brook University. The content is solely the responsibility of the authors and does not necessarily represent the official views of the National Institutes of Health.

References

1. Beig, N., et al.: Perinodular and intranodular radiomic features on lung CT images distinguish adenocarcinomas from granulomas. Radiology **290**(3), 783 (2019)
2. Bertram, R., et al.: Eye movements of radiologists reflect expertise in CT study interpretation: a potential tool to measure resident development. Radiology **281**(3), 805–815 (2016)
3. Bhattacharya, M., et al.: RadioTransformer: a cascaded global-focal transformer for visual attention-guided disease classification. arXiv preprint arXiv:2202.11781 (2022)
4. Bhattacharyya, A.: On a measure of divergence between two statistical populations defined by their probability distributions. Bull. Calcutta Math. Soc. **35**, 99–109 (1943)
5. Bhattacharyya, A.: On a measure of divergence between two multinomial populations. Sankhyā: Indian J. Stat. 401–406 (1946)
6. Carion, N., Massa, F., Synnaeve, G., Usunier, N., Kirillov, A., Zagoruyko, S.: End-to-end object detection with transformers. In: Vedaldi, A., Bischof, H., Brox, T., Frahm, J.-M. (eds.) ECCV 2020. LNCS, vol. 12346, pp. 213–229. Springer, Cham (2020). https://doi.org/10.1007/978-3-030-58452-8_13
7. Dosovitskiy, A., et al.: An image is worth 16×16 words: transformers for image recognition at scale. arXiv preprint arXiv:2010.11929 (2020)
8. Fogel, I., et al.: Gabor filters as texture discriminator. Biol. Cybern. **61**(2), 103–113 (1989)
9. van der Gijp, A., et al.: How visual search relates to visual diagnostic performance: a narrative systematic review of eye-tracking research in radiology. Adv. Health Sci. Educ. **22**(3), 765–787 (2016). https://doi.org/10.1007/s10459-016-9698-1

10. Goldberger, A.L., et al.: PhysioBank, PhysioToolkit, and PhysioNet: components of a new research resource for complex physiologic signals. Circulation **101**(23), e215–e220 (2000)

11. Guo, Z., et al.: A completed modeling of local binary pattern operator for texture classification. IEEE Trans. Image Process. **19**(6), 1657–1663 (2010)

12. Hassani, A., et al.: Escaping the big data paradigm with compact transformers. arXiv preprint arXiv:2104.05704 (2021)

13. Jocher, G., et al.: YOLOv5. Code repository (2020). https://github.com/ultralytics/yolov5

14. Johnson, A.E., et al.: MIMIC-CXR-JPG, a large publicly available database of labeled chest radiographs. arXiv preprint arXiv:1901.07042 (2019)

15. Karargyris, A., et al.: Eye gaze data for chest x-rays. PhysioNet (2020). https://doi.org/10.13026/QFDZ-ZR67

16. Kelahan, L.C., et al.: The radiologist's gaze: mapping three-dimensional visual search in computed tomography of the abdomen and pelvis. J. Digit. Imaging **32**(2), 234–240 (2019)

17. Kelly, B.S., et al.: The development of expertise in radiology: in chest radiograph interpretation, "expert" search pattern may predate "expert" levels of diagnostic accuracy for pneumothorax identification. Radiology **280**(1), 252–260 (2016)

18. Khosravan, N., et al.: A collaborative computer aided diagnosis (C-CAD) system with eye-tracking, sparse attentional model, and deep learning. Med. Image Anal. **51**, 101–115 (2019)

19. Lakhani, P., et al.: The 2021 SIIM-FISABIO-RSNA machine learning COVID-19 challenge: annotation and standard exam classification of COVID-19 chest radiographs (2021)

20. Lanfredi, R.B., et al.: REFLACX: reports and eye-tracking data for localization of abnormalities in chest x-rays (2021)

21. Lanfredi, R.B., et al.: REFLACX, a dataset of reports and eye-tracking data for localization of abnormalities in chest x-rays. arXiv preprint arXiv:2109.14187 (2021)

22. Lee, A., et al.: Identification of gaze pattern and blind spots by upper gastrointestinal endoscopy using an eye-tracking technique. Surg. Endosc. **36**, 1–8 (2021)

23. Li, Y., Shenoy, V., Prasanna, P., Ramakrishnan, I., Ling, H., Gupta, H.: Surgical phase recognition in laparoscopic cholecystectomy. arXiv preprint arXiv:2206.07198 (2022)

24. Lin, T.Y., et al.: Focal loss for dense object detection. In: Proceedings of the IEEE International Conference on Computer Vision, pp. 2980–2988 (2017)

25. Litchfield, D., et al.: Viewing another person's eye movements improves identification of pulmonary nodules in chest x-ray inspection. J. Exp. Psychol. Appl. **16**(3), 251 (2010)

26. Liu, Z., et al.: Swin transformer: hierarchical vision transformer using shifted windows. In: Proceedings of the IEEE/CVF International Conference on Computer Vision, pp. 10012–10022 (2021)

27. Nebbia, G., et al.: Radiomics-informed deep curriculum learning for breast cancer diagnosis. In: de Bruijne, M., et al. (eds.) MICCAI 2021. LNCS, vol. 12905, pp. 634–643. Springer, Cham (2021). https://doi.org/10.1007/978-3-030-87240-3_61

28. Nguyen, H.Q., et al.: VINDR-CXR: an open dataset of chest x-rays with radiologist's annotations. arXiv preprint arXiv:2012.15029 (2020)

29. Parekh, V.S., et al.: Deep learning and radiomics in precision medicine. Expert Rev. Precis. Med. Drug Dev. **4**(2), 59–72 (2019)

30. Prasanna, P., et al.: Radiographic-deformation and textural heterogeneity (r-DepTH): an integrated descriptor for brain tumor prognosis. In: Descoteaux, M., Maier-Hein, L., Franz, A., Jannin, P., Collins, D.L., Duchesne, S. (eds.) MICCAI 2017. LNCS, vol. 10434, pp. 459–467. Springer, Cham (2017). https://doi.org/10.1007/978-3-319-66185-8_52

31. Redmon, J., et al.: YOLOv3: an incremental improvement. arXiv preprint arXiv:1804.02767 (2018)

32. Shih, G., et al.: Augmenting the national institutes of health chest radiograph dataset with expert annotations of possible pneumonia. Radiol.: Artif. Intell. 1(1), e180041 (2019)

33. Singh, G., et al.: Radiomics and radiogenomics in gliomas: a contemporary update. Br. J. Cancer 125(5), 641–657 (2021)

34. Tokuyasu, T., et al.: Development of an artificial intelligence system using deep learning to indicate anatomical landmarks during laparoscopic cholecystectomy. Surg. Endosc. 35(4), 1651–1658 (2020). https://doi.org/10.1007/s00464-020-07548-x

35. Tourassi, G., et al.: Investigating the link between radiologists' gaze, diagnostic decision, and image content. J. Am. Med. Inform. Assoc. 20(6), 1067–1075 (2013)

36. Venjakob, A., et al.: Radiologists' eye gaze when reading cranial CT images. In: Medical Imaging 2012: Image Perception, Observer Performance, and Technology Assessment, vol. 8318, pp. 78–87. SPIE (2012)

37. Waite, S., et al.: Analysis of perceptual expertise in radiology-current knowledge and a new perspective. Front. Hum. Neurosci. 13, 213 (2019)

38. Wang, X., et al.: ChestX-ray8: hospital-scale chest X-ray database and benchmarks on weakly-supervised classification and localization of common thorax diseases. In: Proceedings of the IEEE conference on computer vision and pattern recognition, pp. 2097–2106 (2017)

39. Yoshie, T., et al.: The influence of experience on gazing patterns during endovascular treatment: eye-tracking study. J. Neuroendovascular Ther. oa–2021 (2021)

40. Zhou, X., et al.: Objects as points. arXiv preprint arXiv:1904.07850 (2019)

41. Zimmermann, J.M., et al.: Quantification of avoidable radiation exposure in interventional fluoroscopy with eye tracking technology. Invest. Radiol. 55(7), 457–462 (2020)

Deep Reinforcement Learning for Detection of Inner Ear Abnormal Anatomy in Computed Tomography

Paula López Diez[1](✉), Kristine Sørensen[1], Josefine Vilsbøll Sundgaard[1], Khassan Diab[4], Jan Margeta[3], François Patou[2], and Rasmus R. Paulsen[1]

[1] DTU Compute, Technical University of Denmark, Kongens Lyngby, Denmark
plodi@dtu.dk
[2] Oticon Medical, Research and Technology group, Smørum, Denmark
[3] KardioMe, Research and Development, Nova Dubnica, Slovakia
[4] Tashkent International Clinic, Tashkent, Uzbekistan

Abstract. Detection of abnormalities within the inner ear is a challenging task that, if automated, could provide support for the diagnosis and clinical management of various otological disorders.Inner ear malformations are rare and present great anatomical variation, which challenges the design of deep learning frameworks to automate their detection. We propose a framework for inner ear abnormality detection, based on a deep reinforcement learning model for landmark detection trained in normative data only. We derive two abnormality measurements: the first is based on the variability of the predicted configuration of the landmarks in a subspace formed by the point distribution model of the normative landmarks using Procrustes shape alignment and Principal Component Analysis projection. The second measurement is based on the distribution of the predicted Q-values of the model for the last ten states before the landmarks are located. We demonstrate an outstanding performance for this implementation on both an artificial (0.96 AUC) and a real clinical CT dataset of various malformations of the inner ear (0.87 AUC). Our approach could potentially be used to solve other complex anomaly detection problems.

Keywords: Deep reinforcement learning · Anomaly detection · Inner ear · Congenital malformation

1 Introduction

Sensorineural hearing loss (SNHL) in children is a major cause of disability. Generally SNHL is detected early in many parts of the world, which allows the prescription of interventions that mitigate the risk of abnormal social, emotional and communicative development. Such interventions include Cochlear Implant (CI) therapy which is prescribed each year to about 80,000 infants and toddlers. Congenital SNHL is sometimes the consequence of an abnormal embryonic development. Resulting malformations are generally classified according to

© The Author(s), under exclusive license to Springer Nature Switzerland AG 2022
L. Wang et al. (Eds.): MICCAI 2022, LNCS 13433, pp. 697–706, 2022.
https://doi.org/10.1007/978-3-031-16437-8_67

two categories: membranous malformations, which are not observable in conventional medical scans, and congenital malformations, which can be detected by Computed Tomography (CT) or Magnetic Resonance Imaging (MRI) [4]. These cases raise surgical challenges during surgical planning of the CI therapy and during the surgery itself, often requiring the surgeon to discover and adapt to the anatomy of the malformation during the operation. Anticipating the presence of such malformations from standard imaging modalities is a complex task even for expert clinicians. Categories for these malformations have been described by Sennaroğ L. *et al.* [16], and heuristics have been proposed to help identify them, such as the ones proposed by Dhanasingh A. *et al.* [7]. These heuristics are however of limited use to inexperienced otologists and ear, nose, and throat (ENT) surgeons, who, given the rarity of some of these conditions, cannot easily learn to detect the associated image patterns reliably. We take a first step towards assisting otologists and ENT surgeons in screening or detecting inner ear malformation by proposing the first automated method to detect these anomalies from clinical CT scans.

Different state-of-the-art deep learning methods have shown high performance for automatic detection of anomalies as presented in [5]. Deep learning approaches mostly based in convolutional neural networks used for classification have been used in a clinical context for anatomical anomalies [8]. Training such models requires large amounts of labeled medical data that faithfully represent these anomalies, which is challenging and expensive to acquire, especially because datasets are usually imbalanced because pathological cases are generally rare [17]. We propose a method that is trained uniquely on normative data for landmark location, which makes the approach suitable for adaptation to other anatomies. Knowledge of normal anatomical structural shapes and arrangements acquired during landmark location training brings implicit information for detecting anomalies within that region. Our method is based on multiple landmark location in CT scans of the inner ear. Because we aim to detect abnormalities indirectly by evaluating the output of the model, we define the landmark location as an object search problem and choose to use a deep reinforcement learning (DRL) architecture. We use both the communicative multiple agent reinforcement learning (C-MARL [11]) model and the standard multiple agent reinforcement learning (MARL [19]) model to locate a set of landmarks in the inner ear. We extract two pieces of critical information from these models: First, the variability of the predicted location of a certain landmark across different runs/agents which we evaluate in a subspace defined by the normative data landmarks after they are all aligned using Procrustes, and a principal component analysis (PCA) of the shape variation is performed to define the subspace as presented by López Diez *et al.* in [12]. Second, as a measurement of abnormality, we use the distribution of the predicted Q-values for each agent over the last ten states, including the final position where the landmark is placed. We initially test our approach using a small set of landmarks in a tight crop of the CT images centered on the cochlea versus synthetically generated images of a specific type of inner ear malformation called cochlear aplasia. Furthermore, we tested the approach in real clinical data using a set of twelve landmarks in a bigger crop of the inner ear.

Simpler methods such as PCA can be employed for anomaly detection in physiological measurements [3,10]. Several groups have used models trained on healthy anatomies to derived the detection of anomalies. While conceptually close to the approach we propose here, the methods have relied on spatial autoencoders or CycleGANs, as described by Baur C. *et al.* [1,2], or segmentation models such as the Bayesian UNet used by Seebock *et al.* [15]. These approaches lack the spatial highlighting and interpretability that our landmark-based approach provides by using highly relevant points of interest defined according to the anatomical malformations.

2 Data

We use two different datasets to test our approach. Our first dataset consists of 119 clinical CT scanners from diverse imaging equipment. These images consist of a region of interest (ROI) with a size of (32.1^3 mm^3) with the cochlea in its center and an average voxel resolution of 0.3 mm. To test our approach, we synthetically generated abnormal inner ear CT scans from the original images by removing the cochlea (simulating cochlear aplasia) from the images, thus generating corresponding pairs of normal and abnormal CT scans with the same surrounding structures. The cochlea was segmented using ITK-SNAP software [20] and then replaced by Gaussian noise with mean and standard deviation estimated from the intensities of the tissue surrounding the segmentation [12]. An example of the transformation process as well as the location of the anatomical landmarks we use are shown in Fig. 1. This dataset will be called the **Synthetic Set** from now on.

Our second dataset consists of 300 normal anatomy CT scans from heterogeneous sources and 123 CT scans of inner ears that present diverse congenital malformations. This unique dataset contains full-head CT images of CI patients acquired through different CT scanners. This dataset will be referred as the

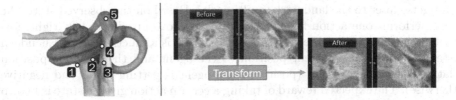

Fig. 1. Left: Set of landmarks used in the Synthetic Set. **1, 2** - Opposite sides of bony cochlear nerve canal in axial view. **3** - Facial Nerve (FN) exiting the Internal Acoustic Canal. **4** - Closest point of FN and cochlea. **5** - Geniculate ganglion of the FN. Edited from [18]. **Right:** Example image of CT scan from test set, before and after the synthetic image generation by inpainting.

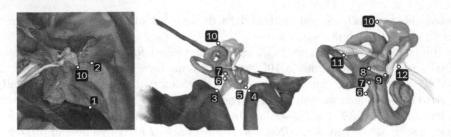

Fig. 2. Set of landmarks annotated in normal anatomy CT images in the Real Abnormality Set. **1** - Sigmoid Sinus (SS) (closest point to EAC). **2** - External Acoustic Canal (EAC) (closest point to SS). **3** - Jugular Bulb (closest point to Round Window (RW)). **4** - Carotid Artery(CA) (closest point to Basal turn of the cochlea). **5** - Basal Turn (closest point to JB). **6, 7** - Anterior and posterior edges of RW. **8, 9** - Anterior and posterior crus of staples. **10** - Short Process of Incus. **11** - Pyramidal Process **12-** Cochleariform Process. Edited from [18].

Real Abnormality Set further on. Out of the 300 normal ears, 175 were manually annotated by an expert with 12 landmarks that define key points of this anatomy. To optimally characterize certain points of interest, these landmarks were designed in close collaboration with our clinical partner, an ENT surgeon specialized in CI therapy in abnormal anatomies. These landmarks are presented in Fig. 2. Simultaneously, the same ROI of 80^3 mm^3 was extracted from the full-head CTs by using the location of the mandible joint and the beginning of the internal acoustic canal for both normal and abnormal anatomies. All images were re-sampled to a 0.5 mm isotropic resolution.

3   Methods

DRL for Landmark Location. Deep-Q-Networks [14] are used to find the optimal strategy for agents to reach their goal. These agents navigate through the 3D image (environment) and observe their state, which is defined as a patch of the image centered on the agent location. This patch becomes smaller as the agent gets closer to the landmark (multi-scale). Based on the observed state, the agent performs one action from the action set (move up, down, left, right, forward, and backward) and receives a reward, which is a function of the Euclidean distance between the current position of the agent and the previous position relative to the target point (positive when agent is getting closer and negative otherwise). The expected reward of taking a certain action given a state is known as the Q-value. In deep reinforcement learning, the Q-value of a certain state associated with each of the possible actions is estimated by the use of a Deep-Q-Network. The architecture of the Deep-Q-Network used for landmark location resembles a typical image classification architecture, but with a set of fully connected layers for each agent. The architecture of the model is shown in Fig. 3. The common convolutional neural network weights among all agents provide implicit

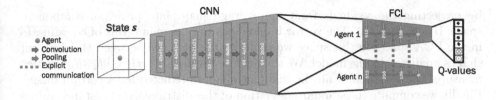

Fig. 3. Diagram of the DRL model used. The explicit communication connections are used in the C-MARL model, but not in the MARL model.

communication between the agents, meaning they share the same layers responsible for extracting the relevant features for their current state. Meanwhile, the shared average weight of the different fully connected layers allows for implicit communication between agents, sharing information of the layers that are used to map the extracted features from the current state to the predicted Q-value of each agent. This setup has been proven especially effective when the different landmarks present a consistent spatial correlation as it is the case with inner ear anatomy [13]. We trained a C-MARL in the normal anatomies of the Synthetic Set and the Real Abnormality Set. Finally we also trained the MARL model on the normal samples of the Real Abnormality Set as we expect that the explicit communication between agents might influence the variability of the model output when facing an abnormal anatomy. The training configuration employed is the same as presented in [13].

PCA Shape Distance Method. The defined landmarks are placed in a spatial configuration which reveals consistency between patients with normal anatomy. We expect that for abnormal cases the landmarks predicted by the model will deviate significantly from this configuration and from one another. In order to test if a case is within the normal configuration, a point distribution model (PDM) is constructed following the approach presented in [6]. We will refer to a full set of landmarks in an image as *shape* where we know there is a point correspondence across all shapes in the training data. The alignment between all the annotated landmarks in normal anatomy is derived using Procrustes analysis [9]. Using this transform, we obtain a PDM that describes the shape variation only and that is invariant to size variation. Once the training shapes are all aligned, a mean shape is computed $\bar{\mathbf{x}}$, followed by a PCA of the shape variation [6]. The outcome of the PCA analysis is a set of principal components concatenated into a matrix Φ, which describes the modes of shape variation. A new shape $\mathbf{x}_{new}$ can be then defined as: $\mathbf{x}_{new} = \bar{\mathbf{x}} + \Phi \mathbf{b}$. The vector $\mathbf{b}$ denotes weights controlling the modes of shape variation and Φ contains the first t principal components. We chose to use a t value such that 90% of the shape variability is contained in the Φ matrix. We found $t = 11$ for the 36-dimension space defined by the twelve 3d-landmarks from Fig. 2 over the 175 annotated normal anatomy images of the Real Abnormality Set and $t = 6$ for the 92 normal anatomy images annotated with five landmarks described in Fig. 1 from the Synthetic Set. A given $\mathbf{x}'$ shape can be aligned to the Procrustes mean and be approximated by the PDM model by projecting the residuals from the average shape into principal component space: $\mathbf{b} = \mathbf{\Phi}^T(\mathbf{x}' - \bar{\mathbf{x}})$. The vector $\mathbf{b}$ describes the shape in terms of coordinates in the PCA space. In this space we evaluate the distance between the different shapes predicted by the model. We then define the distance $d_{ji} = ||b_i - b_j||_2$ which measures the variation of all the different shapes predicted for a certain image. Finally we compute the standard deviation of this distribution of distance values for a certain image, D_{image}, which measures the level of agreement among the multiple predictions computed in the PCA space defined by normative shapes. A sketch of this approach is shown in Fig. 4a.

Q-Value History Distribution Method. Using deep Q-learning means that the network is trained to estimate the Q-value, or estimated reward, of taking a certain action given a certain state. Our hypothesis is that if the current state of the agent that is looking for a certain landmark resembles the normal anatomical configuration of such a region, the Q-values will present a uniform distribution, as the agent should not expect a high reward for moving in a certain direction. On the other hand, when the anatomy of the state does not look like what the agent would expect, the Q-values should be less uniformly distributed, pushing the agent to move in a certain direction. To test this hypothesis, we have computed a measurement of the variability within the distribution of the predicted Q-values of the action set. To compute this abnormality measurement we collect the buffer with the predicted Q-values of the last 10 states of the agent, which have empirically be found sufficient to define the later states of the landmark search procedure. These Q-value vectors are then normalized for each agent and merged together with the Q-values of the different runs for each specific landmark. Then, the standard deviation of the Q-values distribution is computed for each landmark u_n. These uncertainty measurements are then joined together into a single value per image $U_{\text{image}} = \sqrt{\sum_n u_n^2}$. An overview of the process is outlined in Fig. 4b.

The combination of both methods has been computed to evaluate its joint performance. Due to the different magnitude of the measurements of each method, a weighting factor has been included in the combination so both methods have a more balanced contribution. The weighting factor w is defined as $w = \frac{\text{median}(D_{\text{training}})}{\text{median}(U_{\text{training}})}$. Then the combination of both methods is defined as $C_{\text{image}} = \sqrt{D_{\text{image}}^2 + (wU_{\text{image}})^2}$ to analyze the joint performance.

4 Results

Each of our tests has been evaluated over five runs for a more rigorous analysis. Both methods described in the previous section where initially tested in the Synthetic Set. The C-MARL model was trained on 92 images with three agents per landmark for the five landmarks shown in Fig. 1. The average error of landmark location is $0.814 \, \text{mm}$ in the test set. Our method is tested on 27 normal anatomy CT crops and their corresponding artificially created abnormal anatomy scans. The results are shown in Fig. 5a) and d).

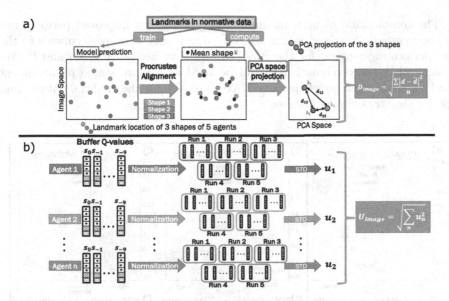

Fig. 4. Diagram of the computation process a) D_{image} b) U_{image} computation process.

To test the performance in the Real Abnormality Set, we trained a C-MARL model with one agent per landmark for the twelve landmarks described in Fig. 2. The model was trained on 150 CT scans of patients with normal inner ear anatomy with an average test error of 1.74 mm for landmark location over the 25 images of the test set. To test our method we used 123 other different CT scans of normal anatomy and 123 of congenital malformations from the Real Abnormality Set, over five runs. The results of our method are shown in Fig. 5b) and e). As was expected, there is a significant drop in performance when comparing the results in the smaller ROI of the Synthetic Set shown in Fig. 5d) and the results with the same architecture (with a different set of landmarks) shown in Fig. 5e). However, we expected our method would benefit from using the MARL model, which does not include the connections that are responsible for the explicit communication between agents. This means that the agents would not share explicit information about their location and search procedure while looking for the landmarks. This makes agents more independent from each other and less tied to the spatial correlation among them. Our hypothesis is that avoiding this communication will derive greater values for the abnormality measurements when facing an anomaly. The MARL model was trained in the exact same configuration as the C-MARL model and obtained an average error of 1.99 mm on landmark location accuracy. The results of applying our method in this configuration can be observed in Fig. 5c) and f). It can be observed that the method does indeed perform better without explicit communication connections in the model.

The combination of both methods C_{image} shows an improved performance for the Synthetic Set as shown in Fig. 5d), and a very close performance to the best-performing method for the Real Abnormality Set see Fig. 5e) and f). We consider that the combination should be used as a more stable measurement which shows an area under the curve (AUC) of 0.96 for the artificial dataset and 0.86 for the large clinical dataset.

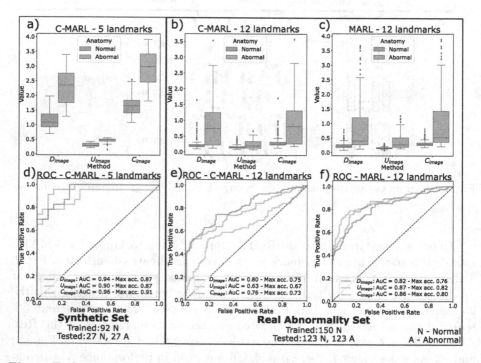

Fig. 5. Evaluation of the different methods over five runs for normal and abnormal anatomies. **Left:** Using 5 landmarks in the Synthetic Set. **Right:** Using 12 landmarks in the Real Abnormality Set. **a), b)** and **c)** Abnormality measurements distribution. **d), e)** and **f)** Derived ROC curves. **a), b), d)** and **e)** C-MARL model. **c)** and **f)** MARL model.

5   Conclusion

We have demonstrated that we can detect abnormal inner ear anatomies by solely training a DRL model on normative data and evaluating the output variability of certain implicit information. This information looks at the relative position of the predicted landmarks over different runs/agents in a subspace defined by the normative annotations as well as the distribution of the Q-values of the last iterations of the agents as a measurement of the uncertainty of the final location. Our MARL model achieved the best performance with an AUC of 0.87 in clinical data which is a high score for such a complex classification problem. We showed how uncertainty information can be derived from a trained model to automatically detect abnormal anatomies, meaning no specific classification model needs to be trained, and therefore annotated abnormal data are not required to build the framework. We proved that the stated methods provide a measurement of the abnormality of the model's output which is linked with the presence of malformations. We examined the approach with good results, not only on artificially generated data, but also in a large dataset of real clinical CT scans of patients with diverse inner ear malformations.

References

1. Baur, C., Graf, R., Wiestler, B., Albarqouni, S., Navab, N.: SteGANomaly: inhibiting CycleGAN steganography for unsupervised anomaly detection in brain MRI. In: Martel, A.L., et al. (eds.) MICCAI 2020. LNCS, vol. 12262, pp. 718–727. Springer, Cham (2020). https://doi.org/10.1007/978-3-030-59713-9_69
2. Baur, C., Wiestler, B., Muehlau, M., Zimmer, C., Navab, N., Albarqouni, S.: Modeling healthy anatomy with artificial intelligence for unsupervised anomaly detection in brain MRI. Radiol. Artif. Intell. **3**(3), e190169 (2021). https://doi.org/10.1148/ryai.2021190169
3. Amor, L. B., Lahyani, I., Jmaiel, M.: PCA-based multivariate anomaly detection in mobile healthcare applications. In: Proceedings of the International Symposium on Distributed Simulation and Real Time Applications (DS-RT), pp. 1–8 (2017). https://doi.org/10.1109/DISTRA.2017.8167682
4. Cairo/EG, R.Z.: Congenital inner ear abnormalities:a practical review. EPOS ECR 2019 / C-1911. https://doi.org/10.26044/ecr2019/C-1911, https://dx.doi.org/10.26044/ecr2019/C-1911
5. Chalapathy, R., Chawla, S.: Deep learning for anomaly detection: a survey (2019), http://arxiv.org/abs/1901.03407
6. Cootes, T.F., Taylor, C.J., Cooper, D.H., Graham, J.: Active shape models-their training and application. Comput. Vis. Image Underst. **61**(1), 38–59 (1995). https://doi.org/10.1006/cviu.1995.1004
7. Dhanasingh, A., et al.: A novel method of identifying inner ear malformation types by pattern recognition in the mid modiolar section. Sci. Rep. **11**(1), 1–9 (2021). https://doi.org/10.1038/s41598-021-00330-6
8. Gill, R.S., et al.: Deep convolutional networks for automated detection of epileptogenic brain malformations. In: Frangi, A.F., Schnabel, J.A., Davatzikos, C., Alberola-López, C., Fichtinger, G. (eds.) MICCAI 2018. LNCS, vol. 11072, pp. 490–497. Springer, Cham (2018). https://doi.org/10.1007/978-3-030-00931-1_56
9. Gower, J.C.: Generalized procrustes analysis. Psychometrika **40**(1), 33–51 (1975). https://doi.org/10.1007/bf02291478
10. Krenn, V.A., Fornai, C., Webb, N.M., Woodert, M.A., Prosch, H., Haeusler, M.: The morphological consequences of segmentation anomalies in the human sacrum. Am. J. Bio. Anthropol. **177**(14), 690–707 (2021). https://doi.org/10.1002/ajpa.24466, https://onlinelibrary.wiley.com/doi/10.1002/ajpa.24466
11. Leroy, G., Rueckert, D., Alansary, A.: Communicative reinforcement learning agents for landmark detection in brain images. In: MLCN/RNO-AI -2020. LNCS, vol. 12449, pp. 177–186. Springer, Cham (2020). https://doi.org/10.1007/978-3-030-66843-3_18

12. Diez, P. L., et al.: Deep reinforcement learning for detection of abnormal anatomies. In: Proceedings of the Northern Lights Deep Learning Workshop, vol. 3 (2022). https://doi.org/10.7557/18.6280

13. López Diez, P., Sundgaard, J.V., Patou, F., Margeta, J., Paulsen, R.R.: Facial and cochlear nerves characterization using deep reinforcement learning for landmark detection. In: MICCAI 2021. LNCS, vol. 12904, pp. 519–528. Springer, Cham (2021). https://doi.org/10.1007/978-3-030-87202-1_50

14. Mnih, V., et al.: Human-level control through deep reinforcement learning. Nature **518**, 529–533 (2015)

15. Seeböck, P., et al.: Exploiting epistemic uncertainty of anatomy segmentation for anomaly detection in retinal oct. IEEE Trans. Med. Imaging **39**(1), 87–98 (2020). https://doi.org/10.1109/TMI.2019.2919951

16. Sennarolu, L., Bajin, M.D.: Classification and current management of inner ear malformations. Balkan Med. J. **34** (2017). https://doi.org/10.4274/balkanmedj.2017.0367

17. Shin, H.-C., et al.: Medical image synthesis for data augmentation and anonymization using generative adversarial networks. In: Gooya, Ali, Goksel, Orcun, Oguz, Ipek, Burgos, Ninon (eds.) SASHIMI 2018. LNCS, vol. 11037, pp. 1–11. Springer, Cham (2018). https://doi.org/10.1007/978-3-030-00536-8_1

18. Trier, P., Noe, K. O., Sørensen, M.S., Mosegaard, J.: The visible ear surgery simulator, vol. 132 (2008)

19. Vlontzos, A., Alansary, A., Kamnitsas, K., Rueckert, D., Kainz, B.: Multiple landmark detection using multi-agent reinforcement learning. In: shen, D., et al. (eds.) MICCAI 2019. LNCS, vol. 11767, pp. 262–270. Springer, Cham (2019). https://doi.org/10.1007/978-3-030-32251-9_29

20. Yushkevich, P.A., et al.: User-guided 3D active contour segmentation of anatomical structures: significantly improved efficiency and reliability. Neuroimage **31**(3), 1116–1128 (2006). https://doi.org/10.1016/j.neuroimage.2006.01.015

Vision-Language Contrastive Learning Approach to Robust Automatic Placenta Analysis Using Photographic Images

Yimu Pan[1(✉)], Alison D. Gernand[1], Jeffery A. Goldstein[2], Leena Mithal[3], Delia Mwinyelle[4], and James Z. Wang[1]

[1] The Pennsylvania State University, University Park, PA, USA
ymp5078@psu.edu
[2] Northwestern University, Chicago, IL, USA
[3] Lurie Children's Hospital, Chicago, IL, USA
[4] The University of Chicago, Chicago, IL, USA

Abstract. The standard placental examination helps identify adverse pregnancy outcomes but is not scalable since it requires hospital-level equipment and expert knowledge. Although the current supervised learning approaches in automatic placenta analysis improved the scalability, those approaches fall short on robustness and generalizability due to the scarcity of labeled training images. In this paper, we propose to use the vision-language contrastive learning (VLC) approach to address the data scarcity problem by incorporating the abundant pathology reports into the training data. Moreover, we address the feature suppression problem in the current VLC approaches to improve generalizability and robustness. The improvements enable us to use a shared image encoder across tasks to boost efficiency. Overall, our approach outperforms the strong baselines for fetal/maternal inflammatory response (FIR/MIR), chorioamnionitis, and sepsis risk classification tasks using the images from a professional photography instrument at the Northwestern Memorial Hospital; it also achieves the highest inference robustness to iPad images for MIR and chorioamnionitis risk classification tasks. It is the first approach to show robustness to placenta images from a mobile platform that is accessible to low-resource communities.

Keywords: Placenta analysis · mHealth · Vision-language pre-training

1 Introduction

The placenta is a temporary organ that forms during pregnancy and acts as fetal life support prior to delivery. Adverse pregnancy outcomes, including chorioam-

This work used the Extreme Science and Engineering Discovery Environment (XSEDE), which is supported by National Science Foundation grant number ACI-1548562.

Supplementary Information The online version contains supplementary material available at https://doi.org/10.1007/978-3-031-16437-8_68.

© The Author(s), under exclusive license to Springer Nature Switzerland AG 2022
L. Wang et al. (Eds.): MICCAI 2022, LNCS 13433, pp. 707–716, 2022.
https://doi.org/10.1007/978-3-031-16437-8_68

nionitis and sepsis (infection) and meconium staining (fetal distress), produce reproducible morphologic changes in the placenta that can be identified by pathologic examination. The current standard placental examination consists of macroscopic examination, production of microscopic slides, manual examination of the slide by a pathologist, and production of a report. This process requires hospital-level equipment and human input at each step, introducing variation and limiting opportunities for scaling. Automatic placenta analysis using a photographic image is more scalable and can benefit low-resource communities with no access to a pathologist.

Related Work. A recent automatic placenta photo analysis approach, the AI-PLAX [4], used a combination of handcrafted features/rules and deep learning methods; A later approach [19] used only deep learning methods. Although these approaches have achieved promising results, their models suffered from data scarcity; a large portion of the collected images was discarded to balance the positive and negative sample ratio and meet certain quality standards; all pathology reports were ignored since their models were not designed to use text data. Recent advances in self-supervised learning [1,8] and vision-language contrastive learning (VLC) [14,18] have shown promising results in pre-training tasks and can potentially benefit the model performance by including the discarded data as part of the pre-training dataset. However, current contrastive loss used in both the self-supervised methods and VLC methods suffered from the feature suppression problem [2]. Although recent work has addressed such a problem in a self-supervised setting [5,12,15,16], to our knowledge, no work has been done in a VLC setting.

Our Contributions. We tackle the data scarcity problem by using an improved VLC technique to train a shared image encoder using the placenta image and the corresponding pathology report. Our technique is designed to learn generalizable placental features that can be applied to many downstream placental analysis tasks without training a separate image encoder for every task. This approach requires less data for the downstream tasks thus alleviating the data scarcity problem. To our knowledge, this is the first work to address the feature suppression problem in VLC. This is also the first automatic placenta analysis approach tested on iPad images. Our work improves both efficiency and robustness over the existing work in automatic placenta analysis.

2 Method

The proposed method is illustrated in Fig. 1. It consists of a pre-training stage and a fine-tuning stage. The pre-trained text encoder is frozen using a stop gradient operation in the pre-training stage. The trained image encoder is frozen using a stop gradient operation and shared for all tasks in the fine-tuning stage.

2.1 Problem Formulation

We have two tasks, the pre-training and the downstream classification. Formally, for the former, we want to learn a function f_v using a learned function f_u such

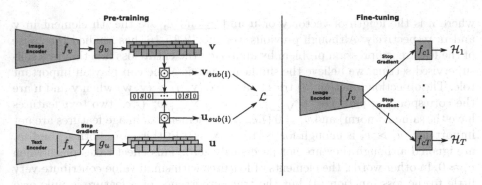

Fig. 1. A diagram illustrating our approach where all the notations correspond to the descriptions in Sect. 2. The inputs are omitted for simplicity.

that, for any pair of input $(\mathbf{x}_i, \mathbf{t}_i)$ and an similarity function sim, we have

$$\texttt{sim}(g_v(f_v(\mathbf{x}_i)), g_u(f_u(\mathbf{t}_i))) > \texttt{sim}(g_v(f_v(\mathbf{x}_i)), g_u(f_u(\mathbf{t}_j))) , \tag{1}$$

where g is a linear projection function to map the output vector to the same shape and $i \neq j$. The details on the objective function for achieving inequality (1) are discussed later.

For the latter task, we want to learn a function f_{ct} using the learned function f_v for each task $t \in [1 : T]$ such that, for a pair of input $(\mathbf{x}_i, l_{ti})$,

$$f_{ct}(f_v(\mathbf{x}_i)) = l_{ti} , \tag{2}$$

which can be achieved by using the cross-entropy loss $\mathcal{H}$ as the objective function.

2.2 Hypothetical Cause of Feature Suppression Problem

Current VLC methods minimize the following contrastive loss:

$$\ell_i^{(v \to u)} = - \log \frac{\exp(\langle \mathbf{v}_i, \mathbf{u}_i \rangle / \tau)}{\sum_{k=1}^{N} \exp(\langle \mathbf{v}_i, \mathbf{u}_k \rangle / \tau)} , \tag{3}$$

where $\langle \mathbf{v}, \mathbf{u} \rangle$ represents the cosine similarity between the two feature vectors $\mathbf{v}, \mathbf{u}$ from $g_v(f_v(x))$ and $g_u(f_u(t))$, respectively, τ is the temperature hyperparameter, and N is the total number of sample is a mini-batch. CLIP [14] and ConVIRT [18] have shown that models trained using this method are more robust. However, as demonstrated in [2], such a contrastive loss suffers from the feature suppression problem where the model only learns the most important feature. This effect is especially problematic in our application since we have multiple tasks for the same image and different tasks may require different features. Current contrastive loss uses cosine similarity $\langle \mathbf{v}, \mathbf{u} \rangle$ which is defined as

$$\langle \mathbf{v}, \mathbf{u} \rangle = \frac{\sum_{i=1}^{n} v_i u_i}{\sqrt{\sum_{i=1}^{n} v_i^2} \sqrt{\sum_{i=1}^{n} u_i^2}} , \tag{4}$$

where n is the length of vector $\mathbf{v}$ or $\mathbf{u}$ and v_i and u_i are the ith element in $\mathbf{v}$ and $\mathbf{u}$, respectively. Although previous work [5,12,15,16] has studied the cause of the feature suppression problem by analyzing the entire loss function in a self-supervised setting, we believe the similarity metric alone can play an important role. The objection function (3) tries to achieve $\langle \mathbf{v}, \mathbf{u} \rangle > \langle \mathbf{v}, \mathbf{w} \rangle$ when $\mathbf{v}$ and $\mathbf{u}$ are the corresponding pair. Given $\sqrt{\sum_{i=1}^{n} u_i^2} = \sqrt{\sum_{i=1}^{n} w_i^2}$ (*i.e.*, two text features have the same L^2 norm) and $v_j = 0$ (*i.e.*, some elements of image features are not important), $u_i > w_i$ is enough for $\langle \mathbf{v}, \mathbf{u} \rangle > \langle \mathbf{v}, \mathbf{w} \rangle$. In this case, both u_j and w_j are ignored although they are not necessarily zero. This effect takes place when $v_j \approx 0$. In other words, the elements of features with small value contribute very little to the loss function (4) but the true importance of a feature is unknown before the downstream tasks. We hypothesize such an effect in the similarity metric is one cause of the feature suppression problem in VLC methods and we can address it by simply replacing the similarity metric.

2.3 Negative Logarithmic Hyperbolic Cosine Similarity

To minimize the feature suppression problem, we propose to use the Negative Logarithmic Hyperbolic Cosine (NegLogCosh) as the similarity metric:

$$\text{NegLogCosh}(\mathbf{v}, \mathbf{u}) = -\frac{1}{n} \sum_{i=1}^{n} \log(\cosh(s(v_i - u_i))), \tag{5}$$

where s is a scaling factor, The advantage of NegLogCosh($\mathbf{v}, \mathbf{u}$) over $\langle \mathbf{v}, \mathbf{u} \rangle$ is that the value change of any v_i or u_i is reflected in the result, thus the trained model tends to focus on more features. Although L1 and L2 loss functions have the same property, NegLogCosh($\mathbf{v}, \mathbf{u}$) has more advantages. First, NegLogCosh has less emphasis than L2 loss when v_i and u_i are very different thus reducing the effect of the dominant feature from either the text side or the image side. Second, NegLogCosh is more stable than L1 loss when $v_i - u_i \approx 0$. The proposed objective function is the following:

$$\tilde{\ell}_i^{(v \to u)} = -\log \frac{\exp(\text{NegLogCosh}(\mathbf{v}_i, \mathbf{u}_i)/\tau)}{\sum_{k=1}^{N} \exp(\text{NegLogCosh}(\mathbf{v}_i, \mathbf{u}_k)/\tau)}. \tag{6}$$

Same as ConVIRT [18], the final loss function given $\lambda \in [0, 1]$ is

$$\mathcal{L} = \frac{1}{N} \sum_{i=1}^{N} \left(\lambda \tilde{\ell}_i^{(u \to v)} + (1 - \lambda) \tilde{\ell}_i^{(v \to u)} \right). \tag{7}$$

2.4 Sub-feature Comparison

Because the similarity metric (5) compares two feature vectors element-wise instead of the angle between the two vectors, we can compare a random subset of the elements in the two features vector to reduce the feature suppression problem further. Inspired by Dropout [17], we can randomly set some element of a feature vector to zero so that we have a sub-feature vector. For a feature vector $\mathbf{v}$ and an index vector $\mathbf{l} = (l_1, l_2, ..., l_k)$ where k is the size of $\mathbf{v}$ and i_j is a sample from a Bernoulli distribution with probability p, a sub-feature $\mathbf{v}_{\text{sub}}$ given $\mathbf{l}$ is defined as:

$$\mathbf{v}_{\text{sub}(\mathbf{l})} = \mathbf{v} \odot \mathbf{l} , \tag{8}$$

where $\odot$ is the element-wise multiplication. Sub-features contain many zero entries determined by $\mathbf{l}$. Replacing $\mathbf{v}$ with $\mathbf{v}_{\text{sub}(\mathbf{l})}$ in the metric (5) produces

$$\text{NegLogCosh}(\mathbf{v}_{\text{sub}(\mathbf{l})}, \mathbf{u}_{\text{sub}(\mathbf{l})}) = -\frac{1}{n} \sum_{i=1}^{n} \log(\cosh(sl_i(v_i - u_i))) , \tag{9}$$

where the index vector $\mathbf{l}$ is shared within the same mini-batch. Sharing the index vector is the main difference between this sub-feature approach and Dropout. Based on inequality (1), once the loss function (7) is minimized, we have:

$$\text{NegLogCosh}(\mathbf{v}_{\text{sub}(\mathbf{l})}, \mathbf{u}_{\text{sub}(\mathbf{l})}) > \text{NegLogCosh}(\mathbf{v}_{\text{sub}(\mathbf{l})}, \mathbf{w}_{\text{sub}(\mathbf{l})}) , \tag{10}$$

for any $\mathbf{v}, \mathbf{u}$ pair and $\mathbf{w}$ that comes from other pairs and for any $\mathbf{l}$. If we construct the corresponding index vector, $1 - \mathbf{l} = (1 - l_1, l - l_2, ..., 1 - l_k)$, which satisfies (10), we obtain

$$\begin{aligned}
&\text{NegLogCosh}(\mathbf{v}, \mathbf{u}) \\
&= \text{NegLogCosh}(\mathbf{v}_{\text{sub}(\mathbf{l})}, \mathbf{u}_{\text{sub}(\mathbf{l})}) + \text{NegLogCosh}(\mathbf{v}_{\text{sub}(1-i)}, \mathbf{u}_{\text{sub}(1-\mathbf{l})}) \\
&> \text{NegLogCosh}(\mathbf{v}_{\text{sub}(\mathbf{l})}, \mathbf{w}_{\text{sub}(\mathbf{l})}) + \text{NegLogCosh}(\mathbf{v}_{\text{sub}(1-i)}, \mathbf{w}_{\text{sub}(1-\mathbf{l})}) \\
&= \text{NegLogCosh}(\mathbf{v}, \mathbf{w})
\end{aligned} \tag{11}$$

from equation (9). Thus, achieving inequality (10) implies achieving inequality (11). However, achieving inequality (11) does not imply inequality (10) because we can select a sub-feature to flip the inequality, and the loss function does not rule out such a possibility. This one-way implication shows that sub-feature comparison enables a VLC model to learn more image and text relationships in the feature space than a traditional approach. In other words, instead of just learning the features presented in the text, we have a chance to learn a more general feature representation. This advantage should both help alleviate the feature suppression problem and reduce over-fitting.

3   Dataset

The primary dataset was collected using a professional photography instrument in the pathology department at the Northwestern Memorial Hospital (Chicago) between 2014 and 2018. After filtering out blurry images and images with sliced placenta, we were left with 13,004 fetal side placenta images and pathology report pairs. We selected 2,811 images from 2017 for fine-tuning and the rest for pre-training.

The final pre-training data set consists of 10,193 image-and-text pairs. Each image contains the fetal side of a placenta, the cord, and a ruler. Each text sequence for the image contains a part of the corresponding pathology report.

The fine-tuning dataset consists of 2,811 images; we first manually checked the images to ensure the placenta is complete and free from obscures. We labeled each image based on the pathology report on four tasks presented in [4,19]: *meconium, fetal inflammatory response* (FIR), *maternal inflammatory response* (MIR), and *chorioamnionitis.* There are different levels or stages for each symptom in the pathology report. We labeled the images as positive for meconium and chorioamnionitis regardless of the level. For FIR and MIR, we labeled the image as negative if the report does not contain any related information or identified the placenta as negative; we labeled the image as positive if the report identifies the placenta as stage 2 or higher; we dropped the image if the stage is higher than 0 but lower than 2 to improve the model's ability to distinguish significant cases. To assess the generalizability, we also labeled 166 images with neonatal sepsis based on the results that are diagnosed by treating physicians using clinical criteria on the infant charts. We then used all the positive examples for each task and uniformly sampled a similar number of negative samples. We then randomly split the data into training, validation, and testing sets with the ratio of 0.25:0.25:0.5 since we do not have the exact test set as in [4,19]. We selected more images for testing to reduce the randomness in the testing result.

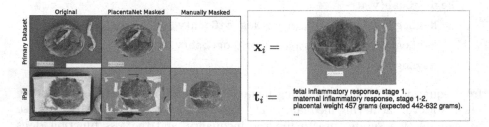

Fig. 2. Left: Example images from the two datasets. There is no manually generated background mask for the primary dataset. The image from the primary dataset and the iPad images have different white balance (see ruler color) and different backgrounds. **Right:** An example input image-and-text pair used in the pre-training. **Best viewed in color.**

To understand the robustness of the proposed method, we collected 52 placenta images at the same hospital in the summer of 2021 using an iPad (2021 model). The placentae were placed on a surgical towel and wiped clean of excess blood, the lighting was adjusted to minimize glare, and the iPad was held near parallel to the bench surface. As shown in Fig. 2, the performance in semantic segmentation from PlacentaNet is sub-optimal, and the white balance is different for the iPad image. We obtained labels for MIR and FIR from microscopical diagnoses [10] by an expert perinatal pathologist and clinical chorioamnionitis from the infant charts. Note that the clinical chorioamnionitis is different from the histologic chorioamnionitis in the primary dataset. Since all data have FIR lower than stage 2, we discarded this label. For the rest of the tasks, we labeled images using the same criteria as the primary dataset. We acknowledge that this dataset is too small to serve as a benchmark, thus we considered a method outperforming others when the difference is significant (*e.g.*, by a few percentage points or higher). We used the iPad images to test the robustness; the main evaluation dataset is the primary dataset. A table containing the detailed breakdown of the data is in the supplementary material.

4   Experiments

4.1   Training and Testing

We used the ResNet50 [9] as our image encoder architecture and a pre-trained BERT [6] as our text encoder. The projection layers for both encoders were one-layer fully-connected neural networks (FC) with no activation. The classifiers were all two-layer FC with ReLU activation in the first layer but no activation in the output layer. For image preprocessing, we masked out the background from each image using PlacentaNet [3] and applied random augmentations. We randomly sampled topics in the text strings with replacements for the text pre-processing. We applied the Adam optimizer [11] and cosine decay learning rate scheduler with warm-up [13]. We selected the hyper-parameters on the baselines and applied them to our method. The details are in the supplementary material.

One independent image encoder was jointly trained with a classifier for each task for the baseline ResNet50. We trained the model on the training set for 100 epochs and saved the model with the highest validation accuracy.

For VLC models, we used the ConVIRT method as the baseline. We changed the projection layer from two-layer FC with ReLU activation to one-layer FC with no activation but kept the essential methodology the same. We trained the model for 400 epochs and saved the encoder in the last epoch. The training procedure for each downstream task was the same as for the baseline ResNet50, but the pre-trained encoder was frozen.

Our proposed method used the NegLogCosh similarity with the sub-feature comparison technique instead of cosine similarity. The training procedure followed the baseline ConVIRT.

We used the same testing procedure for all methods; we used the same pre-possessing steps for all images in the primary dataset but two methods to mask out the background on the iPad images. The first method uses the segmentation map from PlacentaNet, which is sub-optimal (see Fig. 2) due to the difference in image quality. We included manually labeled segmentation maps as the second method to address this issue. In practice, this issue can be minimized by [19].

4.2   Results and Discussion

The mean results and confidence intervals (CIs) for the five experiments on the primary dataset are shown in Table 1. Since we do not have the exact model architectures for all the experiments in [4,19], we are reporting the widely adopted

Table 1. AUC-ROC scores (in %) for placenta analysis tasks. **Top:** The mean and 95% CI of five random splits. The highest means are in bold. **Bottom:** The estimated mean improvements over the baseline ResNet50 and 95% CI using 100 bootstrap samples on the five random splits. The statistically significant improvements (CIs above 0) are underlined.

	Primary					iPad			
	Mecon.	FIR	MIR	H. Chorio.	Sepsis	PlacentaNet		Manual	
						MIR	C. Chorio	MIR	C. Chorio
RN50	77.0±2.9	74.2±3.3	68.5±3.4	67.4±2.7	88.4±2.0	46.7±20.9	42.8±14.0	50.8±21.6	47.0±16.7
ConVIRT	77.5±2.7	76.5±2.6	69.2±2.8	68.0±2.5	89.2±3.6	53.0±8.0	42.4±4.8	52.5±25.7	50.7±6.6
Ours	**79.4±1.3**	**77.4±3.4**	**70.3±4.0**	**68.9±5.0**	**89.8±2.8**	**58.4±7.2**	**45.4±2.7**	**61.9±14.4**	**53.6±4.2**
ConVIRT	0.6±1.8	<u>2.2±1.9</u>	0.8±1.3	0.6±2.3	0.8±1.4	6.4±6.7	-0.7±8.0	1.8±7.5	3.4±10.0
Ours	<u>2.5±1.3</u>	<u>3.1±1.9</u>	<u>1.8±1.3</u>	1.5±2.3	<u>1.4±1.3</u>	<u>11.8±6.5</u>	2.4±7.2	<u>11.3±7.8</u>	6.4±7.7

ResNet50 as the baseline. Our method achieved the highest area under ROC [7] (AUC-ROC) for all tasks in the experiment. Many of the improvements are statistically significant as the 95% CIs estimated using bootstrap samples do not contain 0. Although ConVIRT also outperformed the baseline on all the tasks in the pre-training data, only the improvement on FIR is significant. Moreover, the text features directly correspond to sepsis, which is not in the pre-training tasks, could be suppressed by other features. ConVIRT did not outperform the baseline without addressing the feature suppression problem, even with additional pre-training data. In contrast, our method showed significant improvement.

Additionally, our method also achieved the best when testing on the iPad images regardless of the segmentation map generation method, as shown in Table 1. The performance on clinical chorioamnionitis was much lower for the iPad images because we trained the model using histologic chorioamnionitis. As expected, the manually labeled segmentation maps resulted in higher AUC-ROC scores. Moreover, the proposed method always has smaller CIs, which also confirms the improvement in robustness. Although a larger iPad dataset would be necessary for confirming the improved performance of our approach on the mobile platform, the better robustness of our approach is apparent.

The qualitative examples are in the supplementary materials. Those examples show that all the experimented methods are sensitive to placenta color. We need more control over the lighting when collecting placenta images or better prepossessing to balance the placenta color for better performance.

Moreover, the shared encoder makes our method more efficient than the previous approach as the number of tasks grows.

5   Conclusions and Future Work

We proposed a robust, generalizable, and efficient framework for automatic placenta analysis. We showed that our pre-training method outperformed the popular approaches in almost all the placenta analysis tasks in the experiments. Our approach's robustness on photos taken with an iPad has high clinical value to low-resource communities. We expect our approach to perform better if we have a better image encoder, more data, or a domain-specific text encoder.

In the future, it would be interesting to extend our approach to a zero-shot setting [14] to further reduce the computation cost. More qualitative analysis can be performed to understand the improvement better. Lastly, we can collect a larger clinical dataset to improve the accuracy and robustness.

References

1. Chen, T., Kornblith, S., Norouzi, M., Hinton, G.: A simple framework for contrastive learning of visual representations. In: Proceedings of the International Conference on Machine Learning, pp. 1597–1607. PMLR (2020)
2. Chen, T., Luo, C., Li, L.: Intriguing properties of contrastive losses. In: Advances in Neural Information Processing Systems, vol. 34 (2021)
3. Chen, Y., Wu, C., Zhang, Z., Goldstein, J.A., Gernand, A.D., Wang, J.Z.: PlacentaNet: automatic morphological characterization of placenta photos with deep learning. In: Shen, D., et al. (eds.) MICCAI 2019. LNCS, vol. 11764, pp. 487–495. Springer, Cham (2019). https://doi.org/10.1007/978-3-030-32239-7_54
4. Chen, Y., et al.: AI-PLAX: AI-based placental assessment and examination using photos. Comput. Med. Imaging Graph. **84**(101744), 1–15 (2020)
5. Denize, J., Rabarisoa, J., Orcesi, A., Hérault, R., Canu, S.: Similarity contrastive estimation for self-supervised soft contrastive learning. arXiv preprint arXiv:2111.14585 (2021)
6. Devlin, J., Chang, M.W., Lee, K., Toutanova, K.: BERT: pre-training of deep bidirectional transformers for language understanding. arXiv preprint arXiv:1810.04805 (2018)
7. Fawcett, T.: An introduction to ROC analysis. Pattern Recogn. Lett. **27**(8), 861–874 (2006)
8. He, K., Fan, H., Wu, Y., Xie, S., Girshick, R.: Momentum contrast for unsupervised visual representation learning. In: Proceedings of the IEEE/CVF Conference on Computer Vision and Pattern Recognition, pp. 9729–9738 (2020)
9. He, K., Zhang, X., Ren, S., Sun, J.: Deep residual learning for image recognition. In: Proceedings of the IEEE Conference on Computer Vision and Pattern Recognition, pp. 770–778 (2016)
10. Khong, T.Y., et al.: Sampling and definitions of placental lesions: Amsterdam placental workshop group consensus statement. Archi. Pathol. Lab. Med. **140**(7), 698–713 (2016)
11. Kingma, D.P., Ba, J.: Adam: a method for stochastic optimization. arXiv preprint arXiv:1412.6980 (2014)
12. Li, T., et al.: Addressing feature suppression in unsupervised visual representations. arXiv preprint arXiv:2012.09962 (2020)
13. Loshchilov, I., Hutter, F.: SGDR: stochastic gradient descent with warm restarts. arXiv preprint arXiv:1608.03983 (2016)
14. Radford, A., et al.: Learning transferable visual models from natural language supervision. In: International Conference on Machine Learning, pp. 8748–8763. PMLR (2021)
15. Rezaei, M., Soleymani, F., Bischl, B., Azizi, S.: Deep bregman divergence for contrastive learning of visual representations. arXiv preprint arXiv:2109.07455 (2021)

16. Robinson, J., Sun, L., Yu, K., Batmanghelich, K., Jegelka, S., Sra, S.: Can contrastive learning avoid shortcut solutions? arXiv preprint arXiv:2106.11230 (2021)
17. Srivastava, N., Hinton, G., Krizhevsky, A., Sutskever, I., Salakhutdinov, R.: Dropout: a simple way to prevent neural networks from overfitting. J. Mach. Learn. Res. **15**(56), 1929–1958 (2014)
18. Zhang, Y., Jiang, H., Miura, Y., Manning, C.D., Langlotz, C.P.: Contrastive learning of medical visual representations from paired images and text. arXiv preprint arXiv:2010.00747 (2020)
19. Zhang, Z., Davaasuren, D., Wu, C., Goldstein, J.A., Gernand, A.D., Wang, J.Z.: Multi-region saliency-aware learning for cross-domain placenta image segmentation. Pattern Recogn. Lett. **140**, 165–171 (2020)

Multi-modal Hypergraph Diffusion Network with Dual Prior for Alzheimer Classification

Angelica I. Aviles-Rivero[1]([✉]), Christina Runkel[1], Nicolas Papadakis[2],
Zoe Kourtzi[3], and Carola-Bibiane Schönlieb[1]

[1] DAMTP, University of Cambridge, Cambridge, UK
{ai323,cr661,cbs31}@cam.ac.uk
[2] IMB, Universite de Bordeaux, Bordeaux, France
nicolas.papadakis@math.u-bordeaux.fr
[3] Department of Psychology, University of Cambridge, Cambridge, UK
zk240@cam.ac.uk

Abstract. The automatic early diagnosis of prodromal stages of Alzheimer's disease is of great relevance for patient treatment to improve quality of life. We address this problem as a multi-modal classification task. Multi-modal data provides richer and complementary information. However, existing techniques only consider lower order relations between the data and single/multi-modal imaging data. In this work, we introduce a novel semi-supervised hypergraph learning framework for Alzheimer's disease diagnosis. Our framework allows for higher-order relations among multi-modal imaging and non-imaging data whilst requiring a tiny labelled set. Firstly, we introduce a dual embedding strategy for constructing a robust hypergraph that preserves the data semantics. We achieve this by enforcing perturbation invariance at the image and graph levels using a contrastive based mechanism. Secondly, we present a dynamically adjusted hypergraph diffusion model, via a semi-explicit flow, to improve the predictive uncertainty. We demonstrate, through our experiments, that our framework is able to outperform current techniques for Alzheimer's disease diagnosis.

Keywords: Hypergraph learning · Multi-modal classification · Semi-supervised learning · Perturbation invariance · Alzheimer's disease

1 Introduction

Alzheimer's disease (AD) is an irreversible, neurodegenerative disease impairing memory, language and cognition [9]. It starts slowly but progressively worsens

---

C.-B. Schönlieb—The Alzheimer's Disease Neuroimaging Initiative.

Supplementary Information The online version contains supplementary material available at https://doi.org/10.1007/978-3-031-16437-8_69.

© The Author(s), under exclusive license to Springer Nature Switzerland AG 2022
L. Wang et al. (Eds.): MICCAI 2022, LNCS 13433, pp. 717–727, 2022.
https://doi.org/10.1007/978-3-031-16437-8_69

and causes approximately 60–80% of all cases of dementia [4]. As there is no cure available yet, detecting the disease as early as possible, in prodromal stages, is crucial for slowing down its progression and for patient to improve quality of life. The body of literature has shown the potentials of existing machine learning methods e.g., [20,25,27]. However, there are two major limitations of existing techniques. Firstly, incorporating other relevant data types such as phenotypic data in combination with large-scale imaging data has shown to be beneficial e.g. [19]. However, existing approaches fail to exploit the rich available imaging and non-imaging data. This is mainly due to the inherent problem of how to provide better and higher relations between multi-modal sources. Secondly, whilst vast amount of data is acquired every day at hospitals, labelling is expensive, time-consuming and prone to human bias. Therefore, to develop models that rely on extreme minimal supervision is of a great interest in the medical domain.

To address the aforementioned problems, hypergraph learning has been explored – as it allows going beyond pair-wise data relations. In particular, hypergraphs have already been explored for the task of AD diagnosis e.g. [18,23,24,30]. From the machine learning perspective [6], several works have addressed the problem of hypergraph learning, by generalising the graph Laplacian to hypergraphs e.g. [15,22,29]. The principles from such techniques opened the door to extend the hypergraph Laplacian to graph neural networks (GNNs) e.g. [11,26]. However, the commonality of existing hypergraph techniques for AD diagnosis, and in general in the ML community, is that they consider the clique/star expansion [2], and in particular, follow the hypergraph normalised cut of that [29].

To tackle the aforementioned challenges, we introduce a novel semi-supervised hypergraph learning framework for Alzheimer's disease diagnosis. Our work follows a hybrid perspective, where we propose a new technique based on a dual embedding strategy and a diffusion model. To the best of our knowledge, this is the first work that explores invariance at the image and graph levels, and goes beyond the go-to technique of [29] by introducing a better hypergraph functional.

Our Contributions are as Follows. 1) We introduce a self-supervised dual multi-modal embedding strategy. The framework enforces invariance on two spaces– the manifold that lies the imaging data and the space of the hypergraph structure. Our dual strategy provides better priors on the data distribution, and therefore, we construct a robust graph that offers high generalisation capabilities. 2) In contrast to existing techniques that follow [29], we introduce a more robust diffusion-model. Our model is based on the Rayleigh quotient for hypegraph p-Laplacian and follows a semi-explicit flow.

2   Proposed Framework

This section introduces our semi-supervised hypergraph learning framework (see Fig. 1) highlighting two key parts: i) our dual embedding strategy to construct a robust graph, and ii) our dynamically adaptive hypergraph diffusion model.

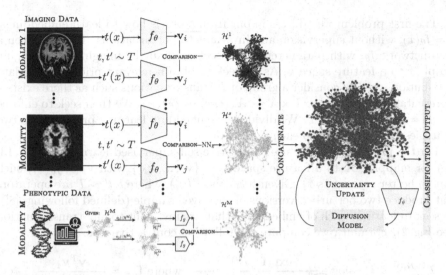

Fig. 1. Visual illustration of our framework. We first use a dual embeddings strategy at the image and hypergraph level. We then introduce a diffusion model for predicting early diagnosis of prodromal stages of Alzheimer's disease.

2.1 Hypergraph Embeddings and Construction

The first part of our framework addresses a major challenge in hypergraph learning – that is, how to extract meaningful embeddings, from the different given modalities, to construct a robust hypergraph. The most recent works for Alzheimer's disease diagnosis are based on extracting embeddings using sparse constraints or directly from a deep network. However, existing works only consider data level features to construct a graph or hypergraph and mainly for imaging data. In contrast to existing works, we allow for higher-order relations between multi-modal imaging and non-imaging data. Secondly, we enforce a dual perturbation invariance strategy at both image and graph levels. With our dual multi-modal embedding strategy, *we seek to provide better priors on the data distribution such as our model is invariant to perturbations, and therefore, we can construct a robust graph that offers high generalisation capabilities.*

We consider a hypergraph as a tuple $\mathcal{G} = (\mathcal{V}, \mathcal{E}, w)$, where $\mathcal{V} = \{v_1, ..., v_n\}$, $|\mathcal{V}| = n$ is a set of nodes and $\mathcal{E} = \{e_1, ..., e_m\}$, $|\mathcal{E}| = m$ the hyperedges. Moreover, $w : \mathcal{E} \to \mathbb{R}_{>0}$ refers to the hyperedges weights, in which each hyperedge e is associated to a subset of nodes. The associated incidence matrix is given by $\mathcal{H}_{i,e} = \begin{cases} 1 & \text{if } i \in e, \\ 0 & \text{otherwise} \end{cases}$, where $i \in \mathcal{V}$ and e is an hyperedge; *i.e.* a subset of nodes of $\mathcal{V}$. In our multi-modal setting, we assume a given $\mathbf{M}$ modalities. We have for each modality N samples $X = \{x_1, ..., x_N\} \in \mathcal{X}$ sampled from a probability distribution $\mathbb{P}$ on $\mathcal{X}$. We then have a multi-modal data collection as $\{X_1, ..., X_s, X_{s+1}, ..., X_{\mathbf{M}}\}$. The first s modalities are imaging data (e.g. PET imaging) whilst the remaining are non-imaging data (e.g. genetics).

The first problem we address in our framework is how to learn embeddings, $\mathbf{v} = f_\theta(x)$, without supervision and invariant to perturbations. That is, given a deep network f_θ, with parameters θ, we seek to map the imaging/non-imaging sample x to a feature space $\mathbf{v}$. As we seek to obtain relevant priors on the data distribution, we also consider a group of T transformations such as there exists a representation function $\phi : T \times \mathcal{X} \to \mathcal{X}, (t, x) \mapsto \phi(t, x)$. We then seek to enforce $\phi(t, x) = \phi(x)$ for all $t \in T$. We divide our embedding learning problem into two strategies corresponding to each type of data.

For the imaging data, we use contrastive self-supervised learning (e.g. [7,13, 14]) for mapping X to a feature space $\mathbf{v} = \{\mathbf{v}_1, ..., \mathbf{v}_N\}$ with $\mathbf{v}_i = f_\theta(x_i)$ such that $\mathbf{v}_i$ better represents x_i. Given $t(x_i)$ and $t'(x_i)$, where $t, t' \sim T$ are operators that produce two perturbed versions of the given sample (defined following [8]), we seek to learn a batch of embeddings that are invariant to any transformation (see Fig. 2). Formally, we compute the following contrastive loss:

$$\mathcal{L}_{visual}^{(i,j),X_i \in s} = -\log \frac{\exp(\mathbf{f}_{i,j}/\tau)}{\sum_{k=1}^{n} \mathbb{1}_{[k \neq i]} \exp(\mathbf{f}_{i,k}/\tau)}, \text{ where } \mathbf{f}_{i,j} = \frac{\mathbf{v}_i^\mathsf{T} \mathbf{v}_j}{||\mathbf{v}_i|| \cdot ||\mathbf{v}_j||}, \quad (1)$$

where $\tau > 0$ is a temperature hyperparameter and $\mathbf{f}_{i,k}$ follows same cosine similarity definition than $\mathbf{f}_{i,j}$. We denote the k-nearest neighbors of an embedding $\mathbf{v}_i$ as $\mathrm{NN}_k(\mathbf{v}_i)$. We then construct for each $X_1, ..., X_s$ a corresponding hypergraph $\mathcal{H}_{ij}^{1,\cdots,s} = [\mathbf{v}_i^\mathsf{T} \mathbf{v}_j]$ if $\mathbf{v}_i \in \mathrm{NN}_k(\mathbf{v}_j)$, otherwise 0.

For the non-imaging data (e.g. genetics and age), we follow different protocol as perturbing directly the data might neglect the data/structure semantics. We then seek to create a subject-phenotypic relation. To do this, we compute the similarity between the subjects $\mathbf{x}$ and the corresponding phenotypic measures, to generate the hypergraphs $\mathcal{H}^{s+1,\cdots,\mathbf{M}}$. Given a set of phenotypic measures, we compute a NN_k graph between subjects given the set of measures $\mathbf{z}$ such as $S(\mathbf{x}, \mathbf{z})$ if $\mathbf{x} \in \mathrm{NN}_k(\mathbf{z})$, otherwise 0; being S a similarity function. That is, we

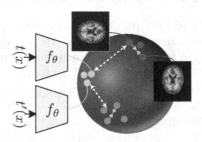

Fig. 2. We seek for distinctiveness. Related samples should have similar features.

enforce the connection of subjects within similar phenotypic measures. We then seek to perturb $\mathcal{H}^{s+1,\cdots,\mathbf{M}}$ such that the transformed versions $\hat{\mathcal{H}}^{s+1,\cdots,\mathbf{M}} \sim \mathrm{q}(\hat{\mathcal{H}}^{s+1,\cdots,\mathbf{M}}|\mathcal{H}^{s+1,\cdots,\mathbf{M}})$ where $\mathrm{q}(\cdot)$ refers to the transformation drawn from the distribution of the given measure. Our set of transformations are node dropping and edge perturbation. We follow, for the group of transformations T, the ratio and dropping probability strategies of [21,28]. To maximise the agreement between the computed transformation, we use the same contrastive loss as defined in (1). We denote the loss as $\mathcal{L}_{pheno}^{(i,j),\mathcal{H}^{s+1,\mathbf{M}}}$ for non-imaging data. The final hypergraph, $\mathcal{H}$, is the result of concatenating all hypergraph structures from all modalities, $\mathcal{H}^1, ..., \mathcal{H}^\mathbf{M}$ given by both imaging and non-imaging data.

2.2   Dynamically Adjusted Hypergraph Diffusion Model

The second key part of our framework is a semi-supervised hypergraph diffusion model. After constructing a robust hypergraph, we now detail how we perform disease classification using only a tiny labelled set.

We consider a small amount of labelled data $X_L = \{(x_i, y_i)\}_{i=1}^l$ with provided labels $\mathcal{L} = \{1, .., L\}$ and $y_i \in \mathcal{L}$ along with a large unlabelled set $X_u = \{x_k\}_{k=l+1}^m$. The entire dataset is then $X = X_L \cup X_U$. The goal is to use the provided labeled data X_L to infer a function, $f : \mathcal{X} \mapsto \mathcal{Y}$, that maps the unlabelled set to class labels. However, to obtain such mapping efficiently and with minimum generalisation error is even more challenging when using multi-modal data e.g. [19]. We cast this problem as a label diffusion process on hypergraphs.

We consider the problem of hypergraph regularisation for semi-supervised classification [11,15,29], where we aim to find a function u^* to infer labels for the unlabelled set and enforce smoothness on the hypergraph. We seek to solve a problem of the form $u^* = \text{argmin}_{u \in \mathbb{R}^{|\mathcal{V}|}} \{\mathcal{L}_{emp}(u) + \gamma \Omega(u)\}$, where u $\mathcal{L}_{emp}(u)$ refers to an empirical loss (e.g., square loss), $\Omega(u)$ denotes a regulariser and $\gamma > 0$ is a weighting parameter balancing the importance of each term. Whilst majority of works consider the clique/star expansion [2] and follow the principles of [29], *we seek to provide a more robust functional* to avoid the inherent disadvantages of such approximations (e.g. bias during the graph diffusion [15]). In particular, we consider the setting of Rayleigh quotient for p−Laplacian. We then seek to estimate, for $p = 1$, the particular family of solutions based on the minimisation of the ratio $\Omega(u) = \frac{\text{TV}_H(u)}{\|u\|}$, where $\text{TV}_H = \sum_{e \in \mathcal{E}} w_e \max_{i,j \in e} |u_i - u_j|$ is the total variation functional on the hypergraph. To do this, we generalise [10] to a hypergraph setting and introduce a dynamic adjustment on the diffusion model, which is based on controlling the predictive uncertainty. Our framework allows for both binary and multi-class classification settings. Given a number of epochs $E \in [1, ..., E]$ and following previous notation, we compute an alternating optimisation process as follows.

Alternating Optimisation for Epoch$\in [1, .., E]$:

Sub-Problem 1: Diffusion Model Minimisation. For a given label, we define a function $u \in \mathbb{R}^n$ over the nodes, and denote its value at node x as $u(x)$. In this binary setting, the objective is to estimate a function u that denotes the presence (resp. absence) of the label for data x. To that end, following [10], we seek to solve $\min \frac{\text{TV}_H(u)}{\|u\|}$ with the following semi-explicit flow, in which u_k is the u value at iteration k:

$$\text{Diffusion Model} \begin{cases} \frac{u_{k+1/2} - u_k}{\delta t} = \frac{\text{TV}_H(u_k)}{\|u_k\|}(q_k - \tilde{q}_k) - \phi_{k+1/2}, \\ u_{k+1} = \frac{u_{k+1/2}}{\|u_{k+1/2}\|_2} \end{cases} \quad (2)$$

where $\phi_{k+1/2} \in \partial \text{TV}_H(u_{k+1/2})$, $q \in \partial \|u_k\|$ (with ∂f the set of possible subdifferentials of a convex function f defined as $\partial f = \{\phi, \text{ s.t. } \exists u, \text{ with } \phi \in \partial f(u)\}$), the scaling $d(x)$ is such that $\tilde{q}_k = \frac{\langle d, q_k \rangle}{\langle d, d \rangle} d$, and δt is a positive time step. Once the PDE has converged, the output of (2) is a bivalued function that can be threshold to perform a binary partition of the hypergraph. In order to realise a multi-class labelling, we consider the generalised model of [5] that includes as prior the available labels of X_L in $\mathcal{L}_{emp}(u)$ to guide the partitioning and estimates L coupled functions $u^L(x)$ $L = 1, \ldots, L$. The final labeling is computed as $\hat{y}_i = \text{argmax}_{i \in \{1, \cdots L\}} u^L(x)$ with $L(x) \in \hat{Y}_i$. *Intuition:* Having our constructed hypergraph as in Subsect. 2.1, we seek to use a tiny labelled set, as we are in a semi-supervised setting, to diffuse the labels for the unlabelled set. A major challenge in semi-supervised learning is how to reduce the prediction bias (e.g. [3]) for the unlabelled set, as it is conditioned solely to the tiny labelled set as prior. We then seek to update the uncertainty of the computed $\hat{y}_i$ via solving Sub-Problem 2.

SUB-PROBLEM 2: UNCERTAINTY HYPERGRAPH MINIMISATION. Following the scheme from Fig. 1 and notation from previous sections, we first initialise f_ψ using the tiny labelled set available, X_L, and then we take $\hat{y}_i$ from Sub-Problem 1 and check the uncertainty of $\hat{y}_i$ by computing the following loss:

$$\mathcal{L}_{DYN}(X, Y, \hat{Y}; \theta) := \min_\theta \sum_{i=1}^{l} \mathcal{L}_{CE}(f_\theta(x_i), y_i) + \sum_{i=l+1}^{m} \gamma_i \mathcal{L}_{CE}(f_\theta(x_i), \hat{y}_i), \quad (3)$$

where $\mathcal{L}_{CE}$ is a cross entropy loss and γ is the measure of uncertainty via entropy [1,17] defined as $\gamma_i = 1 - (H(u_{\hat{y}_i}) / \log(L))$, where H refers to the entropy and $u_{\hat{y}_i}$ is normalised beforehand. *Intuition:* From previous step, we obtain an initial prediction for the unlabelled set on the hypergraph. However, in semi-supervised learning due to the tiny labelled set as prior, the unlabelled data tend to have an asymptotic bias in the prediction. We then ensure that there is a high certainty in the unlabelled predictions since early epochs to avoid the propagation of incorrect predictions.

3   Experimental Results

In this section, we provide all details of our evaluation protocol.

Data Description. We evaluate our semi-supervised hypergraph framework using the Alzheimer's disease Neuroimaging Initiative (ADNI) dataset[1]. ADNI is a multi-centre dataset composed of multi-modal data including imaging and multiple phenotype data. We consider 500 patients using MRI, PET, demographics and Apolipoprotein E (APOE). We included APOE as it is known to be a crucial genetic risk factor for developing Alzheimer's disease. The dataset contains four categories: early mild cognitive impairment (EMCI), late mild cognitive impairment (LMCI), normal control (NC) and Alzheimer's disease (AD).

Evaluation Protocol. We design a three part evaluation scheme. Firstly, we follow the majority of techniques convention for binary classification comparing

---

[1] Data used were obtained from the Alzheimer's Disease Neuroimaging Initiative (ADNI) database (adni.loni.usc.edu).

Table 1. Numerical comparison of our technique and existing (graph) and hypergraph techniques. All comparison are run on same conditions. The best results are highlighted in green colour.

Technique	AD vs NC			EMCI vs LMCI		
	ACC	SEN	PPV	ACC	SEN	PPV
GNNs [19]	81.60±2.81	83.20±3.10	80.62±2.30	75.60±2.50	75.20±3.02	75.80±2.45
HF [23]	87.20±2.10	88.01±2.15	86.60±2.60	80.40±2.02	82.41±2.14	79.23±2.60
HGSCCA [24]	85.60±2.16	87.20±3.11	84.40±2.15	76.01±2.16	75.21±2.01	76.42±2.22
HGNN [11]	88.01±2.60	90.40±2.16	87.59±2.42	80.60±2.05	81.60±2.54	79.60±2.51
DHGNN [16]	89.90±2.40	89.60±2.15	90.21±2.45	80.80±2.47	82.40±2.41	79.80±2.76
Ours	92.11±2.03	92.80±2.16	91.33±2.43	85.22±2.25	86.40±2.11	84.02±2.45

Table 2. Performance comparison of our technique and existing hypergraph models for LMCI vs NC. The results in green colour denotes the highest performance.

Technique	LMCI vs NC		
	ACC	SEN	PPV
GNNs [19]	72.40±2.05	70.40±2.80	73.30±2.04
HF [23]	77.01±2.26	77.6±2.15	78.22±2.51
HGSCCA [24]	74.00±2.10	74.40±2.16	73.80±2.17
HGNN [11]	78.90±3.01	80.01±2.7	78.10±2.67
DHGNN [16]	79.20±2.70	80.03±3.01	78.74±3.22
Ours	82.01±2.16	84.01±2.34	81.80±2.55

AD vs NC, AD vs EMCI, AD vs LMCI, EMCI vs NC, LMCI vs NC and EMCI vs LMCI (see Supplementary Material for extended results). Secondly, we extended the classification problem to a multi-class setting including the four classes AD vs NC vs EMCI vs LMCI. We consider this setting, as one of the major challenges in AD diagnosis is to fully automate the task without pre-selecting classes [12].

To evaluate our model, we performed comparisons with state-of-the-art techniques on hypergraph learning: HF [23], HGSCCA [24], HGNN [11] and DHGNN [16]. We also added the comparison against GNNs [19], which is based on graph neural networks (GNNs). We added this comparison as it also considers imaging and non-imaging data. For a fair comparison in performance, we ran those techniques under same conditions. The quality check is performed following standard convention in the medical domain: accuracy (ACC), sensitivity (SEN), and positive predictive value (PPV) as a trade-off

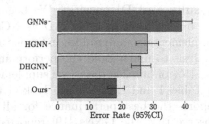

Fig. 3. Performance comparison of ours and SOTA techniques for the four classes case.

Table 3. Error rate comparison of our technique against existing models. The results in green denotes the best performance.

TECHNIQUE	AD vs NC vs MCI ERROR RATE & 95%CI	AD vs NC vs EMCI vs LMCI ERROR RATE & 95%CI
GNNs [19]	36.19±4.45	39.01±3.12
HGNN [11]	26.35±3.20	28.09±3.65
DHGNN [16]	23.10±2.60	26.25±2.55
Ours	16.25±2.22	18.31±2.45

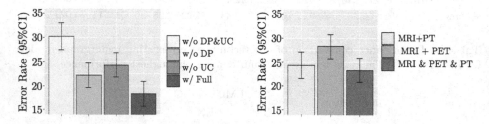

Fig. 4. (left side) Ablation study for the components of our technique. (right side) Performance comparison using different type of multi-modal data.

metric between sensitivity and specificity. Moreover and guided by the field of estimation statistics, we report along with the error rate the confidence intervals (95%) when reporting multi-class results. We set the k-NN neighborhood to $k = 50$, for the alternating optimisation we used a weight decay of 2×10^{-4} and learning rate was set to 5e-2 decreasing with cosine annealing, and use 180 epochs. For the group of transformations T, we use the strategy of that [8] for the imaging data whilst for the non-imaging data the ratio and dropping probability strategies of [21,28]. Following standard protocol in semi-supervised learning, we randomly select the labelled samples over five repeated time and then report the mean of the metrics along with the standard deviation.

Results and Discussion. We began by evaluating our framework following the binary comparison of AD vs NC, EMCI vs LMCI and LMCI vs NC (see supplementary material for extended results). We report a detailed quantitative analysis to understand the performance of ours and compared techniques. We use 15% of labelled data (see supplementary material). The results are reported in Tables 1 and 2. In a closer look at the results, we observe that our technique reports the best performance for all comparison and in all metrics. We also observe that the GNNs [19] reported the worst performance. This is due to the fact that GNNs does not allow for higher-order relations on the graph. We highlight that whilst the other techniques reported a good performance, our technique substantially improved over all methods (with statistical significance, see supplementary material). We underline that all compared techniques follows the approximation of [29], which introduces bias during the hypergraph partition.

We showed that our diffusion model, that follows different principles, mitigates that problem.

To further support our previous results, we also report results in the more challenging setting of multi-class classification for AD diagnosis. The results are reported in Table 3. We observe that our technique performance is consistent with previous results. That is, our technique reported the lowest error rate for both multi-class cases. Our technique decreased the error rate for more than 40% for all techniques. Figure 3 complements these results by reporting the performance with respect to the four classes case.

Finally, to support the design of our technique, we performed two ablations studies regarding our design and the modalities used. We start by evaluating the influence of our full model against without our dual perturbed embedding strategy (displayed as DP) and our uncertainty (denoted as UC) scheme. The results are reported at the left side of Fig. 4 on the four classes case. Whilst our dual strategy substantially improves the performance, we observe that our uncertainty scheme has a major repercussion. The intuition is that our diffusion model ensures, at early epochs, that there is a high certainty in the prediction avoiding incorrect prediction in subsequent epochs. We also include an ablation study regarding the impact of the modalities (PT refers to phenotypic data). From the right side of Fig. 4, we observe that including phenotypic data has a greater positive impact on the performance than using only imaging data.

4 Conclusion

We proposed a novel semi-supervised hypergraph framework for Alzheimer's disease diagnosis. Unlike existing techniques, we introduce a dual embedding strategy. As phenotypic data highly differs from imaging data. Moreover, in contrast to existing techniques that follow the hypergraph approximation of [29]. We introduce a better diffusion model, which solution is provided by a semi-explicit flow in the fra hypergraph learning. From our results, we showed that our technique outperforms other hypergraph techniques. Future work includes a more extensive clinical evaluation with public and in-home datasets.

Acknowledgements. AIAR acknowledges support from CMIH and CCIMI, University of Cambridge. CR acknowledges support from the CCIMI and the EPSRC grant EP/W524141/1 ref. 2602161. ZK acknowledges support from the BBSRC (H012508, BB/P021255/1), Wellcome Trust (205067/Z/16/Z, 221633/Z/20/Z) and Royal Society (INF/R2/202107). CBS acknowledges the Philip Leverhulme Prize, the EPSRC fellowship EP/V029428/1, EPSRC grants EP/T003553/1, EP/N014588/1, Wellcome Trust 215733/Z/19/Z and 221633/Z/20/Z, Horizon 2020 No. 777826 NoMADS and the CCIMI.

References

1. Abdar, M., et al.: A review of uncertainty quantification in deep learning: techniques, applications and challenges. Inf. Fusion **76**, 243–297 (2021)

2. Agarwal, S., Branson, K., Belongie, S.: Higher order learning with graphs. In: Proceedings of the 23rd International Conference on Machine Learning (2006)
3. Arazo, E., Ortego, D., Albert, P., O'Connor, N.E., McGuinness, K.: Pseudo-labeling and confirmation bias in deep semi-supervised learning. In: 2020 International Joint Conference on Neural Networks (IJCNN), pp. 1–8. IEEE (2020)
4. Alzheimer's Association: What is Alzheimer's disease? (2022)
5. Aviles-Rivero, A.I., et al.: GraphXNET - chest X-Ray classification under extreme minimal supervision. In: Shen, D., et al. (eds.) MICCAI 2019. LNCS, vol. 11769, pp. 504–512. Springer, Cham (2019). https://doi.org/10.1007/978-3-030-32226-7_56
6. Berge, C.: Hypergraphs: Combinatorics of Finite Sets, vol. 45. Elsevier, Amsterdam (1984)
7. Chen, T., Kornblith, S., Norouzi, M., Hinton, G.: A simple framework for contrastive learning of visual representations. In: International Conference on Machine Learning, pp. 1597–1607. PMLR (2020)
8. Cubuk, E.D., Zoph, B., Shlens, J., Le, Q.V.: Randaugment: practical automated data augmentation with a reduced search space. In: IEEE/CVF Conference on Computer Vision and Pattern Recognition Workshops, pp. 702–703 (2020)
9. De Strooper, B., Karran, E.: The cellular phase of Alzheimer's disease. Cell $164(4)$, 603–615 (2016)
10. Feld, T., Aujol, J.F., Gilboa, G., Papadakis, N.: Rayleigh quotient minimization for absolutely one-homogeneous functionals. Inverse Prob. $35(6)$, 064003 (2019)
11. Feng, Y., You, H., Zhang, Z., Ji, R., Gao, Y.: Hypergraph neural networks. In: AAAI Conference on Artificial Intelligence, vol. 33, pp. 3558–3565 (2019)
12. Goenka, N., Tiwari, S.: Deep learning for Alzheimer prediction using brain biomarkers. Artif. Intell. Rev. $54(7)$, 4827–4871 (2021)
13. Hadsell, R., Chopra, S., LeCun, Y.: Dimensionality reduction by learning an invariant mapping. In: 2006 IEEE Computer Society Conference on Computer Vision and Pattern Recognition (CVPR 2006), vol. 2, pp. 1735–1742. IEEE (2006)
14. He, K., Fan, H., Wu, Y., Xie, S., Girshick, R.: Momentum contrast for unsupervised visual representation learning. In: Proceedings of the IEEE/CVF Conference on Computer Vision and Pattern Recognition, pp. 9729–9738 (2020)
15. Hein, M., Setzer, S., Jost, L., Rangapuram, S.S.: The total variation on hypergraphs-learning on hypergraphs revisited. In: Advances in Neural Information Processing Systems, vol. 26 (2013)
16. Jiang, J., Wei, Y., Feng, Y., Cao, J., Gao, Y.: Dynamic hypergraph neural networks. In: IJCAI, pp. 2635–2641 (2019)
17. Kendall, A., Gal, Y.: What uncertainties do we need in Bayesian deep learning for computer vision? In: Advances in Neural Information Processing Systems (2017)
18. Pan, J., Lei, B., Shen, Y., Liu, Y., Feng, Z., Wang, S.: Characterization multimodal connectivity of brain network by hypergraph GAN for Alzheimer's disease analysis. In: Ma, H., et al. (eds.) PRCV 2021. LNCS, vol. 13021, pp. 467–478. Springer, Cham (2021). https://doi.org/10.1007/978-3-030-88010-1_39
19. Parisot, S., et al.: Disease prediction using graph convolutional networks: application to autism spectrum disorder and Alzheimer's disease. Med. Image Anal. 48, 117–130 (2018)
20. Pölsterl, S., Aigner, C., Wachinger, C.: Scalable, axiomatic explanations of deep Alzheimer's diagnosis from heterogeneous data. In: International Conference on Medical Image Computing and Computer-Assisted Intervention (2021)
21. Qiu, J., et al.: GCC: graph contrastive coding for graph neural network pre-training. In: Proceedings of the 26th ACM SIGKDD International Conference on Knowledge Discovery & Data Mining, pp. 1150–1160 (2020)

22. Saito, S., Mandic, D., Suzuki, H.: Hypergraph p-Laplacian: a differential geometry view. In: AAAI Conference on Artificial Intelligence, vol. 32 (2018)
23. Shao, W., Peng, Y., Zu, C., Wang, M., Zhang, D., Initiative, A.D.N., et al.: Hypergraph based multi-task feature selection for multimodal classification of Alzheimer's disease. Comput. Med. Imaging Graph. **80**, 101663 (2020)
24. Shao, W., Xiang, S., Zhang, Z., Huang, K., Zhang, J.: Hyper-graph based sparse canonical correlation analysis for the diagnosis of Alzheimer's disease from multi-dimensional genomic data. Methods **189**, 86–94 (2021)
25. Shin, H.-C., et al.: GANDALF: generative adversarial networks with discriminator-adaptive loss fine-tuning for Alzheimer's disease diagnosis from MRI. In: Martel, A.L., et al. (eds.) MICCAI 2020. LNCS, vol. 12262, pp. 688–697. Springer, Cham (2020). https://doi.org/10.1007/978-3-030-59713-9_66
26. Yadati, N., Nimishakavi, M., Yadav, P., Nitin, V., Louis, A., Talukdar, P.: Hyper-GCN: a new method for training graph convolutional networks on hypergraphs. In: Advances in Neural Information Processing Systems, vol. 32 (2019)
27. Yang, F., Meng, R., Cho, H., Wu, G., Kim, W.H.: Disentangled sequential graph autoencoder for preclinical Alzheimer's disease characterizations from ADNI study. In: de Bruijne, M., et al. (eds.) MICCAI 2021. LNCS, vol. 12902, pp. 362–372. Springer, Cham (2021). https://doi.org/10.1007/978-3-030-87196-3_34
28. You, Y., Chen, T., Sui, Y., Chen, T., Wang, Z., Shen, Y.: Graph contrastive learning with augmentations. Adv. Neural. Inf. Process. Syst. **33**, 5812–5823 (2020)
29. Zhou, D., Huang, J., Schölkopf, B.: Learning with hypergraphs: clustering, classification, and embedding. In: Advances in Neural Information Processing Systems (2006)
30. Zuo, Q., Lei, B., Shen, Y., Liu, Y., Feng, Z., Wang, S.: Multimodal representations learning and adversarial hypergraph fusion for early Alzheimer's disease prediction. In: Ma, H., et al. (eds.) PRCV 2021. LNCS, vol. 13021, pp. 479–490. Springer, Cham (2021). https://doi.org/10.1007/978-3-030-88010-1_40

Federated Medical Image Analysis with Virtual Sample Synthesis

Wei Zhu$^{(\boxtimes)}$ and Jiebo Luo

University of Rochester, Rochester, NY, USA
`zwvews@gmail.com`

Abstract. Hospitals and research institutions may not be willing to share their collected medical data due to privacy concerns, transmission cost, and the intrinsic value of the data. Federated medical image analysis is thus explored to obtain a global model without access to the images distributed on isolated clients. However, in real-world applications, the local data from each client are likely non-i.i.d distributed because of the variations in geographic factors, patient demographics, data collection process, and so on. Such heterogeneity in data poses severe challenges to the performance of federated learning. In this paper, we introduce federated medical image analysis with virtual sample synthesis (FedVSS). Our method can improve the generalization ability by adversarially synthesizing virtual training samples with the local models and also learn to align the local models by synthesizing high-confidence samples with regard to the global model. All synthesized data will be further utilized in local model updating. We conduct comprehensive experiments on five medical image datasets retrieved from MedMNIST and Camelyon17, and the experimental results validate the effectiveness of our method. Our code is available at Link.

Keywords: Medical image analysis · Federated learning · Adversarial learning

1 Introduction

Deep-learning-based medical image analysis has received increasing attention in recent years and achieved successes in many applications [16,21,26,34]. However, the successes should be largely attributed to the significant effort to construct large-scale medical image datasets. Nonetheless, due to privacy concerns, regulations, and transmission costs, conducting centralized training by collecting medical images from different hospitals and institutions to a central server is likely infeasible or prohibitive for real-world applications [19,31]. Consequently, there is a real demand for training a generalized medical image analysis model without accessing the raw image.

Supplementary Information The online version contains supplementary material available at https://doi.org/10.1007/978-3-031-16437-8_70.

© The Author(s), under exclusive license to Springer Nature Switzerland AG 2022
L. Wang et al. (Eds.): MICCAI 2022, LNCS 13433, pp. 728–738, 2022.
https://doi.org/10.1007/978-3-031-16437-8_70

We employ federated learning to handle this problem [1,8,11,23,30,31,33]. Federated learning allows the participated hospitals (clients) to train the local models with their private data and only upload the trained models to the central server. The training pipeline for federated medical image analysis is summarized as follows: 1) a central server broadcasts global model to selected clients; 2) clients initialize the local models with the received global model and then conduct training with their private data; 3) the local models are uploaded to the server and further aggregated to update the global model in a data-free manner. As the earliest scheme, Federated Averaging (FedAvg) element-wisely averages the weights of local models [19]. However, since local data are usually non-i.i.d. distributed with inherent heterogeneity, the clients' distribution shifts will make the trained local models inconsistent. Such inconsistency will lead to a sub-optimal global model aggregated from local models in a data-free manner [6,27].

To alleviate the heterogeneity problem, different federated learning methods have been proposed and can be roughly categorized as client-side training improvement methods [5,6,9,10,14,18,24] and server-side aggregation methods [12,15,22,25,27]. In this paper, we focus on client-side methods since these methods can make better use of the local training data and have shown encouraging performance [5,14,24]. For example, FedProx adopts a proximal regularization term to penalize the significant update of local models [24]. FLIT proposed a re-weighted method to handle the heterogenity problem [35]. FedMD assumes a shared standalone dataset and conducts knowledge distillation to facilitate the training of a local model [9]. Contrastive learning is also applied to correct local training [10,29]. Cetinkaya et al. propose to augment the rare classes by image transformation [2]. HarfoFL learns local models with a flat optimum so that the local models would be better aligned [5]. However, on the one hand, the regularization term used by these methods may potentially hinder local training. On the other hand, they also do not fully exploit the rich information in the global model.

In contrast to these methods, we propose Federated Learning with Virtual Sample Synthesis (FedVSS) in this paper. FedVSS jointly uses the local and global models to improve the training on the client side. In particular, FedVSS synthesizes adversarial virtual neighbors by Virtual Adversarial Training (VAT) [20] to enhance the generalization ability of the local model and generate high-confident images with regard to the global model. All synthesized images and the original data will be utilized in updating the local model. In this way, FedVSS can obtain more generalized and consistent local models and the aggregated global model can thus eventually obtain better performance. For simplicity, we adopt FedAvg to aggregate the local models on the server side in this paper but our method can readily work with other model aggregation strategies [12,22,25,27]. We highlight our main contributions as follows:

1. FedVSS applies virtual adversarial training to smooth the local models for better generalization ability;
2. FedVSS makes better use of the global model to synthesize high-confident training samples with regard to the global model, and these samples can promote better alignment between local models;

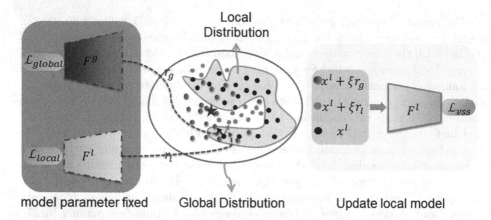

Fig. 1. Illustration of the proposed FedVSS framework. F^g and F^l denote the global and local models, respectively. The star denotes the prototype of the global model. The local models will overfit the local data during training, and the global model will roughly capture the global distribution. Given the training data x^l, we rely on the local model F^l to synthesize adversarial training samples $x^l + \xi r_l$ to smooth the local model and utilize the global model to synthesize high-confident samples $x^g + \xi r_g$ to align the local data distribution to the global data distribution.

3. FedVSS proves its effectiveness in comprehensive experiments on a wide range of medical image datasets.

2   Our Method

In this section, we present the proposed Federated Learning by Virtual Sample Synthesis (FedVSS) scheme for medical image analysis as illustrated in Fig. 1.

2.1   Preliminaries and Notations

We first briefly describe the settings for federated medical image analysis. We assume that L hospitals and research institutions are working on a deep neural network-based automatic disease diagnosis system. We assume that the medical data collected by each hospital are non-i.i.d. distributed, $i.e.$, the heterogeneity problem. Each hospital develops a local neural network, $e.g.$, ResNet [4]. However, the local models trained to inevitably overfit the biased small-scale local data likely suffer from poor generalization performance. The hospitals (clients) intend to collaborate for a global model without sharing the data with the central server or other participants because of privacy concerns and transmission costs.

We apply federated learning to handle this task. Formally, we denote the overall dataset as $D = \{D^l\}_{l=1}^L$, where $D^l = (X^l, Y^l) = \{(x_i^l, y_i^l)\}_{i=1}^{N_l}$ is the local dataset from the l-th hospital/client, x_i^l is the medical image, and y_i^l is the ground-truth label. In this paper, we use a ResNet18 [4] to handle the data on the client-side, and F^l denotes the local model from the l-th client. We also simulate a central server to aggregate the local models for a global model as $F^g = FedAgg(\{F^l\}_{l=1}^L)$, where F^g is the global model. $FedAgg(\cdot)$ is the aggregation function and is implemented with Federated Averaging in this paper for simplicity [19].

2.2 Virtual Samples Synthesis for Federated Learning

Our method can synthesize virtual training samples by jointly using the local models and the global model to mitigate the heterogeneity problem. Generally, on the one hand, we apply virtual adversarial training on a local model to enhance its generalization ability. On the other hand, we utilize the global model to generate high-confident samples to align the local distribution to the global one.

Virtual Adversarial Sample Synthesis with a Local Model. We propose to apply Virtual Adversarial Training (VAT) to improve the generalization performance of a local model for federated learning [20]. Local models will be trained to inevitably overfit the biased small-scale data on the client-side. VAT can smooth the model for better generalization performance by encouraging prediction consistency among neighbors [20]. For detail, given an image x^l from the l-th client, VAT synthesizes adversarial neighbors, which are similar to x^l in terms of the raw pixels but has a different prediction [28]. Concretely, the VAT loss for a local model F^l is formulated as follows [20]:

$$\mathcal{L}_{local} := \min D(F^l(x^l), F^l(x^l + \xi r_l))$$
$$where \; r_l = \arg\max_{r; \|r\| \leq \epsilon} D(F^l(x^l), F^l(x^l + r)), \tag{1}$$

where $\epsilon = 0.001$ is the threshold for the adversarial step size, $\xi = 10$ is the step size to permute the input data, and $D(\cdot)$ denotes KL divergence [20]. Equation (1) first finds a direction r_l that can significantly change the prediction for x^l and then synthesize a virtual sample $x^l + \xi r_l$ by moving towards the direction with step size ξ. Since r is bounded by a small value ϵ, the synthesized sample is believed to be a virtual neighbor for x^l, and we then can minimize the loss $\mathcal{L}_{local}$ for a smooth and generalized local model. r_l can be obtained by applying gradient ascent on the model. We present the method to obtain r_l in supplementary materials and please refer to [20] for more details.

Global Consistent Virtual Sample Synthesis. Existing federated learning methods often only use the global model to initialize the local models [5,6,19]. In this paper, we propose a novel approach to better explore the global model's rich information to alleviate the heterogeneity problem. Our basic assumption is that the global model has roughly learned the global data distribution. We can then synthesize high-confident samples under the guidance of the global model so that the generated samples will be more close to the prototype of the global model as shown in Fig. 1 The distribution of synthesized images will be more consistent with the global data distribution. The local models trained with the synthesized images will then be better aligned with each other. We measure the confidence of the global model for a given image by the loss value, which is simple and shows superior performance [17]. Concretely, following the framework of VAT [20], given a training image x^l with label y^l, a local model F^l, and the global model F^g, our method is formulated as

$$\mathcal{L}_{global} := \min L(y^l, F^l(x^l + \xi r_g))$$
$$where\ r_g = \arg\min_{r;\|r\|\leq\epsilon} L(y^l, F^g(x^l + r)), \tag{2}$$

where $L(\cdot,\cdot)$ is the loss function, i.e. cross-entropy loss, and the hyperparameters are set as Eq. (1). Similar to Eq. (1), Eq. (2) first finds a direction r_g that can minimize the loss (maximize the prediction confidence) of the global model with regarding to x^l and then synthesize a virtual sample $x^l + \xi r_g$ by moving towards the direction with step size ξ. The synthesized images are then used to update the local model. r_g can also be obtained by applying gradient descent on the global model. $\mathcal{L}_{global}$ can facilitate the knowledge transfer and also improves the consistency between the global model to the local models.

2.3   Overall Training Algorithm

FedVSS first synthesizes virtual images jointly with the local and global models as shown in Eq. (1) and Eq. (2) respectively, and then jointly uses all synthesized and original images of the l-th client to train the local model F^l. The overall objective is formulated as

$$\mathcal{L}_{vss} := \min L(y^l, F^l(x^l)) + \alpha\mathcal{L}_{local} + \beta\mathcal{L}_{global}, \tag{3}$$

where α and β are hyperparameters. We summarize the proposed FedVSS in the federated learning settings in Algorithm 1.

3   Experiments

3.1   Datasets and Federated Learning Settings

We conduct experiments on a wide variety of medical image datasets retrieved from Camelyon17 [7,13] and MedMNIST [32] including Camelyon17, OCTM-NIST, BloodMNIST, BreastMNIST, and PenumoniaMNIST. For BloodMNIST, BreastMNIST, and PenumoniaMNIST, we follow the official training, validation, and testing splitting following [32]. For OCTMNIST, we combine the validation and testing sets and then split them as 50:50 since the official testing set is too small. Other MedMNIST datasets show poor federated learning performance according to our experiments e.g. TissueMNIST and RetinaMNIST or are

---

Algorithm 1. Federated Medical Image Analysis with Virtual Sample Synthesis

---

Input: # clients L, # local updates T, # Communication round C, $\epsilon = 0.001$, $\xi = 10$, α, β

Output: Global Model F^g

1: Server initialize a global model F^g ▷ Server init.
2: **while** Communication Round $< C$ **do**
3:     Server broadcasts F^g to clients
4:     **for** $l : 1$ to L in parallel **do**   ▷ Client Update
5:        $F^l \leftarrow F^g$ ▷ Clients initialize the local model F^l with F^g.
6:        **for** $t : 1$ to K **do**   ▷ Update F^l for K steps
7:           Sample a minibatch of local data $\{x_i^l, y_i^l\}_{i=1}^B$
8:           Find r_l with gradient update and obtain $\mathcal{L}_{local}$ by Eq. (1)
9:           Find r_g with gradient update and obtain $\mathcal{L}_{global}$ by Eq. (2)
10:         Update the client model by minimizing $\mathcal{L}_{vss}$ by Eq. (3)
11:        **end for**
12:        Client sends the updated local model F^l to Server
13:     **end for**
14:     Server update the global model $F^g \leftarrow \sum_{l=1}^{L} \frac{|X^l|}{|X|} F^l$ ▷ Server Update
15: **end while**

---

highly imbalanced *e.g.* DermaMNIST, and thus are not suitable for benchmark purposes. We leave federated medical image analysis with problematic datasets as future work. Following the related work [3,27], we adopt Latent Dirichlet Analysis (LDA) to assign the training samples to the local clients for simulating heterogeneous federated learning datasets. We fix the partition parameter of LDA to 0.1 for all datasets except Camelyon17. Camelyon17 is a real-world federated learning dataset and contains 50 patients from 5 hospitals. We randomly select ten patients for testing others for training and validation. We additionally select 20% training samples as the validation set for each hospital regardless of patients to alleviate the discrepancy between validation and testing performance. For other federated learning configurations, we conduct experiments with the number of communication rounds as 10 and 20. We set the number of clients to the number of classes for OCTMNIST, BloodMNIST, BreastMNIST, and PemumoniaMNIST. The number of clients is set to 5 for Camelyon17. We randomly select a subset of clients at each communication round to validate the scalability of federated learning methods [19,27]. Detailed statistics of used datasets are provided in the supplementary material.

3.2 Compared Methods and Evaluation Metrics

To validate the effectiveness of FedVSS, we compare it with Federated Averaging (FedAvg) [19], Federated Proximal (FedProx) [24], and MOON [10]. To show the performance of different components of FedVSS, we also implement two variants as FedVSS-local (FedVSS-l) and FedVSS-global (FedVSS-g). FedVSS-l is formulated as

$$\mathcal{L}_{vss-l} := \min L(y^l, F^l(x^l)) + \alpha \mathcal{L}_{local}, \tag{4}$$

and FedVSS-g is formualted as:

$$\mathcal{L}_{vss-g} := \min L(y^l, F^l(x^l)) + \beta \mathcal{L}_{global}. \tag{5}$$

FedVSS-l and FedVSS-g are developed to validate the effectiveness of the virtual sample synthesis by the local models $\mathcal{L}_{local}$ and the global model $\mathcal{L}_{global}$, respectively. We conduct a grid search on the validation set for hyper-parameter tuning and model selection. We search the hyper-parameter for Fed-Prox from $\{0.001, 0.01, 0.1, 1\}$ and for MOON from $\{0.1, 1, 5, 10\}$ following the original papers [10]. For FedVSS, FedVSS-g, and FedVSS-l, we search the hyper-parameters for α and β from $\{0.1, 1, 2, 5\}$.

We implement all methods with FedML [3] and Pytorch. We adopt ResNet-18 [4] as the backbone network and use momentum SGD with a learning rate of 0.001 and momentum as 0.9 as client optimizer. The batch size is set to 64. We run all experiments three times and report the average accuracy (ACC) and F1 score for all datasets. The experiments are conducted on a Linux machine with 8 NVIDIA RTX 2080 Ti graphics cards.

Table 1. Results on BreastMNIST, BloodMNIST, and PenumoniaMNIST. Best results are in bold and second best are underlined. (%)

Dataset	BreastMNIST				BloodMNIST				PenumoniaMNIST			
Commu round	10	10	20	20	10	10	20	20	10	10	20	20
	ACC	F1	ACC	F1	ACC	F1	ACC	F1	ACC	F1	ACC	F1
FedAvg [19]	70.51	76.73	71.47	78.74	83.46	81.02	85.40	83.39	86.85	90.26	87.17	90.42
FedProx [24]	70.63	76.89	73.72	76.76	83.94	81.95	85.94	84.07	87.42	90.46	87.34	90.49
MOON [10]	66.99	74.83	66.67	74.77	84.81	82.33	**86.35**	83.92	<u>87.66</u>	90.65	87.58	90.76
FedVSS-l (ours)	72.76	78.77	71.79	78.77	<u>85.17</u>	<u>82.75</u>	86.28	<u>84.25</u>	87.58	90.76	**88.38**	<u>91.14</u>
FedVSS-g (ours)	<u>73.28</u>	<u>79.20</u>	<u>74.36</u>	<u>79.36</u>	84.13	81.36	<u>86.54</u>	**84.72**	<u>87.66</u>	<u>90.77</u>	87.81	90.81
FedVSS (ours)	**74.36**	**80.86**	**76.60**	**81.24**	**85.40**	**82.79**	86.25	83.95	**87.90**	**90.89**	<u>88.30</u>	**91.27**

Table 2. Results on OCTMNIST and Camelyon17. Best results are in bold and second best are underlined. (%)

Dataset	OCTMNIST				Camelyon17			
Commu round	10	10	20	20	10	10	20	20
	ACC	F1	ACC	F1	ACC	F1	ACC	F1
FedAvg [19]	69.79	59.00	69.40	57.81	70.77	67.44	71.13	69.94
FedProx [24]	69.82	<u>59.43</u>	<u>71.11</u>	59.45	66.70	65.79	70.94	68.29
MOON [10]	**70.20**	57.53	70.32	<u>59.53</u>	70.58	71.60	69.34	66.36
FedVSS-l (ours)	69.27	58.59	70.69	58.63	69.06	63.16	71.12	73.59
FedVSS-g (ours)	69.05	**59.50**	70.72	58.74	**73.15**	**74.22**	<u>73.65</u>	<u>74.61</u>
FedVSS (ours)	<u>70.16</u>	59.22	**71.32**	**59.68**	<u>72.31</u>	<u>73.82</u>	**75.03**	**76.33**

3.3   Experimental Results

The results are demonstrated in Table 1 and Table 2. We draw several points according to the results. First, our results show that it is feasible to apply federated learning on either synthetic or realistic medical image datasets, and the commonly used FedAvg achieves reasonable performance for these datasets. A large number of communication rounds can benefit all federated learning methods but will induce huge communication costs. Second, methods including FedProx, MOON, and all variants of the proposed FedVSS significantly outperform the baseline FedAvg by considering the heterogeneity problem. For example, the proposed FedVSS achieves 3.85% improvements over FedAvg for BreastMNIST with ten communication rounds. The results suggest that it is necessary to consider the heterogeneity problem for better performance. Third, in general, all variants of FedVSS, especially the proposed FedVSS-g and FedVSS, have significant advantages over other federated learning methods in most cases. For example, FedVSS-g outperforms the best-compared methods MOON around 0.64% for BreastMNIST with 20 communication rounds. Moreover, FedVSS-g leads to better performance than FedVSS-l. FedVSS-g gains 4.09% improvements in accuracy for Camelyon17 with ten communication rounds. This indicates that aligning the local data distribution with the global data distribution is more important than enhancing the model generalization ability by VAT for heterogeneous federated learning. Finally, FedVSS jointly uses the local and global models for virtual sample synthesis and obtains the best overall results. Concretely, FedVSS obtains the top two methods for around 8.5 out of 10 settings. The reasons for the success are intuitively explained as follows. The samples synthesized by the local model make the local model smooth, and those generated by the global model helps align the local data distribution to the global one.

3.4   Sensitivity Studies

We study the parameter sensitivities of FedVSS on the BreadMNIST dataset. FedVSS contains two hyper-parameters as α and β. We fix α (β) and then perform a grid search on β (α). The results are shown in Table 3. According to our experiments, FedVSS is robust to different hyperparameter settings.

Table 3. Sensitivity studies on BreastMNIST.

BreastMNIST		α				β			
Commu round		0.1	1	2	5	0.1	1	2	5
10	ACC	**74.36**	72.12	73.87	71.79	73.40	**74.36**	74.04	72.44
10	F1	**80.86**	79.15	79.60	79.06	79.60	80.86	**81.60**	78.61
20	ACC	75.32	**76.60**	74.68	73.08	75.00	**76.60**	75.00	74.36
20	F1	**81.27**	81.24	81.00	80.64	81.07	**81.24**	81.14	80.68

4 Conclusions

In this paper, we present a novel federated medical image analysis method, namely Federated Learning with Virtual Sample Synthesis (FedVSS), to alleviate the well-known heterogeneity problem. FedVSS jointly uses the local and global models to synthesize virtual training samples. In particular, on one hand, FedVSS applies virtual adversarial training to the local model to enhance its generalization performance. On the other hand, FedVSS synthesizes high-confident virtual samples with regard to the global model to align the local data distributions to the global data distribution. We conduct experiments on five medical image datasets to show the efficacy of our method. We plan to extend our research on federated learning to more problematic data, *e.g.*, noisy labels, highly imbalanced data, and so on.

Acknowledgement. This work was supported in part by NIH 1P50NS108676-01, NIH 1R21DE030251-01 and NSF award 2050842.

References

1. Adnan, M., Kalra, S., Cresswell, J.C., Taylor, G.W., Tizhoosh, H.R.: Federated learning and differential privacy for medical image analysis. Sci. Rep. **12**(1), 1–10 (2022)
2. Cetinkaya, A.E., Akin, M., Sagiroglu, S.: Improving performance of federated learning based medical image analysis in non-IID settings using image augmentation. In: 2021 International Conference on Information Security and Cryptology (ISC-TURKEY), pp. 69–74. IEEE (2021)
3. He, C., et al.: FedML: a research library and benchmark for federated machine learning. arXiv preprint arXiv:2007.13518 (2020)
4. He, K., Zhang, X., Ren, S., Sun, J.: Deep residual learning for image recognition. In: Proceedings of the IEEE Conference on Computer Vision and Pattern Recognition, pp. 770–778 (2016)
5. Jiang, M., Wang, Z., Dou, Q.: HarmoFL: harmonizing local and global drifts in federated learning on heterogeneous medical images. arXiv preprint arXiv:2112.10775 (2021)
6. Karimireddy, S.P., Kale, S., Mohri, M., Reddi, S., Stich, S., Suresh, A.T.: SCAFFOLD: stochastic controlled averaging for federated learning. In: International Conference on Machine Learning, pp. 5132–5143. PMLR (2020)
7. Koh, P.W., et al.: WILDS: a benchmark of in-the-wild distribution shifts. In: International Conference on Machine Learning, pp. 5637–5664. PMLR (2021)
8. Li, D., Kar, A., Ravikumar, N., Frangi, A.F., Fidler, S.: Federated simulation for medical imaging. In: Martel, A.L., et al. (eds.) MICCAI 2020. LNCS, vol. 12261, pp. 159–168. Springer, Cham (2020). https://doi.org/10.1007/978-3-030-59710-8_16
9. Li, D., Wang, J.: FedMD: heterogenous federated learning via model distillation. arXiv preprint arXiv:1910.03581 (2019)
10. Li, Q., He, B., Song, D.: Model-contrastive federated learning. In: Proceedings of the IEEE/CVF Conference on Computer Vision and Pattern Recognition, pp. 10713–10722 (2021)

11. Li, W., et al.: Privacy-preserving federated brain tumour segmentation. In: Suk, H.-I., Liu, M., Yan, P., Lian, C. (eds.) MLMI 2019. LNCS, vol. 11861, pp. 133–141. Springer, Cham (2019). https://doi.org/10.1007/978-3-030-32692-0_16

12. Lin, T., Kong, L., Stich, S.U., Jaggi, M.: Ensemble distillation for robust model fusion in federated learning. arXiv preprint arXiv:2006.07242 (2020)

13. Litjens, G., et al.: 1399 H&E-stained sentinel lymph node sections of breast cancer patients: the CAMELYON dataset. GigaScience **7**(6), giy065 (2018)

14. Liu, Q., Chen, C., Qin, J., Dou, Q., Heng, P.A.: FedDG: federated domain generalization on medical image segmentation via episodic learning in continuous frequency space. In: Proceedings of the IEEE/CVF Conference on Computer Vision and Pattern Recognition, pp. 1013–1023 (2021)

15. Liu, Q., Yang, H., Dou, Q., Heng, P.-A.: Federated semi-supervised medical image classification via inter-client relation matching. In: de Bruijne, M., et al. (eds.) MICCAI 2021. LNCS, vol. 12903, pp. 325–335. Springer, Cham (2021). https://doi.org/10.1007/978-3-030-87199-4_31

16. Liu, T., Siegel, E., Shen, D.: Deep learning and medical image analysis for covid-19 diagnosis and prediction. Ann. Rev. Biomed. Eng. **24** (2022)

17. Liu, W., Wang, X., Owens, J., Li, Y.: Energy-based out-of-distribution detection. Adv. Neural. Inf. Process. Syst. **33**, 21464–21475 (2020)

18. Luo, J., Wu, S.: FedSLD: federated learning with shared label distribution for medical image classification. arXiv preprint arXiv:2110.08378 (2021)

19. McMahan, B., Moore, E., Ramage, D., Hampson, S., y Arcas, B.A.: Communication-efficient learning of deep networks from decentralized data. In: Artificial Intelligence and Statistics, pp. 1273–1282. PMLR (2017)

20. Miyato, T., Maeda, S.I., Koyama, M., Ishii, S.: Virtual adversarial training: a regularization method for supervised and semi-supervised learning. IEEE Trans. Pattern Anal. Mach. Intell. **41**(8), 1979–1993 (2018)

21. Qiu, Z., et al.: A deep learning approach for segmentation, classification, and visualization of 3-D high-frequency ultrasound images of mouse embryos. IEEE Trans. Ultrason. Ferroelectr. Freq. Control **68**(7), 2460–2471 (2021)

22. Reddi, S., et al.: Adaptive federated optimization. arXiv preprint arXiv:2003.00295 (2020)

23. Roth, H.R., et al.: Federated whole prostate segmentation in MRI with personalized neural architectures. In: de Bruijne, M., et al. (eds.) MICCAI 2021. LNCS, vol. 12903, pp. 357–366. Springer, Cham (2021). https://doi.org/10.1007/978-3-030-87199-4_34

24. Sahu, A.K., Li, T., Sanjabi, M., Zaheer, M., Talwalkar, A., Smith, V.: On the convergence of federated optimization in heterogeneous networks. arXiv preprint arXiv:1812.06127 **3** (2018)

25. Seo, H., Park, J., Oh, S., Bennis, M., Kim, S.L.: Federated knowledge distillation. arXiv preprint arXiv:2011.02367 (2020)

26. Taleb, A., Lippert, C., Klein, T., Nabi, M.: Multimodal self-supervised learning for medical image analysis. In: Feragen, A., Sommer, S., Schnabel, J., Nielsen, M. (eds.) IPMI 2021. LNCS, vol. 12729, pp. 661–673. Springer, Cham (2021). https://doi.org/10.1007/978-3-030-78191-0_51

27. Wang, H., Yurochkin, M., Sun, Y., Papailiopoulos, D., Khazaeni, Y.: Federated learning with matched averaging. arXiv preprint arXiv:2002.06440 (2020)

28. Wei, C., Shen, K., Chen, Y., Ma, T.: Theoretical analysis of self-training with deep networks on unlabeled data. arXiv preprint arXiv:2010.03622 (2020)

29. Wu, Y., Zeng, D., Wang, Z., Shi, Y., Hu, J.: Federated contrastive learning for volumetric medical image segmentation. In: de Bruijne, M., et al. (eds.) MICCAI 2021. LNCS, vol. 12903, pp. 367–377. Springer, Cham (2021). https://doi.org/10.1007/978-3-030-87199-4_35

30. Xia, Y., et al.: Auto-FedAvg: learnable federated averaging for multi-institutional medical image segmentation. arXiv preprint arXiv:2104.10195 (2021)

31. Yang, D., et al.: Federated semi-supervised learning for COVID region segmentation in chest CT using multi-national data from China, Italy, Japan. Med. Image Anal. **70**, 101992 (2021)

32. Yang, J., et al.: MedMNIST v2: a large-scale lightweight benchmark for 2D and 3D biomedical image classification. arXiv preprint arXiv:2110.14795 (2021)

33. Yang, Q., Liu, Y., Chen, T., Tong, Y.: Federated machine learning: Concept and applications. ACM Trans. Intell. Syst. Technol. (TIST) **10**(2), 1–19 (2019)

34. Zhu, W., Liao, H., Li, W., Li, W., Luo, J.: Alleviating the incompatibility between cross entropy loss and episode training for few-shot skin disease classification. In: Martel, A.L., et al. (eds.) MICCAI 2020. LNCS, vol. 12266, pp. 330–339. Springer, Cham (2020). https://doi.org/10.1007/978-3-030-59725-2_32

35. Zhu, W., Luo, J., White, A.D.: Federated learning of molecular properties with graph neural networks in a heterogeneous setting. Patterns 100521 (2022)

Unsupervised Cross-Domain Feature Extraction for Single Blood Cell Image Classification

Raheleh Salehi[1,2], Ario Sadafi[1,3], Armin Gruber[1], Peter Lienemann[1],
Nassir Navab[3,4], Shadi Albarqouni[5,6,7], and Carsten Marr[1(✉)]

[1] Institute of AI for Health, Helmholtz Munich - German Research Center
for Environmental Health, Neuherberg, Germany
carsten.marr@helmholtz-muenchen.de

[2] Computer Engineering, Politecnico Di Torino, Turin, Italy

[3] Computer Aided Medical Procedures (CAMP), Technical University of Munich,
Munich, Germany

[4] Computer Aided Medical Procedures, Johns Hopkins University, Baltimore, USA

[5] Clinic for Interventional and Diagnostic Radiology, University Hospital Bonn,
Bonn, Germany

[6] Faculty of Informatics, Technical University Munich, Munich, Germany

[7] Helmholtz AI, Helmholtz Munich - German Research Center for Environmental
Health, Neuherberg, Germany

Abstract. Diagnosing hematological malignancies requires identification and classification of white blood cells in peripheral blood smears.
Domain shifts caused by different lab procedures, staining, illumination,
and microscope settings hamper the re-usability of recently developed
machine learning methods on data collected from different sites. Here,
we propose a cross-domain adapted autoencoder to extract features in an
unsupervised manner on three different datasets of single white blood cells
scanned from peripheral blood smears. The autoencoder is based on an R-
CNN architecture allowing it to focus on the relevant white blood cell and
eliminate artifacts in the image. To evaluate the quality of the extracted
features we use a simple random forest to classify single cells. We show
that thanks to the rich features extracted by the autoencoder trained on
only one of the datasets, the random forest classifier performs satisfacto-
rily on the unseen datasets, and outperforms published oracle networks
in the cross-domain task. Our results suggest the possibility of employing
this unsupervised approach in more complicated diagnosis and prognosis
tasks without the need to add expensive expert labels to unseen data.

Keywords: Unsupervised learning · Feature extraction ·
Autoencoders · Single cell classification · Microscopy · Domain
adaptation

R. Salehi and A. Sadafi—Equal contribution.

Supplementary Information The online version contains supplementary material
available at https://doi.org/10.1007/978-3-031-16437-8_71.

© The Author(s), under exclusive license to Springer Nature Switzerland AG 2022
L. Wang et al. (Eds.): MICCAI 2022, LNCS 13433, pp. 739–748, 2022.
https://doi.org/10.1007/978-3-031-16437-8_71

1   Introduction

Hematopoietic malignancies such as leukemkia are among the deadliest diseases with limited therapeutic options. Cytomorphological evaluation of white blood cells under the microscopic in blood or bone marrow smears is key for proper diagnosis. So far, this morphological analysis has not been automated and is still performed manually by trained experts under the microscope. Recent works demonstrate however the potential in automation of this task. Matek et al. [12] have proposed a highly accurate approach based on ResNext [18] architecture for recognition of white blood cells in blood smears of acute myeloid leukemia patients. In another work [11] we have developed a CNN-based classification method for cell morphologies in bone marrow smears. Boldu et al. [3] have suggested a machine learning approach for diagnosis of acute leukemia by recognition of blast cells in blood smear images. Acevedo et al. [2] suggest a predictive model for automatic recognition of patients suffering from myelodysplastic syndrome, a pre-form of acute myeloid leukemia.

All of these studies have used data provided from a single site. However, many factors in laboratory procedures can affect the data and introduce a domain shift: Different illuminations, microscope settings, camera resolutions, and staining protocols are only some of the parameters differing between laboratories and hospitals. These changes can affect model performance considerably and render established approaches ineffective, requiring re-annotation and re-training of models.

Exposing the optimization to domain shifts can be a solution to align different domains in real-world data. A learning paradigm with dedicated losses is a common way to tackle this problem and has been already applied in many approaches [14]. For instance, Duo et al. [7] propose to learn semantic feature spaces by incorporating global and local constraints in a supervised method, while Chen et al. [5] have developed a method for unsupervised domain adaptation by conducting synergistic alignment of both image and features and applied it to medical image segmentation in bidirectional cross-modality adaptation between MRI and CT.

Here, we present an AutoEncoder-based Cell Feature Extractor (AE-CFE), a simple and economic approach for robust feature extraction of single cells. Our method is based on instance features extracted by a Mask R-CNN [8] architecture that is analyzed by an autoencoder to obtain features of single white blood cells in digitized blood smears. Since the data is coming from different sites, we are introducing a domain adaptation loss to reduce domain shifts. Our method is the first unsupervised two-staged autoencoder approach for cross-domain feature extraction based on instance features of a Mask R-CNN. It outperforms published supervised methods in unseen white blood cell datasets and can thus contribute to the establishment of robust decision support algorithms for diagnosing hematopoietic malignancies. We made our implementation publicly available at https://github.com/marrlab/AE-CFE.

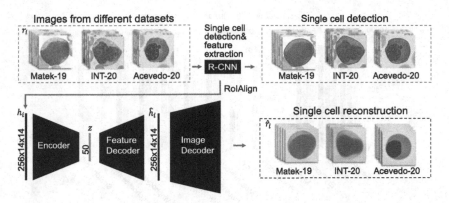

Fig. 1. Overview of the proposed AE-CFE method. A Mask R-CNN detects single cells in images and relevant instance features of the region of interest are extracted. The autoencoder uses the instance features as input and tries to reconstruct (i) instance features and (ii) single cell images. Since features are white blood cell specific, the autoencoder is able to only reconstruct white blood cells and artefacts such as red blood cells are discarded. (Color figure online)

2   Methodology

Our unsupervised feature extraction approach starts with a Mask R-CNN [8] model trained to detect single white blood cells in scanned patient's blood smears. For every detected cell instance-specific features extracted are used for training an autoencoder. This compresses the input to a latent space representation, while a two-staged decoder tries to reconstruct (i) the encoded features and (ii) the single cell images (Fig. 1).

Mask R-CNN is commonly used for instance segmentation. The architecture is based on an underlying feature extractor based on a ResNet-101-FPN [9,10] backbone. It has two stages: (i) A region proposal network (RPN) suggests candidate bounding boxes all over the input image and (ii) different heads of the architecture perform classification, bounding box regression, and segmentation locally only on the region of interest (RoI) based on the features that are extracted for every instance with RoIAlign.

More formally, having an image I_i from dataset D_k,

$$r_{i,j}, h_{i,j} = f_{R-CNN}(I_i) : \forall I_i \in D_k, \tag{1}$$

where $r_{i,j}$ is the jth single cell image cropped out and, $h_{i,j}$ is its corresponding features in i^{th} image of the dataset. For simplicity we assume for now there is only one white blood cell in every image and refer to it with r_i and h_i in the rest of this section.

Our desired feature extraction method can be formulated as

$$z = f_{enc}(h_i; \theta), \tag{2}$$

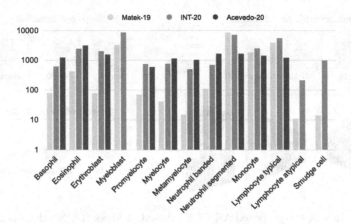

Fig. 2. Number of samples in each of the 13 classes for the three datasets used in our study.

where z is the robust, cross-domain feature vector we get at the bottleneck of the autoencoder and θ are the parameters learned during the training.

The autoencoder consists of three modules: (i) encoder, (ii) feature decoder, and (iii) image decoder. All three parts are trained by optimizing

$$\mathcal{L}(\theta, \gamma, \psi) = \frac{1}{N} \sum_{i=1}^{N} (\hat{h}_i - h_i)^2 + 1 - \text{SSIM}(\hat{r}_i, r_i) + \beta \mathcal{L}_{\text{DA}}, \qquad (3)$$

where $\hat{h}_i = f_{\text{dec}}^{\text{feat}}(z; \gamma)$ is the reconstructed feature vector, $\hat{r}_i = f_{\text{dec}}^{\text{img}}(\hat{h}_i; \psi)$ is the reconstructed image based on the feature reconstruction, γ and ψ are model parameters, and N is number of white blood cells in the dataset. $\mathcal{L}_{\text{DA}}$ is the domain adaptation loss introduced in Sect. 2.1 regulated by constant coefficient β. We use the structural similarity index measure (SSIM) [16] to measure the similarity of the reconstructed image x with the original single cell image y detected by the Mask R-CNN, defined as

$$\text{SSIM}(x, y) = \frac{(2\mu_x \mu_y + c_1)(2\sigma_{xy} + c_2)}{(\mu_x^2 + \mu_y^2 + c_1)(\sigma_x^2 + \sigma_y^2 + c_2)} \qquad (4)$$

where μ and σ are the mean and variance of the images and c_1, c_2 are small constants for numerical stability.

Group normalization (GN) [17] is applied after each layer in the encoder part as an alternative to batch normalization that is not dependent on the batch size. It divides the channels into groups and normalizes the groups with independent group specific mean and variance. In our experiments GN was effective in image generalization.

2.1 Domain Adaptation

When images come from different sites, the latent space can be dominated by a domain shift (see Fig. 4). Domain adaptation with group normalization and distribution-based maximum mean discrepancy has been shown to align the latent space representation of different datasets [15]. We use it to adapt the three datasets, which differ in resolution, size, and color.

With $\mathcal{D} = \{D_1, \ldots D_K\}$ being the datasets we are training on, and a mean matrix μ_k and s_k as softmax of covariance matrix of the embedded features of dataset D_k, our loss is defined by

$$\mathcal{L}_{\mathrm{DA}} = \sum_{k=1}^{K} \{\mathrm{MSE}(\mu_k, \mu_0) + \frac{1}{2}[D_{KL}(s_0\|s_k) + D_{KL}(s_k\|s_0)]\}, \tag{5}$$

where we calculate mean squared error on the mean matrices and a symmetrized Kullback-Leibler (KL) divergence of the covariance matrices for all datasets to bring them closer to the anchor dataset D_0. Any other symmetric divergences, or cosine similarity between the eigenvectors of the covariance matrices would work for this optimization.

3 Evaluation

3.1 Datasets

We are using three different datasets to evaluate our method:

The **Matek-19** dataset consists of over 18,000 annotated white blood cells from 100 acute myeloid leukaemia patients and 100 patients exhibiting no morphological features from the laboratory of leukemia diagnostics at Munich University Hospital between 2014 and 2017. It is publicly available [12] and there are 15 classes in the dataset. Image dimensions are 400×400 pixels or approximately 29×29 µm.

The **INT-20** in-house dataset has around 42,000 images coming from 18 different classes. Images are 288×288 in pixels or 25×25 µm.

The **Acevedo-20** dataset consists of over 17,000 images of individual normal cells acquired in the core laboratory at the Hospital Clinic of Barcelona published by Acevedo et al. [1]. There are 8 classes in the dataset and images are 360×363 pixels or 36×36.3 µm.

Since class definitions of the three datasets are different, we asked a medical expert to categorize different labels into 13 commonly defined classes consisting of: basophil, eosinophil, erythroblast, myeloblast, promyelocyte, myelocyte, metamyelocyte, neutrophil banded, neutrophil segmented, monocyte, lymphocyte typical, lymphocyte atypical, and smudge cells. Figure 2 shows sample distribution between these 13 classes for different datasets.

3.2 Implementation Details

Architecture. The autoencoder is a fully convolutional network. The encoder consists of 6 layers, the feature decoder has 3 layers, and the image decoder has 5 layers. All intermediate layers have ReLU activation functions, and outputs of encoder and feature decoder are regulated by a tanh activation function while

Original
images

Image AE

AE-CFE

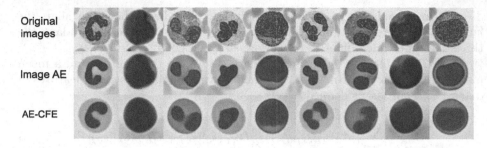

Fig. 3. We are comparing the reconstruction of AE-CFE with an image based autoencoder (Image AE). Red blood cells and artifacts surrounding the white blood cells are eliminated with Mask R-CNN feature extraction. (Color figure online)

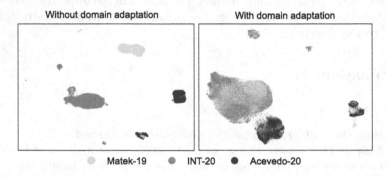

Fig. 4. UMAP embedding of AE-CFE with and without the domain adaptation loss. A more uniformly distributed latent representation is achieved after using the domain adaptation loss.

the image decoder has a sigmoid activation function on the output. To extract richest and least sparse features, we decided to use 50 as the bottleneck size which is the smallest possible.

Training. We performed stratified train and test splits on all datasets keeping 20% of the data for a holdout test set. Training was carried out using an Adam optimizer for 150 epochs with a learning rate of 0.001 on three NVIDIA A100-SXM4-40GB GPUs with a total batch size of 1500 (500 on each). The constant β in Eq. 3 was set to 5.

Random Forest. We used random forest implementation of the scikit-learn library [4] for all of the experiments. Number of estimators was set to 200 and maximum tree depth to 16.

3.3   Single Cell Detection by Mask R-CNN

The Mask R-CNN helps extracting the instance features and eliminate artefacts. To verify this observation, we trained another autoencoder with a similar number of layers in the decoder section on single white blood cell images as a baseline. Figure 3 shows example reconstructions from both methods. The trained autoencoder is affected by the noise from surrounding red blood cells way more than AE-CFE.

The Mask R-CNN model is trained on a small separate dataset of around 1500 images annotated for instance segmentation. The annotation of this dataset does not require any expertise, as white blood cell shapes are annotated with no class information. The Mask R-CNN was trained for 26 epoches reaching a mAP of 0.89. Analysing a pool of 77,363 images coming from three datasets, 65,693 cells were successfully detected (85%).

3.4 Evaluation

To quantitatively compare the quality of the extracted features by AE-CFE, we train a random forest (RF) model on the extracted features trying to classify single white blood cells into one of the 13 defined classes.

Table 1. Comparing the accuracy percentage of a random forest method trained on our proposed AE-CFE approach with 4 other feature extraction methods as baselines: ResNet101 trained on ImageNet, features extracted with Mask R-CNN, an autoencoder trained on instance feature vectors, and an adversarial autoencoder trained on instance feature vector. Mean and standard deviation of accuracy is reported from 5 training runs.

Trained on	Tested on	Reset-RF	R-CNN-RF	AE-RF	AAE-DA	AE-CFE
Matek-19	Matek-19	62.5 ± 1.8	60.5 ± 0.8	86.0 ± 0.04	**87.5 ± 0.8**	83.7 ± 0.5
	INT-20	0	0	46.8 ± 0.2	31.4 ± 0.3	**48.4 ± 0.2**
	Acevedo-20	0	0	20.1 ± 0.1	18.6 ± 0.4	**21.9 ± 0.4**
INT-20	Matek-19	0	0	47.2 ± 3.4	63.9 ± 0.2	**73.2 ± 0.1**
	INT-20	45.2 ± 1.1	46.0 ± 0.4	**69.1 ± 0.4**	66.8 ± 0.4	65.6 ± 0.5
	Acevedo-20	0	0	4.6 ± 0.6	17.7 ± 0.7	**31.8 ± 0.4**
Acevedo-20	Matek-19	0	0	39.5 ± 1.4	39.4 ± 0.6	**45.1 ± 0.5**
	INT-20	0	0	9.7 ± 0.3	17.7 ± 0.7	**21.0 ± 0.5**
	Acevedo-20	37.1 ± 0.8	35.9 ± 1.1	**67.2 ± 0.7**	64.3 ± 0.1	65.2 ± 0.5

In all experiments, the RF is trained on one dataset and tested on the test set of all three datasets. We defined four baselines for our proposed method: (i) ResNet-RF: random forest classification of the features extracted with a ResNet101 [9] architecture trained on ImageNet dataset [6] (ii) R-CNN-RF: random forest classification of the instance features extracted with our trained Mask R-CNN architecture (iii) AE-RF: random forest classification of features extracted by a similar autoencoder trained on all datasets with no domain adaptation. (iv) AAE-DA: trained Adversarial domain adaption on features extracted by a similar autoencoder.

In Table 1 we compare the accuracy of random forest classification of our method with the baselines and report mean and standard deviation of accuracy for 5 runs. For two of the baselines (ResNet-RF & R-CNN-RF) cross-domain evaluations were inaccurate (accuracy close to zero) and random forest was unable to classify any sample correctly.

Next, we compare our method with oracle methods specifically trained for classifying the datasets. Matek et al. [12] have published their ResNext architecture and trained model weights. We trained a similar ResNext model on each of the datasets. In Table 2 and supplementary material, we compare these oracle methods with RF trained on features extracted with our method. We find that both of the oracles are failing on the unseen dataset while a random forest trained on our unsupervised AE-CFE features is performing by far better.

Table 2. Comparing the accuracy percentage of a random forest method trained on our proposed AE-CFE feature extraction approach with 2 other oracle methods specifically trained for each of the datasets. Matek et al.'s published method trained on their dataset, and two ResNext models trained on each of the datasets with a random forest classifying features of our proposed cross-domain autoencoder. Mean and standard deviation of accuracy is reported over 5 runs.

Trained on	Tested on	ResNext	Matek et al.	AE-CFE
Matek-19	Matek-19	–	**96.1**	83.7 ± 0.5
	INT-20	–	29.5	**48.4 ± 0.2**
	Acevedo-20	–	8.1	**21.9 ± 0.4**
INT-20	Matek-19	49.0 ± 6.3	–	**73.2 ± 0.1**
	INT-20	**88.7 ± 1.5**	–	65.6 ± 0.5
	Acevedo-20	16.9 ± 1.6	–	**31.8 ± 0.4**
Acevedo-20	Matek-19	7.3 ± 3.1	–	**45.1 ± 0.5**
	INT-20	8.1 ± 1.4	–	**21.0 ± 0.5**
	Acevedo-20	**85.7 ± 2.4**	–	65.2 ± 0.5

Finally, we are comparing the UMAP [13] embeddings of all feature vectors of the white blood cells from the three datasets with and without our domain adaptation loss. Figure 4 shows that with domain adaptation not only the RF classification results improve but also a more uniform latent distribution is achieved, supporting the results.

4   Discussion and Conclusion

Artefacts in single cell images can greatly affect the performance of a model by falsely overfitting on irrelevant features. For example, the surrounding red blood cells and thus the number of red pixels in images can mislead the model into categorizing samples based on anemic features (i.e. the density of red blood cells) rather than the cytomorphological white blood cell properties. This makes Mask R-CNN an essential element in our design, forcing the algorithm to focus on the instance features cropped out in the region of interest rather than the whole image or features in the background. But what if cells in unseen data are considerably different? The fact that the training dataset for Mask R-CNN was coming from only one of the three datasets used in our study, 85% detection rate for single cells in unseen data is surprisingly high, and obviously good enough for demonstrating the multi-domain applicability of our approach. However, annotation of single white blood cells in images from different datasets to train a better Mask R-CNN is cheap, convenient, and fast and can improve our results even further.

For classification, the random forest model trained on features extracted by our approach is not performing as good as oracle models on the source domains, but its performance in cross-domain scenarios is by far superior. This is partly due to our domain adaptation loss that forces the latent representations from different datasets to be as close as possible to each other. The small feature vectors of only 50 dimensions with minimum sparsity allow usage of these features in many different applications.

Using features from cell nuclei additionally, including the AE-CFE approach in decision support algorithms, and testing our method in continuous training scenarios where datasets are added one by one are just some of the exciting directions we plan to follow in the future works. Our promising results support the quality of cross-domain cell features extracted by AE-CFE and allow expansion of the developed approaches on new data collected from new sites and hospitals.

Acknowledgments. C.M. has received funding from the European Research Council (ERC) under the European Union's Horizon 2020 research and innovation programme (Grant agreement No. 866411).

References

1. Acevedo, A., Merino, A., Alférez, S., Molina, Á., Boldú, L., Rodellar, J.: A dataset of microscopic peripheral blood cell images for development of automatic recognition systems. Data Brief **30** (2020). ISSN 23523409
2. Acevedo, A., Merino, A., Boldú, L., Molina, Á., Alférez, S., Rodellar, J.: A new convolutional neural network predictive model for the automatic recognition of hypogranulated neutrophils in myelodysplastic syndromes. Comput. Biol. Med. **134**, 104479 (2021)
3. Boldú, L., Merino, A., Alférez, S., Molina, A., Acevedo, A., Rodellar, J.: Automatic recognition of different types of acute leukaemia in peripheral blood by image analysis. J. Clin. Pathol. **72**(11), 755–761 (2019)
4. Buitinck, L., et al.: API design for machine learning software: experiences from the scikit-learn project. In: ECML PKDD Workshop: Languages for Data Mining and Machine Learning, pp. 108–122 (2013)
5. Chen, C., Dou, Q., Chen, H., Qin, J., Heng, P.A.: Unsupervised bidirectional cross-modality adaptation via deeply synergistic image and feature alignment for medical image segmentation. IEEE Trans. Med. Imaging **39**(7), 2494–2505 (2020)

6. Deng, J., Dong, W., Socher, R., Li, L.J., Li, K., Fei-Fei, L.: ImageNet: a large-scale hierarchical image database. In: CVPR 2009 (2009)
7. Dou, Q., Coelho de Castro, D., Kamnitsas, K., Glocker, B.: Domain generalization via model-agnostic learning of semantic features. Adv. Neural Inf. Process. Syst. **32** (2019)
8. He, K., Gkioxari, G., Dollár, P., Girshick, R.: Mask R-CNN. In: Proceedings of the IEEE International Conference on Computer Vision, pp. 2961–2969 (2017)
9. He, K., Zhang, X., Ren, S., Sun, J.: Deep residual learning for image recognition. In: Proceedings of the IEEE Conference on Computer Vision and Pattern Recognition, pp. 770–778 (2016)
10. Lin, T.Y., Dollár, P., Girshick, R., He, K., Hariharan, B., Belongie, S.: Feature pyramid networks for object detection. In: Proceedings of the IEEE Conference on Computer Vision and Pattern Recognition, pp. 2117–2125 (2017)
11. Matek, C., Krappe, S., Münzenmayer, C., Haferlach, T., Marr, C.: Highly accurate differentiation of bone marrow cell morphologies using deep neural networks on a large image data set. Blood J. Am. Soc. Hematol. **138**(20), 1917–1927 (2021)
12. Matek, C., Schwarz, S., Spiekermann, K., Marr, C.: Human-level recognition of blast cells in acute myeloid leukaemia with convolutional neural networks. Nat. Mach. Intell. **1**(11), 538–544 (2019)
13. McInnes, L., Healy, J., Melville, J.: UMAP: uniform manifold approximation and projection for dimension reduction. arXiv preprint arXiv:1802.03426 (2018)
14. Tolstikhin, I.O., Sriperumbudur, B.K., Schölkopf, B.: Minimax estimation of maximum mean discrepancy with radial kernels. Adv. Neural Inf. Process. Syst. **29** (2016)
15. Tzeng, E., Hoffman, J., Saenko, K., Darrell, T.: Adversarial discriminative domain adaptation. In: Proceedings of the IEEE conference on computer vision and pattern recognition. pp. 7167–7176 (2017)
16. Wang, Z., Bovik, A.C., Sheikh, H.R., Simoncelli, E.P.: Image quality assessment: from error visibility to structural similarity. IEEE Trans. Image Process. **13**(4), 600–612 (2004)
17. Wu, Y., He, K.: Group normalization. In: Proceedings of the European Conference on Computer Vision (ECCV), pp. 3–19 (2018)
18. Xie, S., Girshick, R., Dollár, P., Tu, Z., He, K.: Aggregated residual transformations for deep neural networks. In: Proceedings of the IEEE Conference on Computer Vision and Pattern Recognition, pp. 1492–1500 (2017)

Attentional Generative Multimodal Network for Neonatal Postoperative Pain Estimation

Md Sirajus Salekin[1], Ghada Zamzmi[1], Dmitry Goldgof[1], Peter R. Mouton[2],
Kanwaljeet J. S. Anand[3], Terri Ashmeade[1], Stephanie Prescott[1],
Yangxin Huang[1], and Yu Sun[1(✉)]

[1] University of South Florida, Tampa, FL, USA
yusun@usf.edu
[2] SRC Biosciences, Tampa, FL, USA
[3] Stanford University, Stanford, CA, USA

Abstract. Artificial Intelligence (AI)-based methods allow for automatic assessment of pain intensity based on continuous monitoring and processing of subtle changes in sensory signals, including facial expression, body movements, and crying frequency. Currently, there is a large and growing need for expanding current AI-based approaches to the assessment of postoperative pain in the neonatal intensive care unit (NICU). In contrast to acute procedural pain in the clinic, the NICU has neonates emerging from postoperative sedation, usually intubated, and with variable energy reserves for manifesting forceful pain responses. Here, we present a novel multi-modal approach designed, developed, and validated for assessment of neonatal postoperative pain in the challenging NICU setting. Our approach includes a robust network capable of efficient reconstruction of missing modalities (e.g., obscured facial expression due to intubation) using an unsupervised spatio-temporal feature learning with a generative model for learning the joint features. Our approach generates the final pain score along with the intensity using an attentional cross-modal feature fusion. Using experimental dataset from postoperative neonates in the NICU, our pain assessment approach achieves superior performance (AUC 0.906, accuracy 0.820) as compared to the state-of-the-art approaches.

Keywords: Generative model · Multimodal learning · Neonatal pain · NICU · Postoperative pain

1 Introduction

It is known that newborns subjected to emergency surgical procedures experience variable levels of pain during the postoperative period. Studies [13] in human and animal neonates reported that postoperative pain leads to long-lasting and likely permanent harm to the normal development of the highly vulnerable nervous system of neonates. Thus, a major challenge for the scientific community

© The Author(s), under exclusive license to Springer Nature Switzerland AG 2022
L. Wang et al. (Eds.): MICCAI 2022, LNCS 13433, pp. 749–759, 2022.
https://doi.org/10.1007/978-3-031-16437-8_72

is to effectively assess, prevent, and mitigate where possible the impact of post-operative pain on neonates. The combination of artificial intelligence (AI)-based methods with continuous monitoring and capture of subtle visual signals have enhanced the capability for continuous assessment of pain behaviors in neonates [14,22,26]. Although these methods achieved promising results, the shortcomings of early methods have limited its applicability for widespread use in real-world clinical practice [13].

Prior approaches, except [14], focused on assessing neonatal acute procedural pain, i.e., short-term distress following brief medical procedure (e.g., immunization) that are routinely experienced by healthy newborns in the presence of caregivers. Given the relatively benign and transient impact of these painful experiences, there is a growing need for expanding pain assessments to help mitigate the long-term and potentially more harmful consequences of postoperative pain in the NICU [13]. Second, prior works were designed for clinical scenarios with full access to visual and audio signals with minimum occlusion and background noise; thus, these pain assessments would be expected to perform poorly or completely fail for intubated neonates, variable light conditions, and ambient sound. A third limitation is that prior works handled failure of signal detection (missing modalities) by ignoring the absent modality and making a final decision based on existing data. The resulting loss of relevant pain information and modality bias could lead to errors since all current manual scales for assessing postoperative pain rely on all modalities to generate the final pain scores.

Our technical contributions are as follows. First, we developed a deep feature extractor followed by an RNN (Recurrent Neural Network) autoencoder network to extract and learn spatio-temporal features from both visual and auditory modalities. Second, we designed a novel generative model that combines all the modalities while learning to reconstruct any missing modalities. Third, instead of using early or late fusion techniques, we used a transformer-based attentional model that learns cross-modal features and generates the final pain label along with its intensity. From an application standpoint, this work presents the first multimodal spatio-temporal approach for neonatal postoperative pain intensity estimation that is designed, developed, and evaluated using a dataset collected in a real-world NICU setting.

2   Related Works

Multimodal Learning: The common approach for multimodal learning involves training different modalities followed by an early or late fusion [4,7,19]. Since these approaches assume that both individual modalities and combined modalities have the same ground truth (GT) labels, it is not suitable for postoperative pain analysis, where modalities have individual GT labels that might differ from the final GT label. For example, the GT label for a specific modality (e.g., sound) could be no-pain while the final assessment of the neonate's state could be pain. To handle this issue, a recent work [14] uses a multimodal Bilinear CNN-LSTM network that was trained using individual modalities' labels and the

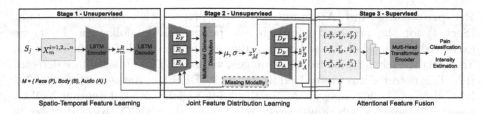

Fig. 1. Proposed approach for postoperative pain assessment (best viewed in color).

final label. The main limitation of this work is that the proposed network can not appropriately handle missing modalities; i.e., it makes the final assessment of pain based on the existing modalities, which can introduce an inherent bias towards available modalities.

Autoencoder (AE): AE [28] is a representational and unsupervised learning technique that learns a compressed feature embedding. In [17], an LSTM-based autoencoder is proposed to reconstruct and predict video sequences. To address the issue of non-regularized latent space in AE, Variational Autoencoder (VAE) [8] generates a probability distribution over the latent space. In multimodal learning, VAE has been extended to utilize multimodal joint learning for generative models such as CMMA [11], JMVAE [18], and MVAE [23].

3   Methodology

Figure 1 shows our novel approach in three stages: spatio-temporal feature learning (Stage 1), joint feature distribution learning (Stage 2), and attentional feature fusion (Stage 3). The following section presents the important notations and pre-processing steps then describes the details of each stage.

Definition and Notations: Let S be the number of visual samples in the video modality, and each sample consists of n number of frames i.e. $S_j = f_1, f_2, f_3, ..., f_n$ where $S_j \in S$. In the case of auditory modality, S_j is just one audio signal. Each pain episode contains three sensory signals $m \in M : face(F)$, $body(B)$, $audio(A)$, i.e. $M = \{F, B, A\}$. For any given sample, we have individual GT labels for F, B and A sensory signals. It is worth mentioning that unlike prior multimodal learning works [6,10,25] in postoperative pain assessment, the individual GT labels are provided based on the observation of the entire modality, not per frame. Finally, a final GT label is provided based on all sensory signals. This final label provides the assessment as pain or no-pain along with an intensity score. As any of these modalities or sensory signals can be missing in the real-world, we aim to detect the pain or no-pain class along with the pain intensity level. If any particular modality is missing, we reconstruct the modality and integrate it into the pain assessment contrary to the previous work [14], which entirely discards the missing modality.

Pre-processing and Augmentation: To prepare the multimodal dataset (Sect. 4) for the proposed approach, we performed the following. First, we extracted the visual (face and body) frames and audio signals from the raw data. To detect the facial region from the images, we used a YOLO-based face detector [12]. This detector was pre-trained using the WIDER face dataset [24] ($\approx$393,703 labeled faces, 32,203 images). As for the body region, we used another YOLO-based [12] detector, which was pre-trained using the COCO object dataset [9] ($\approx$1.5M object instances, 330K images). After detecting the face and body regions, we resized (224×224) all images to provide a consistent data flow in the multimodal network. In the case of the audio modality, we converted all the audio signals to $16K$ mono signals. Due to the partial occlusion of the neonate's face or body in some sequences, some frames were not detected which led to a different number of frames belonging to face and body modalities. To fix this issue and remove repetitive frames, we extracted the salient frames from these sequences with an equal time distribution. Inspired by [21], we divided each sequence into N equal segments. From each segment, we chose F-number of random frames. This has proven to be an efficient frame extraction method in several computer vision tasks [21]. In our experiments, we empirically chose the value of N and F as 10 and 1, respectively. Finally, we performed video augmentation by random rotation (±30) and horizontal flip. This augmentation was applied to all frames of a particular sequence dynamically during the training time.

Spatio-Temporal Feature Learning (Stage 1): We train an LSTM-based AE to capture the spatial and temporal features from the video (F, B) and auditory (A) modalities. Initially, we extracted spatial features from each facial image using FaceNet-based model [16]. This model was pre-trained on the VGGFace2 dataset [2]. For the body region, we used a Resnet18-based model, which was trained on the popular ImageNet [5] dataset. For the auditory modality, we used Google's VGGish model, which was pre-trained with YouTube-8M[1] dataset. Finally, feature sequences of each modality were used to train the LSTM-based AE in an unsupervised manner, where the encoder learns a compressed spatio-temporal feature representation from the deep features. For a spatial feature vector X_m^i with d_m feature-length and n sequence length, this AE maps the sequence as follows:

$$E_R : X_m^{i=1,2,\ldots,n} \rightarrow z_m^R \quad and \quad D_R : z_m^R \rightarrow \hat{X}_m^{i=1,2,\ldots,n} \tag{1}$$

$$L_R = \frac{1}{n} \sum_{i=1}^{n} (X_m^i - \hat{X}_m^i)^2 \tag{2}$$

where $m \in M$, E and D are the RNN encoder and decoder functions, z_m^R is the fixed size latent feature space of the RNN AE, and $\hat{X}$ are the reconstructed features. We used the mean square error (MSE) as the loss function (L_R) to learn the feature reconstruction.

---

[1] http://research.google.com/youtube8m/.

Joint Feature Distribution Learning (Stage 2): After training the LSTM AE, we extracted the latent feature z_m^R for each sensory signal. z_m^R is the feature vector for a particular video (F, B) or audio (A). To learn the joint probability distribution of these vectors, we used VAE [8,11,18,23]. A basic VAE consists of a generative (θ) model and inference (ϕ) model, and it is optimized through Evidence Lower Bound (ELBO). We initially generated a parameterized inference model to estimate the probability distribution (μ, σ) of the latent space for each sensory signal (F, B, A). We used a product of expert approximation (POE) [3] to generate a joint-posterior distribution. This POE acts as a common parameterized inference network to estimate the final probability distribution of the joint latent space. ELBO can be defined based on the combination of the likelihood and Kullback-Leibler (KL) divergence as follows:

$$ELBO(z_m^R) := \mathbb{E}_{q_\phi|z_m^R}[\lambda \log p_\theta(z_m^R|z^V)] - \beta KL[q_\phi(z^V|z_m^R), p(z^V)] \qquad (3)$$

where z_m^R and z^V are the observation and the latent space, respectively; $p_\theta(z_m^R|z^V)$ and $q_\phi(z^V|z_m^R)$ are the generative model and inference network respectively; $p(z^V)$ is the prior; λ and β are the controlled parameters [1,23]. To incorporate the POE over multiple sensory signals, this equation can be extended as:

$$ELBO(z_M^R) := \mathbb{E}_{q_\phi|z_m^R}[\sum_{m \in M} \lambda_m \log p_\theta(z_m^R|z^V)] - \beta KL[q_\phi(z^V|z_m^R), p(z^V)] \qquad (4)$$

Theoretically, training an ELBO consisting of N sensory signals requires 2^N combinations, which is computationally expensive. Therefore, we only optimized ELBO of the joint signals instead of individual signals. We passed *Null* values for the ELBO of the individual signals, and defined the joint learning loss (L_V) from this multimodal AE as follows:

$$L_V = ELBO(z_M^R) + ELBO(z_F^R) + ELBO(z_B^R) + ELBO(z_A^R) \qquad (5)$$

Based on the equation above, the multimodal AE can be trained under different missing data conditions. Specifically, if any signal is missing in the test case, the POE can still create the generative probability distribution, which is used to generate the common latent features (z_M^R) that acts as a common joint feature for all signals. Then, the multimodal AE can reconstruct the individual features $(\hat{z}_m^R)$ again from the common feature space (z_M^R). We used MSE as the loss function for the reconstruction.

Attentional Fusion (Stage 3): After generating the spatio-temporal latent space (z_m^R) and reconstructing missing modalities $(\hat{z}_m^R)$ from the joint probability latent space (z_M^R) in Stage 1 and Stage 2, we stacked the latent features of F, B, and A signals and applied an attentional fusion using the Transformer encoder [20] as follows:

$$Attention(Q, K, V) = softmax(\frac{QK^T}{\sqrt{d_k}})V \qquad (6)$$

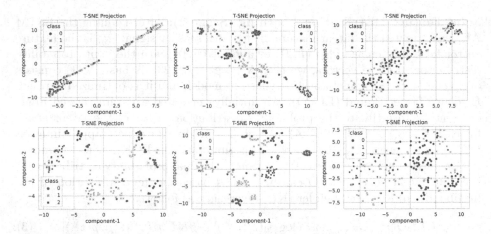

Fig. 2. t-SNE projection of spatio-temporal features using perplexity of 40. Each column represents face, body, and audio (left to right). Top and bottom rows are the baseline and proposed features, respectively (best viewed in color).

where, Q, K, V, and d_k are the query, key, value matrix, and the scaling factor, respectively. As shown in Fig. 1, attentive features were generated using the latent features from all modalities (F, B, A). The generated attentive features were then concatenated and used for pain assessment. Specifically, the pain assessment was produced as follows. The spatio-temporal feature (z_m^R) or reconstructed feature (z_m^V) was selected for each sensory signal. Then, the selected features were stacked followed by performing a multi-head attention to learn the cross-modal relation while focusing on the salient features. Finally, the attentive features were concatenated and used as a final feature vector. This vector was used for assessing pain and estimating its intensity.

4   Experimental Setup and Results

This section presents evaluation of the proposed approach (three stages) and the performance of both pain classification and intensity estimation. We used the accuracy, F-1 score, and AUC to report the performance of binary classification, and MSE and mean absolute error (MAE) to report the performance of intensity estimation. All the models were developed based on PyTorch environment using a GPU machine (Intel core i7-7700K@4.20 GHz, 32 GB RAM, and NVIDIA GV100 TITAN V 12 GB GPU).

Dataset: We used the USF-MNPAD-I [15] neonatal pain dataset, which is the only publicly available neonatal postoperative pain dataset for research use [13]. This dataset has 36 subjects recorded during acute procedural pain, and 9 subjects during postoperative pain. Each subject has videos (face and body) and audios (crying and background noises) recorded in the NICU of a local hospital. Each video and audio contain pain and no-pain segments that are labeled

Table 1. Performance of our approach and previous works when all signals are present.

Approach	Accuracy	Precision	Recall	F1-score	TPR	FPR	AUC
CNN-LSTM [14]	0.7895	0.7913	0.7895	0.7863	0.8761	0.3243	0.8791
EmbraceNet [4]	0.7921	0.7919	0.7921	0.7920	0.8182	0.2405	0.8790
Proposed	0.8202	0.8230	0.8202	0.8207	0.8080	0.1646	0.9055

with two manual pain scales: NIPS scale for procedural pain and N-PASS scale for postoperative pain. We used the procedural part of the dataset to learn the spatio-temporal features. The postoperative part was used to learn the joint feature distribution and reconstruct the missing modalities.

Network Architectures and Training: In Stage 1, we used state-of-the-art models (Sect. 3) to extract spatio-temporal feature vectors with 512-d, 512-d, and 128-d length from F, B, A signals, respectively. For temporal learning, we used an individual LSTM AE with 2 layers, taking the respective spatial feature vector of input sequences to produce a spatio-temporal 128-d latent space. As mentioned above (Sect. 3), the video has a sequence length of ≈ 10 s. In Stage 2, we used MLP encoder-decoder following $128 \to 128 \to 64$ and $64 \to 128 \to 128 \to 128$ encoder and decoder layers for each sensory signal. In Stage 3, a transformer encoder layer with 2 multi-heads had been used to initially perform the scale-dot-product attention. After that, all the features were concatenated $(128 + 128 + 128 = 384)$. Next, an MLP layer following $384 \to 256 \to 128 \to Y$ was used. In case of binary classification, a sigmoid function was used for pain and no-pain classes. As for estimation, $Y = 1$ is just a linear point for pain intensity estimation. A total of 218 postoperative videos (50% pain) were included in our experiments. Following previous approaches [14, 27], we performed a leave-one-subject-out (LOSO) evaluation. For the spatio-temporal training, we used the procedural dataset to learn the spatio-temporal features until convergence. For RNN autoencoder, we used Adam optimizer with 0.001 learning rate and 16 batch size. In the joint learning and attentional feature learning, we followed LOSO and used Adam optimizer with 0.0001 learning rate and batch size of 8.

Visualization of Spatio-temporal Features: Spatio-temporal features were computed using FaceNet (face) [16], ResNet18 (body), and VGGish (sound). To evaluate the quality of the extracted features, we generated the t-SNE projections for all modalities as shown in Fig. 2. Note that all modalities are trained on the procedural pain set (unsupervised) and tested on the postoperative set. From the figure, we can observe that the feature points are scattered in the first row, which shows the baselines for face, body, and sound. The baseline for face and body signals are the raw pixels obtained from the video modality while the baseline for the sound is the mel frequency cepstral coefficients (MFCCs) calculated from the auditory modality. On the contrary, the second row shows the feature points, which are generated by stage 1, grouped into clusters indicating a good differentiation capability of the extracted features.

Table 2. Performance of the proposed approach and [14] when dropping each modality.

Approach	Modalities	Reconstruction?	Accuracy	F1-score	TPR	FPR	AUC
CNN-LSTM [14]	Drop$_{Face}$	No	0.7719	0.7522	0.9897	0.5135	0.8763
	Drop$_{Body}$	No	0.6901	0.6703	0.8866	0.5676	0.8396
	Drop$_{Sound}$	No	0.7076	0.6630	1.0000	0.6757	0.8353
Proposed	Drop$_{Face}$	Yes	0.7921	0.7928	0.7576	0.1646	0.9022
	Drop$_{Body}$	Yes	0.8258	0.8257	0.8485	0.2025	0.9086
	Drop$_{Sound}$	Yes	0.6854	0.6374	0.9899	0.6962	0.8028

Table 3. Ablation study of the attentional feature fusion.

Approach	Accuracy	Precision	Recall	F1-Score	TPR	FPR	AUC
ST + JF	0.5229	0.7559	0.5229	0.3824	0.9999	0.9541	0.5757
ST + JF + AF	0.7890	0.7899	0.7890	0.7888	0.7615	0.1835	0.8870

* ST = Spatio-Temporal, JF = Joint Features, AF = Attentional Fusion

Pain Assessment w/o and w/ Missing Modalities: We compared our proposed classifier with CNN-LSTM approach [14] and another multimodal approach named EmbraceNet [4]. In this experiment, we performed pain assessment in a subset of USF-MNPAD-I [15] that has all the sensory signals present (F, B, A). From Table 1, we can see that the proposed approach outperformed [14] and achieved 0.820 accuracy and 0.906 AUC. Although our approach achieved a lower TPR as compared to [14], it improved the FPR (0.165) by almost 50%. Similarly, our approach significantly outperformed EmbraceNet [4] $(p < 0.01)$.

To evaluate the performance of our approach and the novel reconstruction method, we completely dropped (100%) each sensory signal, reconstructed the features of the dropped signal, combined them with the features of other signals, and reported the performance of multimodal pain classification. We also reported the pain assessment performance using CNN-LSTM as it is the most recent work in the literature that uses USF-MNPAD-I [15] dataset. Recall that this approach [14] discarded missing modalities when making a final assessment. We note that missing a sensory signal is common in clinical practices due to several factors including sensor failure, swaddling, or intbaution, among others. Our model can classify any case with missing modalities as it can reconstruct these modalities and integrate them into the assessment. From Table 2, we observe that reconstructing the features of face and body using our approach improved the performance as compared to CNN-LSTM. The lower performance of sound suggests that sound reconstruction has a higher impact on the final pain/no-pain decision, which is consistent with a similar trend observed in our previous work [14].

Multimodal Assessment with Attentional Feature Fusion: Unlike other approaches, we used an attentional fusion to examine the cross-modal influence on the decision. To evaluate this fusion approach, we performed an ablation study, in which we reported the performance of pain classification with and without attentional fusion. In Table 3, we can observe that the proposed attentional fusion (ST + JF + AF) improved the pain classification performance by a large margin, demonstrating the effectiveness of this fusion approach.

Postoperative Pain Estimation: As the pain intensity in USF-MNPAD-I [15] dataset ranges from 0 to 7, we performed a regression-based training to generate the intensity score. We found an MSE of 3.95 and an MAE of 1.73, which are reasonable for this relatively small and challenging dataset. We further minimized the intensity range and found better results which are 0–4 (MSE 0.75, MAE 0.73) and 0–1 (MSE 0.13, MAE 0.27). We also found that the proposed approach is capable of understanding the no-pain/pain/no-pain transitions while estimating pain intensity with a success rate of 71.15%.

5   Conclusion

This work presents a novel approach for neonatal postoperative pain assessment. Our results demonstrated the efficacy of our novel approach in constructing missing signals, a common situation in NICU settings. Further, our results demonstrated the efficacy of our fusion method in enhancing multimodal pain assessment. These results are promising and suggest the superiority of our approach, which was evaluated on a challenging real-world dataset, as compared to similar works in the literature. In the future, we plan to further evaluate the proposed approach using a large-scale multi-site neonatal multimodal postoperative pain dataset as well as investigate the performance of the proposed approach when two or more modalities are missing.

Acknowledgement. This research is supported by National Institutes of Health (NIH), United States Grant (NIH R21NR018756). Although the second author (G.Z.) is currently affiliated with NIH, this work was conducted while being with the University of South Florida. The opinions expressed in this article are the G.Z.'s own and do not reflect the view of NIH, the Department of Health and Human Services, or the United States government.

References

1. Bowman, S., Vilnis, L., Vinyals, O., Dai, A., Jozefowicz, R., Bengio, S.: Generating sentences from a continuous space. In: Proceedings of The 20th SIGNLL Conference on Computational Natural Language Learning, pp. 10–21 (2016)
2. Cao, Q., Shen, L., Xie, W., Parkhi, O.M., Zisserman, A.: VGGFace2: a dataset for recognising faces across pose and age. In: 2018 13th IEEE International Conference on Automatic Face & Gesture Recognition (FG 2018), pp. 67–74. IEEE (2018)
3. Cao, Y., Fleet, D.J.: Generalized product of experts for automatic and principled fusion of Gaussian process predictions. arXiv preprint arXiv:1410.7827 (2014)
4. Choi, J.H., Lee, J.S.: EmbraceNet: a robust deep learning architecture for multimodal classification. Inf. Fusion **51**, 259–270 (2019)

5. Deng, J., Dong, W., Socher, R., Li, L.J., Li, K., Fei-Fei, L.: ImageNet: a large-scale hierarchical image database. In: 2009 IEEE Conference on Computer Vision and Pattern Recognition, pp. 248–255. IEEE (2009)
6. Haque, M.A., et al.: Deep multimodal pain recognition: a database and comparison of spatio-temporal visual modalities. In: 2018 13th IEEE International Conference on Automatic Face & Gesture Recognition (FG 2018), pp. 250–257. IEEE (2018)
7. Joze, H.R.V., Shaban, A., Iuzzolino, M.L., Koishida, K.: MMTM: multimodal transfer module for CNN fusion. In: Proceedings of the IEEE/CVF Conference on Computer Vision and Pattern Recognition, pp. 13289–13299 (2020)
8. Kingma, D.P., Welling, M.: Auto-encoding variational bayes. In: LeCun, Y.B. (ed.) 2nd International Conference on Learning Representations, ICLR 2014, Banff, AB, Canada, 14–16 April 2014. Conference Track Proceedings (2014)
9. Lin, T.-Y., et al.: Microsoft COCO: common objects in context. In: Fleet, D., Pajdla, T., Schiele, B., Tuytelaars, T. (eds.) ECCV 2014. LNCS, vol. 8693, pp. 740–755. Springer, Cham (2014). https://doi.org/10.1007/978-3-319-10602-1_48
10. Lucey, P., Cohn, J.F., Prkachin, K.M., Solomon, P.E., Matthews, I.: Painful data: the UNBC-McMaster shoulder pain expression archive database. In: 2011 IEEE International Conference on Automatic Face & Gesture Recognition (FG), pp. 57–64. IEEE (2011)
11. Pandey, G., Dukkipati, A.: Variational methods for conditional multimodal deep learning. In: 2017 International Joint Conference on Neural Networks (IJCNN), pp. 308–315. IEEE (2017)
12. Redmon, J., Farhadi, A.: YOLOv3: an incremental improvement. arXiv preprint arXiv:1804.02767 (2018)
13. Salekin, M.S., et al.: Future roles of artificial intelligence in early pain management of newborns. Paediatr. Neonatal Pain 3(3), 134–145 (2021)
14. Salekin, M.S., Zamzmi, G., Goldgof, D., Kasturi, R., Ho, T., Sun, Y.: Multimodal spatio-temporal deep learning approach for neonatal postoperative pain assessment. Comput. Biol. Med. 129, 104150 (2021)
15. Salekin, M.S., et al.: Multimodal neonatal procedural and postoperative pain assessment dataset. Data Brief 35, 106796 (2021)
16. Schroff, F., Kalenichenko, D., Philbin, J.: FaceNet: a unified embedding for face recognition and clustering. In: Proceedings of the IEEE Conference on Computer Vision and Pattern Recognition, pp. 815–823 (2015)
17. Srivastava, N., Mansimov, E., Salakhudinov, R.: Unsupervised learning of video representations using LSTMs. In: International Conference on Machine Learning, pp. 843–852. PMLR (2015)
18. Suzuki, M., Nakayama, K., Matsuo, Y.: Joint multimodal learning with deep generative models. In: 5th International Conference on Learning Representations, ICLR 2017, Toulon, France, 24–26 April 2017. Workshop Track Proceedings. OpenReview.net (2017)
19. Tsiami, A., Koutras, P., Maragos, P.: STAViS: spatio-temporal audiovisual saliency network. In: Proceedings of the IEEE/CVF Conference on Computer Vision and Pattern Recognition, pp. 4766–4776 (2020)
20. Vaswani, A., et al.: Attention is all you need. Adv. Neural Inf. Process. Syst. 30, 5998–6008 (2017)
21. Wang, L., et al.: Temporal segment networks: towards good practices for deep action recognition. In: Leibe, B., Matas, J., Sebe, N., Welling, M. (eds.) ECCV 2016. LNCS, vol. 9912, pp. 20–36. Springer, Cham (2016). https://doi.org/10.1007/978-3-319-46484-8_2

22. Werner, P., Lopez-Martinez, D., Walter, S., Al-Hamadi, A., Gruss, S., Picard, R.: Automatic recognition methods supporting pain assessment: a survey. IEEE Trans. Affect. Comput. **13**(1), 530–552 (2022)

23. Wu, M., Goodman, N.D.: Multimodal generative models for scalable weakly-supervised learning. In: Bengio, S., Wallach, H.M., Larochelle, H., Grauman, K., Cesa-Bianchi, N., Garnett, R. (eds.) Advances in Neural Information Processing Systems 31: Annual Conference on Neural Information Processing Systems 2018, NeurIPS 2018, 3–8 December 2018, Montréal, Canada, pp. 5580–5590 (2018)

24. Yang, S., Luo, P., Loy, C.C., Tang, X.: Wider face: a face detection benchmark. In: Proceedings of the IEEE Conference on Computer Vision and Pattern Recognition, pp. 5525–5533 (2016)

25. Zadeh, A., Pu, P.: Multimodal language analysis in the wild: CMU-MOSEI dataset and interpretable dynamic fusion graph. In: Proceedings of the 56th Annual Meeting of the Association for Computational Linguistics (Long Papers) (2018)

26. Zamzmi, G., Kasturi, R., Goldgof, D., Zhi, R., Ashmeade, T., Sun, Y.: A review of automated pain assessment in infants: features, classification tasks, and databases. IEEE Rev. Biomed. Eng. **11**, 77–96 (2017)

27. Zamzmi, G., Pai, C.Y., Goldgof, D., Kasturi, R., Ashmeade, T., Sun, Y.: A comprehensive and context-sensitive neonatal pain assessment using computer vision. IEEE Trans. Affect. Comput. **13**(1), 28–45 (2022)

28. Zhao, Q., Adeli, E., Honnorat, N., Leng, T., Pohl, K.M.: Variational autoencoder for regression: application to brain aging analysis. In: Shen, D., et al. (eds.) MICCAI 2019. LNCS, vol. 11765, pp. 823–831. Springer, Cham (2019). https://doi.org/10.1007/978-3-030-32245-8_91

Deep Learning Based Modality-Independent Intracranial Aneurysm Detection

Žiga Bizjak[1]([✉]), June Ho Choi[2], Wonhyoung Park[2], and Žiga Špiclin[1]

[1] Laboratory of Imaging Technologies, Faculty of Electrical Engineering,
University of Ljubljana, Ljubljana, Slovenia
ziga.bizjak@fe.uni-lj.si
[2] Department of Neurological Surgery, Asan Medical Center,
University of Ulsan College of Medicine, Seoul, Republic of Korea

Abstract. Early detection of intracranial aneurysms (IAs) allows early treatment and therefore a better outcome for the patient. Deep learning-based models trained and executed on angiographic scans can highlight possible IA locations, which could increase visual detection sensitivity and substantially reduce the assessment time. Thus far methods were mostly trained and tested on single modality, while their reported performances within and across modalities seems insufficient for clinical application. This paper presents a modality-independent method for detection of IAs on MRAs and CTAs. First, the vascular surface meshes were automatically extracted from the CTA and MRA angiograms, using nnUnet approach. For IA detection purpose, the extracted surfaces were randomly parcellated into local patches and then a translation, rotation and scale invariant classifier based on deep neural network (DNN) was trained. Test stage proceeded by mimicking the surface extraction and parcellation, and the results across parcels were aggregated into IA detection heatmap of the entire vascular surface. Using 200 MRAs and 300 CTAs we trained and tested three models, two in cross modality setting (training on MRAs/CTAs and testing on CTAs/MRAs, respectively), while the third was a mixed-modality model, trained and tested on both modalities. The best model resulted in a 96% sensitivity at 0.81 false positive detections per image. Experimental results show that proposed approach not only significantly improved detection sensitivity and specificity compared to state-of-the-art methods, but is also modality agnostic, may aggregate information across modalities and thus seems better suited for clinical application.

Keywords: Cerebral angiograms · Aneurysm detection · Point cloud representation learning · Modality agnostic deep learning · Quantitative validation

1 Introduction

Intracranial aneurysms (IAs) are abnormal vessel wall dilatations in the cerebral vasculature and, according to a study that included 94, 912 subjects [17], have a high 3.2% prevalence in the general population. Most IAs are small and it is estimated that 50–80% do not rupture during a person's lifetime. Still, rupture of IA is one of the most common causes of subarachnoid hemorrhage (SAH) [16], a condition with 50% mortality

© The Author(s), under exclusive license to Springer Nature Switzerland AG 2022
L. Wang et al. (Eds.): MICCAI 2022, LNCS 13433, pp. 760–769, 2022.
https://doi.org/10.1007/978-3-031-16437-8_73

Fig. 1. Vascular 3D surface extracted from CTA *on the left* and MRA *on the right.*

rate [5]. For small IAs with diameter < 5 mm the chances of rupture are below 1%, but increase with ageing and potential IA growth, whereas rupture risk is generally higher for larger IAs. Early detection of IAs is thus necessary to open a window of opportunity to mitigate rupture risk and/or to determine the best time and type of treatment.

Current clinical practice is to search for IAs by visual inspection of 3D angiographic images like CTA, MRA and DSA. Such visual inspection is time consuming (10–15 min per case) and is prone to human error. Even skilled experts achieve a rather low sensitivity of 88% for small IAs on the CTA [19]. This is among the reasons why in recent years many researchers focused extensively on computer-assisted IA detection.

1.1 Background

In recent years, there has been a noticeable increase in the number of publications on the topic of brain aneurysm detection. Deep neural networks have made a significant contribution to this, as it turns out that using deep neural networks can solve aneurysm detection to some extent. A recent review article [9] summarizes most of the methods that are current state-of-the-art in this field, therefore we here describe those that stand out in terms of reported results and number of used cases.

In one of the larger studies conducted by Bo et al. [2] the GLIA-Net was proposed. They used 1476 CTA images (with 1590 aneurysms) and achieved a true positive rate (TPR) of 0.81 on test data and 4.38 false positive per image (FP/image). Yang et al. [18] published CTA aneurysm detection method with sensitivity of 0.975, however the FP/image was 13.8. The authors used 1068 subjects with 1337 aneurysms (688 were assigned to the training set and 649 to the test set). The algorithm also detected eight new aneurysms, which were missed in the baseline radiologic reports. Shahzad et al. [13] used smaller dataset of 215 aneurysms from 185 subjects. With their approach they achieved a TPR of 0.82 with 0.81 FP/image. These results indicate that methods on CTA images generally have either poor TPR or large number of FP/image.

It is noticeable that studies of aneurysm detection on the CTA modality are rare, and that most studies focus on the MRA modality. Recently, Timmins et al. [14] hosted an Aneurysm Detection And segMentation Challenge (ADAM), in which TOF-MRA images from 113 subjects with 129 unruptured IAs were released for training, while test set was comprised of 141 cases with 153 unruptured IAs. The best team achieved a TPR of 0.67 with FP/image of 0.13. Chen et al. [4] used 131 TOF-MRA images with 105 aneurysms and achieved TPR of 0.829 with 0.86 FP/image. Faron et al. [6] used TOF-MRA dataset with 115 IAs and achieved TPR of 0.90 while detecting 6.1 FP/image. Nakao et al. [10] utilized a 6-layer 3D convolutional neural network (CNN) to find IAs in TOF-MRA images. Similarly, Ueda et al. [15] used ResNet-18, pre-trained to detect four vascular pathologies and then fine-tuned to detect IAs in TOF-MRA images. The results of the aforementioned methods are summarized in Table 2.

Deep learning methods seem capable of detecting IAs regardless of their shape and size, but require access to a massive annotated image dataset. Another limitation is the use of intensity information, which renders these methods applicable for the particular modalities they were trained on, while they also are not able to aggregate information from different modalities during training.

In this paper we extended the work of Bizjak et al. [1] who proposed modality independent, deep learning based IA detection approach. However, the method was tested on small dataset of 57 DSA, 5 CTA and 5 MRA images. We further enhance the method by automating the vessel extraction from MRA and CTA images, thus making the IA detection fully automated. Validation is performed on a dataset of 300 CTAs and 200 MRAs.

1.2 Contributions

In this paper we aim to detect aneurysms from 3D meshes automatically obtained from 300 CTA and 200 MRA images. To prove that our approach can be applied and transferred across different modalities, we applied two cross-modality experiments, i.e. learning on MRA and testing on CTA and vice versa, and to demonstrate the potential for aggregating the data on a mixed-modality experiment.

This paper has three major contributions: (1) a novel automated IA detection pipeline that is applicable to different angiographic modalities; (2) our approach achieved state-of-the-art detection sensitivity on large dataset; (3) our approach achieved significantly lower false positive rate per image compared to current state-of-the-art methods and still exhibited high TPR.

2   Data and Methods

For training and validation of IAs detection we used 200 MRA and 300 CTA images. All MRA images were acquired at University Medical Centre Ljubljana, where the institutional ethics committee approved this study. There were 128 subjects (146/54 female/male, median age 53, range 22–74 years) that had one or more unruptured IAs (159 in total), while the remaining 72 were aneurysm-free. Mean diameter of the observed IAs was 5.85 mm, with 46% small (diameter < 5 mm), 47% medium (5 mm $<$ diameter < 10 mm) and 7% large size (diameter > 10 mm). The MRA acquisition protocol was not harmonized, due to the clinical origin of the data, and had in-plane voxel spacing range 0.35–0.45 mm, slice thickness range 0.5–1.29 mm, within-slice sizes from 384×284 to 672×768, and the number of slices in range 160–360.

The CTA scans were of 285 subjects (189/96 female/male, median age 54, range 10–88 years) were acquired at Asan Medical Center, where the institutional ethics committee approved this study. All subjects had at least one aneurysms (430 in total). The mean diameter of IAs was 7.64 mm, with 17% small (diameter < 5 mm), 63% medium (5 mm < diameter < 10 mm) and 20% large size (diameter > 10 mm) aneurysms. The CTA acquisition protocol was not harmonized, due to the clinical origin of the data, and had in-plane voxel spacing range of 0.23–0.57 mm, slice thickness range 0.4–1.0 mm, within-slice size 512 × 512, and the number of slices in range 60–416.

2.1 Vessel Extraction

Cerebrovascular angiographic modalities are 3D images that depict vascular structures with high intensity. Other anatomical structures may also be depicted such as cranial bones in CTA and soft brain tissue in MRA. For vessel segmentation from MRA and CTA images we applied a self adapting framework for U-Net based medical segmentation called nnU-Net [7]. Preprocessing is fully automated and provided by nnU-Net pipeline. Both models (MRA and CTA) were trained on scans with associated manually segmented intracranial vessels. To obtain vessel masks for training the nnU-net, the Slicer3D [11] and its interactive thresholding tools were used. Connected groups with small number of voxels, artifacts and other labeled brain regions were removed using manual segmentation editing. Results from each model were visually inspected by a skilled radiologist. On vascular segmentations we applied marching cubes and smooth non-shrinking algorithms [3,8] in order to obtain surfaces meshes to be input into subsequent aneurysm detection method. Examples of intracranial vessel surface meshes automatically extracted from MRA and CTA images are presented on Fig. 1.

2.2 Aneurysm Detection

This section presents a modality agnostic method for detection of IAs, which involves parcellation of the surface mesh into local patches of unstructured point clouds, learning their point count and geometrically invariant representation using DNN, and then using the DNN to classify each mesh point into either an aneurysm or vessel. In the testing stage, the classification obtained from the mesh patches was aggregated into IA detection heatmap of the entire vascular surface. For training and validating the IA detection method the aneurysm locations were defined by a skilled neurosurgeon, who searched for and segmented each aneurysm on volumetric images. This information was used for annotation of each IA by painting its surface. The flowchart of the proposed method is shown in Fig. 2. The four steps are detailed in the next subsections.

Surface Mesh Parcellation. Surface mesh was transformed into unstructured point cloud by randomly selecting seed points on the whole 3D vascular surface mesh and, for each seed, sampling 4000 closest mesh vertices according to geodesic distance. Hence, each point cloud contained exactly $N = 4000$ points. For prediction purposes, the surface was repeatedly parcellated, with random seed selection, until every point on 3D vascular mesh was sampled at least four times. For training purposes, each 3D mesh was parcellated into 70 point clouds, with ratio 35:35 of clouds containing aneurysm and vessel classes, respectively.

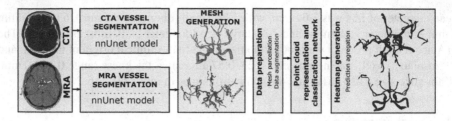

Fig. 2. Flowchart of the proposed aneurysm detection method.

Point Cloud Representation Learning for Classification. We employed the Point-Net architecture [12] to learn ordering, scale and rotation invariant representation of the input point clouds $\{x_i; i = 1, \ldots, N\}, x_i \in \mathbb{R}^n$. The idea of PointNet is to approximate a general function $f(\{x_1, \ldots, x_n\}) \approx g(h(x_1), \ldots, h(h(x_n)))$ by finding two mapping functions, i.e. $h : \mathbb{R}^N \rightarrow \mathbb{R}^K$ and $g : \mathbb{R}^K \times \mathbb{R}^K \times \mathbb{R}^K\} \rightarrow \mathbb{R}$. Functions $h(\cdot)$ and $g(\cdot)$ are modeled as DNNs that take N points as input, apply input and feature transformations, and then aggregate point features by max pooling. In this way, the input shape is summarized by a sparse set of key points K. The network was set to output classification scores for the aneurysm and vessel classes. For DNN training we input the point clouds and manual annotation as output, then use negative log likelihood loss with Adam optimizer run for 100 epochs, learning rate of 0.001, decay rate of 0.5 and decay step 20.

Prediction Aggregration. By extracting the surface mesh from the input image and parcellating into unstructured point clouds, the trained DNN was used to predict IA point labels for each of the point clouds. Then, the obtained soft class prediction probabilities were aggregated across all point clouds and their values normalized based on the number of point predictions. The final output was a heatmap for the aneurysm class as in Fig. 2, *right*, which highlighted potential IA locations.

3   Experiments and Results

The vessel extraction methods were trained and tested using 4-fold cross-validation on 200 MRA and 122 CTA cases and evaluated using the Dice similarity coefficient (DSC) with respect to the manual vessel segmentation masks.

The proposed aneurysm detection method was evaluated by comparing the obtained heatmaps to the manual annotations of IAs made by the neurosurgeon on the same surface mesh. Simple threshold was applied to the heatmap to get binary surface segmentation, then all surface segments larger than 50 connected points were labeled as IA. If a labeled section overlapped with the manually annotated section this was considered a true positive (TP); if not, it was considered as false positive (FP). A false negative (FN) was noted in case no labeled section overlapped with the manual IA annotation. A case was considered a true negative (TN) when vascular mesh didn't include any aneurysms and no sections were labeled.

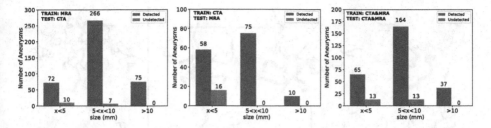

Fig. 3. The number of detected and undetected aneurysms with respect to the aneurysm size in each of the three experiments.

Table 1. Results across the three experiments, differing in the train and test set modality structure.

Experiment	TPR (TP/P)	TNR (TN/N)	FPs/image
Train MRA, Test CTA	0.96 (411/430)	0.40 (2/5)	0.81
Train CTA, Test MRA	0.90 (143/159)	0.58 (42/72)	1.06
Train & Test Mixed	0.91 (266/289)	0.62 (26/42)	0.98

Three distinct experiments were performed: the first two were cross-modality experiments, while the last experiment involved both CTA and MRA scans. We used TPR, TNR and FPs/image as evaluation metrics. We present the overall results, and results with respect to the IA size as obtained from manual annotation.

Vessel Segmentation Performance. The nnU-net models attained DSC values of 0.89 for MRA and 0.9 for CTA when compared to manual vessel segmentation. The average times to compute the segmentations were 26 and 120 s for the MRA and CTA, respectively.

Cross-Modality Detection Performance. We performed two cross-modality experiments, firstly we trained model on MRA images and tested it on CTA images. This model successfully detected 413 out of 430 aneurysms (TPR = 0.96), with 0.81 FP/image. The model successfully detected all large (>10 mm) aneurysms, and high TPR of 0.97% for medium (from 5 to 10 mm) aneurysms. The TPR for small (<5) aneurysms was 0.878. See results are listed in Table 1 and depicted in Fig. 3, *right*.

In the second experiment, we reversed the test and learning set, meaning we trained model on CTA images and tested it on MRA images. This model successfully detected 143 out of 159 aneurysms (TPR = 0.90). The model successfully detected all large and medium intracranial aneurysms, while for small aneurysms the model achieved TPR of 0.78. Out of 72 aneurysm-free subjects our model correctly classified 42 subjects as negative (TNR = 0.58). The number of FP/image was 1.06. Results can be observed in Table 1 and Fig. 3, *middle*.

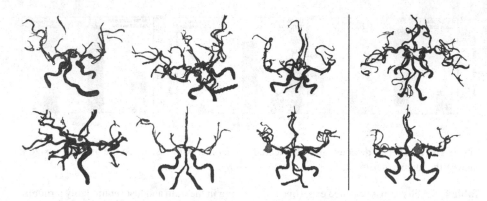

Fig. 4. Aneurysm detections (TPs) are highlighted in red, while the FPs are encircled. (Color figure online)

Mixed-Modality Detection Performance. The last experiment involved vascular meshes extracted from both modalities, using 100 MRAs and 150 CTAs image in train and test sets; i.e. altogether 250 CTAs&MRAs for training and 250 CTAs&MRAs for testing. Model successfully detected 266 out of 292 aneurysms (TPR = 0.91). Once again the model detected all large aneurysms, however it missed 13 small (TPR = 0.83) and 13 medium-sized (TPR = 0.92) aneurysms. Out of 42 aneurysm-free subjects 23 were correctly classified as negative (TNR = 0.65). The model detected 0.98 FP/image. See results in Table 1 and Fig. 3, *left*.

Comparison to State-of-the-Art. Table 2 summarizes the results of state-of-the-art and the proposed method. Compared to the state-of-the-art methods ours achieved a high TPR, while, at the same time, a rather low number of FP/image. Also, our approach was the only one that used both MRA and CTA cases as input in the training and in testing phase.

Table 2. Comparison of state-of-the-art and the proposed methods. Results of the proposed method are highlighted in **bold**.

Method	MRA IAs (#)	CTA IAs (#)	TPR	FPs/image
Bo et al. (2021) [2]	0	1476	0.81	4.36
Yang et al. (2021) [18]	0	1337	0.975	13.8
Shahzad et al. (2020) [13]	0	215	0.82	0.81
Timmins et al. (2021) [14]	282	0	0.67	0.13
Chen et al. (2020) [4]	105	0	0.829	0.86
Faron et al. (2020) [6]	115	0	0.9	6.1
Ueda et al. (2019) [15]	748	0	0.91	10.0
Nakao et al. (2018) [10]	450	0	0.942	2.9
Ours (best)	**200**	**300**	**0.96**	**0.81**

4   Discussion and Conclusion

A practical aneurysm detection method needs to achieve a good sensitivity–specificity balance; namely it should exhibit a high TPR and a low number of FPs/image. Current state-of-the-art methods based on raster image analysis present either a high TPR at the expense of high number of FPs/image [6, 10, 15, 18] or a low number of FPs/image at the expense of low TPR [2, 13, 14]; however, for a method to be considered for clinical application, a high TPR and, at the same time, a low number of FPs/image must be demonstrated on a large dataset including aneurysms of different sizes (but including substantial amount of the small aneurysms) and at different locations.

In this study we focused on extending, improving and extensively evaluating the approach by Bizjak et al. [1], which originally used DSA images. Here we introduced a fully automated vessel surface extraction using nnU-Net for MRA and CTA modality, a prerequisite step to apply their method, and thus made the aneurysm detection process fully automated. We performed quantitative validation using two large datasets, one containing MRA and other CTA images. See the visual outcomes in Fig. 4.

The proposed method is applicable to two major angiographic modalities used for patient screening. To support this claim, we performed the two cross modality experiments. In the first one, training was executed on MRA cases and testing on CTA cases. The result was a state-of-the-art TPR of 0.96 with FPs/image lower then one (0.81). In a reversed experiment, we achieved slightly lower TPR (0.90) with larger number of FPs/image (1.06), which can be attributed to the higher number of small aneurysms present in the MRA dataset; namely, almost 47% were small aneurysms (<5 mm in diameter), which are generally most difficult to detect. Nevertheless, the reported results are still state-of-the-art for the MRA modality (see Table 2). Note that all aneurysms larger than 5 mm were detected in the MRAs (see Fig. 3, *middle*). In the final experiment we trained and tested on both modalities and the resulting model performed better than in CTA–MRA experiment, but slightly worse than in the MRA–CTA experiment. The number of FPs/image was albeit rather low at 0.98.

To the best of our knowledge, our method was the only one tested in the most difficult multi-modal scenario, i.e. training on one angiographic modality and testing on the other, and vice versa. Furthermore, it seems that the proposed method can successfully aggregate information from both modalities as the observed IA detection performance was excellent and consistent with the performance in case of mixed-modality detection.

The current clinical practice is to search for IAs by visual inspection of 3D angiographic images like CTA, MRA and, in the interventional suite also DSA. Such visual inspection is not only time consuming (10–15 min per image), but also prone to human error, especially for small aneurysms (<5 mm in diameter). The high TPR and low number of FPs/image using the proposed method indicates that such method would be feasible as a second reading tool in the clinical workflow.

Alternatively, it may be deployed as a computer-assisted visualization and diagnosis tool, in which the output prediction in form of the 3D heatmap is superimposed on the extracted vascular surface mesh as to highlight potential aneurysm locations and thus assist the neurosurgeon. Based on the presented results we expect that, because the aneurysms are still being visually detected, such a tool to visualize the potential locations would render (small) aneurysm detection more sensitive and/or leave more time for inspection of other potential locations. For instance, the expect visual sensitivity on CTAs was 88% [19]); interestingly, our results for small aneurysms on CTAs indicate the same level of sensitivity (88%). Hence, the added value of the proposed method would be about a 10-fold reduction of inspection time to less then a minute in order to detect and visualize aneurysm locations, thus saving a substantial amount of manual effort and associated costs.

Acknowledgements. This study was supported by the Slovenian Research Agency (Core Research Grant No. P2-0232 and Research Grants Nos. J2-2500 and J2-3059).

References

1. Bizjak, Ž., Likar, B., Pernuš, F., Špiclin, Ž.: Modality agnostic intracranial aneurysm detection through supervised vascular surface classification. In: Medical Imaging 2021: Computer-Aided Diagnosis, vol. 11597, p. 1159700. International Society for Optics and Photonics (2021)
2. Bo, Z.H., et al.: Toward human intervention-free clinical diagnosis of intracranial aneurysm via deep neural network. Patterns **2**(2), 100197 (2021)
3. Cebral, J.R., Löhner, R.: From medical images to anatomically accurate finite element grids. Int. J. Numer. Meth. Eng. **51**(8), 985–1008 (2001)
4. Chen, G., et al.: Automated computer-assisted detection system for cerebral aneurysms in time-of-flight magnetic resonance angiography using fully convolutional network. Biomed. Eng. Online **19**(1), 1–10 (2020)
5. Etminan, N., et al.: Worldwide incidence of aneurysmal subarachnoid hemorrhage according to region, time period, blood pressure, and smoking prevalence in the population: a systematic review and meta-analysis. JAMA Neurol. **76**(5), 588–597 (2019)
6. Faron, A., et al.: Performance of a deep-learning neural network to detect intracranial aneurysms from 3D TOF-MRA compared to human readers. Clin. Neuroradiol. **30**(3), 591–598 (2020)
7. Isensee, F., Jaeger, P.F., Kohl, S.A., Petersen, J., Maier-Hein, K.H.: nnU-Net: a self-configuring method for deep learning-based biomedical image segmentation. Nat. Methods **18**(2), 203–211 (2021)
8. Larrabide, I., et al.: Three-dimensional morphological analysis of intracranial aneurysms: a fully automated method for aneurysm SAC isolation and quantification. Med. Phys. **38**(5), 2439–2449 (2011)
9. Mensah, E., Pringle, C., Roberts, G., Gurusinghe, N., Golash, A., Alalade, A.F.: Deep learning in the management of intracranial aneurysms and cerebrovascular diseases: a review of the current literature. World Neurosurgery (2022)
10. Nakao, T., et al.: Deep neural network-based computer-assisted detection of cerebral aneurysms in MR angiography. J. Magn. Reson. Imaging **47**(4), 948–953 (2018)
11. Pieper, S., Halle, M., Kikinis, R.: 3D slicer. In: 2004 2nd IEEE International Symposium on Biomedical Imaging: Nano to Macro (IEEE Cat No. 04EX821), pp. 632–635. IEEE (2004)
12. Qi, C.R., Su, H., Mo, K., Guibas, L.J.: PointNet: deep learning on point sets for 3D classification and segmentation. In: Proceedings of the IEEE Conference on Computer Vision and Pattern Recognition, pp. 652–660 (2017)
13. Shahzad, R., et al.: Fully automated detection and segmentation of intracranial aneurysms in subarachnoid hemorrhage on CTA using deep learning. Sci. Rep. **10**(1), 1–12 (2020)
14. Timmins, K.M., et al.: Comparing methods of detecting and segmenting unruptured intracranial aneurysms on TOF-MRAS: the ADAM challenge. Neuroimage **238**, 118216 (2021)

15. Ueda, D., et al.: Deep learning for MR angiography: automated detection of cerebral aneurysms. Radiology **290**(1), 187–194 (2019)
16. Van Gijn, J., Kerr, R.S., Rinkel, G.J.: Subarachnoid haemorrhage. Lancet **369**(9558), 306–318 (2007)
17. Vlak, M.H., Algra, A., Brandenburg, R., Rinkel, G.J.: Prevalence of unruptured intracranial aneurysms, with emphasis on sex, age, comorbidity, country, and time period: a systematic review and meta-analysis. Lancet Neurol. **10**(7), 626–636 (2011)
18. Yang, J., et al.: Deep learning for detecting cerebral aneurysms with CT angiography. Radiology **298**(1), 155–163 (2021)
19. Yang, Z.L., et al.: Radiology **285**(3), 941–952 (2017)

LIDP: A Lung Image Dataset with Pathological Information for Lung Cancer Screening

Yanbo Shao[1], Minghao Wang[1], Juanyun Mai[1], Xinliang Fu[1], Mei Li[1], Jiayin Zheng[1], Zhaoqi Diao[1], Airu Yin[1(✉)], Yulong Chen[2], Jianyu Xiao[2], Jian You[2], Yang Yang[3], Xiangcheng Qiu[3], Jinsheng Tao[3], Bo Wang[3], and Hua Ji[1,3(✉)]

[1] Advanced Medical Data Research Center, College of Computer Science, Nankai University, Tianjin, China
{yinar,hua.ji}@nankai.edu.cn
[2] Department of Lung Cancer, Radiology, Tianjin Lung Cancer Center, Tianjin Medical University, Tianjin, China
[3] AnchorDx Medical Co., Guangzhou, China

Abstract. Lung cancer has been one of the greatest lethal cancers worldwide. Computed Tomograph (CT) makes it possible to diagnose lung cancer at an early stage, which can significantly reduce its mortality. In recent years, deep neural networks (DNN) have been widely used to improve the accuracy of benign and malignant pulmonary nodules classification. But the limitation of DNN approach is that AI model's performance and generalization highly depend on the size and quality of the training data. With our best knowledge, almost all existing public lung nodule datasets, e.g., LIDC-IDRI, obtain the crucial benign and malignant labels by radiographic analysis, instead of pathological examination. In this paper, we argue that, without pathology report and hence lack of labels' authenticity, LIDC-IDRI based machine-learning (ML) models are short of generalization. To prove our hypothesis, we introduce a new lung CT image dataset with pathological information (LIDP), for lung cancer screening. LIDP contains 990 samples, including 783 malignant samples and 207 benign samples. More critically, the labels of all samples have been all examined by pathological biopsy. We evaluate various of existing LIDC-based state-of-the-art (SOTA) models on LIDP. Our experimental results show the extreme poor generalization ability of existing SOTA models that are trained on LIDC-IDRI dataset. Our scientific conclusion is striking: the distributions of these datasets are significantly different. We claim that the LIDP dataset is a very valuable addition to the existing datasets like LIDC-IDRI. LIDP can be well used for independent testing or for training new ML models for lung cancer early detection.

Keywords: Lung nodule · Computed tomography (CT) · Computer-aided diagnosis (CAD) · Benign and malignant classification

Supplementary Information The online version contains supplementary material available at https://doi.org/10.1007/978-3-031-16437-8_74.

© The Author(s), under exclusive license to Springer Nature Switzerland AG 2022
L. Wang et al. (Eds.): MICCAI 2022, LNCS 13433, pp. 770–779, 2022.
https://doi.org/10.1007/978-3-031-16437-8_74

1 Introduction

According to the World Health Organization's International Agency for Research on Cancer (IARC) 2020 global cancer statistics [17], there were approximately 2.2 million new cases of lung cancer in 2020, with around 1.8 million fatalities. Lung cancer has the greatest fatality rate of all cancers. Early detection and timely treatment can significantly reduce lung cancer mortality [19]. Imaging examinations, hematological examinations, and pathological examinations are three typical lung cancer screening methods. And the results of pathological examination serving as the gold standard for lung cancer diagnosis [14], while it is performed by puncture or surgery, both of which are harmful to the human body. Low-dose Computed Tomography (LDCT) is a radiographic-based lung cancer screening method that has been shown to significantly reduce the death rate [18]. However, classifying benign and malignant pulmonary nodules based on CT is a difficult task that heavily relies on the doctors' clinical expertise and professional knowledge [5, 19].

In recent years, lots of studies utilize ML, especially, the DNN models to classify pulmonary nodules because of the advantages of automated feature extraction. However, the classification performance and generalization ability of DNN models strongly depend on the underlying training dataset. To our best knowledge, the Lung Image Database Consortium and Image Database Resource Initiative (LIDC-IDRI) [4] has been used as the benchmark for most studies on benign and malignant classification models of pulmonary nodules [1–3, 6, 7, 10, 11, 13, 15, 16, 20–23].

In this paper, we argue that the LIDC-IDRI dataset has significant limitations as a benchmark for benign and malignant lung nodule classification. The benign and malignant labels in LIDC-IDRI dataset are obtained from the diagnose from CT scans by 1–4 radiologists. Based on their imaging characteristics, the radiologist graded the malignancy score of nodules on a scale of 1 to 5. There are two potential problems in LIDC-IDRI dataset. (i) Inaccurate labels. The results of diagnosis are not the gold standard for the judgment of benign and malignant pulmonary nodules [5, 19], so there is the possibility of misdiagnosis. Labels of datasets are used to supervise training and model evaluation in deep learning. As a result, incorrect data labeling can lead to erroneous research findings. (ii) Absence of difficult-to-classify samples. Data with a malignancy score of 3 cannot be classified as positive or negative due to the lack of a precise label, and are thus often eliminated from researches. It results in the loss of the most essential research information related to pulmonary nodules classification, as well as a reduction in the difficulty of the classification task, giving the impression that it is simple to learn.

To address the issues above, we introduce a lung CT dataset, called LIDP. The LIDP dataset has the following two characteristics. (i) **Precise Pathological information.** All of the samples in this dataset were from patients who had undergone lung cancer surgery. As a result, the dataset's benign and malignant labels are gained from pathological examinations, and CT scans were obtained before surgery, which makes it more accurate. **(ii) Challenging difficult-to-classify samples.** Because all of the participants had been identified with a high risk of lung cancer, the benign samples were difficult to identify, which presents new challenges to image-based pulmonary nodules classification. Based on this dataset, we evaluate the existing pulmonary nodules classification model from the perspective of classification performance and generalization, as well as a detailed comparison with baselines built using DCNN models. Our research work can be summarized with the following contributions:

- We build the LIDP dataset with pathological information for early lung cancer screening. The dataset contains 990 samples, including 783 malignant samples and 207 benign samples, which comprises various difficult-to-classify samples. To our best knowledge, LIDP is the largest available dataset with pathological gold standard.
- We use LIDP to evaluate the generalization and classification performance of the LIDC-based SOTA models and establish baseline for classifying lung nodules as benign or malignant. Experimental results clearly show that the LIDC-based SOTA models are extremely lack of generalization and cannot predict the true categories from CT scans accurately. Therefore, there is significant difference in distribution between the two datasets. LIDP can be well used as a good complementary dataset for lung cancer screening study.

2   Related Work

2.1   Dataset

Open datasets are used as benchmarks for comparing the performance of various models. In the field of CAD pulmonary nodules classification, the LIDC-IDRI [4], LUNGx Challenge Dataset [9] and DSB [12] are extensively employed. However, the three datasets have many limitations in terms of lack of pathological information, small amount of data, careless labeling and non-disclosure.

LIDC-IDRI [4] is a dataset for pulmonary nodule detection, benign and malignant classification, and quantitative evaluation. It contains 1018 cases from seven different academic institutions. However, the malignancy scores of LIDC-IDRI are based on visual diagnosis, which may be conflicting with the gold standard, leading to study misdirection. The LUNGx Challenge Dataset was published in the 2014 SPIE-AAPM-NCI Pulmonary Nodule Classification Challenge with pathological information [9]. Unfortunately, there are only 10 nodules in the calibration set and 73 nodules in the test set. It's too small to support DNNs training. Tianchi released a chest CT open dataset (DSB) [12] in 2017, including 2101 samples. However, this dataset is not labeled with nodule levels and no longer accessible at this time.

The existing benign and malignant classification models of pulmonary nodules are all explored using LIDC-IDRI because of the short quantity of data in LUNGx and the unavailability of DSB. As a result, this paper focuses on the comparative analysis of LIDP and LIDC-IDRI.

2.2   LIDC-IDRI Based DNN Models for Benign and Malignant Classification

The study of classification model based on LIDC-IDRI has been carried out from the following perspectives: the fitting of a priori characteristics, the processing of multi-scale and the adaptation of DNNs to small-scale medical datasets. All these DNN models claim to have excellent classification performance on LIDC-IDRI, even if not use pathological information. We summarize the classification performance of related work on LIDC-IDRI in the Table 1.

Table 1. Classification model of pulmonary nodules based on LIDC-IDRI

Model	Year	AUC	Model	Year	AUC
Multi-crop [16]	2017	93%	DenseSharp$^+$(HighAmbig) [24]	2019	95.66%
Fuse-TSD [23]	2018	96.65%	HESAM D_{11c} [11]	2020	–(ACC 99.13%)
MV-KBC [21]	2019	95.70%	Transferable Texture CNN [3]	2020	99.11%
Gated-Dilated [2]	2019	95.14&	MDGAN [10]	2020	94.26%
SSAC [22]	2019	95.81%	NASLung [8]	2021	–(ACC 90.77%)

3   LIDP Dataset

The LIDP dataset is a collection of lung CT images gathered as part of Project Bell[1]. The LIDP dataset comprises 990 CT scans from eight institutions. LIDP comprises pre-operative CT scan, lesion location, lesion contour, and pathological biopsy information, with each pathological information corresponding to just one target nodule in each CT scan. The most crucial element that distinguishes LIDP from other datasets is pathological information.

3.1   Data Collection and Annotation

To ensure the quality and universality of pulmonary nodules, solid nodules, mixed ground glass nodules (mGGN) and pure ground glass nodules (pGGN) with a long diamete of 5 mm to 30 mm can be collected into LIDP dataset. When collecting data in medical institutions, all patients are older than 18 years, excluding pregnant women and lactating women. And images with enlargement of hilar or septal lymph nodes are also excluded to assure the quality of CT scans. Besides, with the exception of three CT scans, all slice resolutions are 512 * 512.

We annotate the nodule position and contour under the guidance of experienced doctors. According to the pulmonary nodules imaging information and pathological information recorded in the original dataset, doctors validate the lesion location, and then instruct the annotators to label all the sections covered

---

[1] http://ncrc.gyfyy.com/index.php?ac=article&at=read&did=509.

by the lesion slice by slice. As part of the labeling procedure, contour labeling for extra non-target nodules larger than or equal to 3 mm contained in CT scan, which may be used for the study of pulmonary nodules detection and segmentation, was done in addition to the target nodules matching to pathological information. The contour was checked twice after annotating. Finally, we annotate 1165 nodules from 990 CT scans in LIDP.

3.2   Dataset Properties

We present the LIDP dataset from the perspectives of pathological information and image characteristics of target nodules. Please refer to supplementary materials (Figure A.1, Table A.1 and Table A.2) for more details on LIDP's data statistic distribution.

Thickness. LIDP only retain CT scans with a thickness of less than or equal to 2 mm during the data collecting procedure. In the LIDP, 1 mm slice thickness CT was the most common, with 674 CT scans, followed by 2 mm slice thickness CT with 188 CT scans.

Pathological Information. Since the CT data in this dataset were gathered from patients having lung cancer surgery, the probability of lung cancer in each instance is high. As a result, there were more malignant instances in this dataset than benign cases, with 783 malignant cases (79.09%) and 207 benign cases (20.91%). Table A.1 depict the distribution of various case subtypes.

Texture. Solid nodules, pGGN, and mGGN are the three categories of pulmonary nodules based on density. Figure A.1(a) shows that in the LIDP dataset, mGGNs were the most common, accounting for 60.81%, followed by pGGNs, which accounts for 25.56%, and solid nodules, which accounts for 13.63%.

Long Diameter. Pulmonary nodule's long diameter is the largest diameter in the plane of a cross section. The number of nodules in the range of (10 mm–20 mm) was the biggest in the LIDP dataset, accounting for 48.69%; the numbers of nodules in the ranges of (0 mm–10 mm) and (20 mm–30 mm) were nearly identical, accounting for 28.38% and 22.93%, respectively.

3.3   Comparison of LIDP and LIDC-IDRI

Sources of Benign and Malignant Labels. The benign and malignant labels of LIDP were determined from pathological examination, whereas LIDC-IDRI collects it by the radiologist's diagnosis based on CT scans. The gold standard for lung cancer diagnosis is pathological examination, hence the accuracy of the benign and malignant labels in LIDP is higher than that in LIDC-IDRI.

Benign and Malignant Distribution. In the LIDC-IDRI dataset, the number of benign and malignant pulmonary nodules were evenly distributed, however in the LIDP dataset, malignant nodules were approximately four times as prevalent as benign nodules. As a result, LIDP provides rich and diverse malignant samples for study which can reduce the missing rate of model.

The Study of Difficult-to-Classify Samples. It is impossible to conduct classification research on difficult-to-classify samples in the LIDC-IDRI dataset. Nodules with a malignancy score of 3 are difficult to classify and are removed from the research due to a lack of precise labels. The benign patients in the LIDP dataset were diagnosed as having a high probability of lung cancer by doctors at the diagnosis stage. Therefore, the LIDP dataset can be used to study difficult-to-classify samples because of the deterministic labeling.

4   Experimental Setup

4.1   Data Preprocessing

Nodules were clipped out from CT scans during classification and then input into the model to limit the effect of other unrelated areas on categorization. The three phases of the data preparation procedure are as follows:

- The pixel spacing in three directions was corrected to 1 mm using cubic spline interpolation.
- The image's HU value less than -1200 was modified to -1200, while the image's HU value more than 600 was adjusted to 600.
- A cube the size of $64 * 64 * 64$ was cut out of the CT using the nodule's center as the midpoint.

4.2   Training Method and Parameter Setting

We split LIDP randomly before the experiment. First, we split an independent test set that accounts for 1/5 of the whole dataset. We aim to balance the benign and malignant samples as much as possible while splitting the testset, and the testset comprises nodules of diverse texture and long diameter. Then, we separated the rest samples into five subsets for 5-fold cross-validation, and all subsets have the same proportion of benign and malignant nodules.

We conducted dynamic sampling of benign and malignant samples in the training data before the training model to mitigate the effects of category imbalance, with a sampling weight ratio of 5:1 for benign and malignant samples. Random data enhancement was performed on the sampled samples. Flipping, cropping, rotation and axis transformation was randomly superimposed to generate various nodule images.

We utilized the cross entropy loss function to supervise the training of the model and the stochastic gradient descent to optimize the model during training. The batchsize was 32. The initial learning rate was0.01 and the learning rate dropped dynamically when the training set loss changes. L2 regularization and early stopping were performed to prevent overfitting. We chose 0.5 as the dividing point for the probability of benign and malignant nodules. Model training and testing is carried out on the PyTorch framework.

5   Evaluation Experiment

Current studies on benign and malignant nodules are mostly based on LIDC-IDRI, the dataset without pathological information. Therefore, we design three sets of experiments to test the SOTA model's generality and classification performance, forethemore, we create a classification baseline of lung nodules based on the LIDP dataset.

Table 2. Classification performance of SOTA benign and malignant classification model on different datasets

Model	Training set	Test set	AUC (%)	Accuracy (%)	Recall (%)	Specific (%)	Precision (%)	F1-score (%)
MC-CNN	LIDC-IDRI	LIDC-IDRI	97.90	92.77	93.14	92.39	91.38	92.22
MC-CNN	LIDC-IDRI	LIDP	55.23	56.94	81.78	22.96	59.27	68.66
MC-CNN	LIDP	LIDP	73.37	67.89	79.33	52.24	69.50	73.99
MV-KBC	LIDC-IDRI	LIDC-IDRI	92.19	79.56	83.55	76.21	7765	79.15
MV-KBC	LIDC-IDRI	LIDP	46.75	56.17	87.98	12.63	57.91	69.72
MV-KBC	LIDP	LIDP	72.51	66.50	86.44	39.21	66.39	74.77
HESAM D_{11c}	LIDC-IDRI	LIDC-IDRI	97.27	91.24	89.85	92.20	91.10	90.34
HESAM D_{11c}	LIDC-IDRI	LIDP	57.66	56.25	87.12	14.01	58.11	69.69
HESAM D_{11c}	LIDP	LIDP	75.59	69.23	75.57	60.53	73.01	73.52

5.1   The Generalization of SOTA Models Trained on LIDC-IDRI

To evaluate the generalization ability of SOTA models, we independently tested three models training on LIDC-IDRI by LIDP. Three SOTA models were chosen from the perspectives of multi-scale, multi-view, and the fitting of prior features, which is multi-crop CNN(MC-CNN) [16], MV-KBC [21], and HESAM [11]. Table 2 show the experimental results.

The AUC of these SOTA models on LIDC-IDRI can achieve 97%, but the testing AUC on LIDP does not exceed 58%. This suggests that the model trained on LIDC-IDRI had no generalization ability on LIDP, and they have been over-fitted to the LIDC-IDRI dataset Therefore, imaging based diagnosis is not a substitute for pathology to guide benign and malignant classification research, which can lead to a large gap between the model and the ground truths. As a result, the pathology data in LIDP is extremely valuable for study.

5.2   The Performance of SOTA Model Trained on LIDP Dataset

To explore the classification performance of SOTA models on LIDP, we train Multi-crop CNN(MC-CNN) [16], MV-KBC [21], and HESAM [11] on LIDP. Table 2 shows the experimental results. The SOTA model's classification AUC on LIDC-IDRI can reach 97%, whereas the classification AUC on LIDP can only reach 72%, and the AUC on various datasets drops by 25%. In short, the model can not predict the pathological results from the CT scans accurately. The LIDP with pathological information can provide a platform for the research of the fitting of pathological findings from CT scans.

Table 3. Performance of pulmonary nodules classification model based on LIDP

Model	AUC (%)	Accuracy (%)	Recall (%)	Specific (%)	Precision (%)	F1-score (%)
ResNet10	73.68	68.17	90.10	38.16	66.71	75.09
ResNet18	71.65	68.56	87.02	43.29	67.93	76.18
ResNet34	71.04	66.28	86.73	38.29	66.07	74.59
ResNet50	71.74	67.12	85.00	43.82	67.80	75.15
VGG11	75.55	69.11	85.48	46.71	69.10	76.14
VGG13	77.33	68.67	85.58	45.53	68.50	75.84
VGG16	75.96	68.50	80.96	51.45	70.82	74.67
VGG19	75.50	69.60	78.66	56.84	71.72	74.35
GoogLeNet	74.48	70.22	87.31	46.85	69.32	77.11

5.3 The Exploration of LIDP Classification Model Baseline

We explore the baseline of classification models on LIDP by training ResNet 18/34/50, VGG 11/13/16/19, and GoogLeNet. The detailed results are presented in Table 3. In short, the optimum AUC model is VGG13, with an AUC of 77.33%. The model trained on LIDP shows high recall and low specificity. This may be related to data distribution, or it may be difficult to identify hard-to-classify samples. The classification result does not improve progressively as the number of network layers increases, but when 9 convolutional layers (ResNet10) or 10 convolutional layers (VGG13) are stacked in the network, the classification performance achieves its top. The reason why VGG series models have higher classification performance might be that VGG series models use several fully connected layers to strengthen the ability of nonlinear features representation.

In addition, compared with the baseline, the classification performance of SOTA model on LIDP did not exceed that of VGG13. Although these SOTA models were proposed particularly for the classification of pulmonary nodules, none of these models show fine classification performance on LIDP, which strongly suggests that previous researches have apparently overstudied LIDC-IDRI.

6   Conclusion

In this paper, we present LIDP, a new lung CT image dataset which contains 783 malignant and 207 benign samples, and all of whose labels were obtained through tissue biopsy. To qualitatively and quantitatively analyze the characteristics between the LIDP and the LIDC-IDRI, we have conducted comparison and evaluation work on multiple SOTA models. According to the experiment results, existing SOTA models cannot learn LIDP pathological information well and show very poor generalization ability. we can clearly conclude two points: **(I)** There is a significant difference in data distribution between LIDP and LIDC-IDRI. **(II)** CAD research on lung cancer benign and malignant classification should pay more efforts on dataset with pathology information rather than radiologist diagnosis. Therefore, LIDP can be a valuable dataset for lung imaging research, assisting in the development, validation of CAD in clinical practice. Researchers interested in LIDP can contact the corresponding author for any non-commercial purpose. The code is available at https://github.com/MHW-NKU/LIDP-model-evaluation.

References

1. Al-Shabi, M., Lan, B.L., Chan, W.Y., Ng, K.H., Tan, M.: Lung nodule classification using deep local-global networks. Int. J. Comput. Assist. Radiol. Surg. **10**, 1815–1819 (2019)
2. Al-Shabi, M., Lee, H.K., Tan, M.: Gated-dilated networks for lung nodule classification in CT scans. IEEE Access **7**, 178827–178838 (2019)
3. Ali, I., Muzammil, M., Haq, I.U., Khaliq, A.A., Abdullah, S.: Efficient lung nodule classification using transferable texture convolutional neural network. IEEE Access **8**, 175859–175870 (2020)
4. Armato, S.G., III., et al.: The lung image database consortium (LIDC) and image database resource initiative (IDRI): a completed reference database of lung nodules on CT scans. Med. Phys. **2**, 915–931 (2011)
5. Del Ciello, A., Franchi, P., Contegiacomo, A., Cicchetti, G., Bonomo, L., Larici, A.R.: Missed lung cancer: when, where, and why? Diagn. Intervent. Radiol. **23**(2), 118 (2017)
6. Dey, R., Lu, Z., Hong, Y.: Diagnostic classification of lung nodules using 3D neural networks. In: 2018 IEEE 15th International Symposium on Biomedical Imaging, pp. 774–778 (2018)
7. Hussein, S., Cao, K., Song, Q., Bagci, U.: Risk stratification of lung nodules using 3D CNN-based multi-task learning. In: Niethammer, M., et al. (eds.) IPMI 2017. LNCS, vol. 10265, pp. 249–260. Springer, Cham (2017). https://doi.org/10.1007/978-3-319-59050-9_20
8. Jiang, H., Shen, F., Gao, F., Han, W.: Learning efficient, explainable and discriminative representations for pulmonary nodules classification. Pattern Recogn. 107825 (2021)
9. Kirby, J.S., et al.: LUNGx challenge for computerized lung nodule classification. J. Med. Imaging (4), 044506 (2016)
10. Kuang, Y., Lan, T., Peng, X., Selasi, G.E., Liu, Q., Zhang, J.: Unsupervised multi-discriminator generative adversarial network for lung nodule malignancy classification. IEEE Access **8**, 77725–77734 (2020)
11. Lei, Y., Tian, Y., Shan, H., Zhang, J., Wang, G., Kalra, M.K.: Shape and margin-aware lung nodule classification in low-dose CT images via soft activation mapping. Med. Image Anal. 101628 (2020)
12. Liao, F., Liang, M., Li, Z., Hu, X., Song, S.: Evaluate the malignancy of pulmonary nodules using the 3-D deep leaky noisy-or network. IEEE Trans. Neural Netw. Learn. Syst. **11**, 3484–3495 (2019)
13. Liu, Y., Hao, P., Zhang, P., Xu, X., Wu, J., Chen, W.: Dense convolutional binary-tree networks for lung nodule classification. IEEE Access **6**, 49080–49088 (2018)
14. Rorke, L.B.: Pathologic diagnosis as the gold standard (1997)

15. Shan, H., Wang, G., Kalra, M.K., de Souza, R., Zhang, J.: Enhancing transferability of features from pretrained deep neural networks for lung nodule classification. In: Proceedings of the 2017 International Conference on Fully Three-Dimensional Image Reconstruction in Radiology and Nuclear Medicine (2017)

16. Shen, W., et al.: Multi-crop convolutional neural networks for lung nodule malignancy suspiciousness classification. Pattern Recogn. **61**, 663–673 (2017)

17. Sung, H., et al.: Global cancer statistics 2020: GLOBOCAN estimates of incidence and mortality worldwide for 36 cancers in 185 countries. CA: Cancer J. Clin. **71**(3), 209–249 (2021)

18. National Lung Screening Trial Research Team: Reduced lung-cancer mortality with low-dose computed tomographic screening. N. Engl. J. Med. **635**(5), 395–409 (2011)

19. Wu, G.X., Raz, D.J.: Lung cancer screening. Lung Cancer 1–23 (2016)

20. Xie, Y., Xia, Y., Zhang, J., Feng, D.D., Fulham, M., Cai, W.: Transferable multimodel ensemble for benign-malignant lung nodule classification on chest CT. In: Descoteaux, M., Maier-Hein, L., Franz, A., Jannin, P., Collins, D.L., Duchesne, S. (eds.) MICCAI 2017. LNCS, vol. 10435, pp. 656–664. Springer, Cham (2017). https://doi.org/10.1007/978-3-319-66179-7_75

21. Xie, Y., et al.: Knowledge-based collaborative deep learning for benign-malignant lung nodule classification on chest CT. IEEE Trans. Med. Imaging **4**, 991–1004 (2018)

22. Xie, Y., Zhang, J., Xia, Y.: Semi-supervised adversarial model for benign-malignant lung nodule classification on chest CT. Med. Image Anal. **57**, 237–248 (2019)

23. Xie, Y., Zhang, J., Xia, Y., Fulham, M., Zhang, Y.: Fusing texture, shape and deep model-learned information at decision level for automated classification of lung nodules on chest CT. Inf. Fusion **42**, 102–110 (2018)

24. Yang, J., Fang, R., Ni, B., Li, Y., Xu, Y., Li, L.: Probabilistic radiomics: ambiguous diagnosis with controllable shape analysis. In: Shen, D., et al. (eds.) MICCAI 2019. LNCS, vol. 11769, pp. 658–666. Springer, Cham (2019). https://doi.org/10.1007/978-3-030-32226-7_73

Moving from 2D to 3D: Volumetric Medical Image Classification for Rectal Cancer Staging

Joohyung Lee[1], Jieun Oh[2], Inkyu Shin[1], You-sung Kim[3], Dae Kyung Sohn[4], Tae-sung Kim[2,3(✉)], and In So Kweon[1]

[1] Korea Advanced Institute of Science and Technology, Daejeon, South Korea
iskweon77@kaist.ac.kr
[2] Healthcare AI Team, National Cancer Center, Goyang, South Korea
tsangel@ncc.re.kr
[3] Department of Radiology, National Cancer Center, Goyang, South Korea
[4] Center for Colorectal Cancer, National Cancer Center, Goyang, South Korea

Abstract. Volumetric images from Magnetic Resonance Imaging (MRI) provide invaluable information in preoperative staging of rectal cancer. Above all, accurate preoperative discrimination between T2 and T3 stages is arguably both the most challenging and clinically significant task for rectal cancer treatment, as chemo-radiotherapy is usually recommended for patients with T3 (or greater) stage cancer. In this study, we present a volumetric convolutional neural network to accurately discriminate T2 from T3 stage rectal cancer with rectal MR volumes. Specifically, we propose 1) a custom ResNet-based volume encoder that models the inter-slice relationship with late fusion (i.e., 3D convolution at the last layer), 2) a bilinear computation that aggregates the resulting features from the encoder to create a volume-wise feature, and 3) a joint minimization of triplet loss and focal loss. With MR volumes of pathologically confirmed T2/T3 rectal cancer, we perform extensive experiments to compare various designs within the framework of residual learning. As a result, our network achieves an AUC of 0.831, which is higher than the reported accuracy of the professional radiologist groups. We believe this method can be extended to other volume analysis tasks.

Keywords: Rectal cancer staging · Medical volume classification · Convolutional neural network · Anisotropy

1 Introduction

Volumetric medical images (VMIs) are widely used image representation in medical field. For example, Magnetic Resonance Images (MRIs) play an essential

---

J. Lee and J. Oh—These authors contributed equally to this work.
T.-s. Kim and I. S. Kweon—These co-corresponding authors contributed equally to this work.

© The Author(s), under exclusive license to Springer Nature Switzerland AG 2022
L. Wang et al. (Eds.): MICCAI 2022, LNCS 13433, pp. 780–790, 2022.
https://doi.org/10.1007/978-3-031-16437-8_75

role in treatment planning for patients as it enables preoperative T-staging [19]. Specifically, the preoperative discrimination of T2 stage from T3 stage rectal cancer is arguably the most difficult yet clinically significant task since it determines whether to use chemo-radiotherapy, a physically burdensome and expensive treatment option [5]. Despite the importance of accurate staging, both radiologists' and surgeons' ability to differentiate between the two stages using MRI vary widely [14].

In our previous study [9], we outlined a 2D convolutional neural network (CNN) model for T2/T3 discrimination that receives 2D MR images. However, it requires radiologists to manually select a representative slice (2D) from each MR volume (3D). In this study, we propose a 3D convolutional neural network (CNN) that classifies rectal MR volumes as T2 or T3 stages.

The decision to use either a 2D or 3D CNN for volume classification remains an open problem [8]. Advocates for 2D convolution address the large anisotropy of volumetric images: for example, the x-y resolution of CT is more than ten times higher than that of its third axis [13,16]. Therefore, they claim that applying 3D CNNs with isotropic kernels to anisotropic medical volume can be problematic [4]. Nonetheless, 3D CNNs can aggregate information from the third axis and is therefore widely used in practice [2,15]. In our work, we investigate ways to include both types of convolutions to create an anisotropic receptive field for VMI analysis.

To capture information from the third axis, most studies in VMI analysis utilize global average pooling [7]. However, in fine-grained image recognition (FGIR) community, bilinear encoding is widely used as an aggregation function [12]. Because VMI classification shares similarity with FGIR, bilinear encoding can be an effective depth aggregation function for VMI classification. Video classification is another area where studies have explored different aggregation functions [21]. Consequently, in this study, we explore various aggregation functions from the field of medical image analysis, FGIR, and video action recognition.

The main contributions of our work are as follows:

– To the best of our knowledge, this is the first reported system that automates rectal cancer staging with 3D MR volumes.
– We introduce a CNN model that classifies rectal MR volumes. Our encoder fuses the third axis information at the last encoding layer, which makes an anisotropic receptive field for an anisotropic MR volume. We also propose aggregating the resulting feature from the encoder using bilinear computation. We train the network with a joint minimization strategy of focal loss and triplet loss.

2   Materials and Methods

Our goal is to solve a clinically challenging problem: to distinguish T2-stage rectal cancer from T3-stage rectal cancer with MRI scans. To this end, we propose a volume classification model for anisotropic medical volumes (Fig. 1). First, we aim to find an effective feature extractor. Second, we compare the performance of various aggregation functions that summarize the extracted features. Finally, we search for supplementary objective functions that improve model performance.

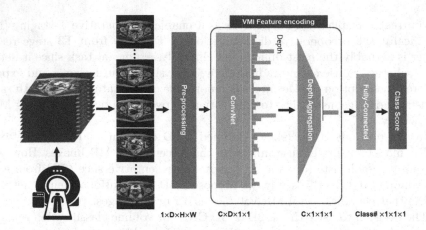

Fig. 1. Proposed volumetric medical image classification architecture. Our work focuses on both the network architecture and depth aggregation function.

Table 1. Subject demographic characteristics and device information. Discovery 750 3.0T, Genesis Signa 1.5T, Signa 1.5T, and Signa HDX 3.0T scanners were manufactured by GE Healthcare ($n = 97$), Achieva 3.0T, Achieva TX 3.0T, and Ingenia CX 3.0T scanners by Philips Healthcare ($n = 468$), and Skyra 3.0T scanners by Siemens ($n = 2$). MFS: magnetic field strength.

	All ($n = 567$)	T2 ($n = 168$)	T3 ($n = 399$)
Age	63.12 ± 11.23	62.95 ± 10.59	63.19 ± 11.48
Sex (Female/Male)	198/369	68/100	130/269
Tumor Volume (cm^3)	7.53 ± 9.64	3.32 ± 5.21	9.3 ± 10.49
N Stage (0/1/2/x)	259/155/133/20	120/27/7/14	139/128/126/6
MFS (1.5T/3T)	77/490	23/145	54/345
Slice Thickness (mm)	3.1 ± 0.42	3.1 ± 0.41	3.11 ± 0.43
Pixel Spacing (mm)	0.439 ± 0.069	0.443 ± 0.069	0.437 ± 0.069

2.1 Dataset and Preprocessing

For this study, we retrospectively (2004–2018) collected 168 T2-stage and 399 T3-stage MR volumes from 567 patients with rectal cancer. The cancer stage labels were pathologically confirmed (determined at surgical resection). Scans were collected with one of seven 3.0T or 1.5T superconducting systems using pelvic phased-array coils. Table 1 contains subject demographic characteristics and device information. For each MR volume, radiologists select one to three representative slices that best reflect the clinical state of the whole volume; we assign the pathologically confirmed cancer stage of an MR volume to all of its representative slice(s), which is our slice-level label. This study was conducted according to the principles of the Declaration of Helsinki, and the protocol was approved by the Institutional Review Board of our institution (NCC2019-0081).

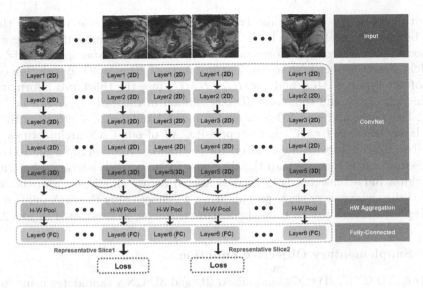

Fig. 2. Selected mixture(f-rMC_5) of 2D CNN and 3D CNN to map volumetric medical image to slice-wise probability.

For pre-processing, we apply N4ITK to correct the intensity inhomogeneity, following the conventions of MRI community [20]. Moreover, we conduct the following training-time augmentations: elastic transformation, optical distortion, grid distortion, random shift, random scale, random rotate, random crop, horizontal flip, CLAHE, random brightness, random contrast, motion blur, median blur, and Gaussian blur. Because localization around rectum region (the union of rectum and tumor) is relatively simple task, and previously reported automatic methods have sufficiently high accuracy (around 94.3% DSC) [10], we start our experiment from the localized setting (around the rectum and tumor) and resize along the first and second axes to create $256 \times 256 \times N$ pixel volumes (where N represents the number of slices in the third axis). We perform 10-fold cross-validation following the procedure of our previous study [10] and report the performance score of hold-out set as described in Sect. 2.5.

2.2   2D CNN vs. 3D CNN for Anisotropic Rectal Volume Analysis

Our first experimental design compares various feature extractors for rectal MR volumes. Here, we address the debate on performance supremacy between 2D CNN and 3D CNN for anisotropic medical volumes. As 2D CNNs can only yield slice-wise scores, we compare the performance of 2D CNN and 3D CNN in slice-level classification by using slice-level labels as described in Sect. 2.1. Therefore, our feature extractor is $f : \mathbb{R}^{H \times W \times D} \rightarrow \mathbb{R}^{D \times 256}$ as illustrated in Fig. 2. This experimental design evaluates whether adjacent slices help slice-wise classification. Motivated by Tran et al. [18], we also compare the performance of various mixtures of 2D and 3D convolutions (f-MC_x, f-rMC_x), which create anisotropic receptive fields. However, unlike Tran *et al.*, we did not stride along the third axis because our goal is to yield slice-wise scores and to examine if training with adjacent slices improves performance. Models with names starting with 'f-' are models that yield slice-wise scores. The mixtures of 2D and 3D convolution consist of two types: $f\text{-}MC_x$ and $f\text{-}rMC_x$. $f\text{-}MC_x$ fuse the third axis information in the early layer and thus have 3D convolution in the early layer (s), i.e., 1st to (x-1)th layer, and have 2D convolution in the late layer, i.e., xth layer to the last layer, whereas $f\text{-}rMC_x$ is the opposite. All of our CNN architectures are based on ResNet-18 with the 32 filters for the first convolution layer [3]. To train our feature extractors, we map the slice-wise features to slice-wise probabilities through a fully-connected layer that is shared across all slices. From the outputted slice-wise cancer stage probabilities, we choose only the probabilities for the representative slices to train the feature extractor as illustrated in Fig. 2.

2.3   Supplementary Objective Function

We train 2D CNN, 3D CNN, and mixed 2D and 3D CNN candidates using focal loss [11] for our feature extractor experiments. This loss function is a generalization of cross-entropy loss with an additional down-weighting parameter for when the prediction is close to the ground truth. The loss can be expressed as follows, where p is the softmax probability:

$$\mathcal{L}_{focal}(p) = -(1-p)^\gamma \log(p) \tag{1}$$

Once we determine the best performing residual block (see Sect. 3.1), we compare the performance of the various supplementary loss functions. These functions are added to focal loss and trained jointly to enhance inter-class separability and intra-class compactness [22]. Specifically, we implement center loss [23] and triplet loss [17]. We use the suggested hyperparameters settings from their respective original manuscripts. The losses are defined as follows:

$$\mathcal{L}_{center} = \frac{1}{2} \sum_{i=1}^{m} \|x_i - c_{y_i}\|_2^2 \tag{2}$$

$$\mathcal{L}_{triplet}(x, x^+, x^-; f) = max(0, \|f - f^+\|_2^2 - \|f - f^-\|_2^2 + m) \tag{3}$$

The y_ith class center of deep features is denoted by $c_{y_i} \in \mathbb{R}^{c'}$. To select both the positive and negative samples for triplet loss, we implement online-hard mining from Schroff *et al.* [17].

2.4   Depth Aggregation Function

Global average pooling is a common choice for aggregation functions in CNNs [3, 6,18]. However, as proper classification of medical volumes is usually contingent on a couple of slices (representative slices), global average pooling may not be the optimal solution. Using the best performing objective function investigated in Sect. 3.2, we evaluate the performance of different aggregation functions on our rectal MR dataset. With the number of extracted feature dimension C, and the number of slices of input VMI D, our depth aggregation function is defined as $f : \mathbb{R}^{C \times D} \rightarrow \mathbb{R}^{C \times 1}$. Specifically, we experiment with four different types of aggregation functions: average pooling, max pooling, attention weighting [21], and bilinear encoding [12]. When i and d are the channel index and depth dimension respectively, attention weighting is defined as follows:

$$g_{i,att} = \sum_{d=1}^{D} A(i,d) f_i^d \tag{4}$$

To design the attention weighting function $A(i, d)$, we did not create an additional embedding space but rather directly applied a softmax function. Therefore, the attention weight vector becomes as follows:

$$A(i,d) = \frac{exp(f_i^d)}{\sum_{d'=1}^{D} exp(f_i^{d'})} \tag{5}$$

Functions other than attention weighting are straightforward. Max-pooling and average-pooling select the maximum and average value along depth per each channel dimension. For bilinear encoding, we used the same features twice to generate the bilinear feature with l_2 normalization scheme [12].

2.5 Training and Testing

To mitigate data imbalance between T2 and T3 stage, we perform oversampling. We train all models using Stochastic Gradient Descent (SGD) with initial learning rate of 0.01 and applied weight decay with ratio (λ) 0.01 to mitigate overfitting. Moreover, to report the final predictions performance, we follow the test-time augmentation procedure from Wu et al. [24]. As a performance metrics, we report mean and standard deviation of average area-under-curve (AUC), accuracy, sensitivity, and specificity over ten folds. We select the best performing module and loss based on mean AUC.

3 Results

3.1 2D CNN vs. 3D CNN for Anisotropic Rectal Volume Analysis

Table 2 contains the performance metrics for models with different mixtures of 2D and 3D convolutions. First, the results show that 2D CNN (f-$R2D$) outperforms 3D CNN (f-$R3D$) for our task. Moreover, f-rMC_x tends to excel f-MC_x, which shows that late fusion (f-rMC_x) along the third axis is more effective to model inter-slice relationship than early fusion (f-MC_x). Among all the models, f-rMC_5, which is illustrated in Fig. 2, scores the highest AUC followed by f-$R2D$; adjacent slices do help learning their center slice. Specifically, the only

Table 2. Performance comparison of different mixture of 2D CNN and 3D CNN.

CNN	AUC	Acc	Recall (T2)	Recall (T3)	# params
f-$R2D$	0.798 ± 0.042	75.0 ± 2.9	60.0 ± 12.6	81.4 ± 7.0	2796001
f-$R3D$	0.763 ± 0.074	74.5 ± 5.3	55.5 ± 12.9	82.5 ± 7.0	8291873
f-$R(2+1)D$	0.758 ± 0.069	74.3 ± 5	50.1 ± 12.8	84.5 ± 5.6	8294563
f-MC_2	0.789 ± 0.061	73.5 ± 3.8	65.5 ± 13.3	76.9 ± 5.1	2799137
f-MC_3	0.746 ± 0.07	70.2 ± 9.4	58.8 ± 16.0	74.9 ± 17.1	2872865
f-MC_4	0.767 ± 0.059	72.5 ± 3.5	55.7 ± 12.8	79.7 ± 6.5	3130913
f-MC_5	0.737 ± 0.042	72.3 ± 4.8	49.7 ± 8.8	81.9 ± 8.0	4163105
f-rMC_2	0.778 ± 0.051	75.1 ± 4.3	55.6 ± 13.9	83.3 ± 6.5	8288737
f-rMC_3	0.796 ± 0.052	75.1 ± 2.2	59.0 ± 15.3	81.8 ± 6.8	8215009
f-rMC_4	0.787 ± 0.055	76.2 ± 4.5	50.0 ± 13.6	87.3 ± 4.6	7956961
f-rMC_5	$\mathbf{0.815 \pm 0.057}$	77.3 ± 4.4	48.5 ± 10.3	89.4 ± 5.0	6924769

difference between f-rMC_5 and f-$R2D$ is that f-rMC_5 makes inter-slice modeling at the last layer of the network. Moreover, f-$R(2+1)D$ doesn't excel unlike in video classification task. We attribute this to the difference between video and medical volume; the third axis of medical volume is spatial as the other two axes whereas that of video is temporal. It has to be noted that f-$R2D$ yields no receptive field along the third axis whereas the receptive field of f-$R3D$ is elongated towards the third axis because the pixel spacing along the third axis is much larger than that along the first and the second.

3.2 Supplementary Objective Function

Table 3 compares the classification performance of different supplementary loss functions. When combined with focal loss, center loss lowers performance whereas the triplet loss boosts the performance of the baseline, i.e., focal loss. Specifically, by adding center loss to focal loss, the performance dropped by 0.012 in AUC. Adding triplet loss to focal loss increases the AUC by 0.019. The addition of the focal loss and triplet loss will be used for all following experiments in this study.

3.3 Evaluation on Aggregation Functions

With CNN module (f-rMC_5) and loss function (focal loss with triplet loss) selected, the performances of various depth-aggregation functions are evaluated and compared. Note that we have already aggregated x-y information by striding and by average pooling as depicted in Fig. 2. The outcome of x-y pooling, which is the slice-level feature, is then summarized into a single volume-level feature by the depth aggregation function. As a result, selecting average pooling as a depth aggregation function equals applying global average pooling after the last

Table 3. Performance comparison among supplementary loss functions.

Loss	AUC	Acc	Acc (T2)	Acc (T3)
Focal	0.815 ± 0.057	77.3 ± 4.4	48.5 ± 10.3	89.4 ± 5.0
+Center	0.803 ± 0.041	74.9 ± 3.9	49.2 ± 13.9	85.8 ± 5.4
+Triplet	$\mathbf{0.834 \pm 0.068}$	78 ± 4.9	49.9 ± 9.7	89.8 ± 4.7

Table 4. Performance comparison of depth aggregation functions to summarize depth-wise feature along depth-axis.

Pool	AUC	Acc	Acc (T2)	Acc (T3)
AVP	0.821 ± 0.055	76 ± 3.8	34 ± 9.6	93.7 ± 4.9
MXP	0.815 ± 0.047	78.1 ± 4.1	47 ± 12.6	91.2 ± 5.9
Bilin	$\mathbf{0.831 \pm 0.062}$	79.3 ± 6.2	62.1 ± 13.5	86.7 ± 6.8
Att	0.817 ± 0.057	77.2 ± 4.8	48.3 ± 14.7	89.5 ± 4.2

convolution layer. Here we have four candidates, including the relatively basic: 1) average pooling, 2) max pooling, and the more complex: 3) bilinear encoding, and 4) attention weighting. Attention weighting is implemented with the hope that attention mechanism may be able to find the representative slices from the volume. No additional parameters are introduced for depth aggregation function except that the feature space is enlarged by bilinear encoding. The results are described in Table 4.

As shown in Table 4, bilinear encoder scores the best performance among four functions to summarize the depth information. Because bilinear encoder has been successful in the field of FGIR, we assume that the ability of bilinear encoder to capture fine-grained detail of tumour can be successful in cancer staging as well [12,25]. Moreover, bilinear encoder notably closes the gap between Acc (T2) and Acc (T3). Note that the performance degradation from the triplet loss of Table 3 and average pooling of Table 4 shows that volume level classification is more challenging than slice-level classification. It may be because volume has more unnecessary areas than a single slice in staging the cancer level.

3.4 Performance Comparison with Radiologists

Two studies have reported the performance of professional radiologists in discriminating T2-stage and T3-stage rectal cancer. Since the datasets are different, direct comparison is not possible. However, performance can still be compared to earn some idea about how competitive our method is (Table 5).

Table 5. Discrimination performance of T2/T3 rectal cancer by radiologists (Ang et al., Maas et al. *et al.*) and by our proposed method

Method	AUC	Acc	Acc (T2)	Acc (T3)
Ang *et al.* [1]	-	68.87	85.5	50.1
Maas *et al.* (1.5T MR) [14]	0.73	66.67	83.33	52.38
Maas *et al.* (3.0T MR) [14]	0.64	56.41	72.22	42.86
f-rMC_5+Bilinear+Triplet (Ours)	**0.831**	79.3	62.1	86.7

4   Conclusion

This paper proposes a CNN model to preoperatively discriminate between T2 and T3 stage rectal cancer from rectal MR volumes. Our network consists of two parts: a CNN feature extractor and a depth aggregation function. The CNN maps medical volumes to slice-wise features while the aggregation function summarizes the slice-wise features into a volume-wise feature. To investigate the best performing CNN for anisotropic rectal VMI, we compared the performance with varying approach to model inter-slice relationship. As a result, f-rMC_5, which represents the late inter-slice interaction, is selected. Moreover, through extensive experimentation, we found that using triplet loss and bilinear encoder as a depth aggregation function and supplementary objective function, respectively, outperformed other functions and losses. To the best of our knowledge, this is the first study to apply the experimental approach from the video action community to address the debate on whether 3D convolutions are an efficient method for anisotropic VMI models. We believe our proposed CNN can be used for other anisotropic VMI-related tasks such as VMI segmentation and detection.

Acknowledgments. The authors would like to thank Young Sang Choi for his helpful feedback. This work was supported by KAIST R&D Program (KI Meta-Convergence Program) 2020 through Korea Advanced Institute of Science and Technology (KAIST), a grant from the National Cancer Center (NCC2010310-1), and the National Research Foundation of Korea (NRF) grant funded by the Korea government (MSIT) (NRF-2020R1C1C1012905).

References

1. Ang, Z.H., De Robles, M.S., Kang, S., Winn, R.: Accuracy of pelvic magnetic resonance imaging in local staging for rectal cancer: a single local health district, real world experience. ANZ J. Surg. **91**(1–2), 111–116 (2021)
2. Çiçek, Ö., Abdulkadir, A., Lienkamp, S.S., Brox, T., Ronneberger, O.: 3D U-Net: learning dense volumetric segmentation from sparse annotation. In: Ourselin, S., Joskowicz, L., Sabuncu, M.R., Unal, G., Wells, W. (eds.) MICCAI 2016. LNCS, vol. 9901, pp. 424–432. Springer, Cham (2016). https://doi.org/10.1007/978-3-319-46723-8_49

3. He, K., Zhang, X., Ren, S., Sun, J.: Deep residual learning for image recognition. In: Proceedings of the IEEE Conference on Computer Vision and Pattern Recognition, pp. 770–778 (2016)
4. Hesamian, M.H., Jia, W., He, X., Kennedy, P.: Deep learning techniques for medical image segmentation: achievements and challenges. J. Digit. Imaging **32**(4), 582–596 (2019). https://doi.org/10.1007/s10278-019-00227-x
5. Horvat, N., Carlos Tavares Rocha, C., Clemente Oliveira, B., Petkovska, I., Gollub, M.J.: MRI of rectal cancer: tumor staging, imaging techniques, and management. Radiographics **39**(2), 367–387 (2019)
6. Howard, A., et al.: Searching for mobilenetv3. In: Proceedings of the IEEE/CVF International Conference on Computer Vision, pp. 1314–1324 (2019)
7. Huang, Z., Zhou, Q., Zhu, X., Zhang, X.: Batch similarity based triplet loss assembled into light-weighted convolutional neural networks for medical image classification. Sensors **21**(3), 764 (2021)
8. Isensee, F., et al.: nnU-Net: self-adapting framework for u-net-based medical image segmentation. arXiv preprint arXiv:1809.10486 (2018)
9. Kim, J., et al.: Rectal cancer: toward fully automatic discrimination of T2 and T3 rectal cancers using deep convolutional neural network. Int. J. Imaging Syst. Technol. **29**(3), 247–259 (2019)
10. Lee, J., Oh, J.E., Kim, M.J., Hur, B.Y., Sohn, D.K.: Reducing the model variance of a rectal cancer segmentation network. IEEE Access **7**, 182725–182733 (2019)
11. Lin, T.Y., Goyal, P., Girshick, R., He, K., Dollár, P.: Focal loss for dense object detection. In: Proceedings of the IEEE International Conference on Computer Vision, pp. 2980–2988 (2017)
12. Lin, T.Y., RoyChowdhury, A., Maji, S.: Bilinear convolutional neural networks for fine-grained visual recognition. IEEE Trans. Pattern Anal. Mach. Intell. **40**(6), 1309–1322 (2017)
13. Liu, S., et al.: 3D anisotropic hybrid network: transferring convolutional features from 2D images to 3D anisotropic volumes. In: Frangi, A.F., Schnabel, J.A., Davatzikos, C., Alberola-López, C., Fichtinger, G. (eds.) MICCAI 2018. LNCS, vol. 11071, pp. 851–858. Springer, Cham (2018). https://doi.org/10.1007/978-3-030-00934-2_94
14. Maas, M., et al.: T-staging of rectal cancer: accuracy of 3.0 tesla MRI compared with 1.5 tesla. Abdom. Imaging **37**(3), 475–481 (2012). DOIurl10.1007/s00261-011-9770-5
15. Milletari, F., Navab, N., Ahmadi, S.A.: V-net: fully convolutional neural networks for volumetric medical image segmentation. In: 2016 Fourth International Conference on 3D Vision (3DV), pp. 565–571. IEEE (2016)
16. Peng, C., Lin, W.A., Liao, H., Chellappa, R., Zhou, S.K.: SAINT: spatially aware interpolation network for medical slice synthesis. In: Proceedings of the IEEE/CVF Conference on Computer Vision and Pattern Recognition, pp. 7750–7759 (2020)
17. Schroff, F., Kalenichenko, D., Philbin, J.: FaceNet: a unified embedding for face recognition and clustering. In: Proceedings of the IEEE Conference on Computer Vision and Pattern Recognition, pp. 815–823 (2015)
18. Tran, D., Wang, H., Torresani, L., Ray, J., LeCun, Y., Paluri, M.: A closer look at spatiotemporal convolutions for action recognition. In: Proceedings of the IEEE Conference on Computer Vision and Pattern Recognition, pp. 6450–6459 (2018)
19. Trebeschi, S., et al.: Deep learning for fully-automated localization and segmentation of rectal cancer on multiparametric MR. Sci. Rep. **7**(1), 1–9 (2017)
20. Tustison, N.J., et al.: N4ITK: improved N3 bias correction. IEEE Trans. Med. Imaging **29**(6), 1310–1320 (2010)

21. Wang, L., et al.: Temporal segment networks for action recognition in videos. IEEE Trans. Pattern Anal. Mach. Intell. **41**(11), 2740–2755 (2018)
22. Weinberger, K.Q., Saul, L.K.: Distance metric learning for large margin nearest neighbor classification. J. Mach. Learn. Res. **10**(2) (2009)
23. Wen, Y., Zhang, K., Li, Z., Qiao, Yu.: A discriminative feature learning approach for deep face recognition. In: Leibe, B., Matas, J., Sebe, N., Welling, M. (eds.) ECCV 2016. LNCS, vol. 9911, pp. 499–515. Springer, Cham (2016). https://doi.org/10.1007/978-3-319-46478-7_31
24. Wu, N., et al.: Deep neural networks improve radiologists' performance in breast cancer screening. IEEE Trans. Med. Imaging **39**(4), 1184–1194 (2019)
25. Zheng, H., Fu, J., Zha, Z.J., Luo, J.: Learning deep bilinear transformation for fine-grained image representation. In: Advances in Neural Information Processing Systems, vol. 32 (2019)

Author Index

Printed in the United States
by Baker & Taylor Publisher Services

Printed in the United States
by Baker & Taylor Publisher Services